Nonlinear Science and Complexity

Nonlinear Science and Complexity

edited by

Albert C J Luo
Southern Illinois University Edwardsville, USA

Liming Dai
University of Regina, Canada

Hamid R Hamidzadeh
Tennessee State University, USA

World Scientific

NEW JERSEY • LONDON • SINGAPORE • BEIJING • SHANGHAI • HONG KONG • TAIPEI • CHENNAI

Published by

World Scientific Publishing Co. Pte. Ltd.

5 Toh Tuck Link, Singapore 596224

USA office: 27 Warren Street, Suite 401-402, Hackensack, NJ 07601

UK office: 57 Shelton Street, Covent Garden, London WC2H 9HE

British Library Cataloguing-in-Publication Data
A catalogue record for this book is available from the British Library.

NONLINEAR SCIENCE AND COMPLEXITY

ISBN-13 978-981-270-436-8
ISBN-10 981-270-436-1

Printed in Singapore by World Scientific Printers (S) Pte Ltd

TRANSACTIONS OF NONLINEAR SCIENCE AND COMPLEXITY

Organizing Committee

Liming Dai (Canada)
Frank Z. Feng (USA)
M.H. Elahinia (USA)
Nail H. Ibragimov (Sweden)
G. Nakhaie Jazar (USA)
Zhongliang Jing (China)
Shijun Liao (China)
Albert C.J. Luo (USA)
Lev A. Ostrovsky (USA)
Sergey Prants (Russia)
Victor I. Shrira (UK)
Subhash Sinha (USA)
Xubin Song (USA)
Gazanfer Unal (Turkey)
Fei-Yue Wang (USA)
Trong Wu (USA)
Pei Yu (Canada)

Conference on
Nonlinear Science and Complexity
August 7-12, 2006
Beijing, China
Conference on

Preface

This volume partially contains the papers presented in the 2006 *International Conference on Nonlinear Science and Complexity*, which held in Beijing, China, August 7-12, 2006. This conference provided a place to exchange recent developments, discoveries and progresses on *Nonlinear Science and Complexity.* The fundamental and frontier theories and techniques for modern science and technology were presented. In addition, this conference provides a platform to exchange the methodology in applied nonlinear science, nonlinear modeling and intelligent computations. The conference organizers believe this conference to stimulate more research in nonlinear science and complexity. The conference focused on the following topics:

- Lie Group Analysis and Applications in Nonlinear Science Complex (Nail H. Ibragimov)
- Nonlinear Wave Dynamics and Patterns in Geophysical Flows (Lev A. Ostrovsky and Victor I. Shrira)
- Chaotic Dynamics and Transport in Classic and Quantum Systems (Sergey Prants)
- Nonlinear Dynamics, Oscillations and Stability (Albert C.J. Luo, Pei Yu, Subhash Sinha)
- Nonlinear Fluid Mechanics (Gazanfer Unal and Shijun Liao)
- Dynamics in Continuous Media and Wave Propagations (Liming Dai)
- Modeling and Nonlinearity in Sensors, Bio-devices, MEMS and Nano-systems (Frank Z. Feng, G. Nakhaie Jazar)
- Nonlinear Modeling and Control of Smart Material Systems (M.H. Elahinia, Xubin Song)
- Nonlinear Modeling and Control and Intelligent Computing (Zhongliang Jing, Trong Wu)

Many papers presented in this conference show excellent achievements in nonlinear science and complexity. The organizers believes this permanent record will make the work last longer and more influence. The conference organizers wish to express their deep appreciation to all the authors and reviewers.

Albert C.J. Luo, Liming Dai, H.R. Hamidzadeh

CONTENTS

Symmetry reductions for an inhomogeneus nonlinear diffusion equation

M.L. Gandarias, M.S. Bruzón

Departamento de Matemáticas, Universidad de Cádiz, PO.BOX 40
11510 Puerto Real, Cádiz, Spain
E-mail marialuz.gandarias@uca.es

Abstract

In this work we derive symmetry reductions and exact solutions for an inhomogeneous equation that model fast diffusion. We find the connection between classes of nonclassical symmetries of the equation and of an associated system. These symmetries allow us to increase the number of solutions. Some of theses solutions are unobtainable by classical symmetries.

Keywords: Partial differential equations; Nonclassical symmetries; Potential symmetries

1 Introduction

The diffusion processes appear in many physics processes such as plasma physics, kinetic theory of gases, solid state, metallurgy and transport in porous medium [1, 12, 15].

In this work we consider a mathematical model for diffusion processes which is the generalised inhomogeneous nonlinear diffusion equation

$$f(x)u_t = [g(x)u^n u_x]_x. \tag{1}$$

In [15] P.Rosenau presented a number of remarkable features of the fast diffusion processes: for $f(x) = 1$, $g(x) = 1$ and $-2 \leq n \leq -1$, the family of fast diffusion (1) coexists with a subclass of superfast diffusions where the whole process terminates within a finite time. The special case with $n = -1$ emerges in plasma physics and reveals a surprising richness of new physic-mathematical phenomena.

In (1) $u(x, t)$ is a function of position x and time t and may represent the temperature, $f(x)$ and $g(x)$ are arbitrary smooth functions of position and may denote the density and the density-dependent part of thermal diffusion, respectively.

There is no existing general theory for solving nonlinear partial differential equations and the methods of point transformations are a powerful tool. One of the most useful point transformations are those which form a continuous group. Lie classical symmetries admitted by nonlinear partial differential equations (PDE's) are useful for finding invariant solutions.

Motivated by the fact the symmetry reductions for many PDE's are known that are not obtained by using the classical Lie method there have been several generalizations of the classical Lie group method for symmetry reductions.

Bluman and Cole [3] developed the nonclassical method to study the symmetry reductions of the heat equation. The basic idea of the method is to require that the N order PDE

$$\Delta = \Delta\left(x,t,u,u^{(1)}(x,t),...,u^{(N)}(x,t)\right) = 0, \tag{2}$$

where $(x,t) \in I\!R^2$ are the independent variables, $u \in I\!R$ is the dependent variable and $u^{(l)}(x,t)$ denote the set of all partial derivatives of l order of u and the invariance surface condition

$$\xi u_x + \tau u_t - \phi = 0, \tag{3}$$

which is associated with the vector field

$$\mathbf{v} = \xi(x,t,u)\partial_x + \tau(x,t,u)\partial_t + \phi(x,t,u)\partial_u, \tag{4}$$

are both invariant under the transformation with infinitesimal generator (4). Since then, a great number of papers have been devoted to the study of nonclassical symmetries of nonlinear PDE's in both one and several dimensions.

In [4, 5] Bluman introduced a method to find a new class of symmetries for a PDE. By writing a given PDE, denoted by $\mathrm{R}\{x,t,u\}$ in a conserved form a related system denoted by $\mathrm{S}\{x,t,u,v\}$ as additional dependent variables is obtained. Any Lie group of point transformations admitted by $\mathrm{S}\{x,t,u,v\}$ induces a symmetry for $\mathrm{R}\{x,t,u\}$; when at least one of the generators of the group depends explicitly of the potential, then the corresponding symmetry is neither a point nor a Lie-Bäcklund symmetry. These symmetries of $\mathrm{R}\{x,t,u\}$ are called *potential* symmetries.

In [14], C. Sophocleous has classified the nonlocal potential symmetries of (1). He obtained that potential symmetries exists only if the parameter n takes the values -2 or $-\frac{2}{3}$ and also certain relations must be satisfied by the functions $f(x)$ and $g(x)$.

In [10], we have derived *nonclassical potential* symmetries for the special case of (1), with $f(x) = 1$ and $g(x) = 1$

$$u_t = [u^{-1}u_x]_x. \tag{5}$$

In [13] connection between classes of nonclassical symmetries of (5), and of nonclassical symmetries of an associated system as well as some new generators have been found.

The aim of this paper is to obtain nonclassical symmetries for (1) and for the associated system given by

$$\begin{aligned} v_x &= f(x)u, \\ v_t &= g(x)u^{-1}u_x, \end{aligned} \tag{6}$$

as well as the connection between these symmetries. These symmetries lead to new solutions, some of these solution exhibit an interesting behaviour.

2 Nonclassical symmetries

2.1 Nonclassical symmetries of the PDE (1)

To obtain nonclassical symmetries of (1), with $n = -1$, we apply the algorithm described in [6, 7] for calculating the determining equations. We can distinguish two different cases:
In the case $\tau \neq 0$, without loss of generality, we may set $\tau(x,t,u) = 1$. The generators that we obtain can be obtained by Lie classical method consequently the nonclassical method, with $\tau \neq 0$ applied to (1) gives only rise to the classical symmetries.

In the case $\tau = 0$, without loss of generality, we may set $\xi = 1$ and we get that the determining equation for the infinitesimal ϕ is

$$\begin{aligned}&u^{(n+2)}\left(fg\phi_{xx} + 2fg'\phi_x - f'g\phi_x + fg\phi^2\phi_{uu} + 2fg\phi\phi_{ux} + fg'\phi\phi_u - f'g\phi_u\right) + fg''\phi - f'g'\phi)\\ &+n\phi u^{(n+1)}\left(3fg\phi_x + fg\phi\phi_u + 2fg'\phi - f'g\phi\right) + fg(n-1)n\phi^3 u^n - f^2\phi_t u^2 = 0.\end{aligned} \tag{7}$$

The complexity of this equation is the reason why we cannot solve (7) in general. Thus we proceed, by making an ansatz on the form of $\phi(x,t,u)$, to solve (7) for $n = -1$. In this way we found, choosing $\phi = \alpha(x,t)u^2 + \beta(x,t)u$, after substituting into the determining equation and splitting with respect to u we obtain an overdetermined system for the functions α and β. So, for equation (1) with $n = -1$, we obtain the infinitesimal generator

$$\mathbf{v} = \partial_x + (\alpha(x,t)u^2 + \beta(x,t)u)\partial_u,$$

where f, g, α and β satisfy the system

$$\begin{aligned}&fg'\alpha^2 - fg\alpha\alpha_x + f^2\alpha_t = 0,\\ &f'g'\alpha - f\,g''\alpha^2 + (fg' + f'g)\alpha\beta + (f'g - 2fg')\alpha_x + fg(\alpha\beta_x - \alpha_x\beta - \alpha_{xx}) + f^2\beta_t = 0,\\ &f\beta\beta_x - fg\beta_{xx} + f'g\beta_x = 0,\\ &\beta(f'g' - fg'') + \beta_x(f'g - 2fg') + \beta^2 fg' + fg(\beta\beta_x - \beta_{xx}) = 0.\end{aligned} \tag{8}$$

2.2 Nonclassical symmetries of the system (6)

We now consider the associated auxiliary system (6) augmented with the invariance surface condition

$$\xi v_x + \tau v_t - \psi = 0, \tag{9}$$

which is associated with the vector field

$$\mathbf{w} = \xi(x,t,u,v)\partial_x + \tau(x,t,u,v)\partial_t + \phi(x,t,u,v)\partial_u + \psi(x,t,u,v)\partial_v. \tag{10}$$

By requiring both (6) and (9) to be invariant under the transformation with infinitesimal generator (10) one obtains an over determined, nonlinear system of equations for the infinitesimals $\xi(x,t,u,v)$, $\tau(x,t,u,v)$, $\phi(x,t,u,v)$, $\psi(x,t,u,v)$. When at least one of the generators of the group depend explicitly of the potential, that is if

$$\xi_v^2 + \tau_v^2 + \phi_v^2 \neq 0 \tag{11}$$

then (10) yields a nonlocal symmetry of (1).

A nonclassical potential symmetry of (1) is a nonclassical symmetry of the associated potential system (6) that satisfies (11).

We are considering $\tau \neq 0$, and without loss of generality, we set $\tau = 1$. The nonclassical method, with $\tau \neq 0$, applied to (6), give rise to nonlinear determining equations for the infinitesimals.

If we require that $\xi_u = \psi_u = 0$, we obtain that

$$\phi = -f\xi_v u^2 + \left(\psi_v - \frac{f_x\xi}{f} - \xi_x\right)u + \frac{\psi_x}{f} \tag{12}$$

and $f(x)$, $g(x)$, $\xi(x,t,v)$ and $\psi(x,t,v)$, must satisfy the following equations:

$$g\xi_{vv} - \xi\xi_v = 0, \tag{13}$$

$$-fg\xi\xi_x + 2fg^2\xi_{vx} + fg\psi\xi_v + 2f'g^2\xi_v - fg\xi_t + fg'\xi^2 - fg\psi_v\xi - fg^2\psi_{vv} = 0, \tag{14}$$

$$\begin{aligned} f^2 g^2\xi_{xx} + f^2g\psi\xi_x + ff'g^2\xi_x - f^2g\psi_x\xi - f^2g'\psi\xi + ff''g^2\xi - f'^2g^2\xi - 2f^2g^2\psi_{vx} \\ +f^2g\psi\psi_v + f^2g\psi_t = 0, \end{aligned} \tag{15}$$

$$-fg\psi_{xx} + f\psi\psi_x + f'g\psi_x = 0. \tag{16}$$

We can distinguish the following cases:
If $\xi_v \neq 0$ by solving (13) and substituting into (14), (15), (16) leads to generators for which (11) is satisfied, consequently they are nonclassical potential generators and have been considered in [11].
If ξ does not depend on v, by substituting $\xi = \xi(x,t)$ in (14) and (15) we obtain that

$$\psi = v\left(-\xi_x - \frac{\xi_t}{\xi} + \frac{g'\xi}{g}\right) + g\left(\frac{\xi_x}{\xi} + \frac{\xi_t}{\xi^2}\right) - g' + \theta(x,t), \tag{17}$$

By substituting into (12) we get that $\phi_v = 0$.
We observe that in this case condition (11) is not satisfied, consequently

$$\mathbf{v} = \xi(x,t)\partial_x + \partial_t + \left((\delta(t) - \frac{f_x\xi}{f} - \xi_x)u + \frac{\psi_x}{f}\right)\partial_u + \psi\partial_v \tag{18}$$

is not a nonclassical potential generator.

2.3 Connection between symmetries of the PDE (1) and of the system (6)

If we assume that ξ and ψ do not depend on v, the system (13-16) becomes

$$\begin{aligned} &-g\xi\xi_x - g\xi_t + g'\xi^2 = 0, \\ &f^2 g^2\xi_{xx} + f^2g\psi\xi_x + ff'g^2\xi_x - f^2g\psi_x\xi - f^2g'\psi\xi + ff''g^2\xi - f'^2g^2\xi + f^2g\psi_t = 0, \\ &-fg\psi_{xx} + f\psi\psi_x + f'g\psi_x = 0. \end{aligned} \tag{19}$$

It is easy to check that denoting $\alpha = -\frac{f}{g}\xi$, $\beta = \frac{\psi}{g}$ systems (8) and (19) coincides. Consequently we can state:

$$\mathbf{w} = \xi(x,t)\partial_x + \partial_t + \psi(x,t)\partial_v - \left(\frac{f'}{f}\xi u + \frac{\psi_x}{f}\right)\partial_u$$

is a generator for system (6) *if and only if*

$$\mathbf{v} = \partial_x + \left(-\frac{f}{g}\xi u^2 + \frac{\psi}{g}u\right)\partial_u$$

is a generator for equation (1).

3 Some exact solutions

In this section we derive some exact solutions by using some generators:

1- From generator

$$\xi = k_4\sqrt{x}, \qquad \psi = 0, \tag{20}$$

for $f = \frac{k_3k_4}{2\sqrt{x}}e^{\frac{k_1k_4\sqrt{x}}{k_2}}$ and $g = \frac{2k_2\sqrt{x}}{k_4}$, we obtain the similarity solution and the ODE that gives rise to the exact solution

$$u = \frac{2\,k_1}{k_3\,k_4\left(k_4\,e^{\frac{k_1\,k_4\,\sqrt{x}}{k_2}} - e^{\frac{k_1\,k_4{}^2\,t}{2\,k_2}+\frac{k_1\,k_4\,k_5}{2\,k_2}}\right)}. \tag{21}$$

We observe that the solution (21), for $x = \frac{(k_4 t + k_5)^2}{4}$, blows up at a parabola.

2- From generator

$$\xi = k, \quad \psi = \frac{-2}{x+kt},$$

for $g = 1$, $f = exp(x)$ and the surface condition we obtain the similarity solution and the ODE that gives rise to the exact solution

$$u = e^{-x}\left(\frac{1}{x-t+k_1} - \frac{1}{x+t}\right). \tag{22}$$

We observe that solution (22) blows up in two straight lines $x - t + k_1 = 0$ and $x + t = 0$.

3- From generator

$$\xi = \tfrac{1}{f}, \quad \psi = -2k_1 \tanh[k_1(t + \textstyle\int f(x)dx)],$$

for $g = \frac{1}{f}$,we obtain the similarity solution and the ODE that gives rise to the exact bounded solution

$$u = -k_1 \tanh(k_1\,t + \int f(x)dx) - k_1 \tanh(k_1\,t - \int f(x)dx). \tag{23}$$

In figure 1 we plot (23) which represents two front waves that evolves changing their shape with opposite velocities with $f(x) = x^2$, $f(x) = sec(x)^2$ respectevely and $k_1 = 1$.

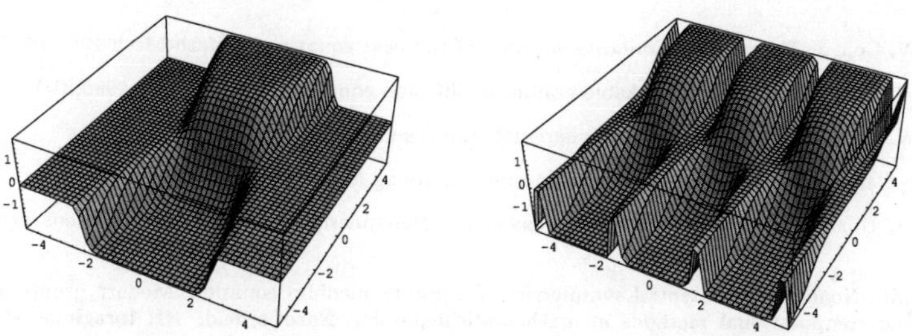

Figure 1: Solution (23) for $g = \frac{1}{f}$, $f(x) = x^2$, $f(x) = sec^2(x)$.

It was pointed out in [14] that the transformation

$$x' = \textstyle\int \frac{dx}{g(x)} = G(x), \quad t' = t, \quad u' = u \tag{24}$$

connects equation (1), and the PDE

$$g(G^{-1}(x'))f(G^{-1}(x'))u'_{t'} = [u'^n u'_{x'}]_{x'}, \tag{25}$$

where G^{-1} is the inverse function of G.

It is clear that we can equivalently use an equation of the form

$$f(x')u'_{t'} = [u'^n u'_{x'}]_{x'}, \tag{26}$$

and then transform the results for equation (1) by using the corresponding point transformation.

We also observe that the transformation

$$x' = x, \quad t' = t, \quad u' = \tfrac{u}{f} \tag{27}$$

connects equation (26) with $f = k_2 e^{k_1 x}$ and $n = -1$, and the PDE (5).

Consequently, it is clear that we can equivalently use for the subsequent analysis an equation of the form (5), which appear in [10] and then transform the results for equation (1) by using the transformation (24) and (27).

Concluding remarks

In this paper, for $n = -1$ we have derived *nonclassical* symmetries for (1) and for (6). We have proved that the nonclassical method with $\tau \neq 0$ applied to (1) gives only rise to classical symmetries. For $\tau = 0$ we have applied an extension [6] of the Bila and Nielsen procedure. We found the connection between classes of nonclassical symmetries of (1) and classes of nonclassical symmetries of (6). We have proved that (1) when the parameter n takes the value -1, that is when the equation model fast diffusion, admits nonclassical symmetries that yield new solutions. Some of theses solutions are unobtainable by classical symmetries and exhibit and interesting behaviour.

References

[1] Berryman JG, Holland CJ. Asymptotic behavior of the nonlinear diffusion equation $n_t = (n^{-1}n_x)_x$. J Math Phys 1982;23:983-987.

[2] Bluman GW. Potential symmetries and linearization. In: Proceedings of NATO Advanced Research Workshop, 1992, Exeter, Kluwer.

[3] Bluman GW, Cole J. The general similarity solution of the heat equation. J Math Mech 1969;18:1025.

[4] Bluman GW, Kumei S. On the remarkable nonlinear diffusion equation. J Math Phys 1980;21:1019-1023.

[5] Bluman GW, Kumei S. Symmetries and Differential Equations. Berlin: Springer; 1989.

[6] Bruzón MS, Gandarias ML. Nonclassical symmetries: applying a new procedure. Preprint.

[7] Clarkson PA. Nonclassical Symmetry Reductions of the Boussinesq Equation. Chaos Solitons Fractals 1995;5:2261-2301.

[8] Gandarias ML. Nonclassical Potential symmetries of a porous medium equation. Modern group analysis : advanced analytical and computational methods in mathematical physics. Nordfjordeid: NH Ibragimov et all, MARS Publ, Trondheim 1997;99-107.

[9] Gandarias ML. Nonclassical Potential symmetries of the Burgers equation. Symmetry in Nonlinear Mathematical Physics. Kiev: Natl Acad Sci Ukraine 1997, Inst Math Publ Kiev 1997;130-137.

[10] Gandarias ML. Phys Lett A 2001;286:153-160.

[11] Gandarias ML, Bruzón MS. Nonclassical potential symmetries for a generalised inhomogeneus nonlinear diffusion equation. Preprint.

[12] Peletier LA. Aplications of Nonlinear Analysis in the Physical Sciences, Pitman: London, 1981.

[13] Popovych RO, Vaneeva OO, Ivanova NM. Potential nonclassical symmetries and solutions of fast diffusion equations. Preprint.

[14] Sophocleous C. Classification of potential symmetries of generalised inhomogeneous nonlinear diffusion equations. Phys A 2003;320:169-183.

[15] Rosenau P. Fast and Superfast Diffusion Processes. Phys Rev Lett 1995;7:1056-1059.

[16] Vinogradov AM. Symmetries of Partial Differential Equations, Dordrecht: Kluver,1989.

Applying a new algorithm to derive nonclassical symmetries

M.S. Bruzón, M.L. Gandarias

Departamento de Matemáticas, Universidad de Cádiz,
Puerto Real, Cádiz, 11510, Spain

Abstract

A new symmetry for the 2+1 dimensional shallow water is determined. This symmetry is obtained extending the procedure described for Bîlă and Niesen in [2] to a different case.

Keywords: Partial differential equations; Symmetries; Nonclassical method

1 Introduction

The most famous and established method for finding exact solutions of differential equations is the classical symmetries method, also called group analysis, which originated in 1881 from the pioneering work of Lie [8].

The nonclassical symmetries method was introduced in 1969 by Bluman and Cole [1] in order to obtain new exact solutions of the linear heat equation. The description of the method can be found in [1, 3, 7]. In [4] Clarkson and Mansfield proposed an algorithm for calculating the determining equations associated with the nonclassical method: the PDE system is augmented with the invariant surface conditions, the nonclassical symmetries are found by seeking the classical symmetries of the augmented system while demanding that the symmetries operator be related to the invariant surface condition.

Bîlă and Niesen in [2] dropped this requirement. Their procedure consists in reducing PDE system to its involutive form and then applying the classical Lie method to the reduced PDE system, but with an arbitrary symmetry operator which is not related anymore to the invariant surface condition.

In this paper we extend the procedure described in [2] to a different case. We apply the procedure to a Cahn-Hilliard equation, to a Boussinesq equation and to a 2+1 dimensional shallow water wave equation.

2 Nonclassical symmetries

The basic idea of the method of Bîlă and Niesen applied to the general nth order PDE, with p independent variables, $x = (x_1, \ldots, x_p)$, and one dependent variables, $u = u(x)$,

$$\Delta \equiv \Delta(x, u, \mathbf{u}^{(1)}(x), \ldots, \mathbf{u}^{(n)}(x)) = 0, \tag{1}$$

where $\mathbf{u}^{(l)}(x)$ denotes the set of all the partial derivatives of order l of u, is the following:
If V,

$$V = \sum_{i=1}^{p} \xi_i(x, u)\frac{\partial}{\partial x_i} + \eta(x, u)\frac{\partial}{\partial u}, \tag{2}$$

is a vector field and $\xi_p \neq 0$ we can multiply V for $\frac{1}{\xi_p}$ and the invariant surface conditions is,

$$\frac{\partial u}{\partial x_p} = \eta(x,u) - \sum_{i=1}^{p-1} \xi_i(x,u)\frac{\partial u}{\partial x_i} = 0. \tag{3}$$

Substituting (3) and its derivatives with respect to x in (1) we obtain a new PDE

$$\Delta' \equiv \Delta'\left(\mathcal{A}_\nu(x,u), \mathbf{u}^{[1]}, \ldots, \mathbf{u}^{[n]}\right) = 0, \tag{4}$$

for the unknown function $u = u(x_1, \ldots, x_{p-1}; x_p)$ of $x_1, \ldots, x_{p-1}$ (here x_p is considered as a parameter); where $\mathcal{A}_\nu(x,u)$ are the coefficients of $u^{[l]}$, and $u^{[l]}$ denotes the set of all the partial derivatives of u with respect to $x = (x_1, x_2, \cdots, x_{p-1})$ up to order N.
Applying the classical Lie method to (4), if (4) is of maximal rank. Invariance of (4) under a Lie group of point transformation, with infinitesimal generator

$$W = \sum_{i=1}^{p} s_i(x,u)\frac{\partial}{\partial x_i} + r(x,u)\frac{\partial}{\partial u}, \tag{5}$$

leads to the determining equations. Substituting $s_p = 1$, $s_i = \xi_i$, $i = 1, \ldots p-2$, $r = \eta$ and the functions $\mathcal{A}_\nu$ into the determining equations we obtain the determining equations of the nonclassical symmetries of the original PDE (1).
The case for $\xi_p = 0$ have not considered in [2] because the authors claim needs to be handled separately. But if $\xi_p = 0$ and at least two $\xi_i \neq 0$, $i = 1, \ldots, p-1$ or if one generator is different to zero and $p-1$ generator are equal zero, without loss of generality, we can procedure in a similar way to the one described in [2].

In the following we consider different PDEs for which we have applied the procedure with $\xi_1 \neq 0$ and $\xi_i = 0$, $i = 2 \ldots, p$.

- Case in which (1) can be written, by using the invariant surface condition

$$u_x = \eta, \tag{6}$$

in the equivalent form

$$u_t = \mathcal{A}(x,t,u), \tag{7}$$

where $\mathcal{A}$ is a arbitrary function which depend of x, t and u.
Invariance of Eq. (7) under a Lie group of point transformations with infinitesimal generator

$$V = \xi(x,t,u)\frac{\partial}{\partial x} + \tau(x,t,u)\frac{\partial}{\partial t} + \phi(x,t,u)\frac{\partial}{\partial u} \tag{8}$$

leads, for $\xi = 1$, $\tau = 0$ and $\phi = \eta$, to the following determining equation

$$\mathcal{A}\eta_u + \eta_t - \mathcal{A}_u\eta - \mathcal{A}_x = 0. \tag{9}$$

- In the case in which (1) can be written, by using (6), in the equivalent form

$$u_{tt} = \mathcal{A}(x,t,u), \tag{10}$$

where $\mathcal{A}$ is a arbitrary function which depend of x, t and u.
Invariance of this equation under a Lie group of point transformations with infinitesimal generator (8) leads, for $\xi = 1$, $\tau = 0$ and $\phi = \eta$, to the following determining equations

$$\begin{aligned} &\eta_{uu} = 0, \\ &\eta_{tu} = 0, \\ &\mathcal{A}\eta_u + \eta_{tt} - \mathcal{A}_u\eta - \mathcal{A}_x = 0. \end{aligned} \tag{11}$$

• Case in which (1) can be written, by using (6), in the equivalent form

$$\mathcal{A}(x,y,t,u)u_y + u_{ty} + \mathcal{B}(x,y,t,u) = 0, \tag{12}$$

where $\mathcal{A}$ and $\mathcal{B}$ are arbitrary functions which depend of x, y, t and u. We apply the classical Lie method to (12). Invariance of this equation under a Lie group of point transformations with infinitesimal generator (5) with $p = 3$ leads, for $s_1 = 1$, $s_2 = s_3 = 0$ and $r = \eta$, to the following determining equations

$$\begin{array}{l} \eta_{uy} = 0, \\ \eta_{uu} = 0, \\ \mathcal{A}_u\eta + \eta_{tu} + \mathcal{A}_x = 0, \\ \mathcal{A}\eta_y - \mathcal{B}\eta_u + \eta_{ty} + \mathcal{B}_u\eta + \mathcal{B}_x = 0. \end{array} \tag{13}$$

We observe that for any equation which can be expressed in the form (7), (10) or (12) the nonclassical determining equations can be derived by substituting the corresponding functions $\mathcal{A}$ and $\mathcal{B}$ into (9), (11) or (13).

3 Some examples

The Cahn-Hilliard equation:

The Cahn-Hilliard equation

$$u_t + ku_{xxxx} - f(u)u_{xx} - f'(u)u_x^2 = 0 \tag{14}$$

describes diffusion for descomposition of a one-dimensional binary solution. In [5, 6] the authors have analyzed the classical and nonclassical symmetries of (14).

In order to apply the nonclassical method to the equation (14) we consider a one-parameter group of transformation generated by the vector field (2), which in our case is,

$$V = \xi_1(x,t,u)\frac{\partial}{\partial x} + \xi_2(x,t,u)\frac{\partial}{\partial t} + \eta(x,t,u)\frac{\partial}{\partial u}. \tag{15}$$

We can distinguish two cases:
Case (i): $\xi_2 \neq 0$, we set $\xi_2(x,t,u) = 1$. The equivalent form of (14), by reducing the initial system, using the invariant surface condition $u_t = \eta - \xi_1 u_x$, is

$$ku_{xxxx} - f(u)u_{xx} - f'(u)u_x^2 + \mathcal{A}_1(x,t,u)u_x + \mathcal{A}_2(x,t,u) = 0, \tag{16}$$

where $\mathcal{A}_1 = -\xi_1$ and $\mathcal{A}_2 = \eta$.
The classical method applied to (16) gives only rise to the classical symmetries of (14).
Case (ii): $\xi_2 = 0$, we may set $\xi_1 = 1$, without loss of generality. Subsequently, we find the equivalent form of (14) using the invariant surface condition (6) and its derivatives. This yields a new equation in the form (7) where

$$\begin{array}{rl} \mathcal{A}(x,t,u) = & -k(\eta_u\eta_{xx} + 3\eta\eta_{uu}\eta_x + 3\eta_{ux}\eta_x + (\eta_u)^2\eta_x + \eta^3\eta_{uuu} + 3\eta^2\eta_{uux} \\ & +4\eta^2\eta_u\eta_{uu} + 3\eta\eta_{uxx} + 5\eta\eta_u\eta_{ux} + \eta(\eta_u)^3 + \eta_{xxx}) + f_u\eta^2 \\ & +f(\eta_x + \eta_u\eta). \end{array} \tag{17}$$

Substituting (17) into (9) we obtain the determining equation for the infinitesimal η derived in [5, 6].

The Boussinesq equation:

We apply the new procedure to the Boussinesq equation

$$u_{tt} + uu_{xx} + u_x^2 + u_{xxxx} = 0. \tag{18}$$

The Boussinesq equation arises in several physical applications: propagation of long waves in shallow water, one-dimensional nonlinear lattice-waves, vibrations in a nonlinear string and ion sound waves in a plasma. In [3], symmetry reductions and exact solutions of this equation using the classical Lie method of infinitesimals, the direct method due Clarkson and Kruskal and the nonclassical method due to Bluman and Cole were derived.

In order to apply the new procedure to equation (18) we consider a one-parameter group of transformation generated by the vector field (15).

We can distinguish two cases:
Case (i): $\xi_2 \neq 0$, we set $\xi_2(x,t,u) = 1$. Substituting

$$u_{tt} = u_x\,\xi_1\,\xi_{1,x} + 2\,(u_x)^2\,\xi_1\,\xi_{1,u} - \eta\,u_x\,\xi_{1,u} - u_x\,\xi_{1,t} + u_{xx}\,\xi_1^2 - 2\,\eta_u\,u_x\,\xi_1 - \eta_x\,\xi_1 + \eta\,\eta_u + \eta_t,$$

which is obtained from the invariant surface condition, in (18) we obtain that the equivalent form is

$$u_{xxxx} + \mathcal{A}(x,t,u)u_{xx} + \mathcal{B}(x,t,u)u_x^2 + \mathcal{C}(x,t,u)u_x + \mathcal{D}(x,t,u) = 0, \tag{19}$$

where $\mathcal{A} = \xi_1^2 + u$ and $\mathcal{B} = 2\xi_1\xi_{1,u} + 1$, $\mathcal{C} = -2\xi_1\eta_u - \eta\xi_{1,u} + \xi_1\xi_{1,x} - \xi_{1,t}$ and $\mathcal{D} = -\xi_1\eta_x + \eta\eta_u + \eta_t$. We apply the classical Lie method to (19). Invariance of Eq. (19) under a Lie group of point transformations with infinitesimal generator (8). Substituting $\xi = \xi_1$, $\tau = 1$, $\phi = \eta$ and $\mathcal{A}$, $\mathcal{B}$, $\mathcal{C}$ and $\mathcal{D}$ we obtain the determining equations derived in ([3], case 2.3.1).
Case (ii): $\xi_2 = 0$, we may set $\xi_1 = 1$, without loss of generality. The equivalent form of (18), using the invariant surface conditions (6) and its derivatives, is given in (10) where

$$\begin{aligned}\mathcal{A}(x,t,u) = \; & -(\eta_x + \eta\,\eta_u)\,u - \eta_{xxx} - \eta_u\eta_{xx} - 3\,\eta\,\eta_{uu}\,\eta_x - 3\,\eta_{ux}\,\eta_x - (\eta_u)^2\,\eta_x \\ & -\eta^3\,\eta_{uuu} - 3\,\eta^2\,\eta_{uux} - 4\,\eta^2\,\eta_u\,\eta_{uu} - 3\,\eta\,\eta_{uxx} - 5\,\eta\,\eta_u\,\eta_{ux} - \eta\,(\eta_u)^3 \\ & -\eta^2.\end{aligned} \tag{20}$$

Substituting (20) in (11) we obtain the determining equations derived in ([3] Case 2.3.2).

The 2+1-dimensional shallow water wave equation:

We apply the procedure described to the 2+1-dimensional shallow water wave equation

$$u_{yt} + \alpha u_x u_{xy} + \beta u_y u_{xx} + u_{xxxy} = 0, \tag{21}$$

where α and β are arbitrary, nonzero, constants. Classical and nonclassical reductions of this equation are classified in [9].

In order to apply the new procedure to equation (21) to obtain the nonclassical symmetries we consider a one-parameter group of transformations generated by the vector field

$$V = \xi_1(x,y,t,u)\frac{\partial}{\partial x} + \xi_2(x,y,t,u)\frac{\partial}{\partial y} + \xi_3(x,y,t,u)\frac{\partial}{\partial t} + \eta(x,y,t,u)\frac{\partial}{\partial u}. \tag{22}$$

There are three cases to consider: (i) $\xi_3 \equiv 1$, (ii) $\xi_3 = 0$ and $\xi_2 \equiv 1$ and (iii) $\xi_2 = \xi_3 = 0$ $\xi_1 \equiv 1$.
Case (i) the infinitesimals are equivalent to the classical infinitesimals [9].
Case (ii) $\xi_3 = 0$ and $\xi_2 \equiv 1$. Substituting $u_{yt} = -\xi_{1,u}u_t u_x - \xi_{1,t}u_x - \xi_1 u_{tx} + \eta_u u_t + \eta_t$ and $u_{xy} = -\xi_{1,x}u_x - \xi_{1,u}\,(u_x)^2 - \xi_1 u_{xx} + \eta_u\,u_x + \eta_x$, which are obtained from the invariant surface condition $u_y = \eta - \xi_1 u_x$, in (21) we obtain

$$\begin{aligned}\mathcal{A}_1 u_{xxxx} + \mathcal{A}_2 u_{xxx} + \mathcal{A}_3 u_x u_{xxx} + \mathcal{A}_4 u_x u_{xx} + \mathcal{A}_5 u_x + \mathcal{A}_6 (u_{xx})^2 + \mathcal{A}_7 (u_x)^2 u_{xx} \\ + \mathcal{A}_8 (u_x)^2 + \mathcal{A}_9 u_{xx} + \mathcal{A}_{10} (u_x)^4 + \mathcal{A}_{11} (u_x)^3 + \mathcal{A}_{12} u_{xt} + \mathcal{A}_{13} u_t + \mathcal{A}_{14} u_x u_t \\ + \mathcal{A}_{15} = 0,\end{aligned} \tag{23}$$

where

$$\begin{array}{ll}\mathcal{A}_1 = \xi_1, \quad \mathcal{A}_2 = 3\xi_{1,x} - \eta_u, & \mathcal{A}_3 = 4\xi_{1,u}, \\ \mathcal{A}_4 = 9\xi_{1,ux} + (\beta+\alpha)\xi - 3\eta_{uu}, & \mathcal{A}_5 = \alpha\eta_x + 3\eta_{uxx} - \xi_{xxx} - \xi_t, \\ \mathcal{A}_6 = 3\xi_u, \quad \mathcal{A}_7 = 6\xi_{1,uu}, & \mathcal{A}_8 = 3\xi_{1,uxx} + \alpha(\xi_{1,x} - \eta_u) - 3\eta_{uux}, \\ \mathcal{A}_9 = \beta\eta - 3\xi_{1,xx} - 3\eta_{ux} \quad \mathcal{A}_{10} = \xi_{uuu}, & \mathcal{A}_{11} = \eta_{uuu} - \alpha\xi_{1,u} - 3\xi_{1,uux}, \\ \mathcal{A}_{12} = -\xi_1, \quad \mathcal{A}_{13} = \eta_u, & \mathcal{A}_{14} = -\xi_{1,u}, \quad \mathcal{A}_{15} = -\eta_{xxx} - \eta_t.\end{array}$$

We apply the classical Lie method to (23). Invariance of Eq. (23) under a Lie group of point transformations with infinitesimal generator (5), $p=3$, leads to the determining equations of the classical method. Substituting $s_1 = \xi_1$, $s_2 = 1$, $s_3 = 0$ and $\mathcal{A}_i$, $i = 1, \ldots, 15$, we deduce the determining equations of the nonclassical method obtained in [9].

Case (iii) $\xi_2 = \xi_3 = 0$, $\xi_1 \equiv 1$, we consider the equivalent form of (21) given in (12), where

$$\begin{aligned}\mathcal{A} = \ & \Big((\eta_{uu} + \beta)\, \eta_x + \eta^2\, \eta_{uuu} + 2\,\eta\, \eta_{uux} + 4\,\eta\, \eta_u \eta_{uu} + \eta_{uxx} + 3\, \eta_u\, \eta_{ux} + (\eta_u)^3 \\ & + (\beta+\alpha)\, \eta\, \eta_u \Big).\end{aligned}$$

$$\begin{aligned}\mathcal{B} = \ & \left(2\,\eta\eta_{uu} + 2\,\eta_{ux} + (\eta_u)^2 + \alpha\,\eta\right) \eta_y + \eta_{xxy} + \eta_u\, \eta_{xy} + \eta_{uy}\eta_x + \eta^2\, \eta_{uuy} \\ & + 2\,\eta\, \eta_u \eta_{uy} + 2\,\eta\, \eta_{uxy}.\end{aligned}$$

Substituting $\mathcal{A}$ and $\mathcal{B}$ in (13) we obtain the determining equations for the infinitesimal η,

$$\begin{aligned}& \eta_{uy} = 0, \\ & \eta_{uu} = 0, \\ & \beta\, \eta_{xx} + (\beta+\alpha)\, \eta_u\, \eta_x + \eta_{uxxx} + 3\, \eta_u\, \eta_{uxx} + 3\, (\eta_{ux})^2 + \eta \left(\beta\, \eta_{ux} + (\beta+\alpha)\, (\eta_u)^2 \right) \\ & \quad + 3\, (\eta_u)^2\, \eta_{ux} + (\beta+\alpha)\, \eta\, \eta_{ux} + \eta_{tu} = 0, \\ & -\eta_u \left(\left(2\, \eta_{ux} + (\eta_u)^2 + \alpha\, \eta \right) \eta_y + \eta_{xxy} + \eta_u \eta_{xy} \right) + \Big(\beta\, \eta_x + \eta_{uxx} + 3\, \eta_u\, \eta_{ux} \\ & \quad + (\eta_u)^3 + (\beta+\alpha)\, \eta\, \eta_u \Big)\, \eta_y + (\alpha\, \eta_x + 2\, \eta_{uxx} + 2\, \eta_u\, \eta_{ux})\, \eta_y + \alpha\, \eta\, \eta_u\, \eta_y \\ & \quad + \eta_{xxxy} + \eta_u\, \eta_{xxy} + \left(2\, \eta_{ux} + (\eta_u)^2 + \alpha\, \eta \right) \eta_{xy} + \eta_{ux}\, \eta_{xy} + \eta_{ty} = 0.\end{aligned} \tag{24}$$

The complexity of this system is the reason why we cannot solve (24) in general. Thus we proceed, by making ansatz on the form of $\eta(x,t,u)$, to solve (24).

If $\alpha + \beta = 0$ one solution of the determining equations is

$$\xi_1 = 1, \qquad \xi_2 = \xi_3 = 0, \qquad \eta = k_1 u + k_2 x + f(y) + g(t).$$

It is easy to check that these generators do not satisfy Lie classical determining equations and that it is a new nonclassical symmetry. Therefore we obtain the nonclassical symmetry reduction

$$z = y, \qquad r = t, \qquad u = e^{k_1 x} h(y,t) - \frac{k_2}{k_1} x - \frac{1}{k_1}(g(t) + f(y)) - \frac{k_2}{k_1^2},$$

where $h(y,t)$ satisfies the PDE

$$h_{yt} + (k_1^3 + \beta k_2) h_y - \beta k_1 f_y h = 0.$$

4 Conclusions

Finding the nonclassical symmetries of PDEs involves a large amount of tedious calculations which can become virtually unmanageable if attempted manually. Some authors have designed computer programs for their obtaining. Bîlă and Niesen [2] developed a procedure to obtain the nonclassical symmetries of a PDEs system. The authors claim that the case $\xi_p = 0$ needs to be handled separately.

In this work we have extended the algorithm described for Bîlă and Niesen to determine the nonclassical symmetries of a PDE for this last case. We observe that for any equation which can be expressed in the form (7), (10) or (12) the nonclassical determining equation can be derived by substituting the corresponding functions $\mathcal{A}$ and $\mathcal{B}$ into (9), (11) or (13).

We also apply the described algorithm to a Cahn-Hilliard equation, to a Boussinesq equation and to a 2+1 dimensional shallow water wave equation. For the 2+1 dimensional shallow water wave equation the method yields a new symmetry reduction which is unobtainable by using Lie classical method.

Acknowledgements

The support of DGICYT project BFM2003-04174, Junta de Andalucía group FQM201 and University of Cádiz are gratefully acknowledged.

References

[1] Bluman GW, Cole JD. The general similarity solutions of the heat equation. J Math Mech 1969;18:1025-1042.

[2] Bîlă N, Niesen J. On a new procedure for finding nonclassical symmetries. J Symb Comp 2004;38:1523-1533.

[3] Clarkson PA. Nonclassical Symmetry Reductions of the Boussinesq Equation. Chaos, Solitons, Fractals 1995;5:2261-2301.

[4] Clarkson PA, Mansfield EL. Algorithms for the nonclassical method of symmetry reductions. SIAM J Appl Math 1994;55:1693-1719.

[5] Gandarias ML, Bruzón MS. Nonclassical symmetries for a family of Cahn-Hilliard equations. Physics Letters A 1999;263:331-337.

[6] Gandarias ML, Bruzón MS. Symmetry Analysis and Solutions for a Family of Cahn-Hilliard Equations. Reports on Mathematical Physics 2000;46:89-97.

[7] Levi D, Winternitz P. Nonclassical symmetry reduction: example of the Boussinesq equation. J Phys A 1989;22:2915-2924.

[8] Lie S. Über die Integration durch bestimmte Integrale von einer Classe linearer partieller Differentialgleichungen, Arch. Math. 1881;6:328-368.

[9] Mansfield EL, Clarkson PA. Symmetries and exact solutions for a 2+1-dimensional shallow water wave equation. Mathematics and Computers in Simulation 1997;43:39-55.

Turbulence and surface gravity waves on the Sun

N. Mole*, R. Erdélyi and A. Kerekes

Space and Atmosphere Research Group, Department of Applied Mathematics, University of Sheffield, The Hicks Building, Hounsfield Road, Sheffield S3 7RH, UK

Abstract

The interior of the Sun cannot be observed directly. Deductions about the structure of the interior are made by observing wave motions at the surface of the Sun ("helioseismology"). Thus, it is desirable to have as full an understanding as possible of these waves, and of how they are influenced by different physical factors. Here we consider only the solar f-mode: an incompressible, buoyancy-driven oscillation confined to the near-surface region. In the classical theory f-modes are described by the standard water wave theory for incompressible, infinitely deep water. Observations of the f-mode show small, but significant, deviations from the classical dispersion relation. In particular, at large wavenumber the observed frequency is smaller by a few percent. One of the physical effects which is neglected in the classical theory is that of turbulent flows in the Sun's photosphere. We examine the impact of such random flows, revisiting the approach of Murawski and Roberts [12]. We assume that the random flow is quasi-steady, and that the turbulent velocity scale is small compared with the phase velocity of f-modes with wavelength comparable to the turbulence length scale — both reasonable assumptions for the parameter regime of interest. For simplicity we also take the flow to be irrotational, which means we cannot capture the scattering of the wave by the turbulence. We then find that, to leading order, the effect of the random flow is to reduce the frequency, by a little more than that observed, and to damp the wave. At large wavenumber we obtain damping identical to that obtained by Sazontov and Shagalov [14]. In contrast to ours, their result was derived purely from the scattering of the wave.

Key words: Surface gravity waves; Turbulence; Dispersion relation; Damping

1 Introduction

The solar f-mode is an incompressible, buoyancy-driven oscillation confined to the near-surface region. In the classical theory f-modes are described by the standard water wave theory for incompressible, infinitely deep water. This gives the dispersion relation

$$\omega^2 = gk, \tag{1}$$

*Corresponding author. *E-mail address:* n.mole@sheffield.ac.uk where ω is the mode frequency, g ($\approx 274\text{ms}^{-2}$) is the gravitational acceleration at the surface of the Sun, and k is the horizontal wavenumber. However, observations of the f-mode show small but significant deviations from (1). At low spherical harmonic degree l the frequency is greater than that given by (1), while at high degree the frequency is a few per cent smaller (Antia and Basu [2]).

Since the f-mode is confined to the region near the surface, the deviation from (1) is expected to be explained by near-surface physical phenomena not included in the classical model. Campbell and Roberts [5], Evans and Roberts [8] and Bogdan *et al.* [4] carried out pioneering studies on atmospheric magnetic field effects, but found that the frequencies were increased for all degrees. Erdélyi *et al.* [7] obtained a similar result for random magnetic field. Murawski and Roberts [12, 13] considered the effects of random photospheric flows, and showed that they could produce a reduction in frequency.

Here we revisit the approach of Murawski and Roberts [12] – henceforth referred to as MR – but correct some errors which appeared in that paper. Our results agree qualitatively with those of MR, but differ in detail. We also examine the damping effect of the turbulent flows, obtaining results in qualitative agreement with the observations (Duvall *et al.* [6], Mędrek *et al.* [11], Antia and Basu [2]).

2 The model

Our main interest is in the behaviour of high-degree f-modes (i.e. $l \gtrsim 1000$), so we ignore spherical effects and consider a simple Cartesian model where $z = 0$ corresponds to the solar surface and the z-axis is directed downwards. The solar interior ($z > 0$) is taken to be an infinitely deep, incompressible fluid of constant density ρ, and the atmosphere ($z < 0$) is taken to have constant pressure p_A.

We assume that in the basic, undisturbed state the flow in the solar interior is steady, horizontal, independent of depth, and random, i.e. we take the velocity $\mathbf{u}$ to be

$$\mathbf{u} = \tilde{\mathbf{U}} \equiv (\tilde{U}_1(x,y), \tilde{U}_2(x,y), 0). \tag{2}$$

Thus we assume that the flow is steady in each realisation, but fluctuates randomly between realisations. If we take the turbulent flow to be associated with granules, then they have a velocity scale $u \sim 2\text{km s}^{-1}$ and a length scale $l_u \sim 1000\text{km}$. Thus $l_u/u \gg 1/\omega$ provided $l \gg 10$, so turbulent fluctuations are slow compared with the f-mode timescales for high-degree modes. Assuming steady flow should therefore give a good approximation.

Furthermore, we assume that the flow is irrotational, both in the basic state and in the perturbed state:

$$\nabla \times \mathbf{u} = \mathbf{0}.$$

Fabrikant and Raevsky [9], assuming quasi-steady flow and small Froude number

$$Fr = \frac{ku^2}{g},$$

and using the Born approximation with only single scattering, showed that the vertical component of vorticity determines the scattering of surface waves by the turbulent flow. With our assumption we will not capture this scattering. We are, however, able to estimate the direct effect of the turbulence on the wave frequency and wave damping, and assuming irrotationality greatly simplifies the calculations.

The governing equations in the interior $z > 0$ are then

$$\nabla \cdot \mathbf{u} = 0$$

$$\frac{\partial \mathbf{u}}{\partial t} + \mathbf{u} \cdot \nabla \mathbf{u} = -\frac{1}{\rho}\nabla p + g\hat{\mathbf{z}} + \mathbf{F},$$

and expressing the velocity in terms of the velocity potential ϕ (i.e. $\mathbf{u} = \nabla\phi$) we have

$$\nabla^2\phi = 0. \tag{3}$$

Here p is the pressure and $\mathbf{F}$ is the body force. $\mathbf{F}$ is assumed to be random and such that it balances the horizontal velocity gradients to produce velocity correlations of the expected form. At the surface we require continuity of pressure and satisfaction of the kinematic boundary condition.

In the basic state (2) we have

$$\tilde{\mathbf{U}} \cdot \nabla\tilde{\mathbf{U}} = -\frac{1}{\rho}\nabla p_0 + g\hat{\mathbf{z}} + \mathbf{F}, \tag{4}$$

where p_0 is the basic state pressure. MR did not include a body force, and took $p_{\mathrm{A}} \equiv 0$. This would imply that $\tilde{\mathbf{U}}$ must be constant, contrary to the assumption of MR. If $\tilde{\mathbf{U}}$ is constant then spatial velocity correlations are independent of separation distance, whereas in a real fluid we expect the correlations to decay with increasing distance. The inclusion of $\mathbf{F}$ avoids this problem. In effect $\mathbf{F}$ represents the processes, not accounted for in this model, which generate and maintain the turbulence.

3 The perturbation equations

We now consider perturbations to this basic flow. We assume that these perturbations are also irrotational, and that $\mathbf{F}$ is unaffected by the perturbations. The boundary conditions at the perturbed surface $z = \eta(x, y, t)$ are

$$p = p_{\mathrm{A}}$$

and

$$\frac{\mathrm{D}}{\mathrm{D}t}(z - \eta) = 0.$$

We let

$$\phi = \phi_0 + \phi',$$

where ϕ_0 and ϕ' are the basic state and perturbation velocity potentials, respectively. Since f-modes are confined to the near-surface region we take

$$\phi' \to 0 \quad \text{as} \quad z \to \infty.$$

Velocities assocated with the f-modes are small compared with the turbulent velocities, so we can take the perturbations to be small, and linearise the equations in the perturbation quantities. If we non-dimensionalise by

$$\mathbf{X} = \frac{\mathbf{x}}{l_u}, \quad T = \sqrt{\frac{g}{l_u}}\,t, \quad \mathbf{U} = \frac{\tilde{\mathbf{U}}}{u},$$

then we obtain (using (4) and Bernoulli's integral in the standard way)

$$M\phi' = \alpha R\phi' + \alpha^2(N\phi' + Q\phi') \quad \text{at} \quad z = 0, \tag{5}$$

where the differential operators M, N, Q and R are given by

$$\begin{aligned}
M &= \partial_{TT} - \partial_Z \\
R &= -2(U_1\partial_{XT} + U_2\partial_{YT}) \\
N &= -(U_1^2\partial_{XX} + 2U_1U_2\partial_{XY} + U_2^2\partial_{YY}) \\
Q &= -\frac{1}{2}\left\{(U^2)_X\partial_X + (U^2)_Y\partial_Y\right\},
\end{aligned}$$

and

$$\begin{aligned}
U^2 &= U_1^2 + U_2^2, \\
\alpha^2 &= \frac{u^2}{gl_u} = \frac{Fr}{kl_u}.
\end{aligned}$$

For granules α^2 is $O(10^{-2})$. In terms of the non-dimensional wavenumber $K = kl_u$ ($\approx (l_u/a)l$ for large spherical harmonic degree l, where a is the solar radius) the condition for the validity of the steady basic flow assumption is $K \gg \alpha^2$, which is thus satisfied provided the waves are not too long, as discussed in the previous section. Alternatively, the condition can be expressed as $K^2 \gg Fr$.

MR missed the N and Q terms. Although it appears from (5) that these terms will only have an impact at higher order in α than will R, we shall see that this is not actually the case.

We now split the perturbation potential ϕ' into mean and fluctuating parts, Φ and ϕ'' respectively:

$$\phi' = \Phi + \phi'', \qquad \Phi = \langle\phi'\rangle, \quad \langle\phi''\rangle = 0,$$

where $\langle\rangle$ denotes the ensemble mean. We assume that the turbulent flow has zero mean velocity:

$$\langle\mathbf{U}\rangle = \mathbf{0} \quad \Rightarrow \langle R\rangle = 0,$$

and is homogeneous so

$$\langle Q \rangle = 0.$$

Taking the ensemble mean of (5) gives

$$M\Phi = \alpha^2 \langle N \rangle \Phi + \alpha \langle R\phi'' \rangle + \alpha^2 \langle N\phi'' \rangle + \alpha^2 \langle Q\phi'' \rangle , \tag{6}$$

and subtracting (6) from (5) gives

$$\begin{aligned} M\phi'' &= \alpha(R\phi'' - \langle R\phi'' \rangle) + \alpha^2(N\phi'' - \langle N\phi'' \rangle) \\ &\quad + \alpha^2(Q\phi'' - \langle Q\phi'' \rangle) \\ &\quad + \alpha R\Phi + \alpha^2(N\Phi - \langle N \rangle \Phi + Q\Phi). \end{aligned} \tag{7}$$

We now assume that $\alpha \ll 1$. (7) shows that

$$\frac{|\phi''|}{|\Phi|} = \mathrm{O}(\alpha),$$

so that, retaining terms required to achieve $\mathrm{O}(\alpha^2)$ accuracy in (6), we have

$$\begin{aligned} M\Phi - \alpha^2 \langle N \rangle \Phi &= \alpha \langle R\phi'' \rangle \\ M\phi'' &= \alpha R\Phi. \end{aligned} \tag{8}$$

MR did not explicitly assume that α was small, but instead assumed that $R\phi'' - \langle R\phi'' \rangle$ could be ignored. Since they did not have any terms nonlinear in the basic flow velocity they obtained (8) without the $\alpha^2 \langle N \rangle \Phi$ term. Since $\alpha \langle R\phi'' \rangle$ also contributes at $\mathrm{O}(\alpha^2)$, their results are incorrect for small α. For small α, assuming that terms of the form $R\phi'' - \langle R\phi'' \rangle$ are negligible is not necessary to make progress at $\mathrm{O}(\alpha^2)$. At $\mathrm{O}(\alpha^3)$, however, the classical turbulence closure problem manifests itself, and some such approximation is needed.

4 The dispersion relation

We solve (8) essentially as in MR, by Fourier transforming the equations with respect to X and Y, and Laplace transforming with respect to T. We let

$$\Phi(\mathbf{X}, T) = \frac{1}{2\pi} \int_{i\gamma-\infty}^{i\gamma+\infty} d\Omega \iint_{-\infty}^{\infty} dK_1 dK_2 \, \phi_{\mathrm{M}}(\mathbf{K}, \Omega) \mathrm{e}^{\mathrm{i}(\mathbf{K}\cdot\mathbf{X} - \Omega T) - KZ} = \mathcal{F}^{-1}\mathcal{L}^{-1} \left\{ \mathrm{e}^{-KZ} \phi_{\mathrm{M}}(\mathbf{K}, \Omega) \right\},$$

where $\mathbf{K} = (K_1, K_2, 0)$, $K = |\mathbf{K}|$, $\mathcal{F}$ denotes the Fourier transform, $\mathcal{L}$ the Laplace transform, and the Z dependence is determined by (3). Similarly, we let

$$\phi''(\mathbf{X}, T) = \mathcal{F}^{-1}\mathcal{L}^{-1} \left\{ \mathrm{e}^{-KZ} \phi_{\mathrm{F}}(\mathbf{K}, \Omega) \right\}.$$

Then Fourier transforming (8) gives

$$\begin{aligned}(\Omega^2 - K)\phi_{\rm M}(\mathbf{K},\Omega) + \alpha^2\phi_{\rm M}(\mathbf{K},\Omega)\left\{\left\langle U_1^2\right\rangle K_1^2 + 2\left\langle U_1 U_2\right\rangle K_1 K_2 + \left\langle U_2^2\right\rangle K_2^2\right\} \\ = 2\alpha\Omega \iint\limits_{-\infty}^{\infty} dk_1 dk_2 \left\langle \phi_{\rm F}(\mathbf{k},\Omega)\left[k_1\hat{U}_1(\mathbf{K}-\mathbf{k}) + k_2\hat{U}_2(\mathbf{K}-\mathbf{k})\right]\right\rangle + I_M(\mathbf{K},\Omega),\end{aligned} \tag{9}$$

and

$$(\Omega^2 - K)\phi_{\rm F}(\mathbf{K},\Omega) = 2\alpha\Omega \iint\limits_{-\infty}^{\infty} dk_1 dk_2\, \phi_{\rm M}(\mathbf{k},\Omega)\left[k_1\hat{U}_1(\mathbf{K}-\mathbf{k}) + k_2\hat{U}_2(\mathbf{K}-\mathbf{k})\right] + I_F(\mathbf{K},\Omega), \tag{10}$$

where

$$I_M(\mathbf{K},\Omega) = -2i\alpha \iint\limits_{-\infty}^{\infty} dk_1 dk_2 \left\langle \hat{\phi}''(\mathbf{k},0)\left[k_1\hat{U}_1(\mathbf{K}-\mathbf{k}) + k_2\hat{U}_2(\mathbf{K}-\mathbf{k})\right]\right\rangle + i\Omega\hat{\Phi}(\mathbf{K},0) - \hat{\Phi}_T(\mathbf{K},0)$$

and

$$I_F(\mathbf{K},\Omega) = -2i\alpha \iint\limits_{-\infty}^{\infty} dk_1 dk_2 \hat{\Phi}(\mathbf{k},0)\left[k_1\hat{U}_1(\mathbf{K}-\mathbf{k}) + k_2\hat{U}_2(\mathbf{K}-\mathbf{k})\right] + i\Omega\hat{\phi}''(\mathbf{K},0) - \hat{\phi}''_T(\mathbf{K},0).$$

Here ^ denotes the Fourier transform, and $\mathbf{k} = (k_1, k_2, 0)$. The random variables in (9) and (10) are $\phi_{\rm F}$, U_1 and U_2.

I_M and I_F are determined by the initial conditions. They are also random, but do not need to be known to derive the dispersion relation. We have, *a priori*, no knowledge of $\phi_{\rm F}$, so we eliminate it by substituting (10) in (9), giving

$$\begin{aligned}D(\mathbf{K},\Omega)\phi_M(\mathbf{K},\Omega) &= I_M(\mathbf{K},\Omega) \\ &+ 2\alpha\Omega \iint\limits_{-\infty}^{\infty} \frac{dk_1 dk_2}{\Omega^2 - k}\left\langle I_F(\mathbf{k},\Omega)\left[k_1\hat{U}_1(\mathbf{K}-\mathbf{k}) + k_2\hat{U}_2(\mathbf{K}-\mathbf{k})\right]\right\rangle\end{aligned} \tag{11}$$

where

$$\begin{aligned}D(\mathbf{K},\Omega) \equiv\; &\Omega^2 - K + \alpha^2\left\{\left\langle U_1^2\right\rangle K_1^2 + 2\left\langle U_1 U_2\right\rangle K_1 K_2 + \left\langle U_2^2\right\rangle K_2^2\right\} \\ &- \frac{4\alpha^2\Omega^2}{\phi_M(\mathbf{K},\Omega)} \iiiint\limits_{-\infty}^{\infty} dk_1 dk_2 dk_1' dk_2' \frac{\phi_{\rm M}(\mathbf{k}',\Omega)}{\Omega^2 - k} \\ &\times \left\langle\left[k_1\hat{U}_1(\mathbf{K}-\mathbf{k}) + k_2\hat{U}_2(\mathbf{K}-\mathbf{k})\right]\left[k_1'\hat{U}_1(\mathbf{k}-\mathbf{k}') + k_2'\hat{U}_2(\mathbf{k}-\mathbf{k}')\right]\right\rangle,\end{aligned} \tag{12}$$

is the dispersion function, and $k = |\mathbf{k}|$, $\mathbf{k}' = (k_1', k_2', 0)$.

We now define the velocity correlation $S_{ij}(\mathbf{R})$ for homogeneous turbulence:

$$S_{ij}(\mathbf{R}) = \langle U_i(\mathbf{X})U_j(\mathbf{X}+\mathbf{R})\rangle .$$

We then find that

$$\left\langle \hat{U}_i(\mathbf{K}-\mathbf{k})\hat{U}_j(\mathbf{k}-\mathbf{k}')\right\rangle = \hat{S}_{ij}(\mathbf{K}-\mathbf{k})\delta(\mathbf{K}-\mathbf{k}'),$$

where δ is the Dirac delta-function. (12) then gives

$$\begin{aligned} D(\mathbf{K},\Omega) &= \Omega^2 - K + \alpha^2\left\{K_1^2 S_{11}(\mathbf{0}) + 2K_1K_2S_{12}(\mathbf{0}) + K_2^2 S_{22}(\mathbf{0})\right\} \\ &\quad - 4\alpha^2\Omega^2 \iint\limits_{-\infty}^{\infty} \frac{dk_1dk_2}{\Omega^2 - k} \\ &\quad \times \left\{K_1\left[k_1\hat{S}_{11}(\mathbf{K}-\mathbf{k}) + k_2\hat{S}_{21}(\mathbf{K}-\mathbf{k})\right] + K_2\left[k_1\hat{S}_{12}(\mathbf{K}-\mathbf{k}) + k_2\hat{S}_{22}(\mathbf{K}-\mathbf{k})\right]\right\}. \end{aligned} \tag{13}$$

The first term proportional to α^2 on the right-hand side was missed by MR because they did not include the terms which are quadratic in the turbulent velocity.

If we assume that the turbulence is isotropic in the horizontal, and that the longitudinal velocity correlations decay with separation distance as a Gaussian function, then we can take (see e.g. sections 3.3, 3.4 of Batchelor [3])

$$S_{ij}(\mathbf{R}) = \mathrm{e}^{-R^2/4}\left\{\frac{1}{2}R_iR_j + \left(1 - \frac{1}{2}R^2\right)\delta_{ij}\right\},$$

where $\mathbf{R} = (R_1, R_2, 0)$, $R = |\mathbf{R}|$. (13) then becomes

$$D(\mathbf{K},\Omega) = \Omega^2 - K + \alpha^2K^2 + \frac{8\alpha^2\Omega^2}{\pi}\iint\limits_{-\infty}^{\infty}\frac{dk_1dk_2}{k-\Omega^2}\mathrm{e}^{-(\mathbf{K}-\mathbf{k})^2}(k_1K_2 - k_2K_1)^2.$$

If we now take the x-direction to be in the wave propagation direction, so $K_1 = K$, $K_2 = 0$, and we change to polar coordinates in the integral by $k_1 = k\cos\theta$, $k_2 = k\sin\theta$, then we have

$$D(\mathbf{K},\Omega) = \Omega^2 - K + \alpha^2K^2 + \frac{8\alpha^2\Omega^2}{\pi}K^2\mathrm{e}^{-K^2}\int_0^{2\pi} d\theta\,\sin^2\theta \int_0^{\infty} dk\,\frac{k^3}{k-\Omega^2}\,e^{-k^2+2Kk\cos\theta}. \tag{14}$$

Setting $D(\mathbf{K},\Omega) = 0$ in (14) would give the corrected version, to $O(\alpha^2)$, of MR (4.7).

(14) can be rewritten (see e.g. 9.6.18 of Abramowitz and Stegun [1] - henceforth referred to as AS) as

$$D(\mathbf{K},\Omega) = \Omega^2 - K + \alpha^2K^2 + 8\alpha^2\Omega^2KE^{-K^2}\int_0^{\infty}\frac{k^2e^{-k^2}I_1(2Kk)}{k-\Omega^2}\,dk, \tag{15}$$

where I_1 is the modified Bessel function.

If we wished to solve the initial value problem we would rearrange (11) and take the inverse Laplace and Fourier transforms to obtain $\Phi(\mathbf{X},T)$. In the Ω-plane this involves integrating along the Bromwich contour, above all singularities of ϕ_M. Here we merely wish to derive the dispersion relation, which is obtained from the zeroes of $D(\mathbf{K},\Omega)$, which also give singularities of ϕ_M. We can deform the Bromwich contour until it lies just above these singularities, which we expect to occur for small $|\mathrm{Im}\,\Omega(\mathbf{K})|$. We can then find the leading order contributions to the integral in (15):

$$\int_0^\infty \frac{k^2 e^{-k^2} I_1(2Kk)}{k-\Omega^2}\,dk \sim \fint_0^\infty \frac{k^2 e^{-k^2} I_1(2Kk)}{k-K}\,dk + i\pi K^2 e^{-K^2} I_1(2K^2).$$

Here $\fint$ denotes the Cauchy principal part of the integral.

Substituting in (15), setting $D(\mathbf{K},\Omega)=0$, and rearranging gives, to $O(\alpha^2)$,

$$\begin{aligned}\mathrm{Re}\ \Omega &= K^{1/2} - \alpha^2\left\{\frac{1}{2}K^{3/2} + 4K^{3/2}e^{-K^2}\fint_0^\infty \frac{k^2 e^{-k^2} I_1(2Kk)}{k-K}\,dk\right\} \\ \mathrm{Im}\ \Omega &= -4\pi\alpha^2 K^{7/2} e^{-2K^2} I_1(2K^2).\end{aligned} \tag{16}$$

5 Asymptotic results

5.1 Small wavenumber

For $z \ll 1$

$$I_1(z) \sim \frac{1}{2}z\left\{1 + O(z^2)\right\}$$

(see e.g. AS 9.6.10), and (16) gives

$$\mathrm{Re}\ \Omega \sim K^{1/2} - \frac{1}{2}\alpha^2 K^{3/2}\left\{1 + 2\sqrt{\pi}K + 4K^2 + O(K^3)\right\}$$

$$\mathrm{Im}\ \Omega \sim -4\pi\alpha^2 K^{11/2}\left\{1 - 2K^2 + O(K^4)\right\}.$$

MR calculated $\mathrm{Re}\,\Omega$ in this limit, but missed the $K^{3/2}$ term.

5.2 Large wavenumber

For $z \gg 1$

$$I_1(z) \sim \frac{e^z}{\sqrt{2\pi z}}\left\{1 + O(z^{-1})\right\}$$

(see e.g. AS 9.7.1), and (16) then gives

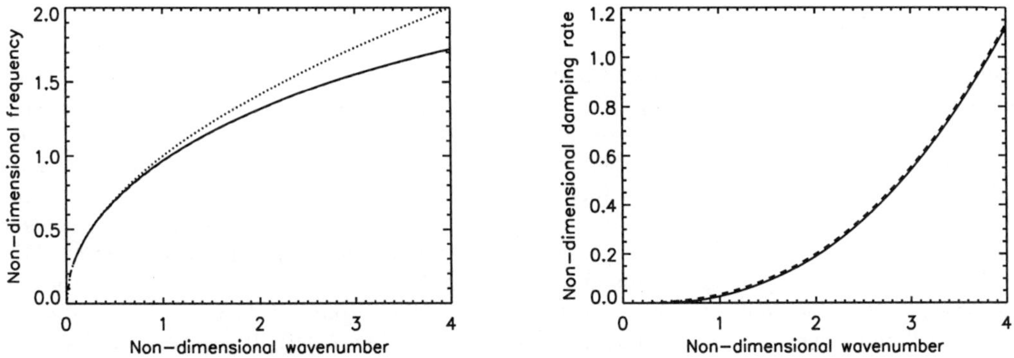

Figure 1. The non-dimensional frequency $\mathrm{Re}\,\Omega$ and damping rate $\mathrm{Im}\,\Omega$ for $\alpha^2 = 0.01$, as functions of the non-dimensional wavenumber K. Full results (solid curves), large wavenumber asymptotic results (dashed curves), and classical result $\Omega^2 = K$ (dotted curve).

$$\begin{aligned} \mathrm{Re}\ \Omega &\sim K^{1/2} - \frac{7}{2}\alpha^2 K^{3/2}\left\{1 + O(K^{-2})\right\} \\ \mathrm{Im}\ \Omega &\sim -2\sqrt{\pi}\alpha^2 K^{5/2}\left\{1 + O(K^{-2})\right\}. \end{aligned} \tag{17}$$

For large wavenumber we can use a heuristic argument to derive the parameter dependence of the leading order frequency shift. Locally we can treat the effect on the real part of the frequency as a Doppler shift by the turbulent velocity. If we take the mean change in the frequency to be weighted by the travel time, then, for turbulent velocities small compared with the phase velocity, we obtain the dependence given in (17).

6 Numerical results

Figure 1 shows the results obtained for the non-dimensional frequency $\mathrm{Re}\,\Omega$ and the non-dimensional damping rate $\mathrm{Im}\,\Omega$, as functions of the non-dimensional wavenumber K. The value of α^2 was taken to be 0.01. The solid lines show the results obtained directly from (16), the dashed lines are obtained from the large wavenumber approximation (17), and the dotted line represents the classical result (1). The large wavenumber results are almost indistinguishable from the full results. At non-dimensional wavenumbers of about 3 there is a 5-6% reduction in the frequency below the classical result. This compares with observed reductions at the solar surface of up to about 3% (Libbrecht *et al.* [10], Antia and Basu [2]). The damping rates shown in Figure 1 are also somewhat larger than those observed (Antia and Basu [2]).

7 Discussion

We find, in agreement with Murawski and Roberts [12], that the direct effect of turbulence causes a significant decrease in the frequency of surface gravity waves below that of the classical result. The magnitude of this decrease is greater than that observed for solar f-modes. In the solar case there are, of course, other physical effects which are likely to be important, and which are ignored in our model. These include vertical sub-surface motions associated with solar convection, atmospheric magnetic field, and a rapid but continuous decrease in density across the surface to a non-zero atmospheric value. Murawski and Roberts [13] and Mędrek *et al.* [11] included vertical velocity and still found a significant decrease in frequency. A more realistic density profile would tend to reduce the frequency, while magnetic field tends to increase it (Campbell and Roberts [5], Evans and Roberts [8], Bogdan *et al.* [4]). It seems likely that the magnetic field plays a role in limiting the decrease produced by the other effects.

We also find that the turbulence has a damping effect on the surface gravity waves. Mędrek *et al.* [11] obtained damping rates from numerical calculations, but did not present any analytical results. Their results were similar to ours. Sazontov and Shagalov [14] analysed the damping of surface gravity waves by turbulent scattering, ignoring the effect of the turbulence on the real part of the frequency. In the large wavenumber approximation their leading order result reduces exactly to that given in (17).

Acknowledgements We wish to thank Misha Ruderman and Victor Shrira for valuable discussions.

References

[1] Abramowitz M, Stegun IA. (eds.) Handbook of Mathematical Functions. New York: Dover; 1965.

[2] Antia HM, Basu S. High-frequency and high-wavenumber solar oscillations. The Astrophysical Journal 1999;519:400–406.

[3] Batchelor GK. The Theory of Homogeneous Turbulence. Cambridge: Cambridge University Press; 1953.

[4] Bogdan TJ, Brown TM, Lites BW, Thomas JH. The absorption of p-modes by sunspots: variations with degree and order. The Astrophysical Journal 1993;406:723–734.

[5] Campbell WR, Roberts B. The influence of a chromospheric magnetic field on the solar p- and f-modes. The Astrophysical Journal 1989;338:538–556.

[6] Duvall Jr TL, Kosovichev AG, Murawski K. Random damping and frequency reduction of the solar f-mode. The Astrophysical Journal 1998;505:L55-L58.

[7] Erdélyi R, Kerekes A, Mole N. Influence of random magnetic field on solar global oscillations: the incompressible f-mode. Astronomy and Astrophysics 2005;431:1083–1088.

[8] Evans DJ, Roberts B. The influence of a chromospheric magnetic field on the solar p- and f-modes. II. Uniform chromospheric field. The Astrophysical Journal 1990;356:704–719.

[9] Fabrikant AL, Raevsky MA. The influence of drift flow turbulence on surface gravity wave propagation. J. Fluid Mech. 1994;262:141-156.

[10] Libbrecht KG, Woodard MF, Kaufman JM. Frequencies of solar oscillations. The Astrophysical Journal Supplement Series 1990;74:1129–1149.

[11] Mędrek M, Murawski K, Roberts B. Damping and frequency reduction of the f-mode due to turbulent motion in the solar convection zone. Astronomy and Astrophysics 1999;349:312–316.

[12] Murawski K, Roberts B. Random velocity field corrections to the f-mode. I. Horizontal flows. Astronomy and Astrophysics 1993;272:595–600.

[13] Murawski K, Roberts B. Random velocity field corrections to the f-mode. II. Vertical and horizontal flow. Astronomy and Astrophysics 1993;272:601–608.

[14] Sazontov AG, Shagalov SV. Scattering of gravity waves by a turbulence of the upper ocean layer. Izvestiya Atmospheric and Oceanic Physics 1986;22:138–143.

The linear stability of interfacial solitary waves in a two-layer fluid

Takeshi Kataoka
Department of Mechanical Engineering, Faculty of Engineering, Kobe University, Rokkodai, Nada, Kobe 657-8501, Japan

Abstract
Linear stability or occurrence of an exchange of stability is studied for interfacial solitary waves propagating in a two-layer fluid. We find that an exchange of stability occurs at every stationary value of the total energy of the solitary waves. Then, we construct a general criterion for the stability of interfacial solitary waves with respect to disturbances that are stationary relative to the basic wave. This criterion is applied to specific interfacial solitary waves and various features on their stability are presented.

PACS: 47.20.Ky; 05.45.Yv; 47.55.Hd
Keywords: Interfacial solitary waves; Two-layer fluid; Exchange of stability

1 Introduction

The stability of finite-amplitude solitary waves was mainly investigated for the surface solitary waves. Tanaka (1986) investigated the linear stability of surface solitary waves numerically and found that an exchange of stability occurs at the amplitude-to-depth ratio of 0.781. This wave amplitude corresponds to the first extremum in the total wave energy of the solitary waves. Tanaka *et al.* (1987) also simulated numerically the time development of unstable solitary waves. The time evolution is largely dependent on the sign of unstable normal mode of disturbances added to the solitary wave. In one case, it leads to a wave breaking. In another case, it undergoes a transition to a lower solitary wave having almost the same total energy. Longuet-Higgins & Tanaka (1997) also presented detailed numerical results of the linear stability analysis. In this way, the stability of surface solitary waves has been elucidated so far. On the other hand, in the case of interfacial waves, the stability has been investigated only for small-amplitude waves and those with strong interfacial tension effects. For small-amplitude waves, linear stability analyses based on the Korteweg de Vries (KdV) equation and its modified version (MKdV equation) were made to show that these waves are stable (Jeffery & Kakutani 1970; Benjamin 1972; Kuznetsov 1984; Weinstein 1986; Bona et al. 1987). For finite-amplitude solitary waves with strong interfacial tension effects, Calvo & Akylas (2003) investigated their linear stability. However, finite-amplitude interfacial solitary waves without interfacial tension effects, which are most fundamental and important, have not been treated so far. We will treat their stability focusing on the exchange of stability.

2 Basic equations

Consider a system under the uniform gravitational acceleration g of two inviscid, incompressible

The corresponding author. *E-mail*: kataoka@mech.kobe-u.ac.jp fluid layers between two horizontal rigid boundaries. The upper fluid layer has depth D_U at rest and constant density ρ_U, and the lower fluid layer has depth D_L at rest and constant density $\rho_L(>\rho_U)$ (see Fig. 1). The effects of interfacial tension are neglected. Hereafter, index U refers to the upper fluid, and index L to the lower. All variables are non-dimensionalized using g, D_L, and ρ_L.

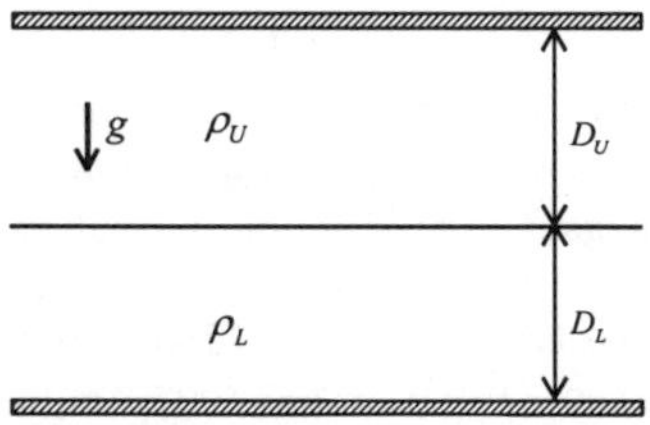

Fig. 1. Schematic of the two-layer fluid system.

Introducing the coordinate system with the x axis horizontal, the y axis vertically upward, and the origin on the undisturbed interface, we obtain the following set of dimensionless governing equations for the velocity potentials $\phi_U(x,y,t)$ (t being the time), $\phi_L(x,y,t)$, and the interfacial elevation $\eta(x,t)$:

$$\nabla^2\phi_U = 0 \quad \text{in the upper-fluid domain,} \tag{1}$$

$$\nabla^2\phi_L = 0 \quad \text{in the lower-fluid domain,} \tag{2}$$

subject to the boundary conditions

$$\frac{\partial\phi_U}{\partial y} = 0 \quad \text{at } y = D, \tag{3}$$

$$\frac{\partial\eta}{\partial t} + \frac{\partial\phi_U}{\partial x}\frac{\partial\eta}{\partial x} = \frac{\partial\phi_U}{\partial y} \quad \text{at } y = \eta, \tag{4}$$

$$\frac{\partial\eta}{\partial t} + \frac{\partial\phi_L}{\partial x}\frac{\partial\eta}{\partial x} = \frac{\partial\phi_L}{\partial y} \quad \text{at } y = \eta, \tag{5}$$

$$\begin{aligned} &-\rho\left\{\frac{\partial\phi_U}{\partial t} + \frac{1}{2}\left[\left(\frac{\partial\phi_U}{\partial x}\right)^2 + \left(\frac{\partial\phi_U}{\partial y}\right)^2\right]\right\} \\ &+\frac{\partial\phi_L}{\partial t} + \frac{1}{2}\left[\left(\frac{\partial\phi_L}{\partial x}\right)^2 + \left(\frac{\partial\phi_L}{\partial y}\right)^2\right] + (1-\rho)\eta = f(t) \end{aligned} \quad \text{at } y = \eta, \tag{6}$$

$$\frac{\partial\phi_L}{\partial y} = 0 \quad \text{at } y = -1, \tag{7}$$

where $y = D$, $y = \eta$, and $y = -1$ represent the upper rigid wall, the interface, and the lower rigid wall, respectively. The linear operator ∇^2 is defined by

$$\nabla^2 = \frac{\partial^2}{\partial x^2} + \frac{\partial^2}{\partial y^2}, \tag{8}$$

and $f(t)$ is the value on the left-hand side of (6) evaluated as $x \to \infty$. ρ and D are, respectively, the density ratio and the depth ratio of the two fluids defined by

$$\rho = \frac{\rho_U}{\rho_L}, \quad D = \frac{D_U}{D_L}. \tag{9}$$

Put a solution of (1)-(7) in the following form:

$$\begin{aligned} \phi_U &= -vx + \Phi_U(x,y), \\ \phi_L &= -vx + \Phi_L(x,y), \\ \eta &= \eta_I(x), \end{aligned} \tag{10}$$

where $\partial\Phi_U/\partial x$, $\partial\Phi_U/\partial y$, $\partial\Phi_L/\partial x$, $\partial\Phi_L/\partial y$, and η_I approach zero as $x \to \pm\infty$, and v is a positive constant. This solution represents a steady propagation of localized wave against a uniform stream of constant speed v in the negative x dierction. We call this solution a solitary wave solution. The existence of the above solitary wave solutions was confirmed numerically for various sets of the parameters ρ, D, and v (Funakoshi & Oikawa 1986; Pullin & Grimshaw 1988; Turner & Vanden-Broeck 1988; Evans & Ford 1996; Laget & Dias 1997; Michallet & Barthēlemy 1998; Grue *et al.* 1999).

Now we make the linear stability analysis of the solitary wave solution (10) on the basis of (1)-(7) and decaying condition as $x \to \pm\infty$. To this end, the solution of (1)-(7) is expressed as sum of the solitary wave solution and its disturbances as

$$\begin{aligned} \phi_U &= -vx + \Phi_U + \hat{\phi}_U(x,y)\exp(\lambda t), \\ \phi_L &= -vx + \Phi_L + \hat{\phi}_L(x,y)\exp(\lambda t), \\ \eta &= \eta_I + \hat{\eta}(x)\exp(\lambda t), \end{aligned} \tag{11}$$

where λ is a complex constant whose value is determined by solving equations for the disturbances. Substituting (11) into (1)-(7), linearizing with respect to $(\hat{\phi}_U, \hat{\phi}_L, \hat{\eta})$, and imposing the decaying condition as $x \to \pm\infty$, we obtain the following set of linear equations for $(\hat{\phi}_U, \hat{\phi}_L, \hat{\eta})$:

$$\nabla^2 \hat{\phi}_U = 0 \quad \text{in the upper-fluid domain,} \tag{12}$$

$$\nabla^2 \hat{\phi}_L = 0 \quad \text{in the lower-fluid domain,} \tag{13}$$

subject to the boundary conditions

$$\frac{\partial \hat{\phi}_U}{\partial y} = 0 \quad \text{at } y = D, \tag{14}$$

$$\mathbf{L_U}[\hat{\phi}_U, \hat{\eta}] = -\lambda \hat{\eta} \cos\theta \quad \text{at } y = \eta_I, \tag{15}$$

$$\mathbf{L_L}[\hat{\phi}_L, \hat{\eta}] = -\lambda \hat{\eta} \cos\theta \quad \text{at } y = \eta_I, \tag{16}$$

$$\mathbf{L_I}[\hat{\phi}_U, \hat{\phi}_L, \hat{\eta}] = \lambda(\rho\hat{\phi}_U - \hat{\phi}_L) \quad \text{at } y = \eta_I, \tag{17}$$

$$\frac{\partial \hat{\phi}_L}{\partial y} = 0 \quad \text{at } y = -1, \tag{18}$$

$$\frac{\partial \hat{\phi}_U}{\partial x} \to 0, \ \frac{\partial \hat{\phi}_U}{\partial y} \to 0, \ \frac{\partial \hat{\phi}_L}{\partial x} \to 0, \ \frac{\partial \hat{\phi}_L}{\partial y} \to 0, \ \hat{\eta} \to 0 \quad \text{as } x \to \pm\infty, \tag{19}$$

where $\mathbf{L_U}$, $\mathbf{L_L}$, and $\mathbf{L_I}$ are the linear operators defined by

$$\mathbf{L_U}[\hat{\phi}_U, \hat{\eta}] = \left\{ \left(-\frac{\partial}{\partial y} + \frac{\mathrm{d}\eta_I}{\mathrm{d}x}\frac{\partial}{\partial x} \right)\hat{\phi}_U + \left[\left(\frac{\partial^2 \Phi_U}{\partial x^2} + \frac{\partial^2 \Phi_U}{\partial x \partial y}\frac{\mathrm{d}\eta_I}{\mathrm{d}x} \right) + \left(-v + \frac{\partial \Phi_U}{\partial x} \right)\frac{\mathrm{d}}{\mathrm{d}x} \right]\hat{\eta} \right\} \cos\theta \tag{20}$$

$$\mathbf{L_L}[\hat{\phi}_L,\hat{\eta}]=\left\{\left(-\frac{\partial}{\partial y}+\frac{\mathrm{d}\eta_I}{\mathrm{d}x}\frac{\partial}{\partial x}\right)\hat{\phi}_L+\left[\left(\frac{\partial^2\Phi_L}{\partial x^2}+\frac{\partial^2\Phi_L}{\partial x\partial y}\frac{\mathrm{d}\eta_I}{\mathrm{d}x}\right)+\left(-v+\frac{\partial\Phi_L}{\partial x}\right)\frac{\mathrm{d}}{\mathrm{d}x}\right]\hat{\eta}\right\}\cos\theta \tag{21}$$

$$\begin{aligned}\mathbf{L_I}[\hat{\phi}_U,\hat{\phi}_L,\hat{\eta}]=&-\rho\left[\left(-v+\frac{\partial\Phi_U}{\partial x}\right)\frac{\partial}{\partial x}+\frac{\partial\Phi_U}{\partial y}\frac{\partial}{\partial y}\right]\hat{\phi}_U+\left[\left(-v+\frac{\partial\Phi_L}{\partial x}\right)\frac{\partial}{\partial x}+\frac{\partial\Phi_L}{\partial y}\frac{\partial}{\partial y}\right]\hat{\phi}_L\\&+\left\{-\rho\left[\left(-v+\frac{\partial\Phi_U}{\partial x}\right)\frac{\partial^2\Phi_U}{\partial x\partial y}+\frac{\partial\Phi_U}{\partial y}\frac{\partial^2\Phi_U}{\partial y^2}\right]+\left(-v+\frac{\partial\Phi_L}{\partial x}\right)\frac{\partial^2\Phi_L}{\partial x\partial y}+\frac{\partial\Phi_L}{\partial y}\frac{\partial^2\Phi_L}{\partial y^2}+1-\rho\right\}\hat{\eta}\end{aligned} \tag{22}$$

Eqs.(12)-(19) constitute an eigenvalue problem for $(\hat{\phi}_U,\hat{\phi}_L,\hat{\eta})$ whose eigenvalue is λ. If it possesses a solution whose λ has a positive real part, the corresponding solitary wave solution is unstable.

Before proceeding to the next section, it is convenient to define the wave amplitude, or the wave height $h(>0)$ of the interfacial solitary wave as the dimensionless maximum interfacial displacement:

$$h\equiv\max|\eta_I|. \tag{23}$$

We also define the dimensionless integral properties of the solitary wave, or the kinetic energy T_U in the upper fluid, the kinetic energy T_L in the lower fluid, the total wave energy E, and the mass M:

$$T_U=\frac{\rho}{2}\iint_{\substack{\text{whole upper-}\\\text{fluid domain}}}\left[\left(\frac{\partial\Phi_U}{\partial x}\right)^2+\left(\frac{\partial\Phi_U}{\partial y}\right)^2\right]\mathrm{d}x\mathrm{d}y,\quad T_L=\frac{1}{2}\iint_{\substack{\text{whole lower-}\\\text{fluid domain}}}\left[\left(\frac{\partial\Phi_L}{\partial x}\right)^2+\left(\frac{\partial\Phi_L}{\partial y}\right)^2\right]\mathrm{d}x\mathrm{d}y,$$

$$E=T_U+T_L+\frac{1-\rho}{2}\int_{-\infty}^{\infty}\eta_I^2\mathrm{d}x,\quad M=\int_{-\infty}^{\infty}\eta_I\mathrm{d}x. \tag{24}$$

3 Asymptotic analysis

Proposition *Suppose that the solitary wave solutions (Φ_U,Φ_L,η_I) do not bifurcate, or they are located on a single solution branch for given ρ and D. Then an exchange of stability occurs only for any stationary value of E ($\mathrm{d}E/\mathrm{d}v=0$). A positive real eigenvalue, which is associated with this exchange of stability, exists on a solution branch that satisfies the following inequality adjacent to the above critical point $\mathrm{d}E/\mathrm{d}v=0$:*

$$\sigma\frac{\mathrm{d}E}{\mathrm{d}v}>0, \tag{25}$$

where

$$\sigma=\frac{\mathrm{d}M}{\mathrm{d}v}\frac{\mathrm{d}\Omega}{\mathrm{d}v}, \tag{26}$$

and

$$\Omega=\frac{2}{v}\left(T_L-\frac{T_U}{D}\right)-\left(1+\frac{\rho}{D}\right)vM. \tag{27}$$

Here *E*, *M*, T_U, and T_L are defined by (24). The bifurcation of the solitary wave solutions implies that a new solution branch bifurcates from the regular solution branch. The above proposition can be proved by an asymptotic analysis of (12)-(19) for small modulus of λ, i.e. as $|\lambda|\to 0$. See Kataoka (2006) for details of the analysis.

4 Criterion for the stability

From the proposition shown in the previous section, we construct a general criterion for the stability of interfacial solitary waves with respect to disturbances that are stationary relative to the basic wave. It is convenient to define two words with quotation mark: 'stable' and 'unstable'. We call the solitary waves 'stable' if they are stable to disturbances that are stationary relative to the basic wave, and 'unstable' if they are unstable to these disturbances. Then the stability of interfacial solitary waves with respect to stationary disturbances is given as follows:

> **<Assumption>**
> The solitary wave solutions do not bifurcate.
>
> **<Stability to stationary disturbances>**
> Follow the branch of solitary wave solutions for fixed ρ and D from small amplitude ($h \ll 1$) to the present one. The solitary waves are 'stable' up to the first point of $\mathrm{d}E/\mathrm{d}v = 0$, and become 'unstable' with respect to a single normal mode after passing through this point. When passing through the subsequent point of $\mathrm{d}E/\mathrm{d}v = 0$, the number of growing modes increases or decreases according as the sign of $\sigma \mathrm{d}E/\mathrm{d}v$ just after passing through this point is positive or negative.

For the validity of <Assumption>, it holds for any solitary waves investigated by us. That is, we calculated the solitary wave solutions using the numerical method of iteration given by Turner & Vanden-Broeck (1988), and the obtained solitary wave solutions did not bifurcate.

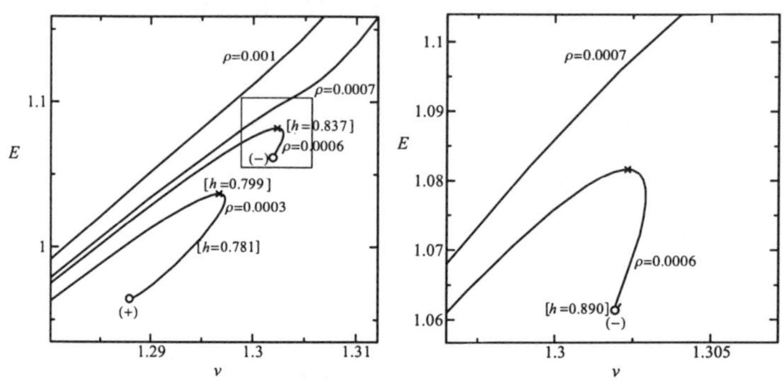

(a) (b)

Fig. 2. Total wave energy E versus the wave speed v of the interfacial solitary waves for $D=1$ and various values of ρ ($\rho = 0.0003$, 0.0006, 0.0007, and 0.001). The crosses indicate the first points of $\mathrm{d}E/\mathrm{d}v = 0$ and the corresponding wave amplitudes h are shown in the square brackets. The circles indicate the second points of $\mathrm{d}E/\mathrm{d}v = 0$ and the corresponding signs of $\sigma \mathrm{d}E/\mathrm{d}v$ just after passing through these points are shown in the parentheses: (a) whole view, (b) blow up of the top-right rectangular region. In Fig. 2(b), the wave amplitude h at the second point of $\mathrm{d}E/\mathrm{d}v = 0$ is shown in the square bracket.

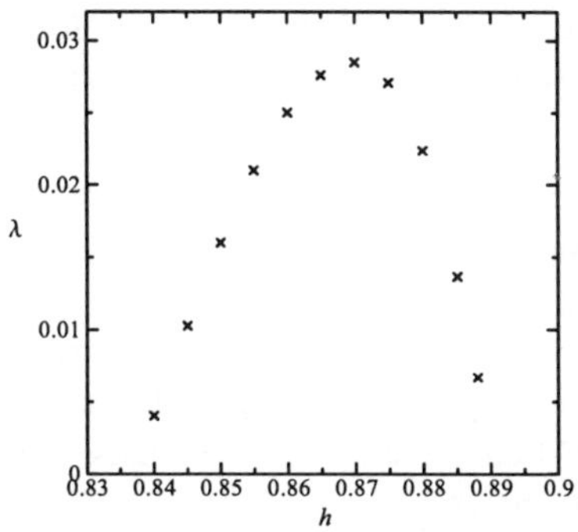

Fig. 3. Growth rate λ of the growing disturbance mode versus wave amplitude h for the interfacial solitary waves with $D=1$ and $\rho=0.0006$. The crosses represent the numerical results of the eigenvalue problem.

5. Stability of specific solitary waves

Here we investigate the stability of specific solitary waves. The solitary wave solutions are numerically calculated using the method described in Turner & Vanden-Broeck (1988). According to the previous studies (Funakoshi & Oikawa 1986; Amick & Turner 1986; Turner & Vanden-Broeck 1988), there are two types of interfacial solitary waves depending on the parameters ρ and D. One is of elevation type ($\rho < D^2$), and the other is of depression type ($\rho > D^2$). The former is treated in Section 5.1 and the latter in Section 5.2.

5.1 Solitary waves of elevation

We treat the cases of $D=1$ and 3 separately.

(i) $D=1$

Figure 2(a) shows E versus v for $\rho=0.0003$, 0.0006, 0.0007, and 0.001. The crosses denote the first points of $\mathrm{d}E/\mathrm{d}v=0$ and the corresponding wave amplitudes h are shown. The circles denote the second points of $\mathrm{d}E/\mathrm{d}v=0$ and the signs (+) or (–) of $\sigma\mathrm{d}E/\mathrm{d}v$ just after passing through these points along the solitary wave solution branches are shown. According to this figure, there appear the points of $\mathrm{d}E/\mathrm{d}v=0$ for $\rho\le 0.0006$, but no $\mathrm{d}E/\mathrm{d}v=0$ occurs for $\rho\ge 0.0007$. In order to see in more detail the above distinction around the density ratios of $0.0006\le\rho\le 0.0007$, blow up of the top-right rectangular region in Fig. 2(a) is shown in Fig. 2(b). One finds that the first points of $\mathrm{d}E/\mathrm{d}v=0$ denoted by the crosses appear for $\rho=0.0006$. On the other hand, there are no points of $\mathrm{d}E/\mathrm{d}v=0$ for $\rho=0.0007$. More interestingly for $\rho=0.0006$, there are the second $\mathrm{d}E/\mathrm{d}v=0$ denoted by the circles at $h=0.890$, and the sign of $\sigma\mathrm{d}E/\mathrm{d}v$ just after passing through this point is negative. According to our criterion, this solitary wave for $\rho=0.0006$ restabilizes at the second point of $\mathrm{d}E/\mathrm{d}v=0$.

In order to confirm this restabilization feature, we also carried out numerical calculation of the eigenvalue problem (12)-(19). The crosses in Fig. 3 denote the numerically calculated growth rates λ of the growing disturbance mode for $\rho=0.0006$. One finds that λ becomes positive from around $h=0.837$ and reaches its maximum at $h=0.87$. Then λ decreases as h becomes larger and seems to intersect the $\lambda=0$ axis at around $h=0.89$. This restabilizing feature is in complete agreement with the stability results based on our criterion mentioned above.

We show in Fig. 4 the critical wave amplitudes h at which an exchange of stability occurs as a function of ρ. The thick solid lines represent an occurrence of exchange of stability at the first points of

$dE/dv = 0$. Multiple solitary wave solutions may exist for given h of course. In such case, the stability given by Fig. 4 applies to the first solitary wave that reaches a given h along its solution branch starting from small amplitude.

(ii) $D = 3$

The total wave energies E versus v for $\rho = 0.03$, 0.05, 0.06, and 0.08 are shown in Fig. 5. The crosses denote the first points of $dE/dv = 0$ and the corresponding wave amplitudes h are shown. The first points of $dE/dv = 0$ appear for $\rho \leq 0.05$, but no points of $dE/dv = 0$ appear for $\rho \geq 0.06$. Thus, according to our criterion, the solitary waves for $\rho \leq 0.05$ are 'stable' up to the first points of $dE/dv = 0$ and become 'unstable' from these critical amplitudes with respect to a single normal mode, and those for $\rho \geq 0.06$ are always 'stable'.

We show in Fig. 6 the critical wave amplitudes h at which an exchange of stability occurs as a function of ρ. The thick solid lines represent an occurrence of exchange of stability at the first points of $dE/dv = 0$.

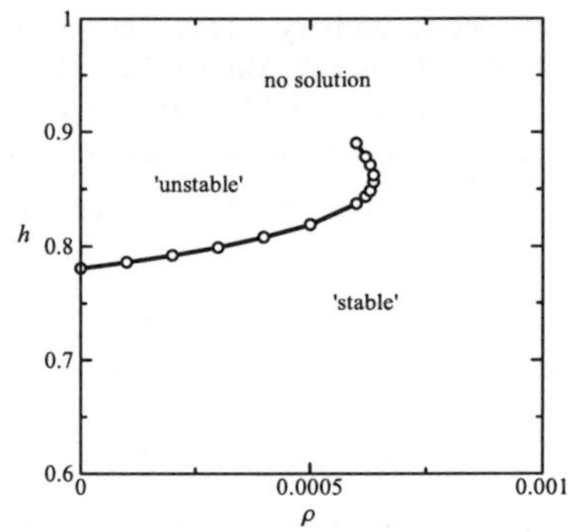

Fig. 4. Critical amplitudes h at which an exchange of stability occurs versus ρ for $D = 1$. The circles represent the recognized positions of the first points of $dE/dv = 0$, and thick solid lines connect them.

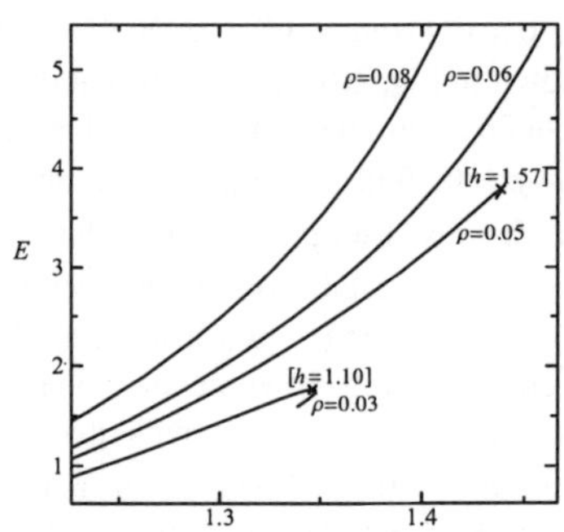

Fig. 5. Total wave energy E versus the wave speed v of the solitary waves for $D = 3$ and various values of ρ (=0.03, 0.05, 0.06, and 0.08). The crosses indicate the first points of $dE/dv = 0$ and the corresponding wave amplitudes h are shown in the square brackets.

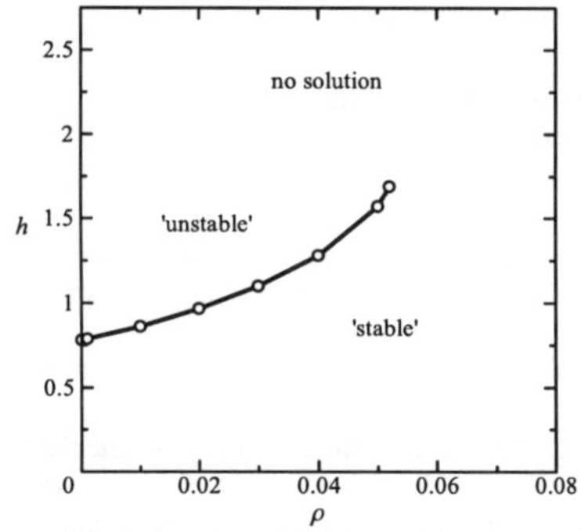

Fig. 6. Critical amplitudes h at which an exchange of stability occurs versus ρ for $D=3$. The circles represent the recognized positions of the first points of $\mathrm{d}E/\mathrm{d}v=0$, and thick solid lines connect them.

5.2 Solitary waves of depression

All the solitary waves of depression investigated by us are 'stable'. That is, the total energy E continues to increase as the wave amplitude h increases, and no points of $\mathrm{d}E/\mathrm{d}v=0$ appear.

6 Concluding remarks

Linear stability or exchange of stability has been investigated for interfacial solitary waves propagating in a two-layer fluid of finite depth. We have constructed a general criterion for the stability of interfacial solitary waves with respect to disturbances that are stationary relative to the basic wave. Application of this criterion to specific solitary wave solutions revealed the following main features on the exchange of stability of interfacial solitary waves:

As for the solitary waves of elevation ($\rho < D^2$), they experience no exchange of stability if the density ratio ρ is larger than some critical value that depends on D, or $\rho > \rho_{cr}(D)$, where $\rho_{cr}(D)$ is an increasing function of D, e.g. $\rho_{cr}(1)=0.0006\sim0.0007$, $\rho_{cr}(3)=0.05\sim0.06$. For the smaller density ratios, or $\rho<\rho_{cr}(D)$, the solitary waves experience an exchange of stability at the first stationary value of their total wave energy E. When ρ is near its critical value $\rho_{cr}(D)$, some solitary waves restabilize. Examples are given for $D=1$ with $\rho=0.0006$.

As for the solitary waves of depression ($\rho > D^2$), they seem to experience no exchange of stability.

References

[1] Amick, C. J. & Turner, R. E. L. 1986 A global theory of internal solitary waves in two-fluid systems. *Trans. Am. Math. Soc.* **298**, 431-484.

[2] Benjamin, T. B. 1972 The stability of solitary waves. *Proc. R. Soc. Lond.* A **328**, 153-183.

[3] Bona, J. L., Souganidis, P. E. & Strauss, W. A. 1987 Stability and instability of solitary waves of Korteweg-de Vries type. *Proc. R. Soc. Lond.* A **411**, 395-412.

[4] Calvo, D. C. & Akylas, T. R. 2003 On interfacial gravity-capillary solitary waves of the Benjamin type and their stability. *Phys. Fluids* **15**, 1261-1270.

[5] Evans, W. A. B. & Ford, M. J. 1996 An integral equation approach to internal (2-layer) solitary waves. *Phys. Fluids* **8**, 2032-2047.

[6] Funakoshi, M. & Oikawa, M. 1986 Long internal waves of large amplitude in a two-layer fluid. *J. Phys. Soc. Jpn.* **55**, 128-144.

[7] Grue, J., Jensen, A., Rusås, P. O., & Sveen, J. K. 1999 Properties of large amplitude internal waves. *J. Fluid Mech.* **380**, 257-278.
[8] Jeffery, A. & Kakutani, T., 1970 Stability of the Burgers shock wave and the Korteweg-de Vries soliton. *Indiana Univ. Math. J.* **20**, 463-468.
[9] Kataoka, T. 2006 The stability of finite-amplitude interfacial solitary waves. *Fluid Dyn. Res.* accepted for publication.
[10] Kuznetsov, E. A. 1984 Soliton stability in equations of the KdV type. *Phys. Lett.* A **101**, 314-316.
[11] Laget, O. & Dias, F. 1997 Numerical computation of capillary-gravity interfacial solitary waves. *J. Fluid Mech.* **349**, 221-251.
[12] Longuet-Higgins, M. S. & Tanaka, M. 1997 On the crest instabilities of steep surface waves. *J. Fluid Mech.* **336**, 51-68.
[13] Michallet, H. & Barthēlemy, E. 1998 Experimental study of interfacial solitary waves. *J. Fluid Mech.* **366**, 159-177.
[14] Pullin, D. I. & Grimshaw, R. H. J. 1988 Finite-amplitude solitary waves at the interface between two homogeneous fluids. *Phys. Fluids* **31**, 3550-3559.
[15] Tanaka, M. 1986 The stability of solitary waves. *Phys. Fluids* **29**, 650-655.
[16] Tanaka, M., Dold, J. W., Lewy, M., & Peregrine, D. H. 1987 Instability and breaking of a solitary wave. *J. Fluid Mech.* **185**, 235-248.
[17] Turner, R. E. L. & Vanden-Broeck, J. -M. 1988 Broadening of interfacial solitary waves. *Phys. Fluids*, **31**, 2486-2490.
[18] Weinstein, M. I. 1986 Lyapunov stability of ground states of nonlinear dispersive evolution equations. *Comm. Pure Appl. Math.* **39**, 51-67.

On the Transition to Diffusion Chaos in the Kuramoto–Tsuzuki equation

Nikolai A. Magnitskii
Institute for Systems Analysis, Russian Academy of Sciences, 9 Prospect 60-let Oktyabrya, Moscow, 117312, Russia

Abstract
It is shown in the present paper that the transition to diffusion or spatiotemporal chaos in the phase space of solutions of the Kuramoto–Tsuzuki diffusion type partial differential equation occurs through the subharmonic and homoclinic cascades of bifurcations of two-dimensional invariant tori with respect to internal as well as external frequencies, and hence this scenario also can be described by universal FSM (Feigenbaum-Sharkovskii-Magnitskii) theory of transition to dynamical chaos in dissipative systems of nonlinear differential equations.

PACS: 05.45; 02.30.Jr
Keywords: nonlinear partial differential equations, dynamical chaos, singular attractor.

1. Introduction

Wide class of physical, chemical, biological, ecological and economic processes is described by reaction-diffusion systems of partial differential equations

$$u_t = D_1 u_{xx} + f(u,v,\mu), \quad v_t = D_2 v_{xx} + g(u,v,\mu),$$

depending on scalar parameter μ. Such system is very complex system. Behavior of its solutions depends on coefficients of diffusion and their ratio, length of space area and edge conditions. As a rule, there exists a value of the parameter μ_0, such that for all $\mu \leq \mu_0$ reaction-diffusion system has a stable stationary and homogeneous solution denoted as thermo dynamical branch. When $\mu > \mu_0$, then thermo dynamical branch loses its stability and after that reaction-diffusion system can have quite different solutions such as periodical oscillations, stationary dissipative structures, spiral waves and nonstationary nonperiodic nonhomogeneous solutions. Last solutions are known as diffusion or spatiotemporal chaos.

It is well-known that any solution of the reaction-diffusion system in a neighbourhood $\mu > \mu_0$ can be approximated by some complex-valued solution $W(r,\tau) = U(r,\tau)+iV(r,\tau)$ of the Kuramoto-Tsuzuki or Time Dependent Ginzburg-Landau equation [1]

$$W_\tau = W + (1+ic_1)W_{rr} - (1+ic_2)W|W|^2, \tag{1}$$

The corresponding author. *Tel.: (495) 931-8731: Fax: (495) 938-2209*
E-mail: nmag@isa.ru; mag@su29.ru

where $r = \varepsilon x$, $\tau = \varepsilon^2 t$, $0 \leq r \leq R$, ε is a small parameter and c_1, c_2 - some real constants.
Clearly that the equation (1) has a homogeneous periodic solution $W(\tau) = \exp(-i(c_2\tau + \varphi))$ for arbitrary phase φ . Such solution is stable in some region of parameters c_1 and c_2. In other region of parameters the equation (1) has a stable automodel solution $W(r,\tau) = F(r)\exp(i(\omega\tau + a(r)))$. But equation (1) has also nonperiodic nonhomogeneous solutions – diffusion or spatiotemporal chaos in some regions of parameters c_1 and c_2. The main question consists in weather or not has spatiotemporal chaos in diffusion type systems the same nature as dynamical chaos in dissipative systems of ordinary differential equations?

It was shown recently in a number of papers (e.g., see [2-5]) that three-dimensional nonlinear dissipative systems of ordinary differential equations have a common universal scenario of transition to chaos through a cascade of period doubling bifurcations of stable cycles (that is, a cascade of Feigenbaum bifurcations [6]) and then through a subharmonic cascade of Sharkovskii bifurcations [7] of stable cycles of all periods up to a cycle of period 3 and then through a homoclinic cascade of Magnitskii bifurcations of stable cycles that tend to the homoclinic contours of singular points [2,3,8]. Some remarkable results lie in the foundation of the theory of this new universal scenario: the Feigenbaum-Sharkovskii theory of bifurcations of stable cycles in one-dimensional mappings and the Magnitskii theory of rotor type singular points of two-dimensional non-autonomous systems of ordinary differential equations [2,8-10] as a bridge between one-dimensional mappings and differential equations. In this new universal scenario of transition to dynamical chaos in dissipative systems of ordinary differential equations, any irregular attractor (which is not a stable cycle or a stable invariant torus) is a singular attractor (closure of semi-stable nonperiodic trajectory) generated only at accumulation points of bifurcation parameter. It was shown also that all classical three-dimensional chaotic systems of ordinary differential equations including the Lorenz hydrodynamic system, the Rossler chemical system, the Chua electro technical system, the Magnitskii macroeconomic system and many others satisfy to this new theory and have the same indicated above universal scenario of transition to chaos. Moreover, it was shown that all classical two-dimensional non-autonomous and many-dimensional autonomous nonlinear dissipative systems of ordinary differential equations and nonlinear differential equations with delay arguments, such as, for instant, Duffing-Holms two-dimensional non-autonomous equation, Mackey-Glass equation with delay argument and many others also satisfy to this theory and have the same universal scenario of transition to dynamical chaos.

However, recently there have been solid argumentation casting doubt on the relevance of the generalization of results valid for few-mode systems of ordinary differential equations to infinite-dimensional systems of partial differential equations, since in systems of higher dimension, the scenario of transition to chaos contains two-dimensional tori and possibly the entire subharmonic cascade of bifurcations of two-dimensional tori with respect to one of the frequencies, just as it is in the five-dimensional system of complex Lorenz equations [11].

In the present paper on the basis of numerical solution, it is shown that the transition to chaos in the space of few-mode approximations for the Kuramoto–Tsuzuki equation occurs through the cascades of Feigenbaum-Sharkovskii-Magnitskii bifurcations of stable cycles, but the transition to spatiotemporal chaos in the phase space of solutions of the Kuramoto–Tsuzuki equation occurs through the cascades of Feigenbaum-Sharkovskii-Magnitskii bifurcations of two-dimensional invariant tori with respect to internal as well as external frequencies, and hence this scenario also can be described by FSM (Feigenbaum-Sharkovskii-Magnitskii) theory.

2. Scenario of transition to chaos in the few-mode approximations

The appearance of diffusion chaos in the second boundary value problem for the Kuramoto–Tsuzuki equation

$$W_t = W + (1+ic_1)W_{xx} - (1+ic_2)W|W|^2,$$
$$W_x(0,t) = W_x(l,t) = 0, \quad W(x,0) = W_0(x), \quad 0 \le x \le l, \quad 0 \le t < \infty, \tag{2}$$

was studied in [12] with the use of Galerkin few-mode approximations

$$W(x,t) \approx \xi^{1/2}(t)\exp(i\theta_1(t)) + \eta^{1/2}(t)\exp(i\theta_2(t))\cos kx, \quad k = \pi / l,$$

which permit one to reduce the infinite-dimensional problem (2) to the simpler three-dimensional system of nonlinear ordinary differential equations

$$\begin{aligned} \dot\xi &= 2\xi - 2\xi(\xi+\eta) - \xi\eta(\cos\theta + c_2\sin\theta), \\ \dot\eta &= 2\eta - 2\eta(2\xi + 3\eta/4) - 2\xi\eta(\cos\theta - c_2\sin\theta) - 2k^2\eta, \\ \dot\theta &= c_2(2\xi - \eta/2) + (2\xi+\eta)\sin\theta + c_2(2\xi-\eta)\cos\theta + 2c_1k^2. \end{aligned} \tag{3}$$

for variables ξ, η and $\theta(t) = \theta_2(t) - \theta_1(t)$, which still has chaotic dynamics.

We show in this item that all irregular attractors of the three-dimensional system (3) are also singular attractors, and that transition to chaos in the system (3) of few-mode approximations also occurs in accordance with FSM (Feigenbaum-Sharkovskii-Magnitskii) theory through the Feigenbaum cascade of period doubling bifurcations for stable cycles, then through the Sharkovskii subharmonic cascade of bifurcations of stable cycles of all periods up to a cycle of period 3 and then through the Magnitskii cascades of bifurcations of stable homoclinic cycles.

Following [12], we consider a small range of the space variable ($l = \pi$). As the Fourier coefficients of solutions are rapidly decreasing with increasing their index, then it seems that there should be a qualitative as well as quantitative correspondence between solutions of the original equation (2) and the few-mode system (3); this pertains to the simplest regular (periodic) as well as irregular attractors and scenarios of their appearance as the parameters c_1 and c_2 change.

The existence domains of stable singular points, simple stable cycles, and double-period cycles of system (3) were found in [12] in the space of parameters (c_1 ,c_2). All more complicated regular (periodic) and irregular attractors of system (3) are simply included in one class corresponding to the remaining domains of the parameter space. Therefore, the approach suggested in [12] does not permit one to explain mechanisms and define scenarios of the onset of chaotic dynamics in the original Kuramoto–Tsuzuki equation (2) as well as in the simplified system (3). Let us show that all irregular attractors of the simplified system (3) are singular in the sense of the definition of [2, 8, 9] and the transition to chaos and the complication of the structure of attractors of system (3) follow the same scenario as in all other known three-dimensional chaotic dissipative systems of ordinary differential equations, including Lorenz, Rossler, Chua systems [2, 5]; i.e., it occurs in accordance with the FSM theory. Further, there may be more complicated cascades of bifurcations of stable cycles in accordance with [2, Chap. 4; 8, 9].

Indeed, set, say, k =1 and c_1 =1.3 and consider the scenario of transition to chaos in system (3) as the parameter c_2 varies. The attractors of system (3) will be observed in three-dimensional phase space with coordinates $x = \xi\cos\theta,\ y = \xi\sin\theta,\ z = \eta$.

By integrating system (3) by a fourth-order Rounge–Kutta method, one can readily show that if the parameter c_2 is negative and decreases, then in system (3), just as in all other chaotic systems there is first a cascade of Feigenbaum period doubling bifurcations for the singular stable limit cycle existing in the system for $c_2 = -5$ and then a subharmonic cascade of bifurcations of stable cycles with arbitrary period in accordance with the Sharkovskii order. For example, a stable cycle of double period is generated in system (3) for $c_2 \approx -6.782$, that of quadruple period for $c_2 \approx -7.92$, of period 8 for $c_2 \approx -8.15$, of period 16 for $c_2 \approx -8.2$, of period 32 for $c_2 \approx -8.21$ and of period 64 for $c_2 \approx -8.211$. The Feigenbaum attractor appears in system (3) for $c_2 \approx -8.2111$. If $c_2 \approx -8.2155$, then in system (3), there is a stable cycle of period 40, which corresponds to the cycle $5*2^3$ in the Sharkovskii order. A cycle of period $20 = 5*2^2$ is generated for $c_2 \approx -8.2509$, a cycle of period $14 = 7*2$ is generated for $c_2 \approx -8.2754$; a cycle of period $10 = 5*2$, for $c_2 \approx -8.2949$; a cycle of period $6 = 3*2$, for $c_2 \approx -8.348$. The generation of the last cycle implies the generation of a stable cycle of period 3 in some two-dimensional plane transversal to the original cycle. For $c_2 \approx -8.564$, in system (3), there appears a stable cycle of period 7; for $c_2 \approx -8.668$, a stable cycle of period 5; and for $c_2 \approx -9.0$, a stable cycle of period 3. All generated cycles undergo their own cascades of Feigenbaum period doubling bifurcations.

It is important to note that, in the three-dimensional phase space (x, y,z) of system (3), for some parameter values, there can simultaneously exist several distinct stable cycles with their attraction domains. Each cycle of this kind can generate its own cascade of bifurcations and its own set of complete or incomplete singular subharmonic attractors. For example, we fix the parameter value $c_2 = 9.0$ and vary the parameter c_1. One can readily show that the stable cycle of period 3 existing in system (3) for $c_1 = 1.3$ results from a subharmonic cascade of bifurcations of the originally stable singular cycle as the parameter c_1 decreases from the value $c_1 = 1.43$. The projection of this original cycle onto the plane (y, z) makes four rotations around some conventional center. A stable cycle of period 8 is generated in the system for $c_1 \approx 1.425$; a stable cycle of period 6, for $c_1 \approx 1.366$; and a stable cycle of period 5, for $c_1 \approx 1.33$. If $c_1 \approx 1.282$, then in system (3), there appears another stable cycle of period 6, which is a double-period cycle for a cycle of period 3. For further decrease of the parameter c_1, a cascade of period doubling bifurcations occurs.

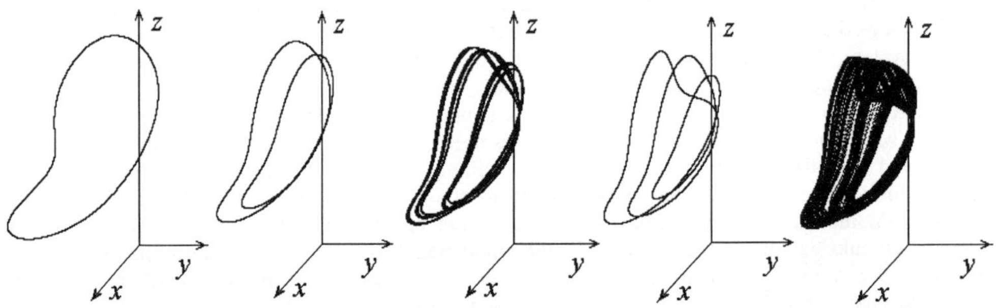

Fig. 1. Singular original cycle, period two cycle, Feigenbaum's attractor, period three cycle and some more complex subharmonic singular attractor in the reduced system (3) for $c_1 = 1.3$ and when c_2 varies.

In addition, as the parameter c_1 increases from the value $c_1 \approx 1.21$, in system (3), there is another subharmonic cascade of bifurcations, which starts from a simple stable singular cycle. The attraction domain of this cycle contains, for example, the initial point $\xi = 0.1$, $\eta = 0.01$, $\theta = -0.2$. The cascade continues with a stable cycle of double period for $c_1 \approx 1.215$, of quadruple period for $c_1 \approx 1.232$, of period 8 for $c_1 \approx 1.237$, of period 16 for $c_1 \approx 1.2375$, and so on. A Feigenbaum attractor is generated here for $c_1 \approx 1.238$. Further, a stable cycle of period 6 is generated in the system for $c_1 \approx 1.2415$; a cycle of period 5, for $c_1 \approx 1.252$; a new stable cycle of period 3, for $c_1 \approx 1.255$; and then a cascade of its period doubling bifurcations occurs as c_1 increases. In the domain $1.26 \leq c_1 \leq 1.28$, there exist numerous singular attractors that are not described by the Sharkovskii theory but are limit attractors for cascades of bifurcations of stable cycles. Some of these cycles of periods 22, 14, and 12 have been detected for the parameter values $c_1 = 1.268$, $c_1 = 1.273$ and $c_1 = 1.279$, respectively. Note that if $c1 = 1.21$, then in a neighborhood of the initial stable cycle of the cascade, there is another stable cycle of double period whose attraction domain contains, for example, the initial point $\xi = 0.1$, $\eta = 0.01$, $\theta = 0$. This cycle does not generate singular attractors.

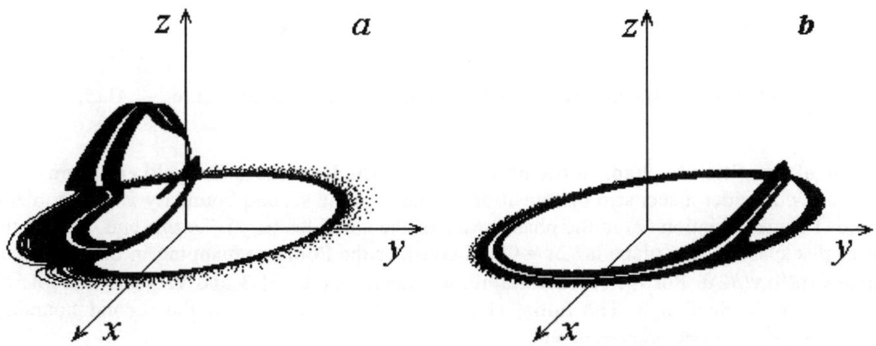

Fig. 2. Singular attractors of system (3) for the following parameter values: $c_2 = -9.0$; (a), $c_1 = 1.765$; (b) $c_1 = 13$.

The third subharmonic cascade of bifurcations generates a stable cycle for $c1 = 1.43$ if the parameter c_1 increases. Here a stable cycle of double period is generated for $c_1 \approx 1.52$, of quadruple period for $c_1 \approx 1.5581$, and so on. The set of irregular subharmonic attractors generated by this cycle is observed approximately in the domain $1.56 \leq c_1 \leq 1.59$. The fourth and fifth subharmonic cascades of bifurcations generate two stable simple cycles for $c_1 = 1.81$ as the parameter c_1 decreases and increases. The attraction domain of the first cycle contains, for example, the initial point $\xi = 0.1$, $\eta = 0.01$, $\theta = 20$. Here a stable cycle of double period is generated for $c_1 \approx 1.802$, of quadruple period for $c_1 \approx 1.7875$, and so on. The set of singular subharmonic attractors generated by this cycle is observed approximately in the domain $1.76 \leq c_1 \leq 1.77$. The attraction domain of the second cycle contains, for example, the point $\xi = 0.1$, $\eta = 0.01$, $\theta = -0.2$. Here a stable cycle of double period is generated for for $c_1 \approx 10.35$, of double period for $c_1 \approx 11.67$ abd so on. The set of singular subharmonic attractors generated by this cycle is observed approximately in the domain $12 \leq c_1 \leq 15$. The singular attractors of system (3) for $c_2 = -9.0$, $c_1 = 1.765$ and $c_1 = 13$ are shown in Fig.2. The singular attractors of system (3) for some other values of the parameters c_1 and c_2 are shown in Fig.3. he above-represented results imply that the scenarios of transition to chaos in the three-dimensional system of few-mode Galerkin approximations for the Kuramoto–Tsuzuki equation do not differ from the scenario considered and theoretically justified in [2, 8, 9]. However, this does not permit one to make a firm conclusion that the chaotic dynamics of the infinite-dimensional system (2) is identical to the chaotic dynamics of its three-dimensional few-mode approximation (3) considered in this section. This problem requires additional investigation to be represented in the forthcoming section of the present paper.

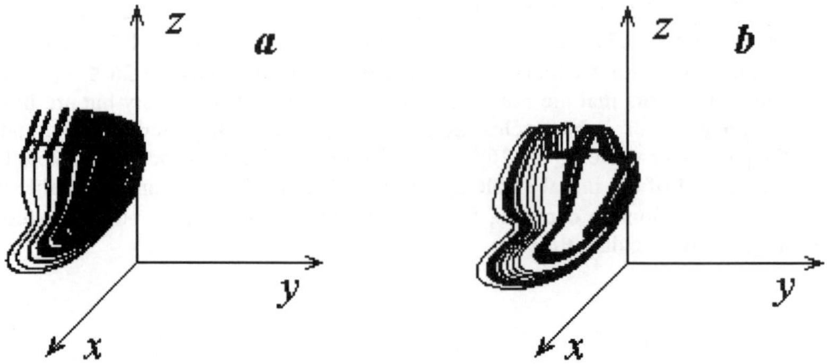

Fig. 3. Singular attractors of system (3) for the following parameter values: (a), c_1 =1.4125, c_2 = -11.5; c_1 =1.512, c_2 = -10.

3. Scenario of transition to chaos in the phase space of the Kuramoto-Tsuzuki equation

Let us now consider a scenario of transition to chaos in the second boundary value problem for the Kuramoto–Tsuzuki equation (2) in the phase space of the variables (u, v). To this end, we use the cross-section of this space by the plane $u(l/2) = 0$ and consider the Poincare mapping in the projection to the coordinates $(u(0),v(l/2))$. For the fixed variable, we again take c_1 =1.3 and vary the variable c_2 in the same domain as in Section 2. The initial conditions for the solution of the second boundary value problem (2) are given to be homogeneous.

We note again that, in the range $c_2 \in [-1.8, 0]$, the second boundary value problem (2) has a homogeneous periodic solution with equal amplitudes of oscillations of variables u(x,t) and v(x,t). For $c_2 \approx -1.81$, this homogeneous solution loses stability, and there appears another stable periodic but inhomogeneous solution, which also has equal amplitudes of oscillations with respect to the variables u(x, t) and v(x, t). For $c_2 \approx -2.66$, the periodic inhomogeneous solution also becomes unstable, and there appears a stable two-dimensional invariant torus, which is justified by the Poincare mapping. For $c_2 \approx -3.549$, there is a period doubling bifurcation of a two-dimensional invariant torus with respect to the basic (internal) frequency.

Note that for the value c_2 = -4.8 corresponding to a cycle with period 4 in the coordinates (ρ_0, ρ_1), the Poincare mapping is represented by a two-dimensional invariant torus with period 2 with respect to both internal and external frequencies (see Fig. 4a). Further, for c_2 = -4.815, one can observe a two-dimensional invariant torus of period 2 with respect to the internal frequency and of period 4 with respect to the external frequency; for c_2 = -4.820, one has a two-dimensional invariant torus of period 2 with respect to the internal frequency and period 8 with respect to the external frequency. Therefore, for problem (2), we have a cascade of period doubling bifurcations with respect to the external frequency for two-dimensional invariant tori of period 2 with respect to the internal frequency. This cascade is finished by the generation of the Feigenbaum attractor for the parameter value $c_2 \approx -4.8225$. The form of the Feigenbaum attractor (which is induced by a cascade of period doubling bifurcations with respect to the external frequency for two-dimensional invariant tori of period 2 with respect to the internal frequency) in the cross-section $u(l/2) = 0$ for the Kuramoto–Tsuzuki equation at the accumulation point (c_1, c_2) = (1.3, 4.8225) is shown in Fig. 4b. Moreover, for the parameter values c_2 = 4.894 and c_2 = 4.955 there are two-dimensional invariant tori of periods 5 and 3 with respect to the external frequency and period 2 with respect to the basic internal frequency (see Fig. 4c). The existence of stable two-dimensional invariant tori of periods 5 and 3 implies that there is a subharmonic cascade of bifurcations of two-dimensional invariant tori in the scenario of transition to chaos in the system of Kuramoto–Tsuzuki equations (2). The form of the singular subharmonic toroidal attractor of period 2 with respect to the internal frequency, that is, the singular attractor resulting from the subharmonic cascade of bifurcations of two-dimensional invariant tori of period 2 with respect to the internal frequency in accordance with the Sharkovskii order, is shown in Fig.4d for $c_2 = -5.05$.

The analysis of solutions of the second boundary value problem (2) for smaller negative values of the parameter c_2 shows that there also exists a cascade of period doubling bifurcations with respect to the basic internal frequency for two-dimensional tori. For $c_2 = -5.79$, the two-dimensional torus in the Poincare mapping has period 4 with respect to the basic (internal) frequency (Fig.5a); and if the parameter c_2 is further decreasing, then the cascade of period doubling bifurcations with respect to the internal frequency starts for this two-dimensional torus. This torus has period 2 with respect to the external frequency for $c_2 = -5.838$, period 4 for $c_2 = -5.840$, and so on (Fig.5b). For the parameter value $c_2 = -5.8589$, this torus in the Poincare mapping has period 3 with respect to the external frequency (Fig. 5c), which again implies that there exists a subharmonic cascade of bifurcations of two-dimensional tori with respect to the external frequency in the scenario of transition to a chaotic regime which appears in the system for the parameter value $c_2 \approx -5.8606$.

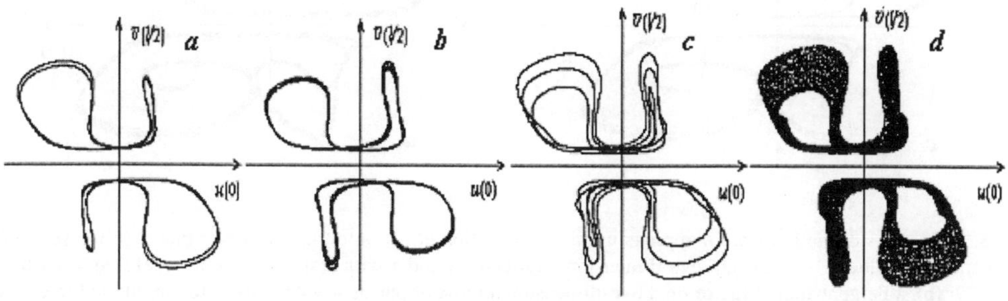

Fig.4. Projections of the Poincare mapping in the cross-section $u(l/2)$ onto the coordinate plane $(u(0),v(l/2))$ for the parameter value $c_1 = 1.3$: (a), a two-dimensional torus of period 2 with respect to the internal frequency and period 2 with respect to the external frequency for $c_2 = -4.8$; (b), a Feigenbaum attractor on a two-dimensional torus of period 2 with respect to the internal frequency for $c_2 = -4.8225$; (c), a two-dimensional torus of period 2 with respect to the internal frequency and period 3 with respect to the external frequency for $c_2 = -4.955$; (d), a singular attractor on a two-dimensional torus of period 2 with respect to the internal frequency for $c_2 = -5.05$.

The above-represented results imply that, in the Kuramoto–Tsuzuki equation (2), there may exist a subharmonic cascade of bifurcations of two-dimensional tori with respect to external as well as basic internal frequency. And such a cascade has been detected for fixed parameter value $c_1 = 2.5$ as the parameter c_2 decreases in the domain of negative values.

In the range $c_2 \in [-1.85,0]$, the second boundary value problem (2) has a homogeneous periodic solutions with equal amplitudes of oscillations of the variables $u(x,t)$ and $v(x,t)$. For the parameter value $c_2 \approx -1.851$, this homogeneous solution loses stability, but there appears another stable solution periodic with respect to time and inhomogeneous with respect to space, which has equal amplitudes of oscillations with respect to the variables $u(x_0,t)$ and $v(x_0,t)$, where $x_0 \in [0,l]$. For $c_2 \approx -2.803$, the periodic inhomogeneous solution also becomes unstable, and in problem (2) there appears a stable two-dimensional invariant torus, which is justified by the Poincare mapping. The period doubling bifurcation for a two-dimensional torus with respect to the basic (internal) frequency occurs for $c_2 \approx -3.134$.

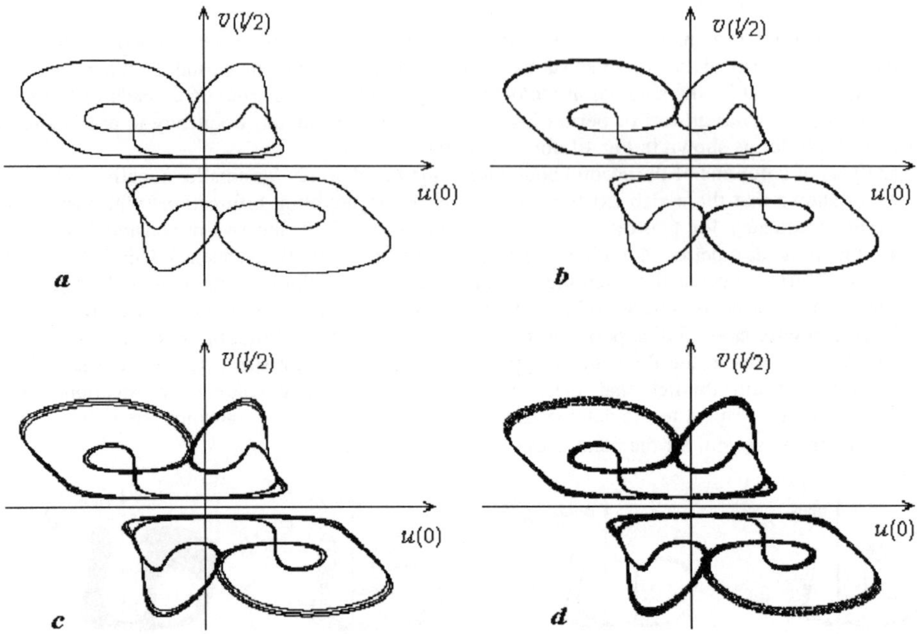

Fig.5. Projections of the Poincare mappings in the cross-section u(l/2) onto the coordinate plane (u(0),v(l/2)) for the parameter value $c_1 = 1.3$: (a), a two-dimensional torus of period 4 with respect to the internal frequency for $c_2 = -5.79$; (b), a Feigenbaum attractor on a two-dimensional torus of period 4 with respect to the internal frequency for $c_2 = -5.843$; (c), a two-dimensional torus of period 4 with respect to the internal frequency and period 3 with respect to the external frequency for $c_2 = -5.85888$; (d), a singular attractor on a two-dimensional torus of period 4 with respect to the internal frequency for $c_2 = -5.8605$.

This bifurcation starts the cascade of period doubling bifurcations for two-dimensional invariant tori with respect to the internal frequency. The solution of problem (2) is given by a two-dimensional invariant torus of quadruple period with respect to the internal frequency for $c_2 \in [-3.6186, -3.537]$, a torus of period 8 with respect to the internal frequency for $c_2 \in [-3.6409, -3.6187]$, a torus of period 16 with respect to the internal frequency for $c_2 \in [-3.64623, -3.6410]$, and so on. The cascade of period doubling bifurcations for two-dimensional tori with respect to the internal frequency finishes by the generation of the Feigenbaum attractor for $c_2 \approx -3.65$ (Fig.6).

For the subsequent decrease of the parameter c_2 , we have a subharmonic cascade of bifurcations of two-dimensional torus with respect to the internal frequency in accordance with the Sharkovskii order with generation of various singular toroidal attractors, one of which is shown in Fig.6d for c2 = -3.75.

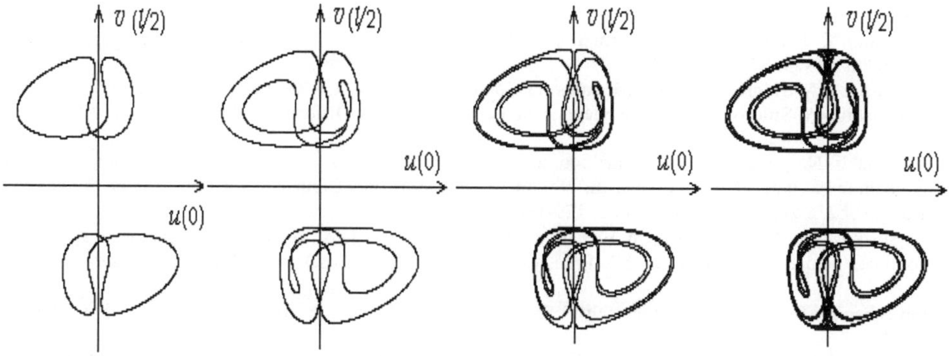

Fig. 6. Projections of the Poincare mappings in the cross-section u(l/2) onto the coordinate plane (u(0),v(l/2)) for the parameter value c_1 =2.5 : (a), a two-dimensional torus of period 4 with respect to the internal frequency for c_2 = -3.6; (b), the two-dimensional torus of period 8 with respect to the internal frequency for c_2 = -3.635; (c), the Feigenbaum attractor with respect to the internal frequency for c_2 = -3.655; (d), a singular attractor on the two-dimensional torus for c_2 = -3.75.

4. Conclusion

The results of numerical analysis of solutions of the Kuramoto–Tsuzuki equation in the space of Fourier coefficients for the few-mode approximation as well as for the original system and in the space of phase variables of this equation permit one to make the following conclusions.

1. The transition to chaotic modes in the Kuramoto–Tsuzuki equation is based on the same mechanisms as in nonlinear dissipative systems of ordinary differential equations [2, 8, 9]; more precisely, for solutions in the space of few-mode approximations these mechanisms include cascades of period doubling bifurcations (Feigenbaum cascades) and subharmonic cascades of cycle bifurcations in accordance with Sharkovskii order. For solutions in the infinite-dimensional phase space, these mechanisms include cascades of Feigenbaum bifurcations and subharmonic cascades of bifurcations of invariant tori; in addition, it was shown that cascades of period multiplication for invariant tori take place with respect to the internal as well as external frequency.

2. It is not adequate to use three-dimensional few-mode approximations to describe diffusion chaos in diffusion-type equations. In three-dimensional few-mode systems, singular attractors are generated only by bifurcations of limit cycles, while in the corresponding diffusion equations, the generation of singular attractors is caused by cascades of bifurcations of at least two-dimensional (no other have been found yet) invariant tori. Moreover, the corresponding bifurcation diagrams of solutions have substantial differences.

3. The following conjecture is likely to be true: the appearance of spatiotemporal chaos in systems of partial differential equations is caused by cascades of bifurcations of two-dimensional invariant tori rather than the destruction of a three-dimensional torus with the generation of some hypothetical strange attractor, as was assumed in modern publications following [13].

Acknowledgments

It is a pleasure to thank S.Sidorov for numerical calculations for this paper. The work was financially supported by the Russian Foundation for Basic Research (project no. 04-01-00225a) and the program OITVS, Russian Acad. of Sciences (project no.1.12).

References

[1] Kuramoto Y, Tsuzuki T. On the formation of dissipative structures in reaction-diffusion systems. Progr. Theor. Phys 1975; 54, 3: 687-699.
[2] Magnitskii NA, Sidorov SV. New Methods for Chaotic Dynamics. URRS: Moscow; 2004.
[3] Magnitskii NA, Sidorov SV. A new view of the Lorenz attractor. Differential Equations 2001; 37, 11: 1568-1579.
[4] Magnitskii NA, Sidorov SV. Transition to chaos in the Lorenz system through a complete double homoclinic cascade of bifurcations. In: Nonlinear Dynamics and Control, S.V.Emel'yanov, S.K.Korovin (Eds.), vol 2. Fizmatlit: Moscow; 2002; 179-194.
[5] Magnitskii NA, Sidorov SV. On scenarios of transition to chaos in nonlinear dynamical systems In: Nonlinear Dynamics and Control, S.V.Emel'yanov, S.K.Korovin (Eds.), vol 3. Fizmatlit: Moscow; 2003;73-98.
[6] Feigenbaum M. Quantitative universality for a class of nonlinear transformations. J. Stat. Phys. 1978; 19: 25-52.
[7] Sharkovskii AN. Cycles coexistence of continuous transformation of line in itself. Ukr. Math. Journal 1964; 26, 1: 61-71.
[8] Magnitskii NA. On nature of chaotic attractors. In: Nonlinear Dynamics and Control, S.V.Emel'yanov, S.K.Korovin (Eds.), vol 4. Fizmatlit: Moscow; 2004 ; 37-58.
[9] Magnitskii NA, Sidorov SV. Rotor type singular points of non-autonomous systems of differential equations and their role in generation of singular attractors of nonlinear autonomous systems. Differential Equations 2004; 40, 11: 1579-1593.
[10] Magnitskii NA., On singular attractors of dissipative systems of nonlinear ordinary differential equations. In: Proc. of ENOC-2005 Int. Conf.; Univ. Technol.: Eindhoven; 2005: 1285-1294.
[11] Magnitskii NA, Sidorov SV. Transition to chaos in nonlinear dynamical systems through a subharmonic cascade of bifurcations of two-dimensional tori. Differential Equations 2002; 38, 12: 1703-1708.
[12] Akhromeeva TS, Kurdyumov SP, Malinetskii GG, Samarskii AA. Nonstationary Structures and Diffusion Chaos. Nauka: Moscow; 1992.
[13] Ruelle D, Takens F. On the nature of turbulence. Commun. Math. Phys. 1971; 20, 3: 167-192.

Fractional calculus for transport in disordered semiconductors

Vladimir V. Uchaikin[1], Renat T. Sibatov

Case Western Reserve University, Cleveland, USA
Ulyanovsk State University, Ulyanovsk, Russia

Abstract

The charge carrier transport in disordered semiconductors is described by making use of the fractional calculus. The physical reasons of introducing fractional derivatives in semiconductor theory are discussed, the process of derivation of fractional differential equations is demonstrated, solutions of fractional differential equations are represented in terms of one-sided stable distributions. The solutions are applied to multiple trapping phenomenon, transient photocurrent and drift mobility, dispersive transport percolation model, transport in bilayer semiconductor. Some numerical results are obtained and their agreement with experimental data is demonstrated.

PACS: 05.40.Fb; 73.50-h; 73.63-b
Keywords: dispersive transport, disordered semiconductor, fractional calculus, stable distribution

1 Introduction

Transport of charge carriers in disordered semiconductors can show anomalous properties contradicting to the Fickian law and Gaussian diffusion in general [1]. In the semiconductor flight experiment, holes are injected near a positive electrode (say, at time $t = 0$), and then transported to a negative electrode where they are absorbed. Moving without trapping with a constant velocity v, they form a rectangular pulse of the photocurrent

$$I(t) \propto \begin{cases} \text{const}, & t < t_T, \\ 0, & t > t_T \end{cases} \tag{1}$$

where the *transient time* t_T is determined by the velocity v and and the size of the sample l: $t_T = l/v$ (because of the high resistivities of amorphous semiconductors the dielectric relaxation time is much longer than the transit time of the carriers). In reality, trapping and thermal release of the carriers cause spreading the packet of carriers. The carrier packet has a symmetrical Gaussian form with mean value $\langle x(t)\rangle \propto t$ and the width $\Delta x(t) \propto \sqrt{t}$ depending on the mean waiting time $\langle\tau\rangle$. Now, $I(t)$ remains constant till the front of the Gaussian packet reaches the end of the sample, then it begins to fall off and disappear during the time $\Delta x/\langle v\rangle$. As a result, we observe a smooth right edge of the photocurrent pulse. Such picture is observed in most ordered material.

However, when determining the drift mobility in some disordered, amorphous and porous semiconductors a specific shape of the transition current is observed. It consists of two time parts of power type with different exponent and some intermediate region between them:

$$I(t) \propto \begin{cases} t^{-1+\alpha}, & t < t_T, \\ t^{-1-\alpha}, & t > t_T, \end{cases} \quad \alpha < 1. \tag{2}$$

[1]Corresponding author. Tel.: +7-8422-320612
E-mail address: uchaikin@sv.uven.ru (V. V. Uchaikin)

The exponent α may depend on temperature and on other control parameters, while the shape of the normalized signal $I(t)/I(t_T)$ versus dimensionless time t/t_T practically does not depend on the applied voltage and size of the sample. This phenomenon was designated *universal*.

The widespread of these features in a variety of disordered materials confirms the universality of the approach A large amount of experimental evidences of the universality has been presented since the first works till the recent ones [1-13].

Several first attempts to explain this phenomenon by involving various aspects of trap limited band transport under conditions where relatively few trapping and release events are experienced by the average carrier during transit have been made without much success [14-16].

A successful interpretation of the anomalous transport called also the *dispersive transport*, with respect to both the time dependence and the pulse form has been provided by Tunaley [2-4] and Scher with Montroll and other collaborators [5-7,10] on the base of the *continuous time random walk* (CTRW) model with an infinite mean waiting time.

2 CTRW model

The basic assumptions of this model called sometimes the Scher-Montroll model are as follows:

(*i*) The transport of charge carriers is one-dimensional jump process that is the continuous time random walk (CTRW) process when the walker changes its position instantaneously at random times.

(*ii*) The carriers jump in the direction of the applied external field independently of each other, intervals between jumps (*waiting* or *trapping times*) are independent identically distributed random variables τ.

(*iii*) The measured current is interpreted as the time derivative of the mean displacement of a carrier $d\langle x(t)\rangle/dt$.

We consider the simplest one-dimensional CTRW model [1] assuming that jumps are performed only in the direction of applied gradient. The lengths of jumps ξ_j and waiting times τ_j are mutually independent random variables with distributions $\mathsf{P}\{\xi > x\} = P(x)$ and $\mathsf{P}\{\tau > t\} = Q(t)$ respectively. The corresponding densities are $p(\xi) = -dP(\xi)/d\xi$ and $q(\xi) = -dQ(\xi)/d\xi$. The particles perform jumps with a constant finite velocity v. Thus, there exists two states for the particle: the state of rest (R) and the state of motion (M).

Let $w(x,t)dx$ be the probability, that the particle starting its history with the transition $M \to R$ at the origin at $t = 0$ will be observed in $(x, x+dx)$ at time $t > 0$. It is known that the Fourier-Laplace transform of this density is expressed through the Fourier-transform $\widetilde{p}(k)$ and Laplace transforms $\widetilde{Q}(\lambda)$ and $\widetilde{q}(\lambda)$ as follows:

$$\widetilde{w}(k,\lambda) = \frac{\widetilde{Q}(\lambda)}{1-\widetilde{q}(\lambda)\widetilde{p}(k)}. \tag{3}$$

3 The normal transport process

On assumption that the mean square of jump length $\langle\xi^2\rangle$ and the mean waiting time $\langle\tau\rangle$ are finite, we can use for the diffusion approximation ($k \to 0, \lambda \to 0$) the expressions

$$\widetilde{p}(k) \sim 1 + ik\langle\xi\rangle - k^2\langle\xi^2\rangle/2,$$

$$\widetilde{q}(\lambda) \sim 1 - \lambda\langle\tau\rangle,$$

and

$$\widetilde{Q}(\lambda) = \frac{1-\widetilde{q}(\lambda)}{\lambda} \sim \langle\tau\rangle.$$

Substituting them in Eq.(3) yields

$$[\lambda - iKk + Dk^2]\widetilde{f}(k,\lambda) = 1, \tag{4}$$

where

$$K = \langle\xi\rangle/\langle\tau\rangle, \; D = \langle\xi^2\rangle/\langle 2\tau\rangle,$$

and $\widetilde{f}(k,\lambda)$ is the Fourier-Laplace transform of the leading asymptotical term $f(x,t)$ of the solution $w(x,t)$.

Taking into account that in the inverse Fourier-Laplace transform process

$$\lambda \mapsto \frac{\partial}{\partial\tau}, \quad -ik\widetilde{f} \mapsto \frac{\partial f}{\partial x}, \quad -k^2\widetilde{f} \mapsto \frac{\partial^2 f}{\partial x^2}, \quad 1 \mapsto \delta(t)\delta(x),$$

we arrive at the normal diffusion equation with drift:

$$\frac{\partial f}{\partial t} + K\frac{\partial f}{\partial x} = D\frac{\partial^2 f}{\partial x^2} + \delta(t)\delta(x), \tag{5}$$

describing the normal transport process in ordered semiconductors. The corresponding equation for disordered semiconductors differs from it by presence of *derivatives of fractional order.*

4 Where fractional derivatives come from

It is very interesting to observe lovely discussion about geometrical and physical meaning of fractional derivatives , although one can doubt in its influence on physical and engineering applications of fractional calculus. Even for derivatives of integer orders, forming a basement of our mathematical thought, we can not formulate a short *common for all integer* n answer to question: what does the nth derivative mean?

Fractional derivatives in consideration of *concrete problems* appear not from mysterious conjuration about complexity or fractality or memory but from *experimental data* or derived from them mathematical models. Introducing the Riemann-Liouville fractional derivative

$$f^{(\alpha)}(t) \equiv \frac{d^\alpha f(t)}{dt^\alpha} = \frac{1}{\Gamma(1-\alpha)}\frac{d}{dt}\int_0^t \frac{f(t')dt'}{(t-t')^\alpha}, \ 0<\alpha<1, \tag{6}$$

we have to prove that the related quantity has the form $\frac{d}{dt}\int_0^t K(t-t')f(t')dt'$ and present the evidence that $K(t) \propto t^{-\alpha}$. Dealing with theoretical models, it is convenient to use the Laplace transform of (6):

$$\widehat{f}^{(\alpha)}(\lambda) \equiv \int_0^\infty e^{-\lambda t} f^{(\alpha)}(t)dt = \lambda^\alpha \widehat{f}(\lambda), \ 0<\alpha<1. \tag{7}$$

This way of derivation of fractional equations is widely used in anomalous transport theory (see, for example, [17],[22]). As one can see below, the mathematical reason for appearance fractional derivatives with respect to time is a long tail distribution

$$Q(t) \equiv \mathsf{P}(\tau > t) \propto t^{-\alpha}, \ 0<\alpha<1, \tag{8}$$

having infinite mean $\langle\tau\rangle$.

The physical grounds of (8) has been discussed in a number of works. The first of interpretation of (8) has been done by Tunaley [2-4] on the base of the following simple jump mechanism. The process goes in an insulator containing randomly distributed point (of small size) traps with exponentially distributed waiting times: $\mathsf{P}\{\tau > t|\theta\}$. Their mean value θ is finite and linked with the random distant δ to the nearest site in the direction of the applied field as follows [23]:

$$\theta = \beta[\exp(\gamma\delta) - 1].$$

Here, γ is a positive constant and β is inversely proportional to the applied potential gradient, both are independent of the temperature of the sample. Taking for δ the exponential distribution with the mean d,

$$\mathsf{P}\{\delta > x\} = e^{-x/d},$$

we obtain the probability density for θ in the following form

$$\mathsf{P}\{\theta > t\} = \mathsf{P}\{\delta > (1/\gamma)\ln(1+t/\beta)\} = \exp[-(1/\gamma d)\ln(1+t/\beta)] = (1+t/\beta)^{-1/(\gamma d)}.$$

Averaging these distribution over d

$$\mathsf{P}\{\tau > t\} = -\int_0^\infty \mathsf{P}\{\tau > t|t'\}d\mathsf{P}\{\theta > t'\} \sim \alpha\Gamma(\alpha)(t/\beta)^{-\alpha}, \ t\to\infty, \ \alpha = 1/(\gamma d)$$

yields (8).

In [10], it has been indicated that the dispersive behavior can be caused also by multiple trapping in a distribution of localized states. On assumption that the localized states below the mobility edge fall off exponentially with energy, one can arrive at Eq.(2) with the exponent α depending on temperature T. In frame of this model, called *random activated energy model*(see also [1,24]) it is assumed, that

(i) the jump rate of a particle hopping over an energy barrier ΔE has the usual quasiclassical form

$$W = \nu e^{-\Delta E/kT};$$

(ii) the conditional waiting time distribution corresponding to a given activation energy $\Delta E = \varepsilon$ is exponential

$$\mathsf{P}\{\tau > t|\varepsilon\} = e^{-W(\varepsilon)t};$$

and

(iii) the activation energy is a random variable with the Boltzman distribution density

$$p(\varepsilon) = (kT_{rmc})^{-1}e^{-\varepsilon/kT_c}.$$

Averaging over activation energy results:

$$\mathsf{P}\{\tau > t\} = \int_0^\infty \mathsf{P}\{\tau > t|\varepsilon\}p(\varepsilon)d\varepsilon = \int_0^\infty \exp[-(\nu e^{-\varepsilon/kT})t]d(e^{-\varepsilon/kT_{rmc}}) = \alpha\Gamma(\alpha)(\nu t)^{-\alpha}$$

with $\alpha = T/T_c$. Here, T_c is the characteristic temperature defining the conduction band tail. For $T < T_c$, thermalization dominates and the photoinjected carriers sink progressively deeper in increasing time; the transport becomes dispersive. For $T > T_c$, the carriers remain concentrated near the mobility edge and the charge transit exhibit non-dispersive behavior. Consequently, the physical meaning of α is that it is representative of disorder: the smaller its value the more dispersive the transport.

The multiple trapping model was successfully used for interpretation of some experimental results in [25]and other works (see [26-30]).

These theoretical results being in accordance with numerous experimental data represent *physical reasons* for using fractional equations when considering transport of charges in amorphous semiconductors.

5 How fractional derivatives come into equations

Taking the waiting time distribution of the inverse power type (8) we obtain that at small λ

$$\widetilde{Q}(\lambda) \sim \lambda^{-1}(\lambda/c)^\alpha, \quad \widetilde{q}(\lambda) \sim 1 - (\lambda/c)^\alpha, \quad \lambda \to 0.$$

Substituting these expressions in Eq.(3) yields

$$\widetilde{w}(k,\lambda) \sim \frac{c^{-\alpha}\lambda^{\alpha-1}}{1 - [1-(\lambda/c)^\alpha]/(1-ik\langle\xi\rangle)}, \quad \lambda \to 0.$$

After some simple algebra, we get for the Fourier-Laplace transform $\widetilde{f}(k,\lambda)$ of asymptotical (in the diffusion sense) part of the solution the following algebraic equation :

$$[\lambda^\alpha - ic^\alpha\langle\xi\rangle k + c^\alpha\langle\xi^2\rangle k^2]\widetilde{f}(k,\lambda) = \lambda^{\alpha-1}, \; 0 < \alpha < 1. \tag{9}$$

Taking into account that in the inverse Fourier-Laplace transform process

$$\lambda^\alpha \widetilde{f} \mapsto \frac{\partial^\alpha f}{\partial t^\alpha}, \quad \text{and} \quad \lambda^{\alpha-1} \mapsto \frac{t^{-\alpha}}{\Gamma(1-\alpha)}\delta(x),$$

we arrive at the following differential equation for spatial distribution of carriers at time t:

$$\frac{\partial^\alpha f}{\partial t^\alpha} + K\frac{\partial f}{\partial x} = D\frac{\partial^2 f}{\partial x^2} + \frac{t^{-\alpha}}{\Gamma(1-\alpha)}\delta(x) \tag{10}$$

where

$$K = c^\alpha\langle\xi\rangle \quad \text{and} \quad D = c^\alpha\langle\xi^2\rangle.$$

The second term on the left hand side results from asymmetry of walking; in the symmetrical case we have (see [17])

$$\frac{\partial^\alpha f}{\partial t^\alpha} = D\frac{\partial^2 f}{\partial x^2} + \frac{t^{-\alpha}}{\Gamma(1-\alpha)}\delta(x), \quad -\infty < x < \infty,\ 0 < t < \infty. \tag{11}$$

However, the model under consideration relates to extreme asymmetrical one-directional walking. In this case, one can neglect the terms with derivatives of orders higher 1 (see Eq.(8)) and use the limit reduced form of the transport equation (the *drift approximation*):

$$\frac{\partial^\alpha f}{\partial t^\alpha} + K\frac{\partial f}{\partial x} = \frac{t^{-\alpha}}{\Gamma(1-\alpha)}\delta(x),\ x \geq 0,\ t > 0. \tag{12}$$

In what follows we shall need the equation

$$\frac{\partial^\alpha f}{\partial t^\alpha} + K\frac{\partial f}{\partial x} = \delta(x)\delta(x),\ x \geq 0,\ t > 0. \tag{13}$$

6 Solutions in terms of α^+-densities

Solutions of fractional equations can be expressed through some special functions, such as Mittag-Leffler functions, Fox H-functions and others. We prefer to represent them in terms of *one-sided alpha-stable densities* (we will call them α^+-densities) which can also be regarded as a kind of special functions. The α^+-density denoted by $g^{(\alpha)}(t)$ is determined by its Laplace transform as follows:

$$\widetilde{g}^{(\alpha)}(\lambda) \equiv \int_0^\infty g^{(\alpha)}(t)e^{-\lambda t}dt = e^{-\lambda^\alpha},\ 0 < \alpha \leq 1.$$

It is equal to 0 on the negative semiaxis including the origin, positive on the positive semiaxis and satisfies the normalization condition

$$\int_0^\infty g^{(\alpha)}(t)dt = 1.$$

Only two densities of the family are expressed in an explicit form:

$$g^{(1)}(x) = \delta(x-1) \quad \text{and} \quad g^{(1/2)}(t) = \frac{1}{2\sqrt{\pi}}t^{-3/2}\exp[-1/(4t)],\ t > 0.$$

Each density $g^{(\alpha)}(x)$ has finite moments only of order $\nu < \alpha$:

$$\int_0^\infty g^{(\alpha)}(t)t^\nu dt = \begin{cases} \Gamma(1-\nu/\alpha)/\Gamma(1-\nu), & -\infty < \nu < \alpha; \\ \infty, & \nu \geq \alpha, \end{cases}$$

The solution to (11) is represented through α^+ density by means of the following integral [17,22]:

$$f(x,t) = \frac{1}{2\sqrt{\pi D t^\alpha}}\int_0^\infty \exp\left(-\frac{x^2\tau^\alpha}{4Dt^\alpha}\right) g^{(\alpha)}(\tau)\tau^{\alpha/2}d\tau. \tag{14}$$

By making use the integral transform method, one can find solutions to Eq's (12)

$$f(x,t) = \frac{t}{\alpha K}\left(\frac{x}{K}\right)^{-1-1/\alpha} g^{(\alpha)}\left(t\left(\frac{x}{K}\right)^{-1/\alpha}\right) \tag{15}$$

and (13)

$$f(x,t) = \frac{1}{K}\left(\frac{x}{K}\right)^{-1/\alpha} g^{(\alpha)}\left(t\left(\frac{x}{K}\right)^{-1/\alpha}\right). \tag{16}$$

When $\alpha = 1$ Eq.(14) describes the normal diffusion process,

$$f(x,t) = \frac{1}{2\sqrt{\pi D t}} \exp\left(-\frac{x^2}{4Dt}\right)$$

while Eq's (15) and (16) represent the same ballistic regime

$$f(x,t) = \delta(x - Kt)$$

(one can easily prove by making use the Laplace transform method, that $t^{-\alpha}/\Gamma(1-\alpha) \to \delta(t)$ when $\alpha \to 1$).

7 Application to some transport problems

7.1 Multiple trapping

The transport equation can be generalized to the case of multiple trapping in a distribution of localized states as follows.

The total concentration of charge carriers $P(\mathbf{r},t)$ is equal to sum of delocalized carriers concentration $P_d(\mathbf{r},t)$ and trapped carriers concentration $P_l(\mathbf{r},t)$. The total current density is given by

$$\mathbf{j} = e\mu P_d \mathbf{E} - eD\nabla P_d, \tag{17}$$

where e is the charge of a carrier, $\mathbf{E}$ is the electric field, μ and D are the mobility and diffusivity of delocalized carriers. The continuity equation is of the form:

$$e\left(\partial P_l/\partial t + \partial P_d/\partial t\right) + \mathrm{div}\mathbf{j} = eS(\mathbf{r},t), \tag{18}$$

where $S(\mathbf{r},t)$ is the source density, that is the number of carriers injected at $(\mathbf{r},t)$ per unit volume and time. In [33], the following interrelation between the Laplace transforms are derived from kinetic equations for trapping-emitting processes:

$$(\lambda/\omega)\widetilde{P}_l(\mathbf{r},\lambda) = A(\lambda/\omega)^{\alpha}\widetilde{P}_d(\mathbf{r},\lambda).$$

Here $\alpha = T/T_c < 1$, $A = \pi\alpha/\sin\pi\alpha$ and ω is the rate of trapping by localized states. The backward transform yields [17,32]

$$\frac{\partial P_l}{\partial t} = A\omega^{1-\alpha}\frac{\partial^\alpha P_d}{\partial t^\alpha}. \tag{19}$$

Gathering Eq's (17)-(19), we arrive at the desired generalization:

$$\frac{\partial P_d}{\partial t} + A\omega^{1-\alpha}\frac{\partial^\alpha P_d}{\partial t^\alpha} + \mu\mathbf{E}\nabla P_d - D\nabla^2 P_d = S(\mathbf{r},t). \tag{20}$$

7.2 Transient photocurrent and drift mobility

For times significantly exceeding a single trapping time we use the drift approximation of Eq.(20). In one-dimensional case, under specific initial conditions when N photoinjecred carriers per unit area appear, we have

$$A\omega^{1-\alpha}\frac{\partial^\alpha P_d}{\partial t^\alpha} + \mu E\frac{\partial P_d}{\partial x} = N\delta(x)\delta(t).$$

The solution of this equation is of the form:

$$P_d = (N/l)(xA/l)^{-1/\alpha} g^{(\alpha)}\left(\omega t (xA/l)^{-1/\alpha}\right)$$

where $l = \langle\xi\rangle = \mu E/\omega$ is the mean path of a carrier between trapping events and x-axis is directed along $\mathbf{E}$.

The transient photocurrent $I(t)$ in a sample of the length L $j(x,t) = e\mu P_d E$ is determined through the conductivity current density as

$$I(t) = (1/L)\int_0^L j(x,t)dx = \frac{e\mu EN\alpha}{AL}(\omega t)^{-1+\alpha}\int_{s_0}^{\infty} s^{-\alpha} g^{(\alpha)}(s)ds, \quad s_0 = \omega t(LA/l)^{-1/\alpha}. \tag{21}$$

For small times $t \ll t_{\mathrm{T}}$, one can replace the lower limit in (21) by 0 and use for calculations the expression (ii) for negative order moments:

$$I(t) \sim \frac{e\mu EN\alpha}{AL}(\omega t)^{-1+\alpha}\int_0^\infty s^{-\alpha}g^{(\alpha)}(s)ds = \frac{e\mu EN\sin\pi\alpha}{\pi L\Gamma(1+\alpha)}(\omega t)^{-1+\alpha}, \quad t < t_{\mathrm{T}}. \tag{22}$$

At large times $t \ll t_{\mathrm{T}}$, we can use for $g^{(\alpha)}(s)$ its first asymptotical term

$$g^{(\alpha)}(s) \sim \frac{\alpha}{\Gamma(1-\alpha)}s^{-\alpha-1},$$

and get

$$I(t) \sim \frac{e\mu ENL\alpha\Gamma(1+\alpha)}{2l^2}(\omega t)^{-1-\alpha}, \quad t > t_{\mathrm{T}}. \tag{23}$$

Equating (22) and (23), we obtain the estimation of the parameter t_T,

$$t_{\mathrm{T}} = \omega^{-1}\left(\frac{L\omega}{\mu E}\sqrt{\frac{A}{2}}\Gamma(1+\alpha)\right)^{1/\alpha}$$

being in agreement with the known relation $t_{\mathrm{T}} \propto (L/E)^{1/\alpha}$[1,10]. The drift mobility is determined by

$$\mu_D = L/Et_{\mathrm{T}}.$$

Results of numerical calculations of the transient photocurrent by means of formula (22) in comparison with experimental data are shown on Fig.1.

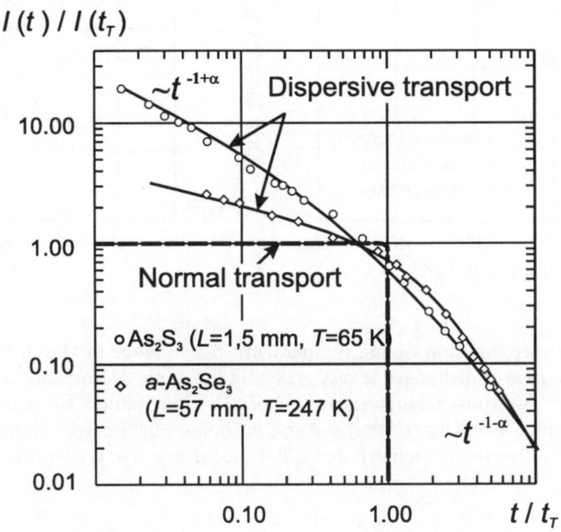

Figure 1: The transient current density (circles represent experimental data for glassy As_2S_3 [34], diamonds show the data for $a-As_2Se_3$ [35], curves give results of our calculation by (19), fitting parameters are α and t_{T}).

7.3 Dispersive transport and percolation

An infinite percolation cluster above the percolation threshold has fractal structure on scales no exceeding some value ξ called the *correlation length*. As reported in [36], transport processes in this region manifest subdiffusion behavior. Fractal model is offered in [37] for description of porous semiconductors: the porous silicon is considered as a set of clusters of silicon atoms surrounded by complexes of SiO_x. Because of dispersion of nanocrystals both in their sizes and in distances between each other the porous silicon is a disordered structure. Investigation of the transient photocurrent in porous silicon samples has shown an essential influence of localized states on the transport process [38]. The values of dispersive transport parameters found in these experiments significantly exceed those predicted by multi-trapping model. By opinion of the authors of [39], the additive spreading of the diffusive packet can be explained by admitting scattering in mobilities caused by inhomogeneity of the material. The temperature dependence of the dispersive parameter experimentally obtained in [11] contradicts to multiple trapping model in frame of which $\alpha \propto T$. Haw can one explain the weak temperature dependence of the dispersive parameter if the tunnelling mechanism does not dominate in this temperature region yet?

In the frame of the percolation model the dispersive parameter depends weakly on temperature and determined mainly by porosity of the material and its structure. The specific features of the dispersive transport are caused by asymptotically power law of time spending by carriers on fractal fragments of percolation clusters. Let us have a look at the data of [11] from point of view of these assumptions. If we approximate the temperature dependence of the dispersive parameter by a linear function $\alpha = c_1 T + c_2$ with small coefficient c_1 (see Fig 2, left panel) and use our formulas (22)-(23) we obtain a good agreement with experiments(Fig 2, right panel).

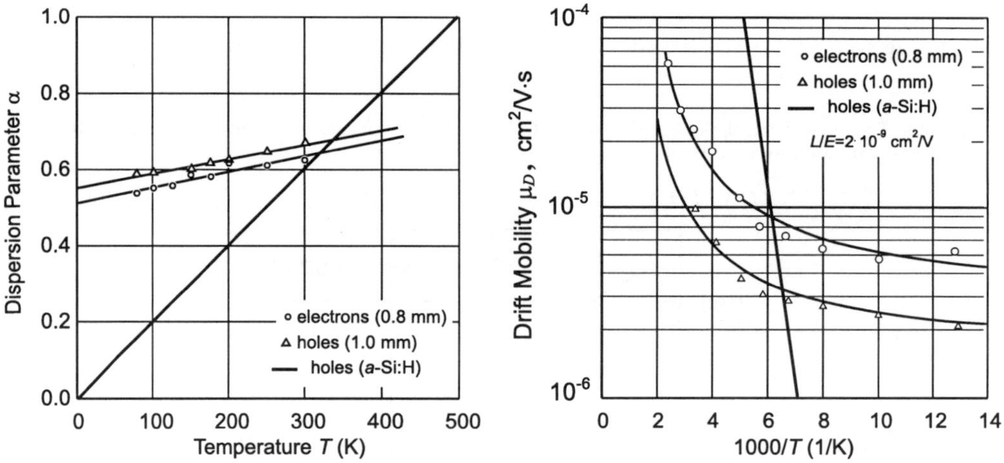

Figure 2: Left panel: approximation of the temperature dependence of the dispersive parameter in the layer of porous silicon in case of dispersive transport (solid line is the linear approximation of experimental data [11], circles are for electrons, triangles are for holes). Right panel: the temperature dependence of drift mobility(circles are for electrons, triangles stand for holes, curve shows results of calculations by formulas (22)-(23), the curve represents experimental data for dispersive transport of holes in hydrogenated amorphous silicon[40]).

7.4 Transient photocurrent in a bilayer semiconductor

Now, we consider transport of charge carriers in the system disordered semiconductor (labelled 1) - crystalline semiconductor (labelled 2). The corresponding equations have the form

$$A\omega^{1-\alpha}\frac{\partial^\alpha P_{d1}}{\partial t^\alpha} + \mu_1 E_1 \frac{\partial P_{d1}}{\partial x} = 0, \quad 0 < x \leq L_1,$$

and

$$\frac{\partial P_{d2}}{\partial t} + \mu_2 E_2 \frac{\partial P_{d2}}{\partial x} = 0, \quad L_1 \leq x \leq L_2,$$

with the initial and boundary conditions

$$P_{d1}(x,0) = N\delta(x), \quad P_{d1}(L_1,t) = P_{d2}(L_1,t).$$

The solution of the first equation is known already:

$$P_{d1}(x,t) = (N/l)(xA/l)^{-1/\alpha} g^{(\alpha)}\left(\omega t (xA/l)^{-1/\alpha}\right), \quad 0 < x \le L_1,$$

the solution of the second one is expressed through the first one taken on the boundary by

$$P_{d2}(x,t) = \int_0^t P_{d1}(L_1, t-\tau)\delta\left(\tau - \frac{x-L_1}{\mu_2 E_2}\right) d\tau = \frac{N}{l}\left(\frac{L_1 A}{l}\right)^{-1/\alpha} g^{(\alpha)}\left(\omega\left(t - \frac{x-L_1}{\mu_2 E_2}\right)\left(\frac{L_1 A}{l}\right)^{-1/\alpha}\right),$$

$$x \in (L_1, L_1+L_2) \bigcap (L_1, L_1 + \mu_2 E_2 t).$$

The transient current in the bilayer system can be written as the sum

$$I(t) = (1/L)\int_0^L j(x,t)dx = (L_1/L)I_1(t) + (L_2/L)I_2(t)$$

where $L = L_1 + L_2$,

$$I_1(t) = \frac{e\mu_1 E_1}{L_1}\int_0^{L_1} P_{d1}(x,t)dx = \frac{e\mu_1 E_1 N\alpha}{AL_1}(\omega t)^{-1+\alpha}\int_\tau^\infty s^{-\alpha} g^{(\alpha)}(s)ds, \quad \tau = \omega t (L_1 A/l)^{-1/\alpha}$$

and

$$I_2(t) = \frac{e\mu_2 E_2}{L_2}\int_{L_1}^{L_1+L_2} P_{d2}(x,t)dx$$

$$= \begin{cases} \frac{N}{l}\frac{e(\mu_2 E_2)^2}{L_2\omega} G^{(\alpha)}\left[\omega t\left(\frac{L_1 A}{l}\right)^{-1/\alpha}\right], & t < \frac{L_2}{\mu_2 E_2}, \\ \frac{N}{l}\frac{e(\mu_2 E_2)^2}{L_2\omega}\left[G^{(\alpha)}\left(\omega t\left(\frac{L_1 A}{l}\right)^{-1/\alpha}\right) - G^{(\alpha)}\left(\omega\left(t - \frac{L_2}{\mu_2 E_2}\right)\left(\frac{L_1 A}{l}\right)^{-1/\alpha}\right)\right], & t > \frac{L_2}{\mu_2 E_2}. \end{cases}$$

Here,

$$G^{(\alpha)}(t) = \int_0^t g^{(\alpha)}(s)ds = 1 - \int_t^\infty g^{(\alpha)}(s)ds$$

is the cumulative distribution function of α^+ random variable. Introducing designations

$$C_1 = (N/l)e\mu_1 E_1 \alpha (L_1 A/l)^{-1/\alpha}, \quad \gamma = (\mu_2 E_2)^2 (L_1 A/l)^{1/\alpha}/L_2 \omega \mu_1 E_1 \alpha,$$

$$\tau = \omega t (L_1 A/l)^{-1/\alpha}, \quad \tau_0 = \omega (L_2/\mu_2 E_2)(L_1 A/l)^{-1/\alpha},$$

we rewrite the expressions for currents components in a more compact form:

$$I_1(t)/C_1 = \tau^{-1+\alpha}\int_\tau^\infty s^{-\alpha} g^{(\alpha,1)}(s)ds,$$

$$I_2(t)/C_1 = \begin{cases} \gamma \; G^{(\alpha)}(\tau), & \tau < \tau_0, \\ \gamma\left[G^{(\alpha)}(\tau) - G^{(\alpha)}(\tau - \tau_0)\right], & \tau > \tau_0. \end{cases}$$

The parameters τ_0 and γ determine position and value of the peak of the transition current curve. The explicit expressions for them are given by

$$\tau_0 = \frac{t_{T2}}{t_{T1}}\left[\frac{\Gamma(1+\alpha)}{\sqrt{2A}}\right]^{1/\alpha}, \qquad \gamma = \frac{\mu_2 E_2}{\mu_1 E_1}\frac{1}{\tau_0}.$$

Fig. 3 displays the comparison of theoretical results obtained within fractional differential model of charge transport with the time-of-flight experimental data for the amorphous semiconductor - crystalline semiconductor (solid lines are the experimental data for the structure a-$Se_{95}As_5$ – c-CdSe from [41], dashed lines are theoretical results).

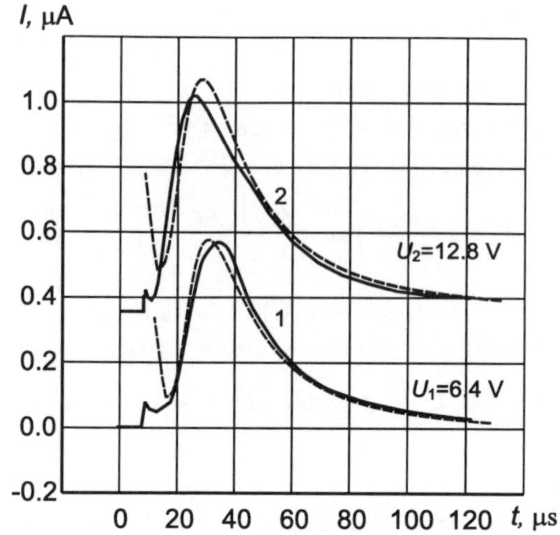

Figure 3: Transient current in the structure amorphous semiconductor – crystalline semiconductor (solid lines are the experimental data for the structure a-$Se_{95}As_5$ – c-CdSe from [41], dashed lines are theoretical results).

8 Concluding remark

We tried to show in this work the physical reasons for introducing fractional derivatives into semiconductor theory, demonstrate derivation of equations with fractional derivatives, their solution, and some numerical results. At the end of the article, we would like to stress the deep link between three aspects of the problem under consideration: disorder, non-Gaussian statistics and non-integer (fractional) differential calculus. A bright example of successful applications of the ideas to engineer practice is given by fractional rheology [42].

References

[1] Madan A., Shaw M. P. The Physics and Applications of Amorphous Semiconductors. Academic Press, Inc, Boston, 1988.

[2] Tunaley J. K. E. J. Appl. Phys. **43**, 1972, p. 3851;

[3] Tunaley J. K. E. J. Appl. Phys. **43**, 1972, p. 4777;

[4] Tunaley J. K. E. J. Appl. Phys. **43**, 1972, p. 4783.

[5] Scher H., Lax M. J.Non-Cryst. Solids **8/10**, 1972, p. 497;

[6] Scher H., Lax M. Phys. Rev. **B7**, 1973, p. 4491;

[7] Scher H., Lax M. Phys. Rev. **B7**, 1973, p. 4502;

[8] Pfister G. Phys. Rev. Lett., 1974, V.33, p.1474.

[9] Sharfe M. E. Bull. Am. Phys. Soc., 1973, V.18, p.454.

[10] Scher H., Montroll E. W. Phys. Rev. **B**, 1975, V.12, p. 2455.

[11] Rao P. N., Schiff E. A., Tsybeskov L., Fauchet P. Chem. Physics., 2002, 284, p.129.

[12] Jonscher A. K. Dielectric Relaxation in Solids, London, Chelsea Dielectric Press, 1983.

[13] Jonscher A. K. Universal Relaxation Law, London, Chelsea Dielectric Press, 1996.

[14] Silver M., Dy K. D., Huang D. L. Phys. Rev. Lett. **27**, 1971.

[15] Marshall J. H., Owen A. E. Phil. Mag. **24**, 1971, p. 1281.

[16] Fox S. J., Locklar H. C. J. Non-Cryst. Solids **8-10**, 1971, p. 552.

[17] Uchaikin V. V. J. of Exper. and Theor. Phys., 1999, V. 115, p. 2113.

[18] Barkai E. Phys. Rev. **E63**, 04618, 2001.

[19] Uchaikin V. V., Sibatov R. T. Review of Appl. and Industr. Math., 2005, V. 12, p. 540.

[20] Sibatov R. T., Uchaikin V. V. Review of Appl. and Industr. Math., 2005, V. 12, p. 1085.

[21] Sibatov R. T. Uch. Zapiski UlGU, 2005, V. 1(17), p. 65.

[22] Saichev A. I., Zaslavsky G. M. Chaos **7**,753, 1997.

[23] Harper W. R. Contact and Frictional Electrification, Oxford Univ. Press, Oxford, England, 1967.

[24] Vlad M. O. Physica **A184**, 1992, p. 303.

[25] Tiedje T., Gebulka J. M., Morel D. L.,Abeles B. Phys. Rev. Lett. **46**, 1981, 1425.

[26] Arhipov V. I., Rudenko A. I., Andriesh A. M. and others. Non-stationary injection currents in solids. Kishinev, 1983, 175 p.

[27] Tiedje T., Rose A. W. Solid State Commun. **37**, 1981, p. 49.

[28] Spear. V. In: Amorphous Silicon and Related materials: – Edited by Frietzsche. – World Scientific, 1991, 544 p.

[29] Arhipov V. I., Kazakova L. P., Lebedev E. A., Rudenko A. I. Semiconductors, 1988, V. 22, p.723.

[30] Arhipov V. I., Nikitenko V. R. Semiconductors, 1989, V. 23, p. 978.

[31] Uchaikin V V, Physica **A 255**, 65, 1998.

[32] Uchaikin V V, Zolotarev V M, Chance and Stability, VSP, Utrecht, tne Netherlands, 1999.

[33] Tiedje T. In.: The Physics of Hydrogenated Amorphous Silicon II. Electronic and Vibrational Properties: – Edited by Joannoulos J. D. and Lucovsky G. – Springer-Verlag, 1984, 448 p.

[34] Shutov S. D., Iovu M. A., Iovu M. S. Semiconductors, 1979, V. 13, p. 956.

[35] Enck R.G., Pfister G. In: Photoconductivity and Related Phenomena, Elseiver, New York, 1976, Ch. 7.

[36] Dyhne A. M., Kondratenko P. S., Matveev L. V. J. of Exper. Theor. Phys. Letters, 2004, p. 464.

[37] Aroutiounian V. M., Ghoolinian M. Zh., Tributsch H. Applied Surface Science, 2000, p. 122.

[38] Averkiev N. S., Kazakova L. P., Piryatinskiy Y. P., Smirnova N. N. Semiconductors, 2003, V. 37, p. 1244.

[39] Averkiev N. S., Kazakova L. P., Smirnova N. N. Semiconductors, 2002, V. 36, p. 355.

[40] Gu Q., Wang Q., Schiff E.A., Li Y.-M., Malone C.T. J. Appl. Phys., 1994, 76, p.2310.

[41] Kazakova L. P., Lebedev E. A. Semiconductors, 1998, V. 32, p. 187.

[42] West B. J., Bologna M., Grigolini P. Physics of Fractal Operators, Springer, New-York, 2003.

Resonant influence of spatial oscillations of a perturbation on motion of a nonlinear oscillator

D.V. Makarov, M.Yu. Uleysky

V.I.Il'ichev Pacific Oceanological Institute of the Russian Academy of Sciences,
690041 Vladivostok, Russia

Abstract

We consider motion of a nonlinear oscillator subject to a perturbation which oscillates in time and in coordinate. Conditions of emergence of chaos are investigated. We found that the resonance between spatial and temporal oscillations of the perturbation induces strong but bounded chaotic diffusion in certain areas of phase space. The model of the Duffing oscillator is used as an example for the numerical simulation.

Keywords: Hamiltonian chaos; Nonlinear resonance; Chaotic diffusion

1 Introduction

An one-dimensional nonlinear oscillator driven by a weak periodic force is one of the simplest dynamical systems exhibiting chaos [1, 2, 3]. The perturbation destroys integrals of motion and causes emergence of unstable area in phase space. If the criterion of Chirikov [1] holds, then the unstable area has form of chaotic sea with fast diffusion therein. A typical chaotic trajectory looks like stochastic oscillations and exact prediction of its long-term behavior is impossible.

Usually chaos is something that we try to overcome. The traditional way of overcoming implies developing some techniques designed to enhance predictability by controlling or suppressing chaotic diffusion [4, 5]. However, in some situations chaos can play a positive role. In particular, it provides amplification of transport processes, that is desirable in various physical applications, such as particle acceleration [6, 7, 8], photoionization of molecules [9], controlling of atom dynamics in optical lattices [10] or ac-induced conductivity [11, 12]. In the present paper we show the way of generating fast chaotic diffusion in certain areas of phase space by means of the perturbation which oscillates in time and in coordinate as well. In this case strong but local chaos arises due to the resonance between spatial and temporal oscillations of the perturbation. The location of the chaotic layer in phase space is essentially determined by the interrelation between the spatial and temporal periods of the perturbation.

The paper is organized as follows. In the next section we give a brief analytical description of the resonance aforementioned. Section 3 contains results of the numerical simulation of the Poincaré map with the model of the double-well Duffing oscillator. In the final section we summarize and discuss the results obtained.

2 Theoretical analysis

Consider a weakly-perturbed nonautonomous oscillator with the Hamiltonian of the generic form

$$H = H_0(x,\, p) + \varepsilon H_1(x,\, p,\, t), \quad \varepsilon \ll 1, \tag{1}$$

where the unperturbed term can be expressed as the sum of kinetic and potential parts

$$H_0 = \frac{p^2}{2} + U(x). \tag{2}$$

In the present paper we study the case of the perturbation of the following form

$$H_1 = \cos(kx + \nu t). \tag{3}$$

A trajectory satisfies the coupled equations of motion

$$\frac{dx}{dt} = p, \tag{4}$$

$$\frac{dp}{dt} = -\frac{\partial U}{\partial x} + \varepsilon k \sin\psi, \tag{5}$$

where

$$\psi = kx + \nu t \tag{6}$$

is the phase of the perturbation. Two types of resonances occur in the system (4), (5). The resonances of the first type satisfy the "usual" resonant condition

$$m\omega = n\nu, \tag{7}$$

where ω is the frequency of the oscillator. The theoretical description of these resonances is given in the framework of the Kolmogorov–Arnold–Moser theory (see for details [1, 2, 3]).

The resonances of the second type correspond to the stationary phase condition for the phase (6)

$$\dot{\psi} = kp + \nu \simeq 0. \tag{8}$$

Hereafter dot denotes differentiation with respect to t. Using (5) and (8) we derive the "pendulum-like" equation [13]

$$\ddot{\psi} + k(U_x' - \varepsilon k \sin\psi) = 0, \tag{9}$$

which describes variation of the phase ψ in the neighborhood of the resonance (8).

Let us assume that k and ν are large enough. Henceforth coordinate x, momentum p and the derivative U_x' can be thought of as slowly-varying quantities along a trajectory $x(t)$. *Capturing* into the resonance (8) means oscillations of ψ near the stable fixed point which appears if

$$U_x' < \varepsilon k. \tag{10}$$

The fixed points can be found from the equation

$$U_x' = \varepsilon k \sin\psi. \tag{11}$$

Confining ψ by the interval $[-\pi : \pi]$ and solving (11) we find the unstable fixed point

$$\psi_1 = \arcsin\left(\frac{U_x'}{\varepsilon k}\right) \tag{12}$$

and the stable one

$$\psi_2 = \pi - \arcsin\left(\frac{U'_x}{\varepsilon k}\right). \tag{13}$$

Equation (9) corresponds to the Hamiltonian with the biased periodic potential

$$\tilde{H} = \frac{1}{2}\dot{\psi}^2 + kU'_x\psi + \varepsilon k^2 \cos\psi. \tag{14}$$

Note that $\tilde{H}$ is not invariant along a trajectory. The model described by this Hamiltonian is known as the pendulum with constant momentum [13]. Its phase portrait consists of oscillating and non-oscillating solutions divided by the separatrix. The value of the Hamiltonian (14) at the separatrix given by the formula

$$\tilde{H}_s = kU'_x\psi_1 + \varepsilon k^2 \cos\psi_1. \tag{15}$$

Passing through the resonance (8) can be qualitatively described in the following way. Let us assume that inequality (10) is fulfilled only within some finite piece of a trajectory, which includes the point where

$$p = p_{\text{res}} = -\frac{\nu}{k}. \tag{16}$$

If the trajectory is far enough from this piece, the phase ψ varies in the ballistic regime, and it is relevant to replace the original equations of motion (4), (5) by the averaged ones

$$\dot{x} = p, \quad \dot{p} = U'_x. \tag{17}$$

Whenever the derivative U'_x becomes smaller than εk, the phase portrait of the pendulum (9) changes, and one enables capturing of the trajectory into the oscillating regime. The capturing takes place if the condition

$$\tilde{H} \leq \tilde{H}_s \tag{18}$$

is satisfied. If the capturing occurs, then the derivative $\dot{\psi}$ remains small for some temporal interval, as long as the inequality (18) holds. The term $\sim \sin\psi$ in the right hand side of equation (5) varies weakly during this interval, that leads to the secular increasing or decreasing of the energy of the oscillator. If the capturing doesn't occur, then the trajectory rushes by the resonance zone without significant changing the energy. This regime is known as the scattering by the resonance [14].

Thus influence of the resonance (8) depends on the duration of the oscillating regime for ψ. One can distinguish regimes of strong and weak capturing. The strong capturing takes place if k is large and the perturbation causes additional minima of the potential. Then the trajectory may be trapped by a microwell which appears near some "new" local minimum. These microwells are not stationary and move along x with the constant velocity $-\nu/k$. When trapped a trajectory performs fast small-scale oscillations inside the microwell. Averaging of these oscillations yields

$$\bar{x} = p_{\text{res}}t = -\frac{\nu}{k}t. \tag{19}$$

The duration of the strong capturing depends on how long the given microwell survives. Roughly speaking, it is proportional to the range in which the microwells appear, which can be evaluated as the maximal distance between two arbitrary local minima of the potential.

The weak capturing takes place if the perturbation doesn't change the number of minima of the potential. In this case a trajectory escapes from the resonant zone during the first cycle around the stable fixed point $\psi = \psi_2$, $\dot{\psi} = 0$. The duration of the weak capturing is given by the following expression

$$\tau = \int_{\psi(t)} \frac{d\psi}{\sqrt{2\tilde{H} - kU'_x - \varepsilon k^2 \cos\psi}}. \tag{20}$$

Here the integral is calculated along the oscillating piece of $\psi(t)$. As it follows from (20), the duration τ decreases with increasing k as $k^{-1/2}$. Taking into account that the resonant term in the right hand side of (5) is proportional to k, the resulting resonant response reads

$$\Delta p \simeq k\tau \sim \sqrt{k}. \tag{21}$$

A similar result was derived in [15] using another approach.

Conditions (10) and (18) imply the maximal duration of the capturing into the resonance (8) if the following condition is satisfied

$$p_{\rm res} \simeq p\big|_{U'_x \simeq 0}\,. \tag{22}$$

Consequently one expects the strongest resonant response near the points of the stable or unstable equilibrium of the oscillator. According to equation (16), near the stable fixed point equation (22) can be written as follows

$$|p_{\rm max}(H)| \simeq \frac{\nu}{k}. \tag{23}$$

In the latter case one yields

$$|p_{\rm min}(H)| \simeq \frac{\nu}{k}. \tag{24}$$

Equations (23) and (24) determine the values of the Hamiltonian (1), related to the trajectories mainly affected by the resonance (8).

3 Numerical simulation

In this section we investigate how the resonance (8) reveals itself in a phase space structure. The double-well Duffing oscillator with the potential

$$U(x) = \frac{x^4}{4} - \frac{x^2}{2} \tag{25}$$

is taken as an example. This potential has two points of the stable equilibrium at $x = \pm 1$. The wells near these points are separated by the potential barrier containing the point of the unstable equilibrium, located at $x = 0$. Oscillations within the wells are separated from the oscillations above the barrier by the separatrix, at which $H = 0$.

Figure 1 represents the Poincaré sections computed with $\nu = 2\pi$ and $k = 20\pi$. For this choice of ν and k, the condition (23) is satisfied only for small-amplitude oscillations near $x = \mp 1$, while the condition (24) holds near the separatrix. As it seen from the Poincaré section, these regions are occupied by the chaotic layers, which are separated from each other at $\varepsilon = 0.001$ (Fig. 1a), and merged at $\varepsilon = 0.01$ (Fig. 1b). Instability near the separatrix should not be surprising because it is the case for a generic periodic perturbation, whereas the chaotic behavior of small-amplitude oscillations indicates directly the role of the resonance (23).

If ν increases, the zones of the strongest influence of the resonance (8) are displaced to the larger values of H, that is followed by the displacement of the chaotic regions in phase space. It is shown in Fig. 2 where the Poincaré sections computed with $\nu = 10\pi$ and $\nu = 20\pi$ are depicted. The chaotic layers are merged for both the cases. In the former case the resulting chaotic sea is located near the separatrix. In the latter case oscillations inside the single wells are almost stable, while strong chaos emerges in the range related to oscillations above the potential barrier.

Thus, we observed that the resonance (8) induces bounded chaotic layers in the ranges of the energy, given by the formulae (23) and (24). It is shown that the layers formed can be conjuncted, that leads to the emergence of the wide chaotic sea.

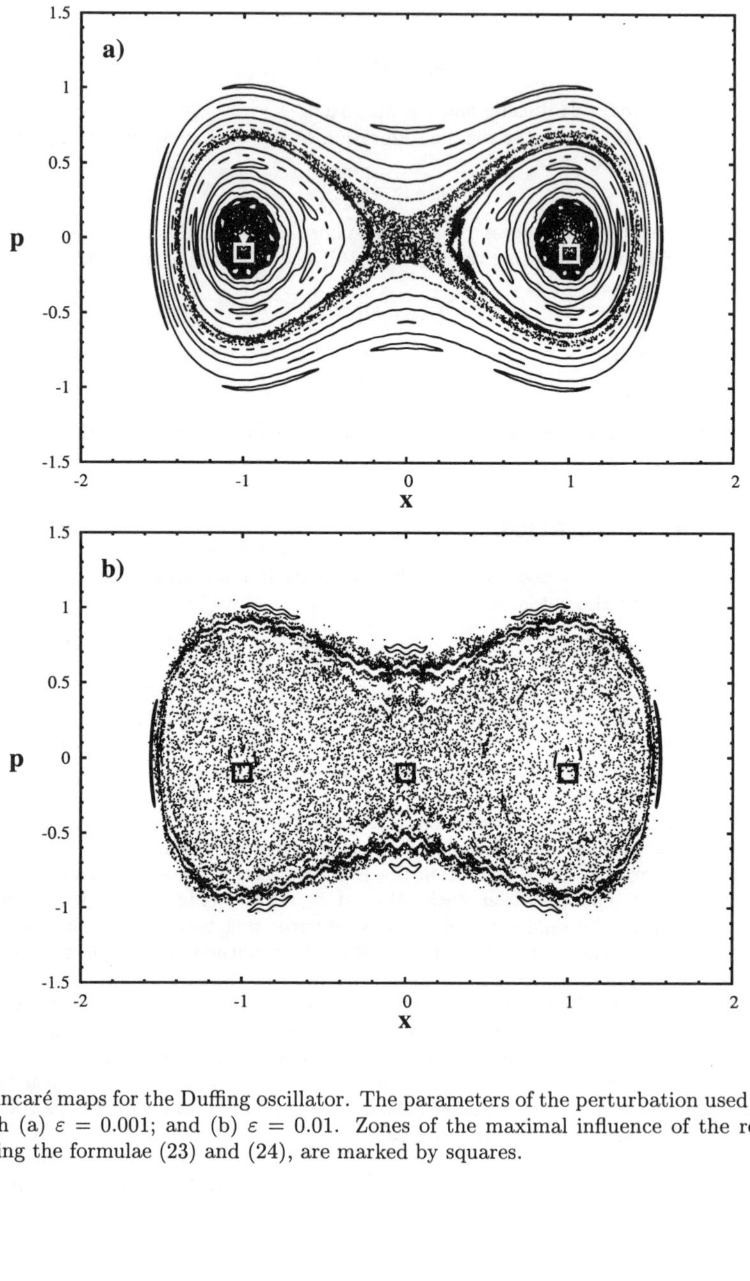

Figure 1: Poincaré maps for the Duffing oscillator. The parameters of the perturbation used were $\nu = 2\pi$, $k = 20\pi$ with (a) $\varepsilon = 0.001$; and (b) $\varepsilon = 0.01$. Zones of the maximal influence of the resonance (8), estimated using the formulae (23) and (24), are marked by squares.

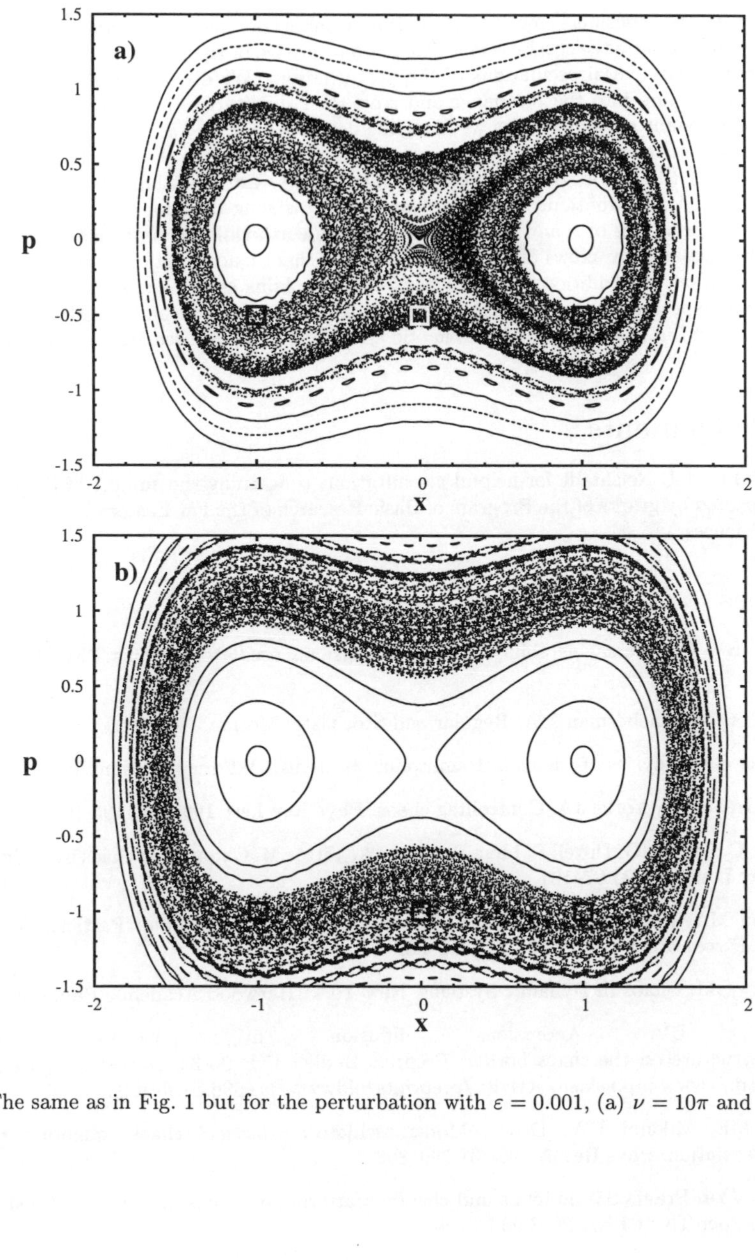

Figure 2: The same as in Fig. 1 but for the perturbation with $\varepsilon = 0.001$, (a) $\nu = 10\pi$ and (b) $\nu = 20\pi$.

4 Summary

The paper presented considers motion of a Hamiltonian nonlinear oscillator under an external perturbation which varies as $\sim \cos(kx+\nu t)$. The main objective of the paper is the resonant interaction between temporal and spatial oscillations of the perturbation. We distinguished the strong and weak regimes of the capturing into the resonance and gave their qualitative description. In particular, it is shown that the resonant response in the weak regime depends on k as $\sqrt{k}$.

The main result of the paper is that the resonance considered leads to the occurrence of chaotic layers in certain ranges of the energy of an oscillator. These ranges are essentially determined by the ratio of the periods of the perturbation. Thus, one can produce strong diffusion in the given area of phase space, using suitable values of ν and k. This effect seems worthwhile for problems related to transport phenomena. For instance, it allows one to achieve efficient chaotic acceleration of particles under a weak external perturbation. In addition, the analogous effect explains strong chaos in the last portion of a received pulse in experiments on long-range sound propagation in the ocean [15]. Manifestations of this effect could be found in the problem of the chaotic advection of passive scalars in open hydrodynamical flows [16].

Acknowledgements

We are grateful to A.I. Neishtadt for helpful consultations concerning the subject of the research. This work was supported by grants of the Program of Basic Research of the Far Eastern Branch of the Russian Academy of Sciences.

References

[1] Chirikov BV. A universal instability of many-dimensional oscillator systems. Phys Rep 1979;52:265–379.

[2] Lichtenberg AJ, Lieberman MA. Regular and Stochastic Motion. New York: Springer; 1992.

[3] Zaslavsky GM. Physics of Chaos in Hamiltonian Systems. Oxford: Academic Press; 1998.

[4] Ott E, Grebogi C, Yorke JA. Controlling chaos. Phys Rev Lett 1990;64:1196–9.

[5] Chandre C, Giraolo G, Doveil F, Lima R, Macor A, Vittot M. Channeling chaos by building barriers. Phys Rev Lett 2005;94:074101.

[6] Liberman MA, Lichtenberg AJ. Stochastic and Adiabatic Behavior of Particles Accelerated by Periodic Forces. Phys Rev A 1972;5:1852–66.

[7] Zaslavsky GM. Chaos in Dynamic Systems. New York: Harwood Academic Publishers; 1985.

[8] Chirikov BV. Anomalous diffusion in microtron and the critical structure on the chaos border, Preprint Budker INP 96-34, Novosibirsk, 1996. Available from ¡http://www.inp.nsk.su/activity/preprints/oldwww/prep96.ru.shtml¿.

[9] Goggin ME, Milonni PW. Driven Morse oscillator: classical chaos, quantum theory, and photodissociation. Phys Rev A 1988;37:796–806.

[10] Argonov VYu, Prants SV. Fractals and chaotic scattering of atoms in the field of a standing light wave. J Exper Theor Phys 2003;96:832–45.

[11] Dupont E, Corcum PB, Liu HC, Buchanan M, Wasilewski ZR. Phase-controlled currents in semiconductors. Phys Rev Lett 1995;74:3596–3599.

[12] Alekseev KN, Cannon EN, McKinney JC, Kusmartsev FV, Campbell DK. Spontaneous dc current generation in a resistively shunted semiconductor superlattice driven by a terahertz field. Phys Rev Lett 1998;80:2669–72.

[13] Neishtadt AI. Capturing into resonance and scattering on resonances in two-frequency systems. Proc of the Steklov Inst of Math 2005;250:183–203.

[14] Vainshtein DL, Neishtadt AI, Mezic I. On passage through resonances in volume-preserving systems. e-print arXiv:nlin.CD/0604047 (2006).

[15] Makarov DV, Uleysky MYu. Sensitivity of ray dynamics in an underwater sound channel to vertical scale of longitudinal sound-speed variations. e-print arXiv:physics/0508226 (2005).

[16] Budyansky MV, Uleysky MYu, Prants SV. Chaotic scattering, transport, and fractals in a simple hydrodynamic flow. J Exper Theor Phys 2004;99:1018–27.

Chaotic transport and fractals in a geophysical jet current

M. V. Budyansky, S. V. Prants

Laboratory of Nonlinear Dynamical Systems, V.I.Il'ichev Pacific Oceanological Institute of the Russian Academy of Sciences, 690041 Vladivostok, Russia

Abstract

We model Lagrangian lateral mixing and transport of passive scalars in meandering oceanic jet currents by two-dimensional advection equations with a kinematic stream function with a time-dependent amplitude of a meander imposed. The advection in such a model is known to be chaotic in a wide range of the meander's characteristics. We study chaotic transport in a stochastic layer and show that it is anomalous. The geometry and topology of mixing are examined and shown to be fractal-like. The scattering characteristics (trapping time of advected particles and the number of their rotations around elliptical points) are found to have a hierarchical fractal structure as functions of initial particle's positions. A correspondence between the evolution of material lines in the flow and elements of the fractal and between dynamical and topological measures of the flow are established.

PACS: 47.52.+j; 47.53.+n; 92.10.Ty
Keywords: Chaotic advection; Meandering jet; Fractals

1 Introduction

Major western boundary currents in the ocean are meandering jets separating water masses with different physical and biogeochemical characteristics. The prominent examples are the Gulf Stream in the Atlantic Ocean and the Kuroshio in the Pacific Ocean. These and similar "heat engines" define the climate in large regions of the planet. Similar jets in the stratosphere play important role in transport and distribution of chemical substances. From the hydrodynamic point of view, they may be considered as jet flows with running waves of different wave lengths and phase velocities imposed. The simplest kinematic model of such a flow is a two-dimensional jet of an ideal fluid with a given velocity profile that is perturbed by an amplitude-modulated wave traveling from the west to the east. The problem of transport and mixing of passive scalars in meandering jets has been considered by many authors in the context of atmospheric and oceanic physics [1, 2, 3, 4, 5, 6, 7, 8]. The typical phase portrait (Fig. 1) consists of two chains of circulations with a zigzag-like jet between them and resembles the phase portrait of a particle in the field of two running waves. In the frame moving with the velocity of one of the waves, the problem is topologically equivalent to the motion of a periodically perturbed nonlinear physical pendulum that is know to demonstrate chaotic oscillations [9, 10]. Different aspects of chaotic mixing of passive particles in meandering jets in the atmosphere and the ocean have been studied in the papers [1, 2, 3, 4, 5, 6, 7, 8].

In the paper we focus on topological and statistical aspects of the chaotic transport and mixing in a specific kinematic model of an eastward meandering jet which has been introduced in Refs. [2, 3] some years ago. We are motivated by the desire to get a more deep insight into the evolution of material lines in the flow and to establish a connection between dynamical, topological and statistical characteristics of the flow. The equations of motion of passive particles advected by a planar incompressible flow is known to have a Hamiltonian form

$$\begin{aligned} \dot{x} &= u(x, y, t) = -\frac{\partial \Psi}{\partial y}, \\ \dot{y} &= v(x, y, t) = \frac{\partial \Psi}{\partial x}, \end{aligned} \tag{1}$$

with the streamfunction Ψ playing the role of a Hamiltonian and the coordinates (x, y) of a particle being canonically conjugated variables. Thus, nonstationary two-dynamical advection is equivalent to a Hamiltonian system with one and half degrees of freedom whose phase space coincides with its configuration space. This property is very useful in visualizing geometric invariant sets in real and numerical experiments.

2 Model flow

To be specific we consider a two-dimensional Bickley jet with the velocity profile $u_0 \operatorname{sech}^2 y$ whose argument is modulated by a zonal running wave [3]. The streamfunction in the fixed frame reference is the following:

$$\psi'(x', y', \tau) = -\psi_0 \tanh\left(\frac{y' - a\cos k(x' - c\tau)}{\lambda\sqrt{1 + k^2 a^2 \sin^2 k(x' - c\tau)}}\right), \tag{2}$$

where a, k and c are amplitude, wavenumber and phase velocity respectively, λ is a measure of the jet's width. After introducing the following notations:

$$x = k(x' - c\tau), \quad x' = \frac{x}{k} + c\tau, \qquad y = ky', \quad y' = \frac{y}{k}, \qquad t = \psi_0 k^2 \tau, \quad \tau = \frac{t}{\psi_0 k^2}. \tag{3}$$

and

$$A = ak, \qquad L = \lambda k, \qquad C = \frac{c}{\psi_0 k}, \tag{4}$$

we get the advection equations (1) in the frame moving with the phase velocity c:

$$\begin{aligned} \dot{x} &= \frac{1}{L\sqrt{1 + A^2 \sin^2 x} \cosh^2 \theta} - C, \\ \dot{y} &= -\frac{A \sin x (1 + A^2 - Ay \cos x)}{L\left(1 + A^2 \sin^2 x\right)^{3/2} \cosh^2 \theta}, \end{aligned} \qquad \theta = \frac{y - A\cos x}{L\sqrt{1 + A^2 \sin^2 x}}. \tag{5}$$

The respective streamfunction

$$\psi(x, y) = -\tanh\left(\frac{y - A\cos x}{L\sqrt{1 + A^2 \sin^2 x}}\right) + Cy \tag{6}$$

has three normalized control parameters: L, A and C are the jet's width, meander's amplitude and its phase velocity. The scaling chosen results in translational invariance of the phase portrait along the x-axis with the period 2π.

The detailed analysis of stationary points and bifurcations of Eqs. (5) has been done in Ref. [11]. Stationary points may exist only under the condition $LC \le 1$. There are four stationary points, two of them are always stable and the other ones are stable under the condition $AL \operatorname{Arcosh} \sqrt{1/LC} > 1$. If the additional condition $C < 1/L \cosh^2(1/AL)$ is fulfilled, there are additional four unstable saddle points. Resuming one gets:

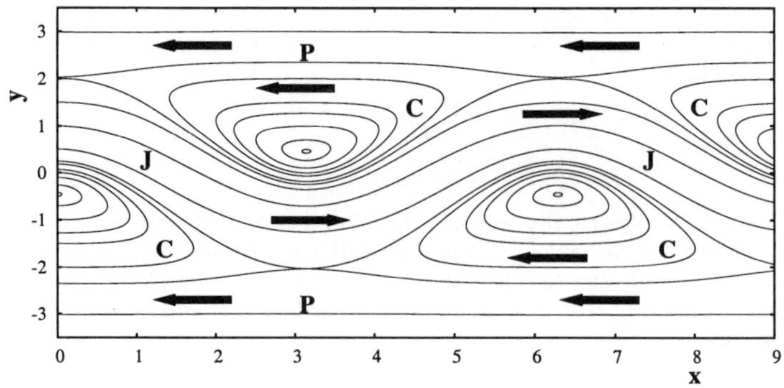

Figure 1: Streamlines of the unperturbed system (5) in the frame moving with the meander's phase velocity c.

1. $C > C_{\rm cr1} = 1/L$, there are no stationary points.

2. $C_{\rm cr1} > C > C_{\rm cr2} = 1/L\cosh^2(1/AL)$ and $C > C_{\rm cr3}$, there are two centers and two saddles with two separatrices connecting the saddles. The jet between the separatrices is westward.

3. $C_{\rm cr1} > C > C_{\rm cr2}$ and $C < C_{\rm cr3}$, the jet is eastward with the same stationary points as in the case 2.

4. $C_{\rm cr2} > C > C_{\rm cr3}$, there are eight stationary points and two separatrices. The jet is westward.

5. $C_{\rm cr2} > C$ and $C < C_{\rm cr3}$, the jet is eastward with the same stationary points as in the case 4.

The respective phase portraits are typical with Hamiltonian systems with running waves in shear flows [6, 12]. In dependence on the values of the phase velocity C one can get different topologies: a homoclinic connection, a heteroclinic connection and a separatrix reconnection. Being motivated by eastward jet currents in the ocean and atmosphere, we deal in this paper with the case 3 (see the phase portrait in Fig. 1).

3 Chaotic mixing and transport

Streamlines with the streamfunction (6), that is time-independent in the moving frame, are shown in Fig. 1. The plot demonstrates three different regions of the flow: a central eastward jet (J), north and south circulations (C) and peripheral westward currents (P). The centers of the circulations are at critical lines to be defined by the condition $u(y_c) = c$, $v(y_c) = 0$, and they are divided by two separatrices connecting saddle points. No exchange between the north and south circulations is possible in the unperturbed system.

Even the simplest periodic perturbation of the meander's amplitude, $A = A_0 + \varepsilon\cos(\omega t + \varphi)$, results in spitting stable and unstable manifolds of the saddle points. There arise stochastic layers with chaotic mixing and transport of water masses fluxes of which depend on the width of a layer, a number of overlapping resonances and other factors. As an example, we plot in Fig. 2 the Poincaré section of the perturbed flow. The control parameters throughout the paper are chosen as follows: $A = 0.785$, $C = 0.1168$, $L = 0.628$, $\varepsilon = 0.0785$, $\omega = 0.2536$ and $\varphi = \pi/2$.

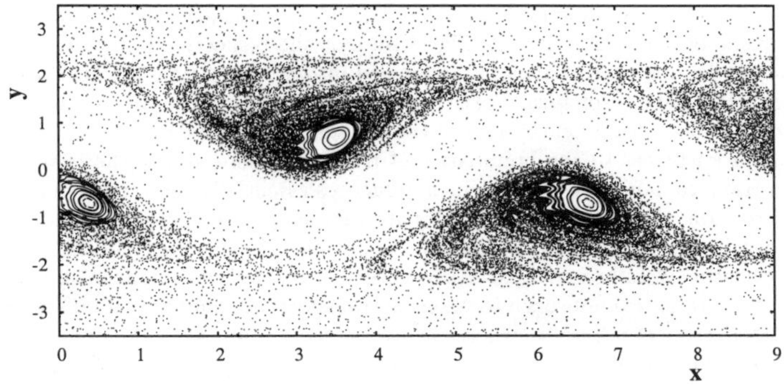

Figure 2: General view of the Poincaré section of the system with a periodically modulated meander's amplitude.

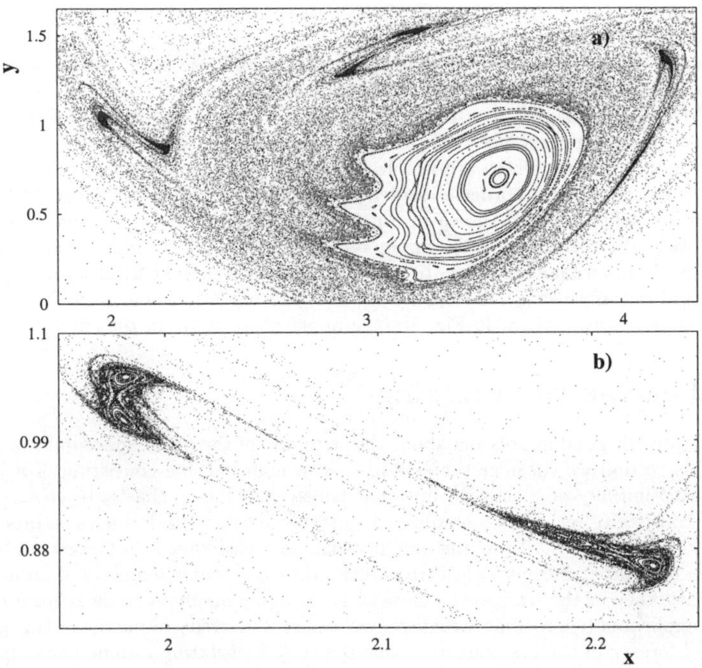

Figure 3: (color online). (a) Poincaré sections in the northern part of the first eastern frame, (b) stickiness to the island's border.

Due to the zonal and meridian symmetries it is enough to consider mixing in the northern part of the first eastern frame only $0 \le x \le 2\pi$. In Fig. 3 we plot the respective Poincaré section. The vortex core, that survives under the perturbation, is immersed into a stochastic sea where one can see 6 small islands belonging to the same resonance. Particles, belonging to these islands, rotate around the elliptic point $x = \pi$. Zoom of the section nearby two of the islands is shown in Fig. 4. The feature we want to pay attention to is a stickiness to the boundaries of the vortex core and of the islands. It is a typical phenomenon with Hamiltonian systems [10] which influences essentially the transport of passive particles.

Without perturbation, the transport properties are very simple: particles either rotate in circulations C or move eastward in the jet J and westward in the peripheral currents P. Under a perturbation, the motion in the stochastic layers become extremely sensitive to small variations in initial conditions, and one is forced to use an statistical approach to describe transport. A commonly used statistical measure of transport is the variance $\sigma^2(t) =< x^2 >$, where the averaging is supposed to be done over an ensemble of particles. Long time behavior of $\sigma^2(t)$ and of the probability density function of particle's displacement $P(x,t)$ may be anomalous with typical chaotic Hamiltonian systems [13]. When $\sigma^2(t) \sim t$ and $P(x,t)$ is a Gaussian distribution, the advection is known to be normal with a well-defined diffusion coefficient $D = \lim_{t\to\infty} \sigma^2/2t$. When $\sigma^2(t) \sim t^\gamma$, with $\gamma \neq 1$, D is either zero or infinite [13], and one gets either subdiffusion ($\gamma < 1$) or super-diffusion ($\gamma > 1$). If the trajectories are dominated by sticking regions nearby boundaries of islands, where particles spend a long time, subdiffusion results. Superdiffusion occurs when particles in the jet travel long distances between sticking events. The respective length and time PDFs are expected to be non Gaussian.

Besides resonant islands with particles moving around the elliptic point in the same frame (Fig. 3), we have found so-called ballistic islands which were situated both in the chaotic sea and in the peripheral jets (Fig. 4a). Particles, belonging to a ballistic island, move with a constant velocity from one frame to another. When mapping their positions at the moments $t = 2k\pi/\omega$ ($k = 1, 2, \ldots$) onto the first frame, we see a chain of islands which are visible in Fig. 4a. In Fig. 4b a zoom of the ballistic island is shown. Stickiness to the island's border is evident. In Fig. 5 we demonstrate two ballistic trajectories corresponding to the ballistic island shown in Fig. 4b. If a particle at $t = 0$ is placed inside the island, it travels to the west in a regular way (the upper blue trajectory in Fig. 5a). Its zonal position x grows linearly with time (the lower blue trajectory in Fig. 5b). The lower green trajectory in Fig. 5a corresponds to a particle placed initially nearby the border of the same ballistic island from outside. Intermittent flight and sticking events are evident in Fig. 5b (the upper green curve in that figure).

4 Fractal geometry of mixing

Poincaré sections provide good impression about the structure of the phase space but not about geometry of mixing. In this section we consider the evolution of a material line consisting of a large number of particles distributed initially on a straight line that transverses the stochastic layer at $x = 0$. A typical stochastic layer consists of an infinite number of unstable periodic and chaotic orbits with islands of regular motion to be imbeded. All the unstable invariant sets are known to possess stable and unstable manifolds. When time progresses particle's trajectories nearby a stable manifold of an invariant set tend to approach the set whereas the trajectories close to an unstable manifold go away from the set. Because of such a very complicated heteroclinic structure, we expect a diversity of particle's trajectories. Some of them are trapped forever in the first eastern frame $0 \le x \le 2\pi$ rotating around the elliptic point $x = \pi$ along heteroclinic orbits. Other ones quit the frame through the lines $x = 0$ or $x = 2\pi$, and then either are trapped there or move to the neighbor frames (including the first one), and so on to infinity.

To get a more deep insight into the geometry of chaotic mixing we follow the methodology of our works [14, 15] and compute the time T, particles spend in the neighbor circulation zones $-2\pi \le x \le 2\pi$ before reaching the critical lines $x = 0, x = \pm 2\pi$, and the number of times $n/2$ they wind around the respective elliptic points $x = \pm\pi$. In the upper panel in Fig. 6 the functions $n(y_0)$ and $T(y_0)$ are shown.

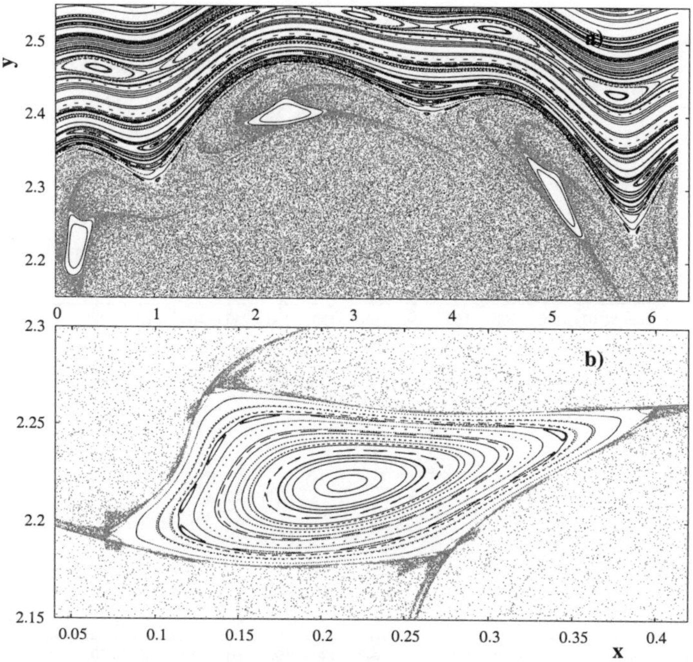

Figure 4: (color online). (a) The northern border between the circulation (C) and the peripheral current (P), (b) stickiness to the border of a ballistic island.

The upper parts of each function (with $n > 0$ and $T > 0$) represent the results for the particles with initial positive zonal velocities which they have simply due to their locations on the material line at $x = 0$. These particles enter the eastern frame and may change the direction of their motion many times before leaving the frame through the lines $x = 0$ or $x = 2\pi$. The time moments of those events we fix for all the particles with $1.9 \leq y_0 \leq 2.045$. The lower "negative" parts of the functions $n(y_0)$ and $T(y_0)$ represent the results for the particles with initial negative zonal velocities ($y_0 \geq 2.045$) which move initially to the first western frame ($-2\pi \leq x \leq 0$). In fact, $T_e(y_0)$ and $T_w(y_0)$ are the time moments when a particle with the initial position y_0 quits the eastern or western frames, respectively. Both the functions consist of a number of smooth U-like segments intermittent with poorly resolved ones. Border points of each U-like segments separate particles belonging to stable and unstable manifolds of the heteroclinic structure. The corresponding initial y-positions is a set (of zero measure) of particles to be trapped forever in the respective frame. A fractal-like structure of chaotic advection in both the frames is shown in the upper panel in Fig. 6, and its fragments for the first levels are shown in the middle panel for the eastern and the western fragments separately. Particles with even values of n quit one the frames through the border $x = 0$, those with odd n – through the border $x = 2\pi$ for the eastern frame and $x = -2\pi$ for the western one.

Let us consider in detail the fractal-like structure in the eastern frame keeping in mind that the results are similar with any other frame. The $n_e(y_0)$-dependence is a complicated hierarchy of sequences

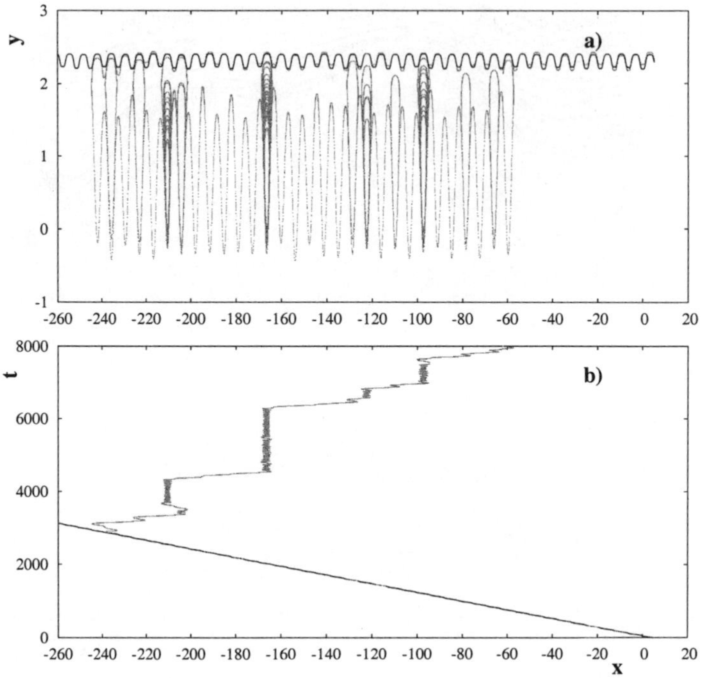

Figure 5: (color online). Examples of ballistic trajectories: the blue and regular trajectory, which is the upper one in (a) and the lower one in (b), is inside a ballistic island, the green and weakly chaotic trajectory, which is the lower one in (a) and the upper one in (b), is just outside the island. (a) $x - y$ plane, (b) dependence of the zonal position on time. Stickiness and flight events are evident with the green chaotic trajectory.

of segments of the material line. Following to the authors of the paper [16], we call as an epistrophe a sequence of segments of the $(n+1)$-level, converging to the ends of a segment of a sequence of the n-th level, whose length decrease in accordance with a law. At $n_e = 1$ we see in Fig. 6 an epistrophe with segment's length A, B, C, D and so on decreasing as $l_m = l_0 q^m$ with $q \approx 0.46$. Letters a and b in Fig. 6 denote the first segments of the epistrophes at the level $n_e = 2$, whereas d and c — the first segments of the epistrophes at the level $n_e = 3$. The respective laws for all those epistrophes are not exponential.

In Fig. 7 we demonstrate fragments of the evolution of the material line in the first eastern frame at the moments indicated in the figure. Letters on the line mark the corresponding segments of the $n_e(y_0)$ and $T_e(y_0)$ functions in Fig. 6. As an example, let us explain formation of the epistrophe ABCD at the level $n_e = 1$. With the period of perturbation $T_0 = 2\pi/\omega \simeq 8\pi$, a portion from the north end of the material line leaves the frame through its eastern border. Look at the segments A and B at $t = 15\pi$ and $t = 23\pi$. They quit the first frame as a fold through the period $T_0 \simeq 8\pi$. The other segments – C and D (not shown in Fig. 7) do the same job. The epistrophe's segments at the odd levels ($n = 2k - 1 > 1$) quit the frame with the period of perturbation T_0 one by one being folded (c and d segments). The folds of the segments of the $(2k-1)$-level are exterior with respects to the folds of the segments of the $(2k+1)$-level.

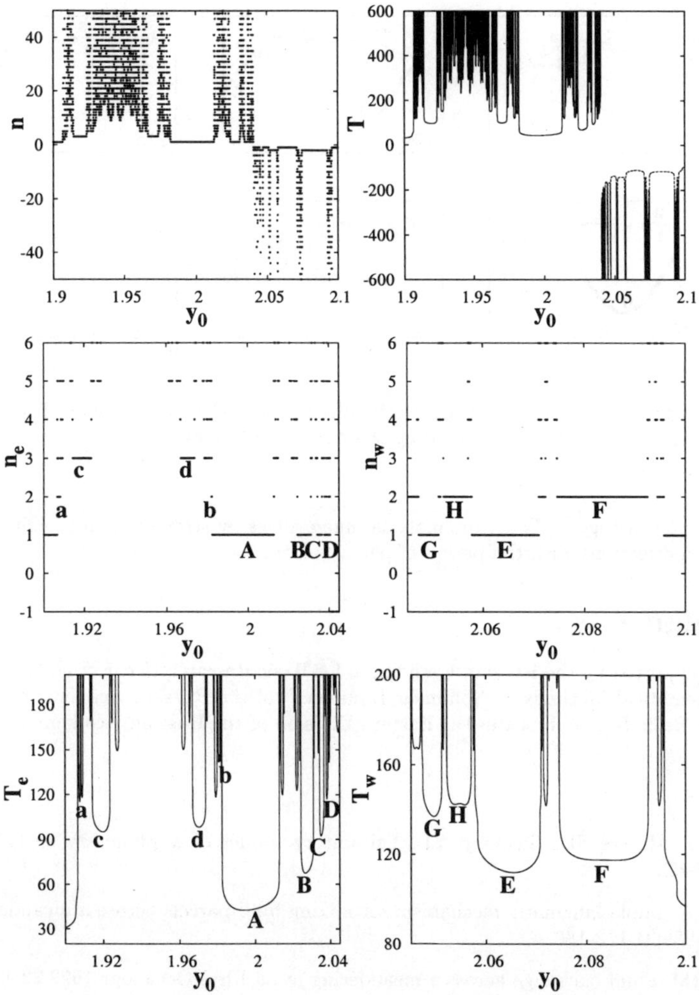

Figure 6: Fractal set of initial positions y_0 of particles that reach the lines $x = 0, \pm 2\pi$ after $n/2$ turns around the elliptic points. T is a time particles need to reach the lines $x = 0, \pm 2\pi$. Indices e and w mean particles moving in the eastward and westward directions, respectively.

The following empirical law is valid: $T_{2k+1} - T_{2k-1} \simeq 2T_0$, where T_{2k-1} is a time when the first segments of the epistrophes at the level $(2k-1)$ (A with $n_e = 1$) reach the line $x = 2\pi$, and T_{2k+1} the respective time for the first segments of the epistrophes at the level $2k+1$ (c and d segments with $n_e = 3$).

Segments of the epistrophes of the even levels ($n = 2k$) leave the frame with the period T_0 as well but through the border $x = 0$ moving to the west. We show the evolution of some of them at the moments

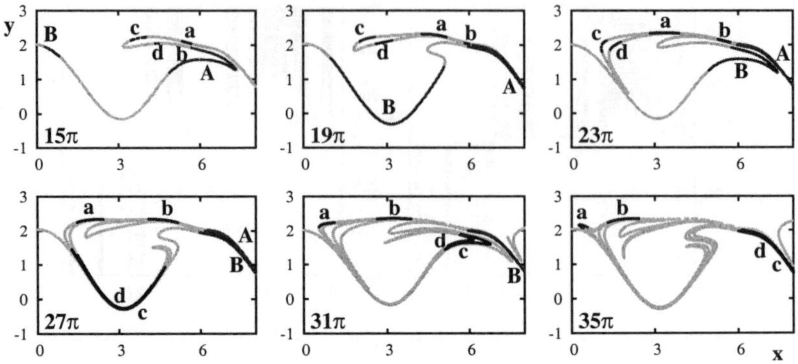

Figure 7: Fragments of the evolution of a material line in the first eastern frame. The fragments of the fractal in Fig. 6 with $n_e = 1, 2, 3$ are marked by the respective letters.

$t = 31\pi$ and $t = 35\pi$ in Fig. 7. Thus, the material line evolves by stretching and folding, and folds quit the frame in both directions with the period of perturbation.

Acknowledgments

This work was supported by the Russian Foundation for Basic Research (Project no.06-05-96032), by the Program "Mathematical Methods in Nonlinear Dynamics" of the Russian Academy of Sciences and by the Program for Basic Research of the Far Eastern Division of the Russian Academy of Sciences.

References

[1] Sommeria J, Meyers SD, Swinney YL. Laboratory model of a planetary eastward jet. Nature 1989;337:58-62.

[2] Bower AS. A simple kinematic mechanism for mixing fluid parcels across a meandering jet. J Phys Oceanogr 1989;21:173-180.

[3] Samelson RM. Fluid exchange across a meandering jet. J Phys Oceanogr 1992;22:431-440.

[4] Solomon TH, Weeks ER, Swinney HL. Observation of anomalous diffusion and Levy flights in a two-dimensional rotating flow. Phys Rev Lett 1993;71:3975-3978.

[5] Del-Castillo-Negrete D, Morrison PJ. Chaotic transport by Rossby waves in shear flow. Phys Fluids A 1993;5:948-965.

[6] Weiss JB, Knobloch E. Mass transport by modulated traveling waves. Phys Rev A 1989;40:2579-2589.

[7] Ngan K, Shepherd T. Chaotic mixing and transport in Rossby-wave critical layers. J Fluid Mech 1997;334:315-351.

[8] Duan JQ, Wiggins S. Fluid exchange across a meandering jet with quasi-periodic time variability. J Phys Oceanogr 1996;26:1176-1188.

[9] Chirikov BV. A universal instability of many-dimensional oscillator systems. Phys Rep 1979;52:263-379.

[10] Zaslavsky GM. Physics of Chaos in Hamiltonian systems. Oxford Academic Press 1998.

[11] Uleysky MYu, Budyansky MV, Prants SV. Chaotic advection in a meandering jet current. Nonlinear Dynamics 2006;1:is.2 [in Russian].

[12] Howard JE, Hohs SM. Stochasticity and reconnection in Hamiltonian systems. Phys Rev A 1984;29:418-421.

[13] Shlesinger MF, Zaslavsky GM, Klafter J. Strange kinetics. Nature 1993;363:31-38.

[14] Budyansky M, Uleysky M, Prants S. Hamiltonian fractals and chaotic scattering by a topographical vortex and an alternating current. Physica D 2004;195:369-378.

[15] Budyansky MV, Uleysky MYu, Prants SV. Chaotic scattering, transport, and fractals in a simple hydrodynamic flow. Zh Eksp Teor Fiz 2004;126:1167-1179 [JETP 2004;99:1018-1027].

[16] Mitchell KA, Handley JP, Tighe B, Delos JB, Knudson SK. Geometry and topology of escape. Chaos 2003;13:880-891.

Experiments on pattern formation in reacting systems with chaotic advection

T. H. Solomon*, M. S. Paoletti† and C. R. Nugent

Department of Physics and Astronomy, Bucknell University, Lewisburg, PA 17837 USA

Abstract

We present the results of experiments on *advection-reaction-diffusion* processes. Two flows are studied: a blinking vortex flow and a chain of alternating vortices. Mixing in both of these flows has been shown to be chaotic in general. The fluid is composed of the chemicals for the Belousov-Zhabotinsky (BZ) chemical reaction. We investigate the effects of chaotic mixing on the patterns that form in this system. Three experiments are described: (a) pattern formation in the oscillatory BZ reaction; (b) front propagation and mode-locking for the excitable BZ reaction in an oscillating vortex chain; and (c) synchronization of a network of fluid oscillators by superdiffusive transport and Lévy flights. The experiments are complemented by numerical simulations that illustrate the chaotic transport of these flows.

Keywords: Chaotic advection, pattern formation

I. Introduction

For several decades, there has been a significant amount of interest in patterns that form in *reaction-diffusion* systems [1],[2], e.g., chemically-reacting or biological systems without any fluid flows. But most fluid systems are not stagnant; rather, there are typically flows, and these flows dramatically affect the mixing properties of the system. This, in turn, has a significant effect on the pattern formation process since reaction of different species in the flow is limited by mixing. The problem is particularly interesting in light of recent studies that have shown that mixing can be chaotic, even for simple, laminar fluid flows. The general *advection-reaction-diffusion* problem has recently begun to receive theoretical and numerical attention, particularly in the regime where mixing is chaotic [3],[4],[5],[6],[7],[8], [9]. However, there have been very few experimental studies of advection-reaction-diffusion systems. In this article, we review some experiments that we have conducted on pattern-formation processes in advection-reaction-diffusion systems with chaotic advection. Three sets of experiments are described: (1) Pattern formation of an oscillatory chemical reaction in a flow with chaotic mixing [10]; (2) front propagation and mode-locking for an advancing chemical reaction in an oscillating vortex chain [11],[12]; and (3) synchronization of chemical oscillators in a fluid flow via superdiffusive transport [13].

* Corresponding author. Tel: +1-570-577-1348; fax +1-570-577-3153
E-mail address: tsolomon@bucknell.edu.
† Current address: Department of Physics, University of Maryland, College Park, MD USA

II. Pattern formation in a blinking vortex flow

In 1984, Aref demonstrated [14] that mixing is typically chaotic in a simple, laminar fluid flow composed of two point vortices that blink on and off periodically (Figure 1a). We have built an experiment that generates this flow in a simple, table-top apparatus. The flow is generated via a magnetohydrodynamic technique [10],[15] (Figure 1b). An electrical current passes radially through a thin layer of an electrolytic solution (either dilute sulfuric acid or the chemicals for the Belousov-Zhabotinsky reaction). This current -- which converges at one of two center electrodes -- interacts with a strong magnetic field produced by a Nd-Fe-Bo magnet below the fluid layer. The result is a flow that circles around that particular electrode. Blinking of the vortices is achieved by alternating periodically between the two center electrodes. The result is a flow that alternates between circling around one vortex and circling around the other.

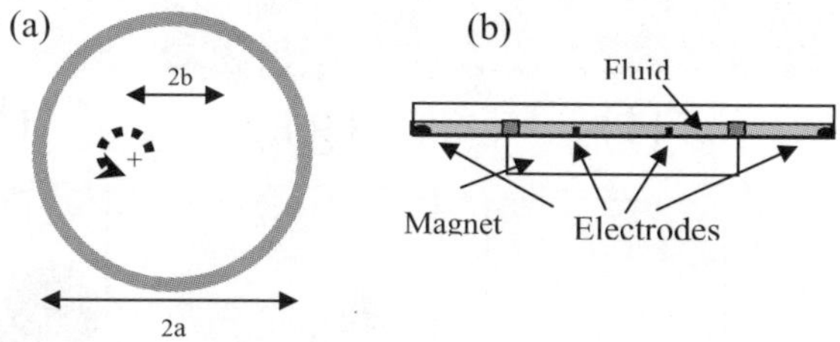

Figure 1. (a) Blinking vortex flow. The fluid alternates between circling around the left vortex and around the right vortex. (b) Apparatus used to generate the flow. An electrical current passes radially through a thin layer of an electrolytic solution, converging on one of two electrodes in the middle. The current interacts with a magnetic field produced by a permanent magnet, resulting in a flow that circles around the electrode

The fluid is composed of the chemicals typically used for the Belousov-Zhabotinsky (BZ) reaction [16],[17], a chemical reaction that is well-known for its oscillatory (and sometimes chaotic) time dependence when well-mixed. (See Reference 10 for details about the chemicals used in this reaction.) In the absence of any fluid flow, the reaction forms a pattern of spirals and/or bulls-eye patterns, typical of reaction-diffusion systems (Figure 2a).

With the flow turned on, the patterns change dramatically (Figures 2b, c and d). For these images, the blinking period T and the flow amplitude A vary; however, the dimensionless blinking period μ – which is the product of A and T, divided by the square of the radius of the system – is held fixed. The dimensionless blinking period determines the mixing patterns that form, but does not determine how *long* it takes for those patterns to form. With decreasing T, the flow mixes faster and faster. In Reference 10, we define a *mixing time* determined by the interplay between advective mixing and molecular diffusion; conceptually, this is the time that it takes for chaotic mixing to stretch elements in the flow into tendrils that are thin enough for molecular diffusion to finish the mixing.

Figures 2 e, f, g and h show mixing fields determined numerically for the same conditions as for Figures 2 a – d. These fields are obtained by simulating the motion of triplets of tracers, initially very close to each other. At each time step, the mixing field at a point is defined as the ratio of the largest separation between the tracers in the triplet divided by the initial separation. It is apparent when comparing Figures 2 b-d with Figures 2 f-h that for the advection-reaction-diffusion case, the mixing fields do a good job of capturing the dominant pattern formation behavior for the oscillatory reaction. (The only exception is for the reaction-diffusion limit for which the mixing fields cannot capture the patterns.)

The implication of these studies is that pattern formation in a reacting system with chaotic mixing can be understood quite well by considering almost purely the mixing behavior of the system without much regard to the details of the reaction itself. We expect, therefore, that mixing fields should be able to capture the dominant pattern formation process for a wide range of advection-reaction-diffusion systems.

More details about these experiments can be found in Reference 10.

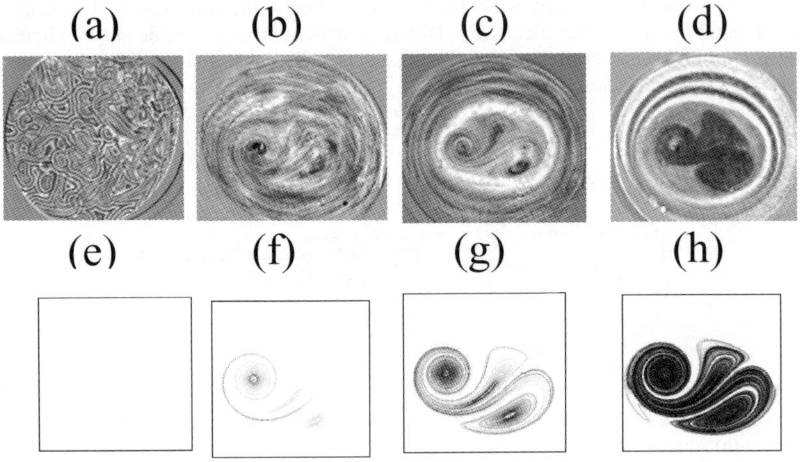

Figure 2. (a)-(d) Images of Belousov-Zhabotinsky patterns in blinking vortex flow. For all cases, the non-dimensional blinking period $\mu = 0.52$ (see Reference 10 for definition). (e)-(h) Mixing fields for the same flows. (a) and (e): no flow. The blinking frequency is 0.010 Hz for (b) and (f), 0.030 Hz for (c) and (g), and 0.050 Hz for (d) and (h).

III. Front propagation and mode-locking

There are numerous processes that are governed by the growth of one species in the system at the expense of another species. The interface between the two regions is often referred to as a *front*, and there has been a significant amount of interest in the manner in which this front moves in reaction-diffusion systems. In fact, there is a well-known theory by Fisher and Kolmogorov [18],[19] that predicts a definitive value for the front propagation speed, given information about the reaction kinetics and the molecular diffusion coefficient. The question arises as to how front propagation is affected by the presence of fluid flows in the system, particularly in situations where the mixing is chaotic. Chaotic mixing processes can often be quantified as enhanced diffusion, so it is natural to predict that the Fisher-Kolmogorov (FK) theory will still work as long as the molecular diffusion coefficient in the
theory is replaced with the enhanced diffusion coefficient. Recent theoretical/numerical studies [5], however, have indicated that a simple extension of the FK theory does not necessarily work for cellular flows with periodic time dependence. Specifically, these studies showed that mode-locking is possible for front propagation in an oscillating chain of vortices (Figure 3), a flow which had been shown previously [15],[20],[21] to display chaotic mixing.

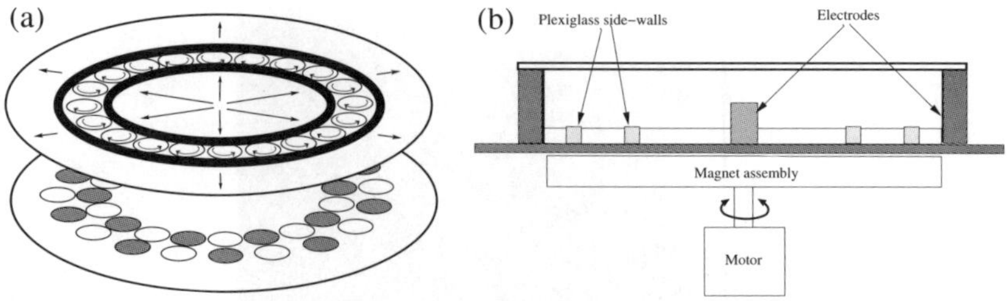

Figure 3. (a) Exploded view of magnetohydrodynamic forcing and resulting flow, composed of an annular chain of 20 counter-rotating vortices. (b) Side view of experimental apparatus. An electrical current passes radially through a thin layer of an electrolytic solution. This current interacts with an alternating magnetic field produced by two rings of magnets mounted in an assembly below the fluid layer. The magnet assembly is mounted coaxially on a motor whose motion can be controlled to oscillate, drift with a constant angular velocity, or move with a combination of both drift and oscillations.

We have reproduced some of these numerical simulations to illustrate locking phenomena [12]. Mode-locking is defined for a propagating front as follows: when mode-locked, the front advances an integer number N of wavelengths in an integer number M of drive periods. Figure 4(a) shows a numerical sequence showing a $(N,M) = (1,1)$ mode such that the front advances two vortex widths (1 wavelength of the flow) in 1 drive period. Figure 4(b) shows a numerical sequence for a (1,2) mode-locked state.

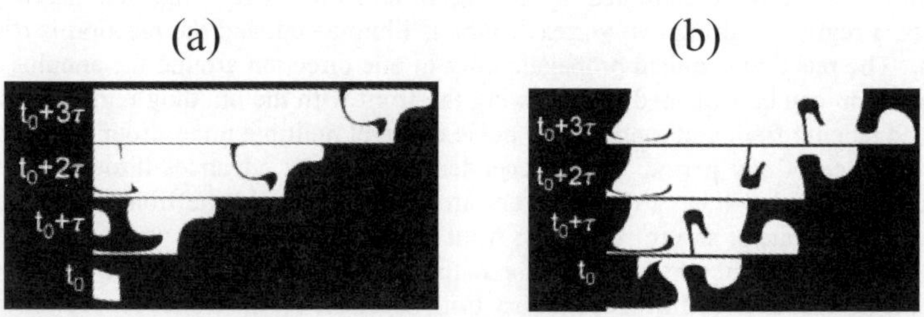

Figure 4. Simulations showing mode-locking with (N,M) = (1,1) for (a) and (1,2) for (b).

We have tested these predictions with experiments on front propagation in an oscillating vortex chain [11],[12]. The flow and apparatus are show in in Figure 3. There are 20 vortices in the vortex chain, which is oriented in an annular configuration. As with the blinking vortex flow described in Section II, the vortex chain is forced magnetohydrodynamically. Two rings of 20 Nd-Fe-Bo magnets are set in a magnet assembly, above which rests a thin layer of an electrolytic solution. The magnets alternate in polarity, as shown in Figure 3. An electrical current passes radially through the fluid layer and interacts with the alternating magnetic field. The result is a chain of 20 alternating vortices in the fluid layer. The magnet assembly is mounted co-axially on a motor which can be programmed to rotate in an oscillatory pattern, with a constant drift velocity, or with a combination of drift and oscillatory terms; the vortex chain itself moves with the magnet assembly. For the experiments discussed in this section, the magnet assembly is programmed to oscillate sinusoidally, although the frequency and amplitude of the oscillations are varied.

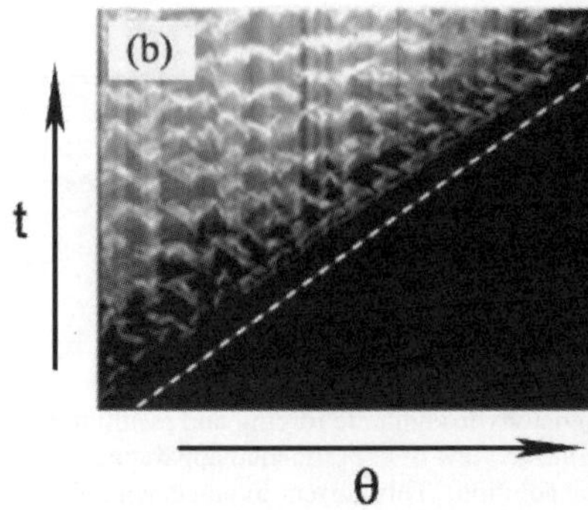

Figure 5. Space-time plot for an experimental run that is mode-locked with (N,M) = (1,2).

The chemical reaction used is an excitable version of the Belousov-Zhabotinsky reaction. (See References 11 and 12 for details of the chemistry.) The chemicals are initially orange in color, but a green front can be triggered by inserting a silver wire into the flow momentarily. Once initiated, the front propagates across the system. The reaction is actually a "pulse" rather than a front – the green reacting zone is followed by a return of the system to the orange state behind the leading edge. In the absence of any flow at all, the result is a green ring that propagates outward with the rest of the system remaining orange. The orange section can be re-triggered multiple times, either if the green front returns due to flows in the system or with the re-insertion of a silver wire. The reaction is also photo-sensitive and can be inhibited by strong illumination with blue and green light. In the experiments, a region covering two vortex widths is illuminated, and the reaction is triggered next to this region. The reaction can then propagate only in one direction around the annulus. The duration of the experiment can be extended by following the front with the blinding region, erasing the region behind the advancing front and enabling the pulse to travel multiple times around the annulus.

In the absence of any periodic time dependence, the front advances through the system due to several factors: (a) advection carries the front around a vortex; (b) the front "burns" its way inward toward the vortex centers; and (c) when the front reaches a corner of a vortex, it "burns" across the separatrix between adjacent vortices and triggers the reaction in the next vortex.

When the system is forced with oscillatory time dependence, the front propagation shows many of the characteristics of chaotic mixing in the oscillating vortex chain. Despite the chaotic behavior, though, the front typically (for large enough amplitude) propagates with a constant velocity, as shown in the spacetime plot of Figure 5. Sequences of images (Figure 6) show that the system is showing mode-locking; compare Figure 6a with Figure 4a and Figure 6b with Figure 4b. The propagation speed can be determined from the slope of the spacetime plot (Figure 5). If plotted as a function of oscillation frequency (Figure 7), two dominant modes are apparent – a (1,1) and (1,2) mode. When locked, the propagation speed grows linearly with frequency.

A parameter space diagram (Figure 8) reveals the conditions for which the fronts are mode-locked. For small amplitudes of oscillation, the system does not mode-lock – the front speed fluctuates in time and there is no repeating pattern. For larger amplitudes, locking has been found for both the (1,1) and (1,2) locking tongues, and there is a region of overlap where the front switches alternately between the (1,1) and (1,2) modes during a single run.

These results agree with the predictions of Reference 5. The fact that the system mode-locks indicates clearly that a simple extension of the FK theory is not valid, and that a new theoretical treatment is needed to predict front propagation in an advection-reaction-diffusion system. In particular, the role of coherent structures (vortices, in this case) needs to be assessed. Our expectation is that deviations from FK predictions will be common, potentially even for turbulent flows, since coherent flow structures are quite common in many natural fluid flows.

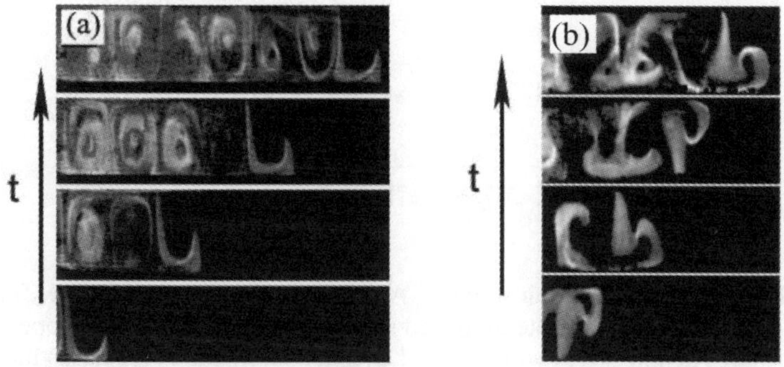

Figure 6. Experimental sequences showing mode-locking with (N,M) = (1,1) for (a) and (1,2) for (b).

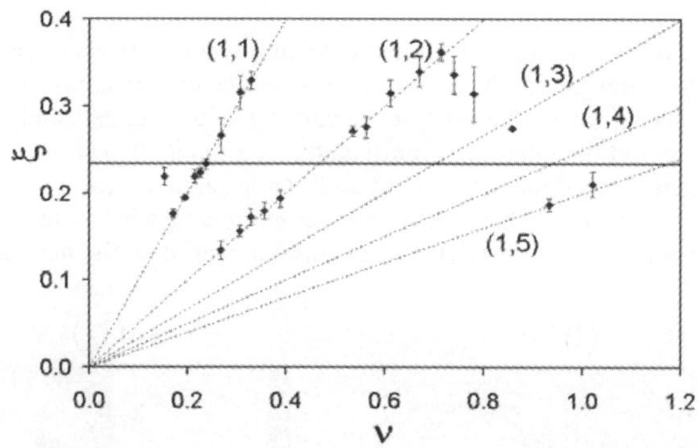

Figure 7. Experimental results showing non-dimensional front speed ξ as a function of non-dimensional frequency ν. The dotted lines show the theoretical predictions (with no fitted parameters) for mode-locked speeds.

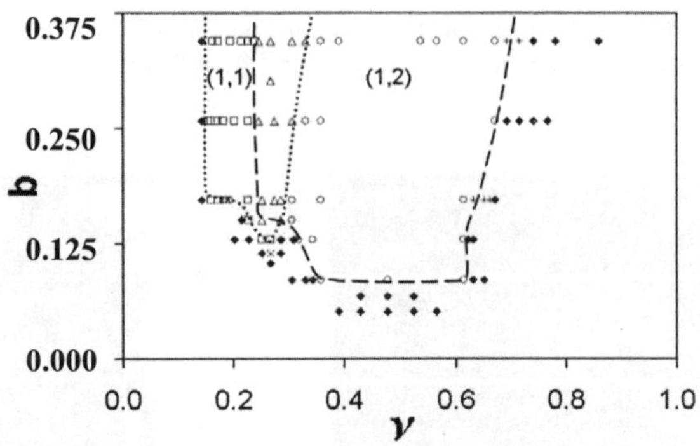

Figure 8. Parameter-space plot showing Arnol'd tongues for (1,1) and (1,2) mode-locked states. Filled diamonds denote unlocked states, whereas open squares, open circles and open triangles denote states with (1,1), (1,2) and combination (1,1)/(1,2) mode-locking, respectively.

More details about these experiments can be found in References 11 and 12.

IV. Synchronization of a continuous network of oscillators

The third set of experiments [13] deals with synchronization of chemical oscillators. There has been a tremendous amount of research recently into networks of oscillators and how they synchronize. Those studies were energized in the past few years by a theoretical study [22] that showed that network connectivity could be enhanced significantly in a *Small-World Network* with random short-cuts that connect distant parts of the network, in addition to regular, nearest-neighbor connections.

Our experiments consider synchronization in a *continuous* (rather than discrete) fluid network in which fluid mixing is the dominant mechanism of communication in the network. The flow is the

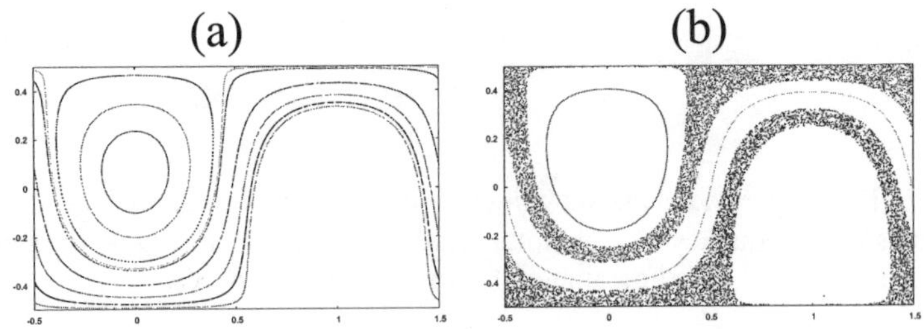

Figure 9. Transport in the drifting vortex chain (a) without oscillations and (b) with oscillations. In the absence of any oscillations, all trajectories are ordered and the flow is divided into two types of regions, one in which tracers rotate within a vortex core, and the other in which tracers move rapidly around and between vortices in a snake-like jet region. If there is oscillatory time dependence as well, chaotic regions forms, denoted by erratic pattern of dots in (b).

same alternating vortex chain discussed in Section III, except that in addition to periodic oscillations, the vortex chain can also drift. If there are no oscillations but there is a drift, a snake-like region forms that winds around and between the vortices (Figure 9a). A tracer in this snake region rapidly moves between vortices and can traverse a long distance in a very short period of time. If the vortex chain has both oscillatory and drifting motion, then there can be a combination of both ordered and chaotic trajectories. If the drift velocity v_d is greater than the maximum oscillation velocity v_o, then the snake region (or portions of it) is maintained (Figure 9b). In this regime, tracers in a chaotic region alternately rotate within a vortex, move between one vortex and the next, or stick to the snake region and undergo rapid motion to a distant vortex.

Trajectories such as this that alternately stay within a localized region and undergo rapid long-range jumps are referred to as Lévy flights [23],[24] (depending on the statistics of the jumps), and transport in this case is typically superdiffusive with a variance that grows faster than linearly in time. This is in contrast to normal, enhanced diffusion with a linear growth in the variance. The result is a system that typically shows normal diffusion for $v_d < v_o$ and superdiffusion for $v_d > v_o$ (Figure 10).

We contend that Lévy flights and superdiffusion play a role in a fluid system similar to the role of short-cuts in a Small World network. We have investigated this experimentally by studying synchronization of the oscillatory BZ reaction (with the same chemistry as in Section I) in the oscillating/drifting vortex chain. In the absence of any time-dependence (i.e., a stationary vortex chain), each vortex acts like an isolated BZ reactor with only minimal communication, since transport between vortices is purely via molecular diffusion. If the flow is time dependent with $v_d < v_o$, there is chaotic mixing between adjacent vortices, and the system spontaneously forms traveling waves (Figure 11a), although the traveling waves evolve in a very complicated manner as a function of time. For time-dependent forcing with $v_d > v_o$, the system typically synchronizes in one of two different ways. In most cases (Figure 11b), *co-rotating* synchronization is observed where the odd vortices synchronize with each other and the even vortices synchronize with each other, but there is an arbitrary phase different between these two sets of vortices. In some cases, however, the system synchronizes globally with all the vortices blinking red/blue in unison (Figure 11c).

The synchronization behavior is summarized with a parameter space plot in Figure 12. Comparing Figure 12 with Figure 10, it can be seen that the system typically synchronizes if the transport is superdiffusive with Lévy flight trajectories, similar to the small-world networks for discrete networks.

The implication of these results is that superdiffusive transport may be a necessary (although probably not sufficient) condition for synchronization in an extended fluid system, i.e., one in which the total system size is appreciably larger than characteristic length scales of the fluid flow. It is also intriguing to consider Lévy flights in these studies as playing a role similar to the "short-cuts" in the Small World networks. But there are differences as well. First, Lévy flights in the vortex chain are not random; rather, they connect every vortex to every other one with a magnitude that decays algebraically (instead of exponentially) with separation. Second, the amount of fluid participating in the flights depends on the sizes of ordered regions in the flow, something that has not been analyzed in detail yet.

We also speculate that techniques used to predict synchronization in a continuous fluid flow could be applied to discrete systems with large numbers of nodes, especially if the nodes are mobile. For a moving population of people, for example, it might be possible to treat the population as a "fluid" and measure its transport properties. Synchronization of various processes in this population (e.g., disease outbreak) might then be predicted by determining whether mixing in the population is diffusive or superdiffusive.

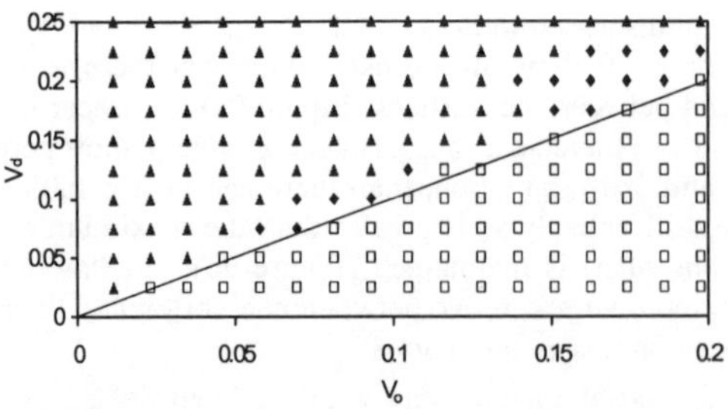

Figure 10. Parameter space plot showing regions of normal (open squares) and superdiffusive transport (filled triangles and diamonds) for the oscillating and drifting vortex chain. Two types of superdiffusive regimes are denoted, one characterized by two chaotic regions separated by an ordered jet region (triangles) and the other denoted by one chaotic region with flight-producing islands (diamonds). The solid line corresponds to $v_d = v_o$.

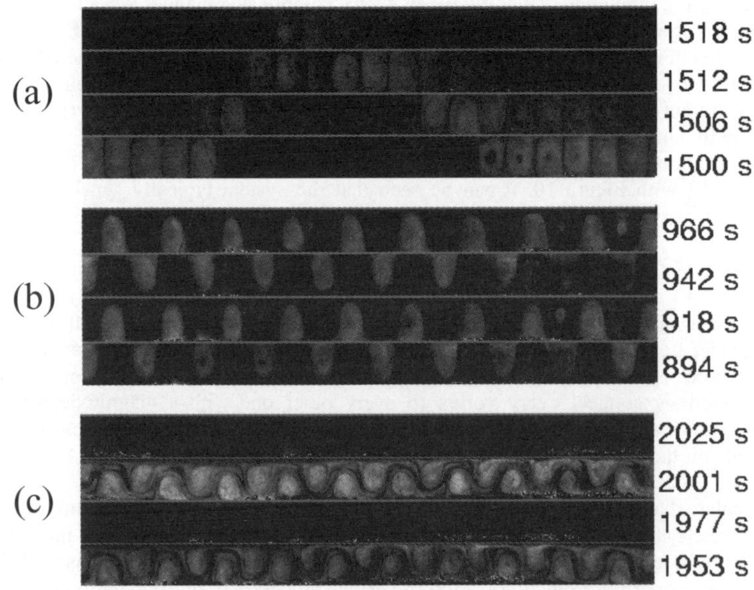

Figure 11. Sequences of images of BZ reaction in oscillating/drifting vortex chain. (a) Wave behavior seen when $v_d < v_o$. (b) Co-rotating and (c) global synchronization for $v_d > v_o$.

V. Continuing work

Numerous issues remain unanswered in advection-reaction-diffusion processes. We are currently conducting several additional experimental investigations on subject: (a) the effects of superdiffusive transport on front propagation processes; (b) the growth of a chemical region in a flow with chaotic mixing; and (c) chaotic traveling waves in an advection-reaction-diffusion system.

Acknowledgements

This work was supported by the US National Science Foundation Grants DMR-0404961 and REU-0097424.

References

[1] Grindrod, P. *The Theory and Applications of Reaction-Diffusion Equations: Patterns and Waves* (Clarendon Press, Oxford, 1996).

[2] Ben-Avraham, D & Havlin, S. *Diffusion and Reactions in Fractals and Disordered Systems* (Cambridge University Press, Cambridge, 2000).

[3] Tel, T., de Moura, A., Grebogi, C. & Karolyi, G. Chemical and biological activity in open flows: A dynamical system approach. *Phys. Rep.* **413**, 91-196 (2005).

[4] Abel, M., Celani, A., Vergni, D. & Vulpiani, A. Front propagation in laminar flows. *Phys. Rev. E* **64**, 046307 (2001).

[5] Cencini, M., Torcini, A., Vergni, D. & Vulpiani, A. Thin front propagation in steady and unsteady cellular flows. *Phys. Fluids* **15**, 679-688 (2003).

[6] Karolyi, G., Pentek, A., Toroczkai, Z., Tel, T. & Grebogi, C. Chemical or biological activity in open chaotic flows. *Phys. Rev. E* **59**, 5468-5481 (1999).

[7] Karolyi, G., Pentek, A., Scheuring, I., Tel, T. & Toroczkai, Z. Chaotic flow: The physics of species coexistence. *Proc. Nat. Acad. Sci. U.S.A.* **97**, 13661-13665 (2000).

[8] Neufeld, Z. Excitable media in a chaotic flow. *Phys. Rev. Lett.* **87**, 108301 (2001).

[9] Neufeld, Z., Kiss, I. Z., Zhou, C. & Kurths, J. Synchronization and oscillator death in oscillatory media with stirring. *Phys. Rev. Lett.* **91**, 084101 (2003).

[10] Nugent, C. R., Quarles, W. M. & Solomon, T. H. Experimental studies of pattern formation in a reaction-advection-diffusion system. *Phys. Rev. Lett.* **93**, 218301 (2004).

[11] Paoletti, M. S. & Solomon, T. H. Experimental studies of front propagation and mode-locking in an advection-reaction-diffusion system. *Europhys. Lett.* **69**, 819-825 (2005).

[12] Paoletti, M. S. & Solomon, T. H. Front propagation and mode-locking in an advection-reaction-diffusion system. *Phys. Rev. E* **72**, 046204 (2005).

[13] Paoletti, M. S., Nugent, C. R. & Solomon, T. H. Synchronization of Oscillating Reactions in an Extended Fluid System. *Phys. Rev. Lett.*, in press (2006).

[14] Aref, H. Stirring by chaotic advection. *J. Fluid Mech.* **143**, 1-21 (1984).

[15] Solomon, T. H., Tomas, S. & Warner, J. L. The Role of Lobes in Chaotic Mixing of Miscible and Immiscible Impurities. *Phys. Rev. Lett.* **77**, 2682-2685 (1996).

[16] Winfree, A.T. Spiral waves of chemical activity. *Science* **175**, 634-636 (1972).

[17] Showalter, K. Pattern-formation in a ferroin-bromate system. *J. Chem. Phys.* **73**, 3735-3742 (1980).

[18] Kolmogorov, A. N., Petrovskii, I. G. & Piskunov, N. S. *Moscow Univ. Math. Bull. (Engl. Transl.)* **1**, 1 (1937).

[19] Fisher, R. A. Proc. Annu. Symp. Eugen. Soc. **7**, 355 (1937).

[20] Solomon, T. H. & Gollub, J. P. Chaotic particle transport in time-dependent Rayleigh-Bénard convection. *Phys. Rev. A* **38**, 6280-6286 (1998).

[21] Camassa, R. & Wiggins, S. Chaotic advection in a Rayleigh-Bénard flow. *Phys. Rev. A* **43**, 774-797 (1991).

[22] Watts, D. J. & Strogatz, S. H. Collective dynamics of 'small-world' networks. *Nature.* **393**, 440-442 (1998).

[23] Shlesinger, M.F., Zaslavsky, G. M. & Klafter, J. Strange kinetics. *Nature* **363**, 31-37 (1993).

[24] Solomon, T. H., Weeks, E. R. & Swinney, H. L. Observation of anomalous diffusion and Lévy flights in a two-dimensional rotating flow. *Phys. Rev. Lett.* **71**, 3975-3978 (1993).

Properties of chaotic advection in a 2-layer model of vortex flow

D.V. Stepanov, K.V. Koshel

Pacific Oceanological Institute, Far East Division, Russian Academy, of Sciences, 690041, Russia

Abstract

The transport properties of a two-dimensional, inviscid incompressible vortex flow using dynamical systems techniques are examined. The flow field induced by the interaction of a bottom topography with a background flow in a 2-layer fluid is considered. Using the concept of background currents, the dynamically consistent stream function of this flow is constructed. If the background flow is steady, then the trajectories of fluid particles coincide with streamlines. The flow field consists of a vortex region (VR) with closed streamlines bounded by a separatrix and a flowing region (FR) with streamlines going to infinity. In the presence of the oscillating perturbation the picture changes dramatically; fluid particles are entrained and detrained from the VR and chaotic particle motion occurs. We focus on how the change in frequency of the perturbation affects the transport and distribution of passive tracers. It's carried out experiment that allows us to estimate the number of fluid particles, initially uniformly filled whole vortex region, leaving the VR. The evolution particle number shows that when the perturbation frequency increases the mean transport decreases. However, the dependence of particle number leaved the VR on perturbation frequency is complicated. It has both global maximum and local extremums. The analyse of evolution Poincaré map from the perturbation frequency showed that local extremums are related to the disappearance and overlap of resonance bands. It is suggested that the disappearance of resonance bands related to the limited of dependence of fluid particle travel time in the VR on the distance from tracer location to vortex center at absent of perturbation.

Keywords: chaotic advection, topographic vortex, Poincaré map

1 Introduction

The role of chaotic advection [1] in the exchange within the topographic vortex structures has been the topic of numerous studies over the past decade. Such studies are aimed at uncovering a mechanism for entrainment and detrainment mass in the vortex structures of oceans and seas without considering turbulence diffusion [2, 3, 4, 5, 6, 7, 8, 9]. The dynamically consistent stream functions [10] of the topographic vortex models [4, 5, 6] have been developed with using the concept of background currents [11] within the barotropic approach. There are reasons to believe, however, that barotropic models are restrictive and so using the baroclinic models are the natural advance of these models. The simplest baroclinic model is a 2-layer model.

The primary purpose of this work is to investigate the chaotic behavior of Lagrangian tracers within the limits of a quasigeostrophic 2-layer model of a topographic vortex in the pulsating flow. In particular, we study the influence of the perturbation frequency on chaotic advection.

This paper is organized as follows. In section 2, we describe an analytical model used in this study. The tracers blob transfer is investigated in section 3 and last, we summarize our results.

2 Model

The flow induced by a topographic bottom elevation with a pulsating incident current is considered. In quasigeostrophic approach of a 2-layer ocean on an f-plane the dynamically consistent stream functions [10] ψ_1–up layer, ψ_2–down layer are [11]

$$\begin{aligned} \psi_1 &= -Uy - \tfrac{f_0\tau_\infty}{2\pi H}\left(\ln k\left(r-r_0\right) + K_0\left(k\left(r-r_0\right)\right)\right), \\ \psi_2 &= -Uy - \tfrac{f_0\tau_\infty}{2\pi H}\left(\ln k\left(r-r_0\right) - (H_1/H_2)K_0\left(k\left(r-r_0\right)\right)\right), \end{aligned} \quad (1)$$

where $r=\sqrt{x^2+y^2}$, r_0 is the seamount location center, $K_0(r)$ is the zero order Macdonald's function, H is the depth of ocean, H_1 and H_2 are the layer widths, f_0 is the Coriolis parameter, U is the incident flow velocity, h^* is the efficient hight of seamount, $\Delta\rho/\rho_2$ is the relative density, $L_d = \sqrt{\frac{H_1H_2g\Delta\rho}{Hf_0^2\rho_2}} = k^{-1}$ is the internal Rossby deformation radius, $\tau_\infty = \pi h^* L_d^2$ is the efficient volume of seamount. The advection equation system has a Hamiltonian form

$$\begin{cases} \dot{x} = -\psi_y, \\ \dot{y} = \psi_x. \end{cases} \quad (2)$$

Taking length scale $L^* = L_d$, velocity scale $U^* = U$, stream function scale $\psi^* = L^*U^*$ and time scale $T^* = \frac{L^*}{U^*}$ and introducing dimensionless variables instead (1) we will have

$$\begin{aligned} \psi_1 &= -Uy - (\ln r + K_0(r)), \\ \psi_2 &= -Uy - (\ln r - (H_1/H_2)K_0(r)). \end{aligned} \quad (3)$$

We shall consider only up layer, where the equation of motion for fluid particles are

$$\begin{cases} \frac{dx}{dt} = U + y/r\left(1/r - K_1\left(r\right)\right), \\ \frac{dy}{dt} = -x/r\left(1/r - K_1\left(r\right)\right). \end{cases}$$

3 Chaotic transport

If the incident flow is steady $U = U_0 = const$, then the streamline picture has the homoclinic structure [4]. Streamlines at the $U_0 = 0.3$ are shown in fig. 1.

A separatrix intersecting itself at the hyperbolic point divides the flow area into two parts: one with closed streamlines surrounding elliptic point inside homoclinic loop is a vortex region (VR) and the other with unbounded trajectories outside it is a flowing region (FR). The fluid particle exchange between VR and FR is absent.

When a small time-dependent perturbation is added the picture of fluid particle motion changes dramatically: fluid is entrained and detrained from the neighborhood of vortex and the chaotic particle motion occurs. In early works [4, 5, 6, 12] have been shown that the perturbation frequency influences on this fluid motion.

Using numerical simulations we investigate the influence of the perturbation frequency and baroclinicity on the exchange of fluid particles between VR and FR. The velocity of incident current is taken:

$$U\left(t\right) = 0.3\left(1 + \delta\sin\left(\omega t\right)\right),$$

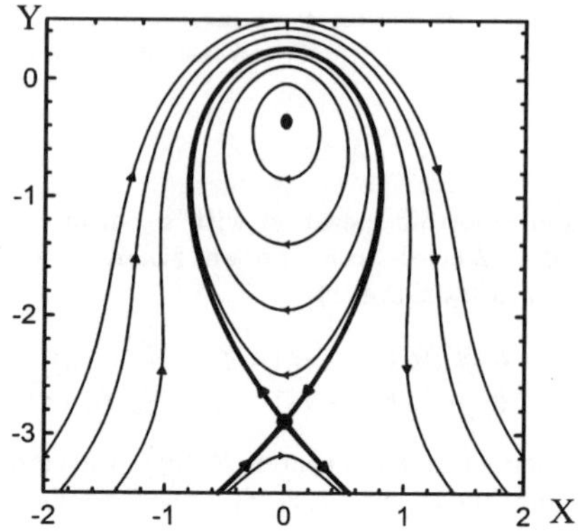

Figure 1: Stream lines at the $U_0 = 0.3$

where $\delta = 0.1$ is the relative perturbation amplitude, ω is the perturbation frequency. We will consider a numerical experiment, which allows us to estimate the number of particles $N(t)$, that have reached the line $x = 3$ from the original, uniformly distributed 8250 tracers in the VR.

Fig. 2 shows the evolution of the leaving particle number $N(t)$ in percentage at the various value of the perturbation frequency.

At first as time goes by the leaving particle number increases dramatically and then is slowly reaching the limiting value. This confirms results of other work [4, 5, 6, 12]. However, though monotonous reduction of the mean transport of tracers from VR to FR, the dependence of the leaving particle number on the perturbation frequency has a non-monotonous character.

Fig. 3 shows the dependence of the limiting particle number $N_\infty(\omega)$ on the perturbation frequency. Along with the presence of the optimal frequency $\omega = 0.59$ [4, 5, 6, 12] at which value of the leaving particle is maximum, series local extremums are observed.

To explain the non-monotonous character of $N_\infty(\omega)$ we used the mean of the Poincaré map. This map takes the position (x, y) of a particle on the plane to its position one oscillation period $T = 2\pi/\omega$ later. The sequence of points (x_i, y_i) obtained by iterating the map on some initial condition is known as an orbit of the Poincare map.

Fig. 4 shows the typical picture of the Poincare map for $\omega = 0.2$. VR consists of the family smooth closed orbits (tori) around central elliptic point is a core and around elliptic point of resonance band is a lateral island [13]. Tracers initially located in the region of core and islands of resonance bands never cannot leave their.

Consider the evolution of VR structure with the change of the perturbation frequency (fig. 4). With the growth latter, the lateral island verges towards core and their total area decreases because of overlap resonance [13]. This results in the increase of leaving particle number (cp. fig. 3 range $\omega = (0.2; 0.25)$). At the $\omega = 0.26$ the degree of overlap of core and the lateral island has the maximal area. For the $N_\infty(\omega)$ this correspondences local maximum (cp. fig. 3 $\omega = 0.26$). With further increase of perturbation frequency, the core is destroyed and the role of core acts the lateral island of disappeared resonance band, which has maximal area. This results in the strong decrease of leaving particle number (cp. fig. 3 range $\omega = (0.26; 0.27)$) and the appearance of local minimum of $N_\infty(\omega)$ (cp. fig. 3, $\omega = 0.27$). Then with the increase of the perturbation frequency, the next resonance band verges towards core and begin to overlap with him. This scenario repeats up to the optimal frequency $\omega = 0.59$. At frequencies $\omega > 0.59$ the

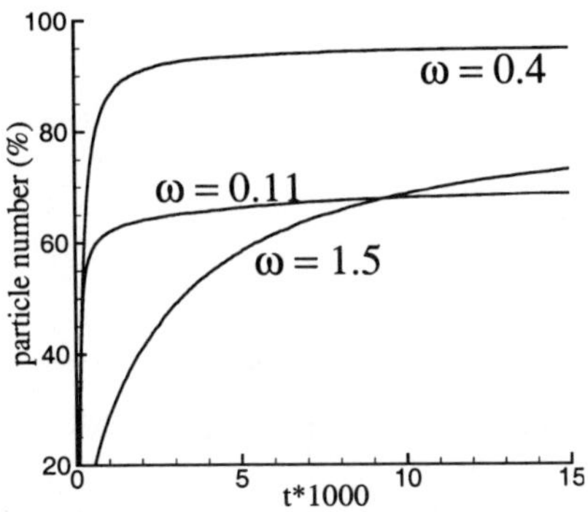

Figure 2: The particle number N(t) (%) evolutions at numerous value of the perturbation frequency.

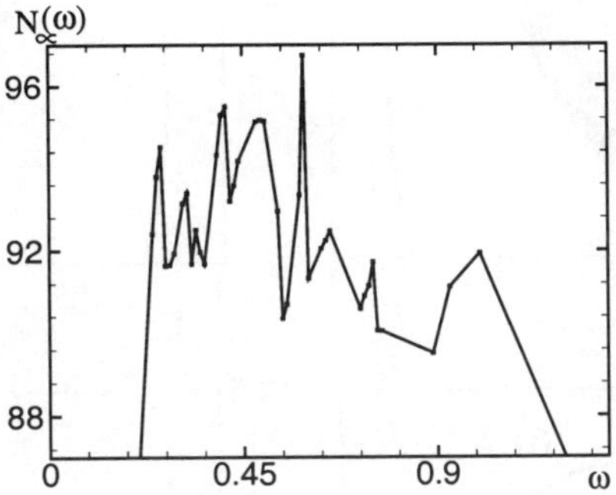

Figure 3: The dependence of limiting particle number $N_\infty(\omega)$ on the perturbation frequency.

process of the destroyed core repeats, but with islands of resonance bands of higher-order. This results in the total decrease of $N_\infty(\omega)$.

Thus, the periodic destroyed of core and the disappearance resonance bands are the primary mechanism of appearance of local maximum and minimum of $N_\infty(\omega)$. We studied the rotating frequency of fluid particle $\omega_0\,(y)$ in the VR when perturbation is absent and found that $\omega_0\,(y)$ is bounded $\omega_{cr} \approx 0.562$ (fig. 5), that is at the perturbation frequency ω in the VR the resonance bands corresponding fractional winding numbers $\frac{n}{m}$, where $n\omega > \omega_{cr}$ are absent. This is the consequence of the baroclinicity.

4 Conclusions and discussion

On based on a quasigeostrophic 2-layer model of topographic vortex, the chaotic advection is investigated. Using the numerical simulations the influence of the perturbation frequency on transfer of the tracer blob from the VR in the FR is studies. It is confirmed the monotonous decrease of mean transport of tracers

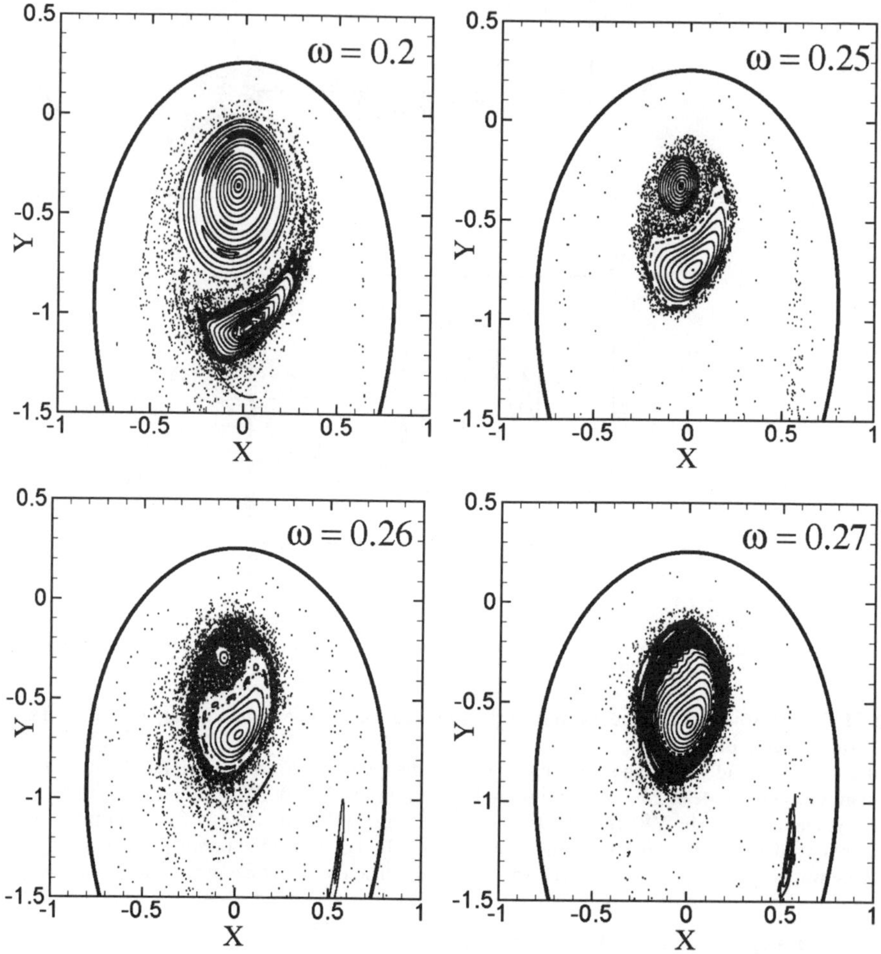

Figure 4: Poincaré maps for corresponding perturbation frequencies.

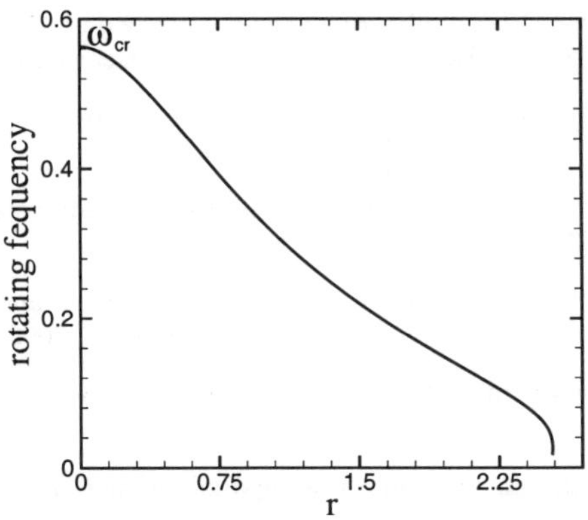

Figure 5: The dependence of tracer rotating frequency in VR from r - the distance from tracer location to vortex center at the $U = U_0 = 0.3$.

with the growth of the perturbation frequency [4, 5, 6, 12]. It was found the non-monotonous dependence of leaving particle number from the perturbation frequency. On based on the analysis of the Poincare map we found that local maximums $N_\infty(\omega)$ related to cases, when degree of overlap of core and islands of resonance bands is maximal and local minimum related to cases of the destroyed of core and the disappearance of resonance bands. It was shown that the destroyed of core related to bounded of $\omega_0\,(y)$, which is the consequence of baroclinicity.

Acknowledgements

The work was completed thanks to financial support of the Russian Fund for Fundamental Studies Grant No. 06-05-96080.

References

[1] Aref H. Chaotic advection of fluid particles. Philos. Trans. R. Soc. London 1990;333:273-288.

[2] Budyansky MV, Prants SV, Uleysky MYu. Fractals and dynamical traps in a simplest model of chaotic advection with a topographic vortex. Doklady Earth Sciences 2002;387:929-932.

[3] Budyansky MV, Uleysky M.Yu, Prants SV. Chaotic scattering, transport, and fractals in a simple hydrodynamic flow. J. Exper. Theor. Phys. 2004;99:1018-1027.

[4] Kozlov VF, Koshel KV. Some features of chaos development in an oscillatory barotropic flow over an axisymmetric submerged obstacle. Izv. Akad. Nauk, Fiz. Atmos. Okeana 2001;37:378-389.

[5] Izrailsky YuG, Kozlov VF, Koshel KV. Some specific features of haotization of the pulsating barotropic flow over elliptic and axisummetric se-mounts. Phys. Fluids 2004;16:3173-3190.

[6] Kozlov VF, Koshel KV, Stepanov DV. Influence of the boundary on chaotic advection in the simplest model of a topographic vortex. Izv. Akad. Nauk, Fiz. Atmos. Okeana 2005;41:99-109.

[7] Waseda T, Mitsudera H. Chaotic advection of the shallow Kuroshio coastal waters. J Oceanogr. 2002;58:627-638.

[8] Gledzer AE. Passive pollutant entrainment and release by ocean eddy structures, Izv. Akad. Nauk. Fiz. Atmos. Okean 1999;35:838-845.

[9] Deese HE, Pratt LJ, Helfrich KL. A laboratory model of exchange and mixing between Western boundary layers and subbasin recirculation gyres. J Phys Oceanogr. 2002;32:1879-1889.

[10] Ngan K, Shepherd T.G. Chaotic mixing and transport in Rossby-wave critical layers. J Fluid Mech 1997;334:315-351.

[11] Kozlov VF. Background currents in geophysical hydrodynamics. Izv. Akad. Nauk, Fiz. Atmos. Okeana 1995;31:245-250.

[12] Rom-Kedar V, Poje AC. Universal properties of chaotic transport in the presence of diffusion. Phys. Fluids. 1999;11:2044-2057.

[13] Lichtenberg AJ, Lieberman MA. Regular and Chaotic dynamics. second ed., Edn. Springer, New York, 1992.

Monte Carlo simulation of atomic transport in a laser field

V. Yu. Argonov, S. V. Prants

Laboratory of Nonlinear Dynamical Systems, V.I.Il'ichev Pacific Oceanological Institute of the Russian Academy of Sciences, 690041 Vladivostok, Russia

Abstract

We treat the motion of two-level atoms in a strong standing-wave laser field taking into account photon recoil effect caused by induced and spontaneous emission. Spontaneous emission is considered as random jumps to be incorporated in the equations of motion by a Monte Carlo method. We derive nonlinear equations with jumps for real-valued renormalized Bloch-like variables composed of complex-valued Schrödinger probability amplitudes and connected to classical atomic position and momentum variables. Well-known dissipative mechanical effects of cooling, heating and velocity grouping of atoms are found numerically with those equations. Correlations between internal and external degrees of freedom of an atom in the optical lattice, i.e. between center-of-mass motion and Rabi oscillations, are found.

PACS: 42.50.Vk; 05.45.Mt; 05.45.Xt
Keywords: Light-induced atomic transport; Synchronization; Quantum jumps

1 Introduction

When emitting and absorbing photons from a standing-wave laser field (optical lattices), atoms change their momentum and position, the effect known as the photon recoil (for reviews of a vast literature on atomic motion in laser fields see, for example, [1, 2, 3, 4]). Nonlinear interplay between atomic internal and external degrees of freedom may cause non-trivial dynamical effects of deterministic chaos, fractality, and anomalous transport of atoms in optical lattices. In recent years we have found and described analytically and numerically effects of Hamiltonian chaos, Lévy flights, correlations between external and internal atomic degrees of freedom, and atomic dynamical fractals both with classical [5, 6] and quantized field in an ideal cavity [7, 8]. In those studies we have neglected all the dissipation phenomena but taken into account a back reaction of atoms on the field. Hamiltonian dynamics in fundamental but simplified. In our recent paper [9] we have studied dissipative dynamics of a two-level atom in an optical lattice, created by two counter-propagating laser waves, and found effects of synchronization of internal and external atomic degrees of freedom, bifurcations, limit cycles, dissipative fractals and strange attractors. Dissipation has been added to the equations of motion phenomenologically, as a smooth exponential decay of atomic internal variables. It is a good approximation with an ensemble of atoms. As demonstrated by various experiments [10, 11], the dynamics of individual atoms is intrinsically stochastic. In practice, spontaneous emission of atoms interrupts the continuous unitary evolution in random time moments and should be modeled with the help of Monte Carlo methods. The main aim of this paper is to study the effect of spontaneous emission on atomic transport in a laser field.

2 Hamilton-Schrödinger equations of motion with quantum jumps

Let us consider a two-level atom with mass m_a, transition frequency ω_a, and the decay rate Γ_a, moving with the momentum P along the axis r through a standing laser wave with the field frequency ω_f, amplitude E, and the wave vector k_f. In the frame, rotating with the field frequency ω_f, we can write the non-Hermitian generalization of the Jaynes-Cummings Hamiltonian [12]:

$$\hat{H} = \frac{\hat{P}^2}{2m_a} + \frac{1}{2}\hbar(\omega_a - \omega_f)\hat{\sigma}_z - \hbar E\,(\hat{\sigma}_- + \hat{\sigma}_+)\cos k_f\hat{r} - i\hbar\frac{\Gamma_a}{2}\hat{\sigma}_+\hat{\sigma}_-. \tag{1}$$

Here $\sigma_{\pm,z}$ are the Pauli operators which describe the transitions between lower, $|1\rangle$, and upper, $|2\rangle$, states. The standing-wave field is considered to be an inexhaustible reservoir (it must be strong enough), so we can neglect a back reaction of atoms on the field. For electronic degree of freedom the simple wavefunction is

$$|\Psi(t)\rangle = a(t)|2\rangle + b(t)|1\rangle, \tag{2}$$

where a and b are the complex-valued probability amplitudes to find the atom in the states $|2\rangle$ and $|1\rangle$, respectively. Using the Hamiltonian (1), we get the Schrödinger equation

$$\begin{aligned} i\frac{da}{dt} &= \frac{\omega_a - \omega_f}{2}a - Eb\cos k_f r \; - i\frac{\Gamma_a}{2}a, \\ i\frac{db}{dt} &= \frac{\omega_f - \omega_a}{2}b - Ea\cos k_f r\;, \end{aligned} \tag{3}$$

where the atomic position r is considered as a parameter, and the norm of the wavefunction, $|a|^2 + |b|^2$, is not conserved.

Schrödinger equation is more appropriate to describe dynamics of single atoms than a master equation for the density matrix. The idea of an quantum trajectory approach [13, 14] is to compute a large number of quantum trajectories with different realizations of a stochastic noise, caused by spontaneous emission, and average over them to get the most probabilistic outcome that can be directly compared with respective experimental outcomes. Quantum trajectories simplify the description of an open quantum system like an atom in an optical lattice in terms of stochastically evolved pure state rather than a density matrix.

The wavefunction Monte Carlo method [13, 14] means that in numerical simulation we have two possible alternatives on every time step : the system evolves coherently according to the Schrödinger equation (including the decay term), or it performs a quantum jump to the lower state ($a = 0$, $|b|^2 = 1$). The common practice is to integrate equations for the probability amplitudes like (3) and to renormalize the wavefunction only just after quantum jumps (see, for example, [15, 16]). Adopting the idea of the wavefunction Monte Carlo method, we propose the following algorithm. To describe the evolution of the internal atomic variables we introduce instead of the complex-valued probability amplitudes a and b the following new real-valued variables:

$$u \equiv \frac{2\,\mathrm{Re}\,(ab^*)}{|a|^2 + |b|^2}, \quad v \equiv \frac{-2\,\mathrm{Im}\,(ab^*)}{|a|^2 + |b|^2}, \quad z \equiv \frac{|a|^2 - |b|^2}{|a|^2 + |b|^2}, \tag{4}$$

which are renormalized quadratures of the atomic dipole moment (u and v) and the atomic population inversion, z. It immediately follows from Eqs.(4) that $u^2 + v^2 + z^2 = 1$, i.e. the length of the Bloch vector is conserved during the coherent evolution in spite of the fact that the norm $|a|^2 + |b|^2$ is not conserved between quantum jumps. Just after spontaneous events, the new internal atomic variables jump to the values: $u \to 0$, $v \to 0$, $z \to -1$.

In the process of emitting and absorbing photons, atoms not only change their internal electronic states but their external translational states change as well due to the photon recoil. If the atomic average momentum is large as compared to the photon momentum $\hbar k_f$, one can describe the translational degree of freedom classically. When spontaneous emission events occur, the momentum changes by the value of the photon recoil $\hbar k_f$. Since we are working with a 1-D geometry, its projection to r axis is chosen to be a random value between $\pm\hbar k_f$. We suppose that spontaneous emission is absolutely isotropic. Considering the translational variables as classical, we do not quantize the recoil projection (in difference from [15]). The dynamics is now governed by the Hamilton-Schrödinger equations

$$
\begin{aligned}
&\dot{\xi} = \omega_r p, \\
&\dot{p} = -\varepsilon u \sin\xi + \sum_{j=1}^{\infty} \tilde{p}_i \delta(\tau - \tau_j), \\
&\dot{u} = \Delta v + 0.5\, uz - u \sum_{j=1}^{\infty} \delta(\tau - \tau_j), \\
&\dot{v} = -\Delta u + 2\varepsilon z \cos\xi + 0.5\, vz - v \sum_{j=1}^{\infty} \delta(\tau - \tau_j), \\
&\dot{z} = -2\varepsilon v \cos\xi - 0.5\,(u^2 + v^2) - (z+1) \sum_{j=1}^{\infty} \delta(\tau - \tau_j),
\end{aligned}
\tag{5}
$$

where $\xi \equiv k_f\langle\hat{r}\rangle$ and $p \equiv \langle\hat{P}\rangle/\hbar k_f$ are classical atomic center-of-mass position and momentum, respectively. Dot denotes differentiation with respect to dimensionless time $\tau \equiv \Gamma_a t$. The normalized recoil frequency, $\omega_r \equiv \hbar k_f^2/m_a\Gamma_a \ll 1$, the normalized Rabi frequency, $\varepsilon \equiv E/\Gamma_a$, and the atom-field detuning, $\Delta \equiv (\omega_f - \omega_a)/\Gamma_a$, are the control parameters, τ_j are random time moments when spontaneous emission occurs, and $\tilde{p}_j$ are the random recoils with values between ± 1. In terms of our normalized time τ the average frequency of spontaneous emission events τ_j is equal to $(z+1)/2$ (the probability of spontaneous emission is maximal when the atom is in the upper state and zero in the lower state). At the moments $\tau = \tau_j$, the atomic momentum jumps to the value $p \to p + \tilde{p}_j$.

When comparing the equations of motion (5) with the equations of motion (6) in our paper [9], which were derived under the assumption of a smooth exponential decay of the excited atomic state, we would like to stress that besides of the stochastisity of Eqs.(5) they are nonlinear in the internal variables in difference from the three Bloch-like equations in the set (6) in Ref.[9]. This nonlinearity is due to a coherent decay of the excited state (see Eqs.(3)) and to the specific choice of the variables u, v and z. The underlying Schrödinger equation (3) is linear, of course. This kind of nonlinearity by itself cannot produce a dynamical instability like chaos because of the existence of the conserved norm of the Bloch vector. Speaking more precisely, we state that in the absence of photon recoils (ξ is a constant) and spontaneous events the deterministic equations for three internal atomic variables with one integral of motion are not chaotic.

3 Numerical simulation of light-induced mechanical effects

It is well-known that light can accelerate and decelerate atoms [1, 2, 3, 4]. Semiclassical theory of the mechanical effects can be found, for example, in Ref.[2]. We have studied laser-induced mechanical effects in [9] considering dissipation as an exponential decay. The approach, based on the method of quantum trajectories [13, 14], makes it possible to describe spontaneous emission much more realistically.

Let us consider the following situation. A collection of non-interacting atoms moves with different velocities in a one-dimensional laser standing wave. What we need to know is a resulting atomic momentum distribution at an arbitrary moment of time after interaction with the laser field. In Fig. 1 we show the result of numerical simulation. The initial momentum distribution along the axis r is shown in Fig. 1a, where 10^4 atoms are distributed over a number of channels with the channel width $40\hbar k_f$. The actual number of atoms per a channel is shown on the y-axis. We have chosen cesium atoms with the decay rate $\Gamma_a \simeq 32$ MHz and the transition wavelength $\lambda_a \simeq 852$ nm. So the respective recoil frequency is $\omega_r = 0.003$. The field is almost resonant, $\omega_f - \omega_a \ll \omega_a$. The initial conditions are: $\xi_0 = 0$, $z_0 = 1$, $u_0 = v_0 = 0$ (atoms are initially in the upper state), and the normalized Rabi frequency is $\varepsilon = 100$.

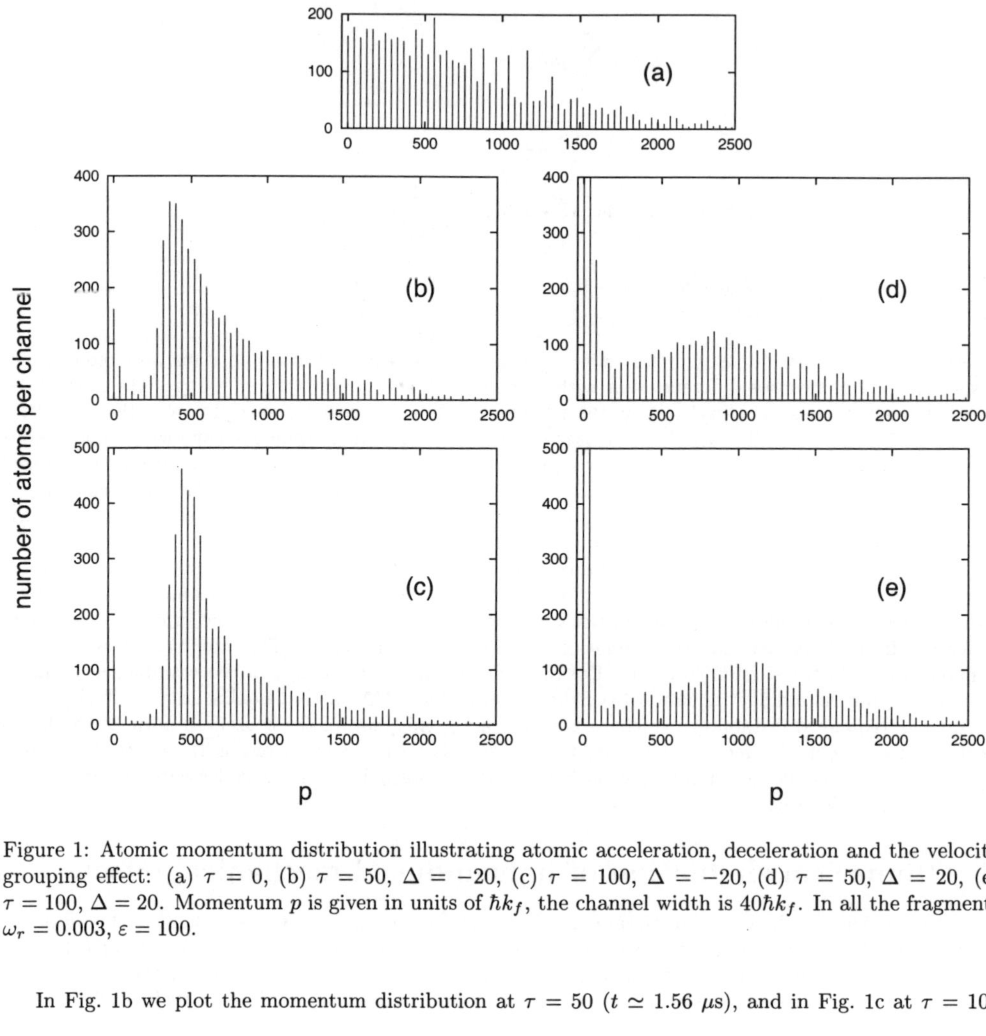

Figure 1: Atomic momentum distribution illustrating atomic acceleration, deceleration and the velocity grouping effect: (a) $\tau = 0$, (b) $\tau = 50$, $\Delta = -20$, (c) $\tau = 100$, $\Delta = -20$, (d) $\tau = 50$, $\Delta = 20$, (e) $\tau = 100$, $\Delta = 20$. Momentum p is given in units of $\hbar k_f$, the channel width is $40\hbar k_f$. In all the fragments $\omega_r = 0.003$, $\varepsilon = 100$.

In Fig. 1b we plot the momentum distribution at $\tau = 50$ ($t \simeq 1.56$ μs), and in Fig. 1c at $\tau = 100$

($t \simeq 3.1\mu s$). In both the cases the detuning is negative, $\Delta = -20$. One can see a distribution with a prominent peak at $p \sim 450$ with a reduced number of fast and slow atoms. The atoms with $p < 100$ in Fig. 1c are those which have been trapped in potential wells up to the time $\tau = 100$. The smaller is the initial momentum of an atom the longer is the time it needs to be accelerated and quits a well. All the slow atoms are accelerated, all the fast ones are decelerated, and we observe at negative detunings the well-known effect of the velocity grouping [1, 2].

A different picture we get with the positive detuning, $\Delta = 20$ (Figs. 1d, e). In this case, slow atoms are decelerated and fast ones are accelerated. The most interesting fact is that the final distribution of slow atoms becomes very narrow with the respective peak consisting of 1708 atoms (total is 10000), so we are forced to truncate its higher part on the figure.

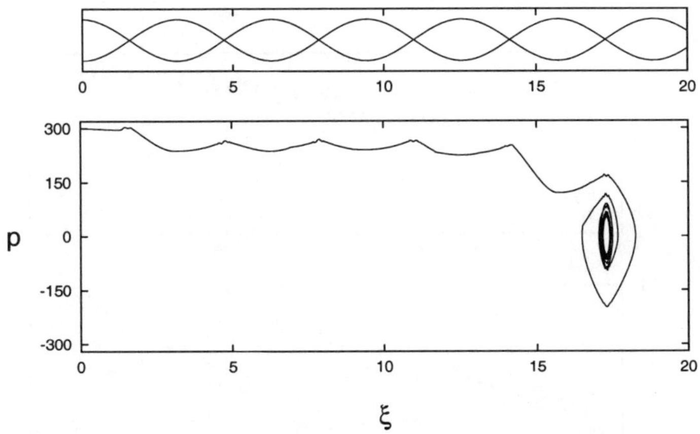

Figure 2: A typical phase trajectory of a single atom decelerated and trapped in the field. The upper fragment shows the respective structure of the standing wave. Atomic momentum p is in units of $\hbar k_f$, position ξ — in units of k_f^{-1}. $\Delta = 20$, $\omega_r = 0.003$, $\varepsilon = 100$, $p_0 = 300$.

Fig. 2 demonstrates the effect of atomic trapping in the optical lattice that occurs at positive detunings. In Fig. 2 we show a typical trajectory of a single atom in the phase plane (ξ, p) for $\Delta = 20$ and the initial momentum $p_0 = 300$. The atom is decelerated and finally is being trapped in the potential well situated at a node of the standing wave. The opposite case, namely, the process of atomic acceleration is shown in Fig. 3. Fig. 3a demonstrates a trajectory of an atom with the same initial momentum $p_0 = 300$ but at the negative detuning $\Delta = -20$. The acceleration is not so prominent as the deceleration in Fig. 2. The atomic motion is highly irregular because of the nonlinear interaction with the field interrupted by spontaneous emission that results in random mechanical recoils. In Fig. 3b the averaged (over 1000 ones) trajectory for the same parameters and initial conditions is shown. It is easy to see a trend. It should be noted that, when averaging over a large number of realizations, we get averaged values of the variables for each values of the atomic position, not time. The saturation momentum is about $p_s \simeq 520$ (not shown in the picture because of a long settling time). In Fig. 3c we plot a trajectory computed at the same conditions but with the equations of motion taken from our paper [9] where atomic relaxation has been treated as an exponential decay. In order to compare our old and new results correctly, we renormalized in the equations (6) in Ref.[9] the variables u and v to be $u/\sqrt{n}$ and $v/\sqrt{n}$, respectively, where n is a saturation number of photons in the standing wave and $\sqrt{n} = \varepsilon$. The comparison with Fig. 3b shows a very good correspondence.

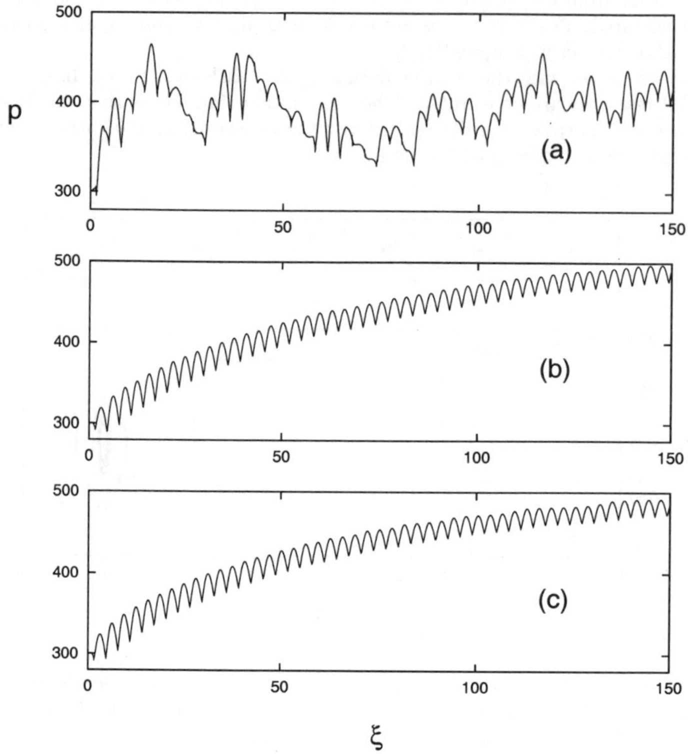

Figure 3: Atomic phase trajectories in the regime of acceleration: (a) a typical trajectory of a single atom, (b) an averaged trajectory with 1000 atoms, (c) a trajectory computed under the same conditions but with the equations with an exponential decay. In all the fragments $\Delta = -20$, $\omega_r = 0.003$, $\varepsilon = 100$, $p_0 = 300$.

It should be stressed that without dissipation the momentum trend does not exist and the average atomic momentum is constant. It is spontaneous emission that causes a dissipative force which can be called a friction force. Whereas in linear systems friction is proportional to momentum, in our case its dependence on momentum is very complicated (see Ref.[9] where we plotted it for the equations with an exponential decay). Friction can be as positive as negative (deceleration and acceleration processes, respectively).

4 Correlations between internal and external degrees of freedom of an atom

Our basic Eqs.(5) describe two coupled oscillators, the external mechanical one (ξ, p) and the internal Rabi oscillator (u, v, z). Without pumping and dissipation, the respective free frequencies may differ in a few orders of magnitude. In a strong field the free frequency of small oscillations of an atom in a well of the optical potential is much smaller then the free Rabi frequency. In Ref.[9] we have shown that synchronization of internal and external degrees of freedom of an atom in a standing-wave field may occur if the atomic excited states decay exponentially in time. The effect consists in equality between the frequency of oscillations of the atomic internal energy and the frequency of variations of the atomic momentum as for trapped and ballistic atoms. The mechanical oscillator forces the Rabi one to oscillate with the same rhythm. A standing wave with the spatial period 2π modulates (with the Doppler shift $\omega_r p_s$) momentum p of an atom, moving ballistically with an average momentum p_s. Under synchronization, all the variables are periodic with the same period. Synchronization manifests itself in the phase space as limit cycles. A limit cycle of period m means a periodicity occurs when an atom transverses $2m$ times the same node (being trapped in a well) or different nodes (a ballistic atom) of the standing wave. We have found in Ref.[9] limit cycles with m=$1, 2, 3, ..., 12$. As an example, we show in Fig. 4b projections of a period-1 limit cycle on the Bloch plane (v, u) and (z, u) and on the plane (p, u) of the internal and external variables at $\Delta = -20$, $\omega_r = 0.003$, $\varepsilon = 100$, $p_0 = 525$ and with the initial conditions $z_0 = 1$, $u_0 = v_0 = 0$. These results have been obtained with the equations (6) in Ref.[9] in which we renormalized the variables u and v as it was described above. The size of a limit cycle is defined by an amplitude of the respective synchronized oscillations, its form — by a spectrum of the oscillations, and the time of circulation of a phase point over the cycle — by a period of oscillations.

In this section we compute Eqs.(5) with quantum jumps under the same conditions as the equations (6) in Ref.[9] in order to find the effect of spontaneous emission on the effect of synchronization. It seems reasonable to expect full destroying of synchronization under random events of spontaneous emission. It is really so with single quantum ballistic trajectories. However, the effect is recovered under averaging over a large number of trajectories. In Fig. 4a we show the same projections in the phase space as in Fig. 4b using 2500 trajectories when averaging. The forms of both limit cycles with the period $T \simeq 4$ ($\simeq 125$ ns) are similar. When dealing with nonlinear oscillations of the atomic center of mass in the wells of the optical lattice, we have found more prominent traces of synchronization between internal and external degrees of freedom even for single trajectories (not shown here). Thus a correlation between internal and external degrees of freedom of an atom in the optical lattice, i.e. between the center-of-mass motion and the Rabi oscillations, is established.

5 Conclusion

In this paper we applied the method of Monte Carlo wavefunction to study atomic transport in a standing laser wave taking into account not only photon recoils caused by induced emission and absorption but spontaneous emission as well. The model is simplified in some relations: in difference from Refs.[6, 7, 8] it does not take into account changes in the field variables when atoms emit and absorb photons, and in difference from Refs.[7, 9] the field is not quantized. It describes the atomic dynamics in terms of real physical values which can be measured in real experiments and take into account the effect of spontaneous emission both on internal and mechanical atomic degrees of freedom. We simulated successfully the main dissipative mechanical effects: atomic acceleration, deceleration, velocity grouping and trapping in potential wells. Correlations between internal and external degrees of freedom, i.e. between center-of-mass motion and Rabi oscillations, are found. The comparison with our previous dissipative model [9] has shown a good correspondence between single trajectories computed with the respective dissipative equations and averaged trajectories computed with the equations with quantum jumps.

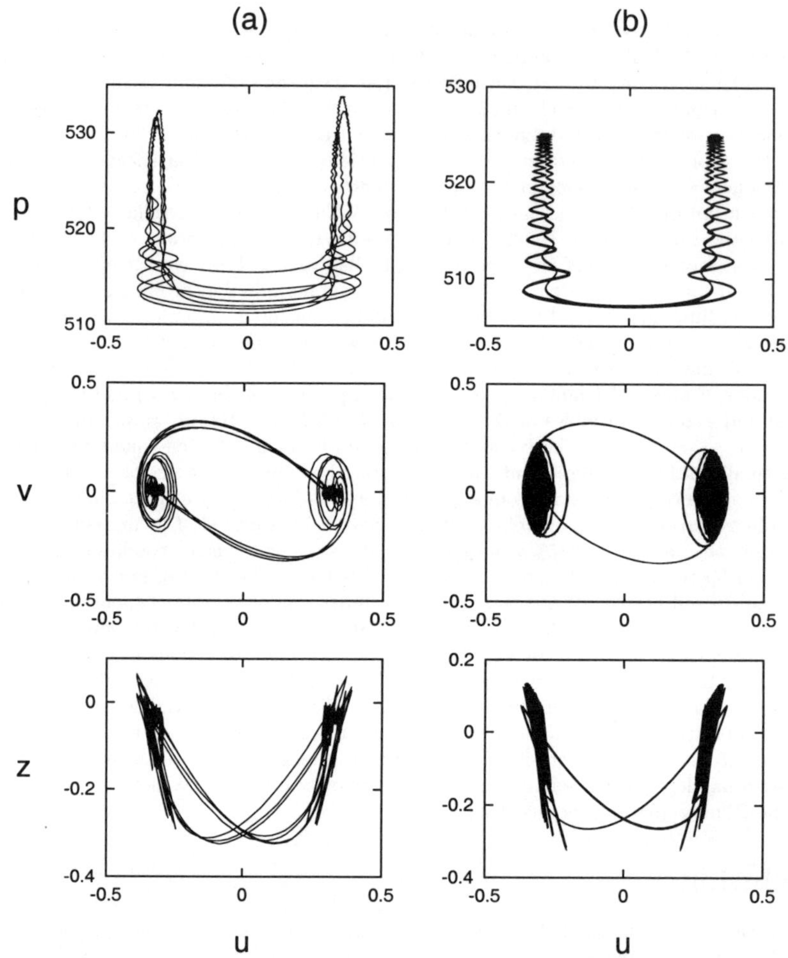

Figure 4: Projections of the phase trajectories of an atom in the optical lattice onto planes (p, u), (v, u) and (z, u): (a) averaging over 2500 trajectories of Eqs.(5) in this paper with quantum jumps, (b) period-1 limit cycle found with equations (6) with exponential decay in Ref.[9]. In all the fragments $\Delta = -20$, $\omega_r = 0.003$, $\varepsilon = 100$.

Acknowledgments

This work was supported by the Russian Foundation for Basic Research (project no. 06-02-16421), by the Program "Mathematical methods in nonlinear dynamics" of the Prezidium of the Russian Academy of Sciences (RAS), and the program of the Prezidium of the RAS Far-Eastern Division.

References

[1] Minogin VG, Letokhov VS. Laser Light Pressure on Atoms. New York: Gordon and Breach; 1987.

[2] Kazantsev AP, Surdutovich GI, Yakovlev VP. Mechanical Action of Light on Atoms. Singapore: World Scientific; 1990.

[3] Stenholm S. The semiclassical theory of laser cooling. Rev Mod Phys 1986;58:699–739.

[4] Chu S. The manipulation of neutral particles. Rev Mod Phys 1998;70:685–706; Cohen-Tannoudji CN. Manipulating atoms with photons. ibid:707–719; Phillips WD. Laser cooling and trapping of neutral atoms. ibid:720–741.

[5] Prants SV. Chaos, fractals and flights of atoms in cavities. JETP Letters 2002;75:651–658 [Pis'ma ZhETF 2002;75:777–785].

[6] Argonov VYu, Prants SV. Fractals and chaotic scattering of atoms in the field of a standing light wave. JETP 2003;96:832–845 [Zh Eksp Teor Fiz 2003;123:946–961].

[7] Prants SV, Uleysky MYu. Atomic fractals in cavity quantum electrodynamics. Phys Lett A 2003;309:357–362.

[8] Prants SV, Uleysky MYu, Argonov VYu. Entanglement, fidelity, and quantum-classical correlations with an atom walking in a quantized cavity field. Phys Rev A 2006;73:023807.

[9] Argonov VYu, Prants SV. Synchronization of internal and external degrees of freedom of atoms in a standing laser. Phys Rev A 2005;71:053408.

[10] Hood CJ, Lynn TW, Doherty AC et al. The atom-cavity microscope: single atoms bound in orbit by single photons. Science 2000;287:1447–1453.

[11] Münstermann P, Fischer T, Maunz P et al. Dynamics of single-atom motion observed in a high-finesse cavity. Phys Rev Lett 1999;82:3791–3794.

[12] Jaynes ET, Cummings FW. Comparison of quantum and semiclassical theories with application to the beam maser. Proc IEEE 1963;51:89–109.

[13] Carmichael H. An Open Systems Approach to Quantum Optics. Berlin: Springer-Verlag; 1993.

[14] Molmer K, Castin Y, Dalibard J. Monte Carlo wave-function method in quantum optics. J Opt Soc Am B 1993;10:524–538.

[15] Doherty AC, Parkins AS, Tan SM, Walls DF. Motion of a two-level atom in an optical cavity. Phys Rev A 1997;56:833–844.

[16] Riedel K, Törmä P, Savichev V, Schleich WP. Control of dynamical localization by an additional quantum degree of freedom. Phys Rev A 1999;59:797–802.

A fuzzy blue sky catastrophe

Ling Hong and J.Q. Sun

Department of Mechanical Engineering
University of Delaware
Newark, DE 19716, USA

Abstract

In this paper, a periodic blue sky catastrophe in autonomous nonlinear systems with fuzzy disturbances is studied by means of the fuzzy generalized cell mapping (FGCM) method. A blue sky catastrophe is caused by the collision of a fuzzy attractor with a fuzzy saddle on the basin boundary, in which the fuzzy attractor together with its basin of attraction suddenly disappears as the intensity of fuzzy noise reaches a critical point. We illustrate this bifurcation event by considering fixed point and limit cycle attractors under additive and multiplicative fuzzy noise. Such a bifurcation is fuzzy noise-induced effects which cannot be seen in deterministic systems.

Keywords: Fuzzy dynamical systems, Fuzzy noise, Fuzzy bifurcation, Cell mapping methods

1 Introduction

Noise is ubiquitous in real-life physical systems and can be usually modeled as a random variable or a fuzzy set dependent on the available information about the noise [1–4]. Noise acting on nonlinear dynamical systems can be a source of new phenomena. It may qualitatively change the system behavior and induce bifurcations. This paper presents a method to analyze the response and bifurcation of nonlinear dynamical systems with fuzzy noise. We are interested in a nonlinear dynamical system whose response is a fuzzy process, and study how the fuzzy response changes as the fuzzy noise intensity varies. Specifically, our attention is focused on the analysis of periodic blue sky catastrophes by fuzzy noise. It should be noted that few works dealing with this problem have been published to date.

In the theory of deterministic dissipative systems, bifurcations can be classified according to the continuity or discontinuity of an attractor path with a control parameter [5,6]. Discontinuous bifurcations of attractors, regular or chaotic, can be divided into two categories, namely catastrophic and explosive bifurcations [7,8]. In the catastrophe, an attractor simply disappears, forcing the system to jump to a remote and entirely new attractor or to the infinity. In the explosion, there is a discontinuous change in the size and form of the attractor.

Bifurcation analysis of noisy (stochastic and fuzzy) nonlinear dynamical systems is in general a difficult subject, partly because even the definition of bifurcation is open to discussion. Take the stochastic system as an example. The commonly accepted definition of the bifurcation is the "qualitative change" of the system response as a control parameter varies. Meunier and Verga studied pitchfork bifurcation of a stochastic dynamical system [9]. They examined the quantities such as invariant measures, Lyapunov exponents, correlation functions, and exit times. It turns out that the behavior of all these quantities near the deterministic bifurcation point changes for different values of the bifurcation parameter, making them a poor indicator of bifurcation in some cases. They proposed an effective potential function of the invariant probability density function of the system response to describe the bifurcation and concluded that corresponding to the bifurcation point of the deterministic system, there is a bifurcation transition region for the stochastic system. Doi, Inoue and Kumagai have also found that the invariant probability density of the system response is not indicative of bifurcation in some cases, and proposed to examine the qualitative changes of the spectrum of a Markov operator [10]. The spectrum of the Markov operator reveals the bifurcation of the stochastic phase lockings of a van der Pol oscillator that does not show up in the invariant probability density of the system response. Other studies phenomenologically define stochastic bifurcation based on the observation of collision of stable attractors with saddle nodes [11].

For fuzzy nonlinear dynamical systems, the subject is even more difficult because the evolution of the membership function of the fuzzy response process is not readily obtained analytically. There is little study in the literature on the bifurcation of fuzzy nonlinear dynamical systems. There are studies of bifurcations of fuzzy control systems where the fuzzy control law leads to a nonlinear and deterministic dynamical system. The bifurcation studies are practically the same as that of deterministic systems [12, 13]. The work in [14] deals with bifurcation of fuzzy dynamical systems having a fuzzy response. Numerical simulations are used to simulate the system response with a given parameter and fuzzy membership grade. The eigenvalues and the membership distribution are both used to describe the bifurcation. For a given membership grade, the bifurcation of the system is defined in the same manner as for the deterministic system. The authors have recently proposed a fuzzy generalized cell mapping (FGCM) method for the bifurcation analysis of fuzzy nonlinear dynamical systems and considered several very interesting scenarios of fuzzy bifurcations [15, 16]. The current paper studies new fuzzy bifurcations of two autonomous nonlinear systems with a fixed point and a limit cycle. In particular, we shall study catastrophic bifurcations involving the collision of fuzzy attractors with a fuzzy saddle.

The remainder of the paper is outlined as follows. In Section 2, we describe the FGCM method, and discuss its properties. In Section 3, we study a blue sky catastrophe of a fixed point and a limit cycle under fuzzy noise. The paper concludes in Section 4.

2 The FGCM Method

2.1 Fuzzy Generalized Cell Mapping

Consider a dynamical system with a fuzzy parameter

$$\dot{\mathbf{x}} = \mathbf{f}(\mathbf{x}, t, S), \quad \mathbf{x} \in \mathbf{D} \tag{1}$$

where $\mathbf{x}$ is the state vector, t the time variable, S a fuzzy set with a membership function $\mu_S(s) \in (0,1]$ where $s \in S$, and $\mathbf{f}$ is a vector-valued nonlinear function of its arguments. It is assumed to be periodic in t with period T for all $s \in S$ and to satisfy the Lipschitz condition for all $s \in S$, and $\mathbf{D}$ is a bounded domain of interest in the state space. A fuzzy Poincaré map can be obtained from Equation (1) as

$$\mathbf{x}(n+1) = \mathbf{G}(\mathbf{x}(n),\ S), \quad n = 0, 1, 2, \cdots \tag{2}$$

The cell mapping method proposes to further discretize the state space in searching for the global solution of the system [17]. In order to apply the cell mapping method, we also need to discretize the fuzzy set S. Suppose that S is a finite interval in R. We divide S into M segments of appropriate length and sample a value $s_k \in S$ $(k = 1, \cdots, M)$ in the middle of each segment. The division of S is such that there is at least one s_k with membership grade equal to one. The domain $\mathbf{D}$ is then discretized into N small cells. Each cell is identified by an integer ranging from 1 to N. For a cell, say cell j, N_p points are uniformly sampled from cell j, $M \times N_p$ fuzzy sample trajectories are computed for one period T, or one mapping step. Each trajectory carries a membership grade determined by that of s_k's. We then find the cells in which the end points of the trajectories fall. Assume that cell i is one of the image cells of cell j, and that there are m $(0 < m \leqslant MN_p)$ trajectories falling in cell i. Define a quantity

$$p_{ij} = \max_{i_k}[\mu_S(s_{i_k})], \quad 0 < p_{ij} \leqslant 1, \tag{3}$$

where i_k $(k = 1, 2, ..., m)$ are referred to the trajectories falling in cell i, and $\mu_S(s_{i_k})$ are the membership grades of the corresponding trajectories. This procedure for computing p_{ij} is known as the sampling point method in the context of generalized cell mapping [17]. Now, assume that the membership grade of the system being in cell j at the nth mapping step is $p_j(n)$ $(0 < p_j(n) \leqslant 1)$. Cell j is mapped in one step to cell i with the membership grade given by

$$\begin{aligned}&\max\{\min[\mu_S(s_{i_1}), p_j(n)], \min[\mu_S(s_{i_2}), p_j(n)], \cdots, \min[\mu_S(s_{i_m}), p_j(n)]\}\\ &= \min[\max_{i_k}(\mu_S(s_{i_k})), p_j(n)] = \min[p_{ij}, p_j(n)].\end{aligned} \tag{4}$$

Considering all possible pre-images of cell i, we have the membership grade of the system being in cell i at the $(n+1)^{th}$ step as

$$p_i(n+1) = \max_j \min[p_{ij}, p_j(n)]. \tag{5}$$

Let $\mathbf{p}(n)$ be a vector with components $p_i(n)$, and $\mathbf{P}$ a matrix with components p_{ij}. Equation (5) can be written in a compact matrix notation

$$\mathbf{p}(n+1) = \mathbf{P} \circ \mathbf{p}(n), \quad \mathbf{p}(n) = \mathbf{P}^n \circ \mathbf{p}(0), \tag{6}$$

where $\mathbf{P}^{n+1} = \mathbf{P} \circ \mathbf{P}^n$ and $\mathbf{P}^0 = \mathbf{I}$. The matrix $\mathbf{P}$ is called the one-step transition membership matrix. The vector $\mathbf{p}(n)$ is called the n-step membership distribution vector, and $\mathbf{p}(0)$ the initial membership distribution vector. The $(i,j)^{th}$ element p_{ij} of the matrix $\mathbf{P}$ is called the one-step transition membership from cell j to cell i. Equation (6) is called a fuzzy generalized cell mapping system, which describes the evolution of the fuzzy solution process $\mathbf{x}(n)$ and its membership function, and is a finite approximation to the mapping (2) in $\mathbf{D}$.

Consider the master equation for the possibility transition of continuous fuzzy processes [18–20],

$$p(\mathbf{x},t) = \sup_{\mathbf{x}_0 \in \mathbf{D}} \left[\min\{p(\mathbf{x},t|\mathbf{x}_0,t_0), p(\mathbf{x}_0,t_0)\}\right], \quad \mathbf{x} \in \mathbf{D} \tag{7}$$

where $\mathbf{x}$ is a fuzzy process, $p(\mathbf{x},t)$ is the membership function of $\mathbf{x}$, and $p(\mathbf{x},t|\mathbf{x}_0,t_0)$ is the transition possibility function, also known as a fuzzy relation [18]. Equation (5) of the FGCM can be viewed as a discrete representation of Equation (7). Friedman and Sandler have derived a partial differential equation from Equation (7) for continuous time processes [19, 20]. This equation is analogous to the Fokker-Planck-Kolmogorov equation for the probability density function of stochastic processes [21]. The solution to this equation is in general very difficult to obtain analytically. Numerically, the FGCM offers a very effective method for solutions to this equation, particularly, for *fuzzy nonlinear* dynamical systems.

2.2 Properties of FGCM

(1) Recall that the min-max operation in Equation (5) really represents the intersection (product) and union (summation) of fuzzy sets in the form of cells in $\mathbf{D}$. Hence, $\mathbf{P} \circ \mathbf{p}$ is an inner product of fuzzy sets. The topological matrix of $\mathbf{P}$, denoted by $[\overline{p}_{ij}]$, and the topological vector of $\mathbf{p}(n)$, denoted by $\{\overline{p}_i(n)\}$, are defined as

$$\overline{p}_{ij} = \begin{cases} 1, & p_{ij} > 0 \\ 0, & p_{ij} = 0 \end{cases}, \quad \overline{p}_i(n) = \begin{cases} 1, & p_i(n) > 0 \\ 0, & p_i(n) = 0 \end{cases}, \quad n \geq 0. \tag{8}$$

Topologically, Equation (6) becomes

$$\overline{\mathbf{p}}(n+1) = \overline{\mathbf{P}} \circ \overline{\mathbf{p}}(n), \quad \overline{\mathbf{p}}(n) = \overline{\mathbf{P}}^n \circ \overline{\mathbf{p}}(0), \tag{9}$$

where $\overline{\mathbf{P}}^{n+1} = \overline{\mathbf{P}} \circ \overline{\mathbf{P}}^n$ and $\overline{\mathbf{P}}^0 = \mathbf{I}$. Note that the min-max operation is equivalent to the logic operations of multiplication $\wedge$ and addition $\vee$ of binary numbers: $0 \wedge 1 = 0$, $1 \wedge 0 = 0$, $0 \wedge 0 = 0$, $1 \wedge 1 = 1$, $0 \vee 1 = 1$, $1 \vee 0 = 1$, $0 \vee 0 = 0$, and $1 \vee 1 = 1$. Hence, the min-max operation in Equation (9) leads to the identical result to that of the topological matrix of the Markov chains. Hence, $\overline{\mathbf{P}}$ forms a digraph and can be partitioned to identify persistent groups of cells representing stable solutions and the transient cells including the unstable solutions [22].

(2) When $\overline{p}_{ij} = 1$, this signifies that cell j is mapped to cell i in one step with a certain possibility. In other words, if the system is found to be in cell i with a possibility, it must be in cell j one step before with another possibility. We introduce a notation $\bar{b}_{ji}$ to represent this backward mapping relationship. When $\bar{b}_{ji} = 1$, cell j is a pre-image of cell i. When $\bar{b}_{ji} = 0$, it is not a pre-image. The matrix defined by these elements such that $\bar{\mathbf{B}} = [\bar{b}_{ij}]$ is called the backward one-step transition topological matrix of the FGCM system. By definition, we have

$$\bar{\mathbf{B}} = [\bar{b}_{ij}] = [\overline{p}_{ji}] = \overline{\mathbf{P}}^T. \tag{10}$$

Hence,

$$\overline{\mathbf{p}}(n-1) = \bar{\mathbf{B}} \circ \overline{\mathbf{p}}(n). \tag{11}$$

Since the matrix $\bar{\mathbf{B}}$ describes the dynamics of the original system backward in time topologically, the stable solutions of the original system appear to be unstable and the unstable solutions appear to be stable in Equation (11). $\bar{\mathbf{B}}$ also forms a digraph and can be used to identify unstable solutions of the system [22].

(**3**) Under the min-max operation in Equation (5), the membership function of the cells is non-increasing in the sense that,

$$||\mathbf{p}(n+1)|| \leq ||\mathbf{p}(n)||, \ \forall n \geq 0. \tag{12}$$

where the norm of the fuzzy membership vector $||\mathbf{p}(n)|| = \max_i\{p_i(n)\}$. Furthermore, the min-max operator does not introduce any new numbers that are not the entries of the matrix $\mathbf{P}$ or the vector $\mathbf{p}(0)$. This implies that $\mathbf{p}(n)$ can assume only finite number of possible values as $n \to \infty$. In the steady state, $\mathbf{p}(n)$ will either converge to a constant vector, or to a set of vectors which form a periodic group and repeat themselves as the iteration goes on. In either case, we consider the system to have converged to the steady state. Because there are only finite number of possible values for $\mathbf{p}(n)$, Equation (5) will converge in finite number of iterations. This is a sharp contrast to the Markov chains, which theoretically converge to steady state in infinite iterations.

3 A Blue Sky Catastrophe

3.1 A Duffing-Van der Pol Oscillator with Fuzzy Additive Noise

A Duffing-Van der Pol (DVP) oscillator

$$\begin{aligned} \dot{x}_1 &= x_2, \\ \dot{x}_2 &= \mu_1 x_1 + \mu_2 x_2 - x_1^3 - x_1^2 x_2, \end{aligned} \tag{13}$$

is one of the most studied systems in nonlinear dynamics. Its local and global bifurcation behavior has been thoroughly studied [23, 24]. The influence of stochastic noise on the Duffing-Van der Pol oscillator exhibiting codimension one and two bifurcations has also been studied [25–27]. For the case of fuzzy noise, to our knowledge, no attempt has been made regarding this problem.

In the present work, we choose $\mu_1 = 0.64$, $\mu_2 = 0.1$ located in the upper right-hand quadrant of Figure 7.3.7 of [23]. When $\mu_1 = 0.64$, $\mu_2 = 0.1$, the DVP equation has two coexistent fixed point attractors and a saddle point on their basin boundary as shown in Figure 1. We consider the DVP equation driven by an additive fuzzy noise

$$\begin{aligned} \dot{x}_1 &= x_2, \\ \dot{x}_2 &= 0.64 x_1 + 0.1 x_2 - x_1^3 - x_1^2 x_2 + S, \end{aligned} \tag{14}$$

where S is a fuzzy parameter with a triangular membership function,

$$\mu_S(s) = \begin{cases} [s - (s_0 - \varepsilon)]/\varepsilon, & s_0 - \varepsilon \leqslant s < s_0 \\ -[s - (s_0 + \varepsilon)]/\varepsilon, & s_0 \leqslant s < s_0 + \varepsilon \\ 0, & \text{otherwise} \end{cases} \tag{15}$$

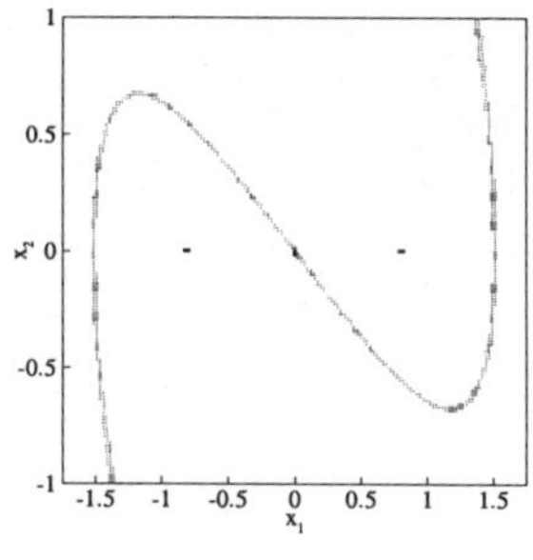

Figure 1: Global phase portrait of the deterministic Duffing-Van der Pol equation (13) with $\mu_1 = 0.64$ and $\mu_2 = 0.1$.

$\varepsilon > 0$ is a parameter characterizing the intensity of fuzziness of S, and s_0 is the nominal value of S with membership grade $\mu_S(s_0) = 1$.

The domain $D = \{-1.75 \le x_1 \le 1.75, \ -1.0 \le x_2 \le 1.0\}$ is discretized into 141×141 cells when applying the FGCM method, 5×5 sampling points are used within each cell. The membership function is discretized into 201 segments ($M = 201$) , hence, out of each cell, there are 5025 trajectories with varying membership grades. These trajectories are then used to compute the transition membership matrix.

We fix $s_0 = 0$ and allow the fuzzy noise intensity ε to vary. As ε increases, two coexistent fuzzy fixed point attractors a_{left} and a_{right} become simultaneously bigger. The global phase portrait is shown in Figure 2 when $\varepsilon = 0.118$. When $s_0 = \mp 0.001$, $\varepsilon = 0.1185$, a blue sky catastrophe, i.e., disappearance of the two fuzzy fixed point attractors a_{right} and a_{left} occurs. In such a case, the fuzzy attractor a_{right} or a_{left} collides with the fuzzy saddle on the basin boundary, and suddenly disappears, leaving behind a fuzzy saddle in the place of the original fuzzy attractor in the phase space after the bifurcation. The global phase portraits are shown in Figure 3.

3.2 A Van der Pol System with Fuzzy Multiplicative Noise

The model system under study is a biased Van der Pol oscillator satisfying an equation of motion,

$$\begin{aligned} \dot{x}_1 &= kx_2 + \mu x_1(b - x_2^2), \\ \dot{x}_2 &= -x_1 + C, \end{aligned} \tag{16}$$

When $C = 0$, Equation (16) reduces to the familiar second-order Van der Pol equation having a limit cycle enclosing a repelling spiral fixed point at the origin. For any $C \neq 0$, the phase

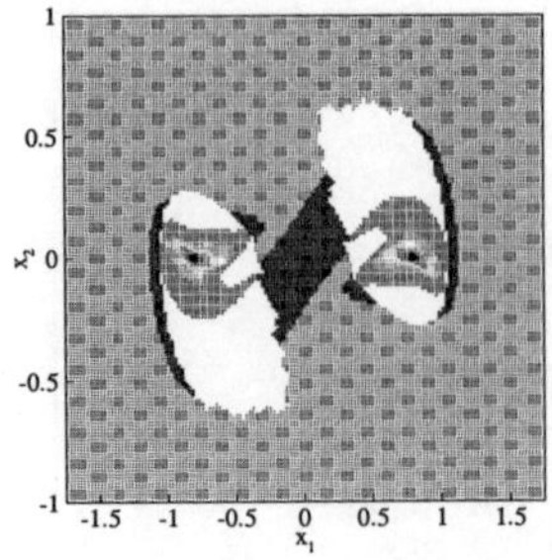

Figure 2: Global phase portrait of the noisy Duffing-Van der Pol equation (14) with $s_0 = 0$ and $\varepsilon = 0.118$. In the figure, the fuzzy attractors are marked by the symbol ■. The membership distribution of fuzzy attractors is color-coded with black=1.0, 0.8<red<1.0, 0.6<green≤0.8, 0.4<yellow≤0.6, 0.2<cyan≤0.4, and 0.0<purple≤0.2. The saddle is marked by the blue symbol ○. The basin boundary is marked by the orange symbol +.

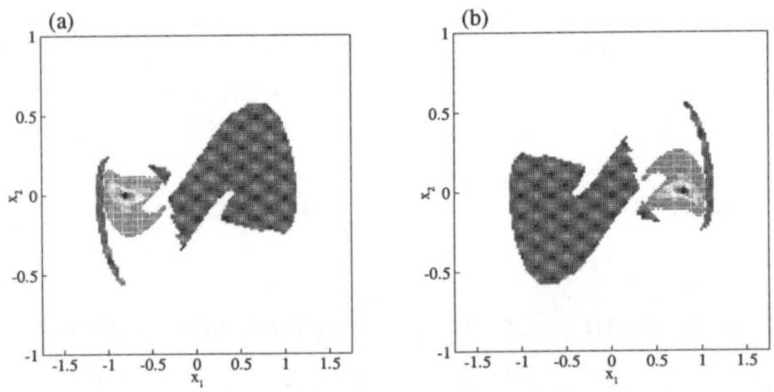

Figure 3: Global phase portrait of the noisy Duffing-Van der Pol equation (14). (a) $s_0 = -0.001$, $\varepsilon = 0.1185$, (b) $s_0 = 0.001$, $\varepsilon = 0.1185$. The color coding is the same as that in Figure 2.

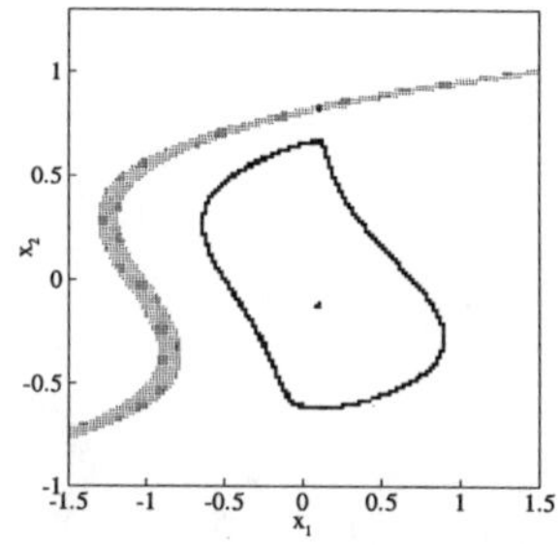

Figure 4: Global phase portrait of the deterministic Van der Pol equation (16) with $k = 0.7$, $\mu = 10$, $b = 0.1$ and $C = 0.10$. The color coding is the same as that in Figure 2.

portrait includes an additional fixed point of saddle type, whose coordinates are found to be

$$x_1 = C, \qquad (17)$$

$$x_2 = \frac{k}{2\mu C} + \sqrt{\left(\frac{k}{2\mu C}\right)^2 + b}.$$

For small C this fixed point lies at a great distance away from the limit cycle, moving to infinity as $C \to 0$. The eigenvalues of this fixed point may be computed algebraically to verify that it is of saddle type. Figure 4 shows a phase portrait of this system for $k = 0.7$, $\mu = 10$, $b = 0.1$ and $C = 0.10$. We consider the Van der Pol system driven by multiplicative fuzzy noise.

$$\dot{x}_1 = 0.7x_2 + Sx_1(0.1 - x_2^2), \qquad (18)$$

$$\dot{x}_2 = -x_1 + 0.1,$$

where S is a fuzzy parameter with the triangular membership function given by Equation (15).

The domain $D = \{-1.5 \le x_1 \le 1.5, \ -1.0 \le x_2 \le 1.3\}$ is discretized into 141×141 cells. 5×5 sampling points are used within each cell. The membership function is discretized into 201 segments ($M = 201$), hence, out of each cell, there are 5025 trajectories with varying membership grades. These trajectories are then used to compute the transition membership matrix. We fix $s_0 = 10$, and allow the fuzzy noise intensity ε to vary. As ε increases, a blue sky catastrophe of a fuzzy limit cycle is discovered in the interval $\varepsilon \in (2.747, 2.748)$ as shown in Figure 5. In this case, the fuzzy limit cycle touches the fuzzy saddle on the basin boundary when $\varepsilon = 2.747$, and vanishes into the blue sky at $\varepsilon = 2.748$ after the collision, leaving behind a repelling fuzzy limit cycle in the place of the original attracting fuzzy limit cycle.

In order to understand this bifurcation, we must consider the global stable and unstable manifolds of the fuzzy saddle referring to Figure 5. As the fuzzy noise intensity ε increases, the fuzzy saddle and its stable manifold move closer to the fuzzy limit cycle. The saddle just touches the limit cycle, as shown in Figure 5(a), the left branch of the stable manifold of the

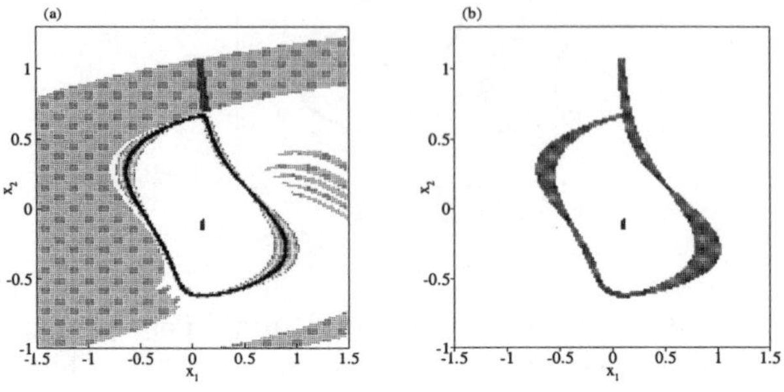

Figure 5: Phase portrait of blue sky catastrophe of a fuzzy limit cycle in the noisy Van der Pol equation (18). (a) $\varepsilon = 2.747$, (b) $\varepsilon = 2.748$. The color coding is the same as that in Figure 2.

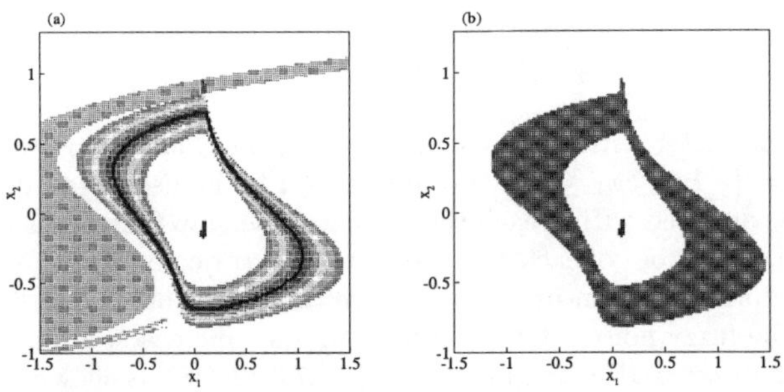

Figure 6: Phase portrait of blue sky catastrophe of a fuzzy limit cycle in the noisy Van der Pol equation (19). (a) $\varepsilon = 0.044$, (b) $\varepsilon = 0.045$. The color coding is the same as that in Figure 2.

saddle and the lower branch of the unstable manifold both coincide with the location of the limit cycle, establishing a *homoclinic connection.* This homoclinic connection may be thought of as a limit cycle of infinite period. By increasing ε further as shown in Figure 5(b), the relative positions of the global stable manifold (left branch) and unstable manifold (lower branch) are interchanged. Trajectories starting below the saddle follow the unstable manifold around the stable manifold and reach the region above the saddle, from which they eventually diverge to infinity. The stable manifold is no longer a separator. The limit cycle no longer exists, and it has vanished into the blue sky. Abraham [7] calls this the *blue sky catastrophe* for a periodic limit cycle. In the terminology of Shilnikov [28] this is a *dangerous boundary*, causing a finite jump to a remote attractor, or the infinity.

As the final example, we consider

$$\begin{aligned}\dot{x}_1 &= 0.7x_2 + 10x_1(S - x_2^2), \\ \dot{x}_2 &= -x_1 + 0.09.\end{aligned} \tag{19}$$

We fix $s_0 = 0.12$, and allow the fuzzy noise intensity ε to vary. As ε increases, a blue sky catastrophe of a fuzzy limit cycle is discovered in the interval $\varepsilon \in (0.044, 0.045)$ as shown in Figure 6. In this case, the fuzzy limit cycle touches the fuzzy saddle on the basin boundary at $\varepsilon = 0.044$ and vanishes into the blue sky at $\varepsilon = 0.045$ after the collision, leaving behind a repelling fuzzy limit cycle in the place of the original attracting fuzzy limit cycle.

4 Concluding Remarks

In this paper, we have introduced the FGCM method and investigated catastrophic bifurcations driven by fuzzy noise where a homoclinic connection leads a fixed point and a limit cycle to disappear into the blue sky. Collision with a fuzzy saddle fixed point is the typical mechanism by which a fuzzy fixed point and a fuzzy limit cycle can abruptly vanish. These fuzzy bifurcations are difficult to analyze with direct numerical simulations or analytical methods. The FGCM method is at present the only effective tool for bifurcation analysis of fuzzy nonlinear dynamical systems.

Acknowledgements

This material is based upon work supported by the National Science Foundation under Grant No. CMS-0219217 and INT-0217453.

References

[1] F. Moss, P. V. E. McClintock, Noise in Nonlinear Dynamical Systems, Cambridge University Press, Cambridge, 1989.

[2] G. J. Klir, T. A. Folger, Fuzzy Sets, Uncertainty, and Information, Prentice-Hall, Englewood Cliffs, New Jersey, 1988.

[3] M. Bucolo, S. Fazzino, M. L. Rosa, L. Fortuna, Small-world networks of fuzzy chaotic oscillators, Chaos Solitons and Fractals 17 (2003) 557–565.

[4] U. Sandler, L. Tsitolovsky, Fuzzy dynamics of brain activity, Fuzzy Sets and Systems 121 (2001) 237–245.

[5] E. C. Zeeman, Bifurcation and catastrophe theory, in: Proceedings of Papers in Algebra, Analysis and Statistics, Providence, Rhode Island, 1982, pp. 207–272.

[6] J. M. T. Thompson, H. B. Stewart, Y. Ueda, Safe, explosive, and dangerous bifurcations in dissipative dynamical systems, Physical Review E 49 (2) (1994) 1019–1027.

[7] R. H. Abraham, H. B. Stewart, A chaotic blue sky catastrophe in forced relaxation oscillations, Physica D 19 (1986) 394–400.

[8] J. M. T. Thompson, H. B. Stewart, Nonlinear Dynamics and Chaos, Wiley, Chichester, New York, 1986.

[9] C. Meunier, A. D. Verga, Noise and bifurcations, Journal of Statistical Physics 50 (1/2) (1988) 345–375.

[10] S. Doi, J. Inoue, S. Kumagai, Spectral analysis of stochastic phase lockings and stochastic bifurcations in the sinusoidally forced van der Pol oscillator with additive noise, Journal of Statistical Physics 90 (5-6) (1998) 1107–1127.

[11] W. Xu, Q. He, T. Fang, H. Rong, Stochastic bifurcation in Duffing system subject to harmonic excitation and in presence of random noise, International Journal of Non-Linear Mechanics 39 (2004) 1473–1479.

[12] Y. Tomonaga, K. Takatsuka, Strange attractors of infinitesimal widths in the bifurcation diagram with an unusual mechanism of onset. Nonlinear dynamics in coupled fuzzy control systems. II, Physica D 111 (1-4) (1998) 51–80.

[13] F. Cuesta, E. Ponce, J. Aracil, Local and global bifurcations in simple Takagi-Sugeno fuzzy systems, IEEE Transactions on Fuzzy Systems 9 (2) (2001) 355–368.

[14] P. K. Satpathy, D. Das, P. B. D. Gupta, A fuzzy approach to handle parameter uncertainties in Hopf bifurcation analysis of electric power systems, International Journal of Electrical Power and Energy System 26 (7) (2004) 531–538.

[15] L. Hong, J. Q. Sun, Bifurcations of fuzzy nonlinear dynamical systems, Communications in Nonlinear Science and Numerical Simulation 11 (1) (2006a) 1–12.

[16] L. Hong, J. Q. Sun, Codimension two bifurcations of nonlinear systems driven by fuzzy noise, Physica D: Nonlinear Phenomena 213 (2) (2006b) 181–189.

[17] C. S. Hsu, Cell-to-Cell Mapping: A Method of Global Analysis for Non-linear Systems, Springer-Verlag, New York, 1987.

[18] Y. Yoshida, A continuous-time dynamic fuzzy system. (I) A limit theorem, Fuzzy Sets and Systems 113 (2000) 453–460.

[19] Y. Friedman, U. Sandler, Evolution of systems under fuzzy dynamic laws, Fuzzy Sets and Systems 84 (1996) 61–74.

[20] Y. Friedman, U. Sandler, Fuzzy dynamics as an altemative to statistical mechanics, Fuzzy Sets and Systems 106 (1999) 61–74.

[21] H. Risken, The Fokker- Planck Equation, Springer-Verlag, New York, 1996.

[22] C. S. Hsu, Global analysis of dynamical systems using posets and digraphs, International Journal of Bifurcation and Chaos 5 (4) (1995) 1085–1118.

[23] J. Guckenheimer, P. J. Holmes, Nonlinear Oscillations, Dynamical Systems and Bifurcations of Vector Fields, Springer-Verlag, New York, 1983.

[24] P. Holmes, D. Rand, Phase portraits and bifurcations of the non-linear oscillator, International Journal of Non-Linear Mechanics 15 (1980) 449–458.

[25] N. S. Namachchivaya, Stochastic bifurcation, Journal of Applied Mathematics and Computation 38 (1990) 101–159.

[26] K. R. Schenk-Hoppe, Bifurcation scenarios of the noisy Duffing-van der Pol oscillator, Nonlinear Dynamics 11 (1996) 255–274.

[27] N. S. Namachchivaya, Co-dimension two bifurcations in the presence of noise, Journal of Applied Mechanics 58 (1991) 259–265.

[28] L. P. Shilnikov, Theory of the bifurcation of dynamical systems and dangerous boundaries, Sov. Phys. Dokl. 20 (1976) 674–676.

Efficient and reliable stability analysis of solutions of delay differential equations

Koen Verheyden[†], Tatyana Luzyanina[‡], Dirk Roose[†]

[†] Department of Computer Science, Katholieke Universiteit Leuven, Celestijnenlaan 200A, B-3001 Heverlee-Leuven, Belgium.
[‡] Institute of Mathematical Problems in Biology, RAS, Pushchino, Moscow region, 142290, Russia.

Abstract

We describe an efficient and reliable method to numerically determine the rightmost, stability-determining, characteristic roots of systems of delay differential equations (DDEs). The method is based on the discretization of the solution operator by a linear multistep (LMS) method. We use a special-purpose LMS method, that aims at preserving the rightmost roots. We discuss a method to determine the steplength in the LMS method to achieve accurate approximations of the roots with real part larger than a user-specified value. The efficiency and the performance of the method is demonstrated for two systems, including a system of DDEs resulting from the space discretization of a hybrid DDE-PDE model. We also compare the method with alternative approaches based on a spectral discretization.

Keywords: Delay differential equations; Stability analysis; Characteristic roots

1 Introduction

The (local) stability of a steady state solution $y(t) \equiv y^\star$ of a (nonlinear) system of delay differential equations (DDEs) with constant delays $\tau_j \geqslant 0$, $j = 1, \ldots, m$, of the form

$$y'(t) = f(y(t), y(t-\tau_1), y(t-\tau_2), \ldots, y(t-\tau_m))$$

with $y(t) \in \mathbb{R}^n$, $f : \mathbb{R}^{n\cdot(m+1)} \to \mathbb{R}^n$, is determined by the stability of (the zero solution of) the variational equation

$$y'(t) = A_0 y(t) + \textstyle\sum_{j=1}^{m} A_j y(t-\tau_j), \tag{1}$$

where $A_j \in \mathbb{R}^{n\times n}$ denotes the partial derivative of f with respect to its $(j+1)$-th argument, evaluated at the steady state solution $y^\star$. The variational equation (1) is asymptotically stable if all roots λ of the characteristic equation

$$\det(\lambda I - A_0 - \textstyle\sum_{j=1}^{m} A_j e^{-\lambda\tau_j}) = 0 \tag{2}$$

lie in the open left half-plane, i.e., $\Re e(\lambda) < 0$, see e.g. [8]. Although (2) has an infinite number of characteristic roots λ, the number of roots satisfying $\Re e(\lambda) \geqslant r$ is finite. Hence, to analyse the stability of a steady state solution, one must determine in a reliable way all roots satisfying $\Re e(\lambda) \geqslant r$, for a given $r < 0$ close to zero.

Most numerical methods to compute the (rightmost) characteristic roots of (2) are based on a discretization of the *infinitesimal generator* [2, 3, 11] or the *solution operator* [6, 1] of the variational equation (1). The solution operator of the variational equation over a time interval t has eigenvalues μ, which are related to the characteristic roots via the equation $\mu = e^{\lambda t}$. Both spectral discretizations and time integration methods can be used to construct a discretization of either operator, that can be represented as a matrix. Computation of the eigenvalues of this matrix yields the characteristic roots. Hence, for the computational efficiency it is important that the size of the resulting matrix eigenvalue problem is as small as possible. Compared to discretization via time integration, spectral discretization may lead to a smaller eigenvalue problem, but in the latter case no strategy is known that guarantees a priori that all characteristic roots with real part larger than r are computed accurately.

The software package DDE-BIFTOOL [4, 5] contains procedures for the continuation, stability analysis and bifurcation analysis of steady state and periodic solutions of DDEs with constant delays. To determine the stability of a steady state solution, the solution operator of the variational equation is discretized by using a *linear multistep* (*LMS*) method. The size of the resulting matrix eigenvalue problem is inversely proportional to the steplength used in the discretization. The *steplength heuristic* proposed in [6] and implemented in DDE-BIFTOOL, aims at a large steplength, while still guaranteeing that all roots with real part larger that a user-specified value are approximated accurately. However, the heuristic is often too conservative, resulting in a small time steplength and thus a high computational cost. In [12] we have developed an improved steplength heuristic, which typically results in a much larger steplength for the LMS method and thus a much smaller computational cost, especially for systems of DDEs arising from the space discretization of parabolic delay partial differential equations. The improvement is based on exploiting theoretical results on the location of the spectrum, before and after the discretization, presented in [10], and on the use of LMS methods that are better suited for our purpose, i.e., the preservation of the rightmost characteristic roots.

In this paper, we summarize the results of [12] and we illustrate the efficiency of the new procedure, also in comparison with the use of a spectral discretization. In Section 2 we describe the computation of the rightmost characteristic roots by an LMS method. In Section 3 we outline the steplength heuristics proposed in [6] and [12] . Section 4 illustrates by two examples the significant reduction in computational cost by using the latter heuristic. In Section 5 we draw conclusions.

2 Computation of the rightmost characteristic roots by an LMS method

In DDE-BIFTOOL, the rightmost characteristic roots λ are approximated by $\tilde{\lambda}$ obtained from the dominant eigenvalues $\tilde{\mu} = e^{\tilde{\lambda}h}$ of the discretized solution operator over one time step of length h. Let $\tau_{\max}$ be the maximal delay. An initial condition $y(\theta)$ on the delay interval $\theta \in [-\tau_{\max}, 0]$ is represented by a discrete set of points $y_i := y(t_i)$ on an equidistant mesh. For the discretization a k-step LMS method with steplength h is used :

$$\sum_{i=0}^{k}\alpha_i y_i = h\sum_{i=0}^{k}\beta_i\Big(A_0 y_i + \sum_{j=1}^{m} A_j\tilde{y}(t_i - \tau_j)\Big), \tag{3}$$

where the approximations $\tilde{y}(t_i - \tau_l)$ (in case $t_i - \tau_l$ does not coincide with a mesh point t_i) are obtained using Lagrange interpolation with s_- and s_+ points to the left and the right, respectively. To avoid the use of future mesh points $y_{k+1}, \ldots$ in (3), we require that

$$h \leqslant h_{\max} := \tau_{\min}/s_+, \tag{4}$$

where $\tau_{\min}$ is the minimal delay. The discretization to the solution operator is the (linear) map between $[y_{L_{min}}, \ldots, y_{k-1}]^T$ and $[y_{L_{min+1}}, \ldots, y_k]^T$ where $L_{min} = -s_- - \lceil \tau_{max}/h \rceil$ and where the mapping is defined by (3) for y_k and by a shift for all variables other than y_k. This map is represented by an $N \times N$ matrix, where

$$N := n(k + \lceil \tau_{\max}/h \rceil + s_-) \approx n\tau_{\max}/h. \tag{5}$$

The eigenvalues $\tilde{\mu}$ of this matrix determine the stability of the discrete scheme. If the eigenvalues all have modulus smaller than one, the trajectories computed using the LMS time integrator converge to zero. If eigenvalues exist with modulus greater than one, trajectories exist which grow unbounded. The eigenvalues can be computed by e.g. the robust QR method, with a computational cost of the order $N^3 \approx n^3(\tau_{\max}/h)^3$.

We now relate the stability properties of the solution of the DDE (1) governed by the characteristic equation, to the stability of the discrete scheme. Note that the characteristic equation (2) can be written in the equivalent form

$$\lambda \in \sigma\big(A_0 + \textstyle\sum_{j=1}^{m} A_j \mathrm{e}^{-\lambda\tau_j}\big), \tag{6}$$

where $\sigma(\cdot)$ denotes the spectrum of a matrix.

We can write a formula similar to (6) for the discrete approximation (3). The eigenvalues $\tilde{\mu}$ of the discrete system are obtained by replacing y_i by $\tilde{\mu}^i$ in (3). Let $z = \tilde{\lambda}h$ and $\tilde{\mu} = \mathrm{e}^z$ with $|\Im m(z)| \leqslant \pi$. As in [6], we characterize the effect of the LMS method by the scalar function

$$\mathrm{LMS}(z) := \frac{\alpha(\mathrm{e}^z)}{\beta(\mathrm{e}^z)}, \quad \text{with} \quad \alpha(\tilde{\mu}) := \sum_{i=0}^{k} \alpha_i \tilde{\mu}^i \quad \text{and} \quad \beta(\tilde{\mu}) := \sum_{i=0}^{k} \beta_i \tilde{\mu}^i. \tag{7}$$

Note that $\alpha(\cdot)$ and $\beta(\cdot)$ in (7) are assumed to be *irreducible*, so that (3) has no roots $\tilde{\lambda}$ that are only caused by the LMS scheme.

The characteristic equation for (3) can be written as

$$\frac{1}{h}\mathrm{LMS}(\tilde{\lambda}h) \in \sigma\Big(A_0 + \sum_{j=1}^{m} A_j \mathrm{e}^{-\frac{\tau_j}{h}\mathrm{Int}_j(\tilde{\lambda}h)}\Big),$$

where $\mathrm{Int}_j(\tilde{\lambda}h) \approx \tilde{\lambda}h$ characterizes the effect of the polynomial interpolation [12]. This formulation of the characteristic equation resembles (6), and is used in the derivation of the steplength heuristics.

3 Steplength heuristics for the LMS method

This section first outlines the steplength heuristic proposed in [6] and then presents the improvements on the heuristic, developed in [12].

3.1 The steplength heuristic in DDE-BIFTOOL

We want to determine the steplength h in the LMS method such that all characteristic roots λ of (2) lying in the right half planes

$$\mathbb{C}^+ := \{\lambda \in \mathbb{C} \,:\, \Re e(\lambda) \geqslant 0\} \quad \text{or} \quad \mathbb{C}^+ + r := \{\lambda \in \mathbb{C} \,:\, \Re e(\lambda) \geqslant r\},$$

with $r \leqslant 0$, are approximated accurately.

We denote the right hand side of (6) by "$\Sigma_\tau(\cdot)$", i.e.,

$$\Sigma_\tau(\lambda) := \sigma(A_0 + \textstyle\sum_{j=1}^{m} A_j \mathrm{e}^{-\lambda\tau_j}), \quad \text{where } \lambda \in \mathbb{C}.$$

Let D be a subset of the complex plane and let $\Sigma_\tau(\mathsf{D}) := \bigcup_{\lambda \in \mathsf{D}} \Sigma_\tau(\lambda)$. Then the roots λ that lie in $\mathbb{C}^+ + r$ are included in $\Sigma_\tau(\mathbb{C}^+ + r)$.

The steplength heuristic implemented in DDE-BIFTOOL, is based on the following properties, cf. [6]:

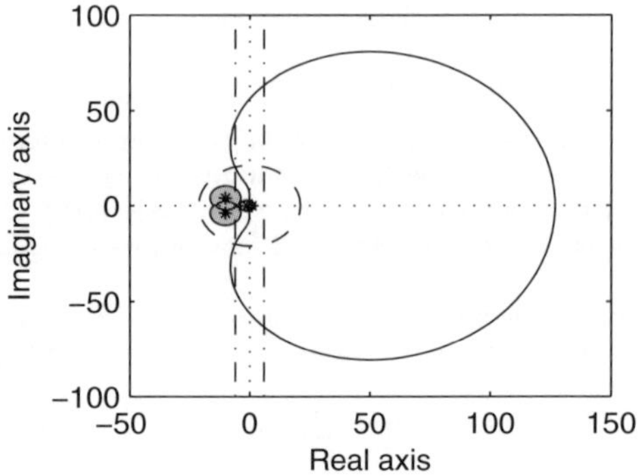

Figure 1: For the DDE system (9) : $\frac{1}{h}$ LMS(i[0, 2π[) for the BDF method of 4th order, where h is given by (8) (solid line); the circle around the origin with radius $\sum_{j=0}^{m} \|A_j\|$ (dashed line); parallel lines $\pm\epsilon/h$ from the imaginary axis (dash-dotted lines) (with a rather large ϵ for visibility).

- $\max|\Sigma_\tau(\mathbb{C}^+)| \leqslant \sum_{j=0}^{m} \|A_j\|$, where $\max|\Sigma_\tau(\mathbb{C}^+)| := \max\{|z| \ : \ z \in \Sigma_\tau(\mathbb{C}^+)\}$,
- LMS(i[0, 2π[) is the boundary of LMS($\mathbb{C}^+$).

The latter statement actually only holds for LMS methods that satisfy the so-called "Property C", cf. [7], and for which LMS($\mathbb{C}^+$) $\cap$ LMS($\mathbb{C}^-$) $= \emptyset$. These conditions are satisfied by e.g. the LMS methods of the BDF class.

The curve LMS(i[0, 2π[) can easily be computed numerically. Part of this curve approximates the imaginary axis. Let $\rho_{\text{LMS},\epsilon}$ the radius of the disc in the complex plane centred at the origin in which the imaginary axis is approximated by LMS(i[0, 2π[) "up to a given tolerance" $\epsilon > 0$. In [6], it is proven that if

$$h = \frac{\rho_{\text{LMS},\epsilon}}{\sum_{j=0}^{m} \|A_j\|}, \tag{8}$$

then the delay-independent stability is preserved by the LMS method up to the tolerance ϵ. This is illustrated in Fig. 1 for a system of four DDEs and one delay with

$$A_0 = \begin{bmatrix} -1 & 0 & 0 & 0 \\ 0 & 1 & 0 & 0 \\ 0 & 0 & -10 & -4 \\ 0 & 0 & 4 & -10 \end{bmatrix}, \qquad A_1 = \begin{bmatrix} 3 & 3 & 3 & 3 \\ 0 & -1.5 & 0 & 0 \\ 0 & 0 & 3 & -5 \\ 0 & 5 & 5 & 5 \end{bmatrix}, \tag{9}$$

and $\tau = 1$. Indeed, looking at LMS(i[0, 2π[), we observe that

a) $\mathbb{C}^-$, that contains all "stable roots" λ of (1), lies in the stability region of the LMS method, up to a strip of width ϵ/h.

b) Since the "unstable roots" λ of (1) lie in the disc with radius $\sum_{j=0}^{m} \|A_j\|$ (see above), they lie in the complement of the stability region of the LMS method, up to a strip of width ϵ/h.

For full details and the proof of the preservation of stability, we refer to [6].

This idea then leads to the following steplength heuristic, implemented in DDE-BIFTOOL, to accurately approximate roots λ in the half-plane $\mathbb{C}^+ + r$:

$$h = 0.9 \frac{\rho_{\text{LMS},\epsilon}}{\|A_0\| + |r| + \sum_{j=1}^{m} \|A_j\| e^{-r\tau_j}}. \tag{10}$$

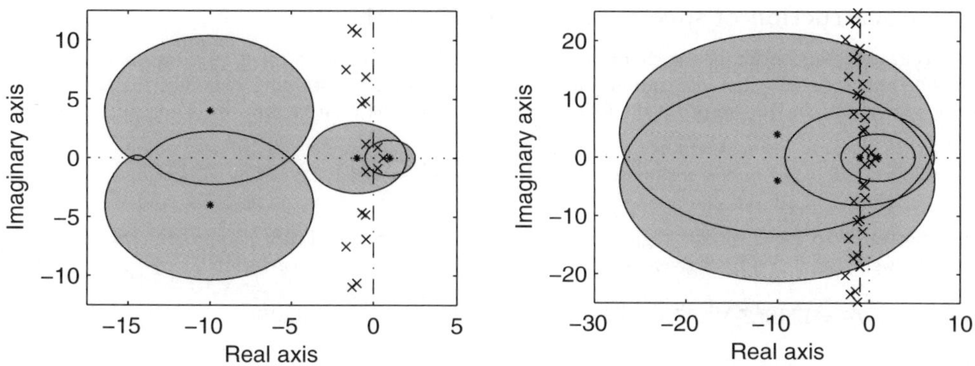

Figure 2: For the DDE system (9) with $n = 4$ and $m = 1$: The characteristic roots λ ($\times$), eigenvalues of A_0 ($*$) and the vertical line $\mathrm{i}\mathbb{R} + r$ (dashed line), $\mathrm{cl}\,\Sigma_\tau(\mathbb{C}^+ + r)$ (colored in gray) and $\Omega(r\vec{\tau})$ (solid line). Left : $r = 0$. Right : $r = -1$.

Here, 0.9 is a safety factor.

However, the denominator of (10) is typically a large overestimation for $\max |\Sigma_\tau(\mathbb{C}^+ + r) \cap (\mathbb{C}^+ + r)|$, especially in case the spectrum has a long "tail" along the negative real axis, as is the case with space discretizations of parabolic partial differential equations. Hence, the heuristic is often very conservative. In [12] we have improved the steplength heuristic by

- locating more precisely $\Sigma_\tau(\mathbb{C}^+ + r)$,
- by modifying the parameters in the LMS method to better suit the purpose of computing the rightmost roots.

We now briefly present these two improvements.

3.2 Locating more precisely $\Sigma_\tau(\mathbb{C}^+ + r)$

To obtain a sharper steplength heuristic, we are particularly interested in bounding the region $\Sigma_\tau(\mathbb{C}^+ + r) \cap (\mathbb{C}^+ + r)$. Let "cl" denote the closure of a set and let $\vec{\tau} := (\tau_1, \ldots, \tau_m)$. It is shown in [10] that the boundary of $\mathrm{cl}\,\Sigma_\tau(\mathbb{C}^+ + r)$ belongs to the set $\Omega(r\vec{\tau})$, where $\Omega(\vec{\gamma})$ is the set-valued function that maps $\vec{\gamma} \in \mathbb{R}^m$ onto

$$\Omega(\vec{\gamma}) := \bigcup_{\vec{\omega} \in [0,2\pi[^m} \sigma(A_0 + \textstyle\sum_{j=1}^m A_j e^{-(\gamma_j + i\omega_j)}). \tag{11}$$

In general, $\Omega(r\vec{\tau})$ is a two-dimensional subset of the complex plane. However, in the case of commensurate delays or a single delay, $\Omega(r\vec{\tau})$ is a union of curves. This is illustrated in Fig. 2 for the system of four DDEs with A_0 and A_1 defined by (9) and one delay, $\tau = 1$. The regions $\mathrm{cl}\,\Sigma_\tau(\mathbb{C}^+ + r)$, for $r = 0$ and $r = -1$ are colored in gray in Fig. 2, and the boundaries of these regions clearly belong to the set $\Omega(r\vec{\tau})$. The latter set can be computed numerically and $\max |\Omega(r\vec{\tau}) \cap (\mathbb{C}^+ + r)|$ gives a much sharper bound than the denominator of (10).

3.3 Construction of special-purpose LMS methods

The accurate computation of the characteristic roots in $\mathbb{C}^+ + r$ imposes other requirements on the LMS method than accurate time integration. In the former context, the stability region of the LMS method is *not* important. In [12] we prove the following estimate for the relative error on a computed root $\tilde{\lambda}$:

$$\frac{|\lambda - \tilde{\lambda}|}{|\tilde{\lambda}|} = \mathcal{O}\left(\left|\frac{\mathrm{LMS}(\tilde{\lambda}h) - \tilde{\lambda}h}{\tilde{\lambda}h}\right|\right). \tag{12}$$

This motivates to focus on the region in the complex plane where LMS(·) approximates the identity mapping well. Let $\delta > 0$ be a given *relative tolerance*. We define the *trust-region* $\mathcal{T}_\delta$ by

$$\mathcal{T}_\delta := \mathrm{LMS}(\mathcal{Z}_\delta), \quad \mathcal{Z}_\delta := \left\{ z \in \mathbb{C}_0^- \cup \mathcal{S}^+ \;:\; |\Im m(z)| < \pi, \quad |\mathrm{LMS}(z) - z| \leqslant \delta|z| \right\}, \tag{13}$$

where

$$\mathcal{S}^+ := \left\{ z \in \mathbb{C}^+ \;:\; \mathrm{LMS}(z) \in \mathbb{C}^+ \right\}.$$

We now explain the meaning of the trust-region $\mathcal{T}_\delta$ in the case that $r = 0$. For a complete discussion, see [10]. As explained in the previous section, "unstable roots" $\lambda \in \mathbb{C}^+$ only occur in subsets of the complex plane bounded by $\Omega(\vec{0}) \cap \mathbb{C}^+$. Moreover, it can also be proven that if a connected component of $\Omega(\vec{0})$ lies entirely in $\frac{1}{h}\,\mathrm{LMS}(\mathcal{Z}_\delta \cap \mathbb{C}^+)$, then it contains characteristic roots λ that are approximated by computed roots $\tilde{\lambda}$ which lie in $\frac{1}{h}\mathcal{Z}_\delta \cap \mathbb{C}^+$. Hence by (12), the relative error on these $\tilde{\lambda}$ is of the order of magnitude of δ. (However it is often much smaller in practice, cf. Section 4.) This theoretical property justifies the heuristic condition that $\Omega(\vec{0})$ should belong to the scaled trust-region $\frac{1}{h}\mathcal{T}_\delta$. In general, for nonzero r, we will impose the condition that $\Omega(r\vec{\tau}) \cap (\mathbb{C}^+ + r)$ belongs to $\frac{1}{h}\mathcal{T}_\delta$. This results in a heuristic for the steplength h.

With the objective of having a large trust-region $\mathcal{T}_\delta$, we have adapted the coefficients α_i and β_i of the LMS method, cf. (3). Also, an additional condition is imposed. Assume that we want to know whether the DDE system (1) is stable or unstable. In order to preserve roots with small real part, it is desirable that

$$\mathrm{LMS}(\mathrm{i}\mathbb{R}) \subseteq \mathrm{i}\mathbb{R}. \tag{14}$$

In [12] it is shown that the LMS methods of *maximal order* $p = 2k$ (with $k \geqslant 1$) have generally small *error constants* C_{err} and result in a large region $\mathcal{T}_\delta$. Furthermore, the construction of the (unique) k-step LMS method of order $p = 2k$ satisfying (14) is described. Fig. 3 shows the trust-regions $\mathcal{T}_\delta$ of the special-purpose methods of 4th and 8th order. These are significantly larger than the $\mathcal{T}_\delta$ of the classical BDF methods. Moreover, by construction, LMS($\mathcal{S}^+$) and $\mathcal{T}_\delta$ contain part of the imaginary axis; a desired property that does not hold for the BDF methods.

3.4 An improved steplength heuristic

We want to determine h such that the part of $\Omega(r\vec{\tau})$ (cf. (11)) which lies in $\mathbb{C}^+ + r$ also belongs to $\frac{1}{h}\mathcal{T}_\delta$ (cf. Fig. 3 for relative tolerance $\delta = 0.1$ and order $p = 4, 8$). In order to obtain a simple formula for the steplength h, that can easily be implemented in software, we replace the trust-region $\mathcal{T}_\delta$ by an ellipse, with the axes aligned with the real and imaginary axes, that is inscribed in $\mathcal{T}_\delta$. For all special-purpose methods and for a given tolerance δ, the length of the semi-axes a_{ell} and b_{ell} of inscribed ellipses can be determined, see [12].

The new steplength heuristic can now be formulated as follows. First, points $q_K \in \Omega(r\vec{\tau})$, for $K = 1, \ldots$, are computed for a given value of r (see below). Ideally, these points are well spread out over $\Omega(r\vec{\tau})$. Next, the points with real parts larger than $r - \Delta r$ for some $\Delta r > 0$ are selected. We use the safety margin Δr because only a limited number of points is computed. Typical values are $r = -1, -2$

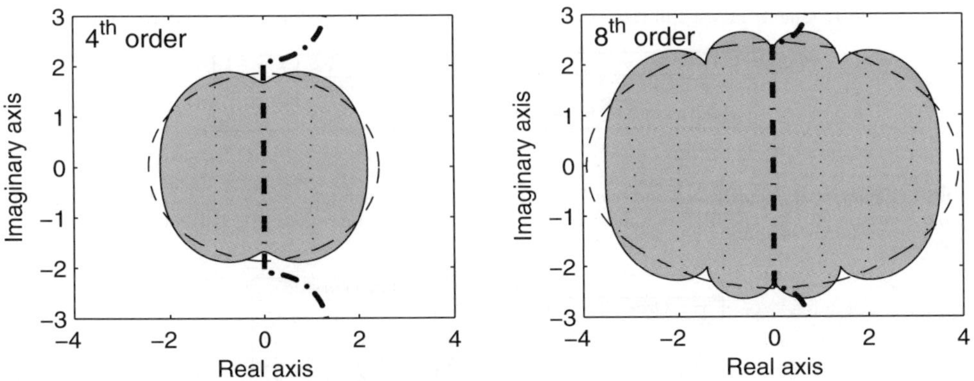

Figure 3: $\mathcal{S}^+$ (bounded by the thick dash-dotted line), $\mathcal{T}_\delta$ for $\delta = 0.1$ (colored in gray and bounded by the solid line), $\text{LMS}^{-1}(\mathcal{T}_\delta)$ (bounded by the dashed line) and (a part of) the image of lines parallel to the imaginary axis (with integer $\Re e(\cdot)$) under $\text{LMS}(\cdot)$ (dotted lines) for the special-purpose methods of 4th and 8th order.

and $\Delta r = -0.1$. Finally, the largest value of h is determined such that the selected points scaled by $h/0.9$ fit into the ellipse inscribed in $\mathcal{T}_\delta$, where 0.9 is a safety factor. Hence h is given by

$$h = \frac{0.9}{\sqrt{\max_{\Re e(q_K) \geqslant r - \Delta r} \left((\Re e(q_K)/a_{\text{ell}})^2 + (\Im m(q_K)/b_{\text{ell}})^2 \right)}}. \tag{15}$$

This novel heuristic always gives a larger steplength than heuristic (10).

The points $q_K \in \Omega(r\vec{\tau})$ (cf. (15)) can be obtained as eigenvalues of the matrix

$$A_0 + \sum_{j=1}^{m} \left(A_j \mathrm{e}^{-r\tau_j}\right) \mathrm{e}^{-\mathrm{i}\omega_j},$$

for a number of m-vectors $\vec{\omega}$ chosen from $[0, \pi[\times[0, 2\pi[^{m-1}$. (By the symmetry of $\Omega(r\vec{\tau})$ w.r.t. the real axis, we do not have to sample the larger set $[0, 2\pi[^m$.) Clearly, the size of these eigenvalue problems, n, is much smaller than N in (5).

4 Examples

This section presents examples to illustrate the efficiency of the novel procedure to compute the rightmost characteristic roots. We consider a small-scale DDE system to illustrate the main features and one large-scale DDE system, obtained after space discretization of a hybrid DDE-PDE system.

4.1 A small-scale system of DDEs with one delay

For a system of four DDEs with A_0 and A_1 defined by (9) and one delay, $\tau = 1$, we computed roots λ with real part larger than $r = 0, -0.5, -1$. We used, for comparison, the special-purpose methods of 4th, 6th and 8th order and the BDF methods of 4th and 6th order. The spectrum and the set $\Omega(r\vec{\tau})$ for

Table 1: Quantities used in the computation of the steplength for different r (and $\Delta r = 0.1$).

	$r = 0$	$r = -0.5$	$r = -1$
$\|A_0\| + \|r\| + \sum_{j=1}^{m} \|A_j\| e^{-r\tau_j}$	21.1	28.3	39.9
$\max \|\Omega(r\vec{\tau}) \cap (\mathbb{C}^+ + r - \Delta r)\|$	2.86	8.58	18.8

Table 2: Values of the steplength h and size N of the eigenvalue problem for $r = 0, -0.5, -1$. Left column : using the old heuristic (10) for the BDF methods. Right column : using the new heuristic (15) for the special-purpose methods.

		old heuristic		new heuristic	
r	Order	h	N	h	N
	4	2.44×10^{-2}	184	5.28×10^{-1} (∗)	20
0	6	2.97×10^{-2}	168	4.84×10^{-1} (∗)	32
	8			6.94×10^{-1} (∗)	44
	4	1.82×10^{-2}	240	1.77×10^{-1}	36
−0.5	6	2.22×10^{-2}	216	2.55×10^{-1}	36
	8			2.32×10^{-1}	48
	4	1.29×10^{-2}	332	8.05×10^{-2}	64
−1	6	1.57×10^{-2}	288	1.12×10^{-1}	56
	8			1.06×10^{-1}	68

$r = 0$ and $r = -1$ are shown in Fig. 2 and the quantities used in the denominator of both steplength heuristics (10) and (15) are given in Table 1.

Table 2 lists the steplength h and the size N of the corresponding eigenvalue problem for the old heuristic (10) and for the novel heuristic (15) with $\Delta r = 0.1$. Obviously, h decreases with r. If $h > h_{\max}$ (cf. (4)), the steplength is set equal to $h_{\max}$. This is indicated by a (∗). In this case, $N = n(3k-1)$ for the special-purpose method of order $p = 2k$. A detailed analysis of this example shows that both adaptions (i.e., the special-purpose LMS methods and the location of $\Omega(r\vec{\tau}) \cap (\mathbb{C}^+ + r)$) contribute to the improvement of the steplength, see [12]. Table 3 gives the ratio of the steplengths and the ratio of the sizes of the eigenvalue problems for the new heuristic (15) and the old heuristic (10). Notice that the ratio $N_{\text{old}}/N_{\text{new}}$ listed in Table 3 decreases with r. However, the novel heuristic remains superior. Fig. 4 shows the approximate roots $\tilde{\lambda}$ with their corrections by Newton's method on the characteristic nonlinear eigenvalue problem, for the special-purpose methods of 4th and 8th order for $r = -1$. Recall that the highest accuracy is achieved for roots $\tilde{\lambda}$ close to the origin.

4.2 A large-scale system of DDEs

This section considers a hybrid DDE-PDE system modelling a semiconductor laser subject to conventional optical feedback and lateral carrier diffusion [9]. The system in the complex scalar variable $A(t)$, representing the electric field, and real $Z(x,t)$, representing the carrier density in the interval $x \in$

Table 3: Ratios of the steplengths $h_{\text{new}}/h_{\text{old}}$ and ratios of the sizes of the eigenvalue problem, $N_{\text{old}}/N_{\text{new}}$, where h_{new} comes from heuristic (15) for special-purpose LMS methods and h_{old} from heuristic (10) for the BDF methods. For $r = 0$, $h_{\text{new}} > h_{\max}$ and h_{new} is set to $h_{\max}$.

Order		$r=0$	$r=-0.5$	$r=-1$
4	$h_{\text{new}}/h_{\text{old}}$	20.5	9.7	6.2
	$N_{\text{old}}/N_{\text{new}}$	9.2	6.7	5.2
6	$h_{\text{new}}/h_{\text{old}}$	11.2	11.5	7.1
	$N_{\text{old}}/N_{\text{new}}$	5.3	6.0	5.1

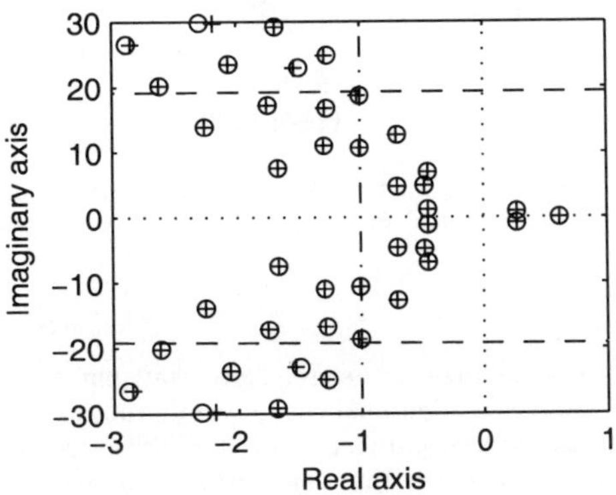

4th order, $h = 8.05 \times 10^{-2}$

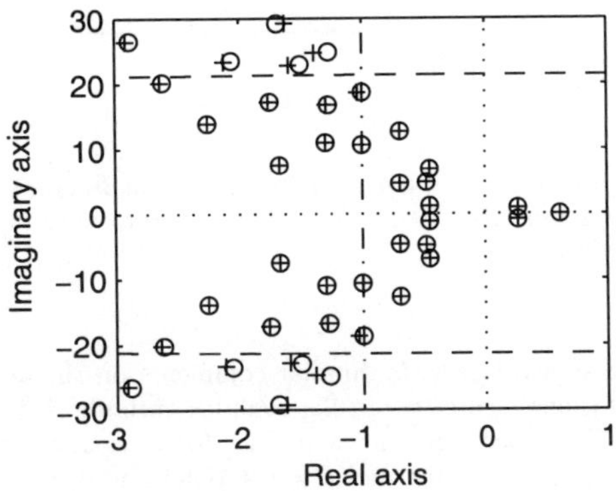

8th order, $h = 1.06 \times 10^{-1}$

Figure 4: For the example of Section 4.1 and $r = -1$: approximate roots $\tilde{\lambda}$ (+), their corrections using Newton iterations (○) and $\frac{0.9}{h}$ LMS$^{-1}(\mathcal{T}_\delta)$ (bounded by the dashed line) for the special-purpose methods of 4th order (left) and 8th order (right).

$[-0.5, 0.5]$, reads as

$$\frac{\mathrm{d}A(t)}{\mathrm{d}t} = (1 - \mathrm{i}\alpha)A(t)\zeta(t) + \eta A(t-\tau)e^{-\mathrm{i}\phi} - \mathrm{i}bA(t), \tag{16}$$

$$T\frac{\partial Z(x,t)}{\partial t} = d\frac{\partial^2 Z(x,t)}{\partial x^2} - Z(x,t) + P(x) - F(x)(1 + 2Z(x,t))|A(t)|^2. \tag{17}$$

The functions $\zeta(t)$, $P(x)$ and $F(x)$ are specified in [9]. Zero Neumann boundary conditions for $Z(x,t)$ are imposed at $x = \pm 0.5$. We fix parameters $\alpha = 3$, $\phi = 0$, $T = 1000$, $d = 1.68 \times 10^{-2}$ and delay $\tau = 1000$. For the numerical computations, the time variable is rescaled as $t \leftarrow 1000\,t$. The symmetry about $x = 0$ is exploited by considering only the interval $[0, 0.5]$. We split (16) into real and imaginary part and discretize (17) in space using a second order central difference formula with constant stepsize $\Delta x = 0.5/128$. Hence the resulting DDE system has size 131. For $\eta \approx 2.5717 \times 10^{-3}$, a steady state Hopf bifurcation arises with $|A| \approx 1.8209$ and $b \approx 1.1119 \times 10^{-3}$ and purely imaginary roots $\lambda = \pm\mathrm{i}47.711$. The system is linearized about this steady state solution. The approximate roots $\tilde{\lambda}$ and their corrections

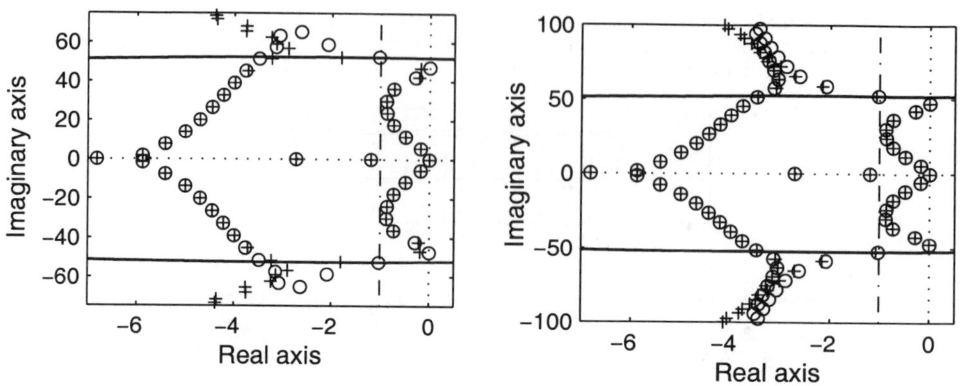

Figure 5: For the large-scale DDE system of Section 4.2 with $r = -1$, using the 6th order special-purpose method : the approximate roots $\tilde{\lambda}$ (+) and their corrections (○) and $\Omega(r\vec{\tau})$ (solid line) for $\delta = 0.1$ (left) and for $\delta = 0.01$ (right).

are shown in Fig. 5. Similar comments on the accuracy can be made as before. Note that, due to the rotational symmetry in the complex variable A, there is always an additional characteristic root at zero.

The spectrum shown in Fig. 5 was computed using the special-purpose method of 6th order and $r = -1$, $\Delta r = 0.1$ and $\delta = 0.1$ (left) and $\delta = 0.01$ (right), where δ is the requested relative accuracy within the trust-region, see (13). For this example, we use a steplength h commensurate with τ. That is, the value of h obtained from heuristic (15) is lowered until τ/h is integer, so that the interpolation is avoided. After this adaption, the novel heuristic (15) gives $h = \tau/24$ ($N = 3\,537$) for $\delta = 0.1$ and $h = \tau/36$ ($N = 5\,109$) for $\delta = 0.01$. The resulting eigenvalue problems are large, but still feasible.

The old heuristic (10), using the BDF method of 6th order, gives $h = \tau/7195$ which would lead to an intractable eigenvalue problem of size $N = 943\,331$ (for $\epsilon = 0.1$). The steplength resulting from heuristic (10) is so small because $\|A_0\| + |r| + \|A_1\|e^{-r\tau} \approx 4\,531.8$ severely overestimates $\max|\Omega(r\vec{\tau}) \cap (\mathbb{C}^+ + r)| \approx 52.5$. The former value is large because the spectrum of this parabolic system extends far to the left in $\mathbb{C}^-$. However, we are only interested in the roots with real part larger than r.

Remark that, by using $\Omega(r\vec{\tau})$, we ensure that the Hopf bifurcation point caused by roots with large imaginary part (shown in Fig. 5), is detected.

Finally, we compare the accuracy and efficiency of the procedure presented in this paper with two alternatives based on a *spectral discretization*. In this type of discretization, the trajectory is not approximated by points on a uniform mesh, but by a high-degree polynomial.

In the first alternative, developed by Breda et al. in [3], the *infinitesimal generator* of the DDE (1) is discretized to obtain a matrix eigenvalue problem for $\tilde{\lambda}$, which approximate the characteristic roots λ. This approach can also be derived by substituting a polynomial approximation for $e^{\lambda t}$ into the formula $(e^{\lambda t})' = \lambda e^{\lambda t}$ and evaluating this expression on a mesh.

In the second alternative, developed in [11], the *solution operator* is discretized to obtain a matrix eigenvalue problem for $e^{\tilde{\lambda} h}$. This approach can analogously be derived from the formula $e^{\lambda(t+h)} = (e^{\lambda h})e^{\lambda t}$. In particular, we consider the so-called "backward" variant of this method where actually the time-reversed system is discretized. For the full details, see [11].

Fig. 6 shows the computed roots for the latter alternative only. However, the accuracy obtained by the spectral method in [3] is similar. In both cases, an approximating polynomial of degree 32 is used,

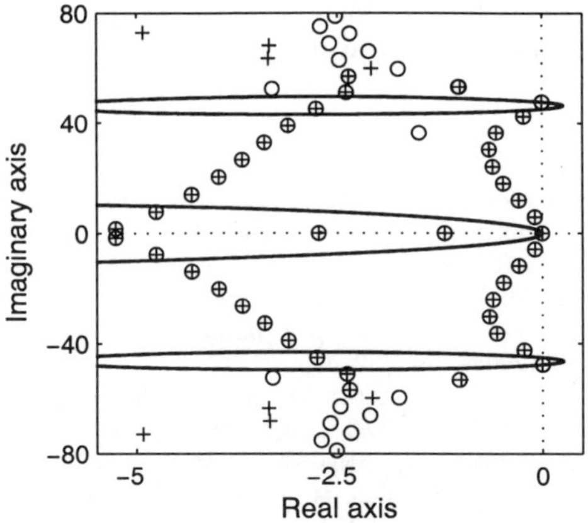

Figure 6: For the large-scale DDE system of Section 4.2 using a spectral method (See the text for some further explanation.) : the approximate roots $\tilde{\lambda}$ (+) and their corrections (○) and $\Omega(\vec{0})$ (solid line).

resulting in a matrix eigenvalue problem of size 4 323. Recall that the size of the eigenvalue problems corresponding to the LMS discretization (cf. Fig. 5, left and right) is 3 537 and 5 109 respectively. It follows from a comparison of Figs. 5–6 that the LMS method and both spectral methods are approximately as efficient in computing the rightmost roots. This is due to the fact that all these methods have difficulties in approximating the complex pair of roots with large imaginary part causing the Hopf bifurcation point. Although the *asymptotic convergence* of the *spectral methods* exceeds the convergence of the LMS methods, both methods are competitive if only a relatively low accuracy is requested. Furthermore, the steplength heuristic for the LMS method, discussed in this paper, allows the automatic selection of an appropriate discretization step.

5 Conclusions

This paper presents a method to efficiently and accurately compute all characteristic roots of a linear DDE system with real part larger than a user-specified value. The solution operator is discretized by a linear multistep (LMS) method combined with polynomial interpolation for the delayed term. This leads to a eigenvalue problem, the size of which is inversely proportional to the steplength used in the LMS method. We have indicated that the steplength heuristic developed in [12] leads to a efficient method, while maintaining the reliability of the numerical results.

We have demonstrated the performance of this method for a small- and a large-scale system of DDEs, and we have compared with methods based on a spectral discretization. Moreover, it was shown that, by using this procedure, no roots with large imaginary parts can be overlooked.

Acknowledgements

This research presents results of the Project IUAP P5/22 funded by the Interuniversity Attraction Poles Programme - Belgian Science Policy. The scientific responsibility rests with the authors. K.V. is a

Research Assistant of the Fund for Scientific Research - Flanders (Belgium). The research leading to this paper was performed during the stay of T.L. at the K.U.Leuven.

References

[1] D. Breda, *Solution operator approximation for delay differential equation characteristic roots computation via Runge-Kutta methods*, Appl. Numer. Math. to appear.

[2] D. Breda, S. Maset, and R. Vermiglio, *Computing the characteristic roots for delay differential equations*, IMA J. Numer. Anal., 24 (2004), pp. 1–19.

[3] ——, *Pseudospectral differencing methods for characteristic roots of delay differential equations*, SIAM J. Sci. Comput., 27 (2005), pp. 482–495.

[4] K. Engelborghs, T. Luzyanina, and D. Roose, *Numerical bifurcation analysis of delay differential equations using DDE-BIFTOOL*, ACM Trans. Math. Softw., 28 (2002), pp. 1–21.

[5] K. Engelborghs, T. Luzyanina, and G. Samaey, *DDE-BIFTOOL v. 2.00: a Matlab package for numerical bifurcation analysis of delay differential equations*, Report TW 330, Department of Computer Science, K.U.Leuven, Leuven, Belgium, 2001. Available from http://www.cs.kuleuven.be/~twr/research/software/delay/ddebiftool.shtml

[6] K. Engelborghs and D. Roose, *On stability of LMS methods and characteristic roots of delay differential equations*, SIAM J. Numer. Anal., 40 (2002), pp. 629–650.

[7] E. Hairer and G. Wanner, *Solving ordinary differential equations. II: Stiff and differential-algebraic problems*, vol. 14 of Springer series in computational mathematics, Springer Berlin, 2nd ed., 1996.

[8] J. Hale and S. M. Verduyn Lunel, *Introduction to Functional Differential Equations*, vol. 99 of Applied Mathematical Sciences, Springer-Verlag, 1993.

[9] K. Verheyden, K. Green, and D. Roose, *Numerical stability analysis of a large-scale delay system modelling a lateral semiconductor laser subject to optical feedback*, Phys. Rev. E 69, 036702 (2004).

[10] K. Verheyden, T. Luzyanina, and D. Roose, *Location and numerical preservation of characteristic roots of delay differential equations by LMS methods.*, Technical Report TW-382, Department of Computer Science, K.U.Leuven, Leuven, Belgium, Dec. 2003. Available from http://www.cs.kuleuven.be/publicaties/rapporten/tw/TW382.abs.html

[11] K. Verheyden and D. Roose, *Efficient numerical stability analysis of delay equations : a spectral method*, in the Proceedings of the IFAC Workshop on Time-Delay Systems 2004, D. Roose and W. Michiels, eds., IFAC Proceedings Volumes, 2004, pp. 209–214.

[12] K. Verheyden, T. Luzyanina, and D. Roose, *Efficient computation of characteristic roots of delay differential equations using LMS methods.*, submitted. Available from http://www.cs.kuleuven.be/~koenv/preprints.html

Grazing Phenomena in a Harmonically Excited Oscillator with Dry-Friction on a Sinusoidally Time-Varying, Traveling Surface

Albert C.J. Luo[1], Brandon Gegg[2] and Steve. S. Suh[2]
[1]*Department of Mechanical and Industrial Engineering, Southern Illinois University Edwardsville, Edwardsville, IL 62026-1805, USA*
[2]*Department of Mechanical Engineering, Texas A & M University, College Station, TX 77840-1942, USA*

Abstract

In this paper, the dynamic mechanism of the grazing phenomena for a dry-friction oscillator interacting with a time varying traveling surface is investigated. The theory of non-smooth dynamical systems for connectable and accessible domains is applied to this oscillator. The stick and non-stick motions are discussed through their corresponding mapping definitions. The grazing motions are presented through the initial and final switching sets, varying with external excitation parameters. The analytical prediction of grazing motion is verified through numerical simulations. This investigation provides a systematic analysis for the grazing motion in such a discontinuous dynamical system.

PACS: 46.30My, 46.30Pa.
Keywords: Friction, Velocity Discontinuity, Grazing, Bifurcation.

1. Introduction

Friction-induced vibrations are widely observed in engineering. Disk brake systems, turbine blades and stringed musical instruments all contain these dynamics. The frictional forces create a discontinuity, which is difficult to solve in any form. Therefore, the friction-induced oscillations have been of great interest for a long time. The dynamic mechanism for the grazing motions of this discontinuous system will be discussed herein as an example to obtain the conditions for the onset and vanishing of such a motion in a non-smooth dynamical system.

In 2004, Luo and Gegg [1] investigated the generalized, linear, mechanical model for the friction-induced oscillator. The mechanism of the stick and non-stick motions for such a generalized model was studied to obtain the conditions for the onset and vanishing of stick motion in such a non-smooth dynamical system. Since friction-induced oscillations exist extensively, this problem has attracted great attention in theory and application. An early study of this non-smooth dynamical system was the topic investigated by Den Hartog [2] in 1931. The periodic motion of the forced linear oscillator with Coulomb and viscous damping was investigated.

From mathematical standpoint, Levitan [3] discussed a friction oscillation model with a periodically driven base in 1960, and the stability of the periodic motion was presented. In 1964, Filippov[4] presented

[1] Corresponding author. Tel: +1-618-650-5389; fax: +1-618-650-2555.
E-mail address: aluo@siue.edu (A.C.J. Luo).

differential equations with discontinuous right-hand sides, which stemmed from the coulomb friction oscillator. A comprehensive discussion of such discontinuous differential equations can be found in reference [5]. In this paper, the grazing motion of the friction-induced oscillator will be investigated from a different point of view. In 1979, Hundal [6] further discussed the dynamical responses of the base driven friction oscillator. In 1986, Shaw [7] investigated the stability for said periodic motion through the Poincare mapping.

In 1995, Luo [8] utilized the mapping structure concept to determine the periodic motion for impact oscillators (also see, [9, 10]). This methodology was used to investigate the periodic and chaotic motions of the periodically driven, piecewise, linear system in [11, 12]. A generalized methodology was expressed through investigation of such a discontinuous dynamical system [13]. On the other hand, Feigin [14] in 1970 investigated the C-bifurcation in piecewise-continuous systems via the Floquet theory of mappings and the motion complexity was classified by the eigenvalues of mappings, which can be referred to recent publications [15,16]. In 1991, Nordmark [17] investigated the non-periodic motion caused by the grazing bifurcation through the discontinuous mapping.

Further, the normal form mapping for such grazing phenomena was developed (e.g., [18-20]). In 2005, Luo [21] developed a general theory for the local singularity of non-smooth dynamical systems on connectable domains. The imaginary, sink and source flows were discussed in [22] to determine the sliding and source motions in non-smooth dynamical systems. In 2006, Luo and Gegg [23] used the local singularity theory to conduct a comprehensive investigation of the generalized, friction-induced oscillator. The complicated motions in parameter space were achieved, and the parameter regions with specific mapping structures were obtained as well for application purpose. In the aforementioned investigation, the grazing phenomena were observed in the periodically forced, friction-induced oscillator.

The mechanism for grazing motion of the model presented herein, for the special case of constant belt velocity, has been studied by Luo and Gegg [24]. The periodic motion was also discussed and studied, for the oscillator presented herein, by Luo and Gegg 25]. The discontinuous mapping techniques will not be used herein. However, the recently developed theory in [21] will be employed to investigate grazing motions in this friction-induced oscillator. In this paper, the necessary and sufficient conditions for grazing in a periodically forced, linear oscillator with dry-friction on a time varying traveling surface will be presented. The generic non-stick and stick mappings for such an oscillator will also be defined. The necessary and sufficient conditions will be expressed through the non-stick mapping. The initial and grazing switching manifolds are introduced. From analytical pre-dictions, illustrations of grazing motions will be illustrated.

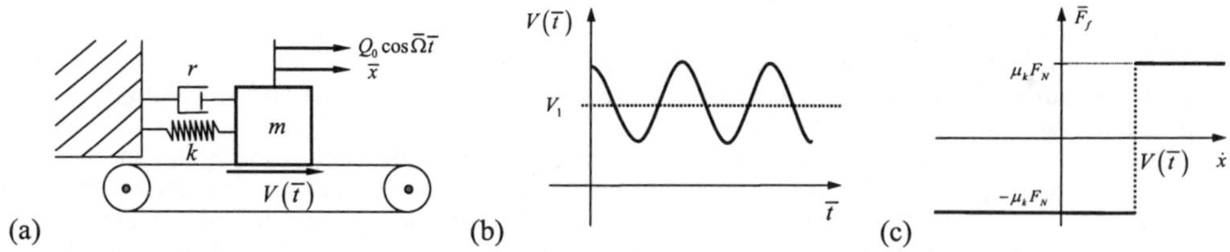

Fig. 1 Mechanical model: (a) schematic model, (b) belt velocity, (c) friction forces.

2. Mechanical Model

Consider a periodically forced oscillator attached to a fixed wall, which consists of a mass *m*, a spring of stiffness *k* and a damper of viscous damping coefficient *r*, as shown in Fig.1(a). The co-ordinate system $(\bar{x},\bar{t})$ is absolute with displacement $\bar{x}$ and time $\bar{t}$. The periodic driving force $\bar{Q}_0\cos\bar{\Omega}\bar{t}$ is exerted on the mass where $\bar{Q}_0$ and $\bar{\Omega}$ are the excitation strength and frequency ratio, respectively. Since the mass contacts the moving belt with friction, the mass can move along or rest on the belt surface. This oscillator rests on the horizontal belt surface which travels with a time varying speed $\bar{V}(\bar{t})$, Fig.1(b).

$$\bar{V}(\bar{t}) = \bar{V}_0\cos(\omega\bar{t}+\beta)+\bar{V}_1 \tag{1}$$

where ω is the oscillatory frequency of the traveling surface, $\bar{V}_0$ is the oscillatory amplitude of the traveling belt surface, and $\bar{V}_1$ is a constant.

Since the mass contacts the moving belt with friction, the mass can move along, or rest on, the belt surface. Further, a kinetic friction force shown in Fig.1(c) is described as

$$\bar{F}_f(\dot{\bar{x}})\begin{cases} = \mu_k F_N, & \dot{\bar{x}} \in [\bar{V}(t), \infty) \\ \in [-\mu_k F_N, \mu_k F_N], & \dot{\bar{x}} = \bar{V}(t) \\ = -\mu_k F_N, & \dot{\bar{x}} \in (-\infty, \bar{V}(t)] \end{cases} \tag{2}$$

where $\dot{\bar{x}} \triangleq d\bar{x}/dt$, and μ_k and F_N are the friction coefficient and a force normal to the contact surface, respectively. For this case, $F_N = mg$ and g is the gravitational acceleration. When the mass adheres to the surface, the non-friction forces acting on the mass in the x-direction during this motion are determined by

$$F_s = \bar{Q}_0 \cos\bar{\Omega}\bar{t} - r\bar{V}(\bar{t}) - k\bar{x}, \quad \text{for } \dot{\bar{x}} = \bar{V}(\bar{t}), \tag{3}$$

where $A_0 = Q_0/m$, $d = r/2m$ and $c = k/m$. This force cannot overcome the friction force during the stick motion, i.e., $|F_s| \le |F_f|$ and $F_f = \bar{F}_f/m$. The mass does not have any relative motion to the belt. Therefore, no acceleration exists, i.e.,

$$\ddot{\bar{x}} = \dot{\bar{V}}(t) = -\bar{V}_0 \omega \sin(\omega\bar{t} + \beta), \qquad \text{for } \dot{\bar{x}} = \bar{V}(\bar{t}). \tag{4}$$

If $|F_s| > |F_f|$, the non-friction force will overcome the static friction force on the mass and the

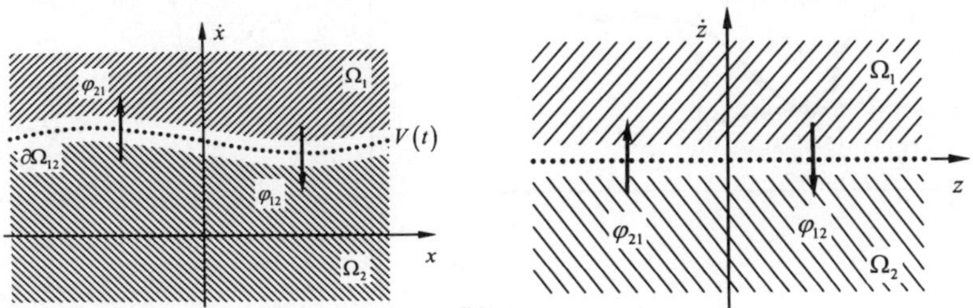

(a) (b)

Fig.2 Phase plane for absolute and relative systems.

non-stick motion will appear. For the non-stick motion, the total force acting on the mass is

$$F = \bar{Q}_0 \cos\bar{\Omega}\bar{t} - \mu_k F_N \operatorname{sgn}(\dot{\bar{x}} - \bar{V}) - r\dot{\bar{x}} - k\bar{x}, \quad \text{for } \dot{\bar{x}} \ne \bar{V}; \tag{5}$$

where $\operatorname{sgn}(\cdot)$ is the sign function. Therefore, the equation of the non-stick motion for this oscillator is

$$\ddot{\bar{x}} + 2\bar{d}\dot{\bar{x}} + \bar{c}\bar{x} = \bar{A}_0 \cos\bar{\Omega}\bar{t} - \tilde{F}_f \operatorname{sgn}(\dot{\bar{x}} - \bar{V}(\bar{t})), \quad \text{for } \dot{\bar{x}} \ne \bar{V}(\bar{t}), \tag{6}$$

where $\bar{A}_0 = \bar{Q}_0/m$, $\bar{d} = r/2m$, $\bar{c} = k/m$ and $\tilde{F}_f = \mu_k F_N/m$. The non-dimensional frequency and time yield,

$$\left.\begin{aligned} &\Omega = \bar{\Omega}/\omega, t = \omega\bar{t}, d = \bar{d}/\omega, c = \bar{c}/\omega^2, A_0 = \bar{A}_0/\omega^2, \\ &F_f = \tilde{F}_f/\omega^2, V_0 = \bar{V}_0/\omega, V_1 = \bar{V}_1/\omega, V = V/\omega, x = \bar{x}; \end{aligned}\right\} \tag{7}$$

$$V(t) = V_0 \cos(t + \beta) + V_1. \tag{8}$$

The phase constant β is used to synchronize the periodic force input with the velocity discontinuity after the modulus of time has been computed. The integration of Eq.(8) gives

$$X(t)=V_0\left[\sin(t+\beta)-\sin(t_i+\beta)\right]+V_1\times(t-t_i)+x_i \text{ for } t>t_i \tag{9}$$

which is the displacement response of the periodically time-varying traveling surface. For $t=t_i$, $X(t_i)=x(t_i)\equiv x_i$.

Introducing the relative displacement, velocity, and acceleration as

$$z(t)=x(t)-X(t),\ \dot{z}(t)=\dot{x}(t)-V(t) \text{ and } \ddot{z}(t)=\ddot{x}(t)-\dot{V}(t), \tag{10}$$

the non-friction force in Eq.(3) becomes

$$F_s=A_0\cos\Omega t-2d\left[V(t)+\dot{z}(t)\right]-c\left[X(t)+z(t)\right]-\dot{V}(t), \quad \text{for } \dot{z}=0. \tag{11}$$

Since the mass does not have any motion relative to the vibrating belt, the relative acceleration is zero, i.e.,

$$\ddot{x}=\dot{V}(t)=-V_0\sin(t+\beta), \quad \text{for } \dot{x}=V(t). \tag{12}$$

$$\ddot{z}=0, \quad \text{for } \dot{z}=0. \tag{13}$$

For non-stick motion, the equation of this oscillator with friction becomes

$$\ddot{x}+2d\dot{x}+cx=A_0\cos\Omega t-F_f\,\text{sgn}\left(\dot{x}-V(t)\right), \text{ for } \dot{x}\neq V(t). \tag{14}$$

or

$$\ddot{z}(t)+2d\dot{z}(t)+cz(t)=A_0\cos\Omega t-F_f\,\text{sgn}\left(\dot{z}(t)\right)-2dV(t)-cX(t)-\dot{V}(t), \text{ for } \dot{z}\neq 0. \tag{15}$$

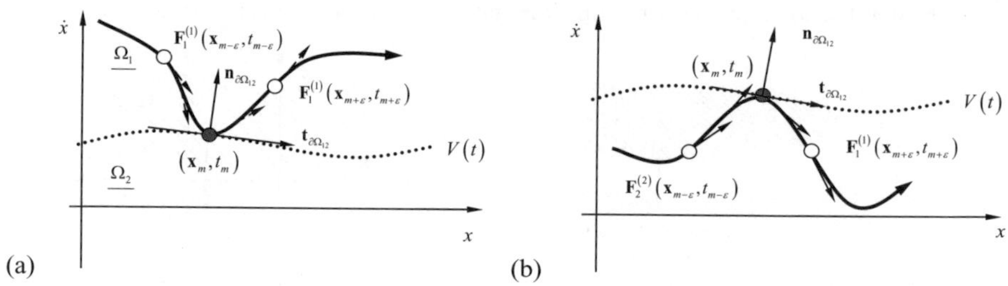

Fig.3 Grazing Motion in (a) domain Ω_1 and (b) domain Ω_2.

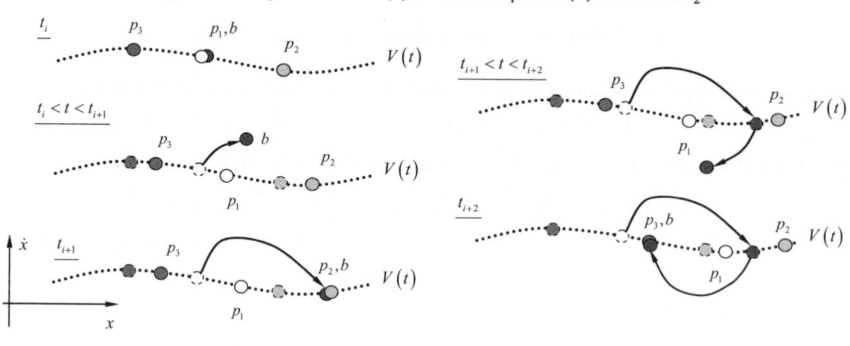

Fig.4 Relative relationship between the mass particle and the belt particle.

3. Grazing Mechanism

As in [1], since the friction force is dependent on the direction of the relative velocity, the phase plane is partitioned into two regions in which the motion is described through the continuous dynamical systems, as shown in Fig.2(a). Because of the discontinuity, the phase partition for this oscillator with friction is shown through the absolute and relative frames, Fig.2(b). A sketch of grazing motions in domain Ω_α ($\alpha=\{1,2\}$) is illustrated in Fig.3(a) and (b). For the absolute frame, the separation boundary is a curve varying with time. However, the discontinuous boundary in the relative frame is constant. From Eq.(9), the belt displacement will increase with increasing time.

However, the oscillator vibrates in the vicinity of equilibrium. The friction is dependent on the relative velocity between the oscillator and belt speed. When the non-stick motion of the oscillator has the same speed of the transport belt, the particle location of the belt has been changed. To understand the stick and non-stick motions, the particle switching on the surface of

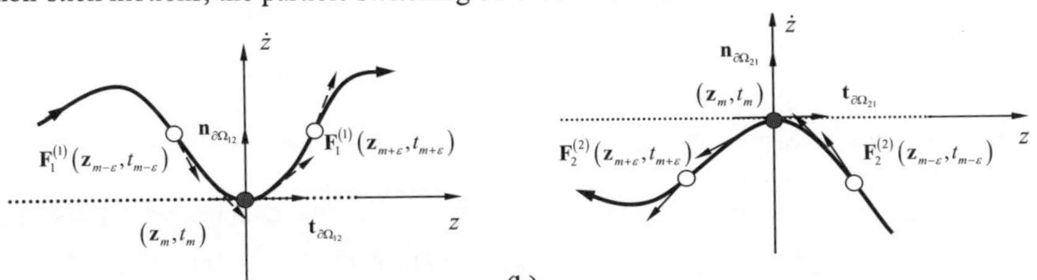

(a) (b)

Fig.5 Vector fields of grazing motions in domains (a) Ω_1 and (b) Ω_2 through the relative frames.

the oscillating belt is shown in absolute phase plane, Fig.4, for the oscillator experiencing the same speed as the oscillating belt at a moment t_i. The particles (p_1, p_2, p_3) on the belts are represented by white, yellow and green circular symbols, respectively. The red circular symbol is the oscillator location.

In the phase plane, we define

$$\mathbf{z} \triangleq (z,\dot{z})^T \equiv (z,v)^T \text{ and } \mathbf{F} \triangleq (v,F)^T. \tag{16}$$

The corresponding regions and boundary are

$$\left.\begin{aligned}\Omega_1 &= \{(z,v) \mid v\in(0,\infty)\},\\ \Omega_2 &= \{(z,v) \mid v\in(-\infty,0)\},\\ \partial\Omega_{\alpha\beta} &= \{(z,v) \mid \varphi_{\alpha\beta}(z,v)=v=0\}.\end{aligned}\right\} \tag{17}$$

The subscript $(\cdot)_{ij}$ defines the boundary from Ω_i to Ω_j. The equations of motion in Eqs.(12) and (14) can be described as

$$\dot{\mathbf{z}} = \mathbf{F}_\lambda^{(\kappa)}(\mathbf{z},t),\ (\kappa,\lambda\in\{0,1,2\}) \tag{18}$$

where

$$\left.\begin{aligned}&\mathbf{F}_\alpha^{(\alpha)}(\mathbf{z},t) = (y,F_\alpha(\mathbf{z},t))^T \text{ in } \Omega_\alpha\ (\alpha\in\{1,2\}),\\ &\mathbf{F}_\alpha^{(\beta)}(\mathbf{z},t) = (y,F_\beta(\mathbf{z},t))^T \text{ in } \Omega_\alpha\ (\alpha\neq\beta\in\{1,2\});\\ &\mathbf{F}_0^{(0)}(\mathbf{z},t) = (0,0)^T \text{ on } \partial\Omega_{\alpha\beta} \text{ for stick},\\ &\mathbf{F}_0^{(0)}(\mathbf{z},t) = \left[\mathbf{F}_\alpha^{(\alpha)}(\mathbf{z},t),\mathbf{F}_\beta^{(\beta)}(\mathbf{z},t)\right] \text{ on } \partial\Omega_{\alpha\beta} \text{ for non-stick.}\end{aligned}\right\} \tag{19}$$

For non-zero values, the subscript and superscript (κ and λ) represent the two adjacent domains for $\alpha,\beta\in\{1,2\}$.

$\mathbf{F}_{\alpha}^{(\alpha)}(\mathbf{z},t)$ is the true (or real) vector field in the α -domain. $\mathbf{F}_{\alpha}^{(\beta)}(\mathbf{z},t)$ is the fictitious (or imaginary) vector field in the α -domain, which is determined by the vector field in the β -domain. $\mathbf{F}_{0}^{(0)}(\mathbf{z},t)$ is the vector field on the separation boundary, and the discontinuity of the vector field for the entire system is presented through such an expression. $F_{\alpha}(\mathbf{z},t)$ is the scalar

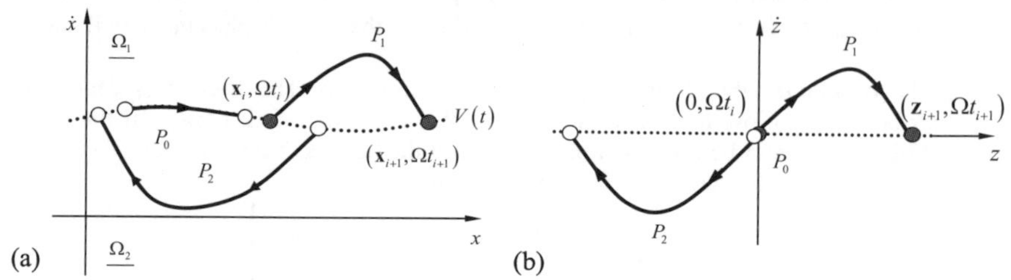

Fig.6 Basic mappings in (a) absolute and (b) relative frames for a linear oscillator with dry-friction.

force in the α -domain. For the system in Eq.(14), the forces in the two domains are for $\alpha \in \{1,2\}$,

$$F_{\alpha}(\mathbf{z},t) = A_0 \cos\Omega t - b_{\alpha} - 2d_{\alpha}\left[V(t) + \dot{z}(t)\right] - c_{\alpha}\left[X(t) + z(t)\right] - \dot{V}(t). \tag{20}$$

Note that $b_1 = \mu g,\ b_2 = -\mu g,\ d_{\alpha} = d$ and $c_{\alpha} = c$ for the model in Fig.1. From Luo [21], the grazing motion is guaranteed by

$$\left.\begin{aligned} &\left[\mathbf{n}_{\partial\Omega_{\alpha\beta}}^{T} \bullet \mathbf{F}_{\alpha}^{(\alpha)}(t_{m\pm})\right] = 0, \\ &\left[\mathbf{n}_{\partial\Omega_{12}}^{T} \bullet D\mathbf{F}_{1}^{(1)}(t_{m\pm})\right] > 0, \quad \left[\mathbf{n}_{\partial\Omega_{21}}^{T} \bullet D\mathbf{F}_{2}^{(2)}(t_{m\pm})\right] < 0 \end{aligned}\right\} \tag{21}$$

where

$$D\mathbf{F}_{\alpha}^{(\alpha)}(\mathbf{z},t) = \left(2F_{\alpha}(\mathbf{z},t), \nabla F_{\alpha}(\mathbf{z},t) \bullet \mathbf{F}_{\alpha}^{(\alpha)}(\mathbf{z},t) + \frac{\partial F_{\alpha}(\mathbf{z},t)}{\partial t}\right)^{T} \tag{22}$$

and

$$\mathbf{n}_{\partial\Omega_{ij}} = \nabla\varphi_{ij} = \left(\frac{\partial\varphi_{ij}}{\partial x}, \frac{\partial\varphi_{ij}}{\partial y}, 0\right)^{T}_{(x_m, y_m)}. \tag{23}$$

where $\nabla = \partial/\partial x\,\mathbf{i} + \partial/\partial y\,\mathbf{j}$ is the Hamiltonian operator. Notice that t_m represents the time for the motion on the velocity boundary and $t_{m\pm} = t_m \pm 0$ reflects the responses on the regions rather than the boundary. Using the third equation of Eq.(16), equation (22) gives

$$\mathbf{n}_{\partial\Omega_{12}} = \mathbf{n}_{\partial\Omega_{21}} = (0,1)^{T}. \tag{24}$$

Therefore, we have

$$\left.\begin{aligned} &\mathbf{n}_{\partial\Omega_{\alpha\beta}}^{T} \bullet \mathbf{F}_{\alpha}^{(\alpha)}(\mathbf{z},t) = F_{\alpha}(\mathbf{z},\Omega t), \quad \alpha \in \{1,2\} \\ &\mathbf{n}_{\partial\Omega_{\alpha\beta}}^{T} \bullet D\mathbf{F}_{\alpha}^{(\alpha)}(\mathbf{z},t) = \nabla F_{\alpha}(\mathbf{z},\Omega t) \bullet \mathbf{F}_{\alpha}^{(\alpha)}(\mathbf{z},t) + \frac{\partial F_{\alpha}(\mathbf{z},\Omega t)}{\partial t}. \end{aligned}\right\} \tag{25}$$

From Eqs.(20) and (21), the conditions for grazing motions are:

$$\left.\begin{aligned} &F_{\alpha}(\mathbf{z}_m, \Omega t_{m\pm}) = 0, \\ &F_{\alpha}(\mathbf{z}_m, \Omega t_{m-\varepsilon}) \times F_{\alpha}(\mathbf{z}_m, \Omega t_{m+\varepsilon}) < 0. \end{aligned}\right\} \tag{26}$$

or

$$F_\alpha(\mathbf{z}_m,\Omega t_{m\pm})=0,\ \nabla F_\alpha(\mathbf{z}_m,\Omega t_{m\pm})\bullet\mathbf{F}_\alpha^{(\alpha)}(\mathbf{z}_m,t_{m\pm})+\frac{\partial F_\alpha(\mathbf{z}_m,\Omega t_{m\pm})}{\partial t}\begin{Bmatrix}>0, \text{ for } \alpha=1,\\<0, \text{ for } \alpha=2.\end{Bmatrix} \tag{27}$$

The grazing conditions are presented in Fig.5 (a) and (b), and the vector fields in Ω_1 and Ω_2 are expressed by the dashed and solid arrow-lines, respectively. The condition in Eq.(20) for the grazing motion in Ω_α is presented through the vector fields of $\mathbf{F}_\alpha^{(\alpha)}(t)$. In addition to $F_\alpha(\mathbf{z}_m,t_{m\pm})=0$, the sufficient condition requires that $F_1(\mathbf{z}_m,\Omega t_{m-\varepsilon})<0$ and $F_1(\mathbf{z}_m,\Omega t_{m+\varepsilon})>0$ in domain Ω_1; and $F_2(\mathbf{z}_m,\Omega t_{m-\varepsilon})>0$ and $F_2(\mathbf{z}_m,\Omega t_{m+\varepsilon})<0$ in domain Ω_2.

4. Grazing Conditions

After grazing motion, the sliding motion will appear. Direct integration of Eq.(8) with the initial condition $(t_i,z_i,V(t_i))$ gives the sliding motion, i.e., Eq.(9). Substitution of Eq.(8, 9) into (14) gives the forces for the small δ - neighborhood of the stick motion ($\delta\to 0$) in the two domains Ω_α ($\alpha\in\{1,2\}$). To produce non-stick motion, select the initial condition on the velocity boundary (i.e., $\dot{x}_i=V(t_i)$), and then select the coefficients of the solution in the Appendix, $C_k^{(\alpha)}(x_i,\dot{x}_i,t_i)\triangleq C_k^{(\alpha)}(x_i,t_i)$ for $k=1,2$. The basic solutions in the Appendix will be used for construction of mappings.

In the phase plane, the trajectories in Ω_α starting and ending at the velocity discontinuity (i.e., from $\partial\Omega_{\beta\alpha}$ to $\partial\Omega_{\alpha\beta}$) are illustrated in Fig.6. The starting and ending points for mappings P_α in Ω_α are (x_i,V,t_i) and (x_{i+1},V,t_{i+1}), respectively. The stick mapping is P_0. The switching planes are defined as,

$$\left.\begin{aligned}\Xi^0&=\{(x_i,\Omega t_i)|\dot{x}_i(t_i)=V(t_i)\},\\ \Xi^1&=\{(x_i,\Omega t_i)|\dot{x}_i(t_i)=V^+(t_i)\},\\ \Xi^2&=\{(x_i,\Omega t_i)|\dot{x}_i(t_i)=V^-(t_i)\};\end{aligned}\right\} \tag{28}$$

where $V^-(t)=\lim_{\delta\to 0}(V(t)-\delta)$ and $V^+(t)=\lim_{\delta\to 0}(V(t)+\delta)$ for arbitrarily small $\delta>0$. Therefore,

$$P_1:\Xi^1\to\Xi^1,\ P_2:\Xi^2\to\Xi^2,\ P_0:\Xi^0\to\Xi^0. \tag{29}$$

From the previous two equations, we have

$$\left.\begin{aligned}P_0&:(x_i,V(t_i),t_i)\to(x_{i+1},V(t_{i+1}),t_{i+1}),\\ P_1&:(x_i,V^+(t_i),t_i)\to(x_{i+1},V^+(t_{i+1}),t_{i+1}),\\ P_2&:(x_i,V^-(t_i),t_i)\to(x_{i+1},V^-(t_{i+1}),t_{i+1}).\end{aligned}\right\} \tag{30}$$

The governing equations for P_0 with $\alpha\in\{1,2\}$ are

$$\left.\begin{aligned}&x_{i+1}-V_0[\sin(t_{i+1}+\beta)-\sin(t_i+\beta)]-V_1\times(t_{i+1}-t_i)-x_i=0,\\ &A_0\cos\Omega t_{i+1}-b_\alpha-2d_\alpha V(t_{i+1})-\dot{V}(t_{i+1})-c_\alpha x_{i+1}=0.\end{aligned}\right\} \tag{31}$$

The mapping P_0 describes the starting and ending of stick motion, where the disappearance of stick motion requires $F_\alpha(\mathbf{z}_{i+1},\Omega t_{i+1})=0$. This paper will not use this mapping, which is presented herein as a generic mapping. For sliding motions, this mapping will be used, and such a discussion is arranged for another paper. From this problem, the two domains Ω_α ($\alpha\in\{1\text{ or }2\}$) are unbounded. However, the flows of the dynamical systems on the corresponding domains should be bounded from Assumptions (A1-A3). Therefore, for non-stick motion, there are three possible stable motions in the two domains Ω_α

($\alpha \in \{1,2\}$). The governing equations of mapping P_α ($\alpha \in \{1,2\}$) are obtained from the displacement and velocity responses for the three cases of motions in the Appendix. Therefore, the governing equations of mapping P_α ($\alpha \in \{0,1,2\}$) can be expressed by

$$\left.\begin{aligned} f_1^{(\alpha)}(x_i, \Omega t_i, x_{i+1}, \Omega t_{i+1}) = 0, \\ f_2^{(\alpha)}(x_i, \Omega t_i, x_{i+1}, \Omega t_{i+1}) = 0. \end{aligned}\right\} \tag{32}$$

If grazing for two non-stick mappings occur at the final state (x_{i+1}, V, t_{i+1}), from Eq.(31), the grazing conditions based on mappings are obtained. With Eq.(27), the grazing condition becomes

$$\left.\begin{aligned} A_0 \cos\Omega t - b_\alpha - 2d_\alpha \dot{x}_{i+1} - c_\alpha x_{i+1} - \dot{V}(t) = 0, \\ -2d_\alpha \dot{V}(t) - c_\alpha V(t) - \ddot{V}(t) - A_0 \Omega \sin\Omega t_{i+1} \begin{cases} > 0 & \text{for } \alpha = 1, \\ < 0 & \text{for } \alpha = 2. \end{cases} \end{aligned}\right\} \tag{33}$$

The grazing conditions for the two non-stick mappings can be illustrated through a range of parameters. The grazing conditions in Eq.(32) are given through the forces. Hence, both the initial and final switching sets of the two non-stick mappings will vary with system parameters. Because the grazing characteristics of the two non-stick mappings are different, illustrations of grazing conditions for the two mappings will be separated. The grazing conditions are computed through Eqs. (31) and (32). Three equations plus an inequality with four unknowns requires one unknown be held constant. In all illustrations, the initial displacement of mapping $P_\alpha(\alpha \in \{1,2\})$ will be fixed to specific values. The resulting three equations describe the initial switching phase, the final switching phase and displacement of mapping $P_\alpha(\alpha \in \{1,2\})$. To ensure the initial switching sets are passable, from Luo [21, 22], the initial switching sets of mapping $P_\alpha(\alpha \in \{1,2\})$ should satisfy the following condition as in [1],

$$F_1(\mathbf{z}_m, \Omega t_{m-\varepsilon}) \times F_2(\mathbf{z}_m, \Omega t_{m+\varepsilon}) > 0. \tag{34}$$

The comprehensive discussion of this condition can be referred to in Luo and Gegg [23, 25]. The condition of Eq.(34) guarantees the initial switching sets of mapping P_α ($\alpha \in \{1,2\}$) are passable on the discontinuous boundary (i.e., $y_i = V(t_i)$). The force product for the initial switching sets is also illustrated to ensure the non-stick mapping exists. The force conditions for the final switching sets of mapping $P_\alpha(\alpha \in \{1,2\})$ is presented in Eq.(14). However, the equivalent grazing conditions based on Eq.(15) give the inequality condition in Eq.(32), which is already embedded in the program for computation of the grazing. Therefore, such a force product of the final switching sets of the two mappings will not be presented.

5. Illustrations

The spring and damper parameters ($d_1 = 1,\ d_2 = 0.1,\ c_1 = c_2 = 30$) are fixed as constants and the external parameters are varied. Consider the grazing variation of mapping P_1 with the belt excitation amplitude (V_0) for the specified parameters ($\Omega = 1,\ A_0 = 100,\ d_1 = 1,\ d_2 = 0.1,\ b_1 = -b_2 = 0.5,\ c_1 = c_2 = 30$). For the initial displacements of $x_i = \{-2, -2.5, \ldots, -5.5\}$, the initial switching force products and phase, and the final switching displacement and phase versus the belt speed are illustrated in Fig.7(a-d), respectively. The initial switching force product on the discontinuous boundary in Fig.7(a) is always positive, which is required in Eq.(33). For $x_i = -2$ and -3, the minimum value of the initial switching force product approaches zero. Also, by reducing the belt excitation amplitude towards zero results in the special case where the belt oscillation is zero. Large values of the belt excitation amplitude produce a wide range of grazing solutions but for practical purposes those illustrated are in the range of $V_0 \in [0,5]$.

The upper bound of the grazing solutions is determined largely by the initial force product, Fig.7(a). The lower bound is the initial and/or final zero-force-product grazing defined by the second Eq. of Eq.(33) is zero. The solution boundary can be verified through Eq.(33). The grazing solutions vanish when the second equation of Eq.(32) is no longer satisfied.

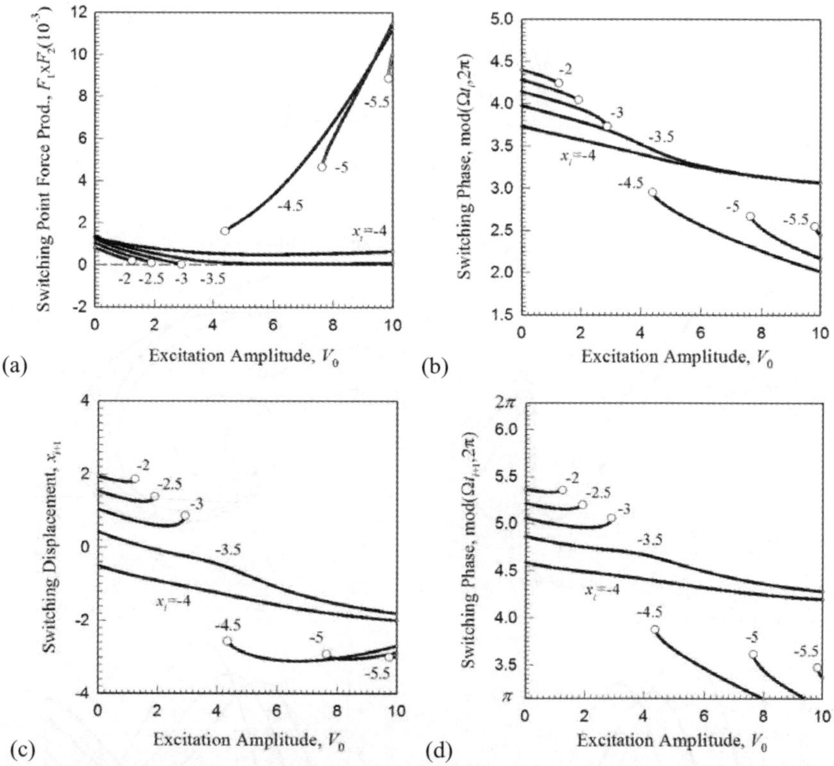

Fig.7 Belt Excitation Amplitude V_0 Illustration: (a) initial switching force product and (b) initial switching phase, (c) final switching displacement, and (d) final switching phase varying with belt excitation amplitude, V_0, ($V_1 = 1$, $A_0 = 100$, $d_1 = 1, d_2 = 0.1$, $b_1 = -b_2 = 0.5$, $c_1 = c_2 = 30$).

6. Numerical Simulations

The grazing motion characteristics of this linear oscillator with dry-friction in parameter space have been systematically investigated. The initial, switching manifolds of grazing and the grazing manifold are presented. To verify the analytical predictions of the grazing motion, the motion responses of the oscillator will be demonstrated through time-history responses and phase space. The grazing strongly depends on the force responses in this discontinuous dynamical system. The force responses will be presented to illustrate the force criteria for the grazing motions in such a friction-induced oscillator. For illustrations, the starting and grazing points of mapping P_α ($\alpha \in \{1,2\}$) are represented by the large, hollow and dark-solid circular symbols, respectively.

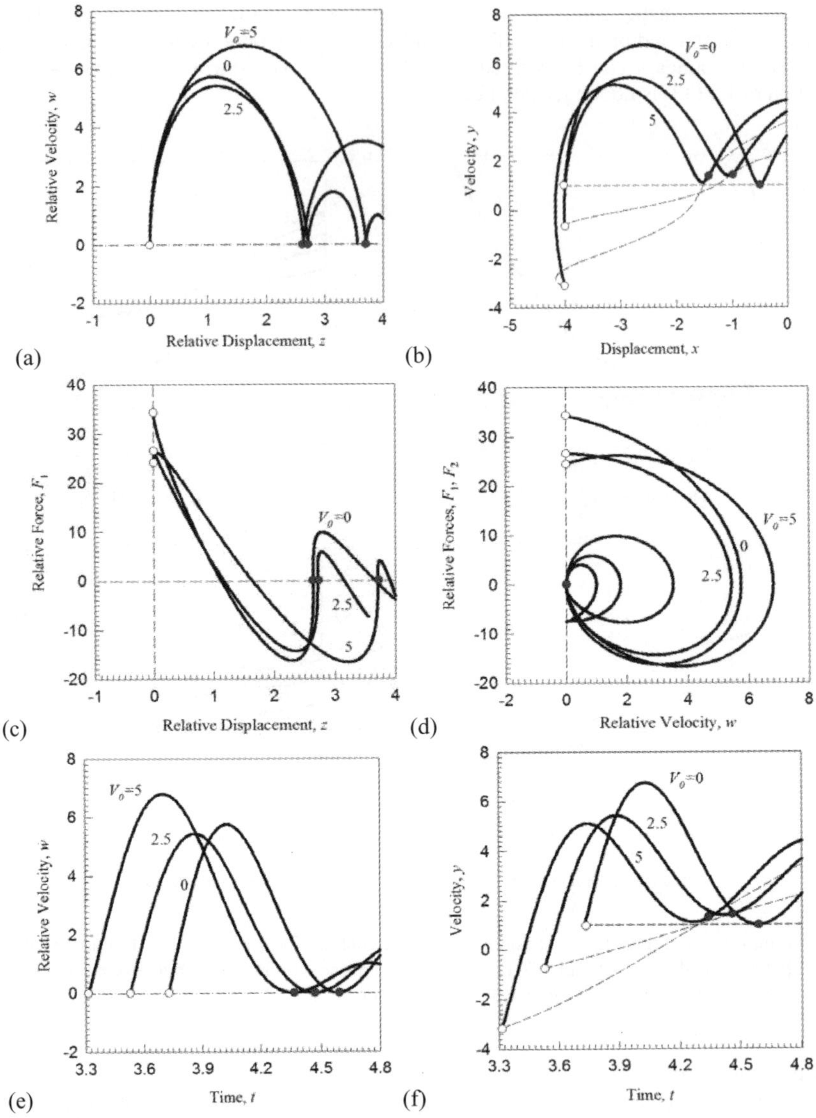

Fig.8 Grazing motion of mapping P_1 for $V_0 = \{0, 2.5, 5\}$: (a) relative phase, (b) absolute phase, (c) force distribution with dis-placement, (d) force distribution along velocity, (e) relative velocity time history, and (f) absolute velocity time history. ($V_1 = 1$, $A_0 = 100$, $d_1 = 1, d_2 = 0.1$, $b_1 = -b_2 = 0.5$, $c_1 = c_2 = 30$).

The switching points from the domain α to the domain β ($\{\alpha, \beta\} \in \{1, 2\}$, $\alpha \neq \beta$) are denoted by the smaller circular symbols. In Fig.8(a-f), phase trajectories, force distribution along displacement, force distribution along velocity, and velocity time-histories are illustrated for the grazing motion of mapping

P_1, respectively. The parameters ($\Omega=1,\ A_0=100,\ V_1=1,\ d_1=1,\ d_2=0.1,\ b_1=-b_2=0.5,\ c_1=c_2=30$) plus the initial conditions $(x_i, y_i)=(-4., V(t_i))$ and $\Omega t_i \approx \{3.7297,\ 3.5288,\ 3.3155\}$ corresponding to $V_0=\{0,2.5,5\}$ are used. In the relative and absolute phase plane, the three grazing trajectories are tangential to the discontinuous boundary ($\dot{z}=w=0$), which can be seen in Fig.8(a,b), respectively. In Fig.8(c), the curve represents the force $F_1(t)$ distribution along displacement. The location denoted by a dark circular symbol is the grazing point where the force $F_1(t)$ has a sign change from negative to positive.

The force distribution along velocity is presented in Fig.8(d), where the grazing point is observed to coincide with the zero relative velocity and zero force, $F_1(t)$. This indicates the first equation of Eq.(33) is satisfied. In the velocity time-history plot, the velocity curves are tangential to the discontinuous boundary (see Fig.8(e,f)). The oscillation amplitude (V_0) of the belt velocity can be observed to impact the grazing motion.

7. Conclusion

The model investigated by Luo and Gegg [1, 23, 24] has been modified with an excitation of the belt velocity. The new model was studied through the local theory of non-smooth dynamical systems on the connectable and accessible domains. The necessary and sufficient conditions for the grazing motions in a dry-friction oscillator were developed through this theory. The sufficient and necessary conditions for the grazing motions were also expressed for grazing at the final point of mappings. The initial and grazing, switching manifolds for grazing mapping were defined. This paper provides a systematic investigation of grazing motion in the dry-friction oscillator. The intuitive analysis of the grazing of discontinuous systems is straightforward, in application, to engineering problems.

Appendix

Solution for Eq.(6) in two regions Ω_j ($j\in\{1,2\}$) are for Case I (i.e., $d_j^2>c_j$):

$$x^{(j)}(t)=C_1^{(j)}(x_i,\dot{x}_i,t_i)e^{\lambda_1^{(j)}(t-t_i)}+C_2^{(j)}(x_i,\dot{x}_i,t_i)e^{\lambda_2^{(j)}(t-t_i)}+A^{(j)}\cos\Omega t+B^{(j)}\sin\Omega t+C^{(j)}, \quad \text{(A1)}$$

$$\dot{x}^{(j)}(t)=\lambda_1 C_1^{(j)}(x_i,\dot{x}_i,t_i)e^{\lambda_1^{(j)}(t-t_i)}+\lambda_2 C_2^{(j)}(x_i,\dot{x}_i,t_i)e^{\lambda_2^{(j)}(t-t_i)}-A^{(j)}\Omega\sin\Omega t+B^{(j)}\Omega\cos\Omega t; \quad \text{(A2)}$$

$$\left.\begin{aligned}
&\lambda_{1,2}^{(j)}=-d_j\pm\sqrt{d_j^2-c_j},\ \omega_d^{(j)}=\sqrt{d_j^2-c_j}\\
&C_1^{(j)}(x_i,\dot{x}_i,t_i)=\frac{1}{2\omega_d^{(j)}}\Big\{-\big[B^{(j)}\Omega+(d_j+\omega_d^{(j)})A^{(j)}\big]\cos\Omega t_i+\big[A^{(j)}\Omega-(d_j+\omega_d^{(j)})B^{(j)}\big]\sin\Omega t_i\\
&\qquad +\dot{x}_i-(d_j+\omega_d^{(j)})(C^{(j)}-x_i)\Big\},\\
&C_2^{(j)}(x_i,\dot{x}_i,t_i)=\frac{1}{2\omega_d^{(j)}}\Big\{\big[B^{(j)}\Omega-(\omega_d^{(j)}-d_j)A^{(j)}\big]\cos\Omega t_i-\big[(\omega_d^{(j)}-d_j)B^{(j)}+A^{(j)}\Omega\big]\sin\Omega t_i\\
&\qquad -\dot{x}_i+(\omega_d^{(j)}-d_j)(x_i-C^{(j)})\Big\}.
\end{aligned}\right\} \quad \text{(A3)}$$

The solution for Case II (i.e., $d_j^2<c_j$)

$$\begin{aligned}x^{(j)}(t)=e^{-d_j(t-t_i)}\big[C_1^{(j)}(x_i,\dot{x}_i,t_i)\cos\omega_d^{(j)}(t-t_i)+C_1^{(j)}(x_i,\dot{x}_i,t_i)\sin\omega_d(t-t_i)\big]\\
+A^{(j)}\cos\Omega t+B^{(j)}\sin\Omega t+C^{(j)},\end{aligned} \quad \text{(A4)}$$

$$\dot{x}^{(j)}(t)=\Big\{\Big[\omega_d C_2^{(j)}(x_i,\dot{x}_i,t_i)-d_j C_1^{(j)}(x_i,\dot{x}_i,t_i)\Big]\cos\omega_d(t-t_i) -\Big[\omega_d C_1^{(j)}(x_i,t_i)+d_j C_2^{(j)}(x_i,\dot{x}_i,t_i)\Big]\sin\omega_d(t-t_i)\Big\}\times e^{-d_j(t-t_i)} -A^{(j)}\Omega\sin\Omega t+B^{(j)}\Omega\cos\Omega t; \tag{A5}$$

$$\left.\begin{aligned} &\omega_d^{(j)}=\sqrt{c_j-d_j^2},\\ &C_1^{(j)}(x_i,\dot{x}_i,t_i)=x_i-A^{(j)}\cos\Omega t_i-B^{(j)}\sin\Omega t_i-C^{(j)},\\ &C_2^{(j)}(x_i,\dot{x}_i,t_i)=\frac{1}{\omega_d^{(j)}}\Big[\dot{x}_i-\big(d_jA+B\Omega\big)\cos\Omega t_i-\big(d_jB^{(j)}-A^{(j)}\Omega\big)\sin\Omega t_i+d_j\big(x_i-C^{(j)}\big)\Big]. \end{aligned}\right\} \tag{A6}$$

The solution for Case III (i.e., $d_j^2=c_j$) is

$$x^{(j)}(t)=\Big[C_1^{(j)}(x_i,\dot{x}_i,t_i)+C_2^{(j)}(x_i,\dot{x}_i,t_i)\times(t-t_i)\Big]e^{\lambda_1^{(j)}(t-t_i)}+A^{(j)}\cos\Omega t+B^{(j)}\sin\Omega t+C^{(j)}, \tag{A7}$$

$$\dot{x}^{(j)}(t)=\Big[\lambda_1^{(j)}C_1^{(j)}(x_i,\dot{x}_i,t_i)+C_2^{(j)}(x_i,\dot{x}_i,t_i)\Big]e^{\lambda_1^{(j)}(t-t_i)}+\lambda_1^{(j)}C_2^{(j)}(x_i,\dot{x}_i,t_i)\times(t-t_i)e^{\lambda_1^{(j)}(t-t_i)} -A^{(j)}\Omega\sin\Omega t+B^{(j)}\Omega\cos\Omega t, \tag{A8}$$

$$\left.\begin{aligned} &\lambda_1^{(j)}=-2d_j,\\ &C_1^{(j)}(x_i,\dot{x}_i,t_i)=x_i-A^{(j)}\cos\Omega t_i-B^{(j)}\sin\Omega t_i-C^{(j)},\\ &C_2^{(j)}(x_i,\dot{x}_i,t_i)=\dot{x}_i+d_jx_i+\big(A^{(j)}\Omega-d_jB^{(j)}\big)\sin\Omega t_i-\big(d_jA^{(j)}+B^{(j)}\Omega\big)\cos\Omega t_i-d_jC^{(j)}. \end{aligned}\right\} \tag{A9}$$

$$\left. A^{(j)}=\frac{A_0\left(c_j-\Omega^2\right)}{\left(c_j-\Omega^2\right)^2+\left(2d_j\Omega\right)^2},\ B^{(j)}=\frac{2d_j\Omega A_0}{\left(c_j-\Omega^2\right)^2+\left(2d_j\Omega\right)^2},\ C^{(j)}=-\frac{b_j}{c_j}.\right\} \tag{A10}$$

References

[1] Luo, A.C.J. and Gegg, B.C., 2004, "On the mechanism of stick and non-stick periodic motion in a forced oscillator including dry-friction", *Proceeding of IMECE 2004*, 2004 ASME International Mechanical Engineering Congress & Exposition, November 13-19, 2004, Anaheim, California. IMECE2004-59218.

[2] Den Hartog, J.P., 1931, "Forced vibrations with Coulomb and viscous damping", *Transactions of the American Society of Mechanical Engineers*, 53, pp.107-115.

[3] Levitan, E.S., 1960, "Forced oscillation of a spring-mass system having combined Coulomb and viscous damping", *Journal of the Acoustical Society of America*, 32, pp.1265-1269.

[4] Filippov, A.F., 1964, "Differential equations with discontinuous right-hand side", *American Mathematical Society Translations, Series 2*, 42, pp. 199-231.

[5] Filippov,A.F., *Differential Equations with Discontinuous Righthand Sides*, Dordrecht: Kluwer Academic Publishers, 1988.

[6] Hundal, M.S. 1979, "Response of a base excited system with Coulomb and viscous friction", *Journal of Sound and Vibration,* 64, pp.371-378.

[7] Shaw, S.W., 1986, "On the dynamic response of a system with dry-friction", *Journal of Sound and Vibration*, 108, pp.305-325.

[8] Luo, A.C.J. *Analytical Modeling of Bifurcations, Chaos and Fractals in Nonlinear Dynamics.* Ph.D. Dissertation, University of Manitoba, Winnipeg, Canada, 1995.

[9] Han, R.P.S., Luo, A.C.J. and Deng, W. 1995, "Chaotic motion of a horizontal impact pair", *Journal of Sound and Vibration*, 181, pp.231-250.

[10] Luo, A.C.J., 2002, "An unsymmetrical motion in a horizontal impact oscillator", *ASME Journal of Vibrations and Acoustics*, 124, pp.420-426.

[11] Menon, S. and Luo, A.C.J., 2005, "An analytical prediction of the global period-1 motion in a periodically forced, piecewise linear system", *International Journal of Bifurcation and Chaos*, 15, pp.1945-1957.

[12] Luo, A.C.J. and Menon, S., 2004, "Global Chaos in a periodically forced, linear system with a dead-zone restoring force", *Chaos, Solitons and Fractals*, 19, pp.1189-1199.

[13] Luo, A.C.J., 2005, "The mapping dynamics of periodic motions for a three-piecewise linear system under a periodic excitation", *Journal of Sound and Vibration*, 283,pp.723-748.

[14] Feigin, M.I., 1970, "Doubling of the oscillation period with C-bifurcation in piecewise-continuous systems", *PMM*, 34, pp.861-869.

[15] Feigin, M.I., 1995, "The increasingly complex structure of the bifurcation tree of a piecewise-smooth system", *Journal of Applied Mathematics and Mechanics*, 59, pp. 853-863.

[16] di Bernardo, M., Feigin, M.I., Hogan, S.J. and Homer, M.E., 1999, "Local analysis of C-bifurcations in *n*-dimensional piecewise-smooth dynamical systems", *Chaos, Solitons & Fractals*, 10, pp.1881-1908.

[17] Nordmark, A.B., 1991, "Non-periodic motion caused by grazing incidence in an impact oscillator", *Journal of Sound and Vibration*, 145, pp. 279-297.

[18] di Bernaedo, M., Budd, C.J. and Champney, A.R., 2001, "Grazing and Border-collision in piecewise-smooth systems: a unified analytical framework", *Physical Review Letters*, 86, pp.2553-2556.

[19] di Bernardo, M., Budd, C.J. and Champneys, A.R., 2001, "Normal form maps for grazing bifurcation in n-dimensional piecewise-smooth dynamical systems", *Physica D*, 160, pp.222-254.

[20] di Bernardo, M., Kowalczyk, P., Nordmark, A.B., 2002, "Bifurcation of dynamical systems with sliding: derivation of normal form mappings" *Physica D*, 170, pp.175-205.

[21] Luo, A.C.J., 2005, "A theory for non-smooth dynamical systems on connectable domains", *Communication in Nonlinear Science and Numerical Simulation*, 10, pp.1-55.

[22] Luo, A.C.J., 2005, "Imaginary, sink and source flows in the vicinity of the separatrix of non-smooth dynamic system", *Journal of Sound and Vibration*, 285,pp.443-456.

[23] Luo, A.C.J. and Gegg. B.C., 2006, "Stick and non-stick periodic motions in a periodically forced, Linear oscillator with dry-friction", *Journal of Sound and Vibration*, 291, pp.132-168. .

[24] Luo, A.C.J. and Gegg. B.C., 2006, "*Grazing Phenomena in a Periodically Forced, Friction-Induced, Linear Oscillator*", *Communications in Nonlinear Science and Numerical Simulation*, 11, pp.777-802.

[25] Luo, A.C.J. and Gegg. B.C., 2006, "Dynamics of a Harmonically Excited Oscillator with Dry-Friction on a Sinusoidally Time-Varying, Traveling Surface", *International Journal of Bifurcation and Chaos*, in press.

Complex dynamics in the trajectory control of redundant manipulators

Maria da Graça Marcos [1], Fernando B. M. Duarte [*], J. A. Tenreiro Machado [2]

[1]*Instituto Politécnico do Porto, Instituto Superior de Engenharia, Dep. Matemática, Rua António Bernardino de Almeida, 4200 Porto, Portugal, mgm@isep.ipp.pt*

[2]*Instituto Politécnico do Porto, Instituto Superior de Engenharia, Dep. Engenharia Electrotécnica, Rua António Bernardino de Almeida, 4200 Porto, Portugal, jtm@.isep.ipp.pt*

Abstract

Redundant manipulators allow the trajectory optimization, the obstacle avoidance, and the resolution of singularities. For this type of manipulators, the kinematic control algorithms adopt generalized inverse matrices that may lead to unpredictable responses. Motivated by these problems this paper studies the complexity revealed by the trajectory planning scheme when controlling redundant manipulators. The results reveal fundamental properties of the chaotic phenomena and give a deeper insight towards the development of superior trajectory control algorithms.

PACS: 02.30.Nw; 03.20.+i

Keywords: Redundant Manipulators; Kinematics; Fourier Transform; Windowed Fourier Transform

1. Introduction

A kinematically redundant manipulator is a robotic arm possessing more degrees of freedom (*dof*) than those required to establish an arbitrary position and orientation of the gripper. Redundant manipulators offer several potential advantages over non-redundant arms. The extra degrees of freedom can be used to move around or between obstacles and, thereby, they can manipulate objects in situations that otherwise would be inaccessible [1-4].

When a manipulator is redundant it is anticipated that the inverse kinematics admits an infinite number of solutions. This implies that, for a given location of the manipulator's gripper, it is possible to induce a self-motion of the structure without changing the location of the end effecter. Therefore, the arm can be reconfigured to find better postures for an assigned set of task requirements.

Several techniques for trajectory planning of redundant manipulators control the gripper through the joint rates using the pseudoinverse of the Jacobian [3, 6]. Nevertheless, these algorithms lead to a kind of chaotic motion with unpredictable arm configurations. Having these ideas in mind, the paper is organized as follows. Section 2 develops the formalism for the matrix generalized inverses. Section 3 introduces the fundamental issues for the kinematics of redundant manipulators. Based on these concepts, section 4

[*]Corresponding author: T*el:*+351232480500;*fax*:+351232424651, *e-mail:* fduarte@mat.estv.ipv.pt

The authors would like to acknowledge FCT, FEDER, POCTI, POSI, POCI and POSC for their support to R&D Projects and GECAD Unit.

presents the trajectory control of the 3 *dof* robot. The results reveal a chaotic behavior that is further analyzed in section 5. Finally, section 6 draws the main conclusions.

2. Generalized Inverses

For $\mathbf{A} \in \Re^{m\times n}$ and $\mathbf{X} \in \Re^{n\times m}$ the Penrose conditions:

$$\mathbf{AXA} = \mathbf{A} \tag{1}$$

$$\mathbf{XAX} = \mathbf{X} \tag{2}$$

$$(\mathbf{AX})^T = \mathbf{AX} \tag{3}$$

$$(\mathbf{XA})^T = \mathbf{XA} \tag{4}$$

lead to the definitions:

- A *generalized inverse* of matrix **A** is a matrix $\mathbf{X} = \mathbf{A}^- \in \Re^{n\times m}$ satisfying condition (1);
- A *reflexive generalized inverse* of matrix **A** is a matrix $\mathbf{X} = \mathbf{A}_r^- \in \Re^{n\times m}$ satisfying conditions (1)-(2);
- A *pseudoinverse* of a matrix **A** (*Moore-Penrose inverse*) is a matrix $\mathbf{X} = \mathbf{A}^{\#} \in \Re^{n\times m}$ satisfying conditions (1)-(4).

For a matrix $\mathbf{A} \in \Re^{m\times n}$:

$$\mathbf{A}^{\#} = \begin{cases} \mathbf{A}^T\left(\mathbf{AA}^T\right)^{-1} \text{ and } \mathbf{AA}^T \text{ is nonsingular} & \Leftarrow m < n \text{ and } r(\mathbf{A}) = m \\ \left(\mathbf{A}^T\mathbf{A}\right)^{-1}\mathbf{A}^T \text{ and } \mathbf{A}^T\mathbf{A} \text{ is nonsingular} & \Leftarrow m > n \text{ and } r(\mathbf{A}) = n \\ (\mathbf{A})^{-1} & \Leftarrow m = n \text{ and } r(\mathbf{A}) = n \end{cases} \tag{5}$$

3. Kinematics of redundant manipulators

A kinematically redundant manipulator has more *dof* than those required to establish an arbitrary position and orientation of the gripper. Fig. 1 depicts a general kinematic structure of a robot with k rotational (R) joints.

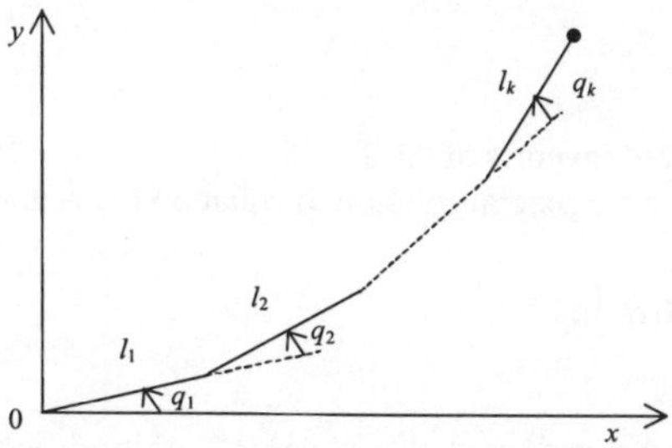

Fig. 1. A *kR* redundant planar manipulator.

When a manipulator is redundant it is anticipated that the inverse kinematics admits an infinite number of solutions. This implies that it is possible to induce a self-motion of the structure without changing the location of the gripper. Therefore, redundant manipulators can be reconfigured to find better postures for an assigned set of task requirements but, on the other hand, have a more complex structure requiring adequate control algorithms.

We consider a manipulator with n degrees of freedom whose joint variables are denoted by $\mathbf{q} = [q_1, q_2, ..., q_n]^T$. We assume that a class of tasks we are interested in can be described by $m < n$ variables, $\mathbf{x} = [x_1, x_2, ..., x_m]^T$ and that the relation between $\mathbf{q}$ and $\mathbf{x}$ is given by the direct kinematics:

$$\mathbf{x} = f(\mathbf{q}) \tag{6}$$

Differentiating (6) with respect to time yields:

$$\dot{\mathbf{x}} = \mathbf{J}(\mathbf{q})\dot{\mathbf{q}} \tag{7}$$

where $\dot{\mathbf{x}} \in \Re^m$, $\dot{\mathbf{q}} \in \Re^n$ and $\mathbf{J}(\mathbf{q}) = \partial f(\mathbf{q})/\partial \mathbf{q} \in \Re^{m \times n}$. Hence, it is possible to calculate a path $\mathbf{q}(t)$ in terms of a prescribed trajectory $\mathbf{x}(t)$ in the operational space. We assume that the following condition is satisfied:

$$\max \operatorname{rank} \{\mathbf{J}(\mathbf{q})\} = m \tag{8}$$

Failing to satisfy this condition usually means that the selection of manipulation variables is redundant and the number of these variables m can be reduced. When condition (8) is satisfied, we say that the degree of redundancy of the manipulator is $n{-}m$. If, for some $\mathbf{q}$

$$\operatorname{rank} \{\mathbf{J}(\mathbf{q})\} < m \tag{9}$$

then the manipulator is in a singular state. This state is not desirable because, in this region of the trajectory, the manipulating ability is very limited. Based on these concepts, to analyze and quantify the problem of object manipulation it was proposed [4] the expression $\mu = \left[det\left(\mathbf{J}\mathbf{J}^T\right)\right]^{1/2}$ as a measure of the manipulability.

Most of the approaches for solving redundancy that have been proposed [5, 8] are based on the inversion of equation (7). A solution in terms of the joint velocities is sought as:

$$\dot{\mathbf{q}} = \mathbf{J}^{\#}(\mathbf{q})\dot{\mathbf{x}} \tag{10}$$

where $\mathbf{J}^{\#}$ is one of the generalized inverses of the $\mathbf{J}$ [8, 9].

It can be easily shown that a more general solution to equation (7) is given by:

$$\dot{\mathbf{q}} = \mathbf{J}^{+}(\mathbf{q})\dot{\mathbf{x}} + \left[\mathbf{I} - \mathbf{J}^{+}(\mathbf{q})\mathbf{J}(\mathbf{q})\right]\dot{\mathbf{q}}_0 \tag{11}$$

where $\mathbf{I}$ is the $n \times n$ identity matrix and $\dot{\mathbf{q}}_0 \in \Re^n$ is a $m \times 1$ arbitrary joint velocity vector and $\mathbf{J}^{+}$ is the pseudoinverse of the $\mathbf{J}$. The solution (11) is composed of two terms. The first term is relative to minimum norm joint velocities. The second term, the *homogeneous solution*, attempts to satisfy the additional constraints specified by $\dot{\mathbf{q}}_0$. Moreover, the matrix $\mathbf{I}-\mathbf{J}^{+}(\mathbf{q})\mathbf{J}(\mathbf{q})$ allows the projection of $\dot{\mathbf{q}}_0$ in the null space of **J**. A direct consequence is that it is possible to generate internal motions that reconfigure the manipulator structure without changing the gripper position and orientation [7-11]. Another aspect revealed by the solution of (10) is that repetitive trajectories in the operational space do not lead to periodic trajectories in the joint space. This is an obstacle for the solution of many tasks because the resultant robot configurations have similarities with those of a chaotic system.

4. Robot trajectory control

The direct kinematics and the Jacobian of a 3-link planar manipulator has a simple recursive nature according with the expressions:

$$\begin{bmatrix} x \\ y \end{bmatrix} = \begin{bmatrix} l_1C_1 + l_2C_{12} + l_3C_{123} \\ l_1S_1 + l_2S_{12} + l_3S_{123} \end{bmatrix} \tag{12.a}$$

$$\mathbf{J} = \begin{bmatrix} -l_1S_1 - \ldots - l_3S_{123} & \cdots & -l_3S_{123} \\ l_1C_1 + \ldots + l_3C_{123} & \cdots & l_3C_{123} \end{bmatrix} \tag{12.b}$$

where l_i is the length of link i, $q_{i\ldots k} = q_i + \ldots + q_k$, $S_{i\ldots k} = Sin(q_{i\ldots k})$ and $C_{i\ldots k} = Cos(q_{i\ldots k})$.

During all the experiments it is considered $\Delta t = 10^{-3}$ sec, $L_{TOT} = l_1 + l_2 + l_3 = 3$, $l_1 = l_2 = l_3$.

In the closed-loop pseudoinverse's method the joint positions can be computed through the time integration of the velocities according with the block diagram depicted in Figure 2.

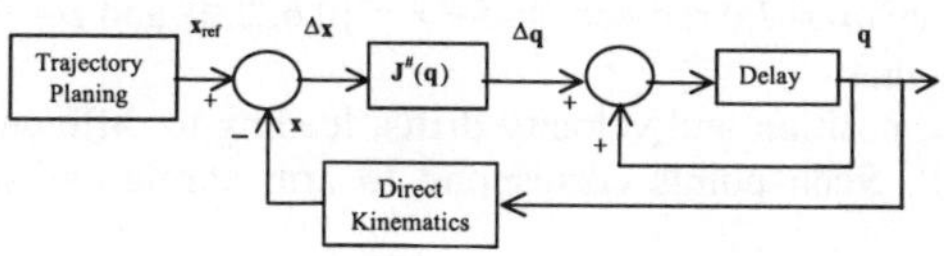

Fig. 2. Block diagram of the closed-loop inverse kinematics algorithm with the pseudoinverse.

Based on equation (12) we analyze the kinematic performances of the 3*R*-robot when repeating a circular motion in the operational space with frequency $\omega_0 = 7.0$ rad sec^{-1}, centre at $r = [x^2+y^2]^{1/2}$ and radius ρ.

Figures 3 show the joint positions and the manipulability μ for the inverse kinematic algorithm (10) for $\rho = \{0.3, 0.5\}$, when $r = 0.6$ and $r = 2.0$, respectively.

We observe that:

- For $r = 0.6$ occur unpredictable motions with severe variations that lead to high joint transients [12]. Moreover, we verify a low frequency signal modulation that depends on the circle being executed. The different fractional order harmonics (*foh*) are visible in the time response but, in order to capture each *foh*, it is required to adopt a specific time window.
- For $r = 2.0$ the motion is periodic with frequency identical to $\omega_0 = 7.0$ rad sec^{-1}.

In what concerns the index of manipulability we conclude that, for $r = 0.6$ it is, during some instants, very close to $\mu = 0$, while for $r = 2.0$ is always $\mu > 2$.

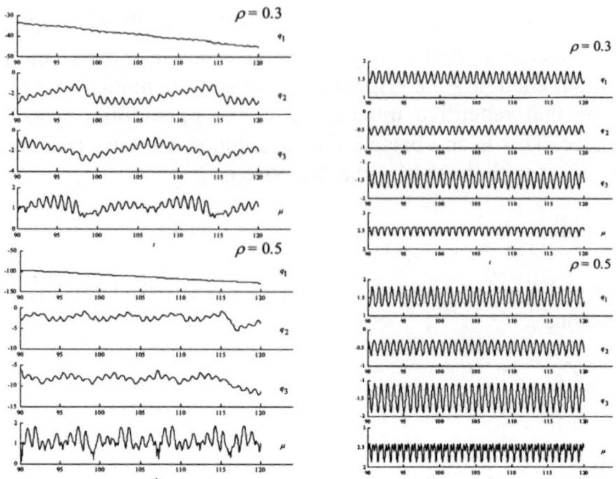

Fig. 3. The 3*R*-robot joint positions and manipulability *versus* time using the pseudoinverse method for $r = \{0.6, 2.0\}$ and $\rho = \{0.3, 0.5\}$.

5. Analyzing the chaotic-like responses of the pseudoinverse algorithm

In the previous section we verified that the pseudoinverse based algorithm leads to unpredictable arm configurations. Bearing these facts in mind, we analyze more deeply the robot joint signals.

Figure 4 depicts the phase-plane of the joint trajectories when repeating a circular motion in the operational space with frequency $\omega_0 = 7.0$ rad sec^{-1}, for $r = \{0.6, 2.0\}$ and $\rho = 0.5$.

From the figures we verify that:

- For $r = 0.6$, besides the position and velocity drifts, leading to different trajectory loops, we have points that are 'avoided'. Such points correspond to arm configurations where several links are aligned;
- For $r = 2.0$ the trajectories are repetitive.

In order to gain further insight into the pseudoinverse nature several distinct experiments are devised in the sequel during a time window of 300 cycles. Therefore, in a first set of experiments we calculate the Fourier transform $F\{\ \}$ of the 3*R*-robot joint velocities for a circular repetitive motion with frequency $\omega_0 =$ 7.0 rad sec^{-1}, radius $\rho = \{0.1, 0.3, 0.5, 0.7\}$ and radial distances $r \in \left]0, L_{TOT} - \rho\right[$.

Figures 5-8 show $\left|F\{\dot{q}_2(t)\}\right|$ *versus* ω_0/ω and *r*.

Induced by the gripper repetitive motion ω_0 an interesting phenomenon is verified, because a large part of the energy is distributed along several sub-harmonics. These *foh* depend on *r* and ρ making a complex pattern with similarities with those revealed by chaotic systems. Furthermore, we observe the existence of several distinct regions depending on *r*.

For example, selecting in Fig. 8 several distinct cases, namely for $r = 0.08$, $r = 0.30$, $r = 0.53$, $r = 1.10$, $r = 1.30$ and $r = 2.00$, we have the different signal Fourier spectra clearly visible in Fig. 9.

Joints 1 and 3 show velocity spectra similar.

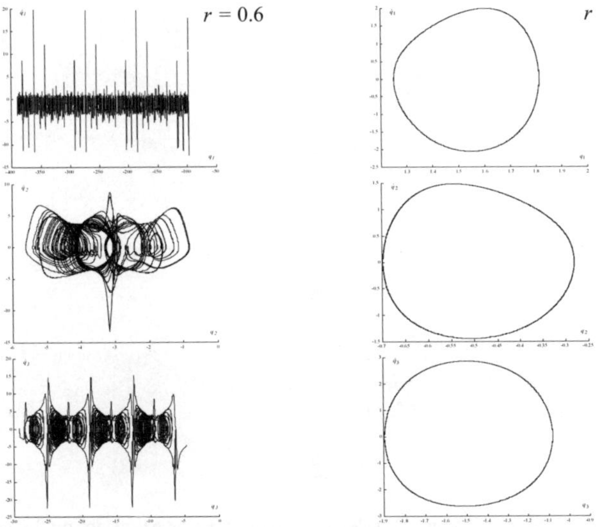

Fig. 4. Phase plane trajectory for the 3R-robot during 300 cycles for $r = \{0.6, 2.0\}$ and $\rho = 0.5$.

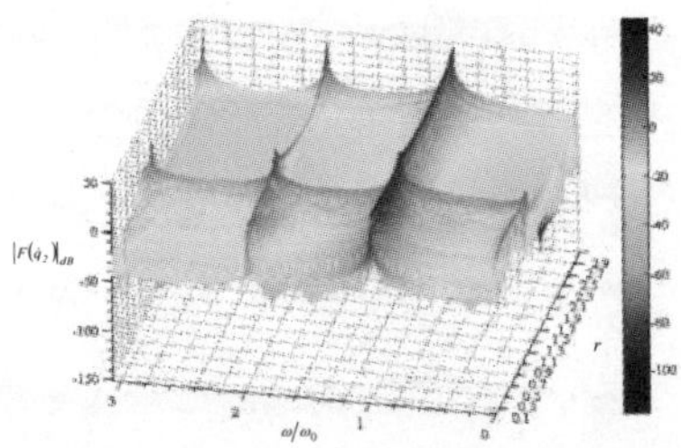

Fig. 5. $\left|F\{\dot{q}_2(t)\}\right|$ of the 3R-robot during 300 cycles, vs r and frequency ratio ω/ω_0, for $\rho = 0.1$, $\omega_0 = 7.0$ rad sec^{-1}.

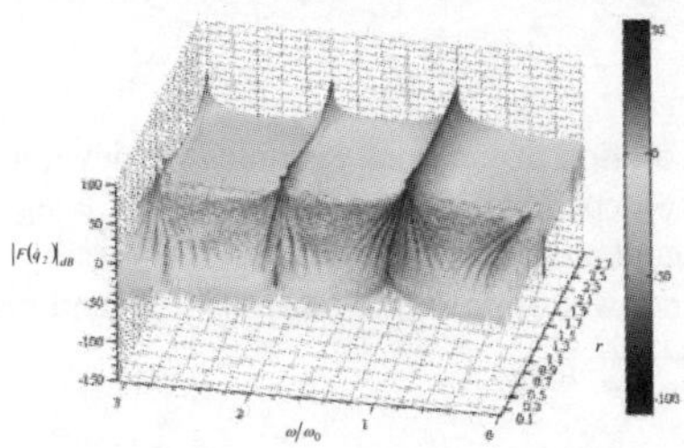

Fig. 6. $\left|F\{\dot{q}_2(t)\}\right|$ of the 3R-robot during 300 cycles, vs r and frequency ratio ω/ω_0, for $\rho = 0.3$, $\omega_0 = 7.0$ rad sec^{-1}.

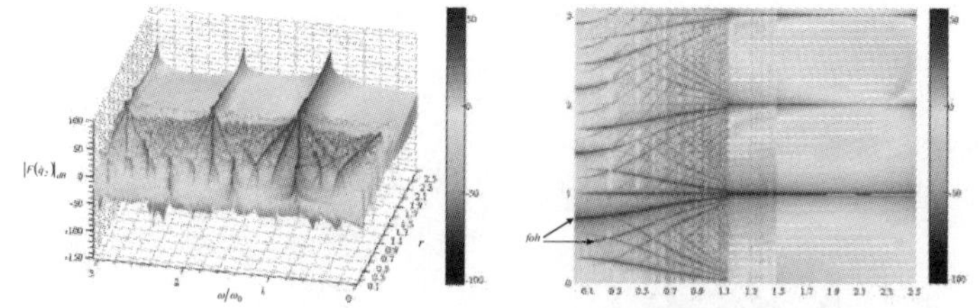

Fig. 7. $\left|F\{\dot{q}_2(t)\}\right|$ of the 3*R*-robot during 300 cycles, vs r and frequency ratio ω/ω_0, for $\rho = 0.5$, $\omega_0 = 7.0$ rad sec^{-1}.

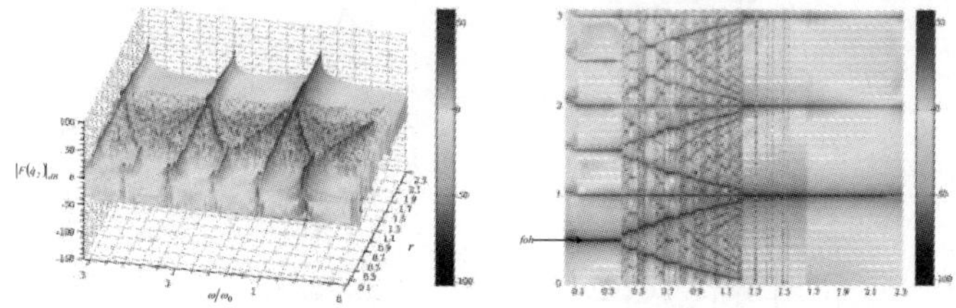

Fig. 8. $\left|F\{\dot{q}_2(t)\}\right|$ of the 3*R*-robot during 300 cycles, vs r and frequency ratio ω/ω_0, for $\rho = 0.7$, $\omega_0 = 7.0$ rad sec^{-1}.

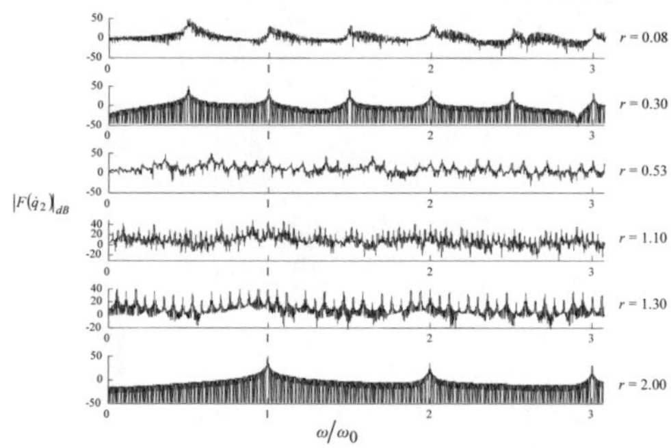

Fig. 9. $\left|F\{\dot{q}_2(t)\}\right|$ for the 3*R*-robot during 300 cycles, vs the frequency ratio ω/ω_0, for r = {0.08, 0.30, 0.53, 1.10, 1.30, 2.00}, $\rho = 0.7$, $\omega_0 = 7.0$ rad sec^{-1}.

In order to capture the time evolution of the joint variables we develop a second set of experiments.

One way of obtaining the time-dependent frequency content of a signal is to take the Fourier transform of a function over an interval around an instant τ, where τ is a variable parameter [13]. This mathematical tool is called the short-time or windowed Fourier transform (*WFT*) and may be defined as follows:

$$\{F_g f\}(\omega,\tau) = \int_{-\infty}^{+\infty} f(t)\overline{g(t-\tau)}e^{-it\omega}dt \qquad (13)$$

where $g(t)$ is the window function and $\tau, \omega \in \Re$. The multiplication by $g(t-\tau)$ localizes the Fourier integral in the neighborhood of $t = \tau$.

The slice of information provided by $\{F_g f\}(\omega,\tau)$ is represented in a time-frequency plane (t,ω) by a region whose location and width depends on the time-frequency spread of $g_{\omega,\tau}(t)=e^{it\omega}g(t-\tau)$. If $\mu(g)$ and $\sigma(g)$ are the centre and the radius, respectively, of the window function $g(t)$, then $\{F_g f\}(\omega,\tau)$ gives information about f and F, essentially in the region $I_t \times I_\omega$ of the time-frequency plane where $\hat{g}=F(g)$, $I_t \equiv [\mu(g)+\tau-\sigma(g), \mu(g)+\tau+\sigma(g)]$ and $I_\omega \equiv [\mu(\hat{g})+\omega-\sigma(\hat{g}), \mu(\hat{g})+\omega+\sigma(\hat{g})]$.

The Heisenberg uncertainty proves that the area of this region is [14]:

$$\sigma(g)\sigma(\hat{g}) \geq 1/2 \tag{14}$$

and this principle states that precise localizations both, on time and frequency, are mutually exclusive. Thus this trade-off between temporal and frequency resolution always exist [14-15]. Moreover, the size of this region is independent of (τ,ω), which means that the *WFT* has the same resolution across the time-frequency plane.

In the experiments we adopt two window functions, $\Psi := \{R_w, G_w\}$, namely the rectangular and Gaussian windows $R_w(t)=1$, and $G_w(t)=e^{-at^2/2}$, $(a=18)$, $t \in W_\Psi$. Moreover, we choose windows that do not overlap in the time domain. In the sequel the corresponding *WFT*s are represented by F_{R_w} and F_{G_w}, respectively.

Figures 10-12 show $\left|F_\Psi\{\dot{q}_2(t)\}\right|$, with window width $W_\Psi=\{5,30,50\}$ cycles, $\Psi := \{R_w, G_w\}$, for $\rho = 0.5$, $r = \{0.6,\ 1.289,\ 2.0\}$. We verify [15] that choosing a shorter (larger) time window W_Ψ increases (decreases) the temporal resolution but, on the other hand, decreases (increases) the frequency resolution.

In Fig. 10 (r = 0.6) we observe that the distribution of the signal energy dependents on the time evolution. In fact, the signal energy of the fundamental harmonic oscillates periodically and we verify that a large amount of the signal energy concentrates at several *foh*.

In Fig. 11 (r = 1.289) we verify that we have two distinct regions: a first one for the leading 60 cycles and a second for the remaining 240 cycles. In the first region we have a signal energy distribution along all frequencies, while in the second the energy is concentrated in the fundamental and multiple higher harmonics.

Finally, in Fig. 12 (r = 2.0) we get a regular behavior and the *WFT*s are invariant with time.

Figures 10-12 reveal also that the phenomena occur independently of the shape $\Psi := \{R_w, G_w\}$ or the width W_Ψ of the time window.

6. Conclusions

This paper discussed several aspects of the phenomena generated by the pseudoinverse-based trajectory control of the 3*R* redundant manipulators.

The closed-loop pseudoinverse's method leads to non-optimal responses, both for the manipulability and the repeatability. Bearing these facts in mind the chaotic responses were analyzed from different point of views, namely, the phase plane and the Fourier Transform. The results revealed the appearance of radial distances for which a large part of the energy is distributed in *fractional order harmonics*. In order to capture the time evolution of the joint variables we develop a set of experiments based on the *WFT*. The results showed that the energy content at the different frequencies of the joint velocity depends also on the time evolution.

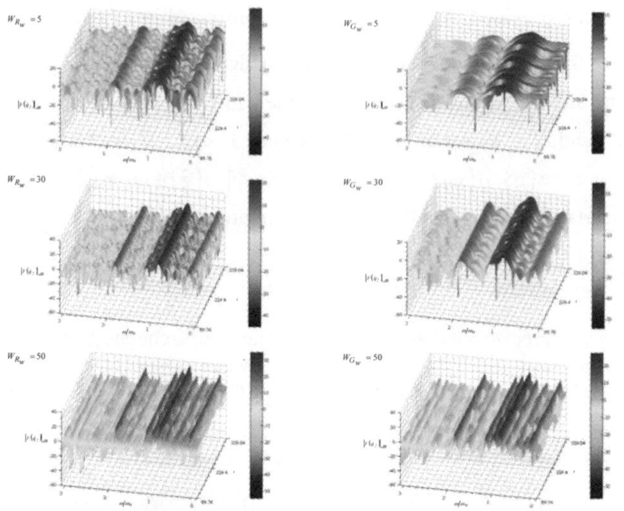

Fig. 10. $\left|F_{R_w}\{\dot{q}_2(t)\}\right|$ of the 3R-robot during 300 cycles, vs time and frequency ratio ω/ω_0, for $\rho = 0.5$, $r = 0.6$, $\omega_0 = 7.0$ rad sec^{-1}, $W_{R_w} = \{5, 30, 50\}$ cycles and $W_{G_w} = \{5, 30, 50\}$ cycles.

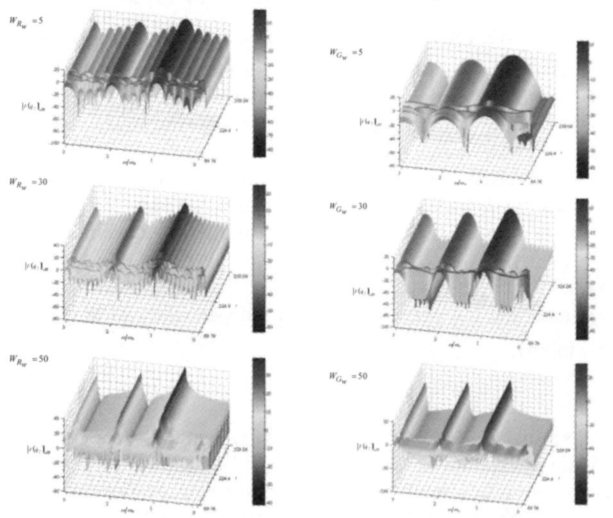

Fig. 11. $\left|F_{R_w}\{\dot{q}_2(t)\}\right|$ of the 3R-robot during 300 cycles, vs time and frequency ratio ω/ω_0, for $\rho = 0.5$, $r = 1.289$, $\omega_0 = 7.0$ rad sec^{-1}, $W_{R_w} = \{5, 30, 50\}$ cycles and $W_{G_w} = \{5, 30, 50\}$ cycles.

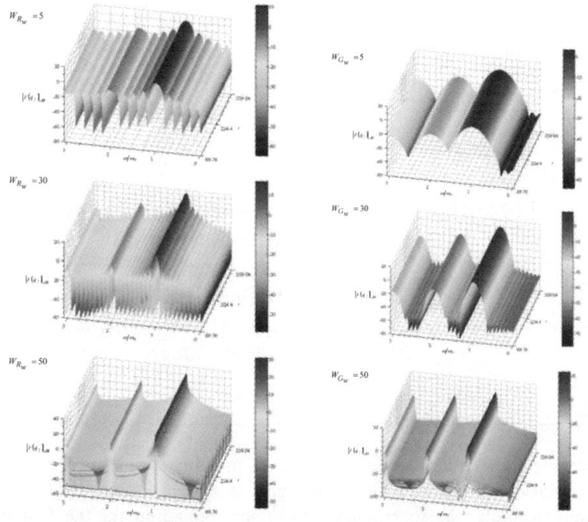

Fig. 12. $\left|F_{R_w}\{\dot{q}_2(t)\}\right|$ of the 3R-robot during 300 cycles, vs time and frequency ratio ω/ω_0, for $\rho = 0.5$, $r = 2.0$, $\omega_0 = 7.0$ rad sec^{-1}, $W_{R_w} = \{5, 30, 50\}$ cycles and $W_{G_w} = \{5, 30, 50\}$ cycles.

References

[1] E. Sahin Conkur, and Rob Buckingham. Clarifying the Definition of Redundancy as Used in Robotics. Robotica 1997; 15:583-586.

[2] S. Chiaverini. Singularity-Robust Task-Priority Redundancy Resolution for Real Time Kinematic Control of Robot Manipulators. IEEE Trans. Robotics Automation 1997; 13:398-410.

[3] C.A Klein, and C. C Huang. Review of Pseudoinverse Control for Use With Kinematically Redundant Manipulators. IEEE Trans. Syst. Man, Cyber 1983; 13:245-250.

[4] Yoshikawa T. Foundations of Robotics: Analysis and Control. MIT Press; 1988.

[5] Rodney Roberts R. and Anthony Maciejewski. Singularities, Stable Surfaces and Repeatable Behavior of Kinematically Redundant manipulators. International Journal of Robotics Research 1994; 13:70-81.

[6] John Bay. Geometry and Prediction of Drift-free trajectories for Redundant Machines Under Pseudoinverse Control. International Journal of Robotics Research 1992; 11: 41-52.

[7] Y. Nakamura. Advanced Robotics: Redundancy and Optimization. Addinson-Wesley; 1991.

[8] Keith L. Doty, C. Melchiorri and C. Bonivento. A Theory of Generalized Inverses Applied to Robotics. International Journal of Robotics Research 1993; 12:1-19.

[9] Bruno Siciliano. Kinematic Control of Redundant Robot Manipulators: A Tutorial. Journal of Intelligent and Robotic Systems 1990; 3:201-212.

[10] W.J.Chung, Y. Youm and W. K. Chung. Inverse Kinematics of Planar Redundant Manipulators via Virtual Links with Configuration Index. J. of Robotic Systems 1994; 11:117-128.

[11] Sanjeev Seereeram and John T. Wen. A Global Approach to Path Planning for Redundant Manipulators. IEEE Trans. Robotics Automation 1995; 11:152-159.

[12] Fernando Duarte, J. A. Tenreiro Machado. Chaotic Phenomena and Fractional-Order Dynamics in the Trajectory Control of Redundant Manipulators. Nonlinear Dynamics, Kluwer, USA 2002; 29:1-4:315-342.

[13] D. Gabor. Theory of communication. J. IEE 1946; 93:429-457.

[14] Stéphane Mallat. A Wavelet Tour of Signal Processing. Academic Press; 1999.

[15] Haldun M. Ozaktas, Zeev Zalevsky, M. Alper Kutay. The Fractional Fourier Transform with Applications in Optics and Signal Processing. John Wiley & Sons Ltd; 2001.

Under-damped Oscillator with Cross-correlated Colored Noises Input Modulated by Periodic Signal

Yanfei Jin*, Haiyan Hu

Institute of Vibration Engineering Research, Nanjing University of Aeronautics and Astronautics, 210016 Nanjing, P. R. China

Abstract

The stochastic resonance of an under-damped linear system with a random natural frequency due to the cross-correlated colored noises modulated by periodic signal is studied in this paper. Using the average method and the Shapiro-Loginov formula, the fourth-order differential equation for the first moment of response is established first. Under the assumption on the form of solution, the expression of the amplitude of the output signal is then obtained. After analyzing the effect of noise on the average coordinate of the oscillator, it is found that the output signal non-monotonically depends on the multiplicative noise intensity, correlation rate and natural frequency. Therefore, the stochastic resonance exists for such an under-damped linear system. Meanwhile, the cross-correlation between the noises can enhance the output signal, while the damping coefficient and the signal frequency can suppress the output signal.

PACS: 05.40.-a

Key Words: Stochastic resonance; Under-damped linear oscillator; Periodically signal-modulated noise

1. Introduction

Stochastic resonance (SR) is a counterintuitive phenomenon in which the response of a dynamical system to an input signal is enhanced by the addition of an optimal amount of noise. This term was firstly proposed by Benzi et al. [1] to explain the periodic switching of the earth's climate between ice ages and warm ages. The SR phenomenon has been investigated both theoretically and experimentally since then [2-7]. The linear-response theory, the perturbation theory and the method of residence-time distribution have been successfully used in studying the SR phenomenon [5-7]. Most of the previous works considered the nonlinearity as an essential ingredient for the presence of SR. However, the SR was also found a few years ago for the over-damped linear systems driven by multiplicative colored noise or dichotomous noise [8-10]. The relevant studies so far have been confined to the over-damped linear systems. When the damping coefficient is quite small, there are few theoretical studies [11-12]. Hence, the dynamical behavior of an under-damped linear system will be discussed in this paper.

Generally, the external noise and the weak periodic force are introduced additively to the study of SR. However, the noise modulated by the periodical signal does occur at the output of amplifiers in optics or radio astronomy and plays an important role [13-14]. For example, a single-mode laser system with periodically signal-modulated noise exhibits the stochastic multi-resonance phenomenon [14], which is absent at all for the case of noise and periodic signal introduced additively. Hence, it is

* Corresponding author: Tel: +86-25-84891672; E-mail: yanfeijin@nuaa.edu.cn (Y. F. Jin).

desirable to gain an insight into the dynamical behavior of the under-damped linear systems subject to periodically signal-modulated noise.

The aim of this paper is to explore the nonlinear phenomenon of the under-damped linear systems driven by periodically signal-modulated noise. In Section 2, the under-damped linear model with random frequency and periodically signal-modulated noise is given. The average method and the Shapiro-Loginov formula are used to establish the fourth-order differential equation for the first moment of response. Then the analytical expression of the first order moment and the amplitude of the output signal are derived by assuming the form of solution. It should be pointed out that the term SR has been applied to the non-monotonic behavior of the output signal amplitude, instead of the usually considered signal-to-noise ratio in this paper. From the discussions about the numerical results, some conclusions are drawn in Section 3.

2. Under-damped linear model

Consider an under-damped linear system with a random natural frequency and a colored noise input modulated by periodic signal, which can be described by the following equation

$$\frac{d^2x}{dt^2}+\gamma\frac{dx}{dt}+\omega^2[1+\xi(t)]x=A_0\sin\Omega t\cdot\eta(t)\,, \tag{1}$$

where A_0 is the amplitude of a periodic signal, Ω is the signal frequency; γ is the damping parameter; $\xi(t)$ and $\eta(t)$ are colored noises [15], and their mean and correlation function can be described by

$$\langle\xi(t)\rangle=0\,,\quad \langle\eta(t)\rangle=0\,,$$
$$\langle\xi(t)\xi(t')\rangle=\sigma_1\exp(-\lambda|t-t'|)\,,$$
$$\langle\eta(t)\eta(t')\rangle=\sigma_2\exp(-\lambda|t-t'|)\,. \tag{2}$$

If $\xi(t)$ and $\eta(t)$ are assumed as the same random source, there exists a correlation between them, i.e.

$$\langle\xi(t)\eta(t')\rangle=\langle\eta(t)\xi(t')\rangle=\sigma_3\exp(-\lambda|t-t'|)\,, \tag{3}$$

where $\sigma_3=\sigma_1 r+\sigma_2 q$, r and q are confined to the interval of [0, 1], measuring the contribution of each individual noise respectively [16].

Equation (1) can be recast as two first-order differential equations

$$\frac{dx}{dt}=y\,, \tag{4}$$

$$\frac{dy}{dt}=-\gamma y-\omega^2[1+\xi(t)]x+A_0\sin\Omega t\cdot\eta(t)\,. \tag{5}$$

After taking average on Equations (4)-(5), one obtains the equations of the first moments as follows

$$\frac{d\langle x\rangle}{dt}=\langle y\rangle\,, \tag{6}$$

$$\frac{d\langle y\rangle}{dt}=-\gamma\langle y\rangle-\omega^2\langle x\rangle-\omega^2\langle x\xi(t)\rangle\,. \tag{7}$$

Using the Shapiro-Loginov formula [17], one has

$$\frac{d\langle \xi x\rangle}{dt}=\left\langle \xi\frac{dx}{dt}\right\rangle-\lambda\langle \xi x\rangle. \tag{8}$$

Multiplying Equation (4) by $\xi(t)$ and taking average, then substituting it into Equation (8) yields

$$\frac{d\langle \xi x\rangle}{dt}=\langle \xi y\rangle-\lambda\langle \xi x\rangle. \tag{9}$$

Multiplying Equation (5) by $\xi(t)$ and taking average, one obtains

$$\left\langle \xi\frac{dy}{dt}\right\rangle=-\gamma\langle \xi y\rangle-\omega^2\langle \xi x\rangle-\omega^2\left\langle x\xi^2(t)\right\rangle+\sigma_3 A_0\sin\Omega t. \tag{10}$$

Equation (10) can be rewritten as the following equation when the Shapiro-Loginov formula is used

$$\frac{d\langle \xi y\rangle}{dt}=-(\gamma+\lambda)\langle \xi y\rangle-\omega^2\langle \xi x\rangle-\omega^2\left\langle x\xi^2(t)\right\rangle+\sigma_3 A_0\sin\Omega t, \tag{11}$$

As ξ^2 is always positive, the term $\left\langle \xi^2 x\right\rangle$ can be replaced by $\sigma_1\langle x\rangle$, which does not affect the asymptotic behavior of the solution of Equation (11) at $t\to\infty$. Hence, Equation (11) can be simplified as

$$\frac{d\langle \xi y\rangle}{dt}=-(\gamma+\lambda)\langle \xi y\rangle-\omega^2\langle \xi x\rangle-\omega^2\sigma_1\langle x\rangle+\sigma_3 A_0\sin\Omega t. \tag{12}$$

By virtue of Equations (6), (7), (9), (12), one obtains the differential equations of unknown functions $\langle x\rangle$, $\langle y\rangle$, $\langle \xi x\rangle$ and $\langle \xi y\rangle$. Solving the above equations, the fourth-order differential equation for $\langle x\rangle$ reads

$$\begin{aligned}&\frac{d^4\langle x\rangle}{dt^4}+2(\gamma+\lambda)\frac{d^3\langle x\rangle}{dt^3}+(\gamma^2+2\omega^2+3\lambda\gamma+\lambda^2)\frac{d^2\langle x\rangle}{dt^2}\\&+(\lambda+\gamma)(2\omega^2+\lambda\gamma)\frac{d\langle x\rangle}{dt}+\omega^2[\omega^2+\lambda\gamma+\lambda^2-\omega^2\sigma_1]\langle x\rangle=-\sigma_3\omega^2A_0\sin\Omega t.\end{aligned} \tag{13}$$

On the basis of the author's previous study [10], the solution of Equation (13) is assumed in the following form

$$\langle x\rangle=A\sin(\Omega t+\phi), \tag{14}$$

Substituting Equation (14) into Equation (13), one has

$$A=\frac{A_0\sigma_3\omega^2}{\sqrt{f_1^2+f_2^2}},\quad \phi=\pi-\arctan(\frac{f_2}{f_1}). \tag{15}$$

where

$$f_1=(\Omega^2-\omega^2)(\Omega^2-\omega^2-\lambda^2)-\Omega^2(\gamma^2+3\gamma\lambda)+\omega^2(\gamma\lambda-\omega^2\sigma_1),$$

$$f_2=\Omega(\lambda+\gamma)[\gamma\lambda-2(\Omega^2-\omega^2)].$$

According to Equation (15), one can easily find the value of multiplicative noise intensity σ_1 at which the amplitude of the output signal A reaches the maximum

$$(\sigma_1)_{\max} = \frac{(\Omega^2-\omega^2)(\Omega^2-\omega^2-\lambda^2)-\Omega^2(\gamma^2+3\gamma\lambda)+\gamma\lambda\omega^2}{\omega^4}. \tag{16}$$

3. Discussions and conclusions

The curves of the amplitude A of a stationary signal versus the multiplicative noise intensity σ_1 are shown in Figures 1 and 2 for different values of damping coefficient γ and the cross-correlated strength σ_3. The amplitude of output signal A is a non-monotonic function of σ_1 and the corresponding maximum of σ_1 can be computed by Equation (16). There is a pronounced single-peak in the curve and SR displays for the case of $\Omega < \omega$, which is absent for the case of noise and periodical signal introduced additively [11]. Meantime, the amplitude of output signal A decreases with an increase of γ but increases with an increase of σ_3. That is, γ suppresses the output signal, while the cross-correlation σ_3 between $\xi(t)$ and $\eta(t)$ enhances the output signal and plays a constructive role.

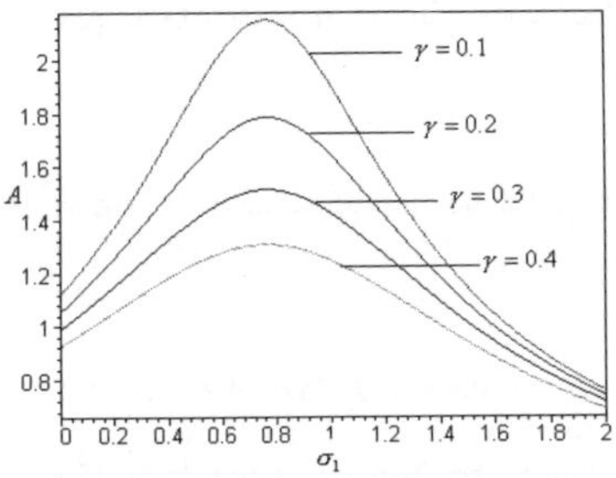

Fig. 1 The amplitude A of a stationary signal as a function of σ_1 for $A_0 = \omega = \sigma_3 = 1$, $\lambda = \Omega = 0.5$ with varied γ.

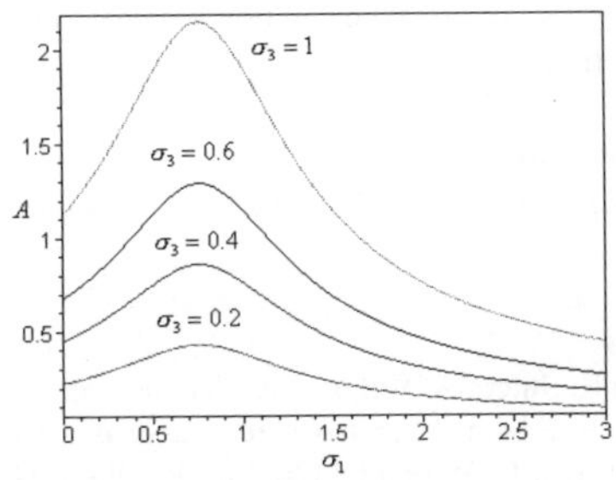

Fig. 2 The amplitude A of a stationary signal as a function of σ_1 for $A_0 = \omega = 1$, $\gamma = 0.1$, $\lambda = \Omega = 0.5$ with varied σ_3.

Figure 3 is a plot of the amplitude A of a stationary signal versus the correlation rate λ with different values of the signal frequency Ω. This graph shows a typical SR non-monotonic behavior for $\gamma = 0$ and $\Omega < \omega$. At the same time, the heights of the maximum monotonically decrease with an increase of Ω, which is opposite to the result in Ref. [11]. The position of peak shifts to the right-hand with an increase of Ω, which can be easily explained from the Equation (16). The effects of the natural frequency ω on the amplitude A of a stationary signal with different values of γ are illustrated in Figure 4. It is seen that the curve exhibits a pronounced single peak, and SR exists. This graph shows a typical SR non-monotonic behavior for the case of ω close to the signal frequency $\omega = \Omega = 0.5$, which is similar to the bona fide SR in a sense.

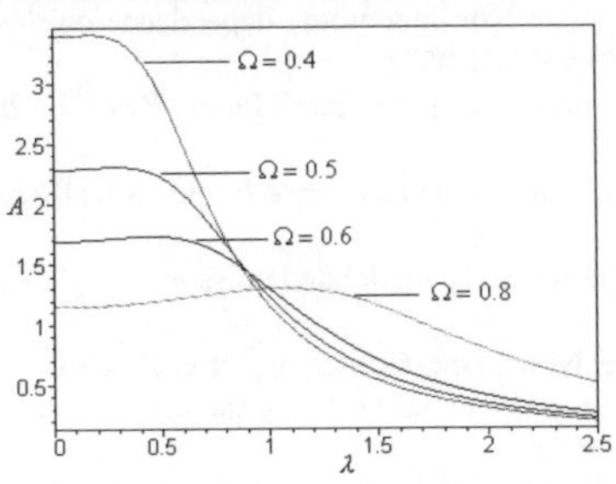

Fig. 3 The amplitude A of a stationary signal as a function of λ for $\sigma_1 = 1$, $\gamma = 0$, $A_0 = \omega = \sigma_3 = 1$ with varied Ω.

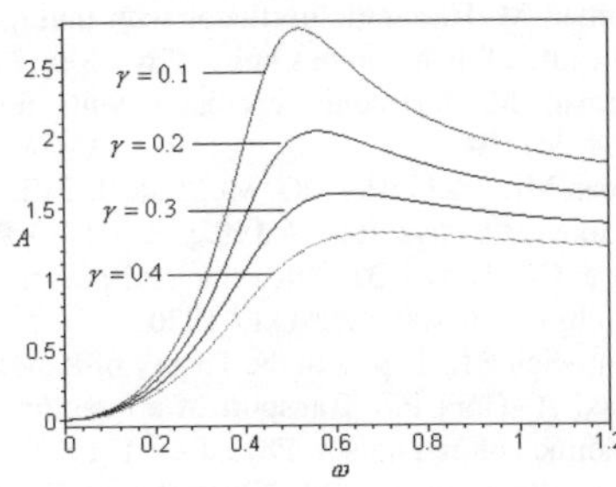

Fig. 4 The amplitude A of a stationary signal as a function of ω for $\sigma_1 = 1$, $A_0 = \sigma_3 = 1$, $\Omega = \lambda = 0.5$ with varied γ.

In this paper, the fluctuations of external parameters are expressed by a multiplicative noise in the equation of motion of an under-damped oscillator with periodically signal-modulated noise. The random natural frequency may cause the instability of the oscillator coordinate (first moment) for a strong noise intensity [18], which is ignored in the paper.

The signal-modulated noise can be used in many fields such as the optical communication and the radio astronomy to describe the fluctuation. And the modulated process is operated in the linear region to avoid the distortion of a modulation signal. Therefore, the investigation of SR for an under-damped linear system with random natural frequency and a colored noise input modulated by periodic signal provides a theoretical basis for the application of signal-modulated noise.

Therefore, the study focuses on SR of an under-damped linear system with a random natural frequency and a colored noise input modulated by periodic signal. The output signal of the under-damped or un-damped oscillator shows the non-monotonic dependence on the multiplicative noise intensity, correlation rate and frequency. And, the SR exists in a broad sense. Moreover, the output signal increases with an increase of cross-correlation between noises, but decreases with an increase of damping coefficient and signal frequency.

Acknowledgments

This work was supported by the China Postdoctoral Science Foundation and Jiangsu Postdoctoral Science Foundation.

References

[1] Benzi R, Sutera A, Vulpiani A. The mechanism of stochastic resonance. J. Phys. A 1981; 14: L453-L457.

[2] Fauve S, Heslot F. Stochastic resonance in a bistable system. Phys. Lett. A 1983; 97A: 5-7.

[3] McNamara B, Wiesenfeld K, Roy R. Observation of stochastic resonance in a ring laser. Phys. Rev. Lett. 1988; 1: 3-4.

[4] McNamara B, Wiesenfeld K. Theory of stochastic resonance. Phys. Rev. A. 1989; 39: 4854-4869.

[5] Dykman MI, Luchinsky D, Mannella R, Stein ND, McClintock PVE, Stocks NG. Stochastic resonance: linear response theory and giant nonlinearity. J Stat Phys 1993; 70: 463-479.

[6] Hu G, Nicolis G, Nicolis C. Periodically forced Fokker-Planck equation and stochastic resonance. Phys. Rev. A. 1990; 42: 2030-2041.

[7] Zhou T, Moss F, Jung P. Escape-time distributions of a periodically modulated bistable system with noise. Phys. Rev. A. 1990; 42: 3161-3169.

[8] Fulinski A. Relaxation, noise-induced transitions, and stochastic resonance driven by non-Markovian dichotomic noise. Phys. Rev. E 1995; 52: 4523-4526.

[9] Berdichevsky V, Gitterman M. Stochastic resonance in linear systems subject to multiplicative and additive noise. Phys. Rev. E 1999; 60: 1494-1499.

[10] Jin YF, Xu W, Xu M, Fang T. Stochastic resonance in linear system due to dichotomous noise modulated by bias signal. J. Phys. A 2005; 38: 3733-3742.

[11] Gitterman M. Harmonic oscillator with multiplicative noise: Non-monotonic dependence on the strength and the rate of dichotomous noise. Phys. Rev. E 2003; 67: 057103-057106.

[12] Gitterman M. Harmonic oscillator with fluctuating damping parameter. Phys. Rev. E 2004; 69: 041104-041104.

[13] Dykman MI, Luchinsky DG, McClintock PVE, Stein ND. Stochastic resonance for periodically modulated noise intensity. Phys. Rev. A 1992; 46: R1713-R1716.

[14] Wang J, Cao L, Wu DJ. Stochastic multi-resonance for periodically modulated noise in a single-mode laser. Chin. Phys. Lett 2003; 20: 1217-1220.

[15] Stratonovich RL. Topics in the Theory of Random Noise. New York: Gordon and Breach; 1963.

[16] Fuliński A, Góra PF. Transport of a quantum particle in a dimer under the influence of two correlated dichotomic colored noises. Phys. Rev. E 1993; 48: 3510-3517.

[17] Shapiro VE, Loginov VM. "Formulae of Differentiation" and their use for solving stochastic equations. Physica A 1978; 91A: 563-574.

[18] Mallick K, Marcq P. Anomalous diffusion in nonlinear oscillators with multiplicative noise. Phys. Rev. E 2002; 66: 041113-041126.

A 3D Turning Model for the Interpretation of Machining Stability and Chatter

Achala V. Dassanayake and C. Steve Suh
Mechanical Engineering Department, Texas A&M University, College Station, Texas 77843-3123, USA.

Abstract
Turning dynamics is investigated using a 3D model that allows for simultaneous workpiece-tool deflections in response to the exertion of nonlinear regenerative force. The workpiece is modeled as a system of three rotors connected by a flexible shaft. Such a configuration enables the motion of the workpiece relative to the tool and tool motion relative to the machining surface to be three-dimensionally established as functions of spindle speed, instantaneous depth-of-cut, material removal rate and whirling. The model is explored along with its 1D counterpart, which considers only tool motions and disregards workpiece vibrations. Different stages of stability for the workpiece and the tool subject to the same cutting conditions are studied.

PACS: 05.45.–a, 46.32.+x
Keywords: Material Removal, Turning, Chatter, Bifurcation

1. Introduction

Machining dynamics along with machine tool chatter has long been studied [1-3]. Stability charts incorporating factors that affect machining stability in general [3-5] and turning in specific [6-8] are widely adopted. A recent literature [9] confirmed the low depth-of-cut instability found experimentally in [10] and reported the observation of multiple stability regions. This is in contrast to the conventional stability lobes. Instead of stability lobes, the use of critical width-of-cut as suggested in [11] for better chatter control was experimental established. However, even following the stability chart to set the ranges of cutting parameters, chatter could still elude one's effort [12].

Most machining operations have two distinct motions: a rotary motion and a translation that is either straight or curvilinear. In operations such as milling and drilling, the tool induces both types of motion. Thus, the tool is the only component that vibrates. In turning operation, on the other hand, the workpiece sees a rotary motion whereas the tool experiences a linear translation. It is obvious that in turning both the tool and workpiece vibrate in material removal. However, only tool vibrations in turning operations received almost exclusively considerations in literature [1, 8-11, 13-15]. As an exception, Ref. [16] investigated workpiece deflections subject to a 3D cutting force to correlate dimensional errors. Machining failures attributable to whirling-induced workpiece vibrations have long been observed [17]. Although workpiece vibrations impact both cutting instability and product quality including surface finishing, interestingly enough, most models developed for surface roughness [18–20] do not consider workpiece vibrations at all. .

Factors affecting machining instability include regenerative effect, nonlinearity and three-dimensionality. The significance of regeneration process in machining instability has long been recognized [3, 21]. Exploring nonlinear models as a major path to understanding machining dynamics is abundant in recent literature [8-10, 13-15, 22-26]. However, many considered either single DOF [13, 14,

Corresponding author. Tel: +1-979-845-1414; fax: +1-979-845-3081.
Email address: ssuh@mengr.tamu.edu (C. Steve Suh)

22-24] or 2 DOF models [9, 11, 15, 26] that admit only tool vibrations. A three-dimensional cutting model was developed by Rao and Shin [10] to investigate stability. But the effect of workpiece vibration on cutting dynamics was neglected.

In response to the aforementioned survey, this paper examines cutting instability using a 3 DOF model that incorporates the followings: (1) regenerative effect, (2) cutting force and structural nonlinearities, (3) simultaneous workpiece-tool vibration, and (4) whirling induced by workpiece material inhomogeneity. Mass and stiffness reduction of the workpiece due to material removal is also considered. Since nonlinearity is a non-negligible parameter in the cutting model, use of a proper characterization tool is crucial in capturing the underlying dynamics. Among popular alternatives to Fourier-based methodologies, discrete wavelet transform is a preferred tool for processing cutting force measurements [27] and the concept of instantaneous frequency [28] is highly effective in characterizing nonlinear rotordynamic responses [29]. In the presentation that follows, instantaneous frequency is used as the characterization tool for simultaneous temporal-spectral domain analysis. As a support for the qualitative analysis using instantaneous frequency, quantitative measure of machining stability is done by determining the largest Lyapunov exponents [30]. One stability state and one instability state of the tool are compared with the experimental results reported in [10] to establish model validation. The equation of motion (EQM) governing the tool motion of the 3D model is uncoupled to formulate a single DOF (SDOF) model. The dynamic responses of the SDOF model are evaluated against the results generated by its coupled 3D model to establish the stability stages of the tool.

2. Model Description

Consider the workpiece given in Fig. 1, which is undergoing longitudinal tuning operation. As a result of the operation, the workpiece has three distinctive sections: unmachined, being machined and machined. The end of the unmachined section is fixed to the spindle, while the other end is pinned to the tailstock,

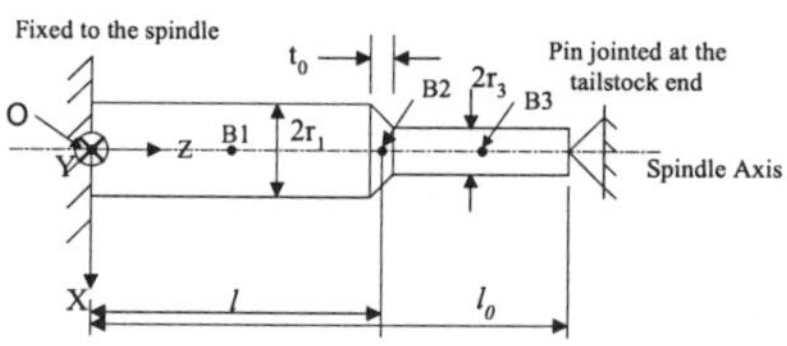

Fig. 1 Workpiece configuration

where l_0 is the full length of the workpiece, l is a variable measuring the distance from the spindle end to the tool position, t_0 is the chip width in feed per revolution, r_1 is the radius of the unmachined section, and r_3 is the radius of the machined section. l is an independent variable of time, which can be written as

$$l = l_i - \dot{l}\,t \tag{1}$$

with $\dot{l}$ being the constant feed rate of the tool and l_i being the value of l at time $t = 0$. For simplicity, the three sections seen in Fig. 1 are assumed to be consisted of 3 rotors connected to each other with a shaft of negligible mass as shown in Fig. 2. Rotor 1 represents the unmachined section having a full length of $(l\text{-}t_0/2)$ measuring from O. The tool is aligned with the spindle axis, ZZ. Rotor 2, having a thickness of t_0, represents the section being machined. It is also where the current position of the tool is. The machined section is represented by Rotor 3, which has a length measuring from $(l + t_0/2)$ to l_0. It is also assumed that the rotors are rigid and remains vertical at all time.

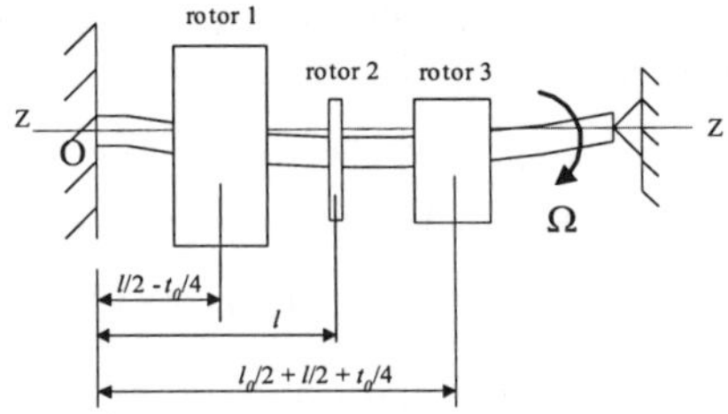

Fig. 2 Workpiece configuration with three rotors

The configuration of Rotor 2 is not cylindrical and hence the center of mass of the section differs from the other two sections even if there is no deflection. The mass of each rotor can be determined using

$$m_1 = \rho \pi r_1^2 (l - t_0 / 2) \tag{2}$$

$$m_2 = \rho \pi r_2^2 t_0 + \rho \pi r_1 t_0 (r_1 - r_2) \tag{3}$$

$$m_3 = \rho \pi r_2^2 (l_0 - l - t_0 / 2) \tag{4}$$

where ρ is the density of the workpiece. Although its section thickness and mass are small compared with those of Rotors 1 and 3, nonetheless, Rotor 2 is where the cutting forces exert and a new machined surface is being generated (Fig. 4). As such, the response of Rotor 2 dominates the dynamics of the 3-rotor system. In the followings, focus will be placed on Rotor 2 to develop an understanding for machining stability subject to the exertion of cutting forces and workpiece whirling.

As depicted in Fig. 3, Rotor 1 when subjected to a spindle speed, Ω, would see an angle $\alpha = \psi_0 + \Omega t$ (where ψ_0 is the angle between the X-axis and $C1G1$ at t = 0) between the mass center (G1) and the geometric center (C1). Note that parameters ε_1, Ω and ψ_0 are all constants. At rest, C1 coincides with B1. While in motion, C1 has x_1 and y_1 as its X- and Y-direction components, respectively. If $C1G1$ coincides with the X-axis at t = 0, then $\psi_0 = 0$.

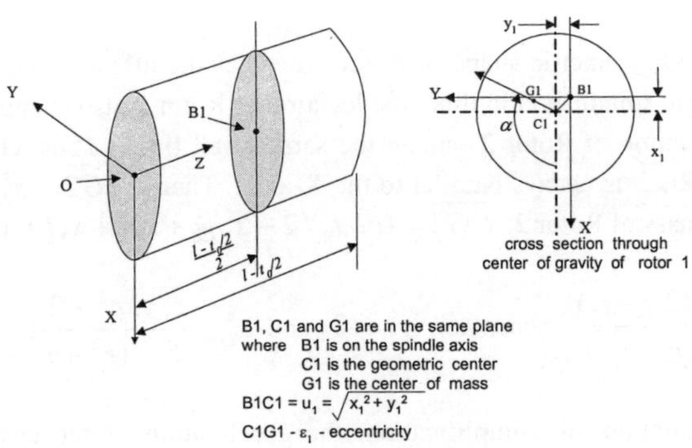

Fig. 3 Rotor 1 with locations of mass and geometric centers

From the figure, the position vector of the center of mass of Rotor 1 is therefore

$$\overrightarrow{OG1} = \overrightarrow{OB1} + \overrightarrow{B1C1} + \overrightarrow{C1G1}$$
$$= 1/2(l - t_0/2)\underline{k} + (x_1 + \varepsilon_1 \cos\alpha)\underline{i} + (y_1 + \varepsilon_1 \sin\alpha)\underline{j} \tag{5}$$

By differentiating Eq. (5) and collecting terms afterwards, the velocity and acceleration of G1 are then

$$v_1 = (\dot{x}_1 - \varepsilon_1 \Omega \sin(\Omega t)\underline{i} + (\dot{y}_1 + \varepsilon_1 \Omega \cos(\Omega t)\underline{j} - \frac{\dot{l}}{2}\underline{k} \tag{6}$$

$$\underline{a}_1 = (\ddot{x}_1 - \varepsilon_1 \Omega^2 \cos\Omega t)\underline{i} + (\ddot{y}_1 - \varepsilon_1 \Omega^2 \sin\Omega t)\underline{j} \tag{7}$$

Note that $\dot{\alpha} = \Omega$ and $\ddot{\alpha} = 0$ for constant spindle speed. Similarly, the velocity and acceleration of the center of mass of Rotor 3 are

$$v_3 = (\dot{x}_3 - \varepsilon_3 \Omega \sin(\Omega t)\underline{i} + (\dot{y}_3 + \varepsilon_3 \Omega \cos(\Omega t)\underline{j} - \frac{\dot{l}}{2}\underline{k} \tag{8}$$

$$\underline{a}_3 = (\ddot{x}_3 - \varepsilon_3 \Omega^2 \cos\Omega t)\underline{i} + (\ddot{y}_3 - \varepsilon_3 \Omega^2 \sin\Omega t)\underline{j} \tag{9}$$

Considering the reduction of eccentricity due to material removal, eccentricity is assumed to be proportional to the radius of each section. Thus,

$$\varepsilon_3 = \frac{r_3}{r_1}\varepsilon_1 \tag{10}$$

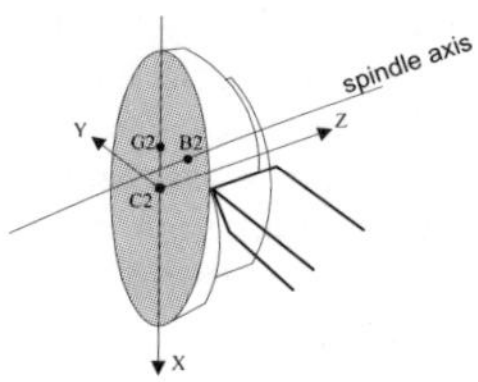

Fig. 4 Rotor 2 with locations of mass and geometric centers

Assume that the asymmetric shape of Rotor 2 negates the effect of material inhomogeneity. As the tool moves toward the spindle end in time, the location of Rotor 2 also changes in time. However, the configuration and orientation of Rotor 2 remain the same at all time. Thus, G2 is always above C2 as shown in Fig. 4 and $\overrightarrow{C2G2}$ is always parallel to the X-axis. Then $\overrightarrow{C2G2} = x_2'\underline{i} + y_2'\underline{j}$ and the position vector of the center of mass of Rotor 2, $\overrightarrow{OG2} = (l - t_0/2 + z_2')\underline{k} + x_2'\underline{i} + x_2\underline{i} + y_2 j + y_2'\underline{j}$, where

$$x_2' = \frac{-r_1(r_1 - r_3)(2r_1 - r_3)}{2\pi(r_3^2 + r_1(r_1 - r_3))} = \text{constant},\ y_2' = 0,\ \text{and}\ z_2' = \frac{t_0}{6}\frac{(3r_3^2 + 2r_1^2 - 2r_1 r_3)}{(r_3^2 + r_1^2 - r_1 r_3)} = \text{constant}.$$

Again, through differentiation and simplification, the acceleration of the center of mass of Rotor 2 becomes

$$\underline{a}_2 = \ddot{x}_2\underline{i} + \ddot{y}_2\underline{j} \tag{11}$$

Note that with respect to the geometric center, *G2* is fixed, meaning that Rotor 2 undergoes only a curvilinear translation, but not a rotation. While the position of Rotor 2 changes continuously with time (since it is a function of *l*), with G2 always holding the same position with respect to C2, Rotor 2 maintains the same shape at all time. Thus, the angular velocity of Rotor 2 is zero. Also note that both the tool and Rotor 2 move towards the spindle end at a constant speed, $\dot{l}$, and that while Rotor 2 has a constant section thickness, t_0, the thicknesses of Rotors 1 and 3 vary with time. .

Since all the depth-of-cuts considered in the study are smaller than 6% of the diameter of the cylindrical workpiece, for the purpose of finding the static deflections at certain locations of interest, the workpiece can be assumed to be a uniform shaft with an average diameter, d_{AV}, as

$$d_{AV} = \frac{1}{l_0}\left[d_1(l - t_0/2) + (\frac{d_1 + d_3}{2})t_0 + d_3(l_0 - l + t_0/2) \right] \tag{12}$$

where d_1 and d_3 are the diameters of Rotors 1 and 3, respectively. The average area moment of inertia of the shaft is therefore

$$I_{AV} = \frac{\pi \, d_{AV}}{64} \tag{13}$$

Considering the shaft as subject to a fixed-pinned boundary condition (Fig. 2) and a concentration load, the static deflection of the center of gravity of each rotor can then be determined as follows. Assume that δ_1, δ_2, and δ_3 are the deflections of Rotors 1, 2 and 3, respectively, and that vibration displacements, u_i, are proportional to static deflections, δ_i . Thus, $u_1/\delta_1 = u_2/\delta_2 = u_3/\delta_3$. Note that u_i is also the radial displacement of the center of gravity of Rotor *i* and the resultant cutting force acting radially on the shaft is the concentration load. Since gravity is negligible compared to the cutting force, one has $x_1/\delta_1 = x_2/\delta_2 = x_3/\delta_3$ and $y_1/\delta_1 = y_2/\delta_2 = y_3/\delta_3$, or equivalently,

$$x_1 = \frac{\delta_1}{\delta_2}x_2 \,,\ x_3 = \frac{\delta_3}{\delta_2}x_2 \,,\ y_1 = \frac{\delta_1}{\delta_2}y_2 \text{ and } y_3 = \frac{\delta_3}{\delta_2}y_2 \tag{14}$$

Note that for particular values for t_0 and l_0, deflections are functions of *l*, the concentration load and the average area moment of inertia.

Forces acting on the three rotors are considered separately. The forces acting on Rotors 1 and 3 are due to the stiffness of the shaft. Rotor 2 sees three more forces in the *X, Y,* and *Z* direction, which together play the role of material removal. The force acting on Rotor 1 due to shaft stiffness is k_1u_1 and its orientation is shown in Fig. 5. Here $u_1 = \overrightarrow{B1C1}$. Consider its component along the *X*- and *Y*-directions, the force acting on Rotor 1 is of the form: $(-k_1u_1\cos\theta)\underline{i} - (k_1u_1\sin\theta)\underline{j}$.

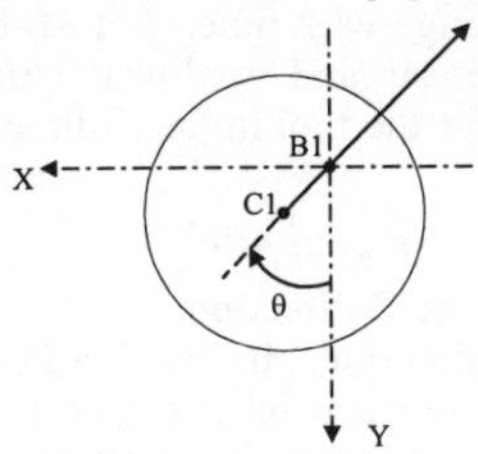

Fig. 5 Force acting on Rotor 1

But since $\cos\theta = x_1 / u_1$ and $\sin\theta = y_1 / u_1$, so the force acting on Rotor 1 is therefore

$$(-k_1x_1)\underline{i} - (k_1y_1)\underline{j} \tag{15}$$

Similarly, the force acting on Rotor 3 is

$$(-k_3x_3)\underline{i} - (k_3y_3)\underline{j} \tag{16}$$

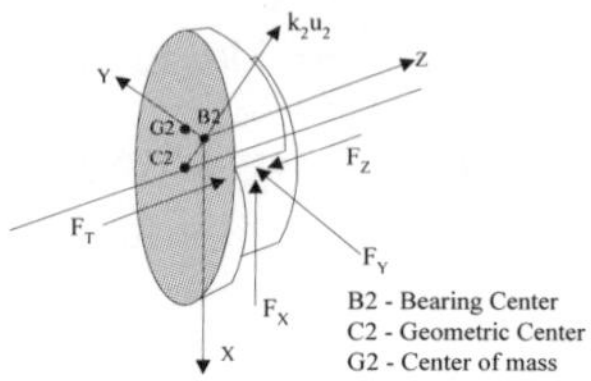

Fig. 6 Components of cutting forces acting on Rotor 2

Other than the forces due to shaft stiffness, a cutting force also acts on Rotor 2. Referring to Fig. 6, Fx, Fy and Fz are three components of the cutting force on Rotor 2 and F_T is the force due to bending stiffness in the *k*-direction. Since the workpiece is not allowed to move along the *Z*-direction, the acceleration of Rotor 2 in the *Z*- direction is therefore zero. In other words, F_T = Fz. As a result, the force acting on Rotor 2 becomes $(-k_2u_2\cos\theta)\underline{i} - (k_2u_2\sin\theta)\underline{j} - F_X\underline{i} + F_Y\underline{j}$, which can be further simplified to be

$$(-F_X - k_2x_2)\underline{i} + (F_Y - k_2y_2)\underline{j} \tag{17}$$

Now applying Newton's 2nd law, $\vec{F} = m\,d\underline{v}/dt + \underline{v}\,dm/dt$, in the *X*- and *Y*-directions and using Eqs. (6)-(9), (11) and (15)-(17), and the time derivatives of Eqs. (2) and (4), the following two equations can be obtained:

$$-k_1x_1 - k_2x_2 - k_3x_3 - F_X = m_1a_{1x} + m_2a_{2x} + m_3a_{3x} + v_{1x}\frac{dm_1}{dt} + v_{3x}\frac{dm_3}{dt} \tag{18}$$

$$-k_1y_1 - k_2y_2 - k_3y_3 + F_Y = m_1a_{1y} + m_2a_{2y} + m_3a_{3y} + v_{1y}\frac{dm_1}{dt} + v_{3y}\frac{dm_3}{dt} \tag{19}$$

Note that the mass of Rotor 2 does not change with time. Let M_Z be the equivalent mass of the machine tool and k_Z and k_{ZC} be the equivalent linear and nonlinear stiffness of the tool in the Z-direction, respectively, then the equation of motion for the tool in the *Z*-direction including nonlinear stiffness term can be written as

$$M_Z\,\ddot{z}_t + k_Z\,z_t + k_{ZC}\,z_t^3 = F_Z \tag{20}$$

where F_Z is the cutting force component in the *Z*-direction.

Consider tool vibrations in the Z-direction. In Fig. 7 a PQ section of Rotor 2 actively engaging with the tool and showing a flatted outer surface on one side is given, where the lowercase z_t is the relative displacement of the tool at time t , and z'_t is the relative displacement of the tool one revolution before. Then the instantaneous chip width, t_i, can be found to be

$$t_i = t_0 - z_t + z_t' \tag{21}$$

Note that if the tool does not vibrate in the Z-direction, the chip width (chip thickness) is equal to the feed rate. It is important to understand that for longitudinal turning, tool feed direction is in the direction along the workpiece. Since the tool is considered infinitely stiff in the X-Y plane and the workpiece is rigidly constrained along the Z-direction, only responses of the tool in the Z-direction and workpiece responses in the X- and Y-directions are considered in the study.

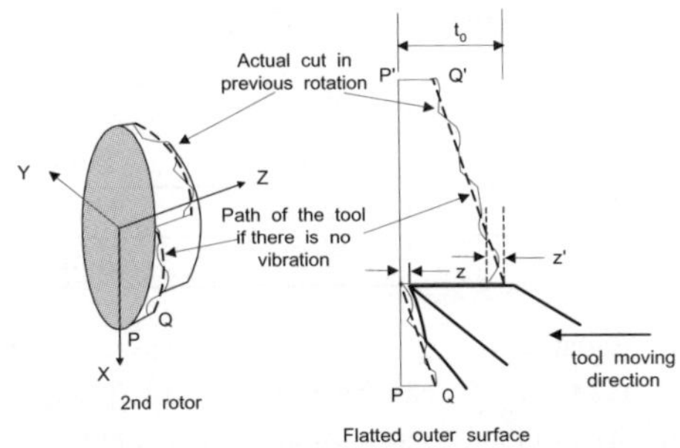

Fig. 7 Instantaneous chip width (instantaneous feed per revolution)

Cutting force components, F_X, F_Y, and F_Z , can be written in the following form [10],

$$F_X = F_f(b_{x1}\sin(\eta_c) + b_{x2}\cos(\eta_c)) + b_{x3}F_n \tag{22}$$

$$F_Y = F_f(b_{y1}\sin(\eta_c) + b_{y2}\cos(\eta_c)) + b_{y3}F_n \tag{23}$$

$$F_Z = F_f(b_{z1}\sin(\eta_c) + b_{z2}\cos(\eta_c)) + b_{z3}F_n \tag{24}$$

where F_f is the friction force, F_n is the normal force and η_c is the chip flow angle. The constants b_{rm}'s (where r = x,y,z and m = 1,2,3) depend on the tool rake angle, side cutting edge angle and inclination angle. In this study, parameter values used in [10] for tool rake angle, side cutting edge angle and inclination angle are adopted for the objective of being able to compare with experimental/physical data. By using a curve fitting for the experimental result found in [10], the chip flow angle can be formulated as a function of DOC that is valid for the range of
0.4mm < DOC < 2.5mm. Thus the chip flow angle is

$$\eta_c = \alpha_1 s^3 + \alpha_2 s^2 + \alpha_3 s + \alpha_4 \tag{25}$$

Here α_i's are constants. The normal force and the friction force are as follows,

$$F_n = k_n A_C \tag{26}$$

$$F_f = k_f A_C \tag{27}$$

in which k_n and k_f are normal and friction pressure components, respectively, and the chip cross sectional area, A_C , is defined using the instantaneous chip width, t_i, and the instantaneous DOC, s, as

$$A_C = t_i s \tag{28}$$

By substituting Eqs. (18) and (19) into (26), the chip cross sectional area becomes

$$A_C = (t_0 - z_t + z_t^{'})(\sqrt{r_1^2 - x_2^2} - r_3 - y_2) \tag{29}$$

In the model presented herein, coefficients k_n and k_f are not constants. But rather they are functions of the instantaneous chip width, t_i . The relationships are modeled as follows

$$k_n = K_{nm} t_i + K_{nn} \tag{30}$$

$$k_f = K_{fm} t_i + K_{ff} \tag{31}$$

Table 1 lists the values along with their units for all K's, which were determined using the experimental data available in [10].

Table 1 Various K values

K_{nm}	-6.0E+12	N/m^2
K_{nn}	2.9E+9	N/m^3
K_{fm}	-8.7E+12	N/m^2
K_{ff}	3.0E+9	N/m^3

Using the equations above the relationships between $\ddot{x}_1, \dot{x}_1, \ddot{x}_2, \dot{x}_2, \ddot{x}_3, \dot{x}_3$ and $\ddot{y}_1, \dot{y}_1, \ddot{y}_2, \dot{y}_2, \ddot{y}_3, \dot{y}_3$ can be established. These established relationships can then be substituted into Eqs. (6)-(9) and (11). The resulted Eqs. (6)-(11), along with Eqs. (2)-(4) and the derived equations for cutting forces, can then be substituted into Eqs. (18)-(20) to obtain three equations of motions that are functions of $\ddot{x}_2, \dot{x}_2, x_2, \ddot{y}_2, \dot{y}_2, y_2, \ddot{z}_t, z_t, z_t'$ and time, t, as follows

$$f_1(t)\ddot{x}_2 + f_2(t)\dot{x}_2 + f_3(t)x_2 = f_4(t, x_2, y_2, z_t, z_t') \tag{32}$$

$$f_5(t)\ddot{y}_2 + f_6(t)\dot{y}_2 + f_7(t)y_2 = f_8(t, x_2, y_2, z_t, z_t') \tag{33}$$

$$M_Z\ddot{z}_t + k_Z z_t + k_{ZC} z_t^3 = f_9(t, x_2, y_2, z_t, z_t') \tag{34}$$

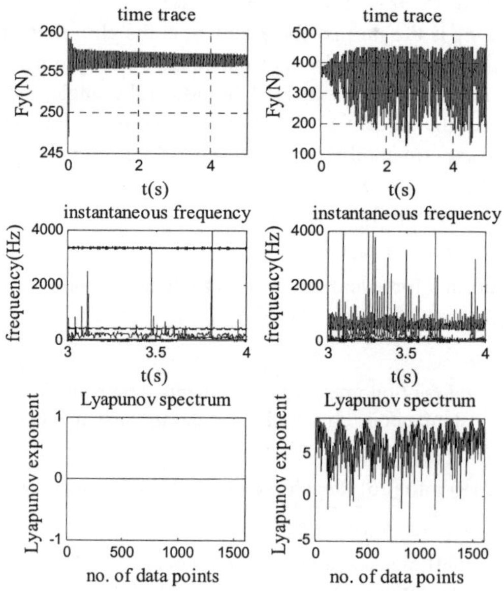

Fig. 8 Y–direction cutting force responses for DOC = 1.62mm (left) and DOC = 2.49mm (right) at Ω = 1250rpm with whirling

3. SDOF Model With Uncoupled EQM For The Tool

Tool motions can be uncoupled from workpiece vibrations by dropping the x and y parameters from the 3D coupled formulation. The resulted equation of motion for tool vibration is a function of z only (thus SDOF) and can be derived from Eq. (20) by eliminating x and y as

$$M_Z\,\ddot{z}_t + k_Z\,z_t + k_{ZC}\,z_t^3 = F_{ZZ} \tag{35}$$

where F_{ZZ} , the resulted/new cutting force function of z only, can be obtained from Eqs. (19) and (20) using Eqs. (36) and (37) below, which define the resulted/new depth-of-cut and new chip cross sectional area. Note that Eqs. (36) and (37) account for motions along the Z-direction only.

$$s_n = r_1 - r_3 \tag{36}$$

$$A_{Cn} = (t_0 - z_t + z_t^{'})(r_1 - r_3) \tag{37}$$

The uncoupled cutting force in the Z-direction can then be derived by modifying Eq. (24) as

$$F_{ZZ} = F_{fn}(b_{z1}\sin(\eta_{cn}) + b_{z2}\cos(\eta_{cn})) + b_{z3}F_{nn} \tag{38}$$

Here η_{cn} , F_{nn} and F_{fn} are the resulted/new friction, normal force and chip flow angle. They can be derived by modifying Eqs. (25)-(27) with s_n and A_{cn} as

$$\eta_{cn} = \alpha_1 s_n^{\,3} + \alpha_2 s_n^{\,2} + \alpha_3 s_n + \alpha_4 \tag{39}$$

$$F_{nn} = k_n A_{Cn} \tag{40}$$

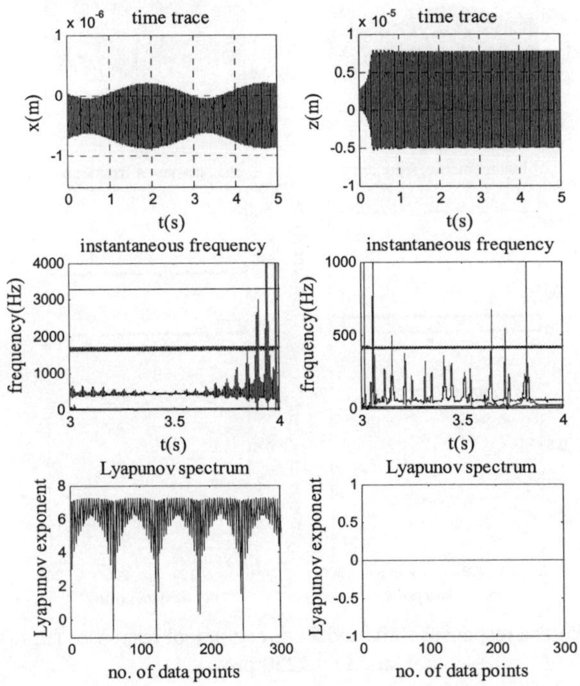

Fig. 9 X–direction workpiece responses (left) and Z- direction tool responses (right) for DOC = 1.00mm and Ω = 1250rpm

$$F_{fn} = k_f A_{Cn} \tag{41}$$

Consequently, the equation governing the uncoupled tool motion is

$$M_Z \ddot{z}_t + k_Z z_t + k_{ZC} z_t^3 = C_1(t_0 - z + z') + C_2(t_0 - z + z')^2 \tag{42}$$

Note that constants C_1 and C_2 depend on tool geometry, chip flow angle and DOC.

4. Numerical Results and Discussion

Both the models are numerically studied to obtain time and frequency domain responses, and Lyapunov spectra. To compare with the experimental results available in [10], a constant spindle speed $\Omega = 1250$rpm, a constant chip width $t_0 = 0.0965$mm, and an eccentricity $\varepsilon_1 = 0.2$mm are considered along with several different DOCs including DOC=1.62mm and 2.49mm. The workpiece considered is of 4140 steel and 0.25m in length (l_0), with the radius of the machined section, $r_3 = 20.095$mm. The starting location of the carbide tool is set at 0.15m from the chuck. There are three types of plots in figures found in the section. The top rows are time history plots, whereas the middle rows show their corresponding time-frequency responses obtained using instantaneous frequency [28]. The last rows give the Lyapunov spectra where the largest Lyapunov exponents are plotted. Instantaneous frequency is employed to realize subtle features that are characteristic of machining instability. Responses of the 3D model (of coupled EQMs) are compared with experimental results available in the literature to validate the model.

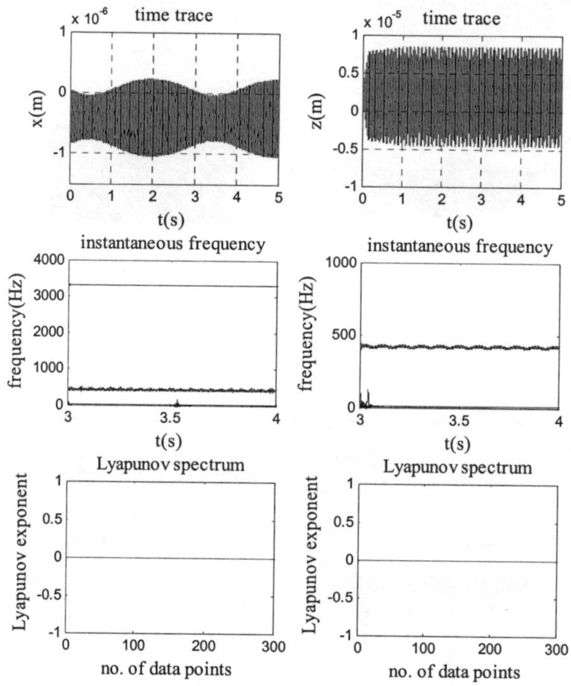

Fig. 10 X–direction workpiece responses (left) and Z- direction tool responses (right) for DOC = 1.25mm and $\Omega = 1250$rpm

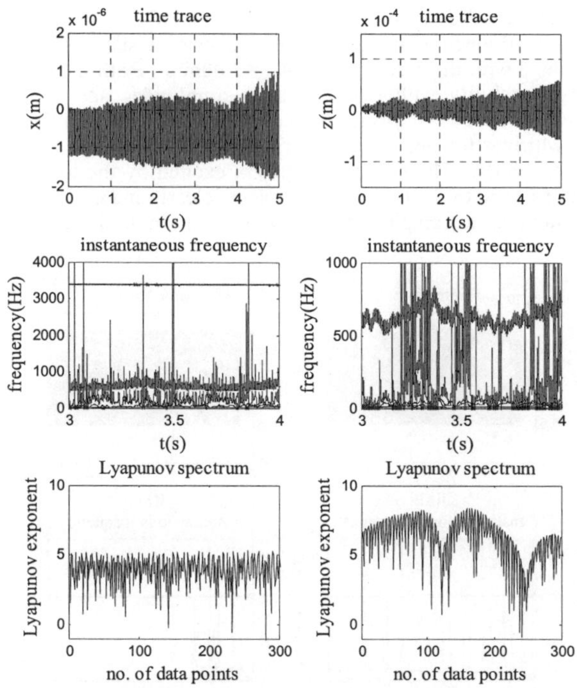

Fig. 11 X–direction workpiece responses (left) and Z- direction tool responses (right) for DOC = 2.00mm and Ω = 1250rpm

The experimental results of Case #4 found in [10] are considered. The paper suggests DOC = 1.75mm as the stable limit and DOC = 1.78mm as the unstable limit. The set of experimental data were taken 9% below the predicted stable limit at DOC = 1.62mm and 40% above the predicted unstable limit at DOC = 2.49mm. To compare with the physical results, the same DOCs, cutting parameter values and tool geometry are adopted in the numerical simulations presented herein. In [10], cutting stability was investigated by analyzing the Y-direction cutting forces in the spectral domain. Fig. 8 shows the time histories, instantaneous frequency and Lyapunov spectra of the Y-direction force components, Fy, for DOC = 1.62mm and DOC = 2.49mm. All responses clearly indicate a stable and a chaotic state of motions, thus agreeing well with the physical data in [10].

Several incremental DOCs are considered to investigate the stability states of the tool and workpiece. All the responses are at a constant spindle speed Ω = 1250rpm. As workpiece behaviors in the X- and Y-directions are similar, only results associated with the X-direction are considered. There are three main frequencies dominate in the workpiece-tool machining system. They are the workpiece natural frequency at around 3300Hz, tool natural frequency at around 420Hz, and whirling frequency at 20.8Hz. Note that the workpiece characteristic frequency varies as its diameter and stiffness vary due to material removal.

Fig. 9 demonstrates the workpiece and tool responses for DOC = 1.00mm. The Lyapunov spectra show that the workpiece is in a motion state different form the tool. The workpiece response is unstable with positive Lyapunov exponents, while the tool response shows a stable condition. The workpiece time-frequency response shows comparably broadband behavior at the tool natural frequency at 420Hz. Moreover, along side the workpiece characteristic frequency at 3280Hz, a 1650Hz frequency is present.

The frequency is a result of period-doubling bifurcation. When DOC is increased to 1.25mm (Fig. 10), all the plots for both the workpiece and tool responses display a stable state of machining. Fig. 11 displays the results associated with the case when DOC = 2.00mm. Both the tool and workpiece see the worst case of machining instability with large vibration amplitudes and broad spectra in the time-frequency plots. This unstable state is confirmed by the positive exponents in the Lyapunov spectra. It is interesting to notice that whirling frequency disappears when the workpiece starts to vibrate chaotically, indicating that it no longer whirls. The broadband spectra excited by the tool natural frequency, along with the aperiodic vibrations of both the tool and workpiece, clearly indicate that the system is in chaos. This would lead the system to a complete failure and result in tool damage and poor workpiece dimensional quality.

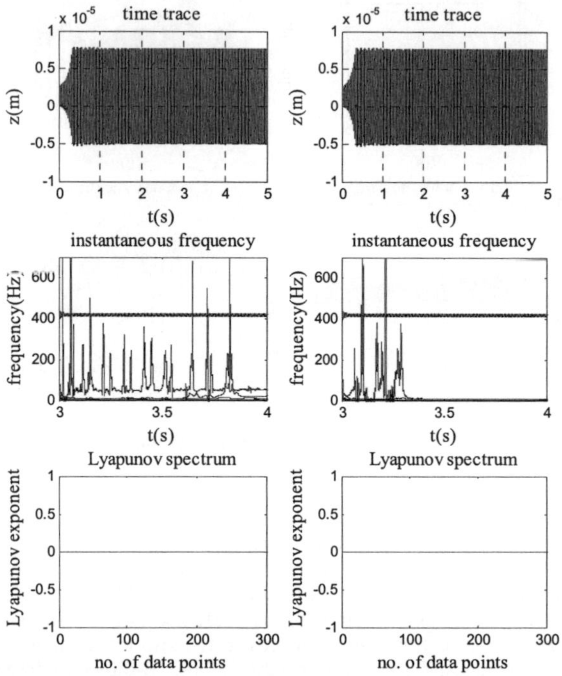

Fig. 12 Z – direction tool responses for coupled equation(left) and Z- direction tool responses for uncoupled equation(right) for DOC = 1.00mm at Ω = 1250rpm

Tool response of the 3D model is compared with the tool response of the 1D model at Ω = 1250rpm. In figures that follow, the left columns represent the tool responses of the coupled model whereas the right columns plot the tool behaviors of the uncoupled SDOF system. In Fig. 12 the instantaneous frequency plots show two different stability stages for the coupled and uncoupled versions. The tool response of the 3D model is of broadband characteristics at 60Hz. Moreover, after 3.7 seconds, another frequency component appears at around 20Hz. The instantaneous frequencies of the SDOF model show that the tool is dynamically stable condition after 3.4 seconds. Recall that the workpiece response for DOC = 1.00mm in Fig. 9 demonstrated unstable situation.

Tool responses corresponding to DOC = 1.75mm indicate a stability state for both the models in Fig. 13. However, the 1D model reveals that the tool vibrates only with its natural frequency while the 3D model has a frequency component other than the tool characteristic frequency at 20Hz. When DOC is increased to 1.83mm in Fig. 14, tool vibration amplitudes are observed to significantly reduce for both models. Even though time domain data exhibit a stable state whose vibration amplitude is only of a few nanometers, the corresponding instantaneous frequencies and Lyapunov exponents depict a completely different situation. The tool response of the 3D model illustrates dynamic instability displaying a broadband component at 3400Hz. In contrast, the 1D model demonstrates a comparably stable state.

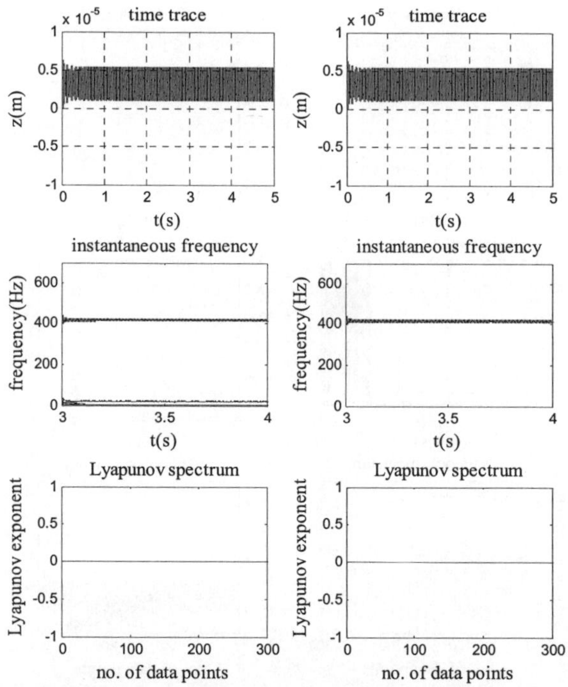

Fig. 13 Z – direction tool responses for coupled equation(left) and Z- direction tool responses for uncoupled equation(right) for DOC = 1.75mm at Ω = 1250rpm

5. Conclusion

Concerns over two major issues were raised and discussed; namely, negligence of workpiece vibration in modeling turning operation and use of SDOF formulation for modeling cutting dynamics. It was demonstrated that the various responses of the presented 3D cutting model agreed well with experimental data found in literature. Ref. [10] reported that chatter was detected at DOC = 2.49mm, but not at DOC = 1.62mm. Model responses shown in Fig. 8 were perfectly in line with the real-world observations. Furthermore, in Fig. 8 the workpiece response demonstrated instability at DOC = 1.00mm, while the tool response was comparably stable. The observations point to a major inconsistency in the dynamic states in which the workpiece and tool are experiencing. They also raise the concern for disregarding workpiece vibrations and for risking instability and ultimate dynamic failure by monitoring only tool vibrations. Chatter at small DOCs has been both predicted theoretically [9] and observed experimentally [10]. Fig. 8 is in agreement with these results and observations. Neglecting workpiece vibrations in modeling fine turning or finish cuts would misinterpret machining dynamics and inevitably impact the surface finish and geometrical tolerance of the final product. The tool responses of the 3D model with coupled EQMs demonstrated more frequency components indicative of instability than those of the SDOF model of a single uncoupled EQM. The SDOF model was shown to be incapable of capturing various instabilities. It was evident that the SDOF model miscomprehended the underlying cutting dynamics. These observations seem suggest against the use of one dimensional equation of motion for studying machining dynamics and for establishing stability limits.

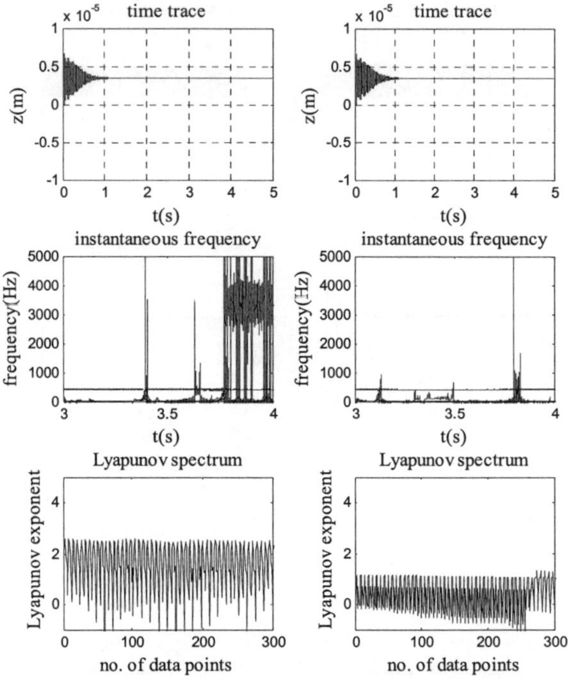

Fig. 14 Z – direction tool responses for coupled equation(left) and Z- direction tool responses for uncoupled equation (right) for DOC = 1.83mm at Ω = 1250rpm

References

[1] R. N. Arnold, 1946, “The mechanism of tool vibration in the cutting of steel,” Proceedings of The Institution of Mechanical Engineering, 154, pp. 267-284.

[2] T. E. Statan and J. H. Hyde., 1925, “An experimental study of the forces exerted on the surface of the cutting tool,” Proceedings of The Institution of Mechanical Engineering, 1(2), pp.175-195.

[3] H. E Merit, 1965, “Theory of self excited machine tool chatter: Contribution to machine tool chatter,” ASME Journals of Engineers for Industry, 84, pp. 447-454.

[4] Minis and R. Yanushevsky, 1993, “A New Theoretical Approach for the Prediction of Machine Tool Chatter in Milling,” ASME Journal of Engineering for Industry, 115 pp. 1-8.

[5] Y. Altintas and E. Budak, 1995, Analytical Prediction of Stability Lobes in Milling, Annals of CIRP, 44(1), pp. 357-362.

[6] Y.S. Chiou, E. S. Chung, and S. Y. Liang, 1995, “Analysis of tool wear effect on chatter stability in turning,” International Journal of Mechanical Sciences, 37(4), pp. 391-404.

[7] T. Schmitz and R. Donaldson, 2000, "Predicting High-Speed Machining Dynamics by Substructure Analysis," CIRP Annals, 49(1), pp. 303-308.
[8] Bason E. et al., 2002, "A Comprehensive Chatter Prediction Model for Face Turning Operation Including Tool Wear Effect," International Journal of Machine Tools and Manufacture, 42, pp. 1035-1044.
[9] M. P. Volger, R. E. DeVor, and S. G. Kapoor, May 2002, "Nonlinear Influence of Effective Lead Angle in Turning Process Stability," Journal of Manufacturing Science and Engineering, 124, pp. 473-475.
[10] Balkrishna C. Rao and Yung C. Shin, 1999, "A Comprehensive Dynamic Cutting Force model for Chatter Prediction in Turning," International Journal of Machine Tools and Manufacture, 9(10), pp. 1631-1654.
[11] Kim J. S. and Lee B. H., 1990, "An Analytical Model of Dynamic Cutting Forces in Chatter Vibration," International Journal of Machine Tools and Manufacture, 31(3), pp. 371-381.
[12] C.S. Suh, P. P. Khurjekar, and B. Yang, 2002 "Characterization and Identification of Dynamic Instability in Milling Operation," Mechanical Systems and Signal Processing, 16(5), pp. 829-848.
[13] Naren Deshpande and M. S. Fofana, 2001 February, "Non-linear Regenerative Chatter in Turning," Robotics and Computer Integrated Manufacturing, 17(1-2), pp 107-112.
[14] Nayfeh et al., Nov 1997, "Perturbation Methods in Non-linear Dynamics- Applications to Machining Dynamics," Journal of Manufacturing Science and Engineering, 119, pp. 485-492.
[15] Igor Grabec, 1988, "Chaotic Dynamics of the cutting process," International Journal of Machine Tools and Manufacture, 28(1), pp. 19-32.
[16] L. Carrino et al., 2002, "Dimensional Errors in Longitudinal Turning Based on the Unified Generalized Mechanics of Cutting Approach," International Journal of Machine Tools and Manufacture, 42(14), pp. 1509-1515.
[17] D. R. H. Jones, 1996, "Whirling Failure in a Woodworking Lathe," Engineering Failure Analysis, 3(1), pp. 71-76.
[18] W. Grzesik, 1996, "A Revised Model for Predicting Surface Roughness in Turning," Wear, 194, pp. 143-148.
[19] Y. Sahin and A.R. Motorcu, 2005, "Surface Roughness Model for Machining Mild Steel with Coated Carbide Tool," Materials and Design, 26(4), pp. 321-326.
[20] M. Thomas et al., 1996, "Effect of Tool Vibration on Surface Roughness during Lathe Dry Turning Process," Computers and Industrial Engineering, 31(3-4), pp. 637-644.
[21] N. H. Hanna and S. A. Tobias, 1974, "Theory of Non-linear Regenerative Chatter," ASME Journal of Engineering for Industry, 96, pp. 247-255.
[22] Grzegorz Litak, 2002, "Chaotic Vibrations in a Regenerative Cutting Process," Chaos, Solutions, and Fractals, 13(7), pp. 1531-1535.
[23] Gouskov et al, 2002, "Nonlinear Dynamics of a machining System with Two Independent Delays," Communications in Nonlinear Science and Numerical Simulations, 7, pp 207-221.
[24] M. Wiercigroch and A.H.D. Cheng, 1997, "Chaotic and Stochastic Dynamics of Orthogonal Metal Cutting," Chaos, Solutions, and Fractals, 8(4), pp 715-726.
[25] X.S. Wang, J. Hu and J.B. Gao, 2005, "Nonlinear Dynamics of regenerative Cutting Process – Comparison of Two Models," Chaos, Solutions, and Fractals, *In Press, Corrected Proof, Available online 11 October 2005*
[26] N. K. Chandiramani and T. Pothala, 2006, "Dynamics of 2DOF Regenerative Chatter during Turning," Journal of Sound and Vibration, 290, pp. 448-464.
[27] Berger et al., 1998, "Wavelet Based Cutting State Identification," Journal of Sound and Vibration, 213(5), pp. 813-827.
[28] N. E. Huang et al., 1998, "The Empirical Mode Decomposition and the Hilbert Spectrum for Nonlinear and Non-stationary Time Series Analysis", Proceedings of the Royal Society of London Series A, 454, pp.903 – 905.
[29] B. Yang and C. S. Suh, 2003, "Interpretation of Crack-Induced Rotor Nonlinear Response Using Instantaneous Frequency," Mechanical Systems and Signal Processing, 18, pp. 491-513.
[30] Wolf et al., 1985, "Determining Lyapunov Exponents from a Time Series Analysis," Physica, 16D, pp. 285-317.

Nonlinear Dynamics and Optimization of Spur Gears

Francesco Pellicano*, Giorgio Bonori, Marcello Faggioni, Giorgio Scagliarini
Department of Mechanical and Civil Engineering University of Modena and Reggio Emilia
Dip. di Ingegneria Meccanica e Civile, V. Vignolese, 905, 41100 Modena

Abstract

In the present study a single degree of freedom oscillator with clearance type non-linearity is considered. Such oscillator represents the simplest model able to analyze a single teeth gear pair, neglecting: bearings and shafts stiffness and multi mesh interactions. One of the test cases considered in the present work represents an actual gear pair that is part of a gear box of an agricultural vehicle; such gear pair gave rise to noise problems. The main gear pair characteristics (mesh stiffness and inertia) are evaluated after an accurate geometrical modelling. The meshing stiffness of the gear pair is piecewise linear and time varying (in particular periodic); it is evaluated numerically using nonlinear finite element analysis (with contact mechanics) for different positions along one mesh cycle, then it is expanded in Fourier series. A direct numerical integration approach and a smoothing technique have been considered to obtain the dynamic scenario. Bifurcation diagrams of Poincaré maps are plotted according to some sample case study from literature. Optimization procedures are proposed, in order to find optimal involute modifications that reduce gears vibration.

Keywords: Gear, stiffness, profile error, backlash, bifurcation, non-smooth, optimization.

Introduction

In the last twenty years, a growing interest was addressed to gear noise problems; i.e. to the parameters which affect the vibration of gears.

A strong interaction between noise and dynamic transmission error (DTE) has been clearly proved; several experiments on gear systems have shown that several nonlinear phenomena occur when the dynamic transmission error is present: multiple coexisting stable motions, sub and super harmonic resonances, fold bifurcations, long period subharmonic and chaotic motions. All these dynamics have been clearly demonstrated experimentally [1,19].

In 1977 Azar and Crossley [5] developed one of first models based on digital computations, in order to study impacts problems in lightly loaded gears, showing the accuracy of simplified models.

In 1983 Sato et al.[8] investigated the role of the contact ratio (the average number of gear teeth in contact during a mesh cycle) and pressure angle error (manufacturing error) on the vibration of gears.

An optimization approach was proposed in [9] in order to find teeth profile modifications able to minimize noise. The effect of parabolic teeth modifications on gear vibrations was analyzed in [10].

In [16-17] an innovative approach for gear modelling was proposed, the method was based on a combined surface integral and a finite element solution; such method allows also dynamic analyses.

General approaches [2] for predicting these phenomena have been developed using time varying parametric excitation and piecewise linear characteristic. Many authors also included impact effects,

* Corresponding author, Tel. +39 059 2056154 Fax +39 059 2056129
E-MAIL: francesco.pellicano@unimore.it

developing specific theories, in order to predict periodic motions in generalized, piecewise linear oscillator with perfectly elastic impacts. Bifurcations of periodic solutions in discontinuous systems have been classified by Leine and van Campen [3]. Luo [4] developed a theory based on generic mapping structures, considering discontinuous boundaries in non-smooth dynamic systems.

In the present study a single degree of freedom oscillator with clearance type non-linearity simulates the dynamics of a simple spur gear pair. The model takes into account a time varying mesh stiffness and a constant viscous damping. Bearing and shaft stiffness are neglected, the main gear pair characteristics (mesh stiffness and inertia) have been evaluated after an accurate geometrical modelling and a finite element analysis, the software CALYX is used for such purpose [16, 17]. The time varying meshing stiffness is periodic (the period is the wheel revolution period times the number of teeth), it is sampled within one mesh cycle, each sample is evaluated through a nonlinear finite element analysis, then a discrete Fourier expansion is carried out in order to obtain an analytical representation.

The backlash is modelled by means of a smoothing technique based on transcendent functions; this approach is able to approximate the piecewise linear function, typical for systems with clearance.

A random technique capable to evaluate composite deviation from a perfect involute profile (theoretical geometry), according to manufacturing process tolerances, is given.

One of the test cases considered in the present work is a gear pair, which is a component of a gear box of an agricultural vehicle. The gear box is affected by noise problems, which cannot be prevented using classical engineering tools, present in the most of Gear Handbooks. The source of noise seems to be the gear pair analyzed in the present work.

A direct numerical integration method is first considered for an actual piecewise linear system. The accuracy of the direct numerical integration of the non-smooth system is checked by means of comparisons with the existent literature.

In the case of smoothing approach, adaptive step-size Runge-Kutta and Adams-Gear algorithms are considered. The accuracy of smoothing technique is checked, and the efficiency of the integration algorithms is evaluated.

Using the smoothing technique the case study is investigated in detail: bifurcation diagrams of Poincaré maps are given in order to show the effect of variable mesh stiffness on the dynamic behaviour: such analysis is able to identify the source of noise.

Optimization procedures are developed in order to obtain micro-geometric gear teeth profile modifications that minimize vibrations within a wide range of operating conditions.

Analytical Model

In the present section general equations of motions are derived. Details of a smoothing technique used for approximating the system are shown. A theoretical approach able to simulate statistically manufacturing errors is developed.

<u>Equation of Motion</u>

The theoretical model developed in this paper considers the spur gear pair as a single-degree-of-freedom lumped system. Each gear is represented by a rigid disk, coupled along the line of action through a time varying mesh stiffness $k(t)$ and a constant mesh damping c (see figure 1). $k(t)$ is evaluated through a nonlinear finite element model in which the geometry take into account typical profile modifications used in technological processes, see figure 2a.

Diameters of the disks are the gear base diameters d_{g1} and d_{g2}. The angular position of the driver tooth wheel (pinion) is θ_{g1}, while the angular position of the driven wheel (gear) is θ_{g2}. The rotary inertia are I_{g1} and I_{g2}. The driver torque is $T_{g1}(t)$, while the breaking torque is $T_{g2}(t)$. Shafts and bearings are supposed to be rigid. A time varying excitation $e(t)$ is included in the model, which considers manufacturing errors. When $e(t)$ is positive a lack of material is considered (see figure 2b).

A backlash function is included in order to simulate clearances:

$$f(t)=f\left(\frac{d_{g1}}{2}\theta_{g1}(t)-\frac{d_{g2}}{2}\theta_{g2}(t)-e(t)\right) \tag{1}$$

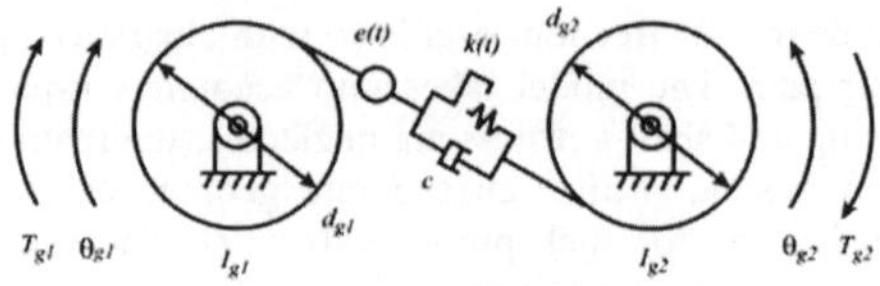

Figure 1. Gear model.

Equations of motions for the gear pair read:

$$I_{g1}\ddot{\theta}_{g1}+c\frac{d_{g1}}{2}\left(\frac{d_{g1}}{2}\dot{\theta}_{g1}-\frac{d_{g2}}{2}\dot{\theta}_{g2}-\dot{e}(t)\right)+\frac{d_{g1}}{2}k(t)f\left(\frac{d_{g1}}{2}\theta_{g1}-\frac{d_{g2}}{2}\theta_{g2}-e(t)\right)=T_{g1}(t) \tag{2}$$

$$I_{g2}\ddot{\theta}_{g2}+c\frac{d_{g2}}{2}\left(-\frac{d_{g1}}{2}\dot{\theta}_{g1}+\frac{d_{g2}}{2}\dot{\theta}_{g2}+\dot{e}(t)\right)+\frac{d_{g2}}{2}k(t)f\left(-\frac{d_{g1}}{2}\theta_{g1}+\frac{d_{g2}}{2}\theta_{g2}+e(t)\right)=-T_{g2}(t) \tag{3}$$

where over-dot means time derivative.

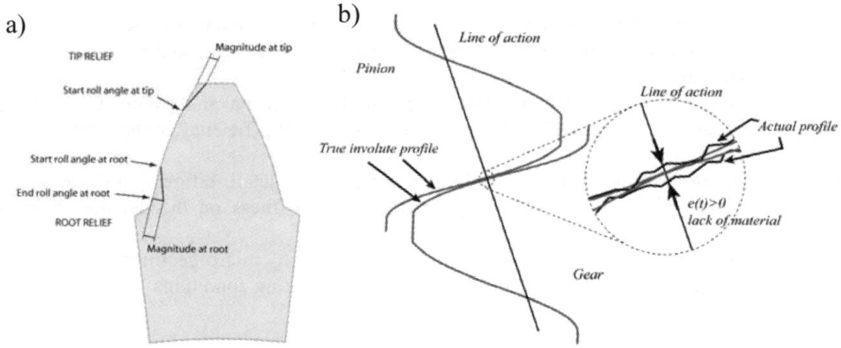

Figure 2. a) profile modification (root and tip relief); b) manufacturing errors: lack of material along the line of action.

The dynamic transmission error x along the line of action is defined as in the following:

$$x(t)=\frac{d_{g1}}{2}\theta_{g1}(t)-\frac{d_{g2}}{2}\theta_{g2}(t) \tag{4}$$

Note that the actual transmission error will be $x(t)$ - $e(t)$.

Using equation (4), after some algebra, equations (1) and (2) can be reduced to the following equation:

$$m_e\ddot{x}(t)+c(\dot{x}(t)-\dot{e}(t))+k(t)f(x(t)-e(t))=T_g(t) \tag{5}$$

where m_e is the equivalent mass:

$$m_e=\frac{1}{\left\{\left(d_{g1}^2/4I_{g1}\right)+\left(d_{g2}^2/4I_{g2}\right)\right\}} \tag{6}$$

T_g is the equivalent applied preload:

$$T_g(t) = m_e\left(\frac{d_{g1}T_{g1}(t)}{2I_{g1}} + \frac{d_{g2}T_{g2}(t)}{2I_{g2}}\right) \tag{7}$$

and we assume $T_{g2}(t) = T_{g1}d_{g2}/d_{g1}$; in the following T_g will be assumed constant. Equation (5) represents the relative dynamics along the line of action. It is to note that T_g is expressed in Newton [N], i.e. it represents a force on the line of action; conversely, T_{g1} and T_{g2} are torques expressed in [N×m].

Backlash

When the clearance is present between two mating teeth, in a gear pair, non-linear phenomena take place [1]. According to literature [5], [6], [7], a non-linear displacement function *f*(*t*) can be used to describe the change of stiffness, which is related to the loosing of contact and the back side low impacts. The gear pair is equivalent to a simple 1dof oscillator (see Figure 3).

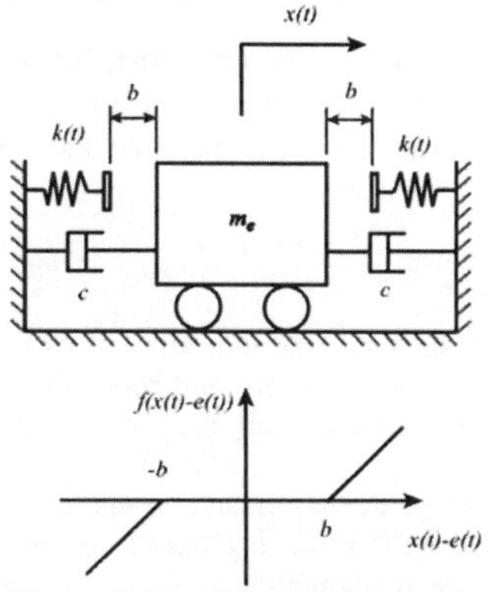

Figure 3. Equivalent gear model.

In this model the gear mesh has a clearance equal to 2*b*; the displacement function, used in describing the restoring force, assumes the following expression:

$$f\left(x(t)-e(t)\right) = \begin{cases} x(t)-e(t)-b & x(t)-e(t) \geq b \\ 0 & \left|x(t)-e(t)\right| \leq b \\ x(t)-e(t)+b & x(t)-e(t) \leq -b \end{cases} \tag{8}$$

Inserting equation (8) into equation (5) a second order piecewise linear time varying differential equation is obtained; for such equation no exact solutions are known.

In order to apply a standard numerical approach to integrate the differential equation (8), a smoothing technique is used. The piecewise linear displacement function, given by equation (8), is approximated by arctangent functions:

$$f(t) = \left(x(t)-e(t)\right) - \frac{1}{\pi}\left\{\left(x(t)-e(t)+b\right)\arctan\left[\alpha\left(x(t)-e(t)+b\right)\right] - \left(x(t)-e(t)-b\right)\arctan\left[\alpha\left(x(t)-e(t)-b\right)\right]\right\} \tag{9}$$

where α is the shape parameter.

In Figure 4 examples of arctangents clarify the quality of the approximation; a simplified form of equation (9) is considered, without loss of generality:

$$y(t) = x(t) - \frac{1}{\pi}\left\{\left(x(t)+1\right)\arctan\left[\alpha\left(x(t)+1\right)\right] - \left(x(t)-1\right)\arctan\left[\alpha\left(x(t)-1\right)\right]\right\} \tag{10}$$

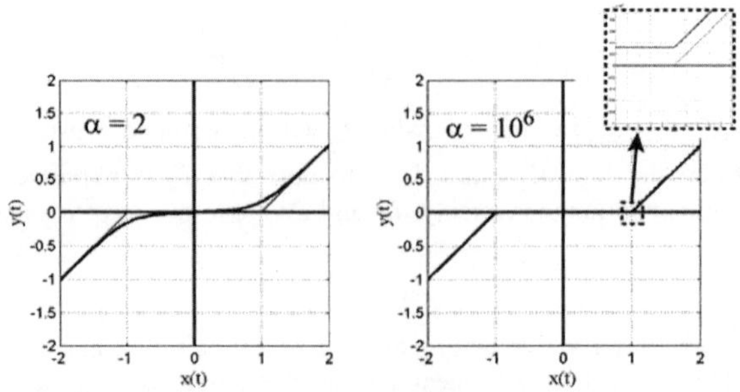

a) b)

Figure 4. Arctangent function: (a) α=10, (b) α=10 zoom, (c) α=10^6.

From the analysis presented in Figure 4 one can argue that, for α>10^6, the approximation is sufficiently accurate. In the following, α=10^8 will be used throughout all computations, such value assures a good approximation of the non-smooth function, at the same time the use of a C^∞ vector field improves the computational efficiency.

Composite Profile Error

Any process used to produce gears results in a certain amount of deviation from the theoretical involute profile. Furthermore, the literature [8-11] suggests tip and root reliefs (profile modifications) to reduce the vibration, which depend on the nominal torque. This profile modification introduces additional deviation from the perfect involute profile.

Many authors, [12-14] identify in shape deviations due to manufacturing errors, a possible source of gear vibration. The estimation of an actual profile error is generally measured using sophisticated and expensive experimental rigs; therefore, few data are available.

In this section a technique is proposed to create a random profile within specified tolerances. A particular manufacturing process quality can be simulated, according to the "K" chart; Figure 5 shows an example of profile fitting tolerance values in a "K" chart, it represents the profile deviation from the involute, projected on the line of action. The roll angle (Figure 5b) is a special coordinates that indicates the position on the tooth flank, see Ref. [15] for details. The deviation from the actual involute profile is provided with respect to normalized roll angle ($\alpha_n = 360(\alpha/\alpha_p)$), where α is the roll angle and α_p is the difference between values of the roll angle at the lowest and the highest point of contact along the tooth profile [15]. Positive values of deviation mean lack of material.

An analytical formulation of the shape of each tooth profile error is developed by means of a Fourier expansion (see Figure 5). The analytical formulation allows to evaluate the composite profile error, during a mesh cycle, as a sum between deviation on meshing teeth, according to the transmission ratio. This approach is repeated $Z_{g1} \times Z_{g2}$ times in order to perform a complete fundamental rotation (Z_{g1} and Z_{g2} are number of teeth on pinion and on gear). The fundamental rotation is the rotation required for the same tooth pair to reach contact in the same position. During such rotation all possible relative teeth contact combinations take place; then, the process is repeated periodically. The composite profile error is approximated by means of a Fourier series:

$$e(t) = \sum_i E_i \cos(i\tilde{\omega}_m t - \gamma_i) \tag{11}$$

where: $\tilde{\omega}_m = \omega_m / (Z_{g1} Z_{g2})$; $\omega_m = Z_{g1}\Omega_{g1}$ is the mesh frequency; Ω_{g1} is the angular velocity of wheel 1 (pinion); E_i and γ_i are amplitude and phase values evaluated through the FFT algorithm.

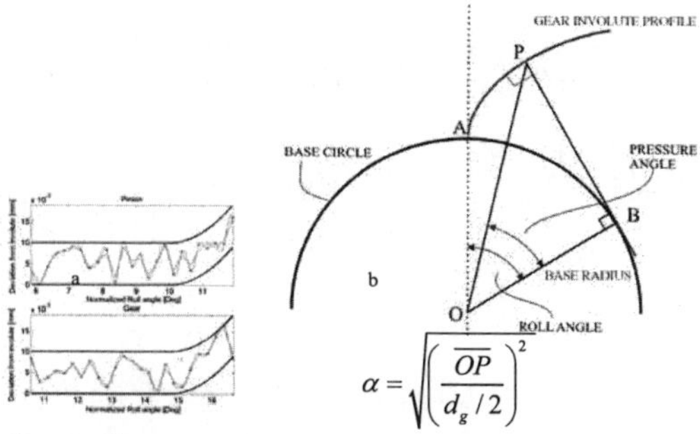

Figure 5. a) Random generation of pinion and gear profile according to "K" chart parabolic tip relief; (blue) random profile; (red) Fourier expansion reconstruction. b) roll angle definition.

Figure 6 shows an example of the spectrum obtained from a composite profile error for a gear pair with profile errors. All frequencies are normalized according to the mesh frequency ω_m. The effect of profile errors results in a periodic excitation with period equal to $2\pi/\omega_m$.

The approach is suitable for simulating manufacturing errors using data regarding production accuracy, but it can be applied on experimental data, when manufacturing errors are measured on a statistical set of actual gears.

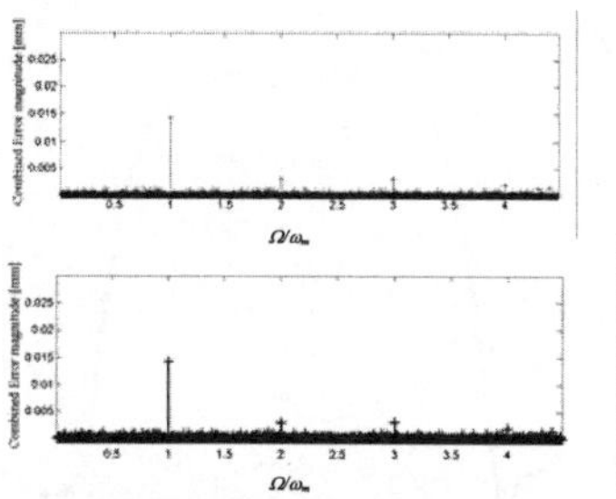

Figure 6. Amplitude of FFT for the composite profile error.

Mesh Stiffness

A commercial software package (CALYX®) is used to evaluate mesh stiffness along one mesh cycle. CALYX® uses a combined surface integral approach and a finite element solution, [16-17]. Furthermore, this technique does not require meshes refinement close to the contact zone to capture the effect of profile modification (see Figure 7).

Twenty gear pair positions, along one mesh cycle, are analyzed for each case study. An analytical formulation of the mesh stiffness is obtained by means of the following Fourier expansion:

$$k(t) = k_0 + \sum_i k_i \cos(i\omega_m t - \varphi_i) \qquad (12)$$

Figure 8 shows a comparison between the mesh stiffness evaluated form finite element analysis and its approximation obtained using equation (12).

Normalization

A dimensionless form of equation (5) is obtained by letting:

$$\omega_n = \sqrt{\frac{k_0}{m_e}};\ \ \zeta = \frac{c}{2m_e\omega_n};\ \ \tau = \omega_n t\ ;\ \ \overline{T}_g = \frac{T_g}{bm_e\omega_n^2}\ ;\ \ \overline{x} = \frac{x}{b};\ \ \overline{e} = \frac{e}{b}$$

$$\overline{f}\left(\frac{x(\tau)-e(\tau)}{b}\right) = \overline{f}\left(\overline{x}(\tau)-\overline{e}(\tau)\right) = \begin{cases} \overline{x}(\tau)-\overline{e}(\tau)-1 & \overline{x}(\tau)-\overline{e}(\tau)\geq 1 \\ 0 & \left|\overline{x}(\tau)-\overline{e}(\tau)\right|\leq 1\ ; \\ \overline{x}(\tau)-\overline{e}(\tau)+1 & \overline{x}(\tau)-\overline{e}(\tau)\leq -1 \end{cases} \tag{13}$$

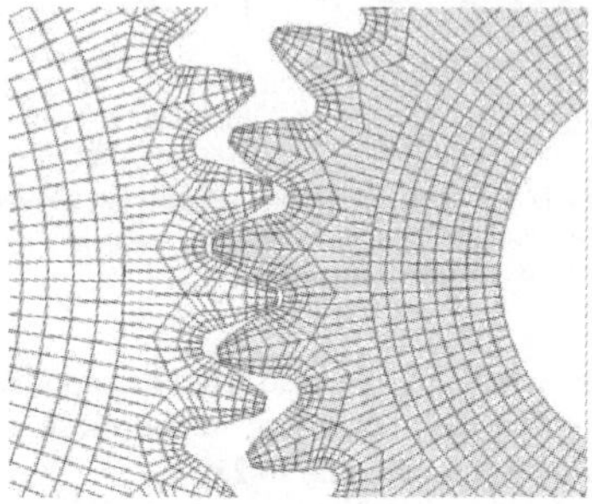

Figure 7. Detail of the gears meshes on the contact area.

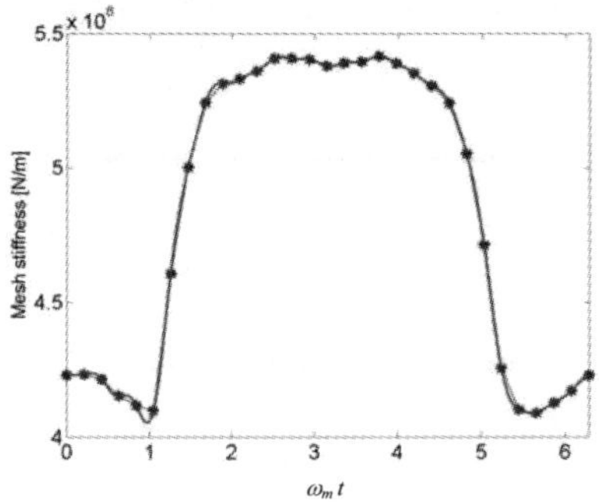

Figure 8. Mesh stiffness; (blue) CALYX®; (red) Fourier series expansion.

and:

$$\overline{k}_i = \frac{k_i}{m_e\omega_n^2}\ ; \qquad \overline{k}(t) = 1+\sum\nolimits_i \overline{k}_i \cos\left(i\frac{\omega_m}{\omega_n}\tau-\varphi_i\right);$$
$$\overline{E}_j = \frac{E_j}{b}\ ; \qquad \overline{e}(\tau) = \sum\nolimits_j \overline{E}_j \cos\left(j\frac{\omega_m}{\omega_n}\tau-\gamma_j\right) \tag{14}$$

Equation (5) assumes the following form:

$$\overline{x}''(\tau)+2\zeta(\overline{x}'(\tau)-\overline{e}'(\tau))+\overline{k}(\tau)\overline{f}(\overline{x}(\tau)-\overline{e}(\tau)) = \overline{T}_g \tag{15}$$

where $(\cdot)' = d(\cdot)/d\tau$.

Comparison with literature

In the present section some numerical tests are carried out on problems studied in literature in order to check the accuracy of the present model.

Trilinear system

Tests are carried out using: *i*) a direct integration of equation (5), i.e. by considering a pure piecewise linear system; *ii*) integration of system (5) approximated by the smoothing function (9). Such test allows to check the accuracy of the smoothing technique.

Let us consider the trilinear system represented by the elastic stiffness represented in Figure 9. Such a system was deeply analyzed in Ref. [18] by means of an analytical procedure. The equation of motion of such system are:

$$m\ddot{x}(t)+c\dot{x}(t)+k(x)=f\sin(\omega t) \tag{16}$$

where

$$k(x)=\begin{cases} k_1 x, & |x|\le x_c \\ k_2 x+(k_1-k_2)x_c\,\mathrm{sgn}(x), & |x|>x_c \end{cases} \tag{17}$$

Note that in such system the boundary x_c plays the role of clearance b present in equation (8).

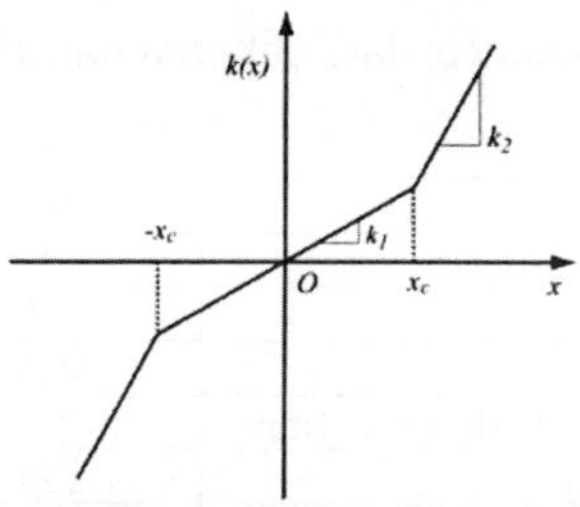

Figure 9. Piecewise linear stiffness Ref. [18].

Equation (16) is now transformed in a nondimensional form $y(t)=x(t)/x_c$

$$\begin{aligned} &\ddot{y}_1+2\zeta\tilde{\omega}\dot{y}_1+\omega_1^2 y_1=f_0\sin(\omega t) && |y|\le 1 \\ &\ddot{y}_2+2\zeta\tilde{\omega}\dot{y}_2+\omega_2^2 y_2=f_0\sin(\omega t)+\mathrm{sgn(y)}\left(\omega_2^2-\omega_1^2\right) && |y|>1 \end{aligned} \tag{18}$$

where $f_0=f/(mx_c)$ and $P=f/(m\omega^2 x_c)$; the latter parameter has no physical meaning, is was used in [18] to parameterize the system.

Simulations are performed by linearyzing equation (16) or integrating it directly without smoothing functions (non-smooth approach). The following parameters are used in simulations: ζ=0.05; k_2=4k_1; $\tilde{\omega}=(\omega_1+\omega_2)/2$; P=0.5.

In Figure 10 the present model is compared with Ref. [18]; in this figure a linearized solution is also included to quantify the nonlinear effect: linearization is accurate for y<1, i.e. below the boundary x_c; the solution is completely wrong for y>1. The direct numerical integration is very close to the solution obtained in [18], similar results are obtained with the smoothing technique.

Comparisons with experiments in literature

In order to investigate the accuracy of the model with respect to real system a comparison with some experimental results was also performed. In 1997 Kahraman and Blankenship [1] presented a number of experiments on a physical system with clearance combining parametric and external forcing excitation [1]. One of the tests concerned a spur gear set with the following geometrical and physical parameters:

A damping value $\zeta = 0.01$ and an external torque T_{gl} of 340 Nm are used. The natural frequency of the system ω_n is 1.983901 10^4 rad/s. Using 20 positions in a mesh cycle the peak to peak value of the mesh stiffness is 6.12079 10^7 N/m. 5 harmonics are considered in the stiffness Fourier expansion: $\bar{k}_1 = 5.882280$ 10^{-2}, $\bar{k}_2 = 4.072213$ 10^{-2}, $\bar{k}_3 = 3.133626$ 10^{-2}, $\bar{k}_4 = 2.176923$ 10^{-2}, $\bar{k}_5 = 1.554604$ 10^{-2}; and no manufacturing errors are included $e(t)=0$. In Figure 11 comparisons between the present model (with smoothing function) and Ref. [1] show that the present theory is accurate in reproducing an actual gear vibration.

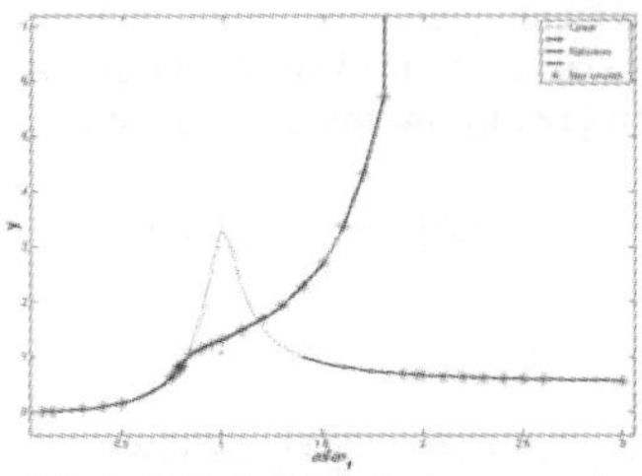

Figure 10. (--) linearized model; (—) Ref. [18]; (--.-) present model, direct integration.

Table 1. Geometrical data of Kahraman's spur gear set.

	Pinion	Gear
Teeth number	50	50
Module [mm]	3	3
Pressure angle [Deg]	20	20
Base diameter [mm]	140.95	140.95
Tooth thickness at pitch diameter [mm]	4.64	4.64
Outer diameter [mm]	156	156
Root diameter [mm]	140.68	140.68
Face width [mm]	20	20
Mass [kg]	2.5161	2.5161
Inertia [kg m^2]	0.0074	0.0074
Young's modulus [MPa]	206000	206000
Poisson's coefficient	0.3	0.3
Center distance [mm]	150	
Backlash [mm]	0.1447	
Profile modifications	None	

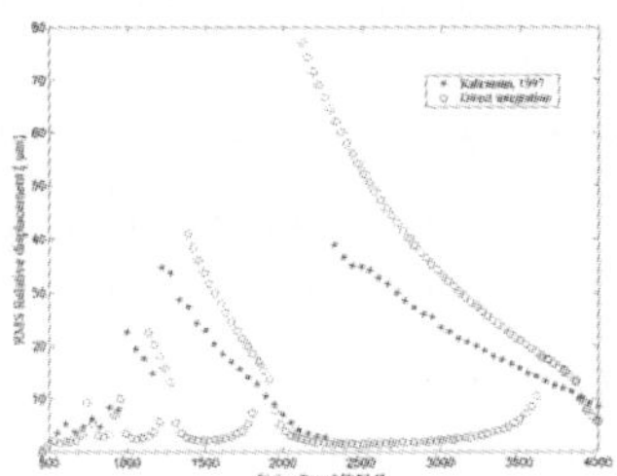

Figure 11. Comparisons with literature: '*' Ref. [1]; 'o' present model.

Optimization

Two optimization strategies are proposed here: the first one is based on the minimization of the static transmission error ($T_g/k(t)$), the second one is based on the dynamic analysis of the system. Both approaches are applied to an actual gear pair.

Table 2 shows the geometrical data of the spur gear pair selected as case study; in this table profile modifications actually used in the gear box production are reported. The goal of the analysis is to verify if such modifications are optimal or the gears could be modified in order to reduce vibration.

Perfect gears are considered, i.e. no manufacturing errors are included in the model: $e(t)$=0 in equation (5).

Table 2. Geometrical data of the spur gear pair.

SISTEM DATA						
Module			3	mm		
Pressure angle			20	deg		
Center distance			111	mm		
Nominal Torque (pinion) T_{g1}			470718.78	Nmm		
GEOMETRIC PROPERTIES			**PINION**		**GEAR**	
Number of teeth			28		43	
Face width			27	mm	22.5	mm
Tooth thickness (on pitch circle)			6.1151	mm	6.7128	mm
Outer diameter			93.1	mm	139.7	mm
Root diameter			79.1	mm	126.2	mm
Inner diameter			40	mm	40	mm
Hob tip radius			0.9	mm	0.9	mm
Addendum modification			1.927	mm	2.74804	mm
Crowning			0.015	mm	0.015	mm
MATERIALS PROPERTIES						
Young Modulus			206000	MPa	206000	MPa
Poisson's ratio			0.3		0.3	
Density			0.00000785	kg/mm^3	0.00000785	kg/mm^3
PROFILE MODIFICATIONS						
Type of modification		1=linear; 2=parabolic	1		1	
Start roll angle at tip			30.15690000	deg	29.21270000	deg
Magnitude at tip			0.01600000	mm	0.01800000	mm
Start roll angle at root			23.47060000	deg	25.20790000	deg
End roll angle at root			14.43340000	deg	20.57640000	deg
Magnitude at root			0.01600000	mm	0.01800000	mm

Static optimization

The first optimization strategy is based on a static approach: the source of vibration is the time varying stiffness, see equation (5), when $e(t)$=0 and $T_g(t)$=constant; in such case a good indicator to consider in noise reduction is the static transmission error (STE) that is simply given by $T_g/k(t)$, see equation (5). Minimizing the amplitude of oscillation (peak to peak) of the static transmission error will result in a reduction of excitation.

Starting from profile modification reported in Table 2, we look for an optimal configuration of (linear) modifications parameters which minimize the peak to peak of static transmission error in order to avoid the dynamic parametric excitation due to the time varying meshing stiffness. Parameters considered for the optimization are:

- Start roll angles at tip on pinion and gear, both are varied between the pitch diameter and the outer diameter;
- Magnitude of modifications at tip on pinion and gear, both are varied between 0 and 40 μm;
- Start roll angles at root on pinion and gear, both are varied between the begin of the involute and the pitch diameter;
- Magnitude of modifications at root on pinion and gear, both are varied between 0 and 40 μm.

The end roll angle at root of the pinion and gear are not considered as optimization parameters, because they are equal to the start of active involute profiles. Therefore, eight parameters are included in the optimization process. In the minimization of the STE peak to peak (PPTE), such parameters will not be considered simultaneously. Conversely, two parameters for time will be varied within a suitable range; therefore, after four steps all parameters will be varied. This is an approximate approach, that doesn't lead to a minimum in a mathematical sense, but allows to obtain 3D surfaces in which we can observe a qualitative influence of the parameters on the PPTE.

In the following the procedure is applied to the test case: the minimum of PPTE versus start roll angles at tip on pinion and gear is found after spanning all possible values, see Figure 12a where a 50×50 grid in the parameters space is considered, the other parameters are left constant. The procedure is repeated for the other parameters pairs: i) the magnitude of modifications at tip on pinion and gear (41×41 grid), Figure 12b; ii) the start roll angles at root on pinion and gear (50×50 grid), Figure 12c; the magnitude of modifications at root (41×41 grid), Figure 12d. It is to note that the PPTE is obtained from the analysis of $k(t)$, which is sampled with 15 samples within a mesh period; therefore, at the end of the optimization process 125430 finite element computations have been carried out.

The result of the optimization process is a new configuration of profile modifications, which reduce the peak to peak of STE from 0.0051 mm to 0.0017 mm.

At each step of optimization we choose a "technological minimum", which doesn't necessarily coincide with the minimum in the 2D surface; indeed, in actual gears the theoretical profile is obtained, because of manufacturing errors. From a practical point of view it is more useful to find a configuration inside a flat region, even though it is not the minimum; the low gradient ensures robustness with respect to perturbations. In Figure 13 is reported an examples where the PPTE surface presents two minima: absolute minimum and technological minimum.

Once the optimization is carried out on static basis, the dynamic analysis must be carried out in order to check the effectiveness of the optimization on gear vibration. In the following amplitude-frequency and bifurcation diagrams are considered in order to carry out the comparison between the initial gears set and the optimized one.

Figure 14 represents results of a simulation in terms of the amplitude-frequency curve; the rotation speed is varied within the range 500-22000 rpm. The initial set of gears (called CNH modifications) presents large amplitude of vibration close to the natural frequency of the system; the response is nonlinear (softening) and loosing of contact is present, as proven from the softening behaviour of the curve. After the optimization, the amplitude of the response is significantly smaller than the initial one; the response is linear. At low frequencies the optimization doesn't necessarily reduce the amplitude of the response, because minimizing the PPTE does not imply a reduction of higher harmonics of $k(t)$, see equation (14).

It is of interest to analyze the gear pair dynamic behaviour in the case of perfectly involute profiles (no profile modifications): this case is quite rare in practical applications, indeed, profile modifications are always present in gears loaded with non negligible torques.

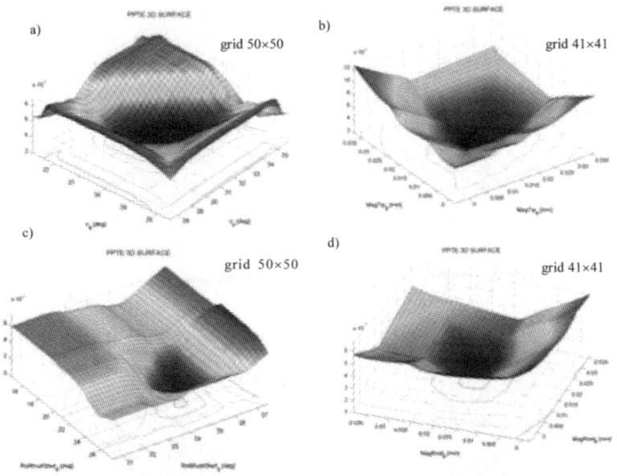

Figure 12. Minimum of STE peak to peak vs.: a) start roll angles at tip on pinion and gear; b) magnitude of modifications at tip on pinion and gear; c) start roll angles at root on pinion and gear; d) magnitude of modifications at root.

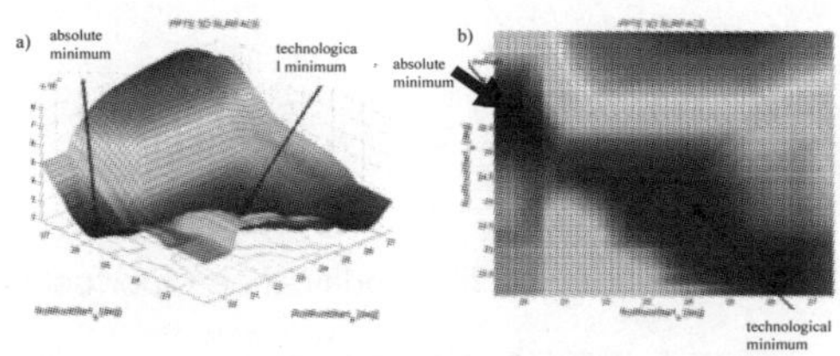

Figure 13. Technological optimum.

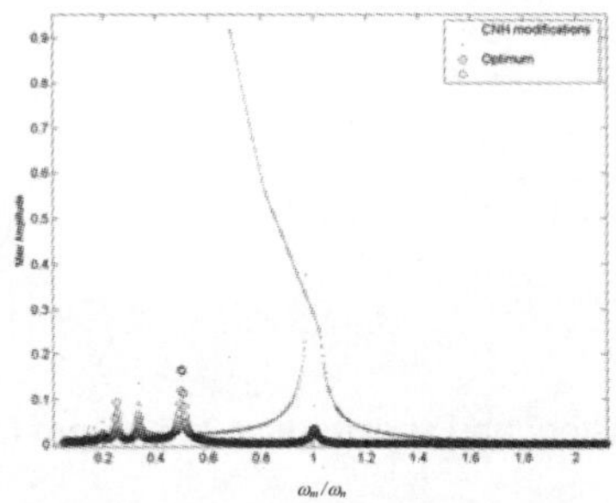

Figure 14. Amplitude-frequency diagram: "·" initial modifications; "o"present optimization.

In absence of profile modifications, gears exhibit the appearance of a parametric resonance, when the normalized frequency is equal to 2. The dynamic instability gives rise to a sudden change of the dynamic behaviour, a jump is clearly visible, see Figure 15. The parametric instability is due to the time varying stiffness that gives rise to a Mathieu type instability and the jump is due to the non-smoothness of the dynamical system. In presence of profile modification the parametric resonance disappears.

The bifurcation diagram, Figure 16, shows that the dynamic scenario, in the case of gears without profile modifications, is particularly rich; the parametric instability leads to very amplitude of vibration when the excitation frequency is decreased. In Figure 17 the time history of the dynamic transmission error (DTE), when the pinion speed is 21535 rpm, is represented. The spectrum confirms that response is 2T subharmonic.

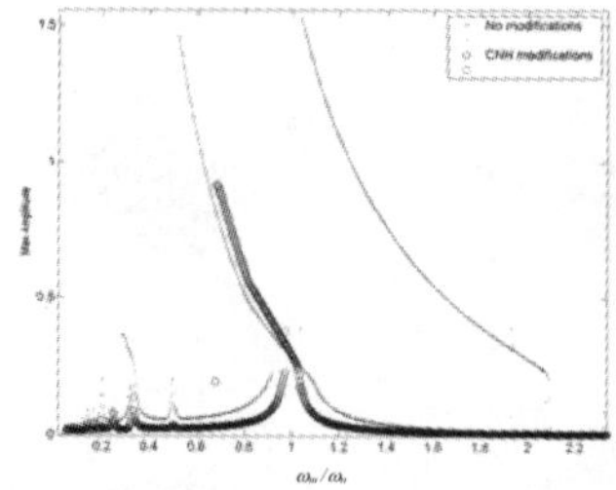

Figure 15. Amplitude-frequency diagram: "·" perfect involute; "o"initial modifications.

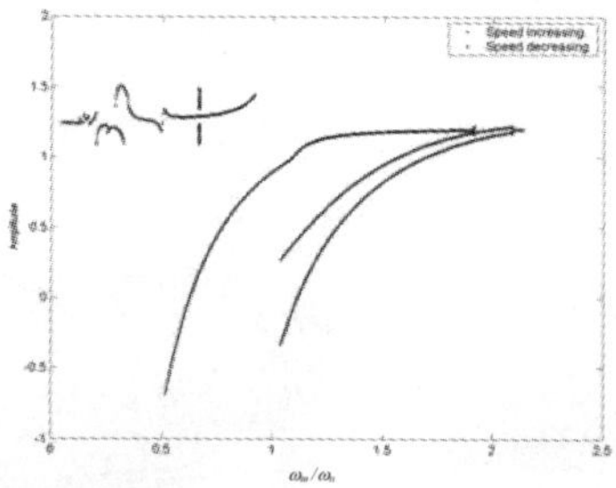

Figure 16. Bifurcation diagram: no profile modifications (perfect involute).

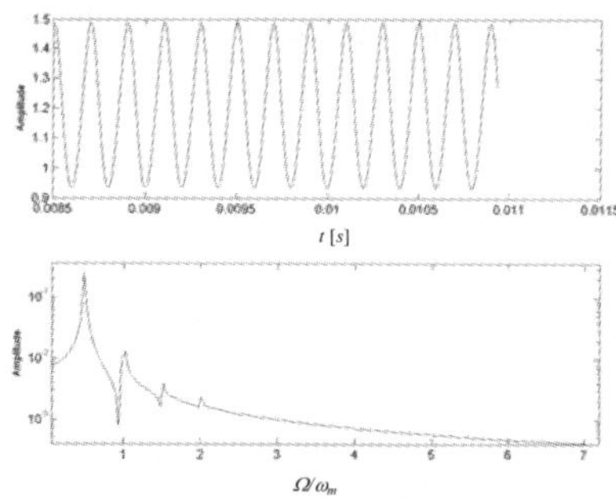

Figure 17. Time history and spectrum at 21535 rpm (perfect involute).

Dynamic optimization

In this section an optimization procedure based on the dynamic response of the system is proposed, the goal is to minimize the maximum amplitude of the response in a range of operating conditions in terms of rotation speed.

The objective function to be minimized is the maximum value of RMS (root mean square) of the gear response (DTE) in a certain range of excitation frequency. It is to note that, in view of the strong nonlinearities present in the system, in the numerical simulations, it is of extreme importance the direction followed in varying the frequency. In the following decreasing excitation frequencies will be considered, because the gear dynamics shows generally a softening behaviour.

Two parameters have been considered in the optimization: the start diameter of tip relief profile modification for both the pinion and the gear. The start tip diameter has been varied between the original value (44.54 and 67.64 mm for pinion and gear respectively, corresponding to roll angles indicated in Table 2) and the outside diameter (see Table 2), the range is discretized using ten samples: a 10x10 grid of values is created. Three different torques are considered: 100%, 66% and 33% of the nominal torque.

In the following the analysis will be carried out on a set of gears without root profile modifications; moreover, the optimization considers also the variation of the static torque. Therefore, the following optimizations are not comparable with the previous section. In order to allow a comparison, a static optimization is carried out using the average peak to peak value of the mesh stiffens over three torque values: 100%, 66% and 33% of the nominal torque; moreover, start tip diameter has been varied both for pinion and gear, similarly to the dynamic optimization.

In Figure 18 the peak to peak of non the dimensional stiffness $k(t)$ for the 3 torque levels is represented versus profile modifications. The optimum is found after averaging the three cases, obtaining an "averaged" optimum.

In Figure 19 the maximum RMS value for the 3 torque levels is represented versus profile modifications. The optimum is found after averaging the three cases, obtaining a dynamic "averaged" optimum.

In Table 3 two possible gears optimizations are proposed on the basis of a static or a dynamic analysis. In order to compare these optima, a dynamic simulation has been performed for the three load cases considered in the optimization. Figure 20 shows that the original configuration (Table 2, without root relief) presents high value of dynamic transmission error (DTE) for all the load cases. After static optimization the dynamic scenario does not change for the nominal torque (Figure 21); however, at 33% the behaviour is improved and at 66% the dynamic response is minimized. In the case of dynamic optimization (Figure 22) the scenario is similar. This is not surprising in view of Table 3, which shows how both procedures lead to similar profile modifications. The scarce quality of the optimization at 100% is expected, because in the optimization the average between 33%, 66% and nominal loads are considered as objective function; this leads to small modifications at the nominal load.

Table3. Optimized configurations

	Pinion	Gear
Start tip diameter [mm] from peak to peak of $k(t)$ optimization	45.88	68.48
Start tip diameter [mm] from RMS optimization	44.99	68.06

Conclusion

In the present paper the vibration of a spur gear pair is studied. A piecewise time varying 1dof model has been developed by taking advantage from a static analysis carried out by means of a finite element code, which include contact mechanics analysis.

A smoothing technique has been developed in order to improve the computational efficiency and reduce the integration time. The accuracy of the smoothing technique has been tested by comparisons with the direct approach.

Static and dynamic optimization techniques are proposed in order to reduce gears vibrations. The static optimization is computationally lighter than the dynamic optimization; therefore, a full optimization can be carried out, greatly improving the gear behaviour.

A preliminary comparison among dynamic and static optimizations show that, both techniques seems to lead to the same optimum; confirming the effectiveness of minimizing the peak to peak of the static transmission error.

Acknowledgments

The authors thank: Case New Holland (Italy) for supporting the present research; Advanced Numerical Solutions (USA) for providing the software CALYX®.

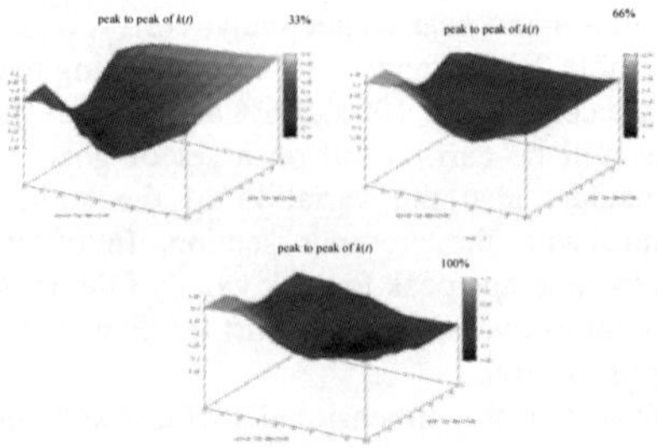

Figure 18. Peak to peak of non dimensional stiffness.

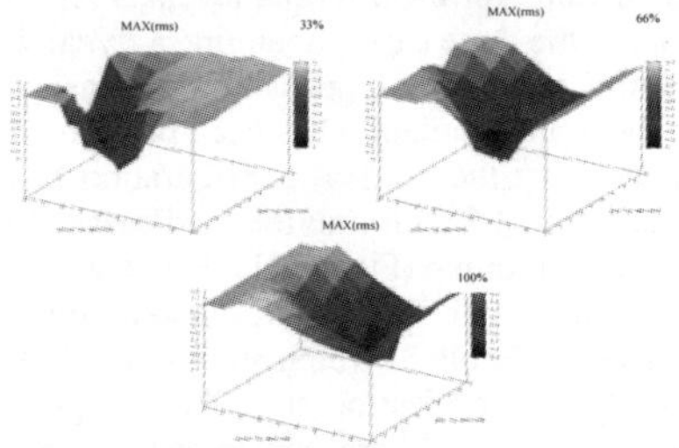

Figure19. MAX(rms) vs. profile modifications for three torque levels: nominal torque; 66%; 33%.

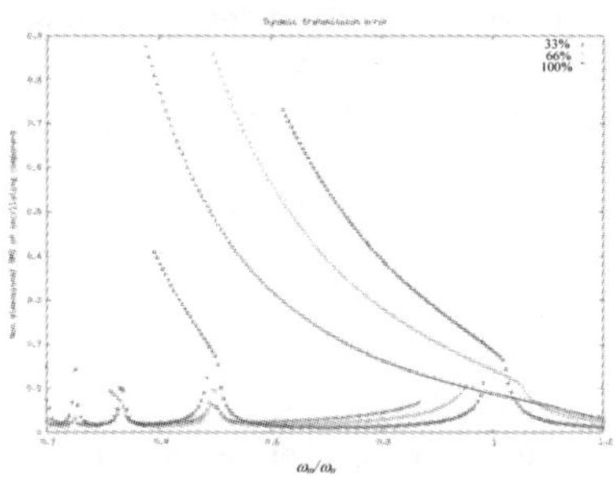

Figure 20. Amplitude frequency curve: initial profile modifications.

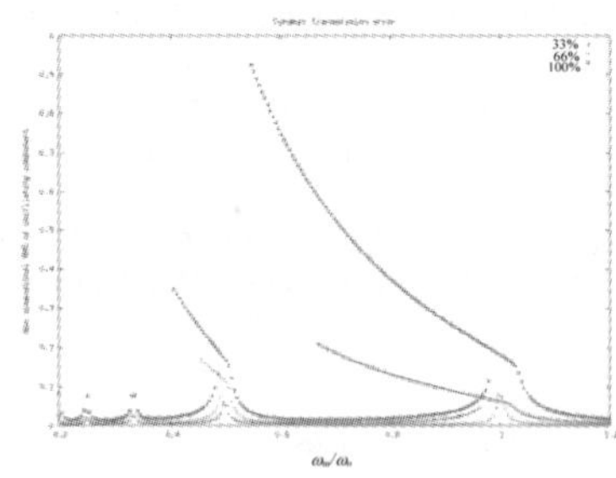

Figure21. Amplitude frequency curve: static optimization

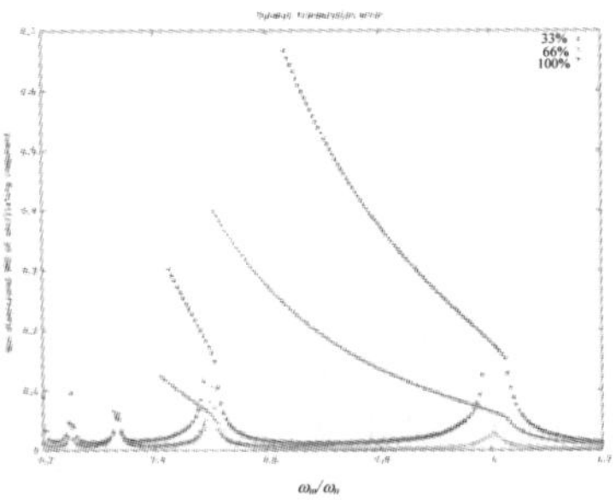

Figure 22. Amplitude frequency curve: dynamic optimization.

References

[1] A. Kahraman, G.W. Blankenship, Experiments on nonlinear dynamic behavior of an oscillator with clearance and periodically time-varying parameters, Journal of Applied Mechanics 64 (1997) 217-226.

[2] Y.B. Kim, S.T. Noah, Stability and bifurcation analysis of oscillators with piecewise linear characteristics: a general approach, Journal of Applied Mechanics 58 (1991) 545-553.

[3] R.I. Leine, D.H. van Campen, Discontinuous Bifurcations of Periodic Solutions, Math. Computer Modelling 36 (2002) 259-273.

[4] A.C.J. Luo, A theory for non-smooth dynamic systems on the connectable domains, Comm. Nonlin. Sci. Num. Sim., 10 (2004), pp. 1-55.

[5] R.C. Azar, F.R.E. Crossley, Digital simulation of impact phenomenon in spur gear system, Journal of Engineering for Industry, (1977) 792-798.

[6] G.R. Tomlinson, J. Lam, Frequency response characteristics of structures with single and multiple clearance-type non-linearity, Journal of Sound and Vibration, 96(1) (1984) 111-125.

[7] G.W. Blankenship, A. Kahraman, Steady state forced response of a mechanical oscillator with combined parametric excitation and clearance type non-linearity, Journal of Sound and Vibration, 185(5) (1995) 743-765.

[8] T. Sato, K. Umezawa, J. Ishikawa, Effect of contact ratio and profile correction on gear rotational vibration, Bulletin of Japanese Society of Mechanical Engineering, 26(221) (1983) 2010-2016.

[9] Y. Cai, T. Hayashi, The optimum modification of tooth profile for a pair of spur gears to make its rotational vibration equal zero, ASME International Power Transmission and Gearing Conference, DE-43(2) (1992) 453-460.

[10] H.H. Lin, F.B. Oswald, D.P. Townsend, Dynamic loading of spur gears with linear or parabolic tooth profile modifications, Mech. Mach. Theory, 29(8) (1994) 1115-1129.

[11] W.S. Rouverol, New modifications eradicate gear noise and dynamic increment at all loads, ASME Power Transmission and Gearing Conference, DE-88 (1996) 17-21.

[12] Munro, R.G., Effect of geometrical errors on the transmission of motion between gears, Proc. Instn. Mech. Engrs., 184(30) (1969) 79-84.

[13] K. Umezawa, T. Sato, Influence of gear errors on rotational vibration of power transmission spur gear, Bulletin of Japanese Society of Mechanical Engineering, 28(243) (1985) 2143-2148.

[14] P. Velex, M. Maatar, A mathematical model for analyzing the influence of shape deviations and mounting errors on gear dynamic behaviour, Journal of Sound and Vibration, 191(5) (1996) 629-660.

[15] D.W. Dudley, D.P. Townsend, Dudley's Gear Handbook, McGraw-Hill, Inc., New York, (1991).

[16] S.M. Vijayakar, A combined surface integral and finite element solution for a three dimensional contact problem, Int. J. Numer. Methods Eng., 31 (1991) 525-545

[17] R. Parker, S.M. Vijayakar, T. Imajo, Nonlinear dynamic response of a spur gear pair: modeling and experimental comparisons, Journal of Sound and Vibration, 237(3) (2000) 435-455.

[18] S. Natsiavas, Periodic Response and Stability of Oscillators With Symmetric Trilinear Restoring Force. Journal of Sound and Vibration. 134(2) (1989) 315-331.

[19] A. Kahraman, R. Singh, Non-Linear Dynamics of a Spur Gear Pair. Journal of Sound and Vibration. 142(1) (1990) 49-75.

Global Bifurcation and Chaotic Behavior Research of a Truncated Conical Shallow Shell Rotating Around a Single Axle

Changping Chen[a], Liming Dai[a]*, [b]Simon Y. Sun
[a]*Industrial Systems Engineering, University of Regina, Regina, Saskatchewan, Canada S4S 0A2*
[b]*DIRECTV Group, Inc., D5 N353, 2230 E. Imperial Hwy., El Segundo, CA 90245 USA*

Abstract
A truncated conical shallow shell rotating around a single axle and subjected to a transverse periodic loading is studied in this research. The Melnikov method is adopted to study the homoclinic bifurcation and subharmonic bifurcation of the dynamic system. Numerical simulations are performed for investigating the system's nonlinear and chaotic behavior. Conditions under which chaotic motions occur are determined.

Key words: Rotationary device, truncated conical shallow shell, nonlinear dynamics, global bifurcation, chaotic motion, numerical simulation

Nomenclatures:

h: thickness of the truncated conical shallow shell
L: generatrix length of the truncated conical shallow shell
R : mean radius of the shell at the truncated en
w : transverse deformation of the shell
φ : semi-vertex angle of the shell
ρ : mass density of the shell
Ω_1 : angular velocity of the shell around axle x
x: axle which cross the centre and parallel the symmetry axis of the shell
$\omega_1, \bar{\mu}, \bar{k}$: constant coefficients of the equation
Q : transverse uniform load of applying on the shell
Ω : periodic variety frequency of the transverse load
ε : perturbation parameter
μ, k : coefficients which have relations with $\omega_1, \bar{\mu}, \bar{k}$ and Ω_1
f : amplitude of transverse uniform load
t, τ : time variable
H : Hamilton action

1. Introduction

Analysis of nonlinear vibration of the plates and shells has been under a considerable amount of research interest in the recent years. However, the literature concerning the nonlinear dynamic analysis of

The corresponding author: Tel: (306)585-4498, Fax: (306)585-4855
Email:Liming.Dai@uregina.ca

shells in rotating reference system is scanty compared to that in static. There also have been respectable developments in the fields of the nonlinear dynamic analysis of the plates and shells under the large overall motion. The dynamic responses of the plates in the non-inertial reference system are firstly studied by Kane [1] by using the Kane equation in 1989. He first extended the Kane equation that suits the motion of a particle system and rigid body to the motion of deformable body and founded a new important approach to research the dynamic problem of a deformable body. Later, Boutaghou [2] and Bolin[3] investigated dynamic problems of beams and plates in large overall motion using Hamilton's principle and finite element method.

As an indispensable rotating component, truncated conical shells have been widely applied in many fields such as space flight, rocket, aviation, and submarine technology. It plays an important role in the construction of the national economy. The dynamic analysis of the truncated conical shells in large overall motion is very important and becoming the study emphasis in this field recently. By using the Rayleigh-Ritz method, Chandraseharan and Ramamnrti [4,5] researched the free vibration of truncated conical thin shell; Dumir and Khatri [6] studied the nonlinear dynamic response of the orthotropic truncated conical thin shells; Xu and Chia[7,8] researched on the nonlinear free oscillations and post bucking of the truncated lamination moderate thickness conical shells and symmetrically laminated moderately thick spherical caps under the condition with consideration of the rotary inertia and transverse shearing deformation. A new approach of the truncated moderate thickness conical shells in the overall motion is established by employed Hamilton theory by Fu and Chen [9] and the nonlinear dynamic vibration analysis is performed in the paper.

The present paper based on the literature research [9], employing the Melnikov method, discussed the Homoclinic orbit bifurcation and subharmonic bifurcation of the dynamic system of the truncated shallow conical shell rotating around an axle, and simulated the system's chaotic motion by numerical method. The simulation figures illustrated that, in the condition of a specific parameter combination, the chaotic motion will occur in the system which the truncated shallow conical shell rotating around a axle under applying of the transverse periodic external load.

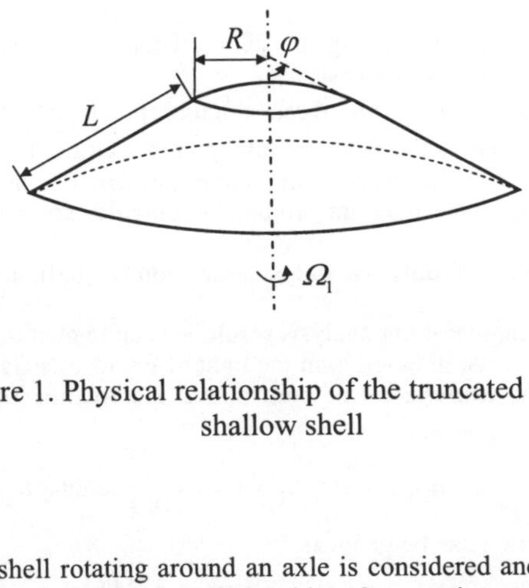

Figure 1. Physical relationship of the truncated conical shallow shell

2. Base equations

A truncated shallow conical shell rotating around an axle is considered and its geometric relationship is shown in figure 1. Figure 1 the Geometric relationship of a rotating truncated conical shallow shell The thickness of the shell is h, generatrix length is L, mean radius of the shell is R at the truncated end, semi-vertex angle is φ and mass density is ρ. The conical shell uniformly rotates around the axle x with an angular velocity Ω_1. According to the literature [9], just by considering the transverse motion of the system, the governing equation of the shell can be written as the following

$$\ddot{w}+(\omega_1^2-\bar{\mu}\Omega_1^2)w+\bar{k}w^3+Q=0 \tag{1}$$

where Q is the transverse uniform load applying on the shell, $\omega_1,\bar{\mu},\bar{k}$ are the constant coefficients, which its expression can be gotten from literature [9]. Transform the style of Eq. (1), suppose $t=\omega_1\tau$, $Q=-\varepsilon f\omega_1^2\cos\Omega\tau$, where Ω is the periodic variety frequency of the transverse load, ε is the perturbation parameter, so Eq. (1) can be expressed by

$$\ddot{w}+(1-\mu)w+kw^3=\varepsilon f\cos\Omega\tau \tag{2}$$

in which, μ is the rotating parameter, $k=\bar{k}/\omega_1^2$ and $\mu=\Omega_1^2/\omega_1^2\,\bar{\mu}$, there, $\mu>0,k>0$.

3. Homoclinic orbit bifurcation

Suppose $x_1=w,x_2=\dot{w}$, the Eq.(2) can be translate into the form as following:

$$\begin{aligned}\dot{x}_1&=x_2\\ \dot{x}_2&=(\mu-1)x_1-kx_1^3+\varepsilon f\cos\Omega t\end{aligned} \tag{3}$$

when $\varepsilon=0$, Eq..(3) can be simplified the following forms:

$$\dot{x}_1=x_2,\ \dot{x}_2=(\mu-1)x_1-kx_1^3 \tag{4}$$

Eq. (4) is the typical Hamilton system, which its Hamilton action is

$$H=\frac{1}{2}x_2^2-\frac{1}{2}(\mu-1)x_1^2+\frac{1}{4}kx_1^4 \tag{5}$$

The next take a singular analysis to Hamilton system (4), suppose $\dot{x}_1=0,\dot{x}_2=0$, then the follow equations can be gotten from eq.(4)

$$x_2=0,(\mu-1)x_1-kx_1^3=0 \tag{6}$$

because $\mu>0$ and $k>0$, so, according to above equations, the Hamilton system's singular distribution may have the following several cases:

1° when $\mu>1$, the system will have three singular points which are point (0, 0), points $(\pm\sqrt{(\mu-1)/k},0)$,where point(0,0) is the saddle point, points $(\pm\sqrt{(\mu-1)/k},0)$ are the center points.

2° when $\mu<1$, the system will only has one singular point: (0, 0), and the singular point is a center point, the path of the phase-plane diagram around the singular are a set of concentric closed loop curves.

3° when $\mu=1$, the system will only has one singular point: (0,0), and the point is higher singular point.

According to the above singular point analysis result, we can known that through saddle point $P(0,0)$, there be Homoclinic orbit Γ exist in case 1°. in the light of Eq. (5), its Hamilton action $H=0$, therefore, from the Eq. (4) and (5), the expressions of undisturbed Homoclinic orbit can be expressed as the following.

$$x_1=\sqrt{\frac{2(\mu-1)}{k}}\operatorname{sech}(\sqrt{\mu-1}t),\ x_2=(\mu-1)\sqrt{\frac{2}{k}}\operatorname{sech}(\sqrt{\mu-1}t)\operatorname{th}(\sqrt{\mu-1}t) \tag{7}$$

Corresponding Melnikov function can be given as.

$$M(t_0)=\int_{-\infty}^{\infty}x_2(t)f\cos\Omega(t+t_0)\mathrm{d}t=\sqrt{\frac{2}{k}}\frac{\Omega\pi\operatorname{ch}\dfrac{\Omega\pi}{2(\mu-1)}}{\operatorname{ch}\dfrac{\Omega\pi}{\mu-1}+1}\sin\Omega t_0,\ [t_0\in(0,T)] \tag{8}$$

From the above expression, it can be seen that:.

1° when all the values of max $M(t_0)$ and min$M(t_0)$ are positive or negative, the single zero point will not exist in expression (8), therefore, the stable manifold W^s and instable manifold W^u through the saddle point P will never intersect.

2° when the value of the product of max $M(t_0)$ and min$M(t_0)$ is negative, the simple zero points will be form in expression (3), consequently, the cross-sections of stable manifold W^s and instable manifold W^u transit saddle point P will intersect at non-degenerate Homoclinic point accordingly, the system (3) will come forth the Homoclinic orbit bifurcation, and the Smale horse-shoe will appear in the phase plane of the system. then the system will be found in the chaotic motion of Smale meaning.

3° As one of the values of max $M(t_0)$ and min$M(t_0)$ is zero, and appear secondary zero point in $M(t_0)$ for some t_0, therefore the stable manifold W^s and instable manifold W^u will bring secondary Homoclinic tangency.

4. Subharmonic bifurcation of resonance periodic orbit

According to the above analysis, it is known that there will appear two sets of parameter periodic orbits in the inside of the Homoclinic orbit Γ, due to expressions (4) and(5), which their orbit can be expressed as.

$$x_{1p} = K\sqrt{\frac{2(\mu-1)}{(1+K^2)k}}\text{sn}[\sqrt{\frac{\mu-1}{1+K^2}}t] \pm \sqrt{\frac{\mu-1}{k}}$$

$$x_{2p} = \frac{K(\mu-1)}{1+K^2}\sqrt{\frac{2}{k}}\text{cn}[\sqrt{\frac{\mu-1}{1+K^2}}t]\text{dn}[\sqrt{\frac{\mu-1}{1+K^2}}t] \qquad (9)$$

where sn, cn and dn are the Jacobi elliptic function, $K(p)$ is the first kind complete elliptic integral, the value of p satisfys the following equation:

$$H_p = \frac{p^2(\mu-1)^2}{(1+p^2)k} \qquad (10)$$

where H_p is the Hamilton action inside the Homoclinic orbit of the system. and

$$T(p) = 4K(p)\sqrt{\frac{1+p^2}{\mu-1}} \qquad (11)$$

is the period of the periodic orbit.

Therefore, the condition of resonating is:

$$T(p) = 4K(p)\sqrt{\frac{1+p^2}{\mu-1}} = \frac{m2\pi}{n\Omega} \qquad (12)$$

where m and n are the integers which can not reduce the fractions to a common. As $\Omega = \sqrt{\mu-1}$, the system will occur primary resonance; when $\Omega = 2\sqrt{\mu-1}$, system will take place $1/2$ subharmonic resonance; when $\Omega = 1/2\sqrt{\mu-1}$, ultraharmonic resonance will occur in the system. The subharmonic Melnikov function corresponding to the expression (9) can be shown as

$$M^{m/n}(t_0)=\int_0^{mT(p)} x_{2p}(t)f\cos\Omega(t+t_0)dt$$

$$=\int_0^{mT(p)}\frac{K(\mu-1)}{1+K^2}\sqrt{\frac{2}{k}}\mathrm{cn}[\sqrt{\frac{\mu-1}{1+K^2}}t]\mathrm{dn}[\sqrt{\frac{\mu-1}{1+K^2}}t]dt \tag{13}$$

$$=\begin{cases}0 & n\neq 1,\ \text{and } m \text{ must be an even number}\\ \dfrac{m\pi^2}{K}\sqrt{\dfrac{2(\mu-1)}{k(1+K^2)}}\cosh\dfrac{m\pi K'}{2K}\cos\Omega t_0 & n=1,\ \text{and } m \text{ must be an odd number}\end{cases}$$

where $K'=K'(p)$, and K' is the secondary elliptic complete integral.

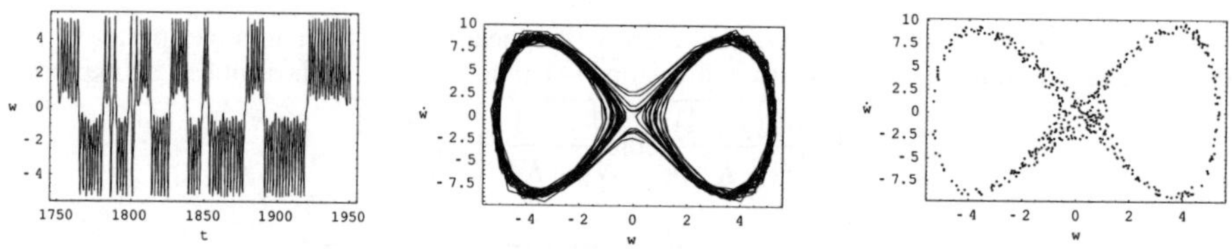

(a) (b) (c)

Figure 2 (a) Time history curves, (b) Phase plane curves, (c) Poincare map

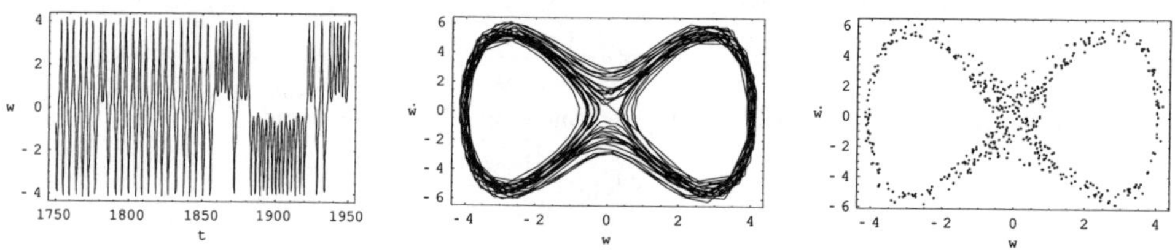

(a) (b) (c)

Figure 3 (a) Time history curves, (b) Phase plane curves, (c) Poincare map

Therefore, according to expression(14), the following conclusion can be obtained. For the case of $n\neq 1$, and m is a even number, $M^{m/n}(t_0)$ will always be zero for any t_0 accordingly, high-order even subharmonic bifurcation will occur in perturbing system (3). For the case $n=1$, and m is odd number, there are three cases shown as following again: when any of the value of $\max M^{m/n}(t_0)$ and $\min M^{m/n}(t_0)$ is positive or negative, so there isn't simple zero point in expression $M^{m/n}(t_0)$ for any t_0, therefore the m/n harmonic will not take place in perturbation system (3); when there is a simple zero point occurs in expression $M^{m/n}(t_0)$, the perturbation system (3) will transit the chaotic state pass by infinite odd subharmonic bifurcation; when one of the value of $\max M^{m/n}(t_0)$ and $\min M^{m/n}(t_0)$ is zero, there will exist secondary zero point in expression $M^{m/n}(t_0)$ for some specification, hereby the subharmonic saddle point bifurcation will occur in perturbation system (3).

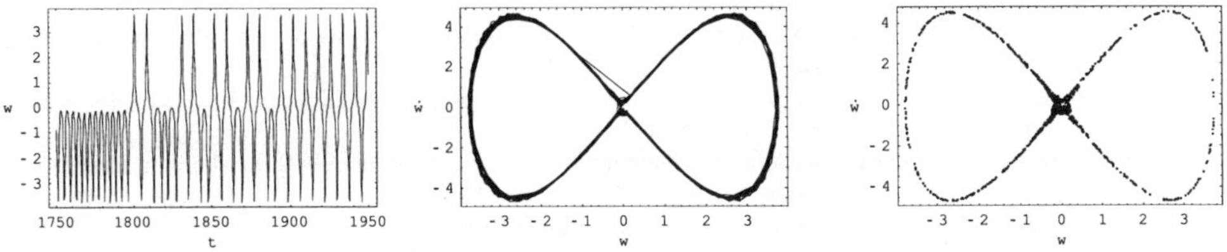

(a) (b) (c)

Figure 4 (a) Time history curves, (b) Phase plane curves, (c) Poincare map

5. Numerical simulation of chaotic motion

As chaotic motion has specific number characteristic, therefore, it is important to open out the character of chaotic motion by numerical simulation, which can be a criterion to identify a steady motion if it can lead to chaos. The following figures are the numerical simulation results of the system (3) by using the fourth order Runge Kutt method, the characteristic of the phase plane figure, time history curves and Poincare map are used to judge the possibility of the chaotic motion that the system will take place. When the parameters $\mu = 13.0, \Omega = 6.0$, the ultra harmonic resonance will happen in the system (3), at this time the value of other parameters are $k = 0.85, \varepsilon = 0.01, f = 100$, $x_{10} = 0.1, x_{20} = 0.4$, As the simulation results are be shown in Figure 2. Figure 3 shows the principle resonance simulating result of the system (3) when the parameters are $\mu = 8.0, \Omega = 7.0$, $k = 0.85, \varepsilon = 0.01, f = 115, x_{10} = 0.01$, $x_{20} = 0.5$. The simulation result of 1/2 subharmonic resonance is shown in Figure 4 with the parameters $\mu = 7.0, \Omega = 12$, $k = 0.85, \varepsilon = 0.01, f = 85$, $x_{10} = 0.2, x_{20} = 0.5$.

6. Conclusive remarks

The truncated conical shallow shell rotating around a single axle and subjected to a transverse periodic loading shows dynamic responses of high complexity. The governing equation for the dynamic system is developed. The homoclinic orbit bifurcation and subharmonic bifurcation of resonance periodic orbit of the system are studied theoretically. Such research has not been seen in the current literature. The numerical simulation results of the present research show a good accordant with that of the theoretical analysis presented in the context of the present paper.

Reference

[1] Kane T. R., Likins P. W and Levinson D.A., Spacecraft Dynamics, McGrand-Hill, New York, 1983.

[2] Boutaghou Z. E., Dynamic of Flexible Beams and Plates in large overall Motion, Journal of Applied Mechanics, Vol. 59, 1992, pp.991~1004.

[3] Chang B. and Shabana A., Finite element formulation for the large displacement analysis of plates, Journal of

Applied Mechanics, Vol.57, 1990, pp.707~717.
[4] Chandrasekharan K., and Ramamurti V., Asymmetric Free Vibrations of Laminated Conical Shell, Journal of Mechanic Desion, Vol.104, 1982, pp.453~462.
[5] Chandrasekharan K., and Ramamurti V., Axisymmetric Free Vibrations of Laminated Conical Shell, Proc Int Symp Mechanical Behaviour Struct, Media Ottawa, 1981, pp.18~21.
[6] Dumir P. C., and Khatri K. N., Axisymmetric Static and Dynamic Buckling of Orthotropic Truncated Shallow Conicels Caps, Comput. and Struct., Vol.22, 1986, pp.335~342.
[7] Xu C. S., Xian Z. Q. and Chia C. Y., Nonlinear theory and vibration analysis of laminated truncated thick conical shells, Int. J. of Nonlinear Mechanics, Vol.16, No.11, 1995.
[8] Xu C. S., Buckling and postbuckling of symmetrically laminated moderately thick spherical caps, Int. Journal of Solids and Structures, Vol.28, No.9, 1991, pp.1171~1184.
[9] Fu Y. M., Chen C. P., Nonlinear Vibration of elastic truncated conical moderately thick shells in large overall motion, Inter. J of Nonlinear Mechanics, Vol.36, No.5, 2001, pp.763~772.
[10] Chia C. Y., Nonlinear analysis of plates, New York: Megraw-Hill, 1980.
[11] Sinharay G. C. and Banerjee B., Large amplitude free vibrations of shallow spherical shell and cylindrical shell – a new approach, Int. J. of Nonlinear Mechanics, V01.20, No.2, 1985, pp.69~78.
[12] Liu R. H. and Li J., Nonlinear vibration of shallow conical sandwich shells, Int. J. of Nonlinear Mechanics, Vol.30, No.2, pp.97~109.
[13] Dumir P. C., Nonlinear axisymmetric response of orthotropic thin truncated conical and spherical caps, Acta Mechanica, Vol.60, 1986, pp.121~132.
[14] Fu Y. M. and Chia C. Y., Nonlinear vibration and postbucking of generally laminated circular cylindrical thick shells with non-uniform boundary conditions, Int. J. Nonlinear Mechanics, Vol28, No.3, 1993, pp.313~327.
[15] Cheung Y. K. and Fu Y. M., Nonlinear static and dynamic analysis for laminated, annular, spherical caps of moderate thickness, Nonlinear Dynamics, Vol.8, 1995, pp.251~268.

Technology of Magnetic Flux Leakage Signal Detection Based on Scale Transformation Stochastic Resonance

Taiyong Wang, Shiguang Hu, Yonggang Leng*, Ying Zhang, Li Zhao

School of Mechanical Engineering, Tianjin University, Tianjin, 300072, China

Abstract

The conception and realization of scale transformation stochastic resonance (STSR) are proposed, which are used to deal with problem of weak signal detection with large parameters that can not be resolved by traditional bistable system based on stochastic resonance. Magnetic flux leakage (MFL) signals inspection for oil well tubing defects is easily destroyed by noise. In order to detect weak MFL signals effectively, the proposed STSR technology is applied to the inspection. The result shows that the STSR technology has the potential application in engineering practice.

PACS: 05.40.-a, 05.45.-a

Key words: Scale transformation stochastic resonance; Magnetic flux leakage signal; Weak signal; Detection

1. Introduction

In the field of oil extraction, non-destructive testing (NDT) of oil well tube is of vital importance for ensuring safety-related working and for decreasing cost. Among technologies of NDT, technology of magnetic flux leakage (MFL) [1] is widely applied to the inspection of oil well tubing. In practice, MFL signal is impure because of noise. It is hard to recognize weak MFL signal from strong noise with normal signal process technologies like wavelet and etc, especially when the data amount is small or when the frequency of noise is close to or the same as that of MFL signal. In the paper, the principle of scale transformation stochastic resonance (STSR) proposed by our group provides a resolution to the problem mentioned above.

2. Theory of stochastic resonance (SR)

With respect to nonlinear system, proper noise could improve the performance of signal output of the system. This phenomenon is called stochastic resonance[2, 3], which provides a resolution to the recognition of weak signals. The equation of Langevin Eq.(1) is the fundamental model for analyzing SR of a nonlinear bistable system.

$$dx/dt = \mu x - x^3 + a\sin(2\pi f_0 t) + n(t) \tag{1}$$

* Corresponding author. Tel.: +86-022-27408118(O); Fax.: +86-022-27404536.

E-mail address: zeyangang@tdme.tju.edu.cn

where $E[n(t)]=0$ and $E[n(t)n(t-\tau)]=2D\delta(\tau)$.

In stead of caring about where the input signal of the bistable system comes from and what the dynamics characteristic of the signal source is, this model pays attention to the signal noise ratio, the characteristic of both of them and whether they can generate SR with bistable system.

Figure 1 shows that SR is generated with signal $Sn(t)$ passing through a bistable system $U(x)$.

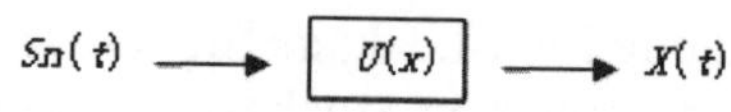

Fig.1 a Bistable system

In the figure, $U(x)$ denotes the reflection-symmetric quartic potential:

$$U(x) = -1/2\mu x^2 + 1/4x^4, \tag{2}$$

Corresponding to the output $X(t)$, $Sn(t)$ is the input and is composed of the weak signal $S(t)$ and noise $n(t)$:

$$Sn(t) = S(t) + n(t), \tag{3}$$

where $S(t) = a\sin(2\pi f_0 t)$ and $n(t) = \sqrt{2D}g(t)$ with D representing noise intensity and $g(t)$ magnifying a zero-mean and unit variance Guassian white noise.
Then equation (1) can be written as

$$dx/dt = \mu x - x^3 + a\sin(2\pi f_0 t) + \sqrt{2D}g(t). \tag{4}$$

With further research on the power spectrum of the system output, obvious spectrum peak is found near f_0, the frequency of the signal $S(t)$.

For example, set parameters corresponding to equation (4) to be $\mu = 1$, $a = 0.3$, $f_0 = 0.01$, $D = 0.31$ and choose a sample frequency $fs = 5$. The input and output waveform and spectra of the bistable system are shown in Fig.2.

The input and output spectra are calculated with 1024 data points, while the display data length in time domain is 4000 points for a better view. As can be seen in the figure, the output is obviously a periodical signal with a frequency of f_0, and the spectrum peak at f_0 is evident.

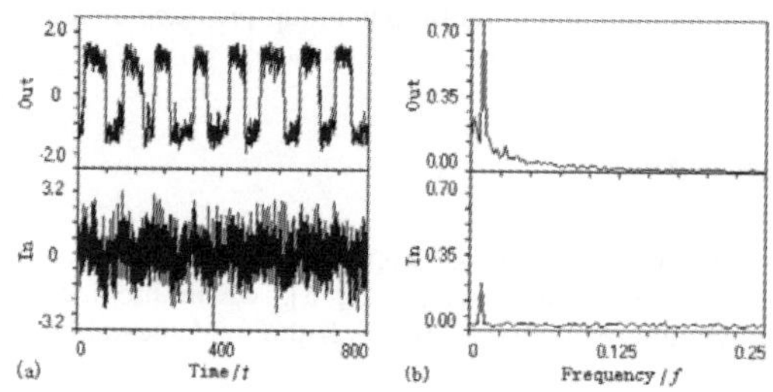

Fig.2 The input and output of the bistable system with small parameters

$\mu = 1$, $a = 0.3$, $f_0 = 0.01$, $D = 0.31$, $fs = 5$. (a) Time domain waveforms. (b) Frequency domain power

3. Scale transformation stochastic resonance (STSR)

The theory of stochastic resonance is generated under the condition of adiabatic approximation with small parameters research objects. In other words, the amplitude and frequency of input signals, as well as noise intensity, of the bistable system are all small than 1. In the case of large parameters, the theory of stochastic resonance based on small parameters will lose efficiency.

Principle of scale transformation stochastic resonance (STSR)[4, 5] is applied to research phenomenon of stochastic resonance with large parameters in the paper. The key point of STSR is by the transformation of frequency from high to low, the frequency of input signal meets or closes to the requirement of small parameter. The frequency of the input signal is renormalized according to newly-defined sampling frequency, so no characteristic of the signal is changed. The detail process of scale transformation is as follows:

(1)Choose a frequency compress scale R.

(2)Define a compressed sampling frequency fsr according to R, that is, $fsr = fs / R$, where fs is the practical sampling frequency.

(3)Calculate the step in accordance with the following equation: $\Delta t = 1 / fsr$.

(4)Get the relevant output of the bistable system.

The compressed sampling frequency fsr decides how f_0 is transferred from high to low, and fs decides whether f_0 is moved into low frequency region on which the noise is concentrated. The potential barrier parameter μ in equation (4) is relative to the amplitude a and plays a role in adjusting resonance spectrum peak at frequency f_0. To generate SR conveniently, a can be, in case that it is quite lower than the threshold, restored to or exceeded threshold by lowering potential barrier ΔU with reducing μ according to the equation $\Delta U = \mu^2 / 4$. In so doing, weak periodical signal submerged by strong noise will be detected.

4. Magnetic flux leakage (MFL) signals detection based on STSR

The MFL signal is a space vector, as is shown in Fig.3(a). Normally this vector is decomposed to horizontal and vertical components which are represented as ϕ_x (Fig.3(b)) and ϕ_y (Fig.3(c)) respectively. By detecting and analyzing ϕ_x and ϕ_y, characteristics of relevant defects are figured out quantitatively. The key process is to detect the MFL signal accurately.

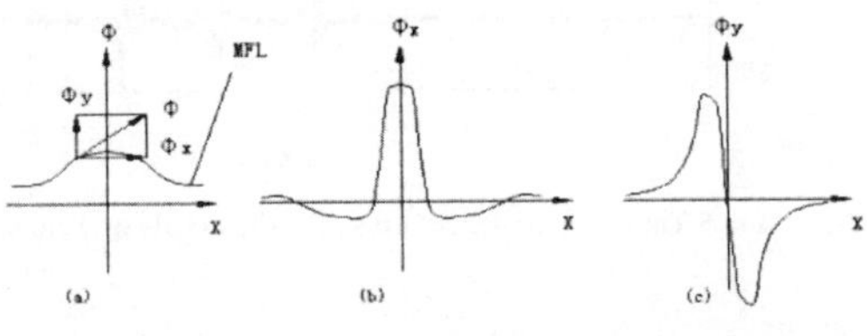

Fig.3 MFL signals and the horizontal and vertical components

The distribution of artificial defects in oil well tubing is show in Fig.4. As can be seen in the figure, 4 through-holes with different diameters are made in the right of the tubing, and 4 different longitudinal cracks are in the left. Also, in the middle of the tubing, there are 3 welding lines, among which small raw-shaped abrasive defects are made.

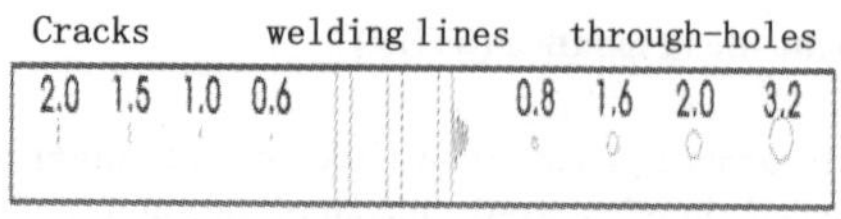

Fig.4 Distribution of artificial defects in experimental oil well tubing

The waveform of MFL signal changes following not only with the size and shape of defects, but also with the speed of detector moving across the tubing which calls for a high sampling frequency. To describe the waveform of MFL signal accurately, the sampling frequency is set to be 4096Hz through lots of experiments. Thus the total sampling length of the whole tubing is 66600 points. With a certain defect, the MFL signal strength sensed by 32 Hall sensors arranged around the tubing is not the same, owing to the difference of their location. So the output signals of several sensors near the possible defect should be considered synthetically.

By comparing the waveforms in Fig.5 and Fig.6, it is found out that, in Fig.6, the amplitudes of MFL signals of large welding lines are lowered and other weak signals are enlarged (referring to the coordinate value). Because of the effect of concentrating energy in the low frequency region by lowering the amplitude of high frequency signals, the waveform of STSR signal is smoother than that of original signal. Refer to the comparison of partial enlarged view in Fig.7.

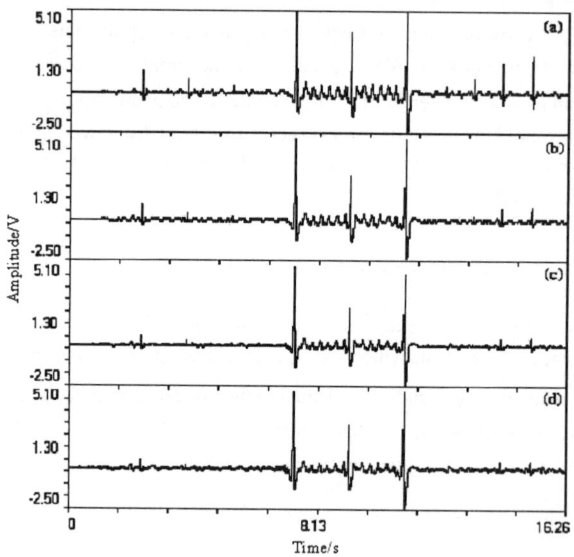

Fig.5 Time domain waveforms of MFL signals of 4 consecutive channels

Fig.5 shows the time domain waveforms of MFL signals of defects in accordance with Fig.4. While the time domain waveforms of relevant STSR signals are shown in Fig.6. The selected parameters in Fig.6 are $R=157.5$ or $fsr = fs/R = 4096/157.5 = 26\,\text{Hz}$ and $\mu = 0.35$.

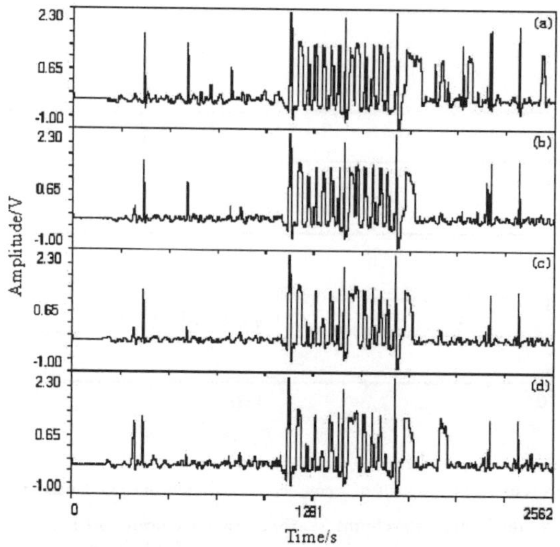

Fig.6 STSR waveforms in according with Fig.5 μ=0.35, fsr=26Hz

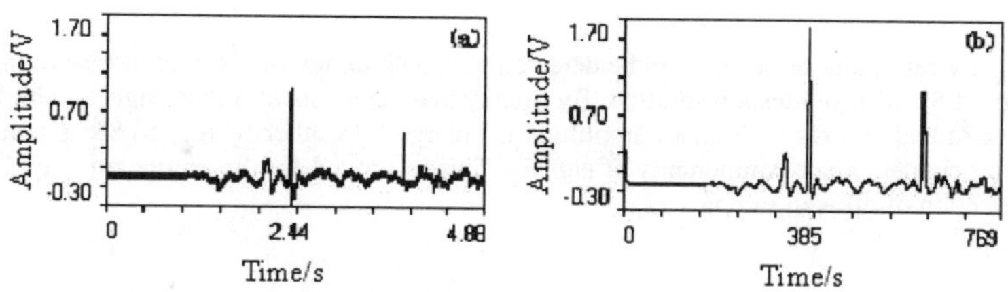

Fig.7 Waveforms of the first 20000 data length in according with Fig.5 and Fig.6 separately

(a) Original waveforms (b) STSR waveforms μ=0.35, fsr=26Hz

Although weak signals of defects are enlarged by STSR, noises or interfering signals with comparable amplitude are enlarged at the same time. So it is not easy to analyze defects only with signals of a single channel. In the paper, signals of 4 consecutive channels are summed and averaged, which is proved by experiments to be able to eliminate interfering components and to highlight signals of defects.

Fig.8 shows the waveforms of average signals of 4 consecutive channels in accordance with Fig.5 and Fig.6 respectively. In terms of convenience, both the coordinate scales in the two figures are set to be the same. Obviously, the original signal waveform which is shown in Fig.8(a) can denote only the longest crack and two largest through-holes. While in Fig.8(b), two cracks and three through-holes are displayed clearly, and the location of the third crack and the forth through-hole can be judged roughly. What is more, weak signals of small raw-shaped abrasion defects are illustrated by the waveform of STSR.

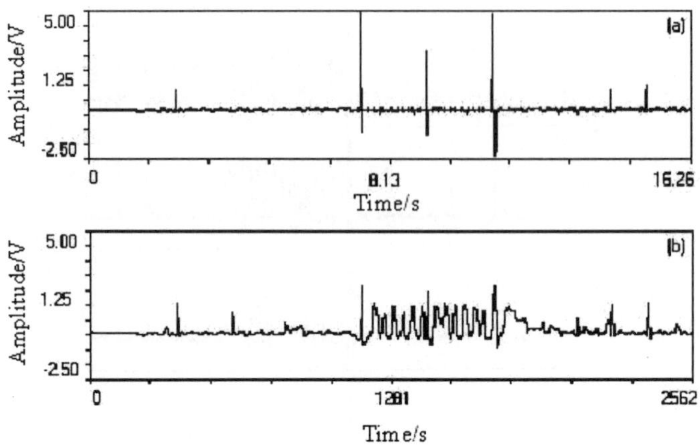

Fig.8 The waveforms of average signals of 4 consecutive channels in accordance with Fig.5 and Fig.6 respectively

(a) Original waveforms (b) STSR waveforms μ=0.35, *fsr*=26Hz

5. Conclusions

Signals of weak faults or defects can be detected with technology of SR, and in case of large parameters, technology of STSR provides a resolution. By time domain analysis of STSR, signals with large amplitude is compressed and signals with small amplitude is enlarged. In other words, STSR is able to lower peak values and heighten weak components of signals. This is proved by the engineering application to MFL signals detection of oil well tubing.

Acknowledgements

This research was supported by the Natural Science Foundation of China (No.50475117) , the Tianjin Municipal Science and Technology Commission (05YFJZJC01800) and the Youth Teacher Foundation of Tianjin University (Project No.5110108).

References

[1]H.Jens, U.Ralf, Evaluation of inverse algorithms in the analysis of magnetic flux leakage data, IEEE Transactions on magnetics. 38(2002)1481-1488.

[2]Benzi R, Sutera A and Vulpiana A, The Mechanism of Stochastic Resonance, Phys. A. 14(1981) L453-457.

[3]Gammaitoni L, Hänggi P, et al, Stochastic resonance, Rev Mod Phys.. 70(1998)223-246.

[4]Y.G.Leng, T.Y.Wang, Numerical research of twice sampling stochastic resonance for the detection of a weak signal submerged in heavy noise, Acta Phys Sin. 52(2003) 2432-2438.

[5]Y.G.Leng, T.Y.Wang, R.X.Li, et al, Scale transformation stochastic resonance for the monitoring and diagnosis of electromotor faults, Proceedings of the Chinese Society of Electrical Engineering. 23(2003) 111-115.

Periodic motions and bifurcations of vibro-impact systems near a strong resonance point

Guanwei Luo [a*], Yanlong Zhang [b], Jiangang Zhang [a], Jianhua Xie [c]
[a] *School of Mathematics, Physics and Software Engineering, Lanzhou Jiaotong University, Lanzhou,* 730070, *China*
[b] *School of Mechanical Engineering, Lanzhou Jiaotong University, Lanzhou* 730070, *China*
[c] *Department of Engineering Mechanics, Southwest Jiaotong University, Chengdu,* 610031, *China*

Abstract

Two typical vibro-impact systems are considered. The periodic-impact motions and Poincaré maps of the vibro-impact systems are derived analytically. A center manifold theorem technique is applied to reduce the Poincaré map to a two-dimensional one, and the normal form map associated with 1:4 strong resonance is obtained. Two-parameter bifurcations of fixed points in the vibro-impact systems, associated with 1:4 strong resonance, are analyzed. The results from simulation illustrate some interesting features of dynamics of vibro-impact systems. Near the bifurcation point for 1:4 strong resonance Neimark-Sacker bifurcations of periodic-impact motions and tangent and fold bifurcations of period-4 orbits are found to exist in the vibro-impact systems.

Keywords: Vibration; Impact; Resonance; Periodic motion; Bifurcation

1. Introduction

Vibrating systems with clearances, gaps or rigid amplitude stops are frequently encountered in technical applications of mechanism. Repeated impacts, i.e., vibro-impacts, usually occur whenever the components of a vibrating system collide with rigid obstacles or with each other. The principle of operation of vibration hammers, impact dampers, inertial shakers, pile drivers, machinery for compacting, milling and forming, etc., is based on the impact action for moving bodies. With other equipment, e.g., mechanisms with clearances, heat exchangers, fuel elements of nuclear reactors, gears, piping systems, wheel-rail interaction of high speed railway coaches, etc., impacts also occur, but they are undesirable as they bring about failures, strain, shorter service life and increased noise levels. Researches into vibro-impact dynamics have important significance on optimization design of machinery with clearances, gaps or rigid amplitude stops, noise suppression and reliability analyses, etc. The physical process during impacts is strongly nonlinear and discontinuous, but it can be described theoretically and numerically by discontinuities in good agreement with reality. Compared with single impact, vibro-impact dynamics is more complicated, and hence, has received great attention.The large interest in analyzing and understanding the performance of such systems is reflected by vast and ever increasing amount of research effort devoted in this area in recent years, Some important problems on vibro-impact dynamics, including global bifurcations [1-5], grazing singularities of impact mapping [6-13], sliding bifurcation[14], chattering impacts [15], quasi-periodic impacts [16,17] and controlling chaos [18], etc., have been studied in the past several years. Along with the theoretical researches into vibro-impact dynamics, the researches into application of such systems are developed, e.g., wheel-rail impacts of railway coaches [19, 20], inertial shaker [21],

* Corresponding author. Tel.: 86-931-4956108; fax: 86-931-4938613
E-mail address: luogw@mail.lzjtu.cn (Guanwei Luo)

vibration hammer [22], Jeffcott rotor with bearing clearance [23], impact damper [24, 25] and gears [26, 27], etc. The purpose of the present study is to focus attention on two-parameter bifurcation of fixed points associated with 1:4 strong resonance. Two typical vibratory systems with repeated impacts are considered in the paper. Periodic motions and bifurcations of the vibro-impact systems, associated with 1:4 strong resonance, are analyzed. Some complicated bifurcations, e.g., T_{in}, T_{on} and T_{out} types of tangent bifurcations of period-4 orbits, are found to exist near the bifurcation point for 1:4 strong resonance.

2. Mechanical models and Poincaré map

The mechanical model for a three-degree-of- freedom vibro-impact system is shown in Fig.1. A rigid body with mass m bounces on the flat horizontal surface of a two-degree-of-freedom vibro-bench with masses M_1 and M_2. Displacements of the masses m, M_1 and M_2 are represented by Y_0, X_1 and X_2, respectively. The masses M_1 and M_2 are connected to linear springs with stiffnesses K_1 and K_2, and linear viscous dashpots with damping constants C_1 and C_2. The excitations on both masses are harmonic with amplitudes P_1 and P_2. Ω is the excitation frequency, and τ the phase angle. The mass M_1 impacts mutually with the bouncing mass m when they are on the same height, so the mass m exhibits the bouncing motion.

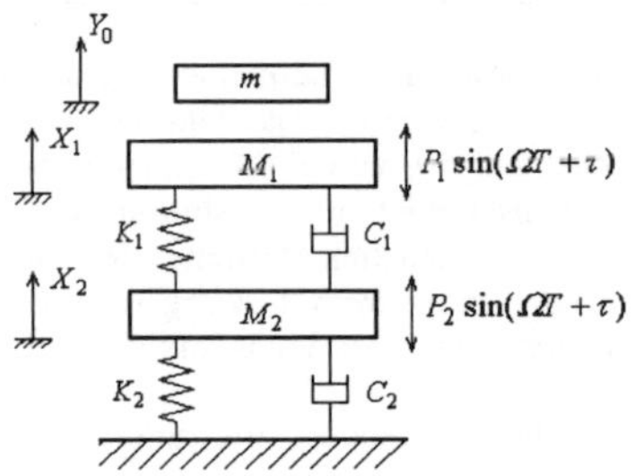

Fig.1. Schematic of the vibratory system impacting an unconstrained rigid body

The motion processes of the system, between any two consecutive impacts, are considered. Between any two consecutive impacts, the time T is always set to zero directly at the instant when the former impact is over, and the phase angle τ is used only to make a suitable choice for the origin of time in the calculation. The state of the vibro-impact system, justimmediately after impact, has become initial conditions in the subsequent process of the motion. Between consecutive impacts, the non-dimensional differential equations of motion are given by

$$\begin{bmatrix} 1 & 0 \\ 0 & \mu_m \end{bmatrix}\begin{Bmatrix} \ddot{x}_1 \\ \ddot{x}_2 \end{Bmatrix}+\begin{bmatrix} 2\zeta & -2\zeta \\ -2\zeta & 2\zeta(1+\mu_c) \end{bmatrix}\begin{Bmatrix} \dot{x}_1 \\ \dot{x}_2 \end{Bmatrix}+\begin{bmatrix} 1 & -1 \\ -1 & 1+\mu_k \end{bmatrix}\begin{Bmatrix} x_1 \\ x_2 \end{Bmatrix}=\begin{Bmatrix} 1-f_{20} \\ f_{20} \end{Bmatrix}\sin(\omega t+\tau),$$

$$\ddot{y}=-\beta, \tag{1}$$

where a dot ($\bullet$) denotes differentiation with to the non-dimensional time t, and the non-dimensional quantities are given by

$$\mu_m=\frac{M_2}{M_1},\quad \mu_k=\frac{K_2}{K_1},\quad \mu_c=\frac{C_2}{C_1},\quad f_{20}=\frac{P_2}{P_1+P_2},\quad \omega=\Omega\sqrt{\frac{M_1}{K_1}},\quad t=T\sqrt{\frac{K_1}{M_1}},$$

$$\zeta=\frac{C_1}{2\sqrt{K_1M_1}},\ x_i=\frac{X_iK_1}{P_1+P_2},\quad \beta=\frac{M_1g}{P_1+P_2},\quad \mu=\frac{m}{M_1},\quad y=\frac{Y_0K_1}{P_1+P_2}. \tag{2}$$

When the impact occurs, for $x_1(t)=y(t)$, the velocities of the masses m and M_1 are changed according to the conservation law of momentum, and the impact equation and the coefficient of restitution R are given by

$$\dot{x}_{1-}+\mu\dot{y}_-=\dot{x}_{1+}+\mu\dot{y}_+,\qquad \dot{x}_{1+}-\dot{y}_+=-R(\dot{x}_{1-}-\dot{y}_-), \tag{3}$$

where the velocities of two masses M_1 and m, immediately before and after impact, are represented by $\dot{x}_{1-}$, $\dot{y}_{-}$, $\dot{x}_{1+}$ and $\dot{y}_{+}$, respectively.

Impacting systems are conveniently studied by use of a map derived from the equations of motion. Each iteration of this map corresponds to the mass M_1 striking the bouncing mass m once. Under suitable system parameter conditions, the system can exhibit periodic impact behavior. The periodic-impact motions of the vibro-impact system can be characterized by using the symbol *n-p*, where *p* is the number of impacts and *n* is the number of the forcing cycles. The Poincaré section associated with the state of the vibro-impact system, just immediately after impact, is chosen, and period *n* single-impact motion and its disturbed map are derived analytically in Ref.[28]. Let $\theta = \omega t$, we can establish Poincaré map of the system by choosing a Poincaré section $\sigma = \{(x_1, \dot{x}_1, x_2, \dot{x}_2, y, \dot{y}, \theta) \in \boldsymbol{R}^6 \times \boldsymbol{S},\ x_1 = y,\ \dot{x}_1 = \dot{x}_{1+},\ \dot{y} = \dot{y}_{+}\}$. The disturbed map of period one single-impact motion is expressed by

$$X' = f(v, X), \tag{4}$$

where $X \in \boldsymbol{R}^6$, v is varying parameter, and $v \in \boldsymbol{R}^1$ or $\boldsymbol{R}^2$; $X = X^* + \Delta X$, $X' = X^* + \Delta X'$, $\Delta X = (\Delta\dot{x}_{1+}, \Delta x_1, \Delta\dot{x}_2, \Delta x_2, \Delta\dot{y}_{+}, \Delta\tau)^{\mathrm{T}}$ and $\Delta X' = (\Delta\dot{x}'_{1+}, \Delta x'_1, \Delta\dot{x}'_2, \Delta x'_2, \Delta\dot{y}'_{+}, \Delta\tau')^{\mathrm{T}}$ are the disturbed vectors of X^*, $X^* = (\dot{x}_{1+}, x_{10}, \dot{x}_2, x_{20}, \dot{y}_{+}, \tau_0)^{\mathrm{T}}$ is the fixed point of period one single-impact motion in the hyperplane σ.

Linearizing the Poincaré map (4), at the fixed point X^*, results in the Jacobian matrix

$$\mathrm{D}f(v, X^*) = \frac{\partial f(v, X)}{\partial X}\bigg|_{(v, X^*)}. \tag{5}$$

The stability of 1-1 motion is determined by computing and analyzing eigenvalues of Jacobian matrix $\mathrm{D}f(v, X^*)$. Variations of the parameters of the system will cause the fixed point and its associated eigenvalues to move. If one of them passes through the unit circle in the complex plane, i.e., $|\lambda_i(v_c) = 1|$ (v_c is a bifurcation value), an instability and an associated bifurcation will occur. In general, bifurcation occurs in various ways according to the numbers of the eigenvalues on the unit circle and their position on the unit circle. Two-parameter bifurcations of fixed points in the vibro-impact system are considered, and dynamics of the system is studied with special attention to the bifurcations for 1:4 strong resonance.

3. Center manifold and normal form map associated with 1:4 resonance case

Let us consider the map $X' = f(v, X)$. $X^*(v)$ is a fixed point for the map for v in some neighborhood of a critical value $v = v_c$ at which $\mathrm{D}f(v, X^*)$ satisfies the following assumptions:

H1. $\mathrm{D}f(v, X^*)$ has a complex-conjugate pair eigenvalues $\lambda_{1,2}(v_c)$ on the unit circle ($|\lambda_{1,2}(v_c)| = 1$), the other eigenvalues $\lambda_3(v_c), \cdots, \lambda_6(v_c)$ stay inside the unit circle;

H2. $\mathrm{d}|\lambda_1(v)|/\mathrm{d}v\big|_{v=v_c} > 0$ and $\lambda_1^{\,4}(v_c) = 1$.

For the map (4), there exists a local center manifold, on which the local behavior of the map (4) can be reduced to a two-dimensional map. By using the center manifold technique and normal form method of maps, we can reduce the map (4) to the normal form map, which is expressed in the complex form by

$$\phi_\mu(\zeta, \bar{\zeta}) = \lambda(\mu)\zeta + C(\mu)\zeta^2\bar{\zeta} + D(\mu)\bar{\zeta}^3 + O(|\zeta|^4), \tag{6}$$

Lemma (Normal form map for 1:4 resonance) [29, 30]: The fourth iterate of the map (6) associated with 1:4 resonance case can be represented, for all sufficiently small $\|\mu\|$, in the form

$$\Gamma_\alpha^4(\zeta) = \varphi_\alpha^1\zeta + O(|\zeta|^4),$$

where φ_α^t is the flow of a planar system

$$\dot{\zeta} = \omega(\mu)\zeta + C_1(\mu)\zeta|\zeta|^2 + D_1(\mu)\bar{\zeta}^3, \tag{7}$$

where $C_1(\mu)$ and $D_1(\mu)$ are smooth complex-valued functions of μ, $\omega(0)=0$, and $C_1(0) = -4iC(0)$, $D_1(0) = -4iD(0)$.

If the complex number $D_1(0) \neq 0$, then we can scale the planar system (7) by taking $\zeta = \gamma(\beta)\eta$. The scaling results in

$$\dot{\eta} = (\beta_1 + i\beta_2)\eta + A(\beta)\eta|\eta|^2 + \bar{\eta}^3, \tag{8}$$

where $A(\beta) = C_1(\mu(\beta)) / |D_1(\mu(\beta))|^2$.

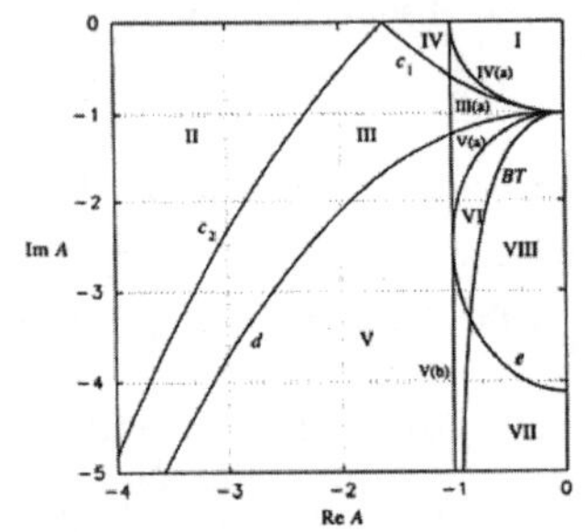

Fig.2. Division of the A-plane into regions with different bifurcation diagrams of (7)

The bifurcation analyses of the system (8) are very complicated and requires analytical and numerical techniques, e.g. Refs.[29, 30]. The bifurcation diagram depends on A=A(0), so the A-plane is divided into eleven regions with different bifurcation diagram. The partitioning of the A-plane into regions with different bifurcation diagrams is given in Fig.2. The boundaries on the A-plane are symmetrical under reflections with respect to the coordinate axes. Thus, it is sufficient to study them in one quadrant of the A-plane. Near 1:4 resonance point there exist very complicated bifurcation phenomena, e.g., Hopf bifurcation of nontrivial equilibria E_k (k=1,2,3,4), T_{in}, T_{on} and T_{out} types of tangent bifurcations, "square" and "clover" heteroclinic cycles, etc. The bifurcation sequences and phase portraits for every region are studied and illustrated in Refs.[29, 30], which make it possible to analyze the complicated bifurcations near the 1:4 resonance point for the multi-degree-of-freedom vibro-impact systems.

4. Numerical analyses of Neimark-Sacker and tangent bifurcations

In this section the analyses developed in the former sections are verified by the presentation of results for the vibro-impact system shown in Fig.1. The existence and stability of 1-1 motion are analyzed explicitly. Also, local bifurcations at the points of change in stability, discussed in the previous section, are considered. The vibro-impact system shown in Fig.1, with non-dimensional parameters: $\beta = 0.5$, $\mu_m = 0.6$, $\mu = 0.3$, $\mu_k = 1$, $f_{20} = 0$ and $R = 0.5$ has been chosen for analyses. The forcing frequency ω and the damping ratio ζ are taken as the control parameters, i.e. $v = (\omega, \zeta)^T$. The eigenvalues of Jacobian matrix $Df(v, X^*)$ are computed with $\omega \in [0.35, 0.6]$ and $\zeta \in [0.02, 0.04]$. All eigenvalues of $Df(v, X^*)$ stay inside the unit circle for v=(0.35, 0.02)^T. By gradually increasing ω and ζ from the point v=(0.35, 0.02)^T to change the control parameter v, we found that a complex conjugate pair of eigenvalues locate at the points (0, ± i) of the unit circle of the complex plane, and the other eigenvalues still stay inside the unit circle as v equals $v_c = (0.4660942, 0.0301314)^T$. Here v_c is a bifurcation point associated with 1:4 resonance, and all eigenvalues of $Df(v, X^*)$, at the bifurcation point v_c, are given as follows

$\lambda_{1,2}(v_c) = -0.00000018 \pm 1.00000178i$, $\lambda_{3,4}(v_c) = -0.09576943 \pm 0.65100270i$,

$\lambda_{5,6}(v_c) = 0.03121547 \pm 0.1269071i$.

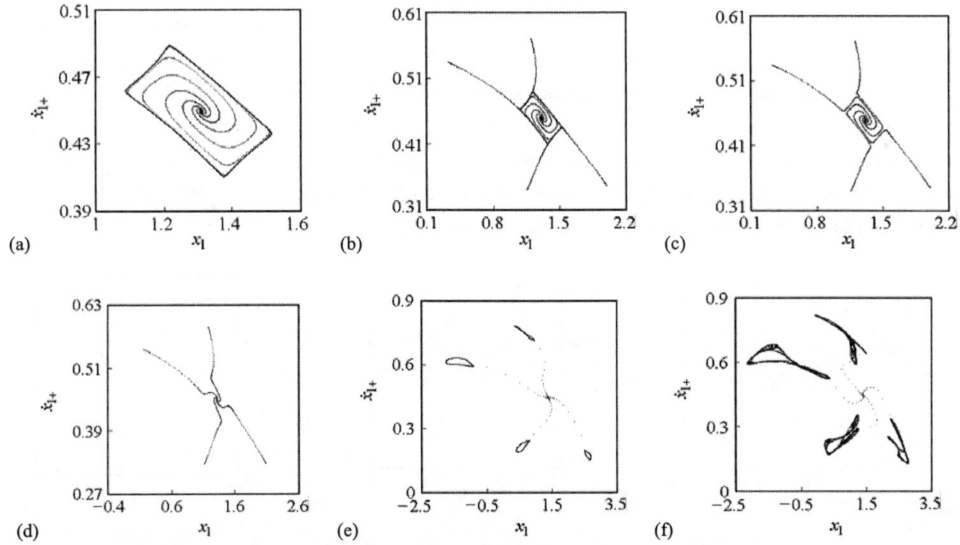

Fig.3. Projected Poincaré sections($\zeta = \zeta_c - \Delta\zeta$): (a) transient points as well as the quasi-periodic attractor associated with 1-1 fixed point, starting from the initial condition near the unstable 1-1 fixed point, $\omega = 0.4661, \zeta = \zeta_c - \Delta\zeta$; (b) 4-4 fixed points generated via T_{out} type of tangent bifurcation of 4-4 fixed points, $\omega = 0.4661221$; (c) transient points, unstable 4-4 fixed points as well as stable 4-4 fixed points, starting from the initial condition near the unstable 1-1 fixed point, $\omega = 0.46613$; (d) 4-4 fixed points, $\omega = 0.4662$; (e) quasi-periodic attractor associated with 4-4 fixed points, $\omega = 0.468$; (f) chaos, $\omega = 0.4682$

Numerical analyses are carried out to unfold dynamic behavior of the vibro-impact system near the 1:4 resonance point. We choose a series of values of damping ratio ζ and change successively the forcing frequency ω in the whole numerical analyses. The results from simulation are shown in Fig.3 in the form of projected portraits of Poincaré map. The Poincaré section is taken in the form σ, which is six-dimensional. The section is projected to the $(x_1, \dot{x}_{1+})$ plane, etc., which is called the projected Poincaré section. The 1-1 fixed point, with the corresponding parameter v, is taken as the initial map point in every numerical analysis. We choose the damping ratio ζ near $\zeta_c = 0.0301314$, e.g., $\zeta = \zeta_c \pm \Delta\zeta$, $\Delta\zeta =$ 0.00043, and change the forcing frequency ω in the numerical analyses. The results from simulation, associated with $\zeta = \zeta_c - \Delta\zeta$, show that the system exhibits stable 1-1 motion with $\omega \in [0.35, 0.4660236]$. As ω passes through $\omega_{c1} = 0.4660236$ increasingly, instability of 1-1 motion occurs, and the motion undergoes Neimark-Sacker bifurcation. The system exhibits quasi-periodic impact motion for $\omega > \omega_{c1}$. The quasi-periodic attractor associated with 1-1 fixed point is represented by an attracting invariant circle in projected Poincaré section as seen in Fig.3(a). With increase in the forcing frequency ω, a T_{out} type of tangent bifurcation of 4-4 fixed points occurs so that it changes the quasi-periodic attractor to two families of 4-4 fixed points, one of which are unstable, the other stabilize; see Figs.3(b) and (c). A full understanding of Figs.3(b) and (c) can be obtained by making a comparison with Fig.9.18(b) of page 442 in Ref.[30]. We can observe the inflection points of four mapping trajectories in Figs.3(b), (c) and (d). The unstable 4-4 fixed points are located near the inflection points of four mapping trajectories. With increase in ω, 4-4 motion changes its stability, and Neimark-Sacker bifurcation of fixed points associated with 4-4 motion occurs so that the system exhibits quasi-periodic impact motion associated with 4-4 points; see Fig.3(e). With further increase in ω, phase locking takes place so that the quasi-periodic motion, associated with 4-4 motion, gets locked into a periodic attractor of higher (than period four) period, which subsequently becomes unstable and chaotic; see Fig.3 (f). For $\zeta = \zeta_c + \Delta\zeta$, the system exhibits stable 1-1 motion with $\omega \in [0.38, 0.4661401]$. As ω passes through $\omega_{c1} = 0.4661401$ increasingly, 1-1 motion undergoes Neimark-Sacker bifurcation; see Fig.4(a).

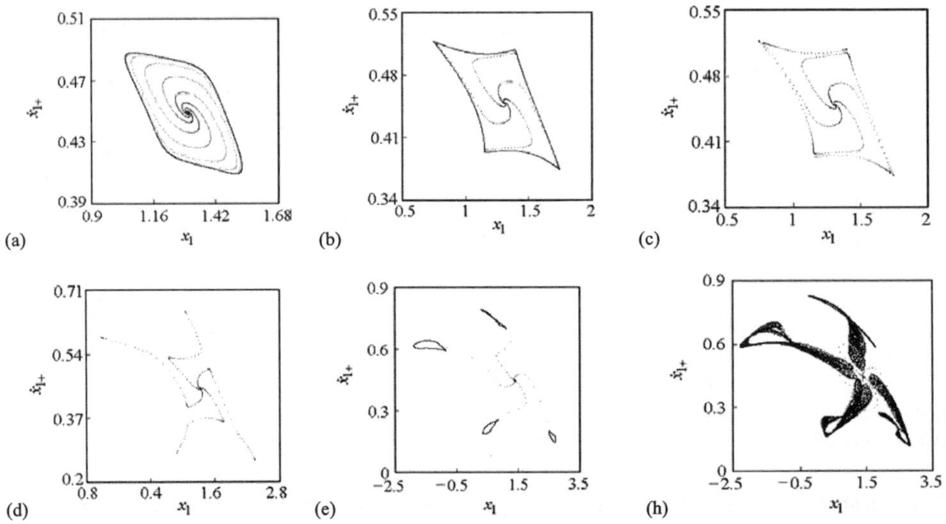

Fig.4. Projected Poincaré sections($\zeta = \zeta_c + \Delta\zeta$): (a) quasi-periodic attractor associated with 1-1 fixed point, $\omega = 0.46616$; (b) quasi-periodic attractor associated with 1-1 fixed point, $\omega = 0.4664833$; (c) 4-4 fixed points generated via T_{on} type of tangent bifurcation of 4-4 fixed points, $\omega = 0.4664834$; (d) 4-4 fixed points(stable nodes), $\omega = 0.4669$; (f) quasi-periodic attractor associated with 4-4 fixed points, $\omega = 0.46825$; (h) chaos, $\omega = 0.4685$.

With increase in the forcing frequency ω, a T_{on} type of tangent bifurcation of 4-4 fixed points occurs so that it changes the quasi-periodic attractor to two families of 4-4 fixed points, one of which are unstable, the other stabilize; see Figs.4(b) and (c). A full understanding of Figs.4(b) and (c) can be obtained by making a comparison with Fig.9.19 of page 443 in Ref.[30]. And then the system exhibits the similar bifurcation sequences and the routes to chaos; see Figs.4(d)~(f).

5. 1:4 strong resonance bifurcations of the vibratory system with symmetrical rigid stops

The mechanical model for a three-degree-of-freedom vibratory system with symmetrical stops is shown in Fig.5. Displacements of the masses M_1, M_2 and M_3 are represented by X_1, X_2 and X_3, respectively. The masses are connected to linear springs with stiffnesses K_1, K_2 and K_3, and linear viscous dashpots with damping constants C_1, C_2 and C_3. The excitations on the masses are harmonic with amplitudes P_1, P_2 and P_3. The excitation frequency Ω and the phase τ are the same for these masses. For small forcing amplitudes the system will undergo simple oscillations and behave as a linear system. As the amplitude is increased, the second mass M_2 eventually begins to hit the stops and the motion becomes nonlinear. The impact is described by a coefficient of restitution R.

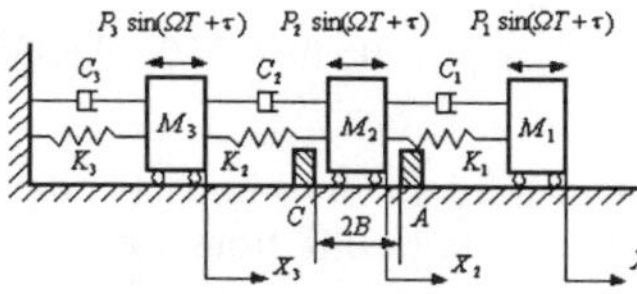

Fig.5. Schematic of a three-degree-of-freedom vibratory system with symmetrical rigid stops

Between the stops, the non-dimensional differential equations of motion can be obtained by introducing the non-dimensional quantities

$$m_i = \frac{M_i}{M_1}, \quad k_i = \frac{K_i}{K_1}, \quad f_{i0} = \frac{P_i}{P_0}, \quad \omega = \Omega\sqrt{\frac{M_1}{K_1}}, \quad t = T\sqrt{\frac{K_1}{M_1}},$$

$$\zeta_i = \frac{C_i}{2\sqrt{K_1 M_1}}, \quad \gamma = \frac{\zeta_i}{k_i}, \quad \delta = \frac{BK_1}{P_0}, \quad x_i = \frac{X_i K_1}{P_0}, \qquad i = 1, 2, 3. \tag{9}$$

When impacts occur, for $|x_2| = \delta$, the velocities of the impacting mass M_2 are changed according to the impact law

$$\dot{x}_{2A+} = -R\dot{x}_{2A-}, \quad (x_2 = \delta), \qquad \dot{x}_{2C+} = -R\dot{x}_{2C-}, \quad (x_2 = -\delta), \tag{10}$$

where $\dot{x}_{2A-}$ and $\dot{x}_{2A+}$ ($\dot{x}_{2C-}$ and $\dot{x}_{2C+}$) represent the impacting mass velocities of approach and departure at the instant of impacting with the stop A (C), respectively.

Periodic-impact motions of the vibro-impact system can be characterized by the symbol *n-p-q*, where *q* and *p* is the number of impacts occurring respectively at the stops *A* and *C*, and *n* is the number of the forcing cycles. Under suitable system parameter conditions, the vibro-impact system can exhibit symmetrical double-impact periodic motion. The Poincaré section associated with the state of the vibro-impact system, just immediately after the impact occurring at the stop *A*, is chosen, and symmetrical double-impact periodic motion and its disturbed map are derived analytically in Ref.[31] .We can choose the Poincaré section $\sigma = \{(x_1, \dot{x}_1, x_2, \dot{x}_2, x_3, \dot{x}_3, \theta) \in \boldsymbol{R}^6 \times S, \ x_2 = \delta, \ \dot{x}_2 = \dot{x}_{2A+}\}$ to establish Poincaré map of the system

$$X' = \tilde{f}(v, X). \tag{11}$$

Under suitable system parameter conditions, the vibratory system with symmetrical rigid stops can exhibit 1-1-1 symmetrical motion due to symmetry of stops and excitation points. In general case, 1-1-1 symmetrical motion undergo pitchfork bifurcation with change in the bifurcation parameters, so that 1-1-1 asymmetrical orbits stabilize. 1-1-1 asymmetrical orbits exhibit two antisymmetrical forms, which may be born by different initial conditions of motion. Near 1:4 resonance point there exist a pair of antisymmetrical 4-4-4 motions due to symmetry of stops. So bifurcation sequences of the vibratory system with symmetrical rigid stops, near 1:4 resonance point, are different from those of the first vibro-impact system, which are more complicated. The system with parameters: m_1=1, m_3=2, k_1=1, k_2=1.4, k_3=1.2, γ=0.03, f_{10}=0, f_{20}=1, f_{30}=0, δ=0.1, R=0.7, has been chosen for analysis. The forcing frequency ω and m_2 are taken as the control parameters, i.e. $v=(\omega, m_2)^{\mathrm{T}}$. The eigenvalues of $D\tilde{f}(v, X^*)$ are computed with $\omega \in [3.5, 4.16]$ and $m_2 \in [0.9, 1.2]$. The moduli of all eigenvalues of $D\tilde{f}(v, X^*)$ are less than one for v=(3.5, 0.9)^{T}. By gradually increasing ω and m_2 from the point v=(3.5, 0.9)^{T} to change the control parameter v, we found that a complex conjugate pair of eigenvalues locate at the points (0, $\pm$ i) of the unit circle of the complex plane, and the other eigenvalues still stay inside the unit circle as v equals $v_c = (3.9295220,\ 1.1151092)^{\mathrm{T}}$. All eigenvalues of $D\tilde{f}(v, X^*)$, at 1:4 resonance point v_c, are given as follows

$\lambda_{1,2}(v_c) = 0.00000022 \pm 1.00000000\mathrm{i}$, $\lambda_{3,4}(v_c) = -0.20879310 \pm 0.93356970\mathrm{i}$,

$\lambda_{5,6}(v_c) = 0.02609331 \pm 0.41288840\mathrm{i}$.

Local behavior of the three-degree-of-freedom vibro-impact system, near 1:4 strong resonance point, are analyzed by taking $m_2 = m_{2c} \pm \Delta m_2$. The results from simulation show that the system exhibits stable period one double-impact symmetrical motion with $\omega \in [3.5, 3.929541]$ and $\Delta m_2 = 0.0001$. As ω passes through $\omega_{c1} = 3.929541$ increasingly, instability of 1-1-1 symmetrical motion occurs, and the motion undergoes Neimark-Sacker bifurcation. The system exhibits quasi-periodic impact motion for $\omega > \omega_{c1}$. The quasi-periodic attractor associated with 1-1-1 fixed point is represented by an attracting invariant circle in projected Poincaré section as seen in Fig.6(a). With increase in the forcing frequency ω, the attracting invariant circle transits into a "tire-like" quasi-periodic attractor; see Figs.6(b) and (c). There exist two quasi-periodic attractors as ω is increased to ω =4.075, one of them is an unstable attracting invariant circle, the other "tire-like" quasi-periodic attractor; see Figs.6(d) and 7(a). With further increase in ω, the unstable attracting invariant circle transits to unstable period four eight-impact fixed points. Two different projected Poincaré sections $(x_3, \dot{x}_3)$ and $(x_1, \dot{x}_1)$ are shown for ω =4.06, 4.075 and 4.077 in Figs.6(d)~(f) and Figs.7(a)~(c), respectively. We can observe, from Figs.6(e) and 7(b), transient map points as well as unstable 4-4-4 fixed point and "tire-like" quasi-periodic attractor corresponding to ω = 4.075, starting from the initial condition near the fixed point of 1-1-1 symmetrical motion. Unstable 4-4-4 fixed points and chaos are shown for ω = 4.077 in Figs.6(f) and 7(c). As ω is further increased, the fixed points associated with long-periodic multi-impact motion, "tire-like" quasi-periodic attractor and attracting invariant circle are generated in turn; see Figs.6(g)~(i).

6. Conclusions

The mechanical models for a two-degree-of-freedom vibratory system impacting a unconstrained rigid body and a three-degree-of-freedom vibratory system impacting two symmetrical rigid stops are considered, respectively. Two-parameter bifurcations of fixed points in the vibro-impact systems, associated with 1:4 strong resonance case, are analyzed. For the first mechanical model, near the bifurcation point for 1:4 strong resonance there exist quasi-periodic impact motion, stable and unstable 4-4 motions, etc. Usually, 4-4 motion is probably produced via a T_{in}, T_{on} or T_{out} type of tangent(fold) bifurcation. The results from simulation show that the routes of 4-4 motions to chaos are multiple due to period doubling or Neimark-Sacker bifurcation of 4-4 fixed points, or the "square heterroclinic" cycle formed by coinciding stable and unstable separatrices of 4-4 fixed points, etc. For the mechanical model of the vibratory system with symmetrical rigid stops, near the bifurcation point for 1:4 strong resonance there exist quasi-periodic impact motions represented by attracting invariant circle and "tire-like" quasi-periodic attractor, stable and unstable 4-4-4 motion, etc. It is to be noted that the bifurcation sequences of the vibratory systems with symmetrical rigid stops, near 1:4 resonance point, are different from those of the vibratory systems impacting an unconstrained rigid body or a single stop, which are more complicated.

The strict conditions of bifurcations, associated with strong resonance cases, are not easy to encounter in practical application of engineering. However, there exist the possibilities that actual nonlinear dynamical systems, with two varying parameters or more, work near the critical values of strong resonance bifurcations due to change of parameters. The change of multi-parameter possibly leads to the results that the vibro-impact systems work near the critical parameters associated with strong resonance cases. It is necessary to study the bifurcations caused by change of multi-parameters and reveal dynamical behavior of nonlinear systems near the strong resonance points.

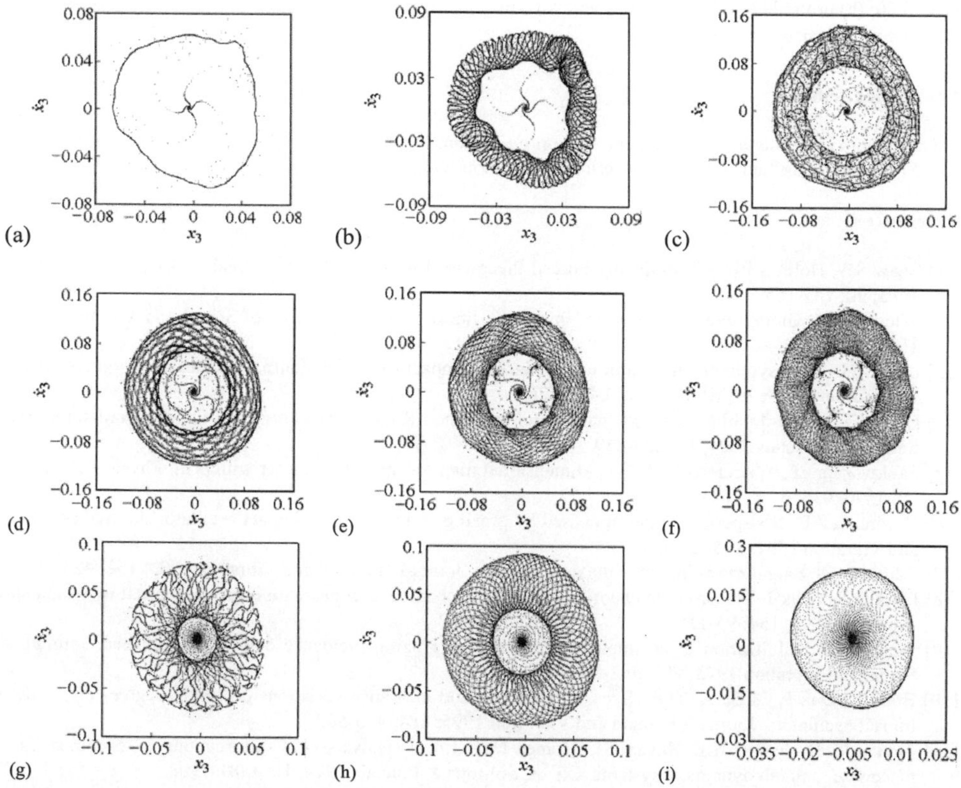

Fig.6. Projected Poincaré sections: (a) quasi-periodic attractor associated with 1-1-1symmetrical fixed point, $\omega = 3.935$; (b) transient map points as well as "tire-like" quasi-periodic attractor, starting from the initial condition near the fixed point of 1-1-1 symmetrical motion, $\omega = 3.992$; (c) "tire-like" quasi-periodic attractor, $\omega = 4.028$; (d) unstable attracting invariant circle and "tire-like" quasi-periodic attractor, $\omega =$ 4.06; (e) unstable 4-4-4 fixed points and "tire-like" quasi-periodic attractor, $\omega = 4.075$; (f) unstable 4-4-4 fixed points and chaos, $\omega = 4.077$; (g) phase locking, $\omega = 4.12$; (h) "tire-like" quasi-periodic attractor, $\omega = 4.11$; (i) quasi-periodic attractor associated with 1-1-1 symmetrical fixed point, $\omega = 4.15$.

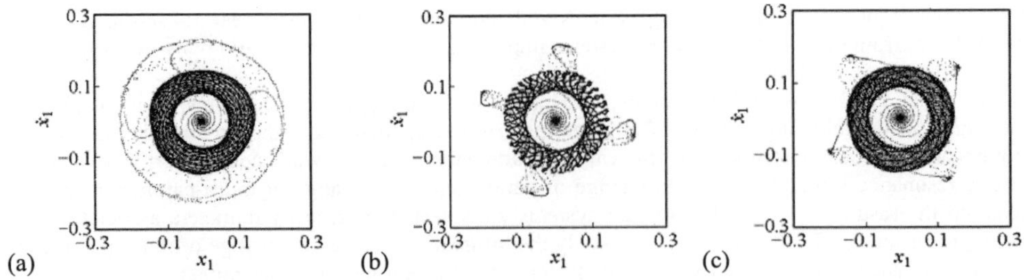

Fig.7. Projected Poincaré sections: (a) unstable attracting invariant circle and "tire-like" quasi-periodic attractor, $\omega =$ 4.06; (b) unstable 4-4-4 fixed points and "tire-like" quasi-periodic attractor, $\omega = 4.075$; (c) unstable 4-4-4 fixed points and chaos $\omega = 4.077$

Acknowledgement

The authors gratefully acknowledge the support by National Natural Science Foundation of China (10572055, 50475109) and 'Qing Lan' Talent Engineering by Lanzhou Jiaotong University.

References

[1] Shaw SW, Holmes PJ. A Periodically Forced Piecewise Linear Oscillator. Journal of Sound and Vibration 1983; 90: 129-155.
[2] Whiston GS. Global dynamics of vibro-impacting linear oscillator. Journal of Sound and Vibration 1987, 118(3), 395-429.
[3] Luo ACJ. An unsymmetrical motion in a horizontal impact oscillator. Journal of Vibration and Acoustics, Transactions of the ASME 2002; 124:420-426.
[4] Luo ACJ. Period-doubling induced chaotic motion in the LR model of a horizontal impact oscillator Chaos, Solitons & Fractals 2004; 19: 823-839.
[5] Pavlovskaia EE, Wiercigroch M. Two-dimensional map for impact oscillator with drift. Physical Review E 2004; 70, 03620.
[6] Nordmark AB. Non-periodic motion caused by grazing incidence in an impact oscillator. Journal of Sound and Vibration 1991; 145: 279-297.
[7] Whiston GS. Singularities in virbo-impact dynamics. Journal of Sound and Vibration 1992; 152: 427-460.
[8] Peterka F, Vacik J. Transition to chaotic motion in mechanical systems with impacts. Journal of Sound and Vibration 1992; 154: 95-115.
[9] Ivanov AP. Stabilization of an impact oscillator near grazing incidence owing to resonance. Journal of Sound and Vibration 1993; 162: 562-565.
[10] Budd C, Dux F, Cliffe A. The Effect of frequency and clearance variations on single-degree- of-freedom impact oscillators. Journal of Sound and Vibration 1995; 184: 475-502.
[11] Bernardo D, Feigin MI, Hogan SJ, Homer ME. Local analysis of C-bifurcations in N-dimensional piecewise- smooth dynamical systems. Chaos, Solitons & Fractals 1999; 10: 1881-1908.
[12] Luo ACJ, Chen LD. Periodic motions and grazing in a harmonically forced, piecewise, linear oscillator with impacts. Chaos, Solitons & Fractals 2005; 24: 567-578.
[13] Luo ACJ. Grazing and chaos in a periodically forced, piecewise linear system. Journal of Vibration and Acoustics 2006; 128: 29-34.
[14] Wagg DJ. Rising phenomena and the multi-sliding bifurcation in a two-degree of freedom impact oscillator, Chaos, Solitons & Fractals 2004; 22: 541-548.
[15] Nguyen DT, Noah ST, Kettleborough CF. Impact behaviour of an oscillator with limiting stops, part I: a parametric study. Journal of Sound and Vibration 1986; 109: 293-307.
[16] Chatterjee S, Mallik AK. Bifurcations and chaos in autonomous self-excited oscillators with impact damping. Journal of Sound and Vibration 1996; 191: 539-562.
[17] Luo GW, Xie JH. Hopf bifurcation of a two-degree-of-freedom vibro-impact system. Journal of Sound and Vibration 1998; 213: 391-408.

[18] Hu HY. Controlling chaos of a periodically forced nonsmooth mechanical system. Acta Mechanica Sinica 1995; 11: 251-258.
[19] Mejaard JP, Pater ADD. Railway vehicle systems dynamics and chaotic vibrations. International Journal of Non-Linear Mechanics 1989; 24: 1-17.
[20] Zeng J, Hu S. Study on frictional impact and derailment for wheel and rail. Journal of Vibration Engineering 2001; 14: 1-6.
[21] Shu ZZ, Shen XZ. Theoretical analysis of complete stability and automatic vibration isolation of impacting and vibrating systems with double masses. Chinese Journal of Mechanical Engineering 1990; 26: 50-57.
[22] Xie JH. The mathematical model for the impact hammer and global bifurcations. Acta Mechanica Sinica 1997; 29: 456-463.
[23] Pavlovskaia EE, Karpenko EV, Wiercigroch M. Non-linear dynamic interactions of a Jeffcott rotor with preloaded snubber ring. Journal of Sound and Vibration 2004; 276: 361–379.
[24] Han PRS, Luo ACJ. Chaotic motion of a horizontal impact pair. Journal of Sound and Vibration 1995; 181: 231-250.
[25] Bapat CN. The general motion of an inclined impact damper with friction. Journal of Sound and Vibration 1995; 184: 417-427.
[26] Kahraman A, Singh R. Non-linear dynamics of a geared rotor-bearing system with multiple clearances. Journal of Sound and Vibration 1991; 144: 469-506.
[27] Dong HJ, Shen YW, Liu MJ, Zhang SH. Research on the dynamical behaviors of rattling in gear system. Chinese Journal of Mechanical Engineering 2004; 40: 136-141.
[28] Zhang YL. Periodic motion and bifurcation of a three-degree-of-freedom vibro-impact system. Journal of Lanzhou Railway University 2003; 22:32-39.
[29] Arnol'd VI. Geometrical methods in the theory of ordinary differential equations. Springer-Verlag; 1983.
[30] Kuznetsov YA. Elements of applied bifurcation theory, Springer-Verlag, New York; 1998.
[31] Luo GW, Zhang YL. Dynamical behavior of a class of vibratory systems with symmetrical rigid stops near the point of codimension two bifurcation. Journal of Sound and Vibration, in press(YJSVI 7818).

On The Nonlinear Dynamic Characteristics of Truck Rear Full-Floating Axle

T.N. Tongele
Department of Mechanical and Industrial Engineering, Southern Illinois University Edwardsville, Edwardsville, IL 62026-1805, USA

Abstract
This paper is essentially concerned with the question of how the rear full-floating axle of a truck behaves in response to road induced excitation. The axle is modeled as a single-degree-of-freedom system excited by a sinusoidal base motion. Linear and nonlinear analytical approaches are used to examine the axle's response characteristics. Linear analysis shows that the axle's response is characterized by its natural frequency and damping ratio. It's worth noticing the occurrence of the jump phenomenon illustrated by the solution of Duffing's equation used to model the axle's nonlinear behavior. Both approaches appear to be complementary in providing more insight into the axle's dynamic behavior.

PACS: 05.40-a
Keywords: Nonlinear vibration, Truck rear full-floating axle, Axle dynamic behavior

1. Introduction

Modeling and simulation of the dynamic response of vehicle to harmonic excitation has been and continue to be subject of numerous studies [1-3]. This paper focuses on the dynamic behavior of a particular part of vehicle, i.e., the rear full-floating axle of a truck. The single-degree-of-freedom used to model the axle's response constitutes the basis of vibration fundamentals [4-5].

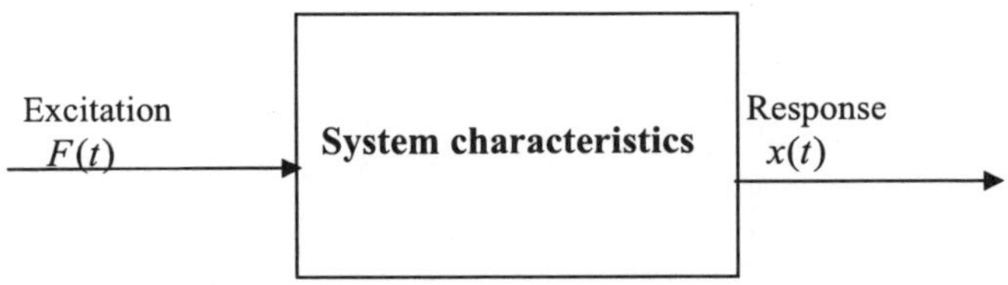

Figure 1 Block diagram relating the excitation and the response

Figure 1 shows that a system subjected to an excitation $F(t)$ exhibits a certain response $x(t)$. The solution to the equation of motion of the system provides answers to how the system responds to the

The corresponding author. *Tel:1-618-650-2820, Fax:1-618-650-2555*
Email: ttealak@siue.edu

excitation. But one of the fundamental questions that arise is whether a system is linear or nonlinear, because the answer has profound implications as far as the solution to the equation of motion is concerned [5]. Physical systems are rather nonlinear. Basically, all the problems in mechanics are nonlinear from the onset. The sources of nonlinearities can be material or constitutive, geometric, inertia, body forces, or friction. The constitutive nonlinearity occurs when the stresses are nonlinear functions of the strains. The geometric nonlinearity is associated with large deformation in solids, such as beam, plates, frames, and shells, resulting in nonlinear strain-displacement relations. The inertia non linearity may be caused by the presence of concentrated or distributed masses. The nonlinear body forces are essentially magnetic and electric forces. The friction nonlinearity occurs because the friction force is a nonlinear function of displacement and velocity, such as dry friction and backlash. The nonlinearities may appear in the governing partial differential equations, or the boundary conditions, or both [6-7]. The modeling of structural systems can therefore be linear modeling, pseudo nonlinear, and nonlinear. The linearizations commonly practiced are approximating devices that are good to the degree of the supporting assumptions. Linear approach yields exact solutions that can be determined analytically. In pseudo nonlinear modeling, the static behavior is described by a nonlinear model, but the dynamic behavior is described by a linear model. Nonlinear modeling makes use of numerical integration to solve equation of motion of the system which cannot be obtained in closed form [7-8].

This paper models the full-floating rear axle of a truck as a single-degree-of-freedom system, uses both linear and nonlinear approaches to solve the governing equation, and compares the dynamic characteristics obtained. After this introduction, the second section will deal with the problem formulation. The third section will present the linear and nonlinear descriptions of the dynamic behavior of the axle. The forth part will discuss the results. And a conclusion will summarize the work and the findings.

2. Problem Statement

(a) Truck drawing

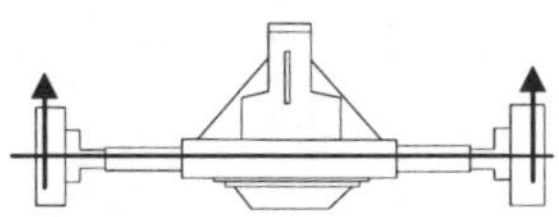

(b) Truck rear axle

Figure 2 A rough sketch of a truck and the real axle

Figure 2 is an illustration of a rear axle and its location in a truck. A full-floating axle transmits engine power, but does not carry any of the vehicle weight. All weight is supported through the outer bearing assembly [9-10]. In here, the axle unit comprises the two wheels bolted to the ends of the shaft, and the differential box in the middle of the axle shaft. The mathematical model of the axle is depicted in figure 3, while the free-body diagram is shown in figure 4. The stiffness and the damping coefficient of the tires are represented by k_1 and c_1; and those of the axle are represented by k_2 and c_2. As the vehicle moves along the road, the vertical displacements at the tires excite the vehicle axle system, whose motion may consist of a translational motion of the center of mass of the axle as well as tilting/rotation about the center of mass. It is assumed that the motion u is the input to the system and the vertical motion y of the axle is the output. The displacement y is measured from the equilibrium position in the absence of the input u. Since the stiffness of the tire and that of the axle are represented in series, the system equivalent stiffness, k, is obtain using the following relation:

$$k = \frac{k_1 k_2}{k_1 + k_2}. \tag{1}$$

Similarly, the equivalent damping coefficient, c, is given by this expression:

$$c = \frac{c_1 c_2}{c_1 + c_2}. \tag{2}$$

The next step consists of deriving and solving the axle's governing equation of motion, which will shed light on how the full-floating axle behaves in response to road induced excitation.

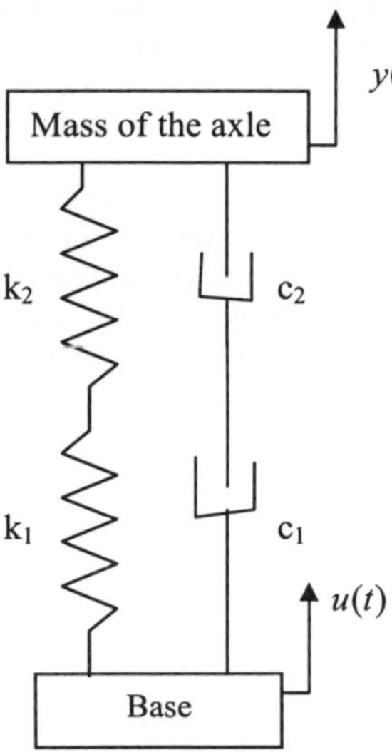

Figure 3 Model of the full-floating axle system

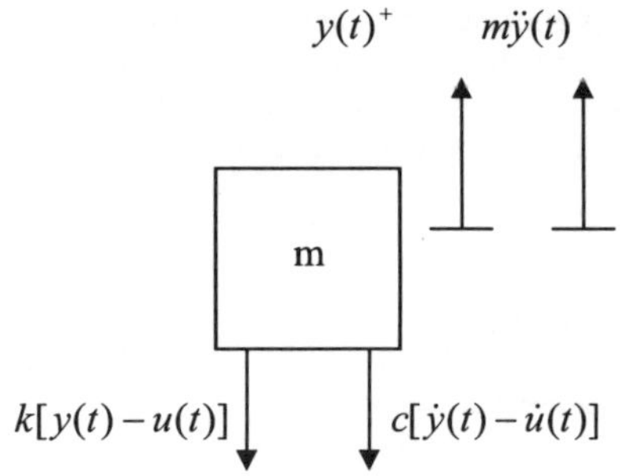

Figure 4 Motion of mass m excited by the base motion

3. Axle's Dynamic Behavior

3.1 Linear Analysis

Summing the relevant forces on the mass, m, as shown in figure 4, (i.e. the inertial force $m\ddot{x}(t)$ is equal to the sum of the two forces acting on the mass m, and the gravitational force is balanced against the static deflection of the spring), the following expression is obtained:

$$.\ddot{m}y(t) + c[\dot{y}(t) - \dot{u}(t)] + k[y(t) - u(t)] = 0. \tag{3}$$

By dropping the time symbol (t), and recognizing that the dot symbol represents the derivative with respect to time, equation (3) can be rewritten as:

$$\ddot{m}y + c(\dot{y} - \dot{u}) + k(y - u) = 0. \tag{4}$$

Since it is assumed that the base motion is harmonic, the input force can be expressed as:

$$u(t) = U \sin \omega t, \tag{5}$$

where U denotes the amplitude of the base motion, and ω represents the frequency of the base oscillation. Substituting equation (5) into equation (4) yields, after rearrangement, the following expression:

$$m\ddot{y} + c\dot{y} + ky = kU \sin \omega t + c\omega U \cos \omega t . \tag{6}$$

Equation (6) is a nonhomogenious, linear, ordinary differential equation governing the motion of the full-floating axle. Dividing equation (6) by m and using the definitions of damping ratio and natural frequency yields

$$\ddot{y} + 2\zeta\omega_n\dot{y} + \omega_n^2 y = \omega_n^2 U \sin \omega t + 2\zeta\omega_n \omega U \cos \omega t , \tag{7}$$

where $\zeta = c/c_c$ is the damping ration, $c_c = 2m\omega_n$ is the critical damping level, and $\omega_n = \sqrt{k/m}$ is the undamped natural frequency.

The general solution of equation (7) has both steady-state and transient parts; i.e., the complete solution to this equation is obtained by combining the general solution of the homogeneous part (left side equation (7)) with the particular solution of the right side of the same equation. Hence,

$$y = y_h + y_p \tag{8}$$

The well know solution to the homogeneous part of equation (7) is:

$$y_h = A\, e^{\left(-\zeta+\sqrt{\zeta^2-1}\right)\omega_n t} + B\, e^{\left(-\zeta-\sqrt{\zeta^2-1}\right)\omega_n t} , \tag{9}$$

where A and B are arbitrary constants to be determined from the initial conditions (values of y and $\dot{y}$ at time $t = 0$) of the system. It can be observed that the behavior of this solution depends upon the magnitude of the damping.

The particular solution y_p can also be expected to be harmonic, and is the linear combination of particular solutions due to the sine and cosine input excitations. The expression is:

$$y_p = \omega_n U \sqrt{\frac{\omega_n^2 + (2\zeta\omega)^2}{\left(\omega_n^2 - \omega^2\right)^2 + (2\zeta\omega_n\omega)^2}} \cos\left(\omega t - \varphi_1 - \varphi_2\right), \tag{10}$$

where U and φ s, are respectively the amplitude and the phase angles of the response to be determined.

$$\varphi_1 = \tan^{-1}\left(\frac{\omega_n}{2\zeta\omega}\right) \quad \text{and} \quad \varphi_2 = \tan^{-1}\left(\frac{2\zeta\omega_n\omega}{\omega_n^2 - \omega^2}\right) . \tag{11}$$

It is convenient to denote the magnitude of the particular solution, y_p, by Y so that

$$Y = U\sqrt{\frac{1+(2\zeta\, r)^2}{\left(1-r^2\right)^2 + (2\zeta\, r)^2}} , \tag{12}$$

where $r = \omega/\omega_n$ is the frequency ratio. Dividing equation (12) by the magnitude of the base motion yield:

$$\frac{Y}{U} = \sqrt{\frac{1+(2\zeta\, r)^2}{\left(1-r^2\right)^2 + (2\zeta\, r)^2}} . \tag{13}$$

Equation (13) expresses the ratio of the amplitude of the response y_p to that of the base motion u, which is called the displacement transmissibility and is used to describe how motion is transmitted from the base to the axle as a function of the frequency ratio r. The displacement transmissibility as a function of the frequency ratio for selected values of the damping ratio are shown in figure 5.

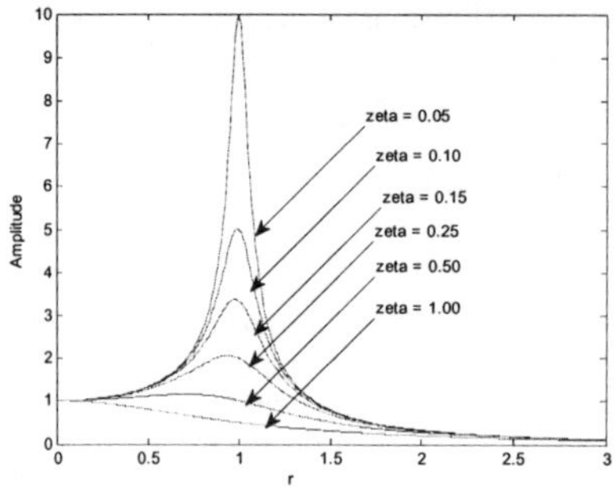

Figure 5 Nondimensional magnitude

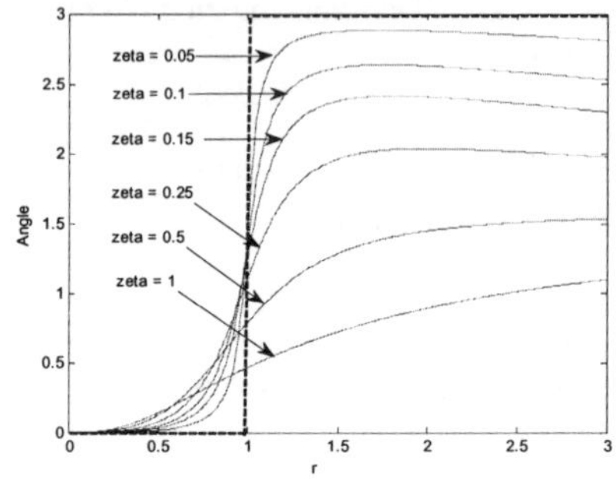

Figure 6 Response phase angle of the response amplitude

The phase angle of the system is obtained by combining the two parts of equation (11) in one expression:

$$\varphi = \tan^{-1}\left(\frac{2\zeta r^3}{1+(4\zeta^2-1)r^2}\right). \tag{14}$$

Figure 6 displays the phase angle variation as a function of the frequency ratio for selected values of the damping ratio. Another quantity of interest is the force transmitted to the axle as the result of the harmonic base excitation. From the free-body diagram (figure 4), the force transmitted is done through the spring and damper, and must balance the inertial force of the mass m. Therefore,

$$F(t) = c(\dot{y}-\dot{u}) + k(y-u) = -m\ddot{y}. \tag{15}$$

In the steady state, the solution for y is given by equation (10). Differentiating twice equation (10) and substituting it into equation (15) yields

$$F(t) = m\omega^2\omega_n U\sqrt{\frac{\omega_n^2+(2\zeta\omega)^2}{\left(\omega_n^2-\omega^2\right)^2+\left(2\zeta\omega_n\omega\right)^2}}\cos(\omega t-\varphi_1-\varphi_2). \tag{16}$$

Introducing the frequency ratio in equation (16) gives

$$F(t) = F_T\cos(\omega t-\varphi_1-\varphi_2), \tag{17}$$

where the magnitude of the transmitted force, F_T, is given by

$$F_T = kUr^2\sqrt{\frac{1+(2\zeta r)^2}{\left(1-r^2\right)^2+(2\zeta r)^2}}. \tag{18}$$

Equation (18) is used to define force transmissibility by forming the following ratio:

$$\frac{F_T}{kU} = r^2 \sqrt{\frac{1+(2\zeta\, r)^2}{\left(1-r^2\right)^2+(2\zeta\, r)^2}}\,. \tag{19}$$

Equation (19) expresses a dimensionless measure of how displacement in the base motion of amplitude U results in a force magnitude applied to the axle. Figure 7 illustrates how the dimensionless force ratio varies as the frequency of the base motion increases.

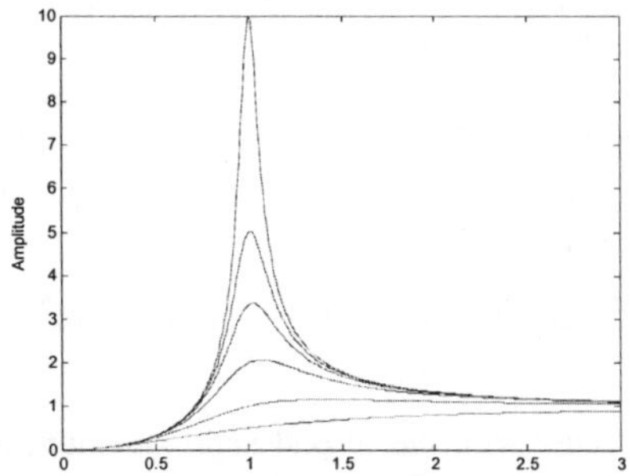

Figure 7 Transmitted Force

It should be remarked that figures 5 and 7 look identical. In fact, the quantity expressed by figure 7, more than depends upon the quantity in figure 5, but is the quantity in figure 5 times the squared ratio of the excitation frequency to the natural frequency.

3.2 Nonlinear Analysis

The use of numerical integration allows consideration of the effect of various nonlinear terms in the equation of motion. The Duffing equation is used to describe the nonlinear vibration of the axle around its static equilibrium [4-5,11-12]. The expression is as follow:

$$\ddot{x} + c\dot{x} + \omega_n^2 x \pm \alpha x^3 = F\cos(\omega t + \varphi)\,. \tag{20}$$

The plus or minus sign is generally used to indicate hardening or softening nonlinearity.
If damping is ignored, equation (20) can be rewritten with a slight modification as

$$\ddot{x} + \omega_n^2 x \pm \alpha x^3 = F\cos(\omega t)\,, \tag{21}$$

or

$$\ddot{x} = -\omega_n^2 x \mp \alpha x^3 + F\cos(\omega t)\,. \tag{22}$$

As a first approximation, the solution to equation (22) can be assumed as:

$$x_1(t) = A\cos(\omega t)\,, \tag{23}$$

where A is unknown. By substituting equation (23) into equation (22), a differential equation is obtained for the second approximation:

$$\ddot{x}_2 = -A\omega_n^2\cos(\omega t) \mp A^3\alpha\cos^3(\omega t) + F\cos(\omega t)\,. \tag{24}$$

Using the following identity

$$\cos^3(\omega t) = \frac{3}{4}\cos(\omega t) + \frac{1}{4}\cos(3\omega t)\,, \tag{25}$$

equation (24) can be expressed as

$$\ddot{x}_2 = -(A\omega_n^2 \pm \frac{3}{4}A^3\alpha - F)\cos(\omega t) \mp \frac{1}{4}A^3\cos(3\omega t). \tag{26}$$

By integrating this equation and setting the constants of integration to zero so as to make the solution harmonic with period $\tau = \frac{2\pi}{\omega}$, a second approximation is obtained as:

$$x_2(t) = \frac{1}{\omega^2}(A\omega_n^2 \pm \frac{3}{4}A^3\alpha - F)\cos(\omega t) \mp \frac{A^3\alpha}{36\varpi^2}\cos(3\omega t). \tag{27}$$

According to Duffing's theory, it $x_1(t)$ and $x_2(t)$ are good approximations to the solution $x(t)$, then the coefficients of $\cos(\omega t)$ in the two equations (23) and (27) should not be very different. Thus by equating these coefficients, the following is obtained:

$$A = \frac{1}{\omega^2}\left(A\omega_n^2 \pm \frac{3}{4}A^3\alpha - F\right) \quad \text{or} \quad \omega^2 = \omega_n^2 \pm \frac{3}{4}A^3\alpha - \frac{F}{A} \tag{28}$$

Stopping the iterative procedure with the second approximation, it can be observed that when $\alpha = 0$, the case becomes linear, with

$$\mathrm{A} = \frac{\mathrm{F}}{\omega_\mathrm{n}^2 - \omega^2}, \tag{29}$$

where A denotes the amplitude of the harmonic response of the linear system. But, when $\alpha \neq 0$, the system is nonlinear with the frequency ω being function of α, A and F. Even though A is only the coefficient of the first term of the system's response, it can be and is commonly taken as the amplitude of the harmonic response of the system. Duffing's equation with dumping as expressed in equation (20) is rewritten in this form:

$$\ddot{x} + c\dot{x} + \omega_n^2 x \pm \alpha x^3 = A_1\cos(\omega t) - A_2\sin(\omega t), \tag{30}$$

where $F = \sqrt{A_1^2 + A_2^2}$ is the amplitude of the applied force, and $\varphi = \arctan\left(A_1/A_2\right)$ is the phase of the applied force. By assuming the first approximation to the solution to be:

$$x_1 = A\cos(\omega t), \tag{31}$$

and substituting equation (31) into equation (30), including the relation of equation (25), the following is obtained:

$$\left[(\omega_n^2 - \omega^2)A \pm \frac{3}{4}A^3\alpha\right]\cos(\omega t) - c\omega A\sin(\omega t) \pm \frac{A^3\alpha}{4}\cos(3\omega t) = A_1\cos(\omega t) - A_2\sin(\omega t) \tag{32}$$

Disregarding terms involving $\cos(3\omega t)$, and equating the coefficients of $\cos(\omega t)$ and $\sin(\omega t)$ on both sides of equation (32), yields the following relations:

$$(\omega_n^2 - \omega^2)A \pm \frac{3}{4}A^3\alpha = A_1 \tag{33}$$

and

$$c\omega A = A_2 \tag{34}$$

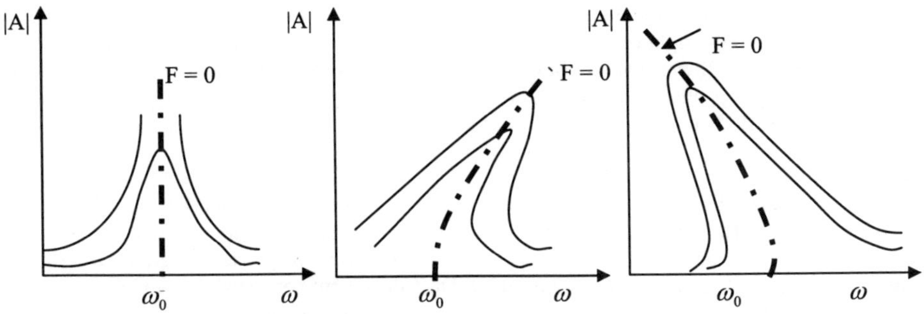

(a) $\alpha. = 0$ (b) $\alpha. > 0$ (c) $\alpha < 0$

Figure 8 Response curves of Duffing's equation

The relation between the amplitude of the applied force and the quantities A and ω is obtained by squaring and adding equations (33) and (34) to obtain:

$$\left[(\omega_n^2 - \omega^2)A \pm \frac{3}{4}A^3\alpha\right]^2 + (c\omega A)^2 = A_1^2 + A_2^2$$

or (35)

$$\left[(\omega_n^2 - \omega^2)A \pm \frac{3}{4}A^3\alpha\right]^2 + (c\omega A)^2 = F^2$$

S.S.Rao [5] presents an excellent illustration of the response curves given by equation (35) as reproduced in Fig.8.

4. Discussions

Looking at figures 5 and 8, it can be observed that while resonance occurs where the excitation frequency and the natural frequency match in the linear approach, resonance in nonlinear analysis occurs at excitation frequencies that are not equal to the linear system's natural frequency. It should also be noted that only the harmonic solution of the Duffing's equation, i.e. solutions for which the frequency is the same as that of the external force $F\cos\omega t$, have been considered in this analysis. When $\alpha \neq 0$, equation (34(b)) shows that the frequency ω is a function of α, A and F. For free vibration of the nonlinear system, $F = 0$, therefore the right side of equation (34b) is equal to zero. Then, it can be shown that the frequency of the response increases with the amplitude $|A|$ for the hardening spring and decreases for the softening spring. For both linear and nonlinear approaches, when $F \neq 0$ (forced vibration), there are two values of the frequency for any given amplitude $|A|$. One of these values of ω is smaller and the other is larger than the corresponding frequency of free vibration at that amplitude. For the smaller value of ω, $A > 0$ and the harmonic response of the system is in phase with the external force. For the larger value of ω, $A < 0$ and the response is 180^0 out of phase with the external force. The amplitude of the applied force is considered fixed $F = \sqrt{A_1^2 + A_2^2}$. It is assumed that c, and α are small. The angular frequency along with the solution varies as a function of the amplitude F. Figure 5 shows the response rising straight towards resonance. But figure 8(b) and (c) show a rise that is not straight, but rather bending to the right or to the left. It is evident that two amplitudes of vibration exist for a given forcing frequency, one lower than the other. This is a jump from one amplitude to the other. The understanding the of the jump phenomenon requires a deeper and more involved analysis of stability which is not part this short study of the nonlinear dynamic characteristic of the rear truck axle.

5. Conclusion

A linear and nonlinear analytical approach has been used to examine the dynamic characteristics of a truck rear full-floating axle. Having modeled the axle as a single-degree-of-freedom system excited by a sinusoidal base motion, it was observed that in linear analysis, the axle's response is characterized by its natural frequency and damping ratio. The solution of Duffing's equation used to model the axle's nonlinear behavior showed that in addition to the natural frequency and the damping ratio, the jump phenomenon occurs, and the period of oscillation depends on the nonlinear element, the amplitude, and the forcing function.

References

[1] Genta, G., *Motor Vehicle Dynamics: Modeling and Simulation*, New Jersey: World Scientific, 2003.

[2] Wong, J.Y., *Theory of Ground Vehicles*, 3rd edition, New York: John Wiley & Son, 2001.

[3] Segel, L. (ed.), *The Dynamics of Vehicles on Road and on Tracks: Proceedings of 14th IAVSD Symposium, Ann Arbor, Michigan, USA, August 21-25, 1995*, Lisse: Swets & Zeitlinger B.V., 1996.

[4] Meirovitch, L., *Fundamentals of Vibrations*, New York, McGraw-Hill, 2001.

[5] Rao, S.S., *Mechanical Vibrations*, 3rd edition, New York, Addison-Wesley, 1995.

[6] Hayashi, C., *Nonlinear Oscillations in Physical Systems*, New York, McGraw-Hill, 1964.

[7] Nayfeh, A.H., and Pai, P.F., *Linear and Nonlinear Structural Mechanics*, Hoboken, New Jersey: Jon Wiley & Sons, 2004.

[8] Inman, D. J., *Engineering Vibration*, 2nd edition, Upper Saddle River, New Jersey: Prentice Hall, 2001.

[9] Stockel, M.W., and al., *Auto Fundamentals*, Tinley Park, Illinois, The Goodheart-Willcox, 2000.

[10] Halderman, J.D., and al, *Automotive Technology*, Upper Saddle River, New Jersey, Prentice Hall, 2003.

[11] Luo, A.C.J., and Han, R.P.S., "A Quantitative Stability and Bifurcation Analyses of the Generalized Duffing Oscillator with Strong Nonlinearity," *J. Franklin Inst.*, (1997), Vol.334B, No.3, pp447-459.

[12] Virgin, L. N., *Introduction to Experimental Nonlinear Dynamics: A Case Study in Mechanical Vibrations*, Cambridge, Cambridge University Press, 2000.

Nonlinear Vibration Analysis of an Unbalanced Rotor on Rolling Element Bearings Due to Cage Run-out

C. Nataraj[a*], S. P. Harsha

Center for Nonlinear Dynamics & Control, Villanova University, Villanova, PA 19085, USA,

Abstract

This paper presents an analytical model to investigate the non-linear dynamic behavior of an unbalanced rotor bearing system due to cage run-out. Due to run-out of the cage, the rolling elements no longer stay equally spaced. The mathematical model takes into account the sources of non-linearity such as Hertzian contact forces and cage run-out, and the resulting transition from a state of no contact to contact between the rolling elements and the races. The contact between the rolling elements and races is treated as nonlinear springs and the system is analyzed for varying number of balls. The results are presented in the form of Fast Fourier Transformations and Poincaré maps. The results show that the ball passage frequency is modulated with the cage frequency. The response falls into three regimes: periodic motion, quasi-periodic oscillations, and chaotic response.

Keywords: Nonlinear dynamics, chaotic vibration, rotor-bearing system, rolling element bearings, ball passage frequency, Poincaré maps.

1. Introduction

Ball bearings are used today in the design of increasingly sophisticated arrangements, involving high speed, high temperature, heavy or unusual loadings and requiring continuous operations; a clear understanding of vibrations associated with them is needed. Rolling element bearings are indeed a non-negligible source of vibration in many types of rotating machinery. With current needs for high precision, many advanced bearing applications now require sound understanding of the dynamic effects and classical quasi-static analysis techniques such as in Jones [11] and Palmgren [18] are inadequate. In addition, since the mathematical underpinnings of linear and nonlinear systems are very different, linear methods cannot even predict *qualitatively* some of the phenomena that are observed in practical industrial systems; multiple frequency peaks is one of innumerable such examples. Hence, nonlinear analysis is essential for accurate analysis and optimal design and to help ensure trouble-free operation of real machinery [17].

Cage failures due to high pocket wear or destructive collision forces between the cage and rolling element or race lands, cage induced audio noise, structural vibration and excessive torque or torque noise are examples of bearing performance characteristics that are significantly affected by cage dynamics. Interactions between the rolling elements and the cage can induce excessive vibrations leading to ball to race skidding and degraded performance. This can cause premature failure in the system.

**The corresponding author, Tel: 1- 610-519-4994*
E-mail: ***nataraj@villanova.edu***

Walters [21] developed an analytical model for ball bearing and cage dynamics with ball raceway slip that was later modified by Gupta [5, 6]. However, the solution of time varying Hertzian contact stress for each ball, along with the integration of each cage impact with the balls or raceways and integration of the ball traction / slip forces at each contact point on the inside and outside raceways, results in long computer run times and can be so costly as to make parametric design studies impractical. In addition, the Walters / Gupta model equations are written in fixed inertial coordinate system which leads to complex equations of motion, excessively long computer times and computational errors due to computer numerical truncation. Kennel and Bupara [12] developed a simplified method for analyzing ball and cage dynamics and assumed that the ball cage only moves in the plane of its major diameter. Meeks et al. [14, 15] have shown that the ball cage motions are far too complex to be modeled with this extreme simplification of cage motion. Meeks [16] developed an analytical model to study and optimize the bearing and cage design parameters. Gad et al. [4] showed that resonance occurs when the ball passage frequency (BPF) coincides with a natural frequency of the system and they also pointed out that for certain speeds, BPF can exhibit its sub and super harmonic vibrations for shaft ball bearing system. Rahnejat and Gohar [19] showed that even in the presence of an elasto-hydrodynamic lubricating film between balls and the races, a peak at the BPF appears in the spectrum. Aktürk et al. [1] performed a theoretical investigation of effect of varying the preload on the vibration characteristics of a shaft bearing system and also suggested that untoward effect of the BPF can be reduced by taking the correct number of balls and the amount of preload in a bearing.

In related work on studies of bearing defects, Harsha et al. [10] developed an analytical model to predict nonlinear dynamic response in a rotor bearing system due to surface waviness. The conclusion of this work shows that for *outer race waviness*, severe vibrations occur when the number of balls and waves are equal. In the case of the *inner race waviness*, the peak amplitude of vibration can be at $q\omega_{wp} \pm p\omega_{cage}$. For the waviness order iN_b, peak amplitude of vibration and super-harmonics appear at the wave passage speed (ω_{wp}). Harsha et al. [8] analyzed the nonlinear behavior of ball bearings due to number of balls and preload effect. Nonlinear dynamic response is found to be associated with the ball passage frequency. The amplitude of the vibration is considerably reduced if the number of balls and preload are correctly selected. Harsha et al. [9] analyzed the nonlinear behavior of a high-speed horizontal balanced rotor supported by ball bearings. The conclusion of this work shows that most severe vibrations occur when the ball passage frequency (BPF) and its harmonics coincide with the natural frequency.

In this paper, a theoretical investigation is conducted to observe the effect of cage run-out on the vibration characteristics of ball bearing system. The results obtained from a large number of numerical integrations are mainly presented in form of Fast Fourier Transformation (FFT) and Poincaré maps.

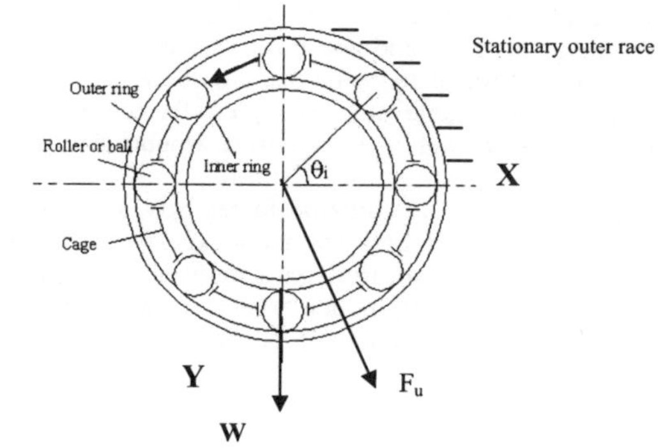

Fig. 1 A Schematic Diagram of a Rolling Element Bearing

2. Modeling of the System

A schematic diagram of a rolling element bearing is shown in Fig. 1. In the mathematical model, the rolling element bearings are considered as a mass-spring system and the contact acts as a nonlinear contact spring. The corresponding spring force appears when the instantaneous spring length is shorter than its unstressed length; otherwise, the separation between ball and race takes place and the resulting force is set to zero. The assumptions made in development of the mathematical model are as follows.

1. The rolling element, the inner and outer races and the cage have motions in the plane of the bearing only. This eliminates any motion in the axial direction.
2. Deformations occur according to the Hertzian theory of elasticity.
3. The cage ensures the constant angular separation (β) between rolling elements; hence there is no interaction between rolling elements. Therefore,

$$\beta = \frac{2\pi}{N_b} \tag{1}$$

2.1 Cage run-out

Due to the run-out of the cage, the rolling elements no longer stay equally spaced, as shown in Fig. 2. The resulting variations of the circumferential angle for a small run-out Γ is:

$$\delta\theta_i = \frac{\Gamma}{R_p}\cos(\theta_i) \tag{2}$$

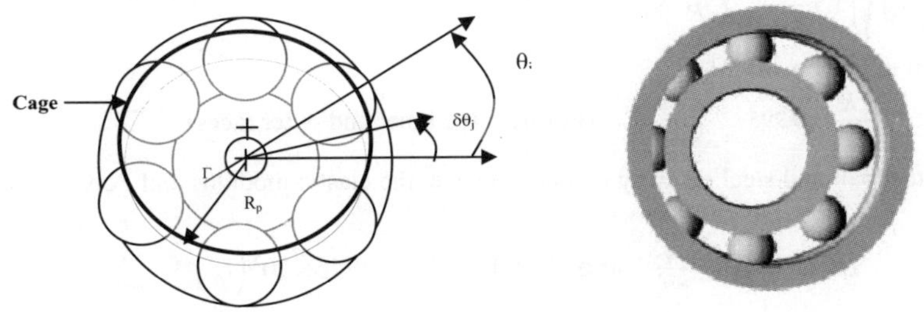

Fig. 2 Non-uniform Ball Spacing due to Cage Run-out

2.2 Calculation of the Restoring Force

The local Hertzian contact force and deflection relationship for the bearing may be written as:

$$F_{\theta_i} = k(r_{\theta_i})^{3/2} \tag{3}$$

The radial displacement (r_{θ_i}) considering the cage run-out, the contact deformation becomes,

$$r_{\theta_i} = x\cos\theta_i + y\sin\theta_i - (\Gamma) \tag{4}$$

Substituting r_{θ_i} in Equation (13), we get

$$F_{\theta_i} = k[(x\cos\theta_i + y\sin\theta_i) - \Gamma]_+^{3/2} \tag{5}$$

2.3 Equations of Motion

The system governing equations accounting for inertia, restoring and damping force and constant vertical force acting on the inner race are,

$$m\ddot{x} + c\dot{x} + \sum_{i=1}^{N_b} k\left[\left(x\cos\theta_i + y\sin\theta_i\right) - \Gamma\right]_+^{3/2} \cos\theta_i = F_u \sin\left(\omega t\right)$$
$$m\ddot{y} + c\dot{y} + \sum_{i=1}^{N_b} k\left[\left(x\cos\theta_i + y\sin\theta_i\right) - \Gamma\right]_+^{3/2} \sin\theta_i = W + F_u \cos\left(\omega t\right) \quad (6)$$

Here, m is the mass of the rotor supported by bearings.

2.4 Contact Stiffness

The Hertz equations for elastic deformation involving ***point*** contact between solid bodies are given by Eschmann [3] as:

$$r_{\theta_i} = 1.5\frac{2K}{\Lambda\mu}\sqrt[3]{\frac{\left(1-\frac{1}{M}\right)^2 \sum\rho}{3E^3}Q^2} \quad (7)$$

Here, Λ, K and μ are Hertz coefficients which depend on the surface properties. E, 1/ M and $\sum\rho$ are the elastic modulus in N/mm^2, Poisson ratio and the sum of curvature of the contacting bodies, respectively. Hence the nonlinear stiffness associated with point contact is:

$$k = \frac{E}{\left(1-\frac{1}{M^2}\right)}\sqrt{\frac{3r_{\theta_i}}{\left(1.5\frac{2K}{\Lambda\mu}\right)^3 \sum\rho}} \quad (8)$$

From the table given in Eschmann [3],

$$\frac{2K}{\Lambda\mu} = 0.995 \qquad \text{(For both the inner and outer races)}$$

Here we consider steel ball and steel raceway contacts; hence, the elastic modulus and Poisson's ratio are as follows:

$$E = 2.0\times10^5 \ \tfrac{N}{mm^2} \ \text{and} \ \tfrac{1}{M} = 0.3,\ k = 1.9845\times10^5\left(r_{\theta_i}\right)^{3/2} \ \tfrac{N}{mm^{1.5}} \quad (9,10)$$

3. Results & Discussion

The equations of motion, Eqs.(6) are solved using the modified Newmark-β method to obtain the radial displacement and velocity of the rolling elements. The longer the time it takes to reach steady state vibrations, the longer the CPU time that is needed and hence the more expensive the computation; we chose a value of c = 200 Ns / m. The initial conditions and step size are very important for successful and economic computational solution. Therefore an optimal choice of the time step needs to be made; for this investigation it was determined to be $\Delta t = 10^{-5}$ sec.

3.1 Cage Run-out

Due to the cage run-out, the rolling elements no longer stay equally spaced which leads to the modulation of the ball passage frequency with the cage frequency. The variation of the circumferential angle for a small run-out Γ is assumed to be 0.001µm and the unbalance force magnitude (Fu) is taken to be 15% of the radial load. The figures show the vibration response for the rolling bearing due to cage run-out and unbalance at several values for the number of balls. For 6 balls, the peak amplitude of vibration appears in the spectrum at the $\omega_{bp} = 200\,Hz$ and $\Omega = 83.33\,Hz$, as shown in Fig. 3. Other major peaks of vibrations appear at $\omega_{bp} + 2\Omega = 366.67\,Hz$, 2 $\omega_{cage} = 100\,Hz, \omega_{cage} = 33.33\,Hz$. The band structure of the frequency spectra and the fractal structure of Poincaré maps show the chaotic nature of system. The chaotic attractor is spread out and the band of frequency in the spectrum formed is also quite prominent.

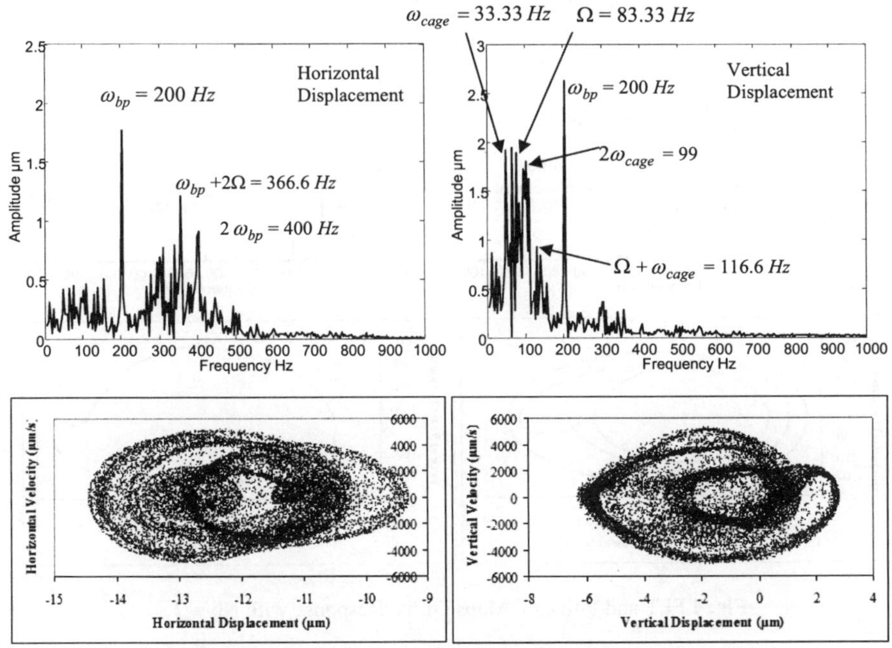

Fig. 3 FFT and Poincaré Maps Of the Response with Nb = 6

When the number of balls is 7, the peak amplitude of vibration appears in the spectrum at the ball passage frequency $\omega_{bp} = 233.33\ Hz$ and at $\Omega = 83.33\ Hz$, as shown in Fig. 4. Other major peaks of vibration appear at $2\Omega = 166.67\ Hz, \omega_{cage} + \Omega = 316.67\ Hz$. The response at ω_{bp} and the super-harmonic character of the frequency spectra is also brought-out by the Poincaré map. As the number of balls increase, the system shows periodic or quasi periodic behavior and the chaotic nature reduces. For 8 balls, the peak amplitude appears at the ball passage frequency $\omega_{bp} = 266.67\ Hz$ as shown in Fig. 5. Other major peaks at super harmonics of vibration appear at rational multiples of the ball passage frequency $2\omega_{bp} = 533.33\ Hz, \Omega = 83.33\ Hz$. As number of balls increase further, the system shows a stable nature. For 13 balls, the peak amplitude of vibration appears at $\omega_{bp} = 433.33\ Hz$ as shown in Fig. 6. Other major peaks at super harmonics appear at $2\omega_{bp} = 866.67\ Hz, \Omega = 83.33\ Hz, 2\Omega = 166.67\ Hz$. The system shows quasi-periodic nature with a closed loop in the Poincaré map. All the major peaks in the vibration spectra are summarized in Table I.

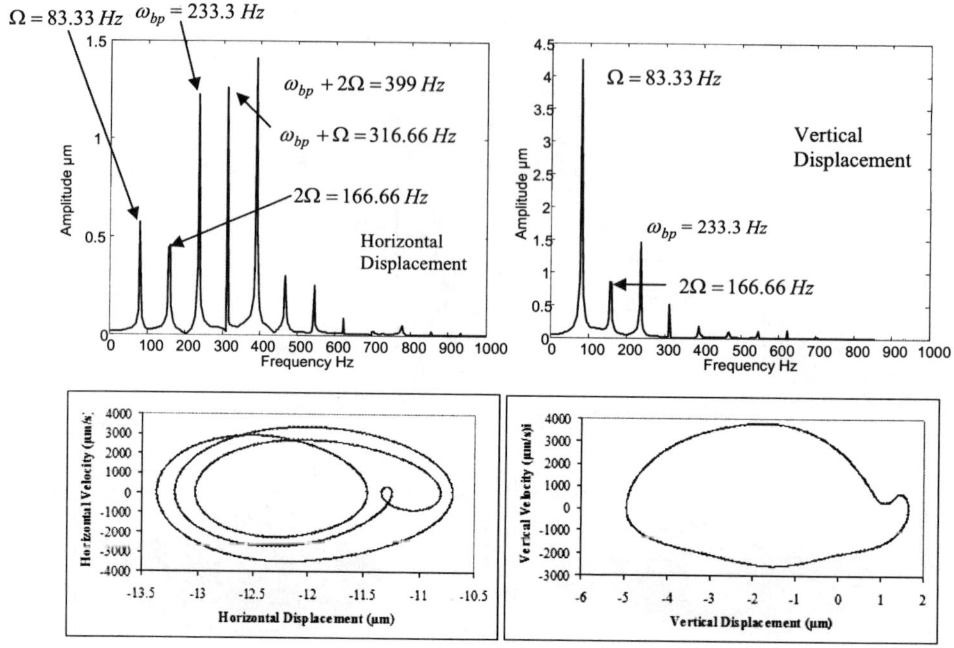

Fig. 4 FFT and Poincaré Maps Of the Response with Nb = 7

4. Conclusion

In the present investigation, an analytical model of a rotor bearing system has been developed to obtain the nonlinear vibration response due to cage run-out in an unbalanced rotor. The system is bi-periodically excited, one due to cage run out, which is at ball passage frequency and the other due to unbalance in the rotor, which is at the rotational frequency. Parametrically excited vibrations occur irrespective of the quality and accuracy of the bearing and are called ball passage vibrations [20] which lead to peak amplitudes of vibrations appearing in the spectrum at the Ball Passage Frequency. These peaks are always present in the results.

The following additional conclusions can be drawn from the obtained results.

1. The highest peaks in the vibrations due to cage run-out are at a frequency of the number of balls times the cage speed; i.e., at the ball passage frequency, as also reported by Gad et al. [4] and Rahnejat and Gohar [19].
2. Increasing the number of balls means the number of balls supporting the rotor, thereby increasing the system stiffness and reducing the vibration amplitude. This effect is exhibited in the results, which shows the Poincaré maps of the displacements for different number of balls. When the number of balls is increased, the vibration reduces drastically implying a stiffer system; this was also reported by Aktürk et al. [1]. From this it can be predicted that increasing the number of balls will reduce the effect of the modulating frequency and because of the cage run-out, ball passage frequency becomes dominant in the vibration spectrum.

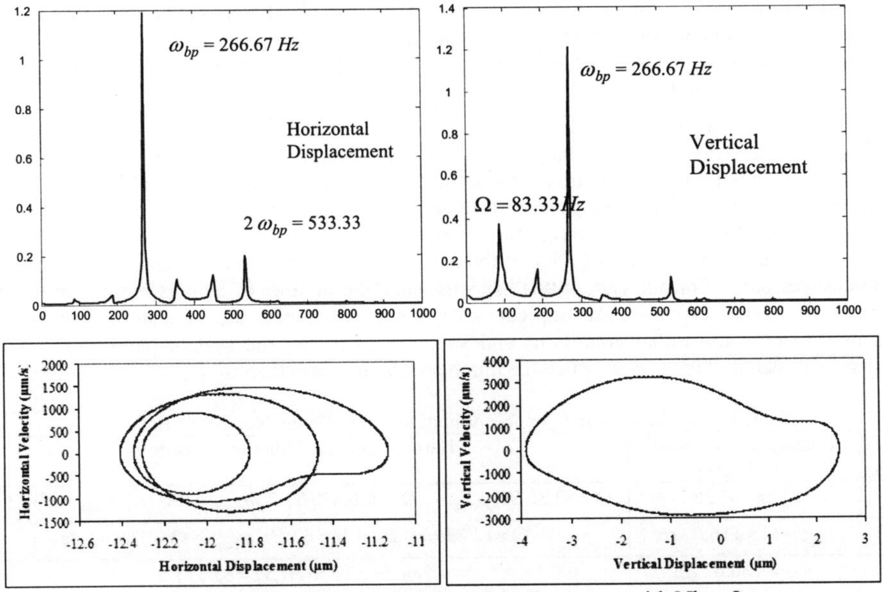

Fig. 5 FFT and Poincaré Maps Of the Response with Nb = 8

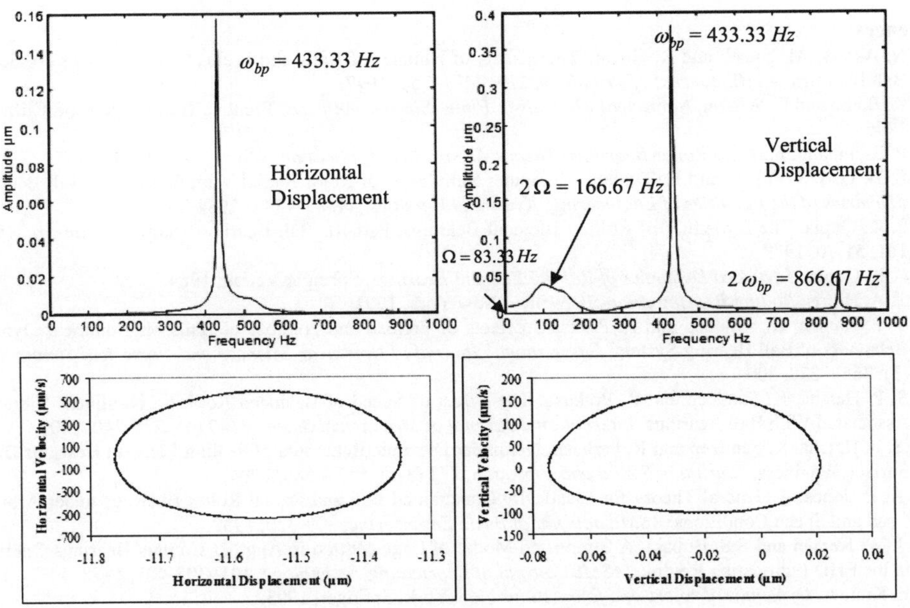

Fig. 6 FFT and Poincaré Maps Of the Response with Nb = 13

3. The number of balls with cage run-out in the bearings can be of importance in the rotor bearing dynamics and should be considered at the design stage.
4. Our analysis predicts that the highest vibrations due to cage run-out for a generic number of balls are at $\left(\omega = q\omega_{bp} \pm k\omega_c Hz\right)$.
5. Based on the characteristics of the dynamic behavior of the system due to cage run-out and unbalanced rotor, the responses may be put in three categories. (i) The system responses are periodic and are not sensitive to initial conditions or small variations of system parameters. This is a well-behaved region, which helps the designer to predict the trends accurately and without ambiguity. For the system considered, this happens when the number of balls is more than 11 and with an unbalanced rotor force of 15% of the radial load. (ii) The system responses are not chaotic but quasi-periodic. For our system this happens when the number of balls is more than 6 but less than 11. (iii) The responses are unpredictable, being either periodic or chaotic and extremely sensitive to both the initial conditions and small variations in the system parameters. For the system considered, this happens when the number of balls is less than 6.

Table I Frequencies of the Possible Vibrations for Different Number of Balls

Number of Balls	Frequencies for Peak Amplitudes	Harmonics in the Vibration Spectrum
6	$\omega_{bp} = 200\,Hz$,	$\Omega = 83.33\,Hz$, $\omega_{bp} + 2\Omega = 366.67\,Hz$, 2 $\omega_{cage} = 100\,Hz, \omega_{cage} = 33.33\,Hz$
7	$\omega_{cage} + \Omega = 116.67\,Hz$,	$\Omega = 83.33\,Hz$, $2\Omega = 166.67\,Hz, \omega_{cage} + \Omega = 316.67\,Hz$
8	$\omega_{bp} = 266.67\,Hz$	$2\omega_{bp} = 533.33\,Hz, \Omega = 83.33\,Hz$
13	$\omega_{bp} = 433.33\,Hz$	$2\omega_{bp} = 866.67\,Hz, \Omega = 83.33\,Hz, 2\Omega = 166.67\,Hz$

References

[1] N. Aktürk, M. Uneeb and R. Gohar. The Effects of Number of Balls and Preload on Vibrations Associated Ball Bearings. *ASME Journal of Tribology*, **119**: 747 – 753, 1997.

[2] K. Bathe and E. Wilson, *Numerical Methods in Finite Element Analysis,* Prentice-Hall Englewood Cliffs, NJ, 1976.

[3] P. Eschmann, *Ball and Roller Bearings- Theory, Design and Application.* Willey, New York, 1985.

[4] E. H. Gad, S. Fukata and H. Tumara. Computer Simulation of Rotor Radial Vibration due to Ball Bearings. *Memories of the Faculties of Engineering, Kyushu universities*, **44**: 83 – 111, 1984 (a).

[5] P. K. Gupta. The Dynamics of Rolling Element Bearings, Part-III: Ball Bearing Analysis. *Trans. of ASME*, **101**: 51-70, 1979.

[6] P. K. Gupta. *Advanced Dynamics of Rolling Element Bearings,* Springer Verlag, 1984.

[7] T. A. Harris. *Rolling Bearing Analysis.* Willey, New York, 1991.

[8] S. P. Harsha, K. Sandeep and R. Prakash. Effects of Preload and Number of Balls on Nonlinear Dynamic Behaviors of Ball Bearing System. *International Journal of Nonlinear Sciences and Numerical Simulation*, **4** (3): 265 – 278, 2003.

[9] S. P. Harsha, K. Sandeep and R. Prakash. The Effect of Speed of Balanced Rotor on Nonlinear Vibrations Associated With Ball Bearings. *International Journal of Mechanical Sciences*, **47** (4): 225-240, 2003.

[10] S. P. Harsha, K. Sandeep and R. Prakash. Nonlinear Dynamic Behaviors of Rolling Element Bearings Due to Surface Waviness. *Journal of Sound and Vibration*, **272** (3-5): 557 – 580, 2004.

[11] A. B. Jones. A General Theory for Elastically Constrained Ball and Radial Roller Bearings under Arbitrary Load and Speed Conditions. *ASME Journal of Basic Engineering*, 309-320, 1959.

[12] J. W. Kennel and S.S. Bupara. A Simplified Model of Cage Motion in Angular Contact Bearings Operating in the EHD Lubricating Regime. *ASME Journal of Lubricating Technology*, **101**: 395-401, 1978.

[13] E. Kramer. *Dynamics of rotors and foundations* New York: Springer, 1993.

[14] C.R. Meeks and L. Tran. Ball Bearing Dynamic Analysis Using Computer Methods, Part-I: Analysis. *ASME Journal of Tribology*, **118**: 52-58 1996.

[15] C.R. Meeks and N. Foster. Computer Simulation of Ball Bearing Dynamics Analytical Predictions and Test Results. *Ball Bearing Symposium and Seminar, Orlando, F*, March 9-12, 1987.

[16] C.R. Meeks. Ball Bearing Dynamic Analysis Using Computer Methods and Correlation with Empirical Data. *International Tribology Conference, Melbourne, Australia*, Dec. 2-4, 1987.

[17] Nataraj, C. and Nelson, H. D. Periodic Oscillations in Rotor Dynamic Systems with Nonlinear Supports, ASME, Journal of Vibration, Acoustics, Stress and Reliability in Design, **111,** 187-193, 1989.

[18] A. Palmgren. *Ball and Roller Bearing Engineering.* SKF Industries Inc., 1959.

[19] H. Rahnejat and R. Gohar. The Vibrations of Radial Ball Bearings. *Proceedings of the Institution of Mechanical Engineers*, **199** (C3): 181-193, 1985.

[20] C. S. Sunnersjo. Varying Compliance Vibrations of Rolling Bearings. Journal of Sound and Vibration, 58 (3), pp. 363 – 373, 1978.

[21] C. T. Walters. The Dynamics of Ball Bearings. *Trans. of ASME, Journal of Lubricating Engineering*, **93**: 1-10, 1971.

Perturbation Analysis of Nip Contact Delay System

L. Yuan *, V.-M. Järvenpää
Department of Mechanical Engineering, Tampere University of Technology
Korkeakoulunkatu 6, P.O.Box 589, 33101, Tampere, Finland

Abstract
This paper presents a perturbation analysis of a nip contact of two paper machine rolls. The two degrees of freedom model of the system is considered. The model includes nonlinear terms of the rolling contact and a time delay term. The multi-scale method has been applied to obtain an approximate solution and limit cycle amplitudes. Some numerical results are given as well.

Keywords: perturbation analysis; nip contact; time delay; nonlinear dynamics

1. Introduction

A roll system with a nip contact is studied in this work. The background of this system is related to the paper manufacturing process in paper mills. Paper machines include multiple roll nips used for finishing the surface of the paper. The vibration behaviours of the nip contact have not been understood completely and the complex solutions are still under investigations [1]. A test nip installation in laboratory environment at Tampere University of Technology has been built for research purposes and it is half scale of a real sized industrial nip unit. The main elements of the system are one polymer covered roll and one pure metal roll in the nip contact. The soft polymer cover enables smooth processing of the paper. But the consequence is that the cover may cause the time delay effect due to the non-recovered polymer material deformations in the rolling contact. The basic data of the nip installation is listed as following. The roll diameters are 550 mm, the polymer cover thickness is 11 mm and the mass of one roll is 3100 kg. The elastic modulus values for the cast iron rolls is $E = 100$GPa and for the polymer cover $E_0 = \sim 50$ MPa. The running speed at the contact is 300-600 m/min, which corresponds to roll rotation frequencies of 3-5 Hz. The lowest natural frequency of the rolls is about 35 Hz, which is the lowest bending mode. There are problems caused by jumps from stable and unstable rolling due to the regenerative chatter effects. The stability of time delay system is more complex and interesting especially when it involves nonlinear elements [2].On the other hand, the dynamic contact force between two rolls has been analyzed and might be nonlinear from the stiffness. So the analysis of this multi degrees of freedom model can be quite complex. From the perturbation theory [3], the multi scale method will be executed for the mathematical model and therefore the amplitude of the limit cycle might be obtained [4].

2. Mathematic Modelling of Nip Contact

The contact pressure distribution in Fig. 1 is unsymmetrical due to viscoelastic material behavior [5]. The static normal force can be described as

* Corresponding author.

$$P = \int_{-a}^{b} p(z)dz \tag{1}$$

$$P = \frac{Ea^2}{R \cdot s}\left[\frac{1}{2}(b+a) - \frac{1}{6a^2}(b^3+a^3) + \beta\xi((b+a) + \frac{1}{2a}(b^2-a^2)) + \beta\xi(1+\xi)(b+a) + a\beta(1+\xi)(e^{\frac{b}{a}\xi^2}-1)\right] \tag{2}$$

Where $R = R_1R_2/(R_1+R_2)$ and $\xi = v_{web}\tau_{relaxation}/a$. R_1 and R_2 are radiuses of the contact surfaces, s is the thickness of the polymer cover layer, v_{web} is the running speed, $\tau_{relaxation}$ is the viscoelastic relaxation time of the polymer material and β is viscoelastic material parameter.

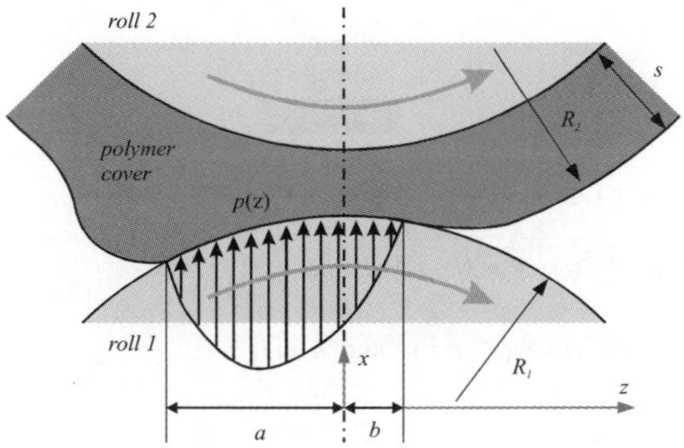

Fig.1. The contact pressure distribution between two rolls.

The compressive strain in an element of the contact at z (horizontal direction) is given by

$$e = -(\delta - z^2/2R)/s\ ,\ -a \le z \le b \tag{3}$$

By assuming $b \approx a$, which is the typical case in practical applications, the a can be solved as the function of x_1 and x_2 as

$$a^2(t) = 2R(\delta_0 + x_1 - x_2) \tag{4}$$

or

$$a(t) = \sqrt{2R\delta_0} \cdot \sqrt{1 + \frac{x_1}{\delta_0} - \frac{x_2}{\delta_0}} \tag{5}$$

where δ_0 is the nominal penetration of the rolls.

Fig. 2 illustrates the equivalent spring-mass system for the nip contact rolls. Let $x_{1\tau} = x_1(t-\tau)$. Symbol τ is the delay time related to the rotational speed of the roll.
The governing equations of motion of the rolling system are expressed as

$$\ddot{x}_1 + 2\zeta_1\dot{x}_1 + p_1{}^2x_1 = -N(x_1, x_{1\tau})/m_1 \tag{6}$$

$$\ddot{x}_2 + 2\zeta_2\dot{x}_2 + p_2{}^2x_2 = N(x_1, x_{1\tau})/m_2 \tag{7}$$

where $2\zeta_i = c_i/m_i$, $p_i^2 = k_i/m_i$, i=1,2.

Considering the time delay effect and Eq. (2) and (5), the dynamic contact force with damping in the contact is

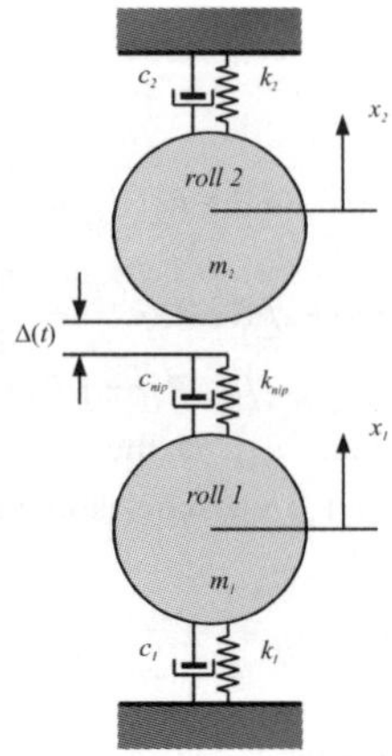

Fig. 2. The equivalent spring-mass system.

$$N = h \cdot F_\Phi \cdot \lfloor a^3(t) - \gamma_e a^3(t-\tau) \rfloor + h \cdot c_{nip} \cdot \dot{\varepsilon}(t) \tag{8}$$

where $F_\Phi = \frac{E}{R}\left[\frac{2}{3} + 2\beta\xi^2 + \beta\xi^2(1+\xi)(1-e^{-\frac{2}{\xi}})\right]$ and $h = 1/s$.

The dynamic penetration function is

$$\varepsilon(t) = x_1(t) - x_2(t) + \Delta(t) \tag{9}$$

where the external excitation $\Delta(t) = \Delta\sin(\Omega t)$, which is assumed as zero here.

The relaxation parameter related to the time delay and polymer material characteristics is

$$\gamma_e = e^{-\tau/\tau_{relaxation}} = e^{-\tau\, k_{nip}/c_{nip}} \tag{10}$$

where the nominal contact stiffness k_{nip} and the damping c_{nip} are given as initial constants.

3. Perturbation Analysis

Before doing perturbation analysis, the nonlinear cubic term $a^3(t)$ can be expanded as a Taylor series

$$\begin{aligned} a^3(t) &= 2R(\delta_0 + x_1 + x_2) \cdot \sqrt{2R\delta_0} \cdot \sqrt{1 + \frac{x_1}{\delta_0} - \frac{x_2}{\delta_0}} \\ &\cong 2R(\delta_0 + x_1 + x_2) \cdot \sqrt{2R\delta_0} \cdot \left[1 + \frac{1}{2}\left(\frac{x_1}{\delta_0} - \frac{x_2}{\delta_0}\right) - \frac{1}{8}\left(\frac{x_1}{\delta_0} - \frac{x_2}{\delta_0}\right)^2 + \frac{1}{16}\left(\frac{x_1}{\delta_0} - \frac{x_2}{\delta_0}\right)^3\right] \end{aligned} \tag{11}$$

where $\left|\frac{x_1}{\delta_0} - \frac{x_2}{\delta_0}\right| \leq 1$ is satisfied.

The perturbation analysis can be performed using the method of multiple scales [6]. In this method, a fast time-scale $T_0 = t$ and a slow time-scale $T_1 = \varepsilon^2 t$.

We seek a second-order expansion for the equations (1 and 2) in the form

$$x_i = \varepsilon x_{i0}(T_0,T_1) + \varepsilon^2 x_{i1}(T_0,T_1) + \varepsilon^3 x_{i2}(T_0,T_1) + O(\varepsilon^4), \qquad i=1,2. \tag{12}$$

The time derivatives as

$$\dot{x}_i = \varepsilon \frac{\partial x_{i0}}{\partial T_0} + \varepsilon^2 \frac{\partial x_{i1}}{\partial T_0} + \varepsilon^3 \frac{\partial x_{i2}}{\partial T_0} + 2\varepsilon^3 \frac{\partial x_{i0}}{\partial T_1} + ... \tag{13}$$

$$\ddot{x}_i = \varepsilon \frac{\partial^2 x_{i0}}{\partial T_0^2} + \varepsilon^2 \frac{\partial^2 x_{i1}}{\partial T_0^2} + \varepsilon^3 \frac{\partial^2 x_{i2}}{\partial T_0^2} + 2\varepsilon^3 \frac{\partial^2 x_{i0}}{\partial T_0 \partial T_1} + ... \tag{14}$$

Then we perturb the contact parameter (effective thickness of the polymer layer), let

$$h = h_c + \varepsilon^2 h_1 \tag{15}$$

where h_c is the inverse of thickness value at the stability boundary.

It is assumed that the term F_Φ is constant. Substituting equations (12-15) into governing equations (6-7), and equating coefficients of like power of ε, we obtain

For the first equation of motion (6),

Order ε^1

$$\frac{\partial^2 x_{10}}{\partial T_0^2} + 2\zeta_1 \frac{\partial x_{10}}{\partial T_0} + p_1^2 x_{10} = F_\Phi\{-h_c \frac{1}{2}(x_{10} - x_{20}) - h_c(x_{10} - x_{20}) + h_c\gamma_e \frac{1}{2}(x_{10} - x_{20})$$
$$- h_c\gamma_e(x_{10} - x_{20})\} - h_c \frac{c_n}{m_1}(\frac{\partial x_{10}}{\partial T_0} - \frac{\partial x_{20}}{\partial T_0}) \tag{16}$$

Order ε^2

$$\frac{\partial^2 x_{11}}{\partial T_0^2} + 2\frac{\partial^2 x_{10}}{\partial T_0 \partial T_1} + 2\zeta_1 \frac{\partial x_{11}}{\partial T_0} + 4\zeta_1 \frac{\partial x_{10}}{\partial T_1} + p_1^2 x_{11}$$
$$= F_\Phi\{-h_c(\frac{1}{2}(x_{11} - x_{21}) - \frac{1}{8\delta_0}(x_{10} - x_{20})^2) - h_c(x_{10} - x_{20} + \frac{1}{2\delta_0}(x_{10} - x_{20})^2)$$
$$+ h_c\gamma_e(\frac{1}{2}(x_{11\tau} - x_{21\tau}) - \frac{1}{8\delta_0}(x_{10\tau} - x_{20\tau})^2) + h_c\gamma_e(x_{11\tau} - x_{21\tau} + \frac{1}{2\delta_0}(x_{10\tau} - x_{20\tau})^2) - h_1\delta_0\}$$
$$- h_c \frac{c_n}{m_1}(\frac{\partial x_{11}}{\partial T_0} + 2\frac{\partial x_{10}}{\partial T_1} - \frac{\partial x_{21}}{\partial T_0} - 2\frac{\partial x_{20}}{\partial T_1}) \tag{17}$$

Order ε^3

$$\frac{\partial^2 x_{12}}{\partial T_0^2} + 2\frac{\partial^2 x_{10}}{\partial T_0 \partial T_1} + 2\zeta_1(\frac{\partial x_{12}}{\partial T_0} + 2\frac{\partial x_{10}}{\partial T_1}) + p_1^2 x_{12}$$
$$= F_\Phi\{-h_c(\frac{1}{2}(x_{12} - x_{22}) - \frac{1}{4\delta_0}(x_{10}x_{11} + x_{20}x_{21} - x_{10}x_{21} - x_{11}x_{20}) + \frac{1}{16\delta_0{}^2}(x_{10}{}^3 - 3x_{10}{}^2 x_{20}$$
$$+ 3x_{20}{}^2 x_{10} - x_{20}{}^3)) - h_c(x_{12} - x_{22} - \frac{1}{\delta_0}(x_{10}x_{11} + x_{20}x_{21} - x_{10}x_{21} - x_{11}x_{20})$$
$$- \frac{1}{8\delta_0{}^2}(x_{10}{}^3 - 3x_{10}{}^2 x_{20} + 3x_{20}{}^2 x_{10} - x_{20}{}^3)) + h_c\gamma_e(\frac{1}{2}(x_{12\tau} - x_{22\tau}) - \frac{1}{4\delta_0}(x_{10\tau}x_{11\tau} + x_{20\tau}x_{21\tau}$$
$$- x_{10\tau}x_{21\tau} - x_{11\tau}x_{20\tau}) + \frac{1}{16\delta_0{}^2}(x_{10\tau}{}^3 - 3x_{10\tau}{}^2 x_{20\tau} + 3x_{20\tau}{}^2 x_{10\tau} - x_{20\tau}{}^3)) + h_c\gamma_e(x_{12\tau} - x_{22\tau}$$
$$- \frac{1}{\delta_0}(x_{10\tau}x_{11\tau} + x_{20\tau}x_{21\tau} - x_{10\tau}x_{21\tau} - x_{11\tau}x_{20\tau})$$
$$- \frac{1}{8\delta_0{}^2}(x_{10\tau}{}^3 - 3x_{10\tau}{}^2 x_{20\tau} + 3x_{20\tau}{}^2 x_{10\tau} - x_{20\tau}{}^3))\} - h_1 \frac{c_n}{m_1}(\frac{\partial x_{10}}{\partial T_0} - \frac{\partial x_{20}}{\partial T_0}) \tag{18}$$

Analogously, one can get similar equations as (16-18) for the second equation of motion, which can be named (16b-18b) but not repeated here.

The solution of equation (16 and 16b) can be expressed as

$$x_{10} = A_1(T_1)e^{i\omega_c T_0} + \overline{A}_1(T_1)e^{-i\omega_c T_0}, x_{20} = A_2(T_1)e^{i\omega_c T_0} + \overline{A}_2(T_1)e^{-i\omega_c T_0}$$

$$x_{10\tau} = A_1(T_1)e^{i\omega_c(T_0-\tau)} + \overline{A}_1(T_1)e^{-i\omega_c(T_0-\tau)}, x_{20\tau} = A_2(T_1)e^{i\omega_c(T_0-\tau)} + \overline{A}_2(T_1)e^{-i\omega_c(T_0-\tau)} \quad (19)$$

where ω_c is the chatter frequency at the Hopf bifurcation

Also let

$$x_{11} = P_1 e^{2i\omega_c T_0} + P_2 e^{-2i\omega_c T_0} + P_3$$
$$x_{21} = R_1 e^{2i\omega_c T_0} + R_2 e^{-2i\omega_c T_0} + R_3 \quad (20)$$

By substituting equations (19-20) into equation 17, we obtain

$$\begin{aligned}
&-4\omega_c^2(P_1e^{2i\omega_c T_0} + P_2e^{-2i\omega_c T_0}) + 2i\omega(A_1'e^{i\omega_c T_0} - \overline{A}_1'e^{-i\omega_c T_0}) + 4\zeta_1 i\omega_c(P_1e^{2i\omega_c T_0} - P_2e^{-2i\omega_c T_0})\\
&+4\zeta_1(A_1'e^{i\omega_c T_0} + \overline{A}_1'e^{-i\omega_c T_0}) + p_1^2(P_1e^{2i\omega_c T_0} + P_2e^{-2i\omega_c T_0} + P_3)\\
&= F_\Phi\{-h_c(\frac{1}{2}(P_1e^{2i\omega_c T_0} + P_2e^{-2i\omega_c T_0} + P_3 - R_1e^{2i\omega_c T_0} - R_2e^{-2i\omega_c T_0} - R_3)\\
&\quad -\frac{1}{8\delta_0}(A_1e^{i\omega_c T_0} + \overline{A}_1e^{-i\omega_c T_0} - A_2e^{i\omega_c T_0} - \overline{A}_2e^{-i\omega_c T_0})^2)\\
&\quad -h_c((P_1e^{2i\omega_c T_0} + P_2e^{-2i\omega_c T_0} + P_3 - R_1e^{2i\omega_c T_0} - R_2e^{-2i\omega_c T_0} - R_3)\\
&\quad -\frac{1}{2\delta_0}(A_1e^{i\omega_c T_0} + \overline{A}_1e^{-i\omega_c T_0} - A_2e^{i\omega_c T_0} - \overline{A}_2e^{-i\omega_c T_0})^2)\\
&\quad +h_c\gamma_e(\frac{1}{2}(P_1e^{2i\omega_c(T_0-\tau)} + P_2e^{-2i\omega_c(T_0-\tau)} + P_3 - R_1e^{2i\omega_c(T_0-\tau)} - R_2e^{-2i\omega_c(T_0-\tau)} - R_3)\\
&\quad -\frac{1}{8\delta_0}(A_1e^{i\omega_c(T_0-\tau)} + \overline{A}_1e^{-i\omega_c(T_0-\tau)} - A_2e^{i\omega_c(T_0-\tau)} - \overline{A}_2e^{-i\omega_c(T_0-\tau)})^2)\\
&\quad +h_c\gamma_e((P_1e^{2i\omega_c(T_0-\tau)} + P_2e^{-2i\omega_c(T_0-\tau)} + P_3 - R_1e^{2i\omega_c(T_0-\tau)} - R_2e^{-2i\omega_c(T_0-\tau)} - R_3)\\
&\quad -\frac{1}{2\delta_0}(A_1e^{i\omega_c(T_0-\tau)} + \overline{A}_1e^{-i\omega_c(T_0-\tau)} - A_2e^{i\omega_c(T_0-\tau)} - \overline{A}_2e^{-i\omega_c(T_0-\tau)})^2) - h_1\delta_0\}\\
&\quad h_c\frac{c_n}{m_1}2(i\omega_c P_1e^{2i\omega_c T_0} - i\omega_c P_2e^{-2i\omega_c T_0} + A_1'e^{i\omega_c T_0} + \overline{A}_1'e^{-i\omega_c T_0} - i\omega_c R_1e^{2i\omega_c T_0}\\
&\quad +i\omega_c R_2e^{-2i\omega_c T_0} - A_2'e^{i\omega_c T_0} - \overline{A}_2'e^{-i\omega_c T_0}
\end{aligned} \quad (21)$$

For (21) and (21b), equating each of the coefficients of exp ($2i\omega_c T_0$), exp ($-2i\omega_c T_0$) and exp (0) on both sides, we obtain

$$P_1 = \phi_1(A_1 - A_2)^2,\ P_2 = \phi_2(A_1 - A_2)^2, P_3 = \phi_3(A_1 - A_2)(\overline{A}_1 - \overline{A}_2) \quad (22)$$

and

$$R_1 = \theta_1(A_1 - A_2)^2, R_2 = \theta_2(A_1 - A_2)^2, R_3 = \theta_3(A_1 - A_2)(\overline{A}_1 - \overline{A}_2) \quad (23)$$

where

$$\phi_1 = \frac{\Gamma_1}{\Delta_1\Gamma_1 - \Delta_2\Gamma_2}(\Delta_2\frac{\Gamma_3}{\Gamma_1} + \Delta_3), \quad \phi_2 = \frac{\Gamma_4}{\Delta_4\Gamma_4 - \Delta_5\Gamma_5}(\Delta_5\frac{\Gamma_6}{\Gamma_4} + \Delta_6),$$
$$\phi_3 = \frac{\Gamma_7}{\Delta_7\Gamma_7 - \Delta_8\Gamma_8}(\Delta_8\frac{\Gamma_9}{\Gamma_7} + \Delta_9), \quad \theta_1 = \frac{\Gamma_2}{\Delta_1\Gamma_1 - \Delta_2\Gamma_2}(\Delta_2\frac{\Gamma_3}{\Gamma_1} + \Delta_3) + \frac{\Gamma_3}{\Gamma_1},$$
$$\theta_2 = \frac{\Gamma_5}{\Delta_4\Gamma_4 - \Delta_5\Gamma_5}(\Delta_5\frac{\Gamma_6}{\Gamma_4} + \Delta_6) + \frac{\Gamma_6}{\Gamma_4}, \quad \theta_3 = \frac{\Gamma_8}{\Delta_7\Gamma_7 - \Delta_8\Gamma_8}(\Delta_8\frac{\Gamma_9}{\Gamma_7} + \Delta_9) + \frac{\Gamma_9}{\Gamma_7} \quad (24)$$

and

$$\Delta_1 = -4\omega_c^2 + 4\zeta_1 i\omega_c + p_1^2 + \frac{3}{2}F_\Phi h_c(1-\gamma_e e^{-2i\omega_c\tau}) + 2i\omega_c\frac{c_n}{m_1}$$
$$\Delta_2 = \frac{3}{2}F_\Phi h_c(1-\gamma_e e^{-2i\omega_c\tau}) + 2i\omega_c\frac{c_n}{m_1}$$
$$\Delta_3 = \frac{5}{8}F_\Phi h_c(1-\gamma_e e^{-2i\omega_c\tau})$$
$$\Delta_4 = -4\omega_c^2 - 4\zeta_1 i\omega_c + p_1^2 + \frac{3}{2}F_\Phi h_c(1-\gamma_e e^{2i\omega_c\tau}) - 2i\omega_c\frac{c_n}{m_1}$$
$$\Delta_5 = \frac{3}{2}F_\Phi h_c(1-\gamma_e e^{2i\omega_c\tau}) - 2i\omega_c\frac{c_n}{m_1}$$
$$\Delta_6 = \frac{5}{8}F_\Phi h_c(1-\gamma_e e^{2i\omega_c\tau}) \quad (25)$$
$$\Gamma_1 = -4\omega_c^2 + 4\zeta_1 i\omega_c + p_1^2 + \frac{3}{2}F_\Phi h_c(1-\gamma_e e^{-2i\omega_c\tau}) + 2i\omega_c\frac{c_n}{m_2}$$
$$\Gamma_2 = \frac{3}{2}F_\Phi h_c(1-\gamma_e e^{-2i\omega_c\tau}) + 2i\omega_c\frac{c_n}{m_2}$$
$$\Gamma_3 = -\frac{5}{8}F_\Phi h_c(1-\gamma_e e^{-2i\omega_c\tau})$$
$$\Gamma_4 = -4\omega_c^2 - 4\zeta_1 i\omega_c + p_1^2 + \frac{3}{2}F_\Phi h_c(1-\gamma_e e^{2i\omega_c\tau}) - 2i\omega_c\frac{c_n}{m_2}$$
$$\Gamma_5 = \frac{3}{2}F_\Phi h_c(1-\gamma_e e^{2i\omega_c\tau}) - 2i\omega_c\frac{c_n}{m_2}$$
$$\Gamma_6 = -\frac{5}{8}F_\Phi h_c(1-\gamma_e e^{2i\omega_c\tau}) \quad (26)$$

Now we can substitute the solutions (20), and (22-23) into equations (18), eliminating the terms that lead to secular terms yields

$$(2i\omega + 4\zeta_1)A_1' = F_\Phi\{-h_c(\frac{1}{4\delta_0}\Lambda_1 + \frac{1}{16\delta_0^2}\Lambda_2) + h_c(\frac{1}{\delta_0}\Lambda_1 + \frac{1}{8\delta_0^2}\Lambda_2)$$
$$+ h_c\gamma_e(-\frac{1}{4\delta_0}\Lambda_1 + \frac{1}{16\delta_0^2}\Lambda_2)e^{-i\omega_c\tau} - h_c\gamma_e(\frac{1}{\delta_0}\Lambda_1 + \frac{1}{8\delta_0^2}\Lambda_2)e^{-i\omega_c\tau}\}e^{i\delta_1} \quad (27)$$
$$- h_1\frac{c_n}{m_1}(1-h_b)A_1 i\omega$$

where

$$\Lambda_1 = (\phi_3 - \theta_3 + \phi_1 - \theta_1)(1-h_b)^3 \cdot \frac{1}{8}\alpha_1^3, \qquad \Lambda_2 = \frac{3}{128\delta_0^2}(1 - h_b + h_b{}^2 - h_b{}^3)\cdot\frac{1}{8}\alpha_1^3$$
$$\Lambda_3 = \frac{1}{4\delta_0}\Lambda_2 - \Lambda_1, \qquad \Lambda_4 = \frac{F_{\phi 1}h_1}{64\delta_0}\Lambda_3$$

Moreover it becomes simpler if we assume

$$A_2 = h_b A_1$$

where the coefficient h_b can be estimated by equation (7) in a linear case. Introducing the polar transformations

$$A_1 = \frac{1}{2}\alpha e^{i\delta}, \; A_1' = \frac{1}{2}\alpha' e^{i\delta} + i\frac{1}{2}\alpha\delta' e^{i\delta} \quad (28)$$

Substituting equation (28) into Equation (27) and separating real and imaginary parts, we obtain the normal form

$$\alpha' = g_1 h_1 \alpha + g_2 \alpha^3 \tag{29}$$

$$\alpha\delta' = g_3 h_1 + g_4 \alpha^3 \tag{30}$$

where

$$g_1 = \frac{-\omega^2 c_n}{(\omega^2 + 4\zeta_1^{\,2}) m_1}(1 - h_b)$$

$$g_2 = (\Lambda_{4i} \frac{\omega}{\omega^2 + 4\zeta_1^{\,2}} \gamma_e \sin(\omega_c \tau) + \Lambda_{4r} \frac{1-\gamma_e \cos(\omega_c \tau)}{2\zeta_1}(1 - \frac{\omega^2}{\omega^2 + 4\zeta_1^{\,2}}))$$

$$g_3 = \frac{-2\zeta_1 \omega c_n}{(\omega^2 + 4\zeta_1^{\,2}) m_1}(1 - h_b)$$

$$g_4 = \frac{2\zeta_1}{(\omega^2 + 4\zeta_1^{\,2})}(\Lambda_{4i} \sin(\omega_c \tau) - \Lambda_{4r} \frac{\omega}{2\zeta_1}(1 - \gamma_e \cos(\omega_c \tau)) \tag{31}$$

where $\Lambda_4 = \Lambda_{4r} + i\Lambda_{4i}$.

The approximate solutions are expressed by (29) and (30). The qualitative behavior near the Hobf bifurcation can be determined by the sign of g_2. When $g_2 < 0$, the bifurcation is supercritical and $g_2 > 0$, it is subcritical.

The amplitude of steady-state limit cycles can be obtained from equation (29) by setting $\alpha' = 0$. It yields the expression of determining the constant amplitude of the limit cycles.

$$\alpha = \sqrt{\frac{g_1}{g_2}(h_c - h)} \tag{32}$$

4. Numerical Results and Discussions

Using the MATLAB® SIMULINK environment the simulation model of the nip installation can be set up for the time domain simulations. To characterize the motions, here we use three plots, namely time traces, phase portraits, and Poincare sections. In practice, the limit cycle might be difficult to find exactly. Some case studies have been carried out instead. For example, Fig. 3 illustrates the characteristics of the motion when speed is 500m/min with small damping. The time delay effects are showed in the time history response and the Poincare section shows some chaotic feature. With moderate damping in Fig. 4 a chaotic fracture emerges in the Poincare section. With large damping in Fig.5 the Poincare section becomes one circle which means the response try to reach the limit cycle.

5. Conclusions

The dynamic behaviors of the rolling nip contact system have been investigated through the two degrees of freedom model considering the nonlinear contact stiffness and time delay effects. The approximation solutions and the amplitudes of the limit cycles are obtained by the perturbation theory. The numerical results show that limit cycles can exist with different time delays and system parameters. Some chaotic behavior can be observed as well

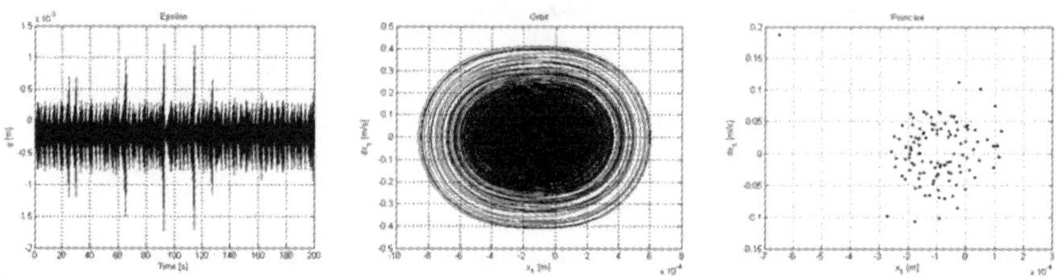

Fig. 3. Speed 500 m/min and small damping. (i) Response (ii) Phase trajectory (iii) Poincare map.

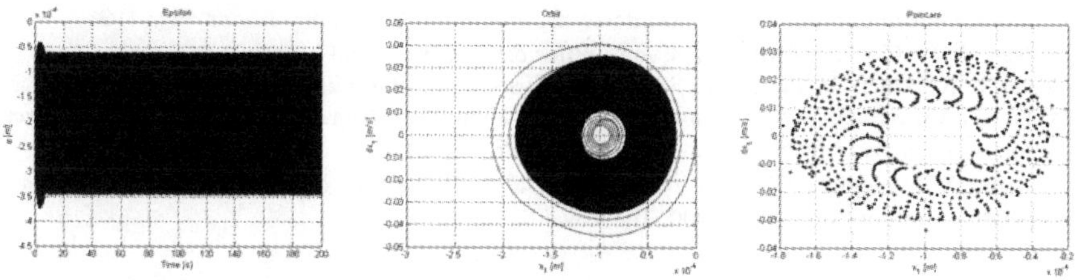

Fig. 4. Speed 500 m/min and moderate damping. (i) Response (ii) Phase trajectory (ii) Poincare map.

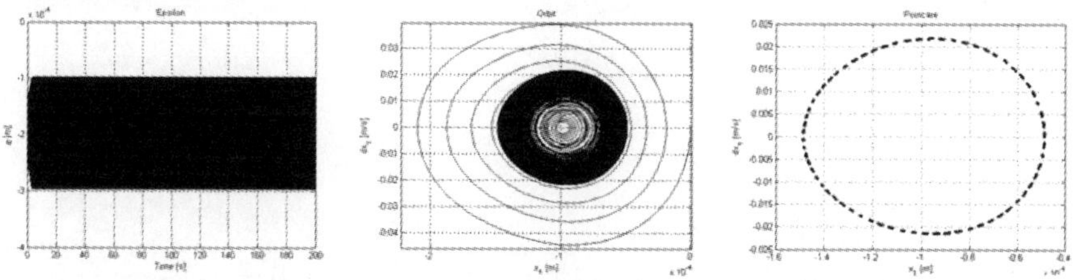

Fig. 5. Speed 500 m/min and large damping. (i) Response (ii) Phase trajectory (ii) Poincare map.

In Fig. 6 the time domain responses with large damping are shown in the speeds of 480m/min, 490m/min and 500m/min. Fig.7 plots the corresponding amplitudes of the responses at a speed range. The lobe curves give the features of the stability of time delay system.

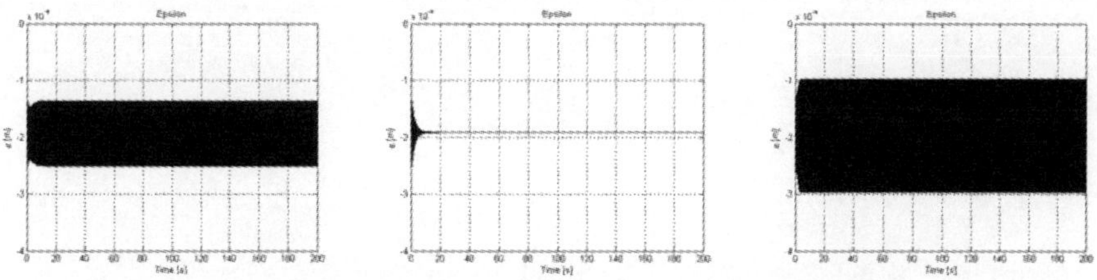

Fig. 6. Large damping, responses at (i) 480 m/min (ii) 490 m/min (iii) 500 m/min

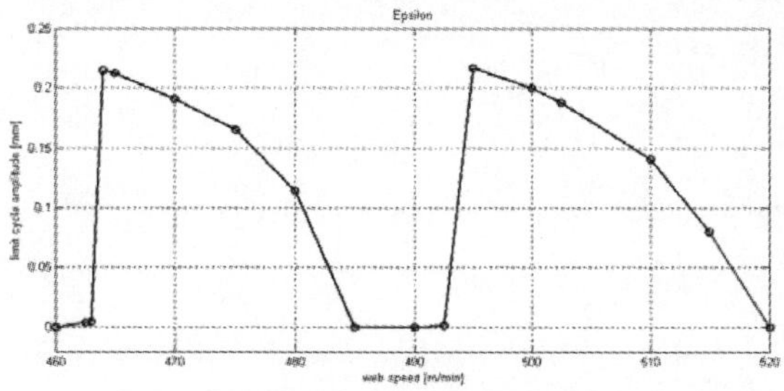

Fig. 7. Amplitude plot versus running speed.

References

[1] E. Keskinen, F. Zardoya, R. Hildebrand, Stochastic Analysis of Rolling Contact, ENOC 2005, The Netherlands.

[2] F.C. Moon, Dynamics and Chaos in Manufacturing Processes, John Wiley &Sons, Inc., 1998
[3] A.H. Nayfeh, Problems in Perturbation, John Wiley &Sons, Inc., 1993
[4] L. Yuan, E. Keskinen, V. Järvenpää, Stability Analysis of Roll Grinding System with Double Time Delay Effects, IUTAM Symposium on Vibration Control of Nonlinear Mechanisms and Structres, Springer, 2005
[5] K.L. Johnson, Contact Mechanics, Cambridge Univ. Press, 2003.
[6] J.R. Pratt, A.H. Nayfeh, Chatter Control and Stability Analysis of a Cantilever Boring Bar under Regenerative Cutting Conditions, 2000, The Royal Society.

Vibration signal analysis and feature extraction based on wavelet energy spectrum

Yongqiang Li, Jie Liu

School of science, Northeastern University, Shenyang 110004, China

Abstract

The FFT is one of the most widely used and well-established methods. Unfortunately, the FFT spectrums are not suitable for non-stationary signal analysis and are also unable to show components with low energy clearly, which make spectrum unable to extract fault features at its early developing stage. In this paper, wavelet energy spectrum is used, which can highlight the components with low energy. This method can be used to extract the fault features when the fault is at an early developing stage. The experimental data of coupling misalignment faults in rotating machinery is analyzed by the wavelet energy spectrum, and the results by FFT analysis are also given. A comparison between two kinds of analysis results is carried out, and comparison results indicate that the wavelet energy spectrum can highlight the components of low energy. Therefore, it is suitable to extract features of fault at an early stage.

Keywords: Signal analysis; FFT; Wavelet energy spectrum

1. Introduction

Fault diagnostics is useful for ensuring the safe running of rotating machines, and vibration signal analysis has been widely used for fault diagnostics. The key problem is how to extract useful features from vibration signals for fault diagnostics. Among many signal analysis methods, the FFT is one of the most widely used and well-established methods. Unfortunately, FFT-based methods are not suitable for non-stationary signal analysis and unable to reveal the inherent information of non-stationary signals. However, various kinds of factors, such as change of the environment and the faults from the machine itself, often make the vibration signal of the running machine contain non-stationary components. Usually, these non-stationary components have abundant fault information, so it is important to analyze the non-stationary signals [1]. Because of disadvantages of the FFT, it is necessary to find supplementary methods for vibration signal analysis.

Hitherto, time-frequency analysis is the most popular method for the analysis of non-stationary vibration signals, such as the Gabor transform [2] (windowed Fourier transform), and the bilinear time-frequency representation [3]. They perform a mapping of one-dimensional signal $x(t)$ to a

*Corresponding author. Tel.: +86 2483684686

E-mail address: lyq525@sohu.com (Li YQ)

two-dimensional function of time and frequency $TFR(x{:}t,\omega)$. But all of the time-frequency analysis methods have some disadvantages. For the Gabor transform, the limitation of Heisenberg-Gabor inequality makes the trade off between time and spectral resolutions unavoidable, and good time and spectral concentration cannot be obtained together on the time-frequency plane. Moreover, the spectrum of the Gabor transform is only a biased estimator of the instantaneous frequency and group delay of the signal. Bilinear time-frequency representations, such as the Wigner-Ville distribution and Margenau-Hill distribution have good concentration on the time-frequency plane. However, when support areas of the signal overlap each other, interference terms will appear on the time-frequency plane. All the disadvantages mentioned above will mislead signal analysis. In order to overcome these disadvantages, many methods of improvement have been proposed. Without exception, however, elimination of one shortcoming will always lead to the loss of other merits. For example, the reduction of interference terms will bring the loss of time-frequency concentration [4].

Over the past 10 years, the wavelet theory [5] has become one of the emerging and fast-evolving mathematical and signal processing tools for its many distinct merits. Different from the Gabor transform, the wavelet transform (WT) can be used for multi-scale analysis of a signal through dilation and translation, so it can extract time-frequency features of a signal electively. In the field of mechanical signal processing, the WT has been used for singularity detection [6], noise reduction [7] and feature extraction [8]. In this paper, the conception and algorithm of wavelet energy spectrum is putted forward, the theory formula and the relationship between wavelet energy spectrum with Fourier transform of original signal are shown. The localization of non-stationary signal and abnormal signal in frequency domain is achieved. The abnormal points or frequency parts which the energy are larger than others may be found by wavelet energy spectrum. It is difficult to do this in usual wavelet decomposition or Fourier analysis.

2 Discrete wavelet transform

We define a square integral function $\psi(t)$ (namely $\psi(t)\in L^2(t)$ as a family of functions, which satisfies the following equation:

$$\int_{-\infty}^{+\infty}\frac{|\hat{\psi}(\omega)|^2}{|\omega|}<\infty \tag{1}$$

Assuming:

$$\psi_{a,b}(t)=\frac{1}{\sqrt{|a|}}\psi\left(\frac{t-b}{a}\right)\quad ,a,b\in R\ ,a\neq 0 \tag{2}$$

$\psi_{a,b}(t)$ is defined as a continuous wavelet, which is derived from a family function $\psi(t)$, a and $b$are the dilation and translation parameters, respectively. a, b represent the family of wavelets obtained from the single function by dilations and translations. The change of parameter a can influence not only the frequency spectrum structure of the continuous wavelet but also the window size and the form.

Assume $f(t)$ to be a finite-energy function, that is, $f(t)\in L^2(R)$. Define a continuous wavelet transform as follows:

$$W_f(a,b;\psi)=\langle f(t),\psi_{a,b}(t)\rangle=a^{-1/2}\int_{-\infty}^{+\infty}f(t)\psi^{*}_{a,b}(t)dt\qquad a>0\ , \tag{3}$$

where the asterisk stands for complex conjugate.Define discrete wavelet transform as follows:

$$W_{j,k}(f)=\int_{-\infty}^{+\infty}f(t)\hat{\psi}_{j.k}(t)dt\quad ,j,k\in R \tag{4}$$

where

$$\psi_{j.k}(t) = 2^{-j/2}\psi\left(\frac{t-2^j k}{2^j}\right) \tag{5}$$

The wavelet coefficient $W_{j,k}(f)$ is taken as the time-frequency map of the original signal $f(t)$. In terms of the relationship between the wavelet function $\psi(t)$ and the scaling function $\phi(t)$, namely:

$$\left|\hat{\phi}(\omega)\right|^2 = \sum_{j=-\infty}^{+\infty}\left|\hat{\psi}(2^j\omega)\right| \tag{6}$$

The discrete scaling function with corresponds to the discrete wavelet function is as follows:

$$\phi_{j.k}(t) = 2^{-j/2}\phi\left(\frac{t-2^j k}{2^j}\right) \tag{7}$$

It is used to discrete the signal, the sampled values are called scaling coefficients Dj,k

$$D_{j,k}(f) = \int_{-\infty}^{+\infty} f(t)\hat{\phi}_{j,k}(t)dt \tag{8}$$

When $j > 1$, the scaling coefficients and the wavelet coefficients are written as follows:

$$D_{j+1,k}(f) = \sum_i h(i-2k)D_{j,k}(f) \tag{9}$$

$$W_{j+1,k}(f) = \sum_i g(i-2k)D_{j,k}(f) \tag{10}$$

where the terms g and h are high-pass and low-pass filters derived from the $\psi(t)$ and $\phi(t)$, the coefficients $Dj1,k$ and $Wj1,k$ represent a decomposition of the (j-1)st scaling coefficient into high and low frequency terms.

3 Discrete wavelet energy spectrum

Assume $f(t)$ to be a finite-energy function, that is, $f(t)\in L^2(R)$. The discrete function norm of $f(t)$ is as follows:

$$\|f\|^2 = \sum_j\sum_k\left\|W_{j,k}(f)\right\|^2 = \sum_j\sum_k\left\|\sum_i g(i-2k)D_{j-1,k}(f)\right\|^2 \tag{11}$$

Define discrete wavelet transform energy as follows:

$$EW_{j,k}(f) = \left\|W_{j,k}(f)\right\|^2 = \left\|\sum_i g(i-2k)D_{j-1,k}(f)\right\|^2 \tag{12}$$

Fourier transform of $EW_{j,k}(f)$ as follows:

$$EW_j\hat{f}(\omega) = \sum_{k=1}^{K} EW_{j,k}(f)\exp(-i\omega k/K) \tag{13}$$

$EW_j\hat{f}(\omega)$ is defined as discrete wavelet energy spectrum.According to equation(11) and equation(12),we have

$$\|f(k)\|^2 = \sum_j EW_{j,k}(f) \tag{14}$$

Considering Parseval identical equation, assume $f(t)$, $g(t)$to be finite-energy function, that is, $f(t)$, $g(t)\in L^2(R)$, yields

$$< f,g > \equiv \int_R f(t)\overline{g(t)}dt = \frac{1}{2\pi} < \hat{f},\hat{g} > \tag{15}$$

When $f = g$,then

$$\|f\|^2 = \langle f, f\rangle = \frac{1}{2\pi}\langle \hat{f}, \hat{f}\rangle = \frac{1}{2\pi}\|\hat{f}\|^2 \quad (16)$$

or

$$\|f\|_2^2 = \frac{1}{2\pi}\|\hat{f}\|_2^2 \quad (17)$$

where

$$\|f\|_2 = \left(\int_{-\infty}^{+\infty}|f(t)|^2 dt\right)^{1/2} \quad (18)$$

According to equation(14) and (16),the relation between signal spectrum and discrete wavelet energy is

$$\|\hat{f}\|^2 = 2\pi\|f\|^2 = 2\pi\sum_j\sum_k EW_{j,k}(f) \quad (19)$$

Considering Parseval identical equation, the relation between signal spectrum and discrete wavelet energy spectrum is given by

$$\|\hat{f}\|^2 = 2\pi\left[\frac{1}{2\pi}\sum_j\sum_{k=1}^{K} EW_j\hat{f}(\omega)\exp(i\omega k/K)\right] = \sum_j\sum_{k=1}^{K} EW_j\hat{f}(\omega)\exp(i\omega k/K) \quad (20)$$

The above equation may be explained that the sum of discrete wavelet energy spectrum equals to discrete signal spectral energy, which is one orthogonal character of signal wavelet decomposition, namely

$$L^2(R) = \bigoplus_{j\in Z} W_j \quad (21)$$

It is feasible to analyze signal character by the wavelet energy spectrum on every decomposition level of wavelet transform. It do not aroused aliasing in time and frequency domain.

4 Numerical experiment

In order to illustrate the characteristics of the discrete wavelet energy spectrum, a signal $f(t)$ will be considered. Figure 1 shows the signal $f(t)$ in the time domain and can be expressed by

$$f(t) = \begin{cases} 2\cos(50\pi t) + 10\cos(0.05\pi t) + \cos(100\pi t) & 25 \le t \le 26 \\ 2\cos(50\pi t) + 10\cos(0.05\pi t) & other \end{cases} \quad (22)$$

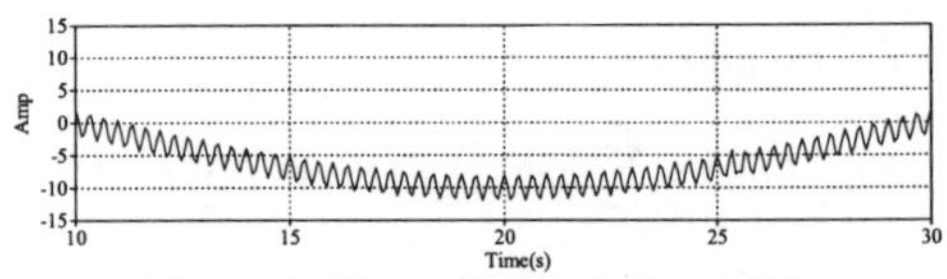

Figure 1. Numerical test signal f(t)

The signal f(t) contains three components whose frequencies are 25,0.025 and 50 Hz respectively. High frequency signal of 50 Hz only exit on time series $25 \leqslant t \leqslant 26$,whose amplitude is under other frequency's. It is find difficulty on the Figure 1. The wavelet transform is used for the signal. The Daubechies wavelet series is used. Singular point is find at the different scales but do not distinctness.

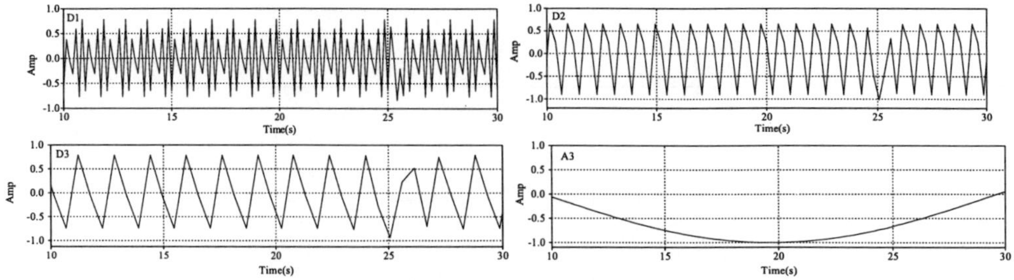

Figure 2 Normalization wavelet decomposition results of $f(t)$

Figure 3 is the wavelet energy of $f(t)$.Figure 3 shows more large amplitude of singular signal than wavelet transform at the different scales. In other words, the wavelet energy can detect easily weak singular signal.

Figure 4 is the spectrum of $f(t)$. The high frequency signal cannot be detected in Figure 4 because its energy is very weakness.

Figure 5 is the wavelet energy spectrum of $f(t)$. The high frequency signal can be detected easily in Figure 5. Figure 4 and Figure 5 show that the wavelet energy spectrum may realize accurately locate on frequency domain for singular signal. It remedy the shortcoming of wavelet transform that can be used to locate on time domain.

Because of good location on frequency domain for singular signal of the wavelet energy spectrum, so it can be used as a useful tool for extracting features of faults at their early stage. In the next section, the wavelet energy spectrum is used to analyze experimental data for coupling misalignment faults of rotating machines.

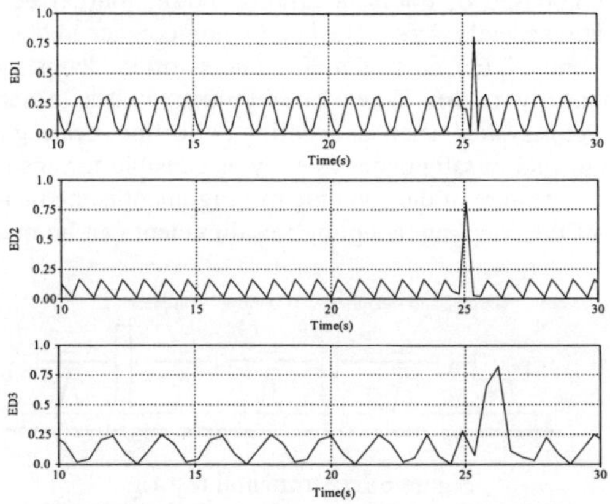

Figure 3 Normalization wavelet energy of $f(t)$

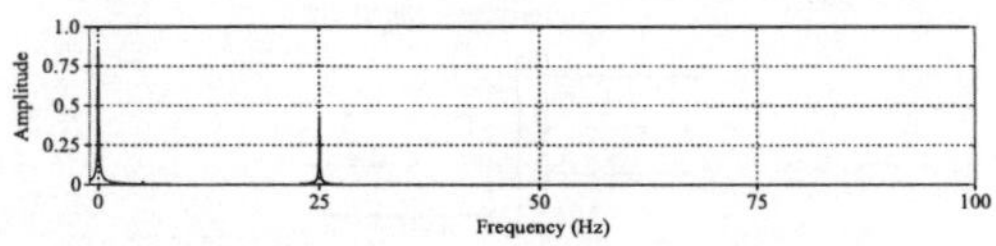

Figure 4 The spectrum of $f(t)$

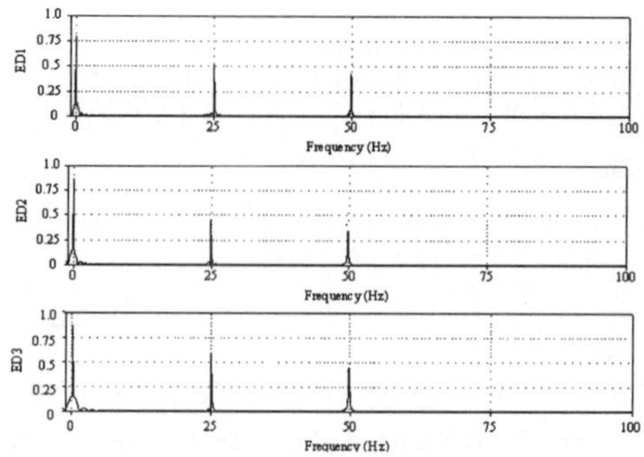

Figure5 Normalization wavelet energy spectrum of $f(t)$

5. Experimental data analysis

Spectrum analysis plays an important role in fault diagnostics of rotating machines, and all faults have corresponding spectral features. During the early developing stage of a fault, however, those features are feeble and often submerged by the noise signal. Therefore, these fault features cannot be recognized clearly from the spectrum. This shortcoming bring some troubles for fault diagnostics at early stage. The wavelet energy spectrum can realize accurately location on frequency domain for singular signal with low energy. Therefore, the wavelet energy spectrum is a useful supplementary method for fault diagnostics at the early developing stage. In this section, the wavelet energy spectrum is used to analyze the experimental data of coupling misalignment faults in rotating machines. Figure 6 shows the experimental test rig, which is mainly composed of rotors, a driving motor, journal bearings and couplings. Both vertical and horizontal vibration signals were picked up by non-contact eddy current transducers.

Coupling misalignment, one of the most familiar faults, often denotes the slant or misalignment between the axes of two nearly rotors. When misalignment exists, a series of dynamic responses undesired will occur in the rotor system, such as coupling deflection, bearing abrasion and oil collapsing, etc. So it is very important to find misalignment as early as possible for ensuring the safe running of the machines. The main frequency feature of the coupling misalignment is the increase of 2X component.

Through the adjustment of the coupling, coupling misalignment can be manually simulated, as shown in Figure 7.

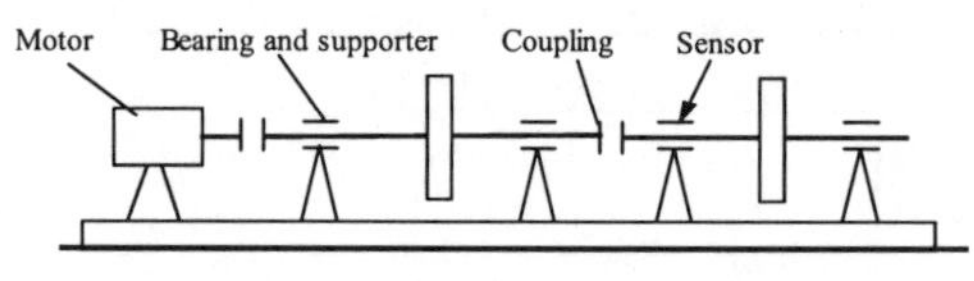

Figure 6 Experimental test rig

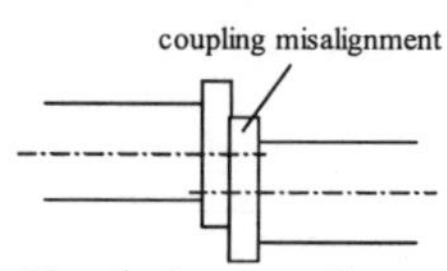

Figure 7. Simulating coupling misalignment

Figure 8 shows a set of horizontal vibration data sampled at coupling misalignment condition. The sampling speed is 1 kHz, and the rotating speed is 4740 r/min.

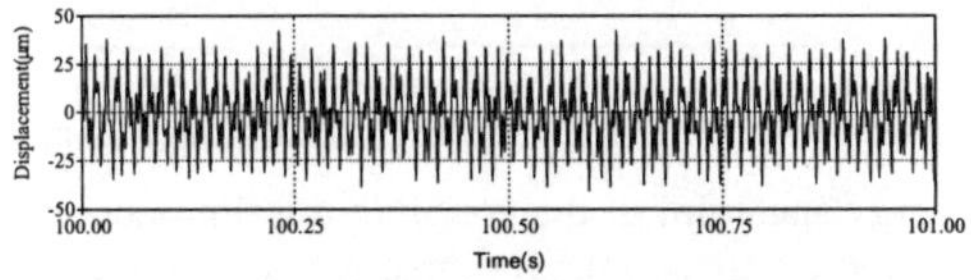

Figure 8 Coupling misalignment signal

The wavelet decomposition and the wavelet energy of coupling misalignment signals are given in Figures 9 and 10.

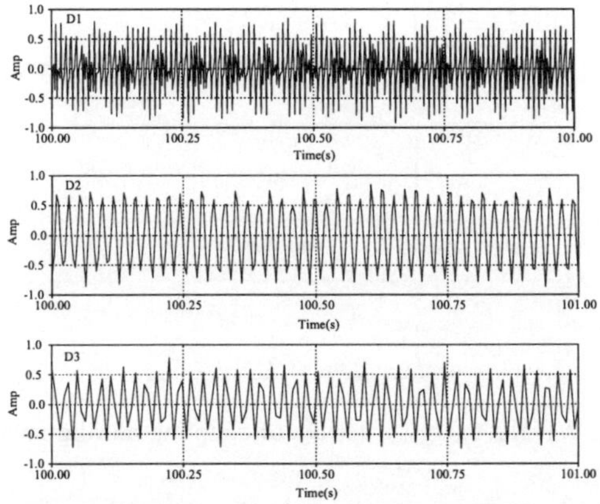

Figure 9 Normalization wavelet decomposition results of coupling misalignment signal

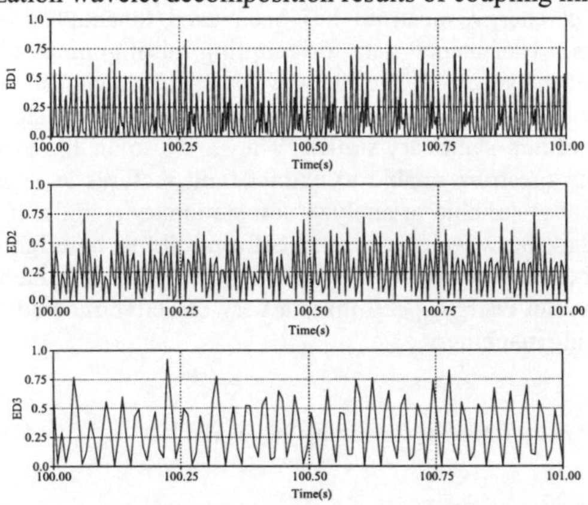

Figure 10 Normalization wavelet energy of coupling misalignment signal

Figure 11 is the spectrum of the coupling misalignment signal, in which 2X component can be recognized. However, the 2X is very small and will be overlooked easily.

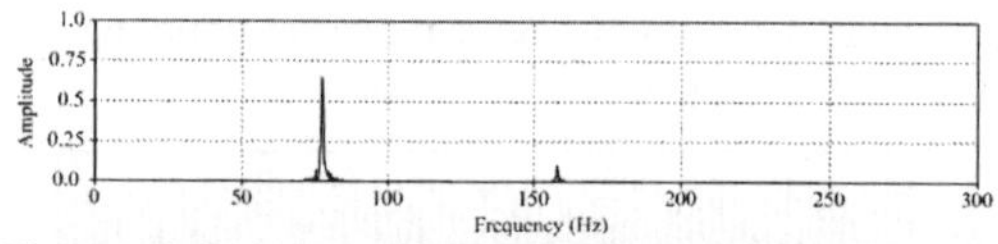

Figure 11 Spectrum of coupling misalignment signal

Figure 12 is the wavelet energy spectrum of the coupling misalignment signal.The 2X component can be seen easily in the wavelet energy spectrum. The results testify once again that the wavelet energy spectrum can give better description than the spectrum in extracting fault features at the early developing stage.

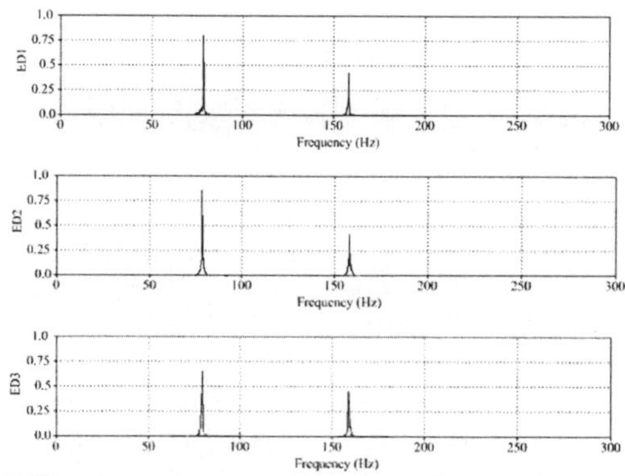

Figure 12 Wavelet energy spectrum of coupling misalignment signal

6. Conclusions

In this paper, the wavelet energy spectrum has been used to analyze vibration signals, including numerical simulation signals, experimental data for coupling misalignment. Comparison was carried out among the analysis results of the spectrum and the wavelet energy spectrum. The results indicate that (1) the FFT spectrums are unable to reveal time-frequency characters of the signal, which make spectrums not suitable for the analysis of non-stationary signals. They are also unable to show components with low energy clearly, which makes spectrum unable to extract fault features at its early developing stage. (2) The wavelet energy spectrum is suitable to analyze non-stationary signals. It is a useful extracting fault features tool. It is shown that the wavelet energy spectrum able to highlight the components with low energy. Therefore, it may be used to extract the fault features when the fault is at an early developing stage. In conclusion, the wavelet energy spectrum is a very effective method of vibration signal analysis for fault diagnostics of rotating machinery.

Acknowledgements

The first author is very grateful to Prof. Liu for supporting this work.

References

[1] M. C. PAN and P. SAS 1996 *IEEE Proceedings of International Conference on Signal Processing* 2,1723-1726. Transient analysis on machinery condition monitoring.

[2] P. C. RUSSELL, J. COSGRAVE, D. TOMTSIS, A. VOURDAS, L. STERGIOULAS and G. R. JONES 1998 *Measurement Science & Technology* 9, 1282-1290. Extraction of information from acoustic vibration signals using Gabor Transform type devices.

[3] I. S. KOO and W. W. KIM 2000 *ISA Transactions* 39, 309-316. Development of reactor coolant pump

vibration monitoring and a diagnostic system in the nuclear power plant.
[4] G. T. ZHENG and P. D. MCFADDEN 1999 *Transactions of the American Society of Mechanical Engineers, Journal of Vibration and Acoustics* 121, 328-333. A time-frequency distribution for analysis of signals with transient components and its application to vibration analysis.
[5] L. A. WONG, J. C. CHEN 2001 *International Journal of Non-near Mechanics* 36, 221-235. Nonlinear and chaotic behavior of structural system investigated by wavelet transform techniques.
[6] Z. CHEN and Y. LU 1997 *Journal of Vibration Engineering* 10, 147-155. Signal singularity detection and its application (in Chinese).
[7] N. TANDON and A. CHOUDHURY 1999 *Tribology International* 32, 469-480. Review of vibration and acoustic measurement methods for the detection of defects in rolling element bearings.
[8] J. LIN and L. QU 2000 *Journal of Sound and Vibration* 234, 135-148. Feature extraction based on Morlet wavelet and its application for mechanical fault diagnosis.

Experimental analysis of cumulants scaling properties in fully developed intermittent turbulence

François G. Schmitt

CNRS, UMR 8013 ELICO, Wimereux Marine Station, University of Lille 1
28 av. Foch, 62930 Wimereux, France
E-mail address: francois.schmitt@univ-lille1.fr

Abstract

We consider velocity structure functions in turbulence through an approach using cumulants, and for a fixed value of the distance ℓ. This allows to consider the cumulant generating function $\Phi_\ell(q) = \log\langle|\Delta V_\ell|^q\rangle$. Using an atmospheric turbulent database, we show that the cumulant generating function is nonanalytic, with a development compatible with a log-stable model of the form $\Phi_\ell(q) = A_\ell q + B_\ell q^\alpha$ and a parameter value of $\alpha = 1.5$. The parameters A_ℓ and B_ℓ are experimentally estimated: they are respectively increasing and decreasing functions of ℓ; their scaling ranges correspond to the scaling range of the velocity fluctuations. The dependence between these two functions is studied in relation to Extended Self Similarity and Generalized Extended Self Similarity properties.

Keywords: Turbulence; Intermittency; equations; Cumulants

1 Introduction

One of the characteristic features of fully developed turbulence is the intermittent nature of velocity fluctuations [1]. Intermittency provides corrections to Kolmogorov's scaling law [2], which are now well established and received considerable attention in the last twenty years. Let us recall how to quantify intermittency effects on scaling laws for Eulerian isotropic turbulence. Denoting $\Delta V_\ell = V(x+\ell) - V(x)$ the longitudinal increments of the Eulerian velocity field at a spatial scale ℓ, their fluctuations are characterized, in the inertial range, using the scale invariant moment function $\zeta(q)$:

$$\langle|\Delta V_\ell|^q\rangle = C_q \ell^{\zeta(q)} \tag{1}$$

where $q > 0$ is the order of moment and C_q is a constant that may depend on q. Kolmogorov's initial proposal, for a non-intermittent constant dissipation, leads to $\zeta(q) = q/3$ [2]. For intermittent turbulence, $\zeta(q)$ is proportional to a cumulant generating function, and is nonlinear and concave; only the third moment has no intermittency correction: $\zeta(3) = 1$. The accuracy of the scaling of Eq.(1) is usually tested for each order of moment, for various values of ℓ in log-log plot, using a least-square regression [3]. The values of $\zeta(q)$ which are then obtained may be compared and fitted to different multifractal models (among many studies, see [4, 5, 6, 7, 8, 9, 10]). This way of estimating $\zeta(q)$ depends on the choice of the scaling range: one usually estimates $\zeta(q)$ for the range of scales where the exact relation $\zeta(3) = 1$ is verified, assuming that the scaling range is the same for each order of moment. One may also more precisely estimate $\zeta(q)$ through Extended Self Similarity (ESS) method [11, 12, 13] (see below). We consider here another approach: instead of studying the scale dependence for each moment, we focus on the moment dependence for each scale. The procedure is the following: we first compute the moments at a fixed scale in order to estimate somes cumulants; in a second step we consider the scaling properties of the coefficients involved in these cumulants. The cumulant approach has already been undertaken in the scaling turbulence framework in a few studies (see e.g. [14, 15, 16]), where the cumulants of the cascade process [15] or a polynomial development of the cumulant generating function [14, 16] have been considered.

2 Cumulants and analycity of the cumulant generating function

We consider here, for a fixed scale ℓ, the cumulant generating function of the generator $g_\ell = \log|\Delta V_\ell|$ defined as [17]:

$$\Phi_\ell(q) = \log\langle|\Delta V_\ell|^q\rangle \tag{2}$$

The function $\Phi_\ell(q)$ is also the second Laplace characteristic function of the generator: $\Phi_\ell(q) = \log\langle e^{qg_\ell}\rangle$. As a second characteristic function, it is convex [18], and can be developed using the cumulants:

$$\Phi_\ell(q) = \sum_{p=1}^{\infty} c_p(\ell)\frac{q^p}{p!} \tag{3}$$

where $c_p(\ell)$ is the nth cumulant. Let us recall that $c_1 = \langle g_\ell\rangle$, $c_2 = \langle g_\ell^2\rangle - c_1^2$, and c_n depends on all moments $\langle g_\ell^p\rangle$ $(1 \leq p \leq n)$. The theorem of Marcienkiewicz states that, if it exists, the development in Eq.(3) is either infinite, or if finite, of degree not higher than 2 [17]. In fact, the development in Eq.(3) may not exist in case of non-analycity of $\Phi_\ell(q)$. This is the case when g_ℓ is a stable process whose second order moment (and hence second order cumulant) diverge [18, 19]. Stable random variables (sometimes also called "Lévy" in the physics literature) correspond to variables that have a domain of attraction and being stable under addition [18, 19, 20]. They have been introduced in the 1930s by Paul Lévy and correspond to a generalisation of the Gaussian law, the main parameter being the index α bounded between 0 and 2. The case $\alpha = 2$ corresponds to the Gaussian law. Log-stable models for turbulent intermittency [21, 22] correspond to a nonanalytic scaling moment function (see also [8]). In this case, we have instead of Eq.(3):

$$\Phi_\ell(q) = A_\ell q + B_\ell q^\alpha \tag{4}$$

where $0 \leq \alpha \leq 2$ is the index of the stable process. When $\alpha = 2$ the generator is a Gaussian process and there are only two cumulants in the development of Eq. (3).

Since $\Phi_\ell(q)$ is a smooth function, it can be numerically differentiated. The first term in the cumulant generating function is linear in q; in order to check the possible nonanalycity of $\Phi_\ell(q)$ and to directly estimate the parameter α, which will be the order of the first nonlinear term in the cumulant generating function, we introduce the following function:

$$\Psi_\ell(q) = q\Phi'_\ell(q) - \Phi_\ell(q) \tag{5}$$

In case of analycity, when the development (3) is infinite, $\Psi_\ell(q)$ is proportional to q^2 for small values of q. When the development (3) is finite, the process is lognormal and $\Psi_\ell(q)$ is proportional to q^2 for a whole range of values of q. For the stable case with $\alpha < 2$, $\Psi_\ell(q)$ is proportional to q^α:

$$\Psi_\ell(q) = (\alpha - 1)B_\ell q^\alpha \tag{6}$$

This is thus an efficient way to estimate the first nonlinear power in the development of $\Phi_\ell(q)$. This is applied in the next section to experimental data.

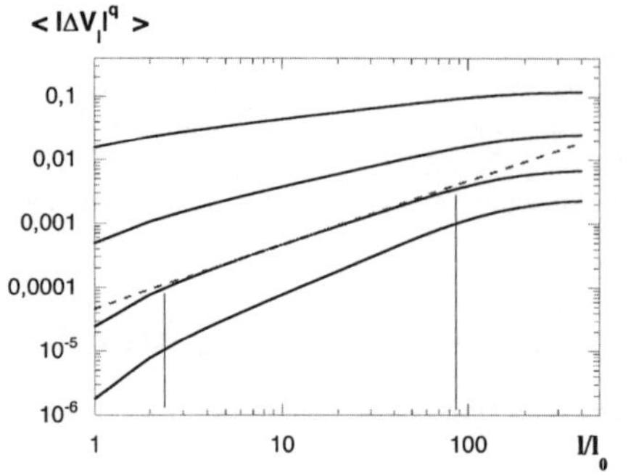

Figure 1: The scaling of the structure functions $\langle|\Delta V_\ell|^q\rangle$ vs. ℓ in log-log plot, for $q = 1, 2, 3$, and 4 (from top to below). The scaling range corresponds here to the scales for which $\langle|\Delta V_\ell|^3\rangle$ is proportional to ℓ (dashed line). Indicated by vertical bars, it corresponds to a scaling range of nearly two decades

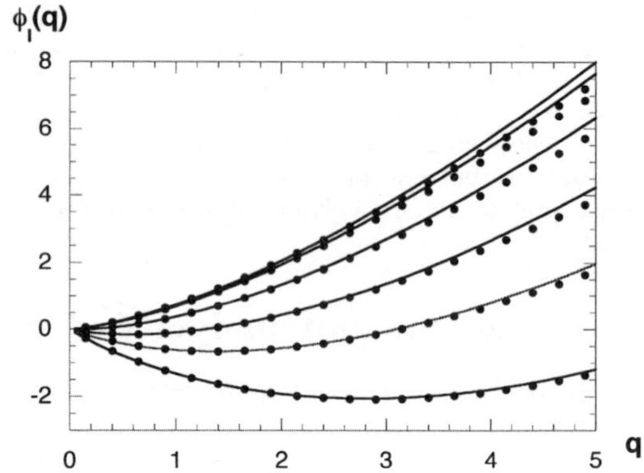

Figure 2: The second characteristic function $\Phi_\ell(q)$ for $\ell/\ell_0 = 1, 4, 16, 64, 256$, and 1024 (from bottom to top): full dots indicate experimental values (plotted only every 5 data point), compared to the fits (continuous lines) obtained through Eq.(4) with the values of A_ℓ and B_ℓ given through Eq.(7). The experimental curves are convex and the fits are better for small values of ℓ.

3 Experimental analysis

3.1 Nonanalycity properties

We use here atmospheric velocity measurements recorded 25 m above ground over a pine forest in south-west France, sampling at 10 Hz. Using an average velocity of 2.7 m/s, we use the usual hypothesis of Taylor to transform time fluctuations into space fluctuations. The elementary length is therefore $\ell_0 = 27cm$. We analysed 22 profiles of duration 55 minutes each, all recorded in near-neutral stability conditions (for more details on the experimental conditions, see [23]). There is a scaling range of nearly two decades, as indicated by the third order structure function (see Fig. 1). Figure 2 represents the second characteristic function $\Phi_\ell(q)$ for several values of ℓ, showing its convexity.

We represent in Fig. 3 $\log \Psi_\ell(q)$ vs. $\log q$ for $0.06 \leq q \leq 6$, for different values of ℓ, from $\ell = \ell_0$ to $\ell = 1024\ell_0$. The straight lines obtained for a wide range of values of q show the range of validity of Eq.(6), and provide the value $\alpha = 1.5$. There is a departure from the straight lines only for vey small values of q, coming from a problem of digits to store the data, and for large orders of q due to sampling limitations. For a wide range of values of q between 0.1 and about 5, $\Psi_\ell(q)$ is proportional to $q^{1.5}$. This result is in agreement with the log-stable model, with $\alpha = 1.5 \pm 0.05$, where the error bars indicate the dispersion found in the different slopes. This value confirms previous results obtained using other approaches [24, 8].

We now turn to the scale dependence of A_ℓ and B_ℓ. These parameters may be estimated for each ℓ, using the first-order moment:

$$\begin{aligned} A_\ell &= \tfrac{1}{\alpha-1}\left(\alpha\Phi_\ell(1) - \Phi'_\ell(1)\right) \\ B_\ell &= \tfrac{1}{\alpha-1}\left(\Phi'_\ell(1) - \Phi_\ell(1)\right) \end{aligned} \tag{7}$$

It can be easily shown that in case of strict scale invariance (Eq. (1)), A_ℓ and B_ℓ are linear functions of

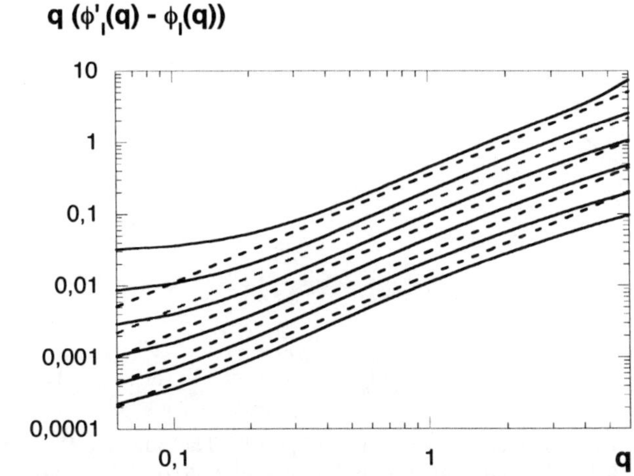

Figure 3: $\Psi_\ell(q)$ vs. q for $\ell/\ell_0 = 1$, 4, 16, 64, 256, and 1024 (from top to bottom). Experimental values are given by continuous lines whereas dotted lines correspond to straight lines with slopes of 1.5. For the clarity of the figure, the empirical curves have been vertically shifted. The wide range of values of q for which $\Psi_\ell(q)$ is proportinal to $q^{1.5}$ confirms the nonanalytic framework applied here.

$\log \ell$:

$$\begin{aligned} A_\ell = &\quad a_0 + a_1 \log(\ell/\ell_0) \\ B_\ell = &\quad b_0 - b_1 \log(\ell/\ell_0) \end{aligned} \tag{8}$$

where $\zeta(q) = a_1 q - b_1 q^\alpha$. Figures 4 and 5 show that this is respected for values of ℓ corresponding to the inertial range, as defined in Fig. 1. These figures show also that A_ℓ is an increasing and B_ℓ a decreasing function of $\log(\ell/\ell_0)$. The value of a_0 clearly depends on the normalization of the data, since dividing the velocity field by the constant value V_0 transforms a_0 into $a_0 - \log V_0$. Thus with the right normalization, a_0 can be put to 0. We obtain here $a_1 = 0.40$, $b_1 = 0.038$ and $b_0 = 0.88$. The fits with these values are given in Figs. 4 and 5. We have $\zeta(1) = a_1 - b_1 = 0.36$ and $\zeta(3) = 3a_1 - 3^{1.5} b_1 = 1$.

3.2 Cumulant relations and extended self similarity

When structure functions are not perfectly scaling, Extended Self Similarity (ESS) has become a classical tool to improve the determination of scaling exponents. The ESS methodology [11, 12, 13] can be translated into the following property for two different moments p and q of velocity increments:

$$\langle |\Delta V_\ell|^q \rangle = E_{p,q} \langle |\Delta V_\ell|^p \rangle^{\zeta(q)/\zeta(p)} \tag{9}$$

where $E_{p,q}$ is independent of ℓ. Relation (9) seems to be widely respected for various turbulent datasets, for a wider range of values of ℓ than strict scaling [11, 12, 13, 25]. In the present cumulant framework, ESS property can be shown to correspond to state that

$$\begin{aligned} A_\ell = &\quad a_0 + a_2 b(\ell) \\ B_\ell = &\quad b_0 - b(\ell) \end{aligned} \tag{10}$$

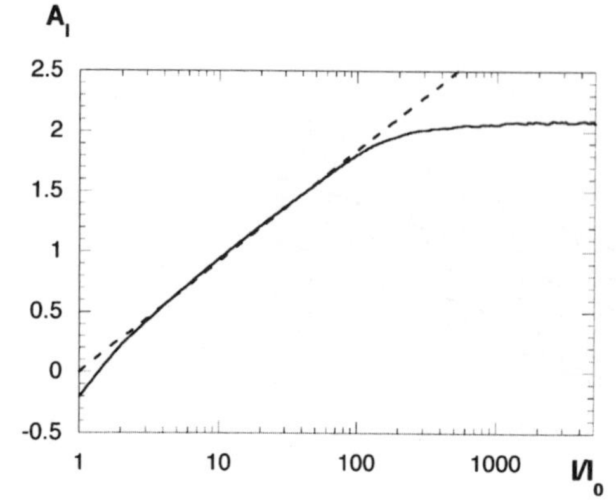

Figure 4: The function A_ℓ estimated for various values of ℓ according to Eq.(7). This function is increasing, and is proportional to $\log(\ell/\ell_0)$ in the inertial range: the dotted line is a fit of equation $A_\ell = 0.4\log(\ell/\ell_0)$. The constant $a_0 = A_{\ell_0}$ has been set to 0 through the right normalisation.

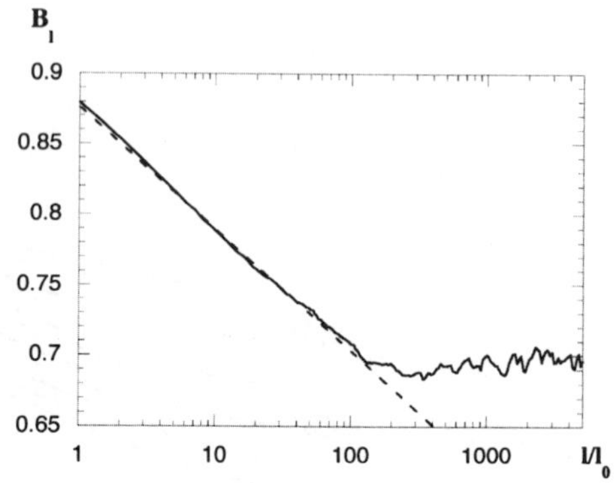

Figure 5: The function B_ℓ estimated for various values of ℓ according to Eq.(7). This function is generally decreasing, and is proportional to $\log(\ell/\ell_0)$ in the inertial range: the dotted line is a fit of equation $B_\ell = 0.88 - 0.0375\log(\ell/\ell_0)$.

where a_0, b_0 and $a_2 = a_1/b_1$ are constants, and $b(\ell)$ is an unknown function of ℓ with $b(\ell_0) = 0$. Equation (10) can be written by elimination of $b(\ell)$, giving $A_\ell = -a_2 B_\ell + a_2 b_0 + a_0$: thus ESS' hypothesis corresponds to state that A_ℓ is a linear function of B_ℓ. We test this in Fig. 6, which shows that this is quite well respected. This may give some hints to better understand the mechanism of ESS, and in particular, help to propose a more precise description of the function $b(\ell)$ which has not been given, up to now, a definite theoretical form.

We may note also that a generalized ESS framework has been also proposed [25], assuming the following form for the cumulant generating function:

$$\Phi_\ell(q) = A_\ell q + B_\ell f(q) \tag{11}$$

This is compatible with the results presented here, taking $f(q) = q^\alpha$. Let us notice finally that the prefactor in Eq.(1) is given here by $C_q = \exp(b_0 q^\alpha)$.

4 Conclusion

The modelling of intermittency is a major issue of turbulence. Many studies have been devoted to this issue. A classical approach is to consider the scaling properties of structure functions of the velocity field. The precise determination of the adequate intermittency model is often contaminated by the non-perfect scale invariance of the statistics. We have considered here another approach: instead of studying the scale dependence for each moment, we focus on the moment dependence for each scale. This way, at a given and fixed scale, the cumulant generating function can be precisely estimated, and its parameters experimentally determined. We have done this here focusing on the non-analytical properties of the cumulant generating function. Using atmospheric experimental data of the turbulent velocity, we have shown that the experimental nonanalycity of the cumulant generating function is compatible with log-stable models of turbulence. We have estimated two cumulant coefficients, and considered their scale

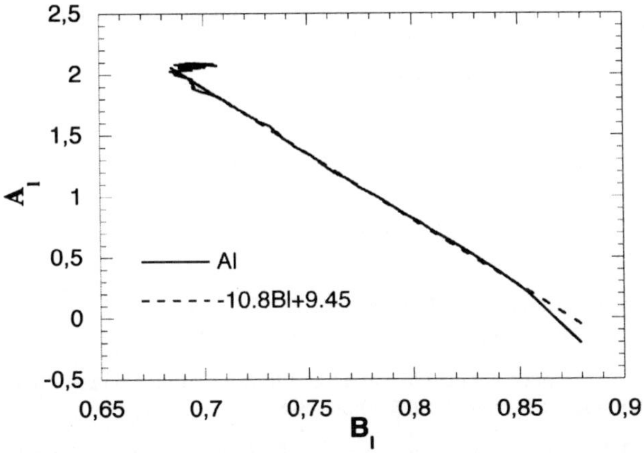

Figure 6: A_ℓ vs. B_ℓ. ESS hypothesis corresponds to state that this function is linear: this is verified for values of ℓ corresponding to the range of A_ℓ values between 0.2 and 2: as shown by Fig.4, this is a wider range than the inertial range

dependence. This also allowed us to revisit in this framework the ESS property. With this study we would like to shed light on the possible nonanalycity of the cumulant generating function for intermittent fields.

Acknowledgements

Y. Brunet is thanked for providing the atmospheric velocity database.

References

[1] Frisch U. Turbulence. Cambridge: Cambridge University Press; 1995.

[2] Kolmogorov AN. The local stucture of turbulence in incompressible viscous fluid for very large Reynolds' numbers. C R Acad Sci USSR 1941;30:301-305.

[3] Anselmet F, Gagne Y, Hopfinger EJ, Antonia RA. High-order velocity structure functions in turbulent shear flows. J Fluid Mech 1984;140:63-89.

[4] She ZS, Leveque E. Universal scaling laws in fully developed turbulence. Phys Rev Lett 1994;72:336-339.

[5] Chen S, Cao N. Inertial range scaling in turbulence. Phys Rev E 1995;52:5757-5759.

[6] Arneodo A, Baudet C, Belin F, Benzi R, et al. Structure functions in turbulence, in various flow configurations, at reynolds number between 30 and 5000, using extend self-similarity. Europhys Lett 1996;34:411-416.

[7] Boratav O. On recent intermittency models of turbulence. Phys Fluids 1997;9:1206-1208.

[8] Schertzer D, Lovejoy S, Schmitt F, Chigirinskaya Y, Marsan D. Multifractal cascade dynamics and turbulent intermittency. Fractals 1997;5:427-471.

[9] van de Water W, Herweijer JA. High-order structure functions of turbulence. J Fluid Mech 1999;387:3-37.

[10] Anselmet F, Antonia RA, Danaila L. Turbulent flows and intermittency in laboratory experiments. Plan Space Sci 2001;49:1177-1191.

[11] Benzi R, Ciliberto S, Tripiccione R, Baudet C, Massaioli F, Succi S. Extended self-similarity in turbulent flows. Phys Rev E 1993;48:29-32.

[12] Benzi R, Ciliberto S, Baudet C, Ruiz Chavarria G, Tripiccione R. Extended self-similarity in the dissipation range of fully developed turbulence. Eur Lett 1993;24:275-279.

[13] Benzi R, Ciliberto S, Baudet C, Ruiz Chavarria G. On the scaling of three-dimensional homogeneous and isotropic turbulence. Physica D 1995;80:385-398.

[14] Delour J, Muzy JF, Arneodo A. Intermittency of 1D velocity spatial profiles in turbulence: a magnitude cumulant analysis. Eur Phys J B 2001;23:243-248.

[15] Eggers HC, Dziekan T, Greiner M. Translationally invariant cumulants in energy cascade models of turbulence. Phys Lett A 2001;281:249-255.

[16] Chevillard L, Roux SG, Leveque E, Mordant N, Pinton JF, Arneodo A. Intermittency of velocity time increments in turbulence. Phys Rev Lett 2005;95:064501.

[17] Gardiner CW. Handbook of stochastic methods. Berlin: Springer, third edition; 2004.

[18] Feller W. An introduction to probability theory and its applications. New York: Wiley; 1971.

[19] Taqqu MS, Samorodnisky G. Stable Non-Gaussian Random Processes. New York: Chapman - Hall; 1994.

[20] Janicki A, Weron A. Simulation and chaotic behavior of α-stable stochastic processes. New York: Marcel Dekker; 1994.

[21] Schertzer D, Lovejoy S. Physical modeling and analysis of rain and clouds by anisotropic scaling multiplicative processes. J Geophys Res 1987;92:9693-9714.

[22] Kida S. Log-stable distribution and intermittency of turbulence. J Phys Soc Japan 1991;60:5-8.

[23] Collineau S, Brunet Y. Detection of Turbulent Coherent Motions in a Forest Canopy. Part II: Time-Scales and Conditional Average. Bound Layer Meteor 1993;66:49.

[24] Schmitt F, Schertzer D, Lovejoy S, Brunet Y. Estimation of universal multifractal indices for atmospheric turbulent velocity fields. Fractals 1997;1:568-578.

[25] Benzi R, Biferale L, Ciliberto S, Struglia MV, Tripiccione R. Scaling property of turbulent flows. Phys Rev E 1996;53:3025-3027.

Exact Solutions of a Second Grade Fluid in a Porous Medium

S. Islam*, C. Y. Zhou[1]
Department of Mechanical Engineering, Harbin Institute of Technology, Shenzhen , 518055, CHINA

Abstract
In this paper, exact solutions are obtained for the equation of class of an unsteady, plane, second grade fluid in a porous medium. This class consists of flows for which the vorticity distribution is proportional to the stream function perturbed by a uniform stream. Expressions for stream lines, velocity components and pressure distributions are derived. The present solutions are compared with the corresponding results of a viscous fluid.

PACS: 47.50.+d; 47.55.Mh
Key words: Second grade fluid, Inverse method, Exact solutions, Porous Medium.

1. Introduction

The fundamental governing equations for fluid mechanics are the Navier-Stokes equations. The inherently non linear set of partial differential equations has no general solution, and only a small number of exact solutions have been found because the non-linear inertial terms do not disappear automatically. Exact solutions are very important not only because they are solutions of some fundamental flows but also because they serve as accuracy checks for experimental, numerical and asymptotic methods. Some exact solutions may be obtained for the Navier Stokes equations in the case where the partial differential equations can be reduced to the ordinary differential equations. In addition, to flows in which the vorticity distribution is chosen in a such way that the governing equations written in terms of the stream function become linear.

The usual technique to assume a particular form of the stream function and apply the so called inverse method to obtain the exact solutions. This technique consists of making certain hypothesis a priori on the velocity field and pressure without making any on the boundaries of the domain occupied by the fluid. These hypothesis are often made on the velocity field and very rarely the pressure. Taking the vorticity to be proportional to the stream function, in 1923 Taylor [1] obtained the solution of the problem of a double infinite array of vortices decaying exponentially with time. After Taylor, these techniques have been discussed by many authors in detail such as Kovaznay [2], Lin and Tobak [3], Wang [4,5], Hui [6], R-Berker [7], Nemenyi [8], Agarwal [9], Hamdan [10,11], Dorrepaal [12], Irmay and Zuzovsky [13], Zuzovsky [14] and others.

* *Corresponding Author:* Tel.: +86-13246656173, fax:+86-755-26033774
Saeed_nihar@yahoo.com.au, saeed@hitsz.edu.cn (S. Islam)

The non linear fluids are increasing being considered to be more important and appropriate in technological applications in comparison with the Newtonian fluids. A large class of real fluids does not exhibit the linear relationship between stress and the rate of strain. Because of the non linear dependence, the analysis of the behavior of the fluid motion on the non Newtonian fluids tends to be much more complicated and subtle in comparison with that of the Newtonian fluids. For the second grade fluids the non-linearities were observed by Rivilin and Erickson [15] in 1955, Colman and Noll [16] in 1960 and they showed that the non-linearties not occur only in the inertial part but also in the viscous part of these equations. Due to these non-linearities the number of such exact solutions are very rare for non Newtonian fluids. By assuming the vorticity to be proportional to the stream function, Rajagopal [17, 18], Rajagopal and Gupta [19], Kaloni and Huchilt [20], Siddiqui and Kaloni [21], Siddiqui [22], Benharbit and Siddiqui [23], Markovitz and Coleman [24], Oku-Ukpong and Chandna [25,26], Truesdell and Noll [27], Ting [28] and others have investigated several flows and obtained several classes of exact solutions.

An understanding of the dynamics of fluids in porous media has practical interest in such disparate fields as petroleum engineering and ground water hydrology; with applications ranging from hydrocarbon migration in reservoirs via paked bed chemical reactors to agriculture drainage and irrigation. In this paper, we study the two dimensional flow of a incompressible second grade fluid in a porous medium by assuming that the vorticity distribution is proportional to the stream function perturbed by a uniform stream that is $\nabla^2\psi = \Lambda(\psi - Uy)$, where Λ is a real constant. The plan of the paper is as follows: In section 2 the basic equations will be developed. Section 3 deals with the problem formulation. In section 4 we discuss the steady flow solutions. Sections 5 deals with the two types of unsteady flow solutions. In section 6, we conclude our results.

2. Basic Equations

The basic equations governing the motion of a second grade in a porous medium are

$$div\, \mathbf{v} = 0 \tag{1}$$

$$\rho\left[\frac{\partial \mathbf{v}}{\partial t} + (\mathbf{v}\cdot\nabla)\mathbf{v}\right] = -\nabla p - \frac{\mu}{k^*}\mathbf{v} + div\mathbf{T} + \rho\mathbf{f} \tag{2}$$

and

$$\mathbf{T} = -P\mathbf{I} + \mu\mathbf{A}_1 + \alpha_1\mathbf{A}_2 + \alpha_2\mathbf{A}_1^{\ 2} \tag{3}$$

where $\mathbf{A}_1 = (grad\,\mathbf{v}) + (grad\,\mathbf{v})^T$ and $\mathbf{A}_2 = \mathbf{A}_{1t} + (grad\,\mathbf{A}_1)\mathbf{v} + (grad\,\mathbf{v})^T\mathbf{A}_1 + \mathbf{A}_1(grad\,\mathbf{v})$, and $\mathbf{v}$ is the velocity vector, p the pressure, $\mathbf{f}$ the body force per unit mass, ρ the constant density, k^* the permeability of porous medium, μ the dynamic viscosity, α_1 and α_2 are the constant stress moduli, $\mathbf{A}_1$ and $\mathbf{A}_2$ are the Rivilin-Erickson tensors.

Dunn and Fosdic found that if an incompressible fluid of a second grade is to have motions which are compatible with thermodynamics in the since that Clausius-Durham inequality is met and the condition that the Helmoltz free energy be a minimum when the fluid is at rest, then the following must satisfy the material constants that are

$$\mu > 0, \qquad \alpha_1 \geq 0, \qquad \alpha_1 + \alpha_2 = 0$$

On the other side, experimental results of tested fluids of a second grade found that $\alpha_1 < 0$ and $\alpha_1 + \alpha_2 \neq 0$ which contradicts with the above conditions and imply that such fluids are unstable. The controversy is discussed in detail in [29] and [30]. However, in our paper we shall assume $\mu > 0$ and $\alpha_1 > 0$.

Let us assume $p = p(x, y, t)$ and the unsteady plane flow where $\mathbf{v}$ is represented by

$$\mathbf{v} = [u(x,y,t),\ v(x,y,t),\ 0] \tag{4}$$

We define the generalized pressure h and vorticity ω functions as

$$h = \frac{\rho}{2}|\mathbf{v}|^2 + p - \left[\alpha_1\left(u\nabla^2 u + v\nabla^2 v\right) + \frac{1}{4}(3\alpha_1 + 2\alpha_2)|A_1|^2\right] \tag{5}$$

$$\omega = \frac{\partial v}{\partial x} - \frac{\partial u}{\partial y} \tag{6}$$

and

$$|A_1|^2 = 4\left(\frac{\partial u}{\partial x}\right)^2 + 4\left(\frac{\partial v}{\partial y}\right)^2 + 2\left(\frac{\partial u}{\partial y} + \frac{\partial v}{\partial x}\right)^2 \tag{7}$$

By using equations (3) to (7) in (1) and (2), and assuming that the body force is null, the equations of motion become

$$\frac{\partial u}{\partial x} + \frac{\partial v}{\partial y} = 0 \tag{8}$$

The x-and y-components of momentum equation are

$$\frac{\partial h}{\partial x} + \rho\left[\frac{\partial u}{\partial t} - v\omega\right] = \mu\nabla^2 u - \frac{\mu}{k^*}u + \alpha_1\nabla^2\frac{\partial u}{\partial t} - \alpha_1 v\nabla^2\omega \tag{9}$$

$$\frac{\partial h}{\partial y} + \rho\left[\frac{\partial v}{\partial t} + u\omega\right] = \mu\nabla^2 v - \frac{\mu}{k^*}v + \alpha_1\nabla^2\frac{\partial v}{\partial t} + \alpha_1 u\nabla^2\omega \tag{10}$$

By introducing the stream function $\psi(x,y,t)$ such that

$$u = \frac{\partial\psi}{\partial y}, \qquad v = -\frac{\partial\psi}{\partial x} \tag{11}$$

Using of (11), the continuity equation is identically satisfied and equations (9) and (10) take the forms

$$\frac{\partial h}{\partial x} + \left[\frac{\partial^2\psi}{\partial t\partial y} - \frac{\partial\psi}{\partial x}(\nabla^2\psi)\right] = \mu\nabla^2\left(\frac{\partial\psi}{\partial y}\right) - \frac{\mu}{k^*}\left(\frac{\partial\psi}{\partial y}\right) + \alpha_1\nabla^2\frac{\partial^2\psi}{\partial t\partial y} - \alpha_1\frac{\partial\psi}{\partial x}\nabla^4\psi \tag{12}$$

$$\frac{\partial h}{\partial y} + \left[-\frac{\partial^2\psi}{\partial t\partial x} - \frac{\partial\psi}{\partial y}(\nabla^2\psi)\right] = -\mu\nabla^2\left(\frac{\partial\psi}{\partial x}\right) + \frac{\mu}{k^*}\left(\frac{\partial\psi}{\partial x}\right) - \alpha_1\nabla^2\frac{\partial^2\psi}{\partial t\partial x} - \alpha_1\frac{\partial\psi}{\partial y}\nabla^4\psi \tag{13}$$

By eliminating the pressure term from equations (12) and (13) by cross differentiation, the above system reduces to a single partial differential equation.

$$\rho\left[\frac{\partial}{\partial t}(\nabla^2\psi) - \frac{\partial(\psi,\nabla^2\psi)}{\partial(x,y)}\right] = \mu\nabla^4\psi - \frac{\mu}{k^*}\nabla^2\psi + \alpha_1\frac{\partial}{\partial t}\left[\nabla^4\psi\right] - \alpha_1\frac{\partial(\psi,\nabla^4\psi)}{\partial(x,y)} \tag{14}$$

Equation (14) represents the planar motion of a second grade fluid in a porous medium.

3. Problem Formulation

Equation (14) is a non-linear differential equation and to linearize we assume that

$$\nabla^2\psi = \Lambda(\psi - Uy) \tag{15}$$

where Λ and U are real constants and $\Lambda \neq 0$. In view of equation (15), equation (14) reduces to the form

$$(\rho - \alpha_1\Lambda)\frac{\partial\psi}{\partial t} + U(\rho - \alpha_1\Lambda)\frac{\partial\psi}{\partial x} = \mu\Lambda(\psi - Uy) - \frac{\mu}{k^*}(\psi - Uy) \tag{16}$$

On setting $\Psi = \psi - Uy$, equations (15) and (16) take the forms

$$\nabla^2\Psi = \Lambda\Psi \tag{17}$$

$$(\rho - \alpha_1\Lambda)\frac{\partial\Psi}{\partial t} + U(\rho - \alpha_1\Lambda)\frac{\partial\Psi}{\partial x} = \frac{\mu(k^*\Lambda - 1)}{k^*}\Psi \tag{18}$$

By setting $U = 0$, $\alpha_1 = 0$ and $k^* \to \infty$ we obtain the Taylor's case [1]. If $\alpha_1 = 0$ and $k^* \to \infty$, we get the Hui case [6]. If $k^* \to \infty$ we recover Benharbit and Siddiqui case [23].

We also observed that for creeping flow we have

$$\frac{\mu(k^*\Lambda - 1)}{k^*}\Psi = 0 \tag{19}$$

Since $\frac{\mu(k^*\Lambda - 1)}{k^*} \neq 0$. We must have $\Psi = 0$, giving the trivial solution $\psi = Uy$ which satisfies identically equation (17).

4. Steady Flow Solutions

For steady flow $\partial\Psi / \partial t = 0$, equation (18) reduces to the form

$$U(\rho - \alpha_1\Lambda)\frac{\partial\Psi}{\partial x} = \frac{\mu(k^*\Lambda - 1)}{k^*}\Psi \tag{20}$$

$$\nabla^2\Psi = \Lambda\Psi \tag{21}$$

If $U = 0$, equation (20) has only the trivial solution, $\Psi = 0$ or $\psi = 0$. If $\rho - \alpha_1\Lambda = 0$, it follows that $\Psi = 0$ or $\psi = Uy$ and equation (21) is identically satisfied.

Let us suppose that $\rho - \alpha_1\Lambda \neq 0$, then (20) becomes

$$\frac{\partial\Psi}{\partial x} = \frac{\mu(k^*\Lambda - 1)}{k^*U(\rho - \alpha_1\Lambda)}\Psi \tag{22}$$

$$\nabla^2\Psi = \Lambda\Psi \tag{23}$$

Solving (22) by separation of variable, we get

$$\Psi = F(y)e^{\alpha x} \tag{24}$$

where $\alpha = \frac{\mu(k^*\Lambda - 1)}{k^*U(\rho - \alpha_1\Lambda)}$. To find $F(y)$, we substitute (24) into (23), and get

$$F''(y) + (\alpha^2 - \Lambda)F(y) = 0 \tag{25}$$

Below the different cases:

a. If $\alpha^2 - \Lambda = \delta^2 > 0$, then the solution for $F(y)$ is

$$F(y) = A\cos(\delta y + B) \tag{26}$$

and the solution for the stream function and velocity components are

$$\psi(x, y) = Uy + Ae^{\alpha x}\cos(\delta y + B) \tag{27}$$

$$u = U - A\delta e^{\alpha x}\sin(\delta y + B), \quad v = -\alpha e^{\alpha x}\cos(\delta y + B) \tag{28}$$

where A and B are arbitrary constants. Solution (27) for the half space $x > 0$ represents a uniform flow with a perturbation part which is periodic in y and decays and grows exponentially as x increases, respectively, when $\alpha > 0$ and $\alpha < 0$. The solution can also be used to describe a flow in $x < 0$.

The pressure distribution for (27) is

$$p = p_\circ - \frac{\rho}{2}U^2 - \frac{\mu}{k^*}Ux - \left[\left\{\frac{\rho}{2}\delta^2 - \alpha_1(\alpha^4 + 5\alpha^2\delta^2) - \frac{\alpha_2}{2}\left(\alpha^4 + 6\alpha^2\delta^2 + \delta^4\right)\right\} - \left\{\left(\alpha_1 + \frac{\alpha_2}{2}\right)(\alpha^2 - \delta^2)^2 \cos 2(\delta y + B)\right\}\right] A^2 e^{2\alpha x} \tag{29}$$

where $p_\circ$ is a constant and $\delta^2 = \left\{\left[\dfrac{\mu(k^*\Lambda - 1)}{k^*U(\rho - \alpha_1\Lambda)}\right]^2 - \Lambda\right\}$.

b. If $\alpha^2 - \Lambda = 0$, then $F(y)$ is

$$F(y) = Cy + D \tag{30}$$

and the solution for the stream function is

$$\psi(x, y) = Uy + e^{\alpha x}(Cy + D) \tag{31}$$

$$u = U + Ce^{\alpha x}, \quad v = -\alpha e^{\alpha x}(Cy + D) \tag{32}$$

where C and D are arbitrary constants. Solution (31) represents a uniform flow with a perturbation part which is neither periodic nor exponential and grows and decreases as x increases, if $\alpha > 0$ and $\alpha < 0$, respectively. The similar description can be given for a flow in $x < 0$.

The pressure distribution to solution (31) is then given by

$$p = p_\circ - \frac{\rho}{2}U^2 - \frac{\mu}{k^*}Ux - \left\{\frac{\rho}{2} - (7\alpha_1 + 4\alpha_2)\alpha^2\right\}C^2 e^{2\alpha x} + \left\{(2\alpha_1 + \alpha_2)\alpha^4\right\} e^{2\alpha x}(Cy + D)^2 \tag{33}$$

where $p_\circ$ is a constant.

c. If $\alpha^2 - \Lambda = -\delta^2 < 0$, then the solution for $F(y)$ is

$$F(y) = Ee^{\delta y} + Fe^{-\delta y} \tag{34}$$

and the solution for the stream function $\psi(x, y)$ and the corresponding velocity components are

$$\psi(x, y) = Uy + e^{\alpha x}\left(Ee^{\delta y} + Fe^{-\delta y}\right) \tag{35}$$

$$u = U + \delta e^{\alpha x}\left(Ee^{\delta y} - Fe^{-\delta y}\right), \quad v = -\alpha e^{\alpha x}\left(Ee^{\delta y} + Fe^{-\delta y}\right) \tag{36}$$

where E and F are arbitrary constants. The solution (35) for the stream function represents a uniform stream with a perturbation part which is not periodic in y and decays and grows exponentially as x increases if $\alpha > 0$ and $\alpha < 0$, respectively. The solution also holds for the region $x < 0$ and the flow is exponential in both cases.

The pressure distribution to (35) is easily given by

$$p = p_\circ - \frac{\rho}{2}U^2 - \frac{\mu}{k^*}Ux + \left[\left\{2\rho\delta^2 - 4\alpha_1(\alpha^4 - 5\alpha^2\delta^2) - 2\alpha_2(\alpha^4 - 4\alpha^2\delta^2 + \delta^4)\right\} EF + \left\{(2\alpha_1 + \alpha_2)(\alpha^2 + \delta^2)^2\right\}\left\{E^2 e^{2\delta y} + F^2 e^{-2\delta y}\right\}\right] e^{2\alpha x} \tag{37}$$

where $p_\circ$ is an arbitrary constant and $\delta^2 = \left\{\Lambda - \left[\dfrac{\mu(k^*K-1)}{k^*U(\rho-\alpha_1 K)}\right]^2\right\}$.

5. Unsteady Flow Solutions (Type A)

By rewriting equation (18) in the form

$$\frac{\partial\Psi}{\partial t} + U\frac{\partial\Psi}{\partial x} = \beta\Psi \tag{38}$$

where $\beta = \dfrac{\mu(k^*\Lambda-1)}{k^*(\rho-\alpha_1\Lambda)}$. We consider the plane wave solution to equation (38) of the form $\Psi = G(X,y)e^{mt}$ with $X = x - Ut$. We substitute $\Psi = G(X,y)e^{mt}$ into equation (38) and find that $m = \beta$. Hence

$$\Psi = G(X,y)e^{\beta t} \tag{39}$$

To find $G(X,y)$ we substitute equation (39) into (17) and obtain

$$G_{XX} + G_{yy} = \Lambda G \tag{40}$$

Plane wave solutions to the Helmholtz equation (40) exist in the form

$$G(X,y) = g(\xi) \tag{41}$$

with $\xi = X\cos\theta + y\sin\theta$, and $-\pi \le \theta < \pi$, which when employed in (40) yields

$$g''(\xi) - \Lambda g(\xi) = 0 \tag{42}$$

Below the different cases:

a. If $\Lambda = -\eta^2 < 0$, then the solution for $h(\zeta)$ is

$$g(\xi) = A(\theta)\cos\eta(\xi + B(\theta)) \tag{43}$$

and the solution for the stream function and velocity components are

$$\psi(x,y,t) = Uy + A(\theta)e^{\frac{-\mu(k^*\eta^2+1)}{k^*(\rho+\alpha_1\eta^2)}t}\cos\eta((x-Ut)\cos\theta + y\sin\theta + B(\theta)) \tag{44}$$

$$u = U - \eta A(\theta)e^{\frac{-\mu(k^*\eta^2+1)}{k^*(\rho+\alpha_1\eta^2)}t}\sin\theta\sin\eta((x-Ut)\cos\theta + y\sin\theta + B(\theta)) \tag{45a}$$

$$v = \eta A(\theta)e^{\frac{-\mu(k^*\eta^2+1)}{k^*(\rho+\alpha_1\eta^2)}t}\cos\theta\sin\eta((x-Ut)\cos\theta + y\sin\theta + B(\theta)) \tag{45b}$$

where $A(\theta)$ and $B(\theta)$ are constants depending on the parameter θ and $-\pi \le \theta < \pi$. Solution (44) represents a uniform stream with a perturbation part and periodic in x and y, and decays exponentially in time at a rate equal to $\dfrac{\mu(k^*\eta^2+1)}{k^*(\rho+\alpha_1\eta^2)}$.

The pressure distribution for the stream function (44) is obtained from the

$$\begin{aligned} p = f(t) - \frac{\rho}{2}U^2 - \frac{\mu}{k^*}Ux + \frac{1}{4}\eta^2\big\{\left[2(\alpha_1+\alpha_2)\eta^2 - 2\rho\right] \\ + \left[\rho + 2(2\alpha_1+\alpha_2)\eta^2\right]\cos 2\eta((x-Ut)\cos\theta + y\sin\theta + B)\big\}A^2 e^{-2\frac{\mu(k^*\eta^2+1)}{k^*(\rho+\alpha_1\eta^2)}} \end{aligned} \tag{46}$$

where $f(t)$ is an arbitrary constant.

b. If $\Lambda = \eta^2 > 0$, then our $h(\zeta)$ is

$$g(\zeta) = Ce^{\eta\zeta} + De^{-\eta\zeta} \tag{47}$$

and the solution for the stream function $\psi(x, y, t)$ and velocity components take the form

$$\psi(x, y, t) = Uy + e^{\frac{\mu(k^*\eta^2-1)}{k^*(\rho-\alpha_1\eta^2)}t}\left(Ce^{\eta((x-Ut)\cos\theta+y\sin\theta)} + De^{-\eta((x-Ut)\cos\theta+y\sin\theta)}\right) \tag{48}$$

$$u = U + \eta\sin\theta e^{\frac{\mu(k^*\eta^2-1)}{k^*(\rho-\alpha_1\eta^2)}t}\left(Ce^{\eta((x-Ut)\cos\theta+y\sin\theta)} - De^{-\eta((x-Ut)\cos\theta+y\sin\theta)}\right) \tag{49a}$$

$$u = -\eta\cos\theta e^{\frac{\mu(k^*\eta^2-1)}{k^*(\rho-\alpha_1\eta^2)}t}\left(Ce^{\eta((x-Ut)\cos\theta+y\sin\theta)} - De^{-\eta((x-Ut)\cos\theta+y\sin\theta)}\right) \tag{49b}$$

where C and D are constants depending on the parameter θ. Plane wave solution (48) is exponential in x and y and t. It decays and grows exponentially in time if $\frac{\mu(k^*\eta^2-1)}{k^*(\rho-\alpha_1\eta^2)} < 0$ and $\frac{\mu(k^*\eta^2-1)}{k^*(\rho-\alpha_1\eta^2)} > 0$, and is meaningful for finite values of time.

The corresponding pressure distribution for the stream function (48) takes the form

$$\begin{aligned} p = f(t) - \frac{\rho}{2}U^2 - \frac{\mu}{k^*}Ux + e^{2\frac{\mu(k^*\eta^2-1)}{k^*(\rho-\alpha_1\eta^2)}t}\eta^2\Big[\{\rho - (\alpha_1 - 2\alpha_2)\eta^2\}CD \\ + (2\alpha_1 + \alpha_2)\eta^2\{C^2e^{2\eta((x-Ut)\cos\theta+y\sin\theta)} + D^2e^{-2\eta((x-Ut)\cos\theta+y\sin\theta)}\}\Big] \end{aligned} \tag{50}$$

where $f(t)$ is an arbitrary constant.

c. If $\Lambda = 0$, then our $h(\zeta)$ is

$$h(\zeta) = E\zeta + F \tag{51}$$

and the solutions for the stream function and velocity components are

$$\psi(x, y, t) = Uy + e^{-\frac{\mu}{k^*\rho}t}(E((x - Ut)\cos\theta + y\sin\theta) + F) \tag{52}$$

$$u = U + E\sin\theta e^{-\frac{\mu}{k^*\rho}t}, \quad v = -E\cos\theta e^{-\frac{\mu}{k^*\rho}t} \tag{53}$$

where E and F are constants depending on the parameter θ.

The pressure distribution for (52) becomes

$$\begin{aligned} p = f(t) - \frac{\rho}{2}U^2 - \frac{\mu}{k^*}Ux - \frac{\rho}{2}E^2e^{-2\frac{\upsilon}{k^*}t} - \rho Ee^{-\frac{\upsilon}{k^*}t}\left(\frac{\upsilon}{k^*}y\cos\theta - \sin\theta(\frac{\upsilon}{k^*}x - U)\right) \\ + \frac{\mu}{k^*}Ee^{-\frac{\upsilon}{k^*}t}(y\cos\theta - x\sin\theta) \end{aligned} \tag{54}$$

where $\upsilon = \mu/\rho$ and $f(t)$ is an arbitrary constant.

Type B

In this section, we give another class of solution to equations (17) and (18) in the form

$$\Psi = H(X, y)e^{mx} \tag{55}$$

where $X = x - Ut$. We need to find m and $H(X, y)$. For this, we use (55) in (18) and we get $m = \gamma$, where $\gamma = \mu(k^*\Lambda - 1)/k^*U(\rho - \alpha_1\Lambda)$, so

$$\Psi = H(X,y)e^{\gamma x} \tag{56}$$

Using (56) in (17), we get

$$\frac{\partial^2 H}{\partial X^2} + \frac{\partial^2 H}{\partial y^2} + 2\gamma\frac{\partial H}{\partial X} + \left(\gamma^2 - \Lambda\right)H = 0 \tag{57}$$

If $X = x$, in (57) the steady state solution is a special case.
There exist plane wave solutions to (57) in the form

$$H(X,y) = h(\zeta) \tag{58}$$

This is substituted into (57) to obtain

$$h''(\zeta) + 2\gamma\cos\theta\, h'(\zeta) + \left(\gamma^2 - \Lambda\right)h(\zeta) = 0 \tag{59}$$

where $\zeta = X\cos\theta + y\sin\theta$. The auxiliary equation of (50) has the roots $m_{1,2} = -\gamma\cos\theta \pm \sqrt{\Lambda - \gamma^2\sin^2\theta}$, depending on the sign of $\Lambda - \gamma^2\sin^2\theta$.

Below the different cases:

a. If $\Lambda = -\lambda^2 < 0$, then the solution for $h(\zeta)$ is

$$h(\zeta) = A(\theta)e^{-\gamma\cos\theta\,\zeta}\cos\left[\zeta\left(\sqrt{\lambda^2 + \gamma^2\sin^2\theta}\right) + B(\theta)\right] \tag{60}$$

and the corresponding stream function and velocity components are

$$\psi(x,y,t) = Uy + A(\theta)e^{-\frac{\mu(k^*\lambda^2+1)}{k^*U(\rho+\alpha_1\lambda^2)}(x-\cos\theta((x-Ut)\cos\theta+y\sin\theta))}$$
$$\times\cos\left[\sqrt{\lambda^2 + \left(\frac{\mu(k^*\lambda^2+1)}{k^*U(\rho+\alpha_1\lambda^2)}\right)^2\sin^2\theta}\left((x-Ut)\cos\theta + y\sin\theta + B(\theta)\right)\right] \tag{61}$$

$$u = U + A(\theta)\gamma\cos\theta\sin\theta\, e^{-\gamma(x-\cos\theta((x-Ut)\cos\theta+y\sin\theta))}\times\cos\left[\Pi\left((x-Ut)\cos\theta + y\sin\theta + B(\theta)\right)\right]$$
$$-\Pi\sin\theta A(\theta)e^{-\gamma(x-\cos\theta((x-Ut)\cos\theta+y\sin\theta))}\times\sin\left[\Pi\left((x-Ut)\cos\theta + y\sin\theta + B(\theta)\right)\right] \tag{62a}$$

$$v = \gamma\sin^2\theta\, A(\theta)e^{-\chi(x-\cos\theta((x-Ut)\cos\theta+y\sin\theta))}\times\cos\left[\Pi\left((x-Ut)\cos\theta + y\sin\theta + B(\theta)\right)\right]$$
$$+\Pi\cos\theta A(\theta)e^{-\chi(x-\cos\theta((x-Ut)\cos\theta+y\sin\theta))}\times\sin\left[\Pi\left((x-Ut)\cos\theta + y\sin\theta + B(\theta)\right)\right] \tag{62b}$$

where A and B are constants depending on the parameter $\theta, \chi = \mu(k^*\lambda^2+1)/k^*U(\rho+\alpha_1\lambda^2)$, and

$$\Pi = \sqrt{\lambda^2 + \left(\frac{\mu(k^*\lambda^2+1)}{k^*U(\rho+\alpha_1\lambda^2)}\right)^2\sin^2\theta}\,.$$

The pressure distribution to (61) is

$$p=f(t)-\frac{\rho}{2}U^2-\frac{\mu}{k^*}Ux+\frac{1}{4}\Big[\{\alpha_1(3(\chi^4\sin^4\theta+\Pi^4+6\chi^2\Pi^2\sin^2\theta)$$
$$-\lambda^2(\lambda^2+2(\chi^2\sin^2\theta+\Pi^2))+2\alpha_2(\chi^4\sin^4\theta+\Pi^4+6\chi^2\Pi^2\sin^2\theta) \tag{63}$$
$$-\rho(\lambda^2+(\chi^2\sin^2+\Pi^2))\}+\{(\alpha_1+2\alpha_2)\lambda^4\cos 2(\Pi((x-Ut)\cos\theta+y\sin\theta+B))\}\Big]$$
$$A^2e^{-2\chi(x-\cos\theta((x-Ut)\cos\theta+y\sin\theta)}$$

where $f(t)$ is an arbitrary constant.

b. If $\Lambda=\lambda^2>\gamma^2>0$, for $-\pi\le\theta<\pi$ and when $\Lambda=\lambda^2<\lambda^2$; where $|\theta|<\theta_\circ$, or $\pi-\theta_\circ<|\theta|\le\pi$, then the solution for $h(\zeta)$ is

$$h(\zeta)=e^{-\gamma\cos\theta\zeta}\left[Ce^{\zeta\left(\sqrt{\lambda^2-\gamma^2\sin^2\theta}\right)}+De^{-\zeta\left(\sqrt{\lambda^2-\gamma^2\sin^2\theta}\right)}\right] \tag{64}$$

and the stream function and velocity components are,

$$\psi(x,y,t)=Uy+e^{\frac{\mu(k^*\lambda^2-1)}{k^*U(\rho-\alpha_1\lambda^2)}(x-\cos\theta((x-Ut)\cos\theta+y\sin\theta))}$$
$$\times\left[Ce^{\left(\sqrt{\lambda^2-\left(\frac{\mu(k^*\lambda^2-1)}{k^*U(\rho-\alpha_1\lambda^2)}\right)^2\sin^2\theta}\right)((x-Ut)\cos\theta+y\sin\theta)}+De^{-\left(\sqrt{\lambda^2-\left(\frac{\mu(k^*\lambda^2-1)}{k^*U(\rho-\alpha_1\lambda^2)}\right)^2\sin^2\theta}\right)((x-Ut)\cos\theta+y\sin\theta)}\right] \tag{65}$$

$$u=U-\chi_1\sin\theta\cos\theta e^{\chi_1(x-\cos\theta((x-Ut)\cos\theta+y\sin\theta))}\times\left[Ce^{\Pi_1((x-Ut)\cos\theta+y\sin\theta)}+De^{-\Pi_1((x-Ut)\cos\theta+y\sin\theta)}\right]$$
$$+\Pi_1\sin\theta e^{\chi_1(x-\cos\theta((x-Ut)\cos\theta+y\sin\theta))}\times\left[Ce^{\Pi_1((x-Ut)\cos\theta+y\sin\theta)}-De^{-\Pi_1((x-Ut)\cos\theta+y\sin\theta)}\right] \tag{66a}$$

$$v=-\Big[\chi_1\sin^2\theta e^{\chi_1(x-\cos\theta((x-Ut)\cos\theta+y\sin\theta))}\times\left[Ce^{\Pi_1((x-Ut)\cos\theta+y\sin\theta)}+De^{-\Pi_1((x-Ut)\cos\theta+y\sin\theta)}\right]$$
$$+\Pi_1\cos\theta e^{\chi_1(x-\cos\theta((x-Ut)\cos\theta+y\sin\theta))}\times\left[Ce^{\Pi_1((x-Ut)\cos\theta+y\sin\theta)}-De^{-\Pi_1((x-Ut)\cos\theta+y\sin\theta)}\right]\Big] \tag{66b}$$

where C and D are constants depending on the parameter θ, $\chi_1=\mu(k^*\lambda^2-1)/k^*U(\rho-\alpha_1\lambda^2)$ and $\Pi_1=\left(\sqrt{\lambda^2-\left(\frac{\mu(k^*\lambda^2-1)}{k^*U(\rho-\alpha_1\lambda^2)}\right)^2\sin^2\theta}\right)$. The pressure distribution for (65) is

$$p=f(t)-\frac{\rho}{2}U^2-\frac{\mu}{k^*}Ux+\Big[\{(2\alpha_1+\alpha_2)\lambda^4\}\times\{C^2e^{2\Pi_1((x-Ut)\cos\theta+y\sin\theta)}+D^2e^{-2\Pi_1((x-Ut)\cos\theta+y\sin\theta)}\}$$
$$+\{\{\lambda^2-(\chi_1^2\lambda^2\sin^2\theta-\Pi_1^2)\}\rho+\alpha_1\{\lambda^2(2(\chi_1^2\sin^2\theta-\Pi_1^2)-\lambda^2)+3(\chi_1^4\sin^4\theta+\Pi_1^4 \tag{67}$$
$$-6\chi_1^2\Pi_1^2\sin^2\theta)\}+\alpha_2\{(\chi_1^4\sin^4\theta+\Pi_1^4-6\chi_1^2\Pi_1^2\sin^2\theta)\}\}CD\Big]e^{2\chi_1(x-\cos\theta((x-Ut)\cos\theta+y\sin\theta))}$$

where $f(t)$ is an arbitrary constant.

c. If $\Lambda=\gamma^2$, and $\theta=\pi/2$, then the solution for $h(\zeta)$ is

$$h(\zeta)=E\zeta+F \tag{68}$$

The solution for the stream and velocity components are

$$\psi(x,y,t)=Uy+e^{\gamma x}(Ey+F) \tag{69}$$

$$u = U + Ee^{\gamma x},\ v = -\gamma e^{\gamma x}(Ey + F) \tag{70}$$

where E and F are constants depending on the parameter θ and γ is defined above.

The pressure distribution for (69) takes the form

$$p = f(t) - \frac{\rho}{2}U^2 - \frac{\mu}{k^*}Ux - \left\{\frac{\rho}{2} - (7\alpha_1 + 4\alpha_2)\gamma^2\right\}E^2e^{2\gamma x} + \left\{(2\alpha_1 + \alpha_2)\gamma^4\right\}e^{2\gamma x}(Ey + F)^2 \tag{71}$$

where $\gamma = \dfrac{\mu(k^*\Lambda - 1)}{k^*U(\rho - \alpha_1\Lambda)}$ and $f(t)$ is an arbitrary constant.

d. If $\Lambda = \lambda^2 < \gamma^2,\ \theta_o \le |\theta| \le \pi - \theta_o$, where $\sin\theta_o = \dfrac{\lambda}{\gamma}$, then the solution to $h(\zeta)$ is

$$h(\zeta) = e^{-\gamma\cos\theta\zeta}[Me^{\zeta\left(\sqrt{\gamma^2\sin^2\theta - \Lambda}\right)} + Ne^{-\zeta\left(\sqrt{\gamma^2\sin^2\theta - \Lambda}\right)}] \tag{72}$$

The stream function and velocity components are found to be

$$\psi(x, y, t) = Uy + A(\theta)e^{-\frac{\mu(k^*\lambda^2 - 1)}{k^*U(\rho - \alpha_1\lambda^2)}(x - \cos\theta((x - Ut)\cos\theta + y\sin\theta))} \times \cos\left[\sqrt{\left(\frac{\mu(k^*\lambda^2 - 1)}{k^*U(\rho - \alpha_1\lambda^2)}\right)^2 \sin^2\theta - \lambda^2}\left((x - Ut)\cos\theta + y\sin\theta + B(\theta)\right)\right] \tag{73}$$

$$u = U - A(\theta)\chi_1\cos\theta\sin\theta\, e^{\chi_1(x - \cos\theta((x - Ut)\cos\theta + y\sin\theta))} \times \cos[\Pi_2((x - Ut)\cos\theta + y\sin\theta + B(\theta))] - \Pi_2\sin\theta\, A(\theta)e^{\chi_1(x - \cos\theta((x - Ut)\cos\theta + y\sin\theta))} \times \sin[\Pi_2((x - Ut)\cos\theta + y\sin\theta + B(\theta))] \tag{74a}$$

$$v = \Pi_2\cos\theta A(\theta)e^{\chi_1(x - \cos\theta((x - Ut)\cos\theta + y\sin\theta))} \times \sin[\Pi_2((x - Ut)\cos\theta + y\sin\theta + B(\theta))] - \chi_1\sin^2\theta\, A(\theta)e^{\chi_1(x - \cos\theta((x - Ut)\cos\theta + y\sin\theta))} \times \cos[\Pi_2((x - Ut)\cos\theta + y\sin\theta + B(\theta))] \tag{74b}$$

where M and N are arbitrary constants, depending on the parameter θ, χ_1 is defined above and

$$\Pi_2 = \sqrt{\left(\frac{\mu(k^*\lambda^2 - 1)}{k^*U(\rho - \alpha_1\lambda^2)}\right)^2 \sin^2\theta - \lambda^2}\ .$$

The corresponding pressure distribution to (73) is

$$p = f(t) - \frac{\rho}{2}U^2 - \frac{\mu}{k^*}Ux + \frac{1}{4}\Big[\{\alpha_1(3(\chi_1^4\sin^4\theta + \Pi_2^4 + 6\chi_1^2\Pi_2^2\sin^2\theta) + \lambda^2(2 - \lambda^2)) + 2\alpha_2(\chi_1^4\sin^4\theta + \Pi_2^4 + 6\chi_1^2\Pi_2^2\sin^2\theta) + \rho(\lambda_1^2 - (\chi_1^2\sin^2 + \Pi_2^2))\} + \{(\alpha_1 + 2\alpha_2)\lambda^4\cos 2(\Pi_2((x - Ut)\cos\theta + y\sin\theta + B))\}\Big]A^2e^{2\chi_1(x - \cos\theta((x - Ut)\cos\theta + y\sin\theta)} \tag{75}$$

where $f(t)$ is an arbitrary constant. We observed that if $\Psi = h(\zeta)e^{\gamma x}$ cannot be recovered from the solution $\Psi = h(\zeta)e^{\beta t}$ and vice versa, except when $\theta = 0$.

6. Conclusion

Exact solutions are obtained for incompressible second grade fluid in a porous medium where the vorticity distribution is proportional to the stream function perturbed by a uniform stream. We discuss the two different sections of problems (section 4) and (section 5), respectively. In section 4 we discuss the possible solutions for the steady case. If $k^* \to \infty$, we recover the solution of Benharbit and Siddiqui [23] case from (27). If $\alpha_1 = 0$ and $k^* \to \infty$, the solution (27) reduce to the Kovaznay [2] grid flow and Hui [6] case solutions. For the solution (31), if $k^* \to \infty$, we get the Benharbit and Siddiqui case [23] solution and if $\alpha_1 = 0$ and $k^* \to \infty$, the solution (31), reduce to the Hui case [6] solution. If $k^* \to \infty$, the solution (35) reduce to the Benharbit and Siddiqui [23] case solution. Also for the solution (35), if $\alpha_1 = 0$ and $k^* \to \infty$, our results are compatible with the Lin and Tobak [3] revered flow with suction and Hui [6] case solutions. Turning to section 5, we discuss the two different cases. In case 1; we obtain plane wave solutions. In solutions (44), (48) and (52), if $k^* \to \infty$, we get the Benharbit and Siddiqui [23] case solutions and if $\alpha_1 = 0$ and $k^* \to \infty$, our results are compatible with the Hui [6] case solutions. In case 2, the solutions (61), (65), (69) and (73) are compatible with those of Hui [6] case solutions if $\alpha_1 = 0$ and $k^* \to \infty$, and Benharbit and Siddiqui [23] case solutions if $k^* \to \infty$. The results for pressure distributions in each case do not have parallel in any of the works of Hui [6] and Benharbit and Siddiqui [23]. Also our results are more general and several results of various authors as already mentioned in the text can be recovered in the limiting cases. The problem can be extended to the MHD aligned fluid flows and also combined with hall current effects for porous medium channel.

Acknowledgement
The authors would like to thank Albert C. J. Luo and anonymous reviewer for helpful comments on earlier draft of this work.

References

[1] Taylor GI. On the decay of vortices in a viscous fluid. Phil. Mag. 46, 671-674,1923.
[2] Kovaznay LIG. Laminar flow behind a two dimensional grid. Proc. Cambridge, phil. Soc. 44, 58-62,1948.
[3] Lin SP, Tobak M. Reversed flow above a plate with suction. AIAAJ. 24, 334-335, 1986.
[4] Wang CY. On a class of exact solutions of the Navier-Stokes equations. Journal Applied. 696-698.
[5] Wang CY. Exact solutions of the steady state Navier-Stokes equations. Ann. Rev. Fluid Mech. 23, (1991), 159-177.
[6] Hui WH. Exact solutions of the 2-dim Navier-Stokes equations. ZAMP 38, 689-702, 1987.
[7] Berker R. Integration des equations de mouvement d'un fluide visques incompressible. Vol.VIII/2, stromungsmechanic II, hand buck der physics. 1963.
[8] Nemenyi PF. Recent developments in inverse and semi inverse methods in the mechanics of continua. Adv. Appl. Mech. 11.123-151, 1951.
[9] Agarwal HL. A new exact solution of the equations of viscous motion with axial symmetry. Q. Journal Mech. Applied Math., 10, 42-44. 1957.
[10] Hamdan HM. Single-phase flow through porous channels. A-Review: Flow models and entry Conditions. J. Appl. Math. Comp. 62(2) (1994), 203-222.
[11] Hamdan HM. An alternative approach to exact solutions of a special class of Navier Stokes flows. Journal of Appl. Math. Computation. 93, 1998, 83-90.
[12] Dorrepaal JM. An exact solution of the Navier-Stokes equations which describes non orthogonal stagnation-point flow in two dimensions. J. Fluid Mech. 163 (1986) 141-147.
[13] Irmay S, Zuzovsky M. Exact solutions of the Navier Stokes equations in two-way flows. Israel Journal of Technology. Vol. 8, No. 4, 1970, 307-315.
[14] Zuzovsky M. Two-way flows-exact solutions of the Navier Stokes equations, M.Sc thesis, Publ. No. 133, Fac. Civil Enggng., Technion-Israel Institute of Technology, Haifa, 35-iv pp. 1969.
[15] Rivilin RS, Erikson IL. Stress deformation relations for isotropic materials. J. Rat. Mech. Anal.4, 323-425, 1955.
[16] Coleman BD, Noll W. An approximation theorem for functionals with application in continuum Mechanics. Arch. Rat. Mech. Anal. 6, 355-370, 1960.

[17] Rajagopal KR. On the decay of vortices of a second grade fluid. Mechanics, 9, 185-188, 1980.

[18] Rajagopal KR. A note on unsteady unidirectional flows of a non-Newtonian fluid. Int. J. Non-Linear. Mech. 17, 369-373, 1982.

[19] Rajagopal KR, Gupta AS. On a class of exact solutions to the equation of motion of a second grade fluid. Int. Journal of Eng. Sci. 19, 1009-1014 (1981).

[20] Kaloni PN, Hushchilt K. Semi-Inverse solutions of non-Newtonian fluid. Int. Journal of Non-Linear Mech. 19, 373-381, 1984.

[21] Siddiqui AM, Kaloni PN. Certain inverse solutions of non-Newtonian fluid. Int. Journal of Non-Linear Mech. 21, 459-473, 1986.

[22] Siddiqui AM. Some inverse solutions of a non-Newtonian fluid. Mech. Res. Comm. 17, 157-163, 1986.

[23] Benharbit AM, Siddiqui AM. Certain solutions of the equations of the planar motion of a second grade fluid for steady and unsteady cases. Acta Mechanica, 94, 85-96 (1992).

[24] Markovitz H, Coleman BD. Incompressible second order fluids. Adv. Appl. Mech. 8, 69-101, 1964.

[25] Ting TW. Certain non-steady flows of a second order fluids. Arch. Rat. Mech. Anal. 14, 1-26, 1963.

[26] Oku- Ukpong EO, Chandna OP. Some reversed and non reversed flows of a second grade fluid. Mech. Res. Comm. 20, 73-84, 1993.

[27] Chandna OP, Oku-Ukpong EO. Unsteady second grade aligned MHD fluid. Acta Mechanica, 107, 77-91, 1994.

[28] Truesdell C, Noll W. The non linear field theories of mechanics, in: hand buck der physic, III/3, Springer, Berlin, 1955.

[29] Dunn JE, Fosdic RL. thermodynamics, stability and boundedness of fluids complexity 2 and fluids of second grade. Arch. Rat. Mech. Anal. 3, 191-252 (1974).

[30] Fosdic RL, Rajagopal KR. Anomalous features in the model of second order fluids. Arch. Rat. Mech. Anal. 70, 145-152 (1979).

Approximate analytic solutions of stagnation-point flows in a porous medium

V. Kumaran *, R. Tamizharasi
Department of Mathematics, National Institute of Technology, Tiruchirappalli – 620 015, India

Abstract
An efficient analytical technique is used to obtain approximate analytical series solutions of Brinkman equation for the two-dimensional and axi-symmetric stagnation point flows in a porous medium. Analytical approximate solutions of the classical two-dimensional and axi-symmetric stagnation point flows are also obtained as limiting cases. The obtained zeroth order series solutions agree well with the computed numerical solutions.

PACS: 47.15.Cb; 47.56.+r.
Keywords: Stagnation point; Laminar boundary layer; Porous medium.

1. Introduction

The classical viscous stagnation point flows towards a rigid plane and rigid axi-symmetric surface are well known as exact solutions of Navier-Stokes equations in the literature which were analyzed by Hiemenz and Homann respectively and were discussed in the text book by Schlichting [1]. Similar flows in a porous medium is governed by Brinkman flow when the inertial effects are negligible and the Reynolds number based on the pore diameter is less than one. An excellent review of existing theoretical and experimental work on this subject can be found in the recent monographs by Nield and Bejan [2], Ingham and Pop [3]. Recently Wu et al [4] proposed a new parameter β which is the ratio of square of the two lengths, the classical boundary layer thickness and fibre interaction layer thickness to explore the transition from Brinkman stagnation point flows ($\beta >> 1$) to the classical Hiemenz and Homann flows ($\beta = 0$). They have obtained asymptotic series solution for the large β limit in powers of β^{-1}, which fails to give solution for $\beta = 0$ where as it need many terms to get a better solution for moderate values of β. Hence in this paper, we introduce an artificial non-negative parameter A and we seek for a series solution in powers of $\varepsilon = (\beta + A)^{-1}$. It is shown that the present zeroth order series solutions are fairly accurate for all values of β, in particular it gives more accurate results for moderate and large values of the parameter β.

2. Governing equations

Following the same notation of Wu et al [4], except for the two variables defined as follows:

* Corresponding author. Tel.: +91 (431) 2501801; fax: +91 (431) 2500133.
E-mail address: sugunakumaran@yahoo.com (V. Kumaran).

$$\zeta = \eta\varepsilon^{-1/2}, \quad G(\zeta) = g(\eta)\varepsilon^{-1/2} \tag{1}$$

where $\varepsilon = (\beta + A)^{-1}$ and A is a non-negative constant to be determined, Eqn. (37) and (20) of Wu et al [4] take the form

$$G'''-G'+1+\varepsilon[GG''+\frac{1}{n}(1-G'^2)-A(1-G')]=0 \tag{2}$$

subject to

$$G(0)=0, \; G'(0)=0, \; G'(\infty)=1 \tag{3}$$

where $n=1$ refers to two-dimensional case and $n=2$ refers to axi-symmetric case.

3. Numerical solutions

Solving the Eqn. (37) subject to Eqn. (20) of Wu et al [4] for $n=1, 2$ and β values ranging from 0 to 1000, using a shooting method combined with Runge-Kutta method, we obtain the numerical solutions for both the cases of two-dimensional and axi-symmetric viscous stagnation point flows in a porous medium. The results are presented and discussed through graphically in section 5. An analysis on the asymptotic behavior as $\eta \to \infty$ of $g(\eta)$ from Eqs. (37) and (20) shows that

$$(g(\eta)-\eta) \to A_1 \quad \text{as} \quad \eta \to \infty, \text{ when } n=1 \tag{4}$$

and

$$(g(\eta)-\eta) \to A_2 \quad \text{as} \quad \eta \to \infty, \text{ when } n=2 \tag{5}$$

where A_1 & A_2 are two constants.

In particular from the numerical computations we observed the following values for case of $\beta = 0$, i.e., for the cases of classical two-dimensional and axi-symmetric incompressible viscous stagnation point flows:

$$A_1 = -0.647901 \text{ and } A_2 = -0.804549. \tag{6}$$

4. Series solutions

Let us search for a series solution of the form

$$G(\zeta,\varepsilon) = G_0(\zeta) + \varepsilon G_1(\zeta) + \varepsilon^2 G_2(\zeta) + \varepsilon^3 G_3(\zeta) + \ldots \tag{7}$$

of the boundary value problem (2) and (3), we obtain the *zeroth-order* problem:

$$G_0'''-G_0'=-1 \tag{8}$$

subject to

$$G_0(0)=G_0'(0)=1-G_0'(\infty)=0 \tag{9}$$

Zeroth order solutions: Solving the Eqs. (8) - (9), we get the simple analytic solution

$$G_0(\zeta) = e^{-\zeta} + \zeta - 1 \tag{10}$$

Hence the zeroth order solution takes the form

$$g(\eta) \approx \eta + \varepsilon^{1/2}(e^{-\eta/\sqrt{\varepsilon}} - 1) \tag{11}$$

It is important to note that the zeroth order solution is independent of n. But still we find different zeroth order solutions for both the cases of $n = 1$ and $n = 2$ by finding different values of the unknown parameter A in Eqn. (11). From Eqn. (11),

$$(g(\eta) - \eta) \to -\varepsilon^{1/2} \quad \text{as} \quad \eta \to \infty \tag{12}$$

Now we obtain A value in such a way that the zeroth order solution for the worst case $\beta = 0$ of the series (Eqn. (11)) and the numerical solution of $\beta = 0$ (given in section 3) have the same asymptotic behavior. i.e., from Eqn. (12) for $\beta = 0$ and Eqs. (4) - (5), we get

$$A = \begin{cases} 2.382225 & for \quad n = 1 \\ 1.54488 & for \quad n = 2 \end{cases} \tag{13}$$

Thus Eqs. (11) and (13) gives the zeroth order solutions of both two-dimensional and axi-symmetric stagnation point flows in a porous medium. The quantities of interest based on the zeroth order solutions are:

i) $g''(0) \approx \varepsilon^{-1/2}$ (14)

ii) The asymptotic value as $\eta \to \infty$ of $(g(\eta) - \eta) \approx -\varepsilon^{1/2}$ (15)

iii) The boundary layer thickness $\eta_\delta \approx \varepsilon^{1/2} \log(100)$ (16)

(which is defined as $g'(\eta_\delta) = 0.99$).

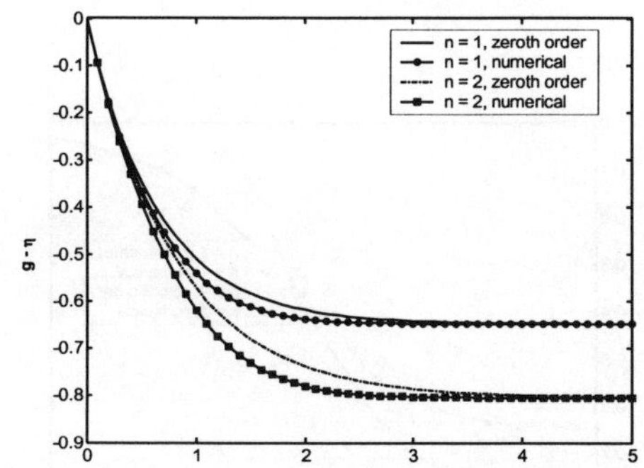

Fig. 1. Zeroth order and numerical profiles of $g(\eta) - \eta$ for $\beta = 0$

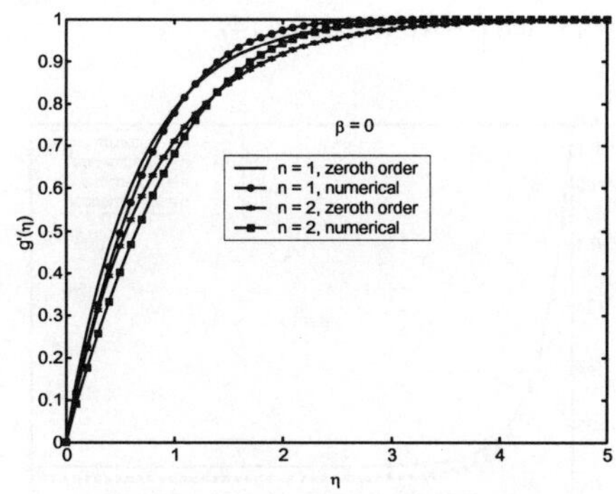

Fig. 2. Zeroth order and numerical profiles of $g'(\eta)$ for $\beta = 0$

5. Results and discussion

In this section, comparison of numerical solutions obtained in section 3 and zeroth order series solutions for both the cases of two-dimensional ($n = 1$) and axi-symmetric ($n = 2$) stagnation point flows in a porous medium are presented graphically. Figs. (1) - (2) show the zeroth order and numerical profiles of $g(\eta) - \eta$ and $g'(\eta)$ for $\beta = 0$. Even for the worst case $\beta = 0$, the maximum error is about five percent to eight percent only. Fig. (3) shows the comparison of the zeroth order solutions and numerical solutions of $g(\eta) - \eta$ for $\beta = 0,1$ and 5.

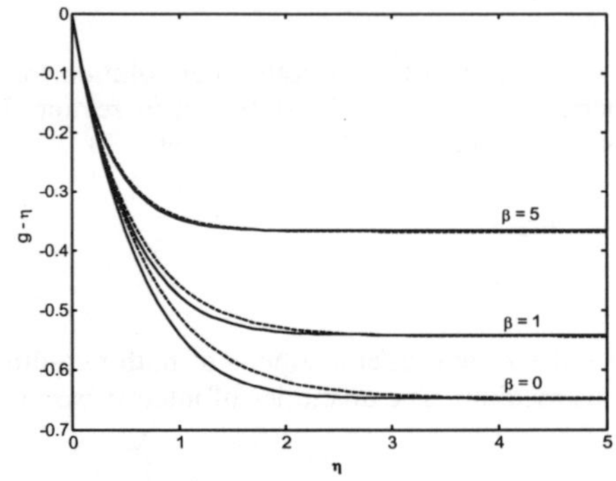

Fig. 3. Two-dimensional stagnation-point flow profiles of $g(\eta) - \eta$ for different values of β (continuous lines -numerical, broken lines - zeroth order)

The maximum error between the zeroth order and numerical solutions reduce gradually with increasing βvalues. In fact, the deviation between the approximate and numerical solutions are not visible on the scale shown when β = 10 from Fig. (4).

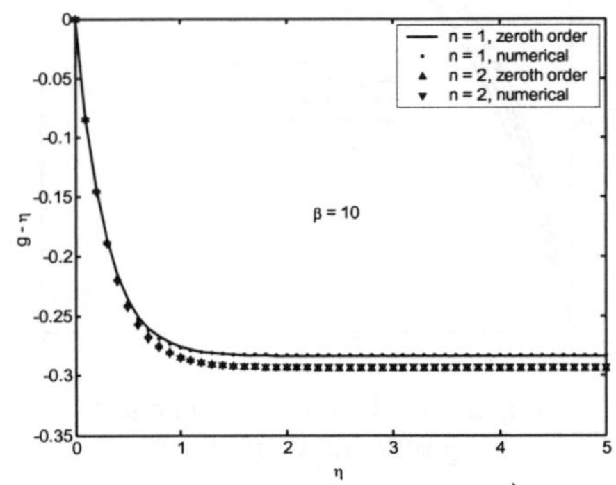

Fig. 4. Zeroth order and numerical profiles of $g(\eta) - \eta$ for $\beta = 10$

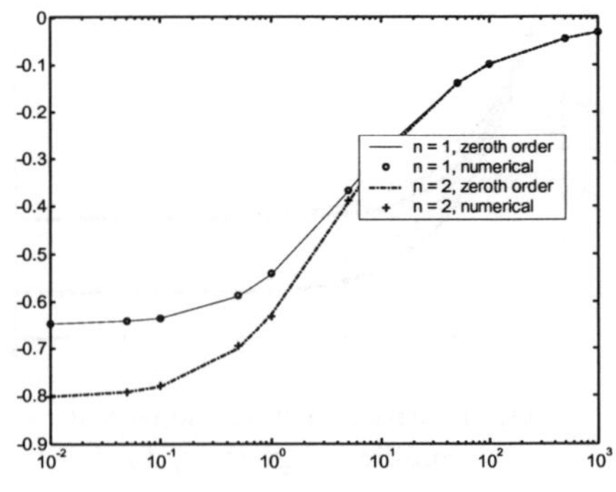

Fig. 5. Zeroth order and numerical values of $g(\eta) - \eta$ as $\eta \to \infty$ versus β

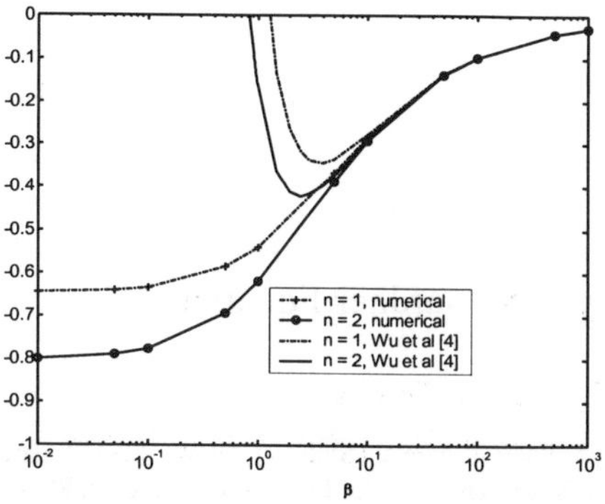

Fig. 6. Wu et al [4] values and numerical values of $g(\eta)-\eta$ as $\eta \to \infty$ versus β

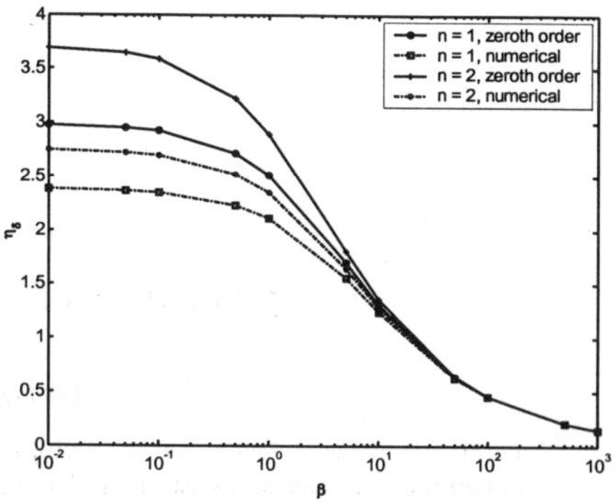

Fig. 7. Zeroth order and numerical values of η_δ versus β

As measures of accuracy of the zeroth order approximate solutions, the asymptotic values as $\eta \to \infty$ of $(g(\eta)-\eta)$ and the boundary layer thickness η_δ are plotted against β values ranging from 10^{-2} to 10^{3} in Figs. (5), (6) and (7) respectively. These figures confirm that the present zeroth order solutions are very good approximations (maximum error is less than one percent) for large β values, good approximations (maximum error is less than two percent) for moderate values of β and fair approximations (maximum error is less than five to eight percent) for small β values which includes the case $\beta = 0$.

6. Conclusions

The advantages of the new parameter $\varepsilon = (\beta + A)^{-1}$ valid for all values of β when compare to the Wu et al [4] parameter β^{-1} valid for large values of β only are as follows.

1. The power series expansion in powers of ε is valid and approximates fairly even for the worst case $\beta = 0$. Thus provide new analytical approximate solutions for both the classical Hiemenz flow and Homann flow.
2. When $0 \le \beta < \infty$, the parameter ε varies from A^{-1} to 0. Hence enable to get a fast converging series solutions in powers of ε, valid for all values of β, in fact zeroth order solutions agree well with the numerical solutions even for moderate values of β.
3. The two term asymptotic solution for large β limit of Wu et al [4] is recovered as a special case, when A = 0. The present zeroth order solutions are better than Wu et al [4] two term series solutions for all values of the parameter β.
4. Two different zeroth order solutions are obtained for two-dimensional ($n = 1$) and axi-symmetric ($n = 2$) cases respectively, where as Wu et al [4] zeroth order solution is the same for both the cases.

Obtaining more accurate approximate analytical solutions by including higher order terms of the series is in progress.

References

[1] Schlichting H. Boundary layer theory, 6th ed. McGraw Hill;1960.
[2] Nield DA, Bejan A. Convection in porous media, 2nd ed. New-York:Springer;1999.
[3] Ingham DB, Pop I. (1998, 2002) Transport phenomena in porous media. Oxford:Pergaman; 1998, 2002.
[4] Wu Q, Weinbaum S, Andreopoulos Y. Stagnation-point flows in a porous medium. Chemical Engineering Science 2005;60:123-34.

Lump Solutions of 2D Generalized Gardner Equation

Y.A. Stepanyants[1*], I.K. Ten[2], H. Tomita[3]

[1*]Australian Nuclear Science and Technology Organisation, PMB 1, Menai (Sydney), NSW, 2234, Australia.
[2] Ocean and Ice Engineering Department, National Maritime Research Institute, 181-0004, Tokyo-to, Mitaka-shi, Shinkawa 6-38-1, Japan.
[3] National Maritime Research Institute, 6-38-1, Shinkawa, Mitaka, Tokyo 181-004, Japan.

Abstract

Results of numerical study of lump solutions (2D solitons) of a generalised 2D Gardner equation are presented. In the moving coordinate frame the equation is:

$$(u_t + \alpha u u_x + \alpha_1 u^2 u_x - \beta u_{xxx})_x = \gamma u - c u_{yy}/2.$$

To construct such solutions, the Petviashvili method is further developed for the evolution equations with the non-power nonlinearity. Solution obtained for different relationships between quadratic and cubic nonlinearity as well as between small- and large-scale dispersions (the β- and γ-terms in the equation) are compared with the known lump solution for the classical Kadomtsev–Petviashvili equation with positive dispersion. The structure of constructed solutions is analysed in terms of two dimensionless parameters characterising the cubic nonlinearity and large-scale dispersion.

PACS: 05.45, 52.35, 02.60
Keywords: nonlinear waves, soliton, lump, Petviashvili method, numerical study, Gardner equation

1. Introduction

In recent years researchers study more and more complex problems of nonlinear wave propagation in different media. There are already a plethora of well-studied integrable models of nonlinear waves (see, e.g., [1, 4, 6, 7] and references therein). In addition to that, new advanced models permanently appear; most of them are not integrable analytically. Therefore, a development of approximate and numerical methods is one of the topical problems in the nonlinear wave theory.

One of the powerful numerical methods for construction of steady-state solutions of nonlinear evolution equations is the Petviashvili method [14, 15]. The method was heuristically suggested by V.I. Petviashvili [14] in 1976; it allows one to construct stationary solutions for a wide class of evolution equations both in one-dimensional and multi-dimensional cases. Strong mathematical justification of the method for a certain class of nonlinear evolution equations with power-type nonlinearity was done in [13]; the range of method convergence was also obtained.

In this paper we present a further development of the Petviashvili method and disseminate it on two-dimensional (2D) problems with the non-power nonlinearity. The basic equation to be studied is a generalised 2D-Gardner equation (gen2D-Gardner):

Corresponding author, E-mail: Yury.Stepanyants@ansto.gov.au

$$\frac{\partial}{\partial x}\left(\frac{\partial u}{\partial t}+c\frac{\partial u}{\partial x}+\alpha u\frac{\partial u}{\partial x}+\alpha_1 u^2\frac{\partial u}{\partial x}-\beta\frac{\partial^3 u}{\partial x^3}\right)=\gamma u-\frac{c}{2}\frac{\partial^2 u}{\partial y^2}. \quad (1)$$

The coefficients of this equation depend on the concrete physical application; $c > 0$ stands for some characteristic velocity of linear perturbations, α, α_1, β and γ may be arbitrary in general but we will consider the case when β and γ are positive constants. Such equation may appear in different physical areas where cubic nonlinearity should be taken into account along with the quadratic one (this can happen, e.g., for internal waves in two-layer fluid [9, 11]). Medium rotation leads to the appearing of the first term in the right-hand side of Eq. (1) with positive γ (see, e.g., [8, 10, 5]), however in some other physical problems this term can appear with the negative coefficient γ. The existence of such term in the equation reflects the presence of a large-scale dispersion in the physical system, whereas the β-term in the equation represents a small-scale dispersion (the former dispersion manifests itself at large wavelengths, and the later one – at small wavelengths). In the particular case, $\alpha_1 = \gamma = 0$, Eq. (1) reduces to the well-known Kadomtsev–Petviashvili (KP) equation (see, e.g., [1, 3, 15]). This equation is completely integrable and in the case $\beta > 0$ (KP1 equation in contrast to KP2 equation when $\beta < 0$) it possesses 2D soliton solutions which call lumps. Similar solutions also exist when KP1 equation is augmented by the γ-term but so that $\beta\gamma > 0$ (2D version of the Ostrovsky equation) [10]. They were constructed only numerically and their analytical form is unknown so far. For the modified KP1 equation ($\alpha_1 \neq 0$, $\alpha = \gamma = 0$) with positive small-scale dispersion, $\beta > 0$, this equation also possesses lump solutions which were constructed numerically in [2, 13].

2. Modification of the Petviashvili numerical method

Let us consider gen2D-Gardner equation (1). It combines quadratic and cubic nonlinearity as well as small- and large-scale dispersion. All these effects in physical systems usually are small. The dispersion relation for infinitesimal perturbations of the form $u \sim \exp[i(\omega t - k_x x - k_y y)]$ may be presented in the form of phase velocity dependency $V_{ph} \equiv \omega / k_x$ on wavenumber k: (Fig. 1):

$$V_{ph} = c + \beta k_x^2 + \left(\gamma + c k_y^2/2\right)/k_x^2 \quad (2)$$

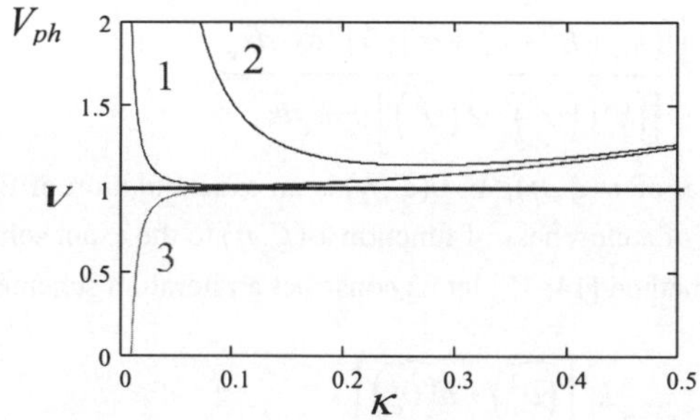

Fig. 1. Normalised phase velocity V_{ph}/c versus normalised wavenumber κ_x for two values of κ_y.
Line 1 – $\kappa_y = 0$, $B = 10^{-4}$; line 2 – $\kappa_y = 0.1$, $B = 10^{-4}$; line 3 – $\kappa_y = 0$, $B = -10^{-4}$.

where $\kappa_x = k_x\sqrt{\beta/c}$, $\kappa_y = k_y\sqrt{\beta/c}$, $B = \beta\gamma/c^2$. As follows from Eq. (2), phase velocity has a minimum if $\beta\gamma > 0$, $\left(V_{ph}\right)_{\min} = c + 2\sqrt{\beta\gamma}$, which occurs at $\left(k_x\right)_{\min} = \sqrt[4]{\gamma/\beta}$, $k_y = 0$. The presence of the velocity minimum is important for the construction of 2D soliton solutions because such stationary objects can exist when their velocities is less than $\left(V_{ph}\right)_{\min}$. Otherwise, there is a resonance of a stationary moving object, whose velocity is V_o, with linear perturbations as shown in Fig. 1 (intersection of line V_0 and line 3). Due to this resonance, the object generates wave perturbations in the medium and losses its energy, therefore it cannot move stationary without action of any external force.

Considering further stationary solutions of Eq. (1), i.e., solutions depending on $\xi = x - Vt$ and y, one can rewrite it as

$$\frac{\partial^2}{\partial \xi^2}\left[(c-V)u + \frac{\alpha u^2}{2} + \frac{\alpha_1 u^3}{3} - \beta\frac{\partial^2 u}{\partial \xi^2}\right] = \gamma u - \frac{c}{2}\frac{\partial^2 u}{\partial y^2}. \quad (3)$$

Let us make a transformation of variables to reduce this equation to the dimensionless form:

$$\varsigma = \xi\sqrt{\frac{c-V}{\beta}}, \quad \eta = y(c-V)\sqrt{\frac{2}{\beta c}}, \quad \upsilon = -\frac{\alpha u}{2(c-V)}. \quad (4)$$

Eq. (3) in new variables becomes

$$\frac{\partial^2}{\partial \zeta^2}\left(\upsilon - \upsilon^2 + \delta\upsilon^3 - \frac{\partial^2 \upsilon}{\partial \zeta^2}\right) = \sigma\upsilon - \frac{\partial^2 \upsilon}{\partial \eta^2}, \quad (5)$$

where $\delta = 4\alpha_1(c-V)/(3\alpha^2)$, $\sigma = \beta\gamma/(c-V)^2$.

Let us make a Fourier transformation of this equation on variables ζ and η denoting the Fourier transformation of function $\upsilon(\zeta,\eta)$ as $\tilde{\upsilon}(k_x,k_y) = \frac{1}{2\pi}\iint \upsilon(\zeta,\eta)e^{i(k_x\zeta + k_y\eta)}d\zeta d\eta$ (the integration here is taken over whole ranges of axes ζ and η from minus to plus infinity).

$$\left(k_x^2 + k_y^2 + k_x^4 + \sigma\right)\tilde{\upsilon}\left(k_x,k_y\right) = k_x^2\left[\widetilde{\left(\upsilon^2\right)} - \delta\widetilde{\left(\upsilon^3\right)}\right]. \quad (6)$$

Multiply now this equation by $\tilde{\upsilon}(k_x,k_y)$ and integrate it over k_x and k_y from minus to plus infinity assuming that function $\tilde{\upsilon}(k_x,k_y)$ is real (this corresponds to the assumption that function $\upsilon(\zeta,\eta)$ is even both on ζ and on η). By dividing left-hand side of the resultant equation on the right-hand side, let us form a ratio

$$M\left[\upsilon(\zeta,\eta)\right] = \frac{\iint\left(k_x^2 + k_y^2 + k_x^4 + \sigma\right)\left(\tilde{\upsilon}\right)^2 dk_x dk_y}{\iint k_x^2\left[\widetilde{\left(\upsilon^2\right)} - \delta\widetilde{\left(\upsilon^3\right)}\right]\tilde{\upsilon}dk_x dk_y}, \quad (7)$$

which is obviously a functional of $\upsilon(\zeta,\eta)$. If $\upsilon(\zeta,\eta)$ is an exact solution of Eq. (5), then $M = 1$. Thus, M can be treated as a measure of a closeness of function $\upsilon(\zeta,\eta)$ to the exact solution.

In a spirit of the Petviashvili method [14, 15] let us construct an iteration scheme presenting Eq. (6) in the form

$$\tilde{\upsilon}_{n+1}\left(k_x,k_y\right) = M^p\left[\upsilon_n\right]\frac{k_x^2\left[\widetilde{\left(\upsilon_n^2\right)} - \delta\widetilde{\left(\upsilon_n^3\right)}\right]}{k_x^2 + k_y^2 + k_x^4 + \sigma}. \quad (8)$$

Here index $n = 0, 1, 2, \ldots$ stands for a number of the iteration step; M^p with p being some real value plays a role of a stabilizing factor which provides a convergence of the iterative scheme (8) for any pulse type start function $\upsilon(\zeta,\eta)$. Without this factor or, equivalently, with $p = 0$, the iterative scheme diverges and does not provide any solution. In relatively simple cases when there is only one power-type function υ in the right-hand side of Eq. (8), i.e., $\widetilde{\left(\upsilon_n^q\right)}$, where q is an arbitrary positive number, a strong mathematical criterion can be obtained for the exponent p of the stabilising factor which provides a convergence of the iterative scheme, $1 < p < (q + 1)/(q - 1)$, and the fastest convergence occurs at $p = q/(q - 1)$ [13]. As a criterion that the iterations yield to the solution, one usually uses the inequality $\left|1 - M\right| < \varepsilon$, where $\varepsilon << 1$ is some small parameter. Note that the above assumption about the positiveness of a product $\beta\gamma > 0$ implies that so is σ, which guaranties the absence of singularity in Eq. (8).

In our case with a complex non-power nonlinearity we choose the value of $p = 5/4$ to be within the convergence interval both for quadratic and cubic nonlinearity. Our practice showed that in the majority of cases the iteration procedure converges to some solutions indeed. The solution shape depends on the parameters of Eq. (5), δ and σ. For $\delta = \sigma = 0$ Eq. (5) reduces to the classical KP equation with positive dispersion. Its particular solution in the form of lump-soliton is well known (see, e.g., [1]):

$$\upsilon(\zeta,\eta) = 12\frac{3+\eta^2-\zeta^2}{\left(3+\eta^2+\zeta^2\right)^2}. \qquad (9)$$

In the dimensionless variables this solution does not contain any parameter; it can be used for testing of a numerical scheme. In the original variables this solution is

$$u(x,y,t) = -24(c-V)\frac{\beta c}{\alpha}\frac{3\beta c+2(c-V)^2 y^2 - c(c-V)(x-Vt)^2}{\left[3\beta c+2(c-V)^2 y^2 + c(c-V)(x-Vt)^2\right]^2} \qquad (10)$$

with $\beta > 0$ and $c > V$ (see Eqs. (4)). Depending on the sign of the coefficient α it may have positive or negative polarity with the amplitude proportional to the speed of propagation $U = -8(c - V)/\alpha$. Characteristic sizes of the lump are different in x and y directions: $\Delta_x = \sqrt{3\beta/(c-V)} = \sqrt{-24\beta/\alpha U}$, $\Delta_y = \sqrt{3\beta c/2(c-V)^2} = -8\sqrt{3\beta c}/\sqrt{2}\alpha U$ (note that in the plane case the relationship between the KdV-soliton width and amplitude is very similar: $\Delta_x = \sqrt{\beta/(c-V)} = \sqrt{-12\beta/\alpha U}$).

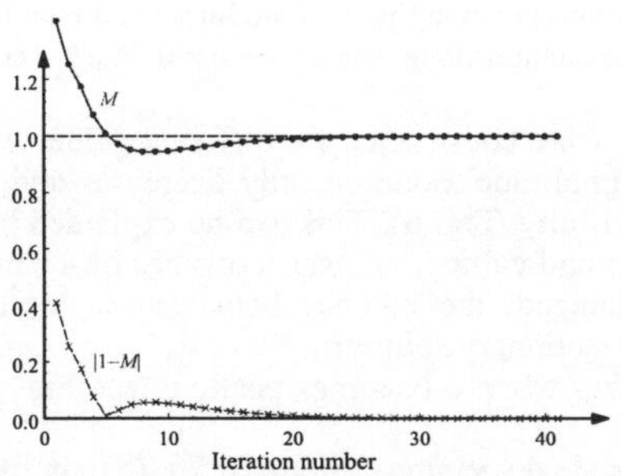

Fig. 2. Convergence of a stabilising factor M and a complementary factor $|1 - M|$ versus iteration number n for the KP equation (Eq. (5) with $\delta = \sigma = 0$ and $\varepsilon = 10^{-4}$).

3.Results of numerical calculations

As a first step, the numerical scheme was tested by comparison of a numerical solution against the analytical one for the KP lump. With the particular choice of parameter values $\delta = \sigma = 0$, solution of Eq.

(5) was readily obtained by means of iteration scheme (8) with the Gaussian start function $\upsilon(\zeta,\eta)=4\exp\left(-\zeta^2-\eta^2\right)$. A typical numerical domain was 50×50 in ζ,η-plane but for some cases (depending on the characteristic width of the solution) it varied from 20×20 to 200×200; it was discretized by a mesh of 1024×1024 points. The numerical scheme convergence with $\varepsilon = 10^{-4}$ was achieved after 41 iterations. Figure 2 shows the scheme convergence in terms of M and $|1 - M|$ versus iteration number n. The corresponding lump solution is presented in Fig. 3 and its main cross-sections are shown in Fig. 4. As one can see, the correspondence between the numerical and analytical solutions is very good. Only a small discrepancy between these solutions occurs at the lump maximum, the numerical value is a bit higher than the analytical one (for the case presented in Fig. 4 the discrepancy is less than 0.2%). This discrepancy however decreases when the domain size increases. Thus, the comparison provides us with a confidence in construction other numerical solutions with non-zero values of δ and σ.

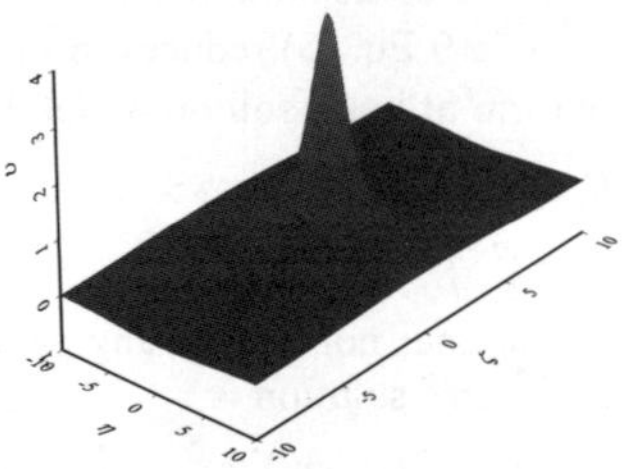

Fig. 3. Lump solution of the KP equation (only a fragment of a whole domain is shown).

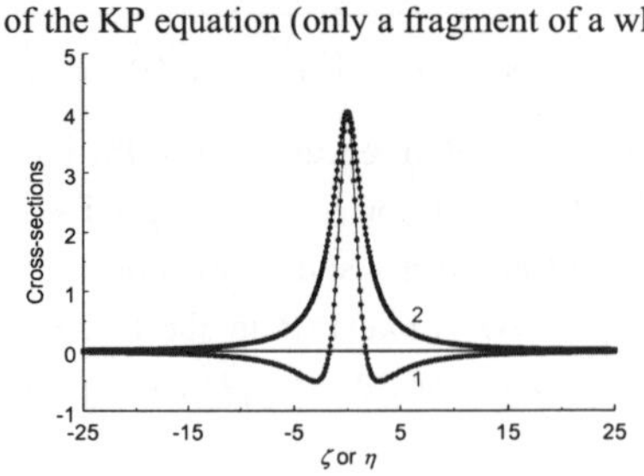

Fig. 4. Main cross-sections of KP lump shown in Fig. 3. Solid lines are drown from the analytical solution, Eq. (9), dots – numerical data. Line 1 – cross-section along axis ζ with $\eta = 0$; line 2 – cross-section along axis η with $\zeta = 0$.

Next, a family of lump solutions were constructed for different values of δ with $\sigma = 0$. In Fig. 5 one can see their cross-sections. Lump amplitude monotonically decreases and, apparently, vanishes eventually when δ varies from 0 to minus infinity (Fig. 6). This can be explained by increase of a total nonlinearity in Eq. (5) because both quadratic and cubic nonlinear terms are of a similar sign. As the dispersion term (small-scale dispersion) is unchanged, the balance between the total nonlinearity and dispersion is possible at smaller amplitudes of stationary solutions.

Lump amplitude abruptly increases when δ becomes positive (see Fig. 6) so that we failed constructing lump solution for $\delta > 0.1$.

Then, an influence of large-scale dispersion (σ-term in Eq. (5)) on the structure of lumps was studied. Again the lumps of qualitatively the same shapes were obtained for different values of σ. In Fig. 7 their cross-sections are shown for fixed value of $\delta = 0$. Lumps become more symmetrical and compact with almost equal characteristic width in ζ and η directions, and their amplitudes monotonically increase when σ increases. Apparently it is caused now by increase of a total dispersion (small-scale and large-scale) in Eq. (5). As the total dispersion becomes greater, the balance between the nonlinearity and total dispersion can be achieved at higher amplitudes of stationary solutions. The dependency of lump amplitudes on parameter σ is shown in Fig. 8.

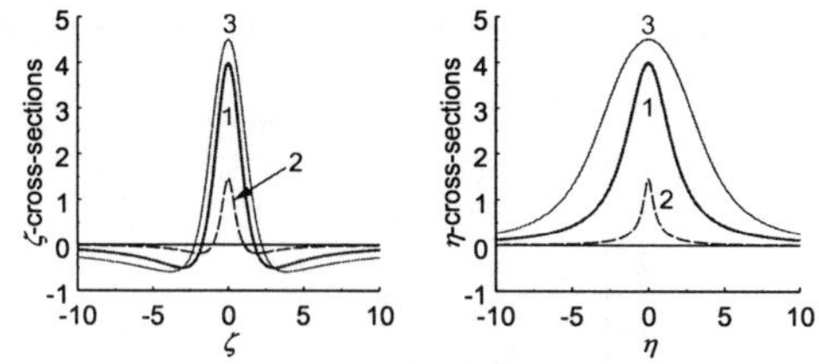

Fig. 5. ζ- and η-cross-sections of lumps numerically constructed for different values of δ with $\sigma = 0$: 1 – $\delta = 0$; 2 – $\delta = -5.0$; 3 – $\delta = 0.1$.

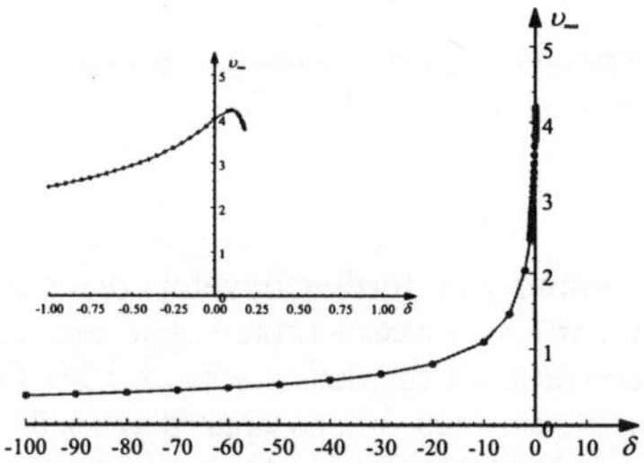

Fig. 6. Lump amplitude versus δ with $\sigma = 0$. The insertion shows the plot on the stretched interval $-1 < \delta < 1$.

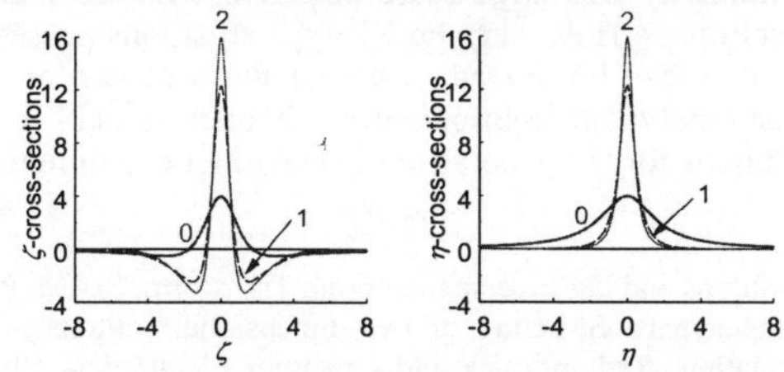

Fig. 7. ζ- and η-cross-sections of lumps numerically constructed for different values of σ with $\delta = 0$: 0 – $\sigma = 0$; 1 – $\sigma = 1.0$; 2 – $\sigma = 2$.

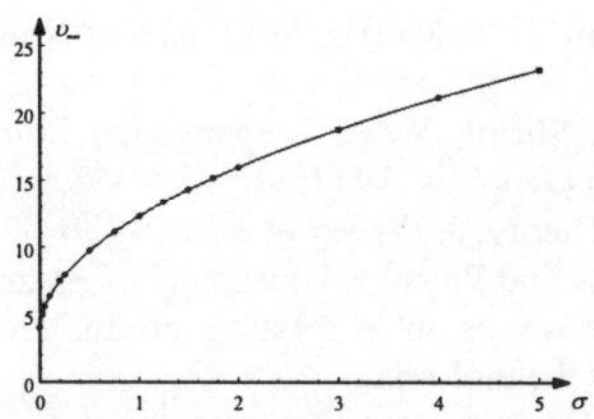

Fig. 8. Lump amplitude versus σ with $\delta = 0$.

In our last series of numerical experiments the cooperative effect of cubic nonlinearity and large-scale dispersion (both δ- and σ-terms in Eq. (5)) was studied. When the parameters δ and σ are of opposite signs ($\delta < 0$ and $\sigma > 0$), their influence on the lump amplitude is opposite, the δ-term leads to decreasing amplitude, whereas the σ-term leads to its increasing. As a result, at certain relationships between parameters δ and σ lump amplitude remains constant despite of their variation. One particular case of cooperative action of cubic nonlinearity ($\delta = -1$) and large-scale dispersion ($\sigma = 1$) is shown in Fig. 9 in comparison with the standard KP lump. As one can see, lump amplitudes are close to each other, whereas their characteristic sizes are different (the parameter σ may be chosen more accurately to make soliton amplitudes exactly equal).

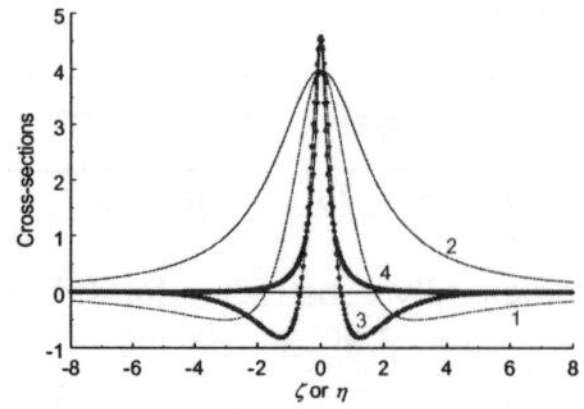

Fig. 9. Main cross-sections of KP lump (line 1 – cross-section along axis ζ with $\eta = 0$; line 2 – cross-section along axis η with $\zeta = 0$) in comparison with the lump with of the following parameters $\delta = -1$, $\sigma = 1$ (corresponding dotted lines).

4. Conclusion

Thus, the Petviashvili numerical method was further developed for the solution of evolution equations with the non-power nonliearity of a particular case – mixed quadratic and cubic nonlinearity. By means of this method lump solutions were constructed for the generalised 2D Gardner equation containing small- and large-scale dispersion. Obtained solutions are compared with the known lump soliton of the KP1 equation. The structure of constructed solutions is analysed in terms of two dimensionless parameters characterising the cubic nonlinearity and large-scale dispersion. In this paper we focused only on the construction of single lump solutions. However, the KP-type equations possess also multilump stationary solutions (see, e.g., [2, 15]); they can be treated as bound states of two or several lumps. For the KP1 equation such multilump exact analytical solutions were obtained in [12]. It could of interest to study a family of such multilump solutions for the generalised 2D Gardner equation too.

References

1. M.J. Ablowitz, H. Segur, Solitons and the Inverse Scattering Transform, SIAM, Philadelphia, 1981.
2. L.A. Abramyan, Yu.A. Stepanyants, Structure of two-dimensional solitons in the context of a generalised Kadomtsev–Petviashvili equation, Radiophysics and Quantum Electronics, 1987, 30, n. 10 (1987) 861–865 (English translation from the Russian journal Izvestia VUZov, Radiofizika).
3. V.Yu. Belashov, The KP Equation and its Generalization. Theory and Applications, NEISRI FEB RAS, Magadan, 1997 (in Russian).
4. R.K. Dodd, J.C. Eilbeck, J.D. Gibbon, H.C. Morris, Solitons and Nonlinear Wave Equations, Academic Press, 1984.
5. R. Grimshaw, L.A. Ostrovsky, V.I. Shrira, Yu.A. Stepanyants, Long nonlinear surface and internal gravity waves in a rotating ocean, Surveys in Geophys. 19 (1998) 289–338.
6. G.L. Lamb Jr., Elements of Soliton Theory, J. Willey & Sons, 1980.
7. A.C. Newell, Solitons in Mathematics and Physics, University of Arizona, SIAM, 1985.
8. L.A. Ostrovsky, Nonlinear internal waves in a rotating ocean, Oceanology, 18 (1978) 119–125 (English translation from the Russian journal Okeanologia).
9. L.A. Ostrovsky, Yu.A. Stepanyants, Do internal solitons exist in the ocean?, Rev. Geophys. 27 (1989) 293–310.
10. L.A. Ostrovsky, Yu.A. Stepanyants, Nonlinear surface and internal waves in rotating fluids, in: A.V. Gaponov-Grekhov, M.I. Rabinovich, J. Engelbrecht (Eds.), Nonlinear Waves 3. Proc. 1989 Gorky School on Nonlinear Waves, Springer-Verlag, Berlin–Heidelberg 1990, pp. 106–128.
11. L.A. Ostrovsky, Y.A. Stepanyants, Internal solitons in laboratory experiments: Comparison with theoretical models, Chaos, 15 (2005) 037111, 28 p.
12. D.E. Pelinovsky, Y.A. Stepanyants, New multisoliton solutions of the Kadomtsev–Petviashvili equation, JETP Lett., 57 (1993) 24–28 (English translation from the Russian journal Pis'ma v ZhETF).

13. D.E. Pelinovsky, Y.A. Stepanyants, Convergence of Petviashvili's iteration method for numerical approximation of stationary solutions of nonlinear wave equations, SIAM J. Numerical Analysis, 42 (2004) 1110–1127.
14. V.I. Petviashvili, Equation of an extraordinary soliton, Soviet J. Plasma Phys., 2 (1976) 247 (English translation from the Russian journal Fizika Plazmy).
15. V.I. Petviashvili, O.V. Pokhotelov, Solitary Waves in Plasmas and in the Atmosphere, Gordon and Breach, Philadelphia, 1992.

Wood fracture behavior simulation using stochastic FEM

Mingbao Li[a,*], Jun Cao[b] and Shiqiang Zheng[b]

[a]*School of Civil Engineering, Northeast Forestry University*
[b]*School of Electromechanical Engineering, Northeast Forestry University*
No. 26 Hexing Road, Harbin, 150040, China

Abstract

Stochastic finite element method (FEM) is introduced into wood fracture mechanics in this paper. Constitutive equations with the simplified penalty functions are used to develop the power-law nonlinear stochastic finite element method. Considering the randomness of crack size, *J*-integral is calculated and analyzed. According to the variation ratio of *J*-integral and the *J-K* relationship, the change curve of wood strength factor is given. To prove the validity of numerical results, test species machined from *Maor Mountain* larch is tightened and the testing curve of wood strength factor is got. The two curves have the same change trend, and errors between them are within a standard deviation. It is concluded that adopting stochastic finite element method to analyze and predict the wood fracture performance is feasible and practical.

PACS: 46.50.+a; 83.80.Mc
Keywords: Wood; Stochastic finite element method; Fracture mechanics

1. Introduction

Wood has been used as a material for construction, tools, furniture and decoration for thousands of years. However, many details of wood fracture behavior are not yet understood. Knowledge of the fracture behavior has relevance not only to the structural use of wood, but has also importance to processes like cutting and machining. A fundamental requirement for effective utilization of wood as a competitive structural material is the accurate knowledge of its mechanical properties. However, the experimental identification and the analytical modeling of mechanical behavior of wood remains an open problem, due to its natural variability, inhomogeneity and anisotropy. Wood is a natural macromolecule composite material which is porous, anisotropic, multilayer, and inhomogeneous. It is a

* Corresponding author. Tel.: +86-451-82191615; fax: +86-451-82191406.
E-mail address: zsqrq@yahoo.com.cn (M.B. Li).
This work is supported by National Natural Science Foundation of China (30371126) and by Natural Science Foundation of Heilongjiang Province of China (C0308).

combination of wood fiber in the different thicknesses at the microstructure level and has an elastic behavior of unidirectional fiber reinforced material.

Wood is approximately considered as the orthotropic material, where R, T and L refer to radial, tangential and longitudinal directions of growth rings respectively. Supposed that the first symbol represents the normal direction of crack plane and the second represents the extension direction, six basic crack extension forms are got, that is: TL, RL, LT, LR, TR, RT. This paper focuses on the TL crack extension (seen from Fig. 1). The toughness, deformation and damage performance of wood is related to its internal structure, crack extension and fracture parameters. The random distribution of micro-crack controls the macrostructure performance to a certain extent, such as tensile toughness, compressive toughness, *etc.* Wood damage is mainly caused by various kinds of potential defects inside it. So the investigation on the wood fracture parameters can help to know the wood fracture emergence mechanism and provide theory foundation for the prediction of wood structure failure and rational utilization of wood timber.

Relations between wood fiber structure and mechanical behaviors are very complex. Wood fracture mechanics performances include the size and shape of crack, the fracture resistance parameters, crack growth factor, stress strength factors, *etc.* The process of calculating and analyzing these parameters possesses randomness. In view of that, the paper proposes a stochastic finite element method (FEM) to calculate the wood fracture parameters and makes a quantitative determination on the randomness of the parameters. Welson *et al.* put forward a linear elasticity perturbation stochastic finite element method to calculate the ratio of stress strength K by setting up the displacement expression of singular elements. This method first calculates the displacements with stochastic elements, and then gets ratio of K by relations between the displacements and K values. The precision of this method is relatively low under the influence of singularity. The calculations of J-integral can reduce the singularity effect at crack tip. K can be calculated by J-integral that uses the relations between J-G and G-K in the linear elasticity condition. However, the expression of J-integral is not a form of numerical calculation and can not be programmed by computer. This study simplifies constitutive equations using the penalty function method and develops the power-law linear stochastic finite element method (SFEM) on the basis of the previous works by others.

2. Power-law nonlinear stochastic finite element method

2.1. General remarks

At the middle of 1980s, Kumar V. *et al.* from General Electric (GE) proposed a method of plastic fracture analysis for Electric Power Research Institute (EPRI). The core of this method is transforming a complex elastic-plastic crack problem into an addition of elasticity and plasticity. Then the solution of elastic-plastic problem is given by numerical method. This paper is aim to develop stochastic finite element method of elastic-plastic fracture mechanics based on the work of EPRI. Therefore simplified penalty function is presented to construct power-law nonlinear SFEM.

Incompressible volume problem exists in pure plastic solution. That cannot produce much difficulty in the plane stress condition. In the plane strain or other 3-D stress condition, however, it will result in the singularity of finite element stiffness matrix. In addition, the power-law constitutive relationship will make material nonlinearity for numerical calculations.

Goldman *et al.* used the corrected constitutive equation [2]

$$\frac{\varepsilon_{ij}}{\varepsilon_0}=\frac{3}{2}\alpha\left[\frac{\sigma_e}{\sigma_0}\right]^{n-1}\frac{s_{ij}}{\sigma_0}+\frac{1}{3}\beta\left[\frac{\sigma_h}{\sigma_0}\right]^{n-1}\frac{\sigma_{kk}}{\sigma_0}\delta_{ij} \tag{1}$$

The pure power-law and full plastic solutions of the center crack plate is got by $\beta\to 0$ in the plane strain. Needleman and Shih (1978) proposed a finite element solution method that realizes volume incompressible by removing node freedom. Shih and Needleman (1984) [3] proposed a method that uses penalty function to solve the full plastic crack equation. In the implement of finite element, however, constitutive equations are also corrected: stress-strain relationship with $n>1$ and $\varepsilon<\varepsilon_0$ in the uniaxial stress condition can be expressed as

$$\frac{\varepsilon}{\varepsilon_0}=\frac{n}{n-1}-\left\{\frac{1+n}{(n-1)^2}-\left[\frac{\sigma}{\sigma_0}+\frac{1}{n-1}\right]^2\right\}^{1/2} \tag{2}$$

The above methods establish the foundation of full plastic crack solution. The shortage lies in that the adopted formulas are complicated and inconvenient for the element finite formula deduction. Therefore, with the former work of elastic-plastic stochastic finite element, the paper introduces a newly simple constitutive equation with power-law nonlinear penalty function and finite element formulations are given based on the nonlinear continuous medium mechanics.

2.2. Penalty function method for solving incompressible problem

Assumed that the deformation of volume is elastic and the drive force of deformation is provided by hydrostatic pressure in the continuous medium mechanics, we can get:

$$\varepsilon_v=\sigma_m/\beta \tag{3}$$

where ε_v is volume strain; σ_m is hydrostatic pressure; β is penalty factor meaning volume elastic modulus in mechanics.

While σ_m is finite value, we get

$$\lim_{\beta\to\infty}\varepsilon_v=\lim_{\beta\to\infty}\sigma_m/\beta \tag{4}$$

Eq.(4) is the mechanics principle of solving incompressible problem using penalty function FEM. Being consistent with usual expressing form of finite element, the formulation is given by

$$\sigma=D\varepsilon+\beta m^T m\varepsilon \tag{5}$$

where D is the performance matrix of incompressible constitutive material. $m=(1, 1, 1, 0, 0, 0)^T$ at 3-D stress.

$$\varepsilon/\varepsilon_0=\alpha\left(\sigma/\sigma_0\right)^N \tag{6}$$

where α is material constant; N is power-law exponent; ε_0 and σ_0 are reference values of strain and stress; N is the hardening exponent with $N=1$ (linear elastic state) and $N\to\infty$ (ideal plastic state).

Apply unitary curve postulate of complex stress and generalize it into 3-D, we get

$$\sigma=\frac{2}{3}\alpha^{-1/N}\left[\frac{\varepsilon}{\varepsilon_0}\right]^{(1-N)/N}\left[\frac{\sigma}{\sigma_0}\right]\varepsilon+\beta m^T m\varepsilon \tag{7}$$

Substitute Eq.(7) into virtual work equation in the form of finite element

$$\int_{\Omega} b^T \sigma d\Omega + f = 0 \tag{8}$$

Finite element formulation is given as

$$kd + f = 0 \tag{9}$$

where b is strain matrix; k is stiffness matrix; d is node displacement vector; f is node strength vector.

Perturbation method is a powerful tool that analyzes the nonlinear problem. It is widely applied in the mechanics and other engineering fields. Handa [4] first developed perturbation SFEM and applied it into structure statics analysis. Hisada *et al.*[5-7] took a further study on this method. Erik Vanmarcke [8] introduces local average theory of random field into stochastic finite element, which is made more available for application.

Assumed that one parameter of general structure, z, is stochastic perturbation quantity. z can be expressed as the sum of definitive component and stochastic component, namely

$$z = z_0 (1 + \alpha) \tag{10}$$

where z_0 is mean value; α is the stochastic quantity with mean value being zero, representing the stochastic performance of parameters.

Assume small perturbation and apply the perturbation method, we can get

$$k = k_0 + \sum_{i=1}^{n} \frac{\partial k}{\partial \alpha_i} \alpha_i + \frac{1}{2} \sum_{i=1}^{n} \sum_{j=1}^{n} \frac{\partial^2 k}{\partial \alpha_i \partial \alpha_j} \alpha_i \alpha_j \tag{11}$$

$$F = F_0 + \sum_{i=1}^{n} \frac{\partial F}{\partial \alpha_i} \alpha_i + \frac{1}{2} \sum_{i=1}^{n} \sum_{j=1}^{n} \frac{\partial^2 F}{\partial \alpha_i \partial \alpha_j} \alpha_i \alpha_j \tag{12}$$

$$\delta = \delta_0 + \sum_{i=1}^{n} \frac{\partial \delta}{\partial \alpha_i} \alpha_i + \frac{1}{2} \sum_{i=1}^{n} \sum_{j=1}^{n} \frac{\partial^2 \delta}{\partial \alpha_i \partial \alpha_j} \alpha_i \alpha_j \tag{13}$$

where K_0, F_0, δ_0 are the value of each random variable, K, F, δ, at mean value. $\alpha_i = X_i - m_{xi}$ is minor perturbation quantity while random variable X_i is at the mean value m_x.

According to the second-order perturbation method, Eq.(1) can be expressed as the following equations:

$$\delta_0 = k_0^{-1} F_0 \tag{14}$$

$$\frac{\partial \delta}{\partial \alpha_i} = k_0^{-1} \left(\frac{\partial F}{\partial \alpha_i} - \delta_0 \frac{\partial k}{\partial \alpha_i} \right) \tag{15}$$

$$\frac{\partial^2 \delta}{\partial \alpha_i \partial \alpha_j} = k_0^{-1} \left(\frac{\partial^2 F}{\partial \alpha_i \partial \alpha_j} - \delta_0 \frac{\partial^2 k}{\partial \alpha_i \partial \alpha_j} - \frac{\partial \delta \partial k}{\partial \alpha_i \partial \alpha_j} - \frac{\partial k \partial \delta}{\partial \alpha_i \partial \alpha_j} \right) \tag{16}$$

Expand Eq.(13) by second-order Tailor series at the mean value, and the second-order means and variances of node displacement can be approximately given as

$$E[\delta] \cong \delta_0 + \frac{1}{2}\sum_{i=1}^{n}\sum_{j=1}^{n}\frac{\partial^2 \delta}{\partial\alpha_i\partial\alpha_j}\mathrm{cov}\left[\alpha_i\alpha_j\right] \tag{17}$$

$$Var[\delta] \cong \sum_{i=1}^{n}\sum_{j=1}^{n}\frac{\partial^2 \delta}{\partial\alpha_i\partial\alpha_j}\mathrm{cov}\left[\alpha_i\alpha_j\right] \tag{18}$$

As the same principle, stress can be extended at the mean value as follows:

$$E[\sigma] \cong \sigma_0 + \frac{1}{2}\sum_{i=1}^{n}\sum_{j=1}^{n}\frac{\partial^2 \sigma}{\partial\alpha_i\partial\alpha_j}\mathrm{cov}\left[\alpha_i\alpha_j\right] \tag{19}$$

$$Var[\sigma] \cong \sum_{i=1}^{n}\sum_{j=1}^{n}\frac{\partial^2 \sigma}{\partial\alpha_i\partial\alpha_j}\mathrm{cov}\left[\alpha_i\alpha_j\right] \tag{20}$$

3. Wood fracture parameter and its ratio

J-integral is an important parameter in the elastic-plastic fracture mechanics. It can describe the strength of stress-strain field at crack tip, as well as be determined conveniently by experiments. Through the relation of *J-K*, stress strength factor *K* is obtained as fracture parameter, namely

$$k = \sqrt{E'J} \tag{21}$$

In the above equation, E' can be expressed as $E' = E/(1-v)$ for the plane strain problem.

$$\frac{\partial k}{\partial \alpha_j} = \frac{1}{2k}\left| J\frac{\partial E'}{\partial \alpha_j} + E\frac{\partial J}{\partial \alpha_j}\right| \tag{22}$$

For 2-D problem, *J*-integral is defined to describe the mechanics state near the plane crack tip by Rice [9]. That is

$$J = \int_\Gamma \left(w dy - T_i \frac{\partial u}{\partial x} ds \right) \tag{23}$$

where w is the strain energy density in the uniform monotonic load state; T represents the outside stress vector operated at the micro-arc ds of curve $\boldsymbol{\Gamma}$ and u is the displacement vector there. Expand *J*-integral by Taylor series at the mean value and means and variances are given by

$$E[J] \cong J_0 + \frac{1}{2}\sum_{i=1}^{n}\sum_{j=1}^{n}\frac{\partial^2 J}{\partial\alpha_i\partial\alpha_j}\mathrm{cov}\left[\alpha_i\alpha_j\right] \tag{24}$$

$$Var[J] \cong \sum_{i=1}^{n}\sum_{j=1}^{n}\frac{\partial^2 J}{\partial\alpha_i\partial\alpha_j}\mathrm{cov}\left[\alpha_i\alpha_j\right] \tag{25}$$

Seen from Eq.(25), the linchpin of random analysis for crack volume is the calculation of $\partial J/\partial\alpha_j$. Perform partial derivative for *J*-integral of Rice on each side.

$$\frac{\partial J}{\partial \alpha_j} = \int_\Gamma \left[\frac{\partial w}{\partial \alpha_j} dL + w \frac{\partial dy}{\partial \alpha_j} - \frac{\partial}{\partial \alpha_j} \left(T_i \frac{\partial u}{\partial T} \right) ds - T_i \frac{\partial u}{\partial T} \frac{\partial ds}{\partial \alpha_j} \right] \tag{26}$$

where

$$\begin{aligned} \frac{\partial w}{\partial \alpha_j} = \frac{1}{2} \Bigg\{ & \frac{\partial \sigma_{TT}}{\partial \alpha_j} \frac{\partial u}{\partial T} + \frac{\partial \sigma_{TL}}{\partial \alpha_j} \left(\frac{\partial u}{\partial L} + \frac{\partial v}{\partial T} \right) + \frac{\partial \sigma_{TL}}{\partial \alpha_j} \frac{\partial v}{\partial L} + \sigma_{TT} \frac{\partial}{\partial T} \left(\frac{\partial u}{\partial \alpha_j} \right) \\ & + \sigma_{TL} \left[\frac{\partial}{\partial \alpha_j} \left(\frac{\partial u}{\partial \alpha_j} \right) + \frac{\partial}{\partial T} \left(\frac{\partial L}{\partial \alpha_j} \right) \right] + \sigma_{yy} \frac{\partial}{\partial y} \left(\frac{\partial L}{\partial \alpha_j} \right) \Bigg\} \end{aligned} \tag{27}$$

$$\begin{aligned} \frac{\partial}{\partial \alpha_j} \left(T_i \frac{\partial u}{\partial T} \right) = & (\sigma_{TT} n_1 + \sigma_{TL} n_2) \frac{\partial}{\partial T} \left(\frac{\partial u}{\partial \alpha_j} \right) + (\sigma_{TL} n_1 + \sigma_{LL} n_2) \frac{\partial}{\partial T} \left(\frac{\partial v}{\partial \alpha_j} \right) \\ & + \left[\frac{\partial \sigma_{TT}}{\partial \alpha_j} n_1 + \frac{\partial \sigma_{TL}}{\partial \alpha_j} n_2 + \sigma_{TT} \frac{\partial n_1}{\partial \alpha_j} + \sigma_{TL} \frac{\partial n_2}{\partial \alpha_j} \right] \frac{\partial u}{\partial T} + \left[\frac{\partial \sigma_{TL}}{\partial \alpha_j} n_1 + \frac{\partial \sigma_{TL}}{\partial \alpha_j} n_2 + \sigma_{TL} \frac{\partial n_1}{\partial \alpha_j} + \sigma_{RL} \frac{\partial n_2}{\partial \alpha_j} \right] \frac{\partial v}{\partial T} \end{aligned} \tag{28}$$

In the condition of power-law full plastic, w is given as:

$$w = \alpha \sigma_0 \varepsilon_0 \left| \frac{n}{n+1} \right| \left| \frac{\sigma_e}{\sigma_0} \right|^{n+1} \tag{29}$$

According to single curve hypothesis of total theory, only if $\sigma_0 = E\sigma_0$, we get

$$w = \alpha^{1/n} E \varepsilon_0^2 \left| \frac{n}{n+1} \right| \left| \frac{\varepsilon_e}{\varepsilon_0} \right|^{(n+1)/n} \tag{30}$$

While random variable $\alpha_i = \zeta$ (here ζ is the random quantity of crack length), we can get

$$\frac{\partial w}{\partial \alpha_i} = \alpha^{1/n} E \varepsilon_0^2 \frac{1}{\varepsilon_e} \left| \frac{\varepsilon_e}{\varepsilon_0} \right|^{(n+1)/n} \frac{\partial \varepsilon_e}{\partial \alpha_i} \tag{31}$$

As the same principle, random variables E, n, ε_0 can be calculated by SFEM.

4. Fracture parameter ratio calculated by SFEM and difference check

For the analysis of elastic-plastic fracture, the calculation of fracture parameter adopts the definitive FEM. Its ratio can be get approximately by difference method, namely finite difference method in the fracture analysis. The accuracy of this method is concerned with step length of increment. It can be checked mainly by the SFEM calculation result.

For each basic random variable α_i, apply difference formula, we can get

$$\frac{\partial f}{\partial \alpha_i} = \frac{f(\alpha_i + \Delta \alpha_i) - f(\alpha_i)}{\Delta \alpha_i} \tag{32}$$

where $f(\alpha_i+\Delta\alpha_i)$, $f(\alpha_i)$ can be got by definitive FEM. For example, compact tension (CT) plate with single edge crack as Fig. 1 shown is made as the test specimen. The elasticity modulus E, Poisson ratio μ, power-law exponent n is chosen as basic random variables. The result is seen from Table 1. Here variation ratio of J-integral versus the basic random variable, J/α_i, is get by SFEM, while $\Delta J/\Delta\alpha_i$ is get by difference method.

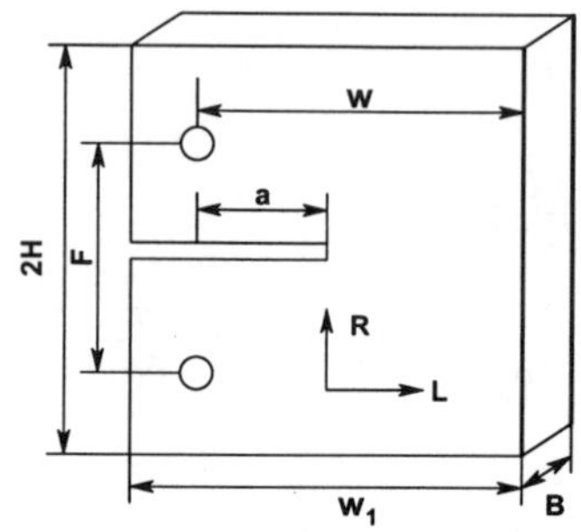

Fig. 1. Compact tension specimen

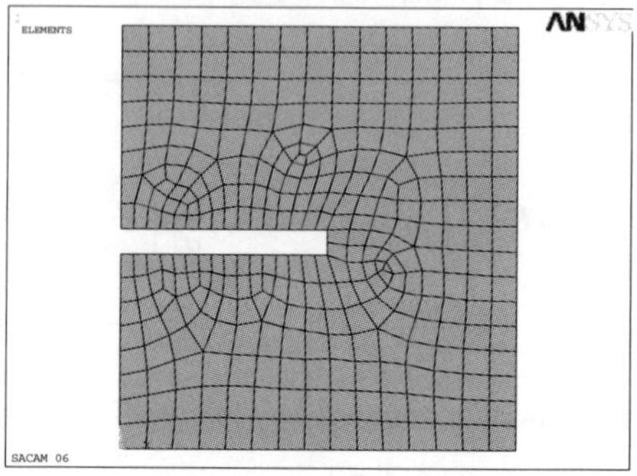

Fig. 2. Finite element meshing

Table 1. Variation rate of J-integral and its test result

	SFEM: $\partial J/\partial\alpha_i$	Difference Method: $\Delta J/\Delta\alpha_i$	Error
N=1 J_0=0.05907			
E	0.3757E-06	0.3804E-06	1.25%
ϵ	0.4326E+00	0.4098E+00	5.27%
n=1 J_0=0.05907			
E	0.9806E-06	0.9473E-06	3.40%
ϵ	0.8315E+00	0.8027E+00	3.46%

Table 2. Calculating values of fracture parameters (J- integral, δ- crack opening displacement (COD), a- crack size chosen as basic stochastic parameter)

	J/M Pa-m	δ/m	$\partial J/\partial a$	$\partial\delta/\partial a$
σ_0 = 100MPa	4.52E01	5.0845-E04	0.2385	3.82E-01
σ_0 = 130MPa	5.48E01	6.1325-E04	0.2457	4.68E-01
σ_0 = 150MPa	7.53E01	9.0386-E04	0.2981	6.91E-01

5. Stochastic analysis example of fracture

The work of stochastic analysis is to calculate the means and variances. The mean value can be got by the definitive FEM. Variances can be calculated with parameter variation ratio. According to the standard GB4161-84, compact tension specimen with single edge crack is machined. And its corresponding parameter is in the following: elastic modulus E=1.627×10^{4}MPa; parameter coefficient α=1; Poisson ratio μ_{LT}= 0.49; yield limit σ_0 is selected as 100MPa, 130Mpa, 150MPa, respectively; hardening exponent N=0.5; tension load P=1000N. Since the test specimen thickness t/w is more than 0.25, it can be treated as the plane strain state. The result of nonlinear SFEM in the condition of plane strain is listed in Table 2.

6. Wood fracture strength factor and its variation ratio

After fracture parameter J and its variation ratio versus basic stochastic derived, we can get the statistic characteristic of fracture parameter K_I based on basic stochastic variables. Loads are applied to the test specimen, and we can get gradient vector contour of the strain variety by simulation software ANSYS as Fig. 3 shown. From the crack tip, we define the paths on the strain variation rate in the x direction and y direction, respectively. Fig. 4 shows the curve of strain variation rate. With the calculation of finite element, the change curve of K_I is fitted. Fig. 4 illustrates the variation regulation of K_I followed different original crack size. Then the transformed graphics can reflect directly the variation curves of K_I along with different original crack. In order to prove the validity of SFEM results, a softwood species, Maor-Mountain larch (*Larix*) from the Forestry Research Center of Heilongjiang Province, is selected for the study. The tree growth age is 30 years with a height of 16 meter, a breast width of 32 centimeter and a density of 485±40 kg·m^{-1}. Discs are taken at breast height and one sector is sawed out. There is no standardized test procedure or test specimen configuration for mode I fracture testing of wood material. In this study, the geometry of the specimen cut out along the longitudinal direction is similar to a CT-specimen. CT specimens are cut are indicated in Fig. 2. Tests are performed referring to metal material plane fracture test methods (GB4161-84). All tests were carried out using *Maor Mountain Larch*, previously kiln dried and stored until the stabilization of the moisture content at 12%. Specimens were conditioned at 20±2°C and 65±5% RH.

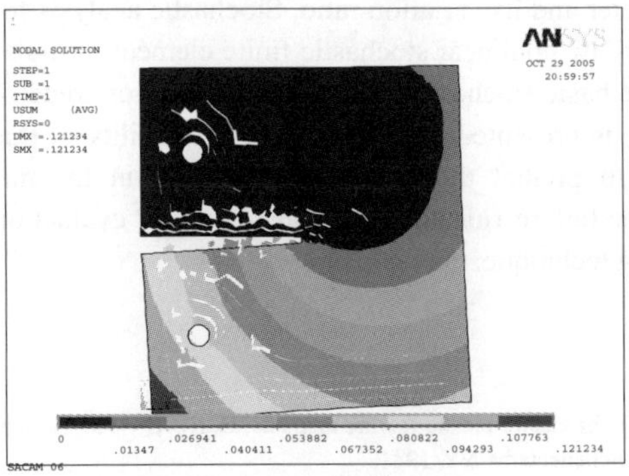

Fig. 3. Contour of wood fracture in displacement

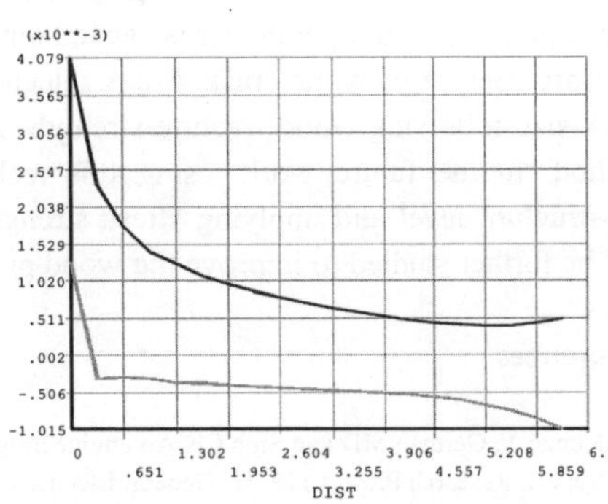

Fig. 4. the curve of strain variation rate

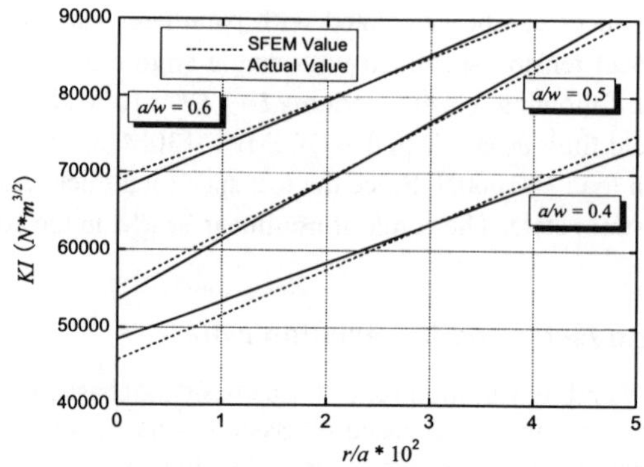

Fig. 5. The curve of stress strength factors

Four groups of test specimen are performed for the corresponding mechanics tension experiments. Variation curve graphics is drawn in the three different cases $a/w = 0.4$, $a/w = 0.5$ and $a/w = 0.6$. Compared the curve from the two methods, variance of the two results is within one standard deviation. We also conclude from Fig. 3 that while σ_0 is invariable, K_I will increase along with the crack size growing, and the ability of macro crack instable-extension is better. For this reason, the investigations adopting stochastic finite element method to analyze and predict the wood fracture performance are feasible and practical.

7. Conclusions

The paper simplifies constitutive equations with penalty function on the basis of stochastic finite element formulations, and develops power-law nonlinear SFEM. SFEM is presented for the wood fracture mechanics for calculating the fracture parameter and its variation ratio. Stochastic analysis for elastic-plastic fracture parameter is carried out. Power-law nonlinear stochastic finite element formulations are considered while crack size is selected as the basic stochastic variable. The variation trend of crack size following wood fracture strength factor K_I is presented. So it proves the feasibility of this method. In the future work, using this technique to predict the location of fracture in the micro-structure level and applying stress strength factor failure rule to the wood reliability evaluation will be further studied to improve the wood processing technique.

References

[1] Kumar V, German MD and Shih CF. An engineering approach for elastic-plastic facture analysis. EPRI NP-1931 Topical Report. Research Project 1237-1, General Electric Company, Schenectady, NY, 1981.
[2] Goldman NLand Hutchinson JW. Fully plastic crack problems: the centre crack tip under plane strain. Int. J. Solids Structures 1975; 11: 575–591.
[3] Shih CF, *et al.* Fully plastic crack problems part I: solutions by a penalty method. Journal of Applied Mechanics 1984; 51: 48–56.

[4] Handa K and Anderson K. Application of finite element methods in statistical analysis of structures. In: Proc 3rd Int Conf on Structure Safety and Reliability. Nroway: Trondheim. 409–417, 1981.
[5] Hisada T and Nakagiri S. Stochastic finite element method developed for structure. Safety and reliability. In: Proc 3rd Int Conf on Structure Safety and Reliability. Nroway: Trondheim. 395–408, 1981.
[6] Nakagiri S and Hisada T. Stochastic finite element method applied to structure analysis with uncertain parameter. In: Proc Int Conf on FEM. Australia. 206–211, 1982.
[7] Hisada T and Nakagiri S. Role of the stochastic finite element method in structures. Safety and reliability. In: Proc 4th Int Conf on Struct Satety and Reliability. Japan: Kobe 213–219, 1985.
[8] Erik Vanmarcke. Random field: Analysis and Synthesis. Cambridge: The MIT Press 1983.
[9] J. R. Rice, A path independent integral and approximate analysis of strain concentration by notches and cracks, J. Appl. Mech. 1968; 35: 379–386.
[10] Wilson WK. Finite element methods for elastic bodies containing cracks. Mech. Of Fract. (G. C. Sihed.), 1973.
[11] Zao J-P, *et al.* The application of 2D elastic stochastic finite element method in the field of fracture mechanics. The International Journal of Pressure Vessels and Piping 1997; 71: 169–174.
[12] Li Jiankang, *et al.* A penalty finite element method for the full plastic problems of pure power-hardning materials. Chinese Journal of Computational Mechanics 1999; 16: 379–389,.
[13] Kumar V, German MD and Wilkening WW. Further development in the engineering approach for elastic-plastic fracture analysis. EPRI NP-3607 Topical Report. Research Project 1237–1, General Electric Company, Schenectady, NY, 1983.

A novel numerical computation based on FEM for wood temperature distribution field

Liping Sun[a], Mingbao Li[b,*] and Shiqiang Zheng[a]

[a]*School of Electromechanical Engineering, Northeast Forestry University*
[b]*School of Civil Engineering, Northeast Forestry University*
No. 26 Hexing Road, Harbin, 150040, China

Abstract

A novel numerical computation based on finite element method (FEM) for temperature distribution field of anisotropic wood is proposed in this paper. The basic theory for the analysis of wood steady heat conduction is presented. Based on finite element theory, numerical method analysis and calculation for wood temperature field is performed, and also the related experiments are performed to prove the validity of computer simulation result. Through comparison numerical simulation solutions with testing experiment results, the simulation results is very consistent with the actual values, and the errors between them lie in the range of −3.5%~ 4.2% (full range). This performance can meet the requirements of engineering calculation precision. It is also proved that this method on wood heat conductivity is feasible and practical.

PACS: 02.70.D; 83.80.Mc

Keywords: Wood temperature field; Finite element; Numerical simulations

1. Introduction

Wood is an anisotropic, porous material with complicated cellular and macro scale structure features and material properties. Heat stress and deformation of wood structural level in the wood can be controlled effectively by modeling temperature field and predicting accurately the temperature distributing gradient during the heating process. Heat transfer on a wood board scale, which has resulted in numerous empirical and theoretical models, has been studied extensively. Above these, we can carry on in-depth investigations on the other wood applications, such as wood drying, wood preservation, insecticide, etc. Development of these models has been reviewed by Kamke and Vanek [1]. Most heat

---

* Corresponding author. Tel.: +86-451-82191615; fax: +86-451-82191406.
E-mail address: zsqrq@yahoo.com.cn (M.B. Li).

This work is supported by Natural Science Funds for Distinguished Young Scholar of Heilongjiang Province, China.

transfer models for wood are 2-Dimensional and ignore the longitudinal heat transfer effects because of its relatively long path compared to the transverse heat transfer in the board [2,3]. Most models also do not differentiate heat properties with respect to the ideal axial symmetry, radial, or tangential directions, but assume averaged heat transfer characteristics (Forest Products Laboratory, US) based only on an average species-density and moisture relationship. In some boards, however, the radial and tangential orientation of the rings can curve such that the radial directions could act both horizontally and vertically in the same board that could have a significant effects on the heat properties. Two 2-D finite element models have been developed that take these parameters into accout by Hongmei Gu and John F. Hunt [4].

The present study focuses on numerical simulations of wood 2-D temperature field. Calculations of wood steady heat conduction are carried out. While wood is applied to heat treatment, the temperature field along the longitudinal direction has the same change and the fields vary gradually along both the radial and tangential directions. So it can be processed by 2-D steady temperature field. Finite difference method is difficult to solve the complex geometric boundary condition for anisotropic material. For this reason, finite element method (FEM) is adopted. Assumed that the approximation function representing each element and the boundary among elements is continuous, complete solutions are calculated combined with each single solution. In order to analyze the wood heat transfer, the paper proposes a finite element algorithm that uses a rectangle element with a bilinear basis function to solve heat diffusion equation. This method is based on that the thermo-physical property of wood does not vary with temperature, namely focuses on the linear heat conduction problem. According to the above mentioned theories, the numerical work is performed to develop the temperature field distribution of the heated wood using computer program by ANSYS, which is a general purpose finite element modeling package for numerically solving a wide variety of mechanical problems.

2. Basic theory of wood steady heat conduction analysis

2.1. Wood heat diffusion equation

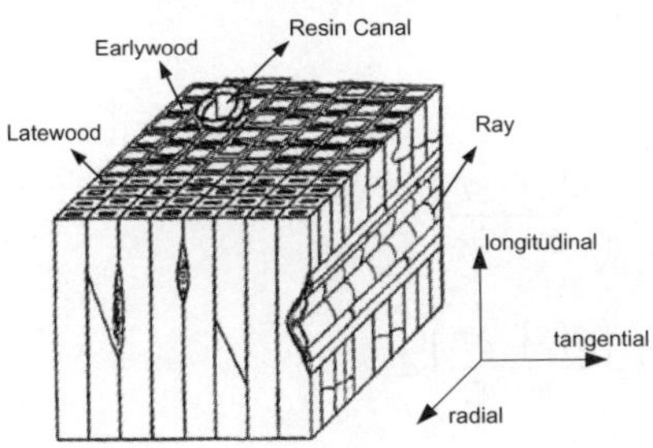

Fig. 1. 2-D wood temperature field

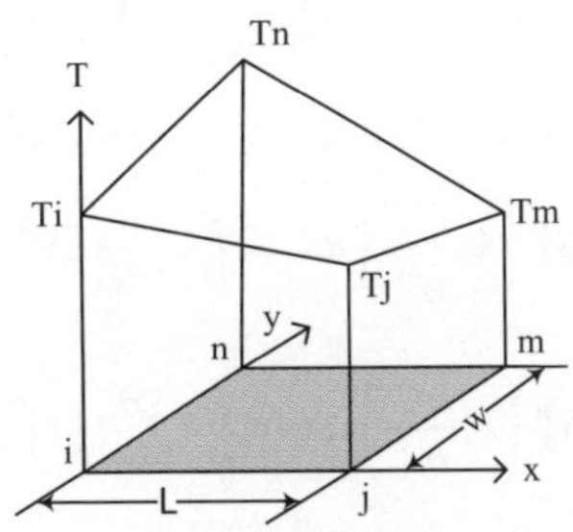

Fig. 2. Typical rectangle unit

Wood is usually applied in the form of rectangle board, so it is considered that the temperature field of each cross section along the longitudinal direction transfers similarly. Therefore, it can be treated approximately as 2-D temperature field. Applied energy conservation law of thermodynamics [6], heat diffusion equation is set up in the Cartesian coordinates system as follows:

$$k_x \frac{\partial^2 T}{\partial x^2} + k_y \frac{\partial^2 T}{\partial x^2} + \dot{q} = 0 .$$

where $\dot{q}$ is the unit heat volume; k_x and k_y are the effective heat conduction coefficient in the x (tangential) and y (radial) directions, respectively; T is wood temperature.

2.2. Equation deducing of rectangle element

2-D bilinear rectangle element can be expressed by the node temperature and shape functions, that is

$$T^{(e)} = \begin{bmatrix} S_i & S_j & S_m & S_n \end{bmatrix} = [S]\,[T]^T .$$

where $[T]$ is the node temperature; $[S]$ is the shape function.

$$[T] = \{T_i \quad T_j \quad T_m \quad T_n\};\ [S] = \{S_i \quad S_j \quad S_m \quad S_n\} = \left\{\left(1-\frac{x}{l}\right)\left(1-\frac{y}{w}\right) \quad \frac{x}{l}\left(1-\frac{y}{w}\right) \quad \frac{xy}{lw} \quad \frac{y}{w}\left(1-\frac{x}{l}\right)\right\}.$$

Applying *Galerkin* method to the heat diffusion equation, we can get

$$\int_A [S]^T \left(k_x \frac{\partial^2 T}{\partial x^2} + k_y \frac{\partial^2 T}{\partial y^2} + \dot{q} \right) dA = 0 .$$

The equation consists of three major integrals

$$\int_A [S]^T \left(k_x \frac{\partial^2 T}{\partial x^2} \right) dA + \int_A [S]^T \left(k_y \frac{\partial^2 T}{\partial y^2} \right) dA + \int_A [S]^T \dot{q}\, dA = 0 . \tag{1}$$

Assumed that the second derivative representing temperature linear function is zero, second order can be reduced to first order. Chain rule is applied in the form

$$[S]^T \frac{\partial^2 T}{\partial x^2} = \frac{\partial}{\partial x}\left([S]^T \frac{\partial T}{\partial x} \right) - \frac{\partial [S]^T}{\partial x^2} \frac{\partial T}{\partial x} . \tag{2}$$

Substitute Eq.(2) into the first two terms of Eq.(1):

$$\int_A [S]^T \left(k_x \frac{\partial^2 T}{\partial x^2} \right) dA = \int_A k_x \frac{\partial}{\partial x}\left([S]^T \frac{\partial T}{\partial x} \right) dA - \int_A k_x \left(\frac{\partial [S]^T}{\partial x} \frac{\partial T}{\partial x} \right) dA \tag{3}$$

$$\int_A [S]^T \left(k_y \frac{\partial^2 T}{\partial y^2} \right) dA = \int_A k_y \frac{\partial}{\partial y}\left([S]^T \frac{\partial T}{\partial y} \right) dA - \int_A k_y \left(\frac{\partial [S]^T}{\partial y} \frac{\partial T}{\partial y} \right) dA \tag{4}$$

where:

$$\frac{\partial T}{\partial x} = \frac{\partial}{\partial x}[S]\,[T]^T = \frac{1}{lw}[(-w+y) \quad (w-y) \quad y \quad -y]\,[T]^T \tag{5}$$

$$\frac{\partial[S]^T}{\partial x}=\frac{1}{lw}\begin{Bmatrix}-w+y\\ w-y\\ y\\ -y\end{Bmatrix}=\frac{1}{lw}[B]. \tag{6}$$

Combined with Eq.(5), Eq.(6), the latter term of Eq.(3) can be expressed as

$$-\int_A k_x\left(\frac{\partial[S]^T}{\partial x}\frac{\partial T}{\partial x}\right)dA=-k_x\int_A\frac{1}{(lw)^2}[B]\,[B]^T[T]^T\,dA=-\frac{k_x w}{6l}\begin{bmatrix}2&-2&-1&1\\-2&2&1&-1\\-1&1&2&-2\\1&-1&-2&2\end{bmatrix}[T]^T \quad . \tag{7}$$

Also Eq.(3) can be given in the form

$$-\int_A k_y\left(\frac{\partial[S]^T}{\partial y}\frac{\partial T}{\partial y}\right)dA=-\frac{k_y l}{6w}\begin{bmatrix}2&-2&-1&1\\-2&2&1&-1\\-1&1&2&-2\\1&-1&-2&2\end{bmatrix}[T]^T. \tag{8}$$

Then the latter term in Eq.(1), namely thermal load term, is calculated:

$$-\int_A[S]^T\dot{q}\,dA=\dot{q}\int_A\begin{Bmatrix}S_i\\S_j\\S_m\\S_n\end{Bmatrix}dA=\frac{1}{4}\dot{q}\,A\begin{Bmatrix}1\\1\\1\\1\end{Bmatrix}. \tag{9}$$

And the integral terms $\int_A k_x\frac{\partial}{\partial x}\left([S]^T\frac{\partial T}{\partial x}\right)dA$ and $\int_A k_y\frac{\partial}{\partial y}\left([S]^T\frac{\partial T}{\partial y}\right)dA$ can be expressed by unit boundary integral using Green theory as follows:

$$\int_A k_x\frac{\partial}{\partial x}\left([S]^T\frac{\partial T}{\partial x}\right)dA=\int_\tau k_x[S]^T\frac{\partial T}{\partial x}\cos\theta\,d\tau,\quad \int_A k_y\frac{\partial}{\partial y}\left([S]^T\frac{\partial T}{\partial y}\right)dA=\int_\tau k_y[S]^T\frac{\partial T}{\partial y}\sin\theta\,d\tau \tag{10}$$

where τ is the unit boundary, θ is the angle of unit normal. Conduction boundary condition and anisotropic behavior of wood into are considered:

$$-k_x\frac{\partial T}{\partial x}=h_x\left(T-T_f\right),\qquad -k_y\frac{\partial T}{\partial y}=h_y\left(T-T_f\right)\quad . \tag{11}$$

where h_x, h_y are the coefficients of heat transfer in the x and y direction of temperature field, respectively; T_f is the ambient air temperature; $\frac{\partial T}{\partial y}$ is the temperature gradient.

Take Eq.(11) into Eq.(10), respectively:

$$\int_{\tau} k_x [S]^T \frac{\partial T}{\partial x}\cos\theta \, d\tau = -\int_{\tau} h_x [S]^T (T - T_f)\cos\theta \, d\tau , \quad \int_{\tau} k_y [S]^T \frac{\partial T}{\partial y}\sin\theta \, d\tau = -\int_{\tau} h_y [S]^T (T - T_f)\sin\theta \, d\tau \quad (12)$$

$$[K]^{(e)} = \frac{h_x l_{ij}}{6} = \begin{bmatrix} 2 & 1 & 0 & 0 \\ 1 & 2 & 0 & 0 \\ 0 & 0 & 0 & 0 \\ 0 & 0 & 0 & 0 \end{bmatrix}, \quad [K]^{(e)} = \frac{h_y l_{jm}}{6} = \begin{bmatrix} 0 & 0 & 0 & 0 \\ 0 & 2 & 1 & 0 \\ 0 & 1 & 2 & 0 \\ 0 & 0 & 0 & 0 \end{bmatrix},$$

$$[K]^{(e)} = \frac{h_x l_{mn}}{6} = \begin{bmatrix} 0 & 0 & 0 & 0 \\ 0 & 0 & 0 & 0 \\ 0 & 0 & 2 & 1 \\ 0 & 0 & 1 & 2 \end{bmatrix}, \quad [K]^{(e)} = \frac{h_y l_{ni}}{6} = \begin{bmatrix} 2 & 0 & 0 & 1 \\ 0 & 0 & 0 & 0 \\ 0 & 0 & 0 & 0 \\ 1 & 0 & 0 & 2 \end{bmatrix}.$$

From Fig. 2, $l_{ij} = l_{mn} = l$, $l_{jm} = l_{in} = w$ can be got. Integrate these terms in Eq.(12) along the different border of conduction boundary condition and the load matrix is given as

$$\{F\}^{(e)} = \frac{h_x T_f l_{ij}}{2}\begin{Bmatrix} 1 \\ 1 \\ 0 \\ 0 \end{Bmatrix}, \quad \{F\}^{(e)} = \frac{h_y T_f l_{jm}}{2}\begin{Bmatrix} 0 \\ 1 \\ 1 \\ 0 \end{Bmatrix}, \quad \{F\}^{(e)} = \frac{h_x T_f l_{mn}}{2}\begin{Bmatrix} 0 \\ 0 \\ 1 \\ 1 \end{Bmatrix}, \quad \{F\}^{(e)} = \frac{h_y T_f l_{ni}}{2}\begin{Bmatrix} 1 \\ 0 \\ 0 \\ 1 \end{Bmatrix}.$$

So the conduction matrix is rearranged as the form:

$$[K]^{(e)} = \frac{k_x w}{6l}\begin{bmatrix} 2 & -2 & -1 & 1 \\ -2 & 2 & 1 & -1 \\ -1 & 1 & 2 & -2 \\ 1 & -1 & -2 & 2 \end{bmatrix} + \frac{k_y l}{6w}\begin{bmatrix} 2 & 1 & -1 & -2 \\ 1 & 2 & -2 & -1 \\ -1 & -2 & 2 & 1 \\ -2 & -1 & 1 & 2 \end{bmatrix}.$$

Finally, combined with element matrix, global matrix is obtained and the equation $[K]\ \{T\} = \{F\}$ is solved. Node temperature will be got with the relation.

3. Numerical computation

3.1. Equation deducing of rectangle element

In this paper, the heat conductivity property related to wood anatomical structures will be studied in the radial and tangential direction to examine the possible influence on the heat transfer in the two directions and to estimate theoretically the heat conductivity values. Before numerical simulations, it is necessary to determine the heat conduction coefficient and heat transfer coefficient of test species since these two coefficients are critical parameters in the analysis of wood temperature field. The test instrument in this study is Heat-Flow Meter of HFM436 series adopting heat flux method by NETZSCH Inc. The fixed size (6cm×6cm×1cm) test sample is inserted into the middle of two boards, and certain gradation of wood temperature is set to the sample. Heat flows through the sample is measured by the calibrated heat flux sensor. The sensor is placed closed to the sample between board and sample. Heat conduction coefficient is obtained by the measuring of sample thickness, temperature gradation and heat flow through sample. *Maor Mountain larch*, which is air-dried (its moisture content is approximately at 12%), is selected as the test species. After several valid measurements, the tangential heat conductivity k_x=0.1289W/m·K and the radial heat conductivity k_y= 0.1399W/m·K are got. With heating source set to 100ºC and exposed air temperature being 30ºC, the corresponding nature heat transfer coefficients in the tangential and radial directions are h_x =14.1W/m^2·K and h_y =16.7W/m^2·K, respectively. Based on these parameters, temperature distributions in the wood species will be studied further.

3.2. Numerical computation based on FEM

In this study, the cross section of the selected species which is cut along the longitudinal direction is square. Since the square is of symmetry, we select a quarter of the section as the object to analysis such that temperature field distribution of whole cross section is got. The section analyzed is then divided into 8 elements with 15 nodes. The solution can be inquired from Table 1.

Table 1. Relation between element and corresponding nodes

Element	i	j	m	n
(1)	1	2	6	5
(2)	2	3	7	6
(3)	6	7	11	10
(4)	5	6	10	9
(5)	4	5	8	9
(6)	8	9	13	12
(7)	9	10	14	13
(8)	10	11	15	14

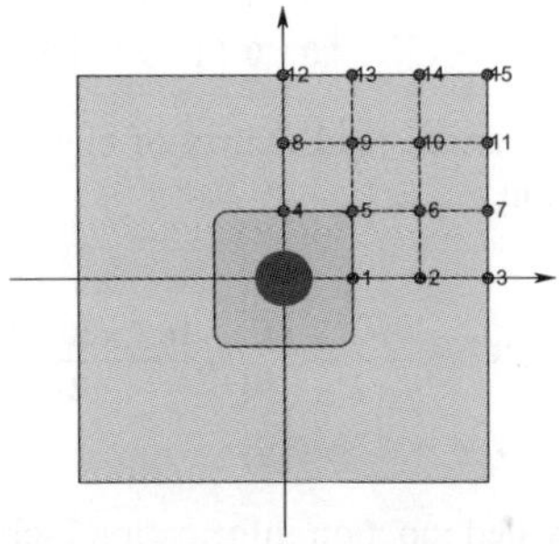

Fig. 3. Finite elements of board cross section

Conduction matrix of the element is:

$$[K]^e = \frac{k_x w}{6l}\begin{bmatrix} 2 & -2 & -1 & 1 \\ -2 & 2 & 1 & -1 \\ -1 & 1 & 2 & -2 \\ 1 & -1 & -2 & 2 \end{bmatrix} + \frac{k_y l}{6w}\begin{bmatrix} 2 & 1 & -1 & -2 \\ 1 & 2 & -2 & -1 \\ -1 & -2 & 2 & 1 \\ -2 & -1 & 1 & 2 \end{bmatrix} = \begin{bmatrix} 0.0896 & -0.0197 & -0.0448 & -0.0251 \\ -0.0197 & 0.0896 & -0.0251 & -0.0448 \\ -0.0448 & -0.0251 & 0.0896 & -0.0197 \\ -0.0251 & -0.0448 & -0.0197 & 0.0896 \end{bmatrix} = \begin{bmatrix} M \end{bmatrix}$$

where $l = w = 0.01$m.

In order to combine the elements conveniently, serial numbers are indicated on the up-side and right-side of each matrix.

$$[K]^{(1)} = \begin{bmatrix} M \end{bmatrix}\begin{matrix} 1 \\ 2 \\ 6 \\ 5 \end{matrix},\ [K]^{(2)} = \begin{bmatrix} M \end{bmatrix}\begin{matrix} 2 \\ 3 \\ 6 \\ 7 \end{matrix},\ [K]^{(3)} = \begin{bmatrix} M \end{bmatrix}\begin{matrix} 6 \\ 7 \\ 11 \\ 10 \end{matrix},\ [K]^{(4)} = \begin{bmatrix} M \end{bmatrix}\begin{matrix} 5 \\ 6 \\ 10 \\ 9 \end{matrix}$$

$$[K]^{(5)} = \begin{bmatrix} M \end{bmatrix}\begin{matrix} 4 \\ 5 \\ 9 \\ 8 \end{matrix},\ [K]^{(6)} = \begin{bmatrix} M \end{bmatrix}\begin{matrix} 8 \\ 9 \\ 12 \\ 13 \end{matrix},\ [K]^{(7)} = \begin{bmatrix} M \end{bmatrix}\begin{matrix} 9 \\ 10 \\ 14 \\ 13 \end{matrix},\ [K]^{(8)} = \begin{bmatrix} M \end{bmatrix}\begin{matrix} 10 \\ 11 \\ 15 \\ 14 \end{matrix}.$$

As already stated, conduction matrix and thermal load matrix will be affected by conduction boundary condition. Elements (2), (3) and (8) are affected by h_x as follows:

$$[K]^{(2)}=\frac{h_y l_{jm}}{6}\begin{bmatrix}0&0&0&0\\0&2&1&0\\0&1&2&0\\0&0&0&0\end{bmatrix}=\begin{bmatrix}0&0&0&0\\0&0.0577&0.0278&0\\0&0.0278&0.0577&0\\0&0&0&0\end{bmatrix}\begin{matrix}2\\3\\7\\6\end{matrix}=\begin{bmatrix}\\N\\\\\end{bmatrix},\quad [K]^{(3)}=\begin{bmatrix}\\N\\\\\end{bmatrix}\begin{matrix}6\\7\\11\\10\end{matrix},\quad [K]^{(8)}=\begin{bmatrix}\\N\\\\\end{bmatrix}\begin{matrix}10\\11\\15\\14\end{matrix}$$

Elements (6), (7) and (8) are affected by h_y as follows:

$$[K]^{(6)}=\frac{h_x l_{mn}}{6}\begin{bmatrix}0&0&0&0\\0&0&0&0\\0&0&2&1\\0&0&1&2\end{bmatrix}=\begin{bmatrix}0&0&0&0\\0&0&0&0\\0&0&0.0470&0.0235\\0&0&0.0235&0.0470\end{bmatrix}\begin{matrix}8\\9\\13\\12\end{matrix}=\begin{bmatrix}\\P\\\\\end{bmatrix}\begin{matrix}8\\9\\13\\12\end{matrix},\quad [K]^{(7)}=\begin{bmatrix}\\P\\\\\end{bmatrix}\begin{matrix}9\\10\\14\\13\end{matrix},\quad [K]^{(8)}=\begin{bmatrix}\\P\\\\\end{bmatrix}\begin{matrix}10\\11\\15\\14\end{matrix}$$

The thermal load matrix of elements (2) (3) (8) along *jm* side is affected by conduction boundary condition.

$$\{F\}^{(e)}=\frac{h_y T_f l_{jm}}{2}\begin{Bmatrix}0\\1\\1\\0\end{Bmatrix}=\frac{16.7\times 30\times 0.01}{2}\begin{Bmatrix}0\\1\\1\\0\end{Bmatrix}=\begin{Bmatrix}0\\2.5050\\2.5050\\0\end{Bmatrix}=\begin{Bmatrix}\\I\\\\\end{Bmatrix}$$

Provided junction information included are given as

$$\{F\}^{(2)}=\begin{Bmatrix}\\I\\\\\end{Bmatrix}\begin{matrix}2\\3\\7\\6\end{matrix},\quad \{F\}^{(3)}=\begin{Bmatrix}\\I\\\\\end{Bmatrix}\begin{matrix}6\\7\\11\\10\end{matrix},\quad \{F\}^{(8)}=\begin{Bmatrix}\\I\\\\\end{Bmatrix}\begin{matrix}10\\11\\15\\14\end{matrix}$$

Elements (6) (7) (8) effected by the boundary condition are developed as

$$\{F\}^{(e)}=\frac{h_x T_f l_{mn}}{2}\begin{Bmatrix}0\\0\\1\\1\end{Bmatrix}=\begin{Bmatrix}0\\0\\2.1150\\2.1150\end{Bmatrix}=\begin{Bmatrix}\\R\\\\\end{Bmatrix},\quad \{F\}^{(6)}=\begin{Bmatrix}\\R\\\\\end{Bmatrix}\begin{matrix}8\\9\\13\\12\end{matrix},\quad \{F\}^{(7)}=\begin{Bmatrix}\\R\\\\\end{Bmatrix}\begin{matrix}9\\10\\14\\13\end{matrix},\quad \{F\}^{(8)}=\begin{Bmatrix}\\R\\\\\end{Bmatrix}\begin{matrix}10\\11\\15\\14\end{matrix}$$

Collecting all elements matrixes together, apply element junction information given by adjacent others, then the total stiffness matrix will be

$$[K]=\begin{bmatrix}
0.0896&-0.0197&0&0&-0.0251&-0.0448&0&0&0&0&0&0&0&0&0\\
-0.0197&0.1792&-0.0197&0&-0.0448&-0.0699&-0.0251&0&0&0&0&0&0&0&0\\
0&-0.0197&0.1453&0&0&0.0027&-0.0448&0&0&0&0&0&0&0&0\\
0&0&0&0.0896&-0.0197&0&0&-0.0251&-0.0448&0&0&0&0&0&0\\
-0.0251&-0.0448&0&-0.0197&0.2688&-0.394&0&-0.0448&-0.0502&-0.0448&0&0&0&0&0\\
-0.0448&-0.0699&0.0027&0&-0.0394&0.4141&-0.0394&0&-0.0448&-0.0502&-0.0448&0&0&0&0\\
0&-0.0251&-0.0448&0&0&-0.0394&0.2349&0&0&-0.0448&0.0027&0&0&0&0\\
0&0&0&-0.0251&-0.0446&0&0&0.1792&-0.0394&0&0&-0.0448&-0.0251&0&0\\
0&0&0&-0.0448&-0.0502&-0.0448&0&0.0394&0.3584&-0.0394&0&-0.0251&-0.0448&-0.0448&-0.0251\\
0&0&0&0&-0.0448&0.0502&-0.0446&0&-0.0394&0.3584&0.0163&0&-0.0448&-0.0502&-0.0448\\
0&0&0&0&0&-0.0448&-0.0027&0&0&-0.0394&0.2906&0&0&-0.0448&0.0027\\
0&0&0&0&0&0&0&-0.0448&-0.0251&0&0&0.1366&0.0038&0&0\\
0&0&0&0&0&0&0&-0.0251&0.0699&-0.0448&0&0.0038&0.2732&0.0038&0\\
0&0&0&0&0&0&0&0&-0.0446&-0.0502&-0.0448&0&0.0038&0.2732&0.0038\\
0&0&0&0&0&0&0&0&0&-0.0448&0.0027&0&0&0.0038&0.1923
\end{bmatrix} \triangleright$$

Combined all the thermal load matrixes together, it gives:

$$\left[\{F\}^{(G)}\right]^{\mathrm{T}}=\{0\quad 0\quad 2.505\quad 0\quad 0\quad 0\quad 5.01\quad 0\quad 0\quad 0\quad 5.01\quad 2.115\quad 4.23\quad 4.23\quad 4.62\}$$

The final thermal load matrix is expressed with normal temperature boundary condition applied at junction (1), (5), (4):

$$\left[\{F\}^{(G)}\right]^{\mathrm{T}}=\{7.05\quad 0\quad 2.505\quad 8.35\quad 10.9281\quad 0\quad 5.1\quad 0\quad 0\quad 0\quad 5.1\quad 2.115\quad 4.23\quad 4.23\quad 4.62\}$$

Finally, according to $[K]\ [T]=[F]$, linear equations are solved as:

$$[T]^{T}=[100\quad 70.1\quad 45.8\quad 100\quad 100\quad 66.8\quad 43.1\quad 70.83\quad 67.2\quad 51.56\quad 39.64\quad 48.11\quad 43.67\quad 40.1\quad 32.73]$$

4. Experimental tests and analysis

4.1. Measurement experiment

To prove the validity of temperature numerical simulation result, we have conducted a number of experiments to verify the theory. The test species is the same as Section 3.1. The ambient temperature is 30°C. A heating source, remaining as 100°C, is installed in the center of the specimen. After heated seven minutes, the nodes 1, 4, 5 arrive near 100°C. When the temperature error remains in the range of ±1.5%, it coincides with the assumed condition. As stated above, a quarter of the cross section which can represent the whole cross section is analyzed. The section analyzed is then divided into 8 elements with 15 nodes where Pt-resistance temperature sensors are embedded (location as Fig. 3 shown). In order to measure each node temperatures, automatic scanning circuit controlled by microprocessor is designed. When the temperature of each node arrives at some steady value, it is termed as the steady temperature. Experiments are performed four times, and comparisons between the experimental values and theoretical values are listed in Table 2.

Compared with actual values (measured by automatic scanning circuit) from Table 2, the simulation results are consistent with the actual measurement values, and the error ranges from –3.5% to 4.2% (full range) which can meet the requirement of engineering calculation.

4.2. Computer simulation

The present finite element analysis based on software consists of five essential parts:

- a geometric description of the tile, which is in this case is completely parametic. Geometrically similar tiles can be modeled;
- with an optimized element mesh for each load case;
- the definition of material properties;
- the thermal solution;
- the stress and strain solution;
- the extraction and presentation of relevant results.

A finite element model for wood temperature field is developed using ANSYS finite element software. PLANE55 can be used as a plane element or as an axisymmetric ring element with a 2-D heat conduction capability. The element has four nodes with a single degree of freedom, temperature, at each node. After model creating, meshing, loading and solving, we can get the contour of temperature field. From Fig. 4, the temperature distribution of the board can be displayed visibly, and its values coincide with the numerical solution in section 3.2. We also can conclude from the contour that temperature gradient in the x (tangential) direction is a little smaller than that in the y (radial) directions. It is because tangential conductivity k_x is smaller than radial conductivity k_y.

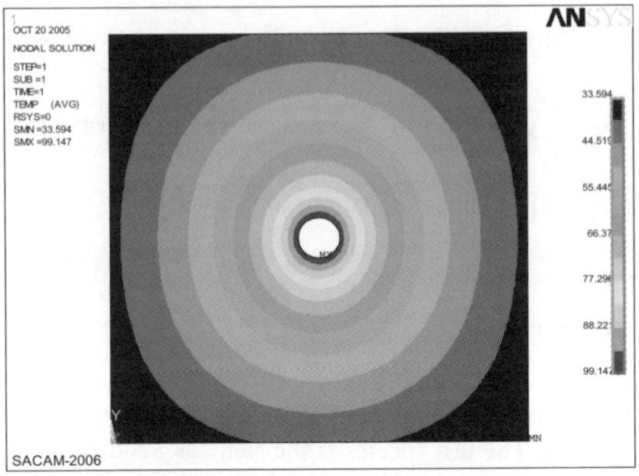

Fig. 4. The contour of wood temperature field

Table 2. Temperature distribution comparisons between experimental values and simulation results

Node	Numercial value (/ °C)	Experimental Value (/°C)			
		1	2	3	4
1	100	98.6	99.6	99.2	98.4
2	70.1	68.8	72	71.5	71.6
3	49.01	49.8	49.9	48	47.8
4	100	99.7	98.7	99.7	97.8
5	100	98.3	99.2	98.1	97.9
6	66.8	64.9	67.9	65.3	64.5
7	43.1	42.1	44.9	44.8	43.6
8	70.83	71.2	70.7	70.8	70
9	67.02	66.4	68.5	67.6	65.2
10	51.56	50.1	51.2	51.6	50.3
11	39.64	40.3	41.3	38.5	38.5
12	45.88	43.5	44.5	46.7	46
13	43.67	44.5	41.7	42.9	42.5
14	40.1	39.4	38.7	41.6	41.5
15	32.73	33.5	32.1	33.7	31.8

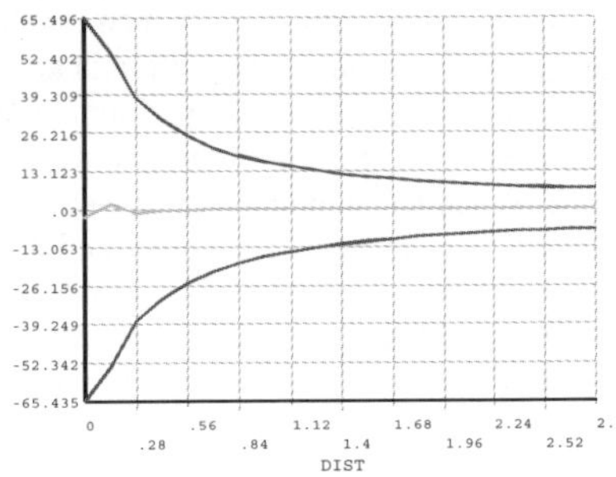

A. Radial Direction (x)

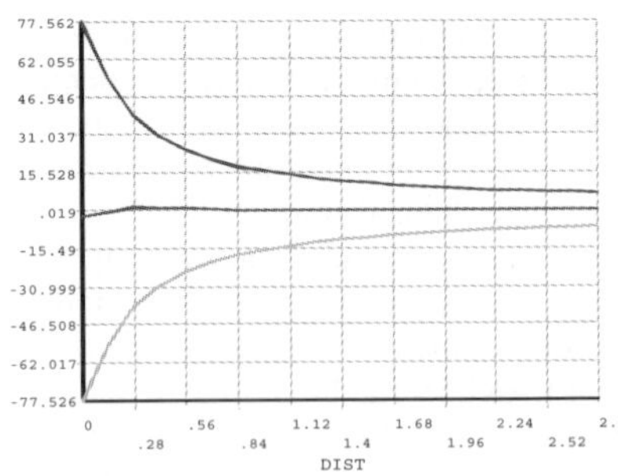

B. Tangential Direction (y)

Fig. 5. Temperature gradient changes

Fig. 5 proves the above conclusion. Also it shows that $\frac{\partial T}{\partial X}=0$ along radial direction and $\frac{\partial T}{\partial Y}$ is equal to the vector sum in the Fig. 5 (A). The similar case is in the Fig. 5 (B). All the experiment results show that it is feasible and effective to analyze wood temperature distribution fields using finite element method.

5. Conclusions

The paper performs a primary investigation on the temperature field of anisotropic wood, deduces the finite element algorithm for wood temperature field and also carries out the related experiments. From the validity experiments and numerical analysis, the simulation results are consistent with the actual measurement values, and the error ranges from -3.5% to 4.2% (full range) which can meet the requirement of engineering calculation. It shows that the investigations on wood heat conductivity adopting finite element method are feasible and practical. In addition, the anisotropic material property affects on the temperature field in radial and tangential directions is discussed, and studies show that heat conduction differences between the radial direction and tangential direction impact on wood temperature distributions. However, many factors, such as moisture content, density, earlywood and latewood differentiation, etc., are not taken into account in this paper. So in the future research, many other wood heat performances influenced by these factors need to be further studied.

References

[1] Kamke FA. and Vanek M. Comparison of wood drying models, Proceedings of the 4th IUFRO International Wood Drying Conference, Rotorua, New Zealand, 1994.

[2] Hukka A.. Evaluation of parameter values for a high-temperature drying simulation model using direct drying experiments, Drying Technology1997; 15:1213–1229.

[3] Pang S. Relationship between a diffusion model and a transport model for softwood drying, Wood and Fiber Science 1997; 29: 58–67.

[4] Gu HM, John F and Hunt PE. Two dimensional finite element heat transfer models for softwood, Proceedings of the 7th Pacific Rim Bio-based Composites Symposium, Nanjing Forestry University, Nanjing China November 1–4, 2004.

[5] Couturier MF, George K and Schneider MH. Thermophysical properties of wood-polymer composite, Wood Science and Technology 1996; 30: 179–196.

[6] Suleiman BM, Largeldt JB and Gustavsson M. Thermal conductivity and diffusivity of wood, Wood Science and Tehnology 1999; 33: 465–473.

[7] Harada T, Hata T and Ishihara S. Thermal constant of wood during the heating process measured with the laser flash method, J. Wood Science 1998; 44: 425–431,.

[8] Luikov AV. Analytical Heat Diffusion Theory, Academic Press. Ine(London) LTD, 1968.

[9] Siau JF. Wood: Influence of Moisture on Physical Properties, Department of Wood Science and Forest Products, Virginia Tech, 227, 1995.

[10] Gu HM. Ph.D dissertation, Structure Based, Two-dimensional Anisotropic, Transient Heat Conduction Model for Wood, Virginia Polytechniq Institute and State University (Virginia Tech), Department of Wood Science & Forest Products, Blacksburg, 2001.

Elastic wave field in a porous medium fully saturated with a Newtonian viscous fluid

Liming Dai* Guoqing Wang

Industrial Systems Engineering, University of Regina, Regina, Saskatchewan, Canada S4S 0A2

Abstract

This is an investigation on the elastic wave field in a homogeneous and isotropic porous medium which is fully saturated with a Newtonian viscous fluid. The wave field in the porous medium can be described by a new methodology with consideration of the excitations of multi-energy sources. The model of the wave motion with multiple cylindrical wave sources is established with the utilization of Hankel function and a new developed moving-coordinate method. Numerical simulations of the wave field with the multiple energy sources in the porous media saturated with viscous fluid are performed.

Key words: Porous medium; multi-source wave model; wave fluid; wave superposition

1. Introduction

Wave velocities and attenuation are two important aspects of the waves in porous media, since they are important in analyzing the dynamic response of the media with respect to the properties of the media and the wave sources, such as viscosity, frequency and porosity. By introducing the assumptions that the solid skeleton of the porous medium obeys the laws of homogeneous linear elasticity and the fluid obeys Darcy's laws, Biot (Biot, 1956a, b) formulated the governing equations for wave propagation in a fluid-saturated medium. The second compressional wave is usually named as slow wave that has a strongly dispersive characteristic, for which the displacements of fluid and solid are out of phase. Following Biot's theory, Vardoulakis and Beskos (1986) developed a theory describing wave propagation in a three-phase porous medium which is applicable to partially-saturated materials. Recently, numerous descent research works are performed to improve Biot's theory and to broaden the applications of Biot's theory. Gurevich *et al.* (1999) used experiment and simulation methods to verify Biot's theory. Pham *et al.* (2002) obtained the wave velocities and quality factors of clay-bearing sandstones as a function of pore pressure, frequency and partial saturation. Fluid in a porous medium has viscosity, thus, to investigate the dynamic response of the medium, the viscosity of the fluid needs to be taken into account. However, very few studies focusing on investigating the relative displacements between the fluid and solid in a porous medium fully-saturated with Newtonian fluid are found in the literature. Almost all the research works in this field assumed a single energy source in the field, except the recent study made by the authors (Dai and Wang, 2005). Obviously, the investigations on the propagation, superposition, dispersion and interactions of the waves generated by multiple energy sources are more significant.

The aim of this research is to develop a methodology to describe the wave field in a porous medium

* Corresponding author: Dr. Liming Dai, P.Eng., Industrial Systems Engineering, University of Regina, Regina, SK S4S 0A2, Canada; Phone: (306) 585-4498; E-mail: liminig.dai@uregina.ca

fully-saturated with a viscous fluid under the excitation of multiple energy sources. The results of the research will allow a quantitative description of the wave field in the fluid-saturated porous media subjected to the excitations of multiple energy sources.

2. Mathematical model establishment

Generally, the stresses acting on a porous medium can be separated into two parts; one is on the solid frame which can be written as σ_{ij}^s, the other is on the fluid represented by $s_{ij} = -\phi p \delta_{ij}$, thus the total stresses are expressed by: $\sigma_{ij} = \sigma_{ij}^s + s_{ij}$. Where ϕ is the porosity of the medium; p is the fluid pressure; δ_{ij} is Kronecker symbol; the negative sign existing in the equation is for the association of directions between fluid pressure and stress. Starting with the above stress expressions of a porous medium and by employing the force equilibrium relation, the dynamics equations of a porous medium can be written as:

$$\begin{cases} N\nabla^2\mathbf{u} + \nabla[(A+N)e + Q\varepsilon] = \dfrac{\partial^2}{\partial t^2}(\rho_{11}\mathbf{u} + \rho_{12}\mathbf{U}) + b\dfrac{\partial}{\partial t}(\mathbf{u} - \mathbf{U}) \\ \nabla[Qe + R\varepsilon] = \dfrac{\partial^2}{\partial t^2}(\rho_{12}\mathbf{u} + \rho_{22}\mathbf{U}) - b\dfrac{\partial}{\partial t}(\mathbf{u} - \mathbf{U}) \end{cases} \quad (1a, b)$$

The coefficient b is related to Darcy's coefficient of permeability k by

$$b = \frac{\mu\phi^2}{k} \quad (2)$$

where, μ is the fluid viscosity and ϕ is the porosity of the medium.

In Eq. (1), $\mathbf{u}$ and $\mathbf{U}$ are the displacement vectors of fluid and solid respectively, while e and ε are the volume strains of the solid and fluid respectively with the expressions: $e = \nabla\cdot\mathbf{u}$, $\varepsilon = \nabla\cdot\mathbf{U}$. ρ_{11}, ρ_{12} and ρ_{22} are density terms, which can be expressed as: $\rho_{11} = (1-\phi)\rho_s$, $\rho_{22} = \phi\rho_f$, $\rho_{12} = -(\alpha - 1)\phi\rho_f$, while ρ_s is the mass density of the solid grains, ρ_f is the mass density of the fluid in pores, $\alpha = (1/2)[\phi^{-1} + 1]$, ϕ is the porosity of the medium. *A*, *N*, *Q* and *R* are the physical parameters of the medium. *A* and *N* are similar as Lame coefficients in elastic theory. *N* represents the shear modulus of the medium; *R* is a measure of pressure on the fluid required to drive a unit volume of fluid into the porous medium. *Q* describes the coupling between the volume change of solid and that of fluid. The expressions for *A*, *N*, *Q* and *R* will be given in following section.

Applying Helmholtz decomposition to the displacement vectors of solid and fluid:

$$\begin{cases} \mathbf{u} = \mathrm{grad}(\varphi_s) + \mathrm{curl}(\boldsymbol{\psi}_s) \\ \mathbf{U} = \mathrm{grad}(\varphi_f) + \mathrm{curl}(\boldsymbol{\psi}_f) \end{cases} \quad (3a, b)$$

where φ_s and φ_f are scalar potentials of solid and fluid respectively, $\boldsymbol{\psi}_s$ and $\boldsymbol{\psi}_f$ are vector potentials for the displacements of solid and fluid. $\boldsymbol{\psi}_s$ and $\boldsymbol{\psi}_f$ also satisfy the conditions: $\nabla\cdot\boldsymbol{\psi}_s = 0$ and $\nabla\cdot\boldsymbol{\psi}_f = 0$.

For P-wave, also named compressible wave, the displacement is corresponding to the scalar potentials, without rotation, that implies, $\nabla\times\mathbf{u} = 0$. For S-wave, also known as rotational wave or shear wave, the displacement is due to vector potentials, $\nabla\cdot\mathbf{u} = 0$. Substituting Eq. (3) into Eq. (1), and rearranging the terms according to the scalar and vector potentials, as Lin *et al.* (2001) did in their research, two sets of equations can be obtained corresponding to scalar potentials and vector potentials of the fluid and solid. Thus, the expressions for P- and S-waves can be given as:

For P-wave:

$$\begin{cases} \nabla^2(P\varphi_s + Q\varphi_f) = \dfrac{\partial^2}{\partial t^2}(\rho_{11}\varphi_s + \rho_{12}\varphi_f) + b\dfrac{\partial}{\partial t}(\varphi_s - \varphi_f) \\ \nabla^2[Q\varphi_s + R\varphi_f] = \dfrac{\partial^2}{\partial t^2}(\rho_{12}\varphi_s + \rho_{22}\varphi_f) - b\dfrac{\partial}{\partial t}(\varphi_s - \varphi_f) \end{cases} \tag{4a, b}$$

For S-wave:

$$\begin{cases} N\nabla^2\boldsymbol{\psi}_s = \dfrac{\partial^2}{\partial t^2}(\rho_{11}\boldsymbol{\psi}_s + \rho_{12}\boldsymbol{\psi}_f) + b\dfrac{\partial}{\partial t}(\boldsymbol{\psi}_s - \boldsymbol{\psi}_f) \\ 0 = \dfrac{\partial^2}{\partial t^2}(\rho_{12}\boldsymbol{\psi}_s + \rho_{22}\boldsymbol{\psi}_f) - b\dfrac{\partial}{\partial t}(\boldsymbol{\psi}_s - \boldsymbol{\psi}_f) \end{cases} \tag{5a, b}$$

Eqs. (4) and (5) are the governing equations of waves propagating in porous media in terms of displacement potentials.

Let φ be a general displacement scalar potential and **u** a general displacement vector. For P-wave, the displacement vector **u** is just related to the scalar potential φ by:

$$\mathbf{u} = \nabla\varphi \tag{6}$$

The scalar potential φ also has the following property:

$$\nabla(\nabla^2\varphi) = \nabla[\nabla\cdot(\nabla\varphi)] = \nabla\times[\nabla\times(\nabla\varphi)] + \nabla^2(\nabla\varphi) = \nabla^2(\nabla\varphi) \tag{7}$$

Therefore, with equations of Eqs. (6) and (7), the governing equations of Eq. (4) for the dilatation waves in the form of displacements can be written as:

$$\begin{cases} \nabla^2(P\mathbf{u}_{sp} + Q\mathbf{U}_{fp}) = \dfrac{\partial^2}{\partial t^2}(\rho_{11}\mathbf{u}_{sp} + \rho_{12}\mathbf{U}_{fp}) + b\dfrac{\partial}{\partial t}(\mathbf{u}_{sp} - \mathbf{U}_{fp}) \\ \nabla^2[Q\mathbf{u}_{sp} + R\mathbf{U}_{fp}] = \dfrac{\partial^2}{\partial t^2}(\rho_{12}\mathbf{u}_{sp} + \rho_{22}\mathbf{U}_{fp}) - b\dfrac{\partial}{\partial t}(\mathbf{u}_{sp} - \mathbf{U}_{fp}) \end{cases} \tag{8a, b}$$

in which, the subscript '*s*' represents the displacement of solid, '*f*' represents the displacement of the fluid, '*p*' represents the displacement due to the P-wave. In Eq. (8), the parameters of material, *P, Q, R* can be determined by the processes described by Lin *et al.* (2001). Eq. (8) is the governing equations for P-wave propagating in the porous medium.

For analyzing the wave propagations generated by the vertical wells and with the concentration on the combined wave motion with multiple energy sources, the waves are assumed to be generated by vertical cylindrical energy sources. In other words, the waves considered in the present research are cylindrical compressible waves. Furthermore, the wave sources are assumed to be uniformly distributed along the cylindrical axis in a 3D domain. The material properties of the porous medium are also assumed to be isotropic and homogeneous. With these assumptions, a 2D model can be used to simulate the real field, and it is convenient for the governing equations and corresponding solutions to be expressed in polar coordinates. In polar coordinates, the operators, ∇ and ∇^2 are given as:

$$\begin{cases} \nabla = (\frac{\partial}{\partial r} + \frac{1}{r})\vec{r} \\ \nabla^2 = \frac{\partial^2}{\partial r^2} + \frac{1}{r}\frac{\partial}{\partial r} \end{cases} \tag{9a, b}$$

The equations for dilatational waves can be written as:

$$\begin{cases} (\frac{d^2}{dr^2} + \frac{1}{r}\frac{d}{dr})(\sigma_{11}\mathbf{u} + \sigma_{12}\mathbf{U}) = \frac{1}{V_c^2}\frac{\partial^2}{\partial t^2}(\gamma_{11}\mathbf{u} + \gamma_{12}\mathbf{U}) + \frac{b}{H}\frac{\partial}{\partial t}(\mathbf{u} - \mathbf{U}) \\ (\frac{d^2}{dr^2} + \frac{1}{r}\frac{d}{dr})(\sigma_{12}\mathbf{u} + \sigma_{22}\mathbf{U}) = \frac{1}{V_c^2}\frac{\partial^2}{\partial t^2}(\gamma_{12}\mathbf{u} + \gamma_{22}\mathbf{U}) - \frac{b}{H}\frac{\partial}{\partial t}(\mathbf{u} - \mathbf{U}) \end{cases} \tag{10a, b}$$

For the sake of convenience of derivation process, the following parameters are used as introduced by Biot (1956a),

$$V_c^2 = H/\rho \tag{11}$$

$$\begin{cases} \sigma_{11} = \frac{P}{H}, \quad \sigma_{12} = \frac{Q}{H}, \quad \sigma_{22} = \frac{R}{H} \\ \gamma_{11} = \frac{\rho_{11}}{\rho}, \quad \gamma_{12} = \frac{\rho_{12}}{\rho}, \quad \gamma_{22} = \frac{\rho_{22}}{\rho} \end{cases} \tag{12}$$

in which

$$H = P + R + 2Q, \qquad \rho = \rho_{11} + \rho_{22} + 2\rho_{12} \tag{13}$$

According to Sommerfeld Radiation Condition (Pao, 1973), the wave propagating from a cylindrical hole can be assumed as:

$$\begin{cases} \mathbf{u} = C_1 H_0^{(1)}(lr)\exp(-i\omega t) \\ \mathbf{U} = C_2 H_0^{(1)}(lr)\exp(-i\omega t) \end{cases} \tag{14a, b}$$

C_1 and C_2 are the amplitudes of the waves propagating in solid and fluid respectively; l is wave number; r is the distance from the considered point to the source. $H_0^{(1)}(\cdot)$ is the zero-order Hankel function of the first kind. The subscript '0' represents zero order, in the following equations these subscripts have the same meaning; the superscript '(1)' means the function is the first kind. $\exp(-i\omega t)$ is the time factor of the harmonic wave; $i = \sqrt{-1}$ is the complex unit; ω is the frequency of wave.

Rearrange the equations, the following quadric equation can be obtained:

$$(\sigma_{11}\sigma_{12} - \sigma_{12}^2)\xi^2 - (\sigma_{11}\gamma_{12} + \sigma_{22}\gamma_{11} - 2\sigma_{12}\gamma_{12})\xi + (\gamma_{11}\gamma_{22} - \gamma_{12}^2) + \frac{ib}{\alpha\rho}(\xi - 1) = 0 \tag{15}$$

In this case l and ξ are complex variables. Denoting ξ_I and ξ_{II} are the roots of Eq. (15), which correspond to the velocities of the purely elastic waves as given by Eq. (1), and assume that ξ_I is the root which corresponds to the first compressible wave, while ξ_{II} is that corresponds to waves of the second kind.

By solving the quadratic equations of Eq. (15) about the velocities, two complex roots can be obtained; the image part reflects the attenuation; while the real part designates the propagation velocity of the wave. It should be noted that this is not the speed of the particle vibration. The ratio of the image part to the real part is important since it describes the degree of damping of the wave.

3. Consideration of multi-energy sources

It is more significant and practically meaningful to study the dynamic response of porous media and the relative displacement between solid and fluid when the domain is excited by multiple energy sources. A model with multiple sources needs be built for a more accurate analysis of wave field of a porous medium. A newly developed moving coordinate method can be employed in building such model and describing the displacement field excited by multiple wave sources.

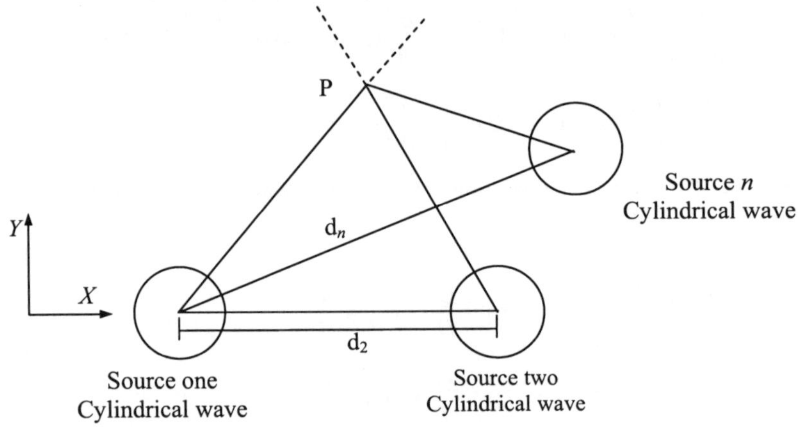

Figure 1. Multi-Source Model

As shown in Fig. 1, if the global coordinates are located at one source, then, the coordinates of other source locations can be expressed by

$$d_j = r_{j0}(\cos\theta_{j0} + i\sin\theta_{j0}) \tag{16}$$

All the energy sources considered in this research are supposed to be continuous and harmonic cylindrical waves generated by multiple cylindrical sources. Furthermore, only steady state is considered. The waves can therefore have the following expressions if they are expressed in their own local coordinates with the origins locating at the sources:

$$\begin{cases} \mathbf{u}_r = u_0 \,\mathrm{Re}[H_0^{(1)}(lr)\exp(-i\omega t)](\cos\theta + i\sin\theta) \\ \mathbf{U}_r = U_0 \,\mathrm{Re}[H_0^{(1)}(lr)\exp(-i\omega t)](\cos\theta + i\sin\theta) \end{cases} \tag{17}$$

in which, the term $(\cos\theta + i\sin\theta)$ is introduced to represent the direction of the displacement. Consequently, this term can be replaced by $[z/|z|]$. z has the expression, $z = x + iy$, with $x = r\cos\theta$ and $y = r\sin\theta$ in the polar coordinate system.

Thus, the waves propagating from each of the sources can be determined.

In order to investigate the superposed action of multiple waves conveniently, the expression for each wave is to be written in a common coordinate system by using the moving-coordinate method (Wang, 2002).

Expressing wave j in the xoy-coordinates as shown in Fig. 1, $z_j = z - d_j$:

$$\begin{cases} u_{rj} = u_{0j}\,\mathrm{Re}[H_0^{(1)}(l_j r_j)\exp(-i\omega_j t)][\frac{z_j}{|z_j|}] = u_{0j}\,\mathrm{Re}[H_0^{(1)}(l_j|z-d_j|)\exp(-i\omega_j t)][\frac{z-d_j}{|z-d_j|}] \\ U_{rj} = U_{0j}\,\mathrm{Re}[H_0^{(1)}(l_j r_j)\exp(-i\omega_j t)][\frac{z_j}{|z_j|}] = U_{0j}\,\mathrm{Re}[H_0^{(1)}(l_j|z-d_j|)\exp(-i\omega_j t)][\frac{z-d_j}{|z-d_j|}] \end{cases} \tag{18a,b}$$

d_j are the coordinates of the jth wave source in the common coordinates.

With the equations developed, the total displacements of any given point, P, in the domain considered can be described in a common coordinate system. *xoy*-coordinates can be considered as the common coordinates (also named global coordinates). This implies that $d_1 = 0$. The combined displacements can now be presented by:

$$\begin{cases} \mathbf{u}_r = \sum_{j=1}^{n} \mathbf{u}_{rj} = u_{01}\,\mathrm{Re}[H_0^{(1)}(l_1|z_1|)\exp(-i\omega_1 t)][\frac{z_1}{|z_1|}] + \cdots + u_{0n}\,\mathrm{Re}[H_0^{(1)}(l_n|z-d_n|)\exp(-i\omega_n t)][\frac{z-d_n}{|z-d_n|}] \\ \mathbf{U}_r = \sum_{j=1}^{n} \mathbf{U}_{rj} = U_{01}\,\mathrm{Re}[H_0^{(1)}(l_1|z_1|)\exp(-i\omega_1 t)][\frac{z_1}{|z_1|}] + \cdots + U_{0n}\,\mathrm{Re}[H_0^{(1)}(l_n|z-d_n|)\exp(-i\omega_n t)][\frac{z-d_n}{|z-d_n|}] \end{cases} \tag{19a,b}$$

The displacement wave field excited by multiple cylindrical sources can be quantified by using the model provided above. The characteristics of the wave field can be analyzed quantitatively when the parameters of material and the sources or the locations of the sources are specified.

4. Numerical Simulation

To demonstrate the application of the model established, numerical simulations are performed as the basis of the wave model and the solutions developed. A numerical simulation for the wave generated by two energy sources is performed.

The distance between the two sources is noted as *d*, the position of point P in the field is expressed as: $z = x + iy$, the frequencies of the two source waves are ω_1 and ω_2 respectively. For the sake of simplification, it is assumed that the solid skeleton system is formed by spherical solid particles as the assumption made by the other researches conventionally. The particles' compressibility can be neglect.

Once the physical parameters are given, the coefficient values of waves can be determined by the wave model established. The phase velocity of the wave and the relative displacements of a random point P in the wave field are computed. The comparison between the results with the consideration of fluid viscosity and result without the concern of viscosity of the fluid is also performed.

Fig. 2 shows the non-dimensional relative displacement amplitudes along the line connecting the two sources. The non-dimensional relative displacement used in Fig. 2 is defined by *(U-u)/u*. The locations of the two sources are at $x = 0, y = 0$ and $x = 1600, y = 0$ respectively. It should be noted that, for each of the waves, the amplitudes of the wave decrease in general with the increasing distance from the energy source. Moreover, the amplitude of the combined wave at steady state is not simply the summation of the amplitudes of the two waves.

As can be seen from Fig. 2, when the porous medium is excited by two energy sources, the wave response (maximum amplitudes of the displacements) is totally different from that of the single source (represented by the curves of "left effect" and "right effect" respectively). For some areas, the amplitude of the combined wave is smaller than that of single source, while for some other areas the amplitude is larger than that of the single source. One may also find from the figure that the amplitude of the wave can be zero at a certain location between the two sources. The reason is because the frequencies of the two sources are equal, the phenomena of standing wave appears. It is also noted that the frequency of the resulting wave generated by the two energy sources are varied from the frequencies of the two energy sources.

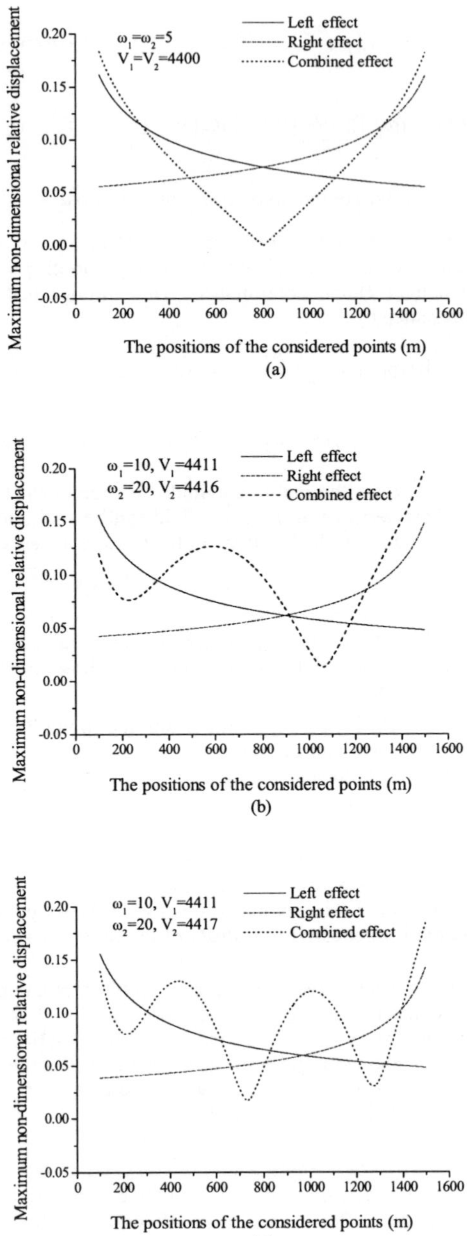

Figure 2 The maximum non-dimensional relative displacement changes vs. location of the concerned point

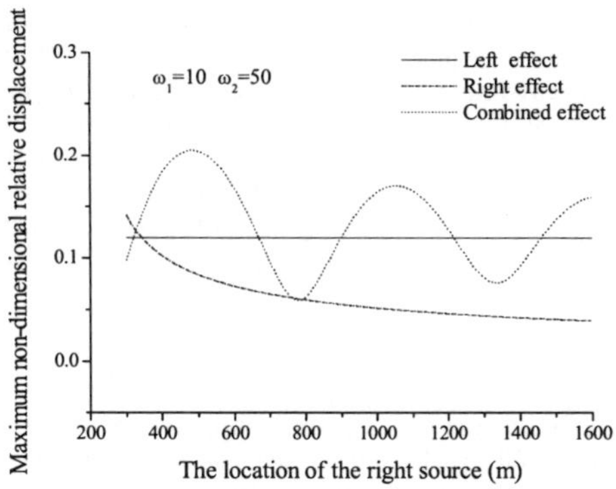

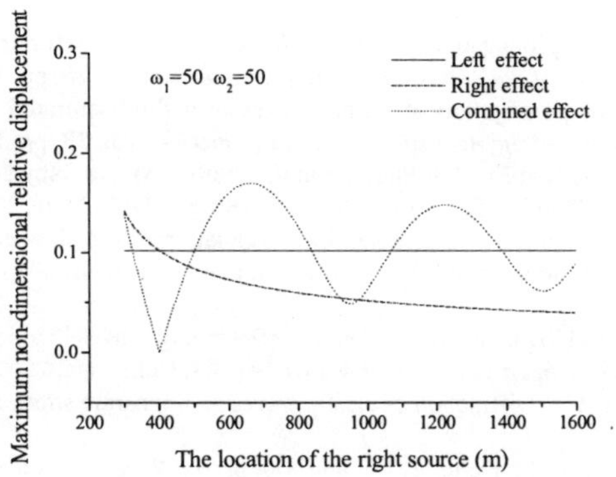

Figure 3 Maximum relative displacements v.s. location of the right source with respect to the location of the left source

Effects of distance between the two sources on the wave motion of the porous medium are also evaluated in the present research. Fig. 3 shows the relative displacement of a point at $x = 200, y = 0$, with respect to the excitations of the left source with a constant distance from the point and the right source with a varying distance from the point. As exhibited in the figures that the effect of the right source decreases as the distance between the concerned point and the right source increases. It may also be observed from the figure that the peak value of the relative displacement varies periodically with the increase of the distance between the right source and the point considered. Although only one poine is considered in the numerical calculation, the wave motion of all the particles in the domain considered can be conveniently determined and plotted by the formulas developed, if desired.

5. Conclusion

A new methodology for analyzing the behavior of wave field in elastic porous medium saturated with a viscous fluid is developed in the present research. The wave motion generated by multiple energy sources can be quantified with the mathematical model established. Such a research has not been found in current literature. The expressions of combined waves from multiple cylindrical sources are constructed with the utilization of Hankel function based on polar coordinates. Making use of the model established, the behavior of any specified point in the considered domain of the porous medium can be quantified, and the relative displacement between the fluid and solid of the medium can be conveniently determined. The wave field of the considered porous medium can thus be determined for any given time. The analysis of the wave motions in the medium is then readily available. The numerical simulations of this research show the efficiency of applying the model established in quantifying the effects of the waves generated by different energy sources on the motions of the fluid and solid of a porous medium. A directly advantage from implementing multi-energy sources is that the energy can be saved. This makes a great significance on controlling the displacements of the particles in a wave field desired.

Acknowledgments

The authors wish to acknowledge the financial supports to this research from the Petroleum Technology Research Centre (PTRC) of Canada, the Natural Sciences and Engineering Research Council of Canada (NSERC), and Canada Foundation for Innovation (CFI).

Reference

[1] Biot, M.A., "Theory of propagation of elastic waves in a fluid-saturated porous solid, part I: low frequency range", *Journal of the Acoustical Society of America*, Vol. **28**, pp. 168-178, 1956.

[2] Biot, M.A., "Theory of propagation of elastic waves in a fluid-saturated porous solid, part II: higher frequency range", *Journal of the Acoustical Society of America*, Vol. **28**, pp. 179-191, 1956.

[3] Dai, L. and Wang, G., "Wave Field in Fluid Saturated Porous Media Subjected to Excitations of Multiple Energy Sources," DETC2005-85601, Proceedings of ASME IDETC/CIE 2005, Long Beach, CA, 2005.

[4] Gurevich, B., Kelder, O., and Smeulders, D.M.J., "Validation of the slow compressional wave in porous media: Comparison of experiments and numerical simulations", *Transport in Porous Media*, Vol. **36**, pp. 149-160, 1999.

[5] Lin, C.H., Lee, V.W., and Trifunac, M.D., "On the reflection of waves in a poroelastic half-space saturated with non-viscous fluid", Report No. CE 01-04, Los Angeles, California, 2001.

[6] Pao, Y.H., and Mow, C.C., "Diffraction of elastic waves and dynamic stress concentration", Crane-Russak Inc., New York, 1973.

[7] Pham, N.H., Carcione, J.M., Helle, H.B., and Ursin, B., Wave velocities and attenuation of shaley sandstones as a function of pore pressure and partial saturation, Geophysical prospecting, Vol. 50, pp. 615-627, 2002.

[8] Vardoulakis, I. and Beskos, D., "Dynamic behavior of nearly saturated porous media", *Mechanics of Composite Materials*, Vol. 5, 87–108, 1986.

[9] Wang, G., and Liu, D., "Scattering of SH-wave by multiple circular cavities in half space", *Journal of Earthquake Engineering and Engineering Vibration*, Vol. **1**(1), pp. 36-44, 2002.

Ball Bearing Remnant Life Prediction of Induction Motors – Impact Inspection Approach

Longfu Luo[a], Liming Dai[b], Mingzhe Dong[c], Lisa Fan[d]
[a]*College of Electrical & Information Engineering, Hunan University, hangsha, 410082, P.R. China*
[b]*Industrial Systems Engineering, University of Regina, Regina, Saskatchewan, anada S4S 0A2*
[c]*Petroleum Systems Engineering, University of Regina, Regina, Saskatchewan, Canada S4S 0A2*
[d]*Computer Science, University of Regina, Regina, Saskatchewan, Canada S4S 0A2*

Abstract
This research intends to establish a random processes model of hitch degree development of induction motor ball bearings based on impact inspections. The variance function and the mean function of the random process are studied for establishing an approximate folding line method and an extrapolation method which can be used to predict the remnant life of the ball bearings. The methods established are verified with the data collected from the petroleum industries utilizing the motor bearings.

Keywords
Induction motor; random process, ball bearings; impact inspection; remnant life prediction

1. Introduction

Remnant life prediction or remnant life diagnosis is a technique of assessing the remnant life of a device on the basis of status inspection and prediction theory, with which one may deduce the remnant life of the device that has already been running under operation conditions for a certain period of time.

In large-scale petroleum or chemical industries, for example, thousands of electric motors are employed to drive oil pumps, water pumps and fan machineries which may be expected to work continuously for a long period of time. Among all electromotor failures in the industries, 35% of them are caused by hitches or defects of ball bearings. Diagnosis for the defects with utilization of impact pulse meters can diagnose the hitch level and the defect types of the bearing. Therefore, diagnosis using impact pulse meters is a widely utilized technique at petroleum industries. In the diagnosis practices, if the hitch level of a ball bearing arrives at $A_{dB} = 60$ dB, the bearing is considered to reach its end of the operational life and has to be replaced. However, people may not necessarily know how long it will take the running ball bearing to last on the basis of the current status of bearing's hitch level. (For example, how long a bearing with $A_{dB} = 40$ dB may last or reach the hitch level of $A_{dB} = 60$ dB under operating conditions). A reliable technique is therefore needed for predicting the remnant life of a ball bearing on the basis of the bearing's current status quantified with its hitch level. The objective of the present research is to establish such a technique on a theoretically and practically sound basis.

2 The hitch inspection method and the simple principles

The most effective method to diagnose the hitch level of a ball bearing relies on the utilization of impact pulse meters. Comparing with that of the vibrations of the motor itself and surroundings, the spectrum character of the impact signal is that it contains much richer high frequency components. The impact pulse meter catches this impact signal by making use of resonance demodulation technology and converts it into hitch level A_{dB}. The advantages of the method are that resonance demodulation wave is corresponding with the hitch of the bearing, the amplitude of the wave increases with the impact degree of the hitch so that obtaining the amplitude is getting the level of the hitch, and the hitch level A_{dB} includes the amplitude of impact after conversion with consideration of the rotate speed and the diameter of the bearing, which makes the A_{dB} becomes comparable.

According to the magnitude of A_{dB}, the status of the bearing can be estimated approximately, as shown in Table 1.

Table 1: The status standard of the general bearing in mechanical industry

Hitch level A_{dB}	Status	Damage phase
A_{dB} <20 dB	Good condition	No damage or very tiny damage period
20 dB≤A_{dB} ≤35 dB	Alarm	Damage developing period
35 dB≤A_{dB} ≤50 dB	Severe damage	Visible damage period
50 dB≤A_{dB} ≤60 dB	Severe damage	Breakage period

The type of the hitches can be judged by making further frequency table analysis, which is the advanced hitch diagnose technology.

3. The sample data

To study the developing process of the bearing hitch, the writer collected the sample data according to the following descriptions:

1. The data acquisition for the bearing hitches detection is utilized with the instrument of a handhold impact pulse meter. The measurement operation fully complies with operation manual;
2. The working condition of the ball bearing are: fixed in middling or minitype motor; well maintained; driving oil pump, water pump and fan continuously;
3. The bearing material is general bearing steel. For example: GCr15;
4. During the past several years, the writer got 171 groups of data, which cover the whole process of the bearing hitch development from A_{dB}=20dB to A_{dB}=60dB, with measurement frequency once a week. All the data cited in this paper belong to the bearing disfigurement.

4. Data analysis and mathematical modeling

Fig.1 is the hitch level developing curves of the bearings No. 1, 41, 81, 121 and 161, in which horizontal coordinate represents time by weeks. The writer reaches the following conclusions after analysis:

1 All the curves show the increasing behavior after a certain phase;

2 The longer the lifetime of the bearing is, the later the curve starts to display increasing behavior;

3 There are no obvious rules among the hitch level development of the bearings.

Fig.2 is another set of curves derived from the same data, in which horizontal coordinate is the relative time by %, where A_{dB}=60dB is corresponding T=100%.

We can find that the hitch level development process of all bearings is an unsteady random processes, in which the horizontal coordinate is the relative time by %, the mathematic model is

$$A_{dB} = (t\%,\omega), t \in T\%(0 \sim 100\%), \omega \in \Omega = 171$$

Apparently, the random processes have two variance curves, which cannot be obtained by

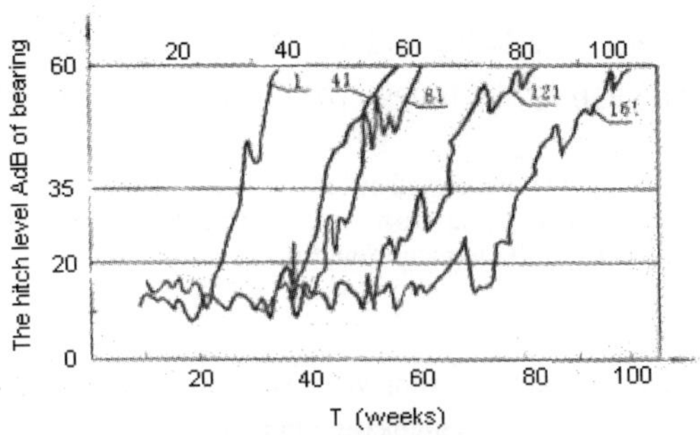

Fig.1 The developing trend of the hitch level of bearing

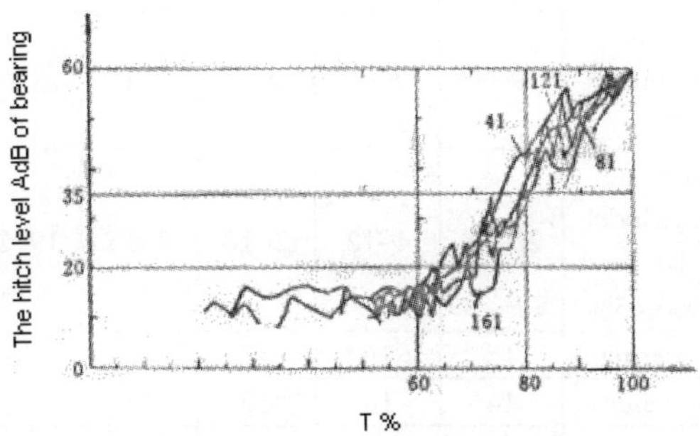

Fig.2 The developing trend of the hitch level of bearing

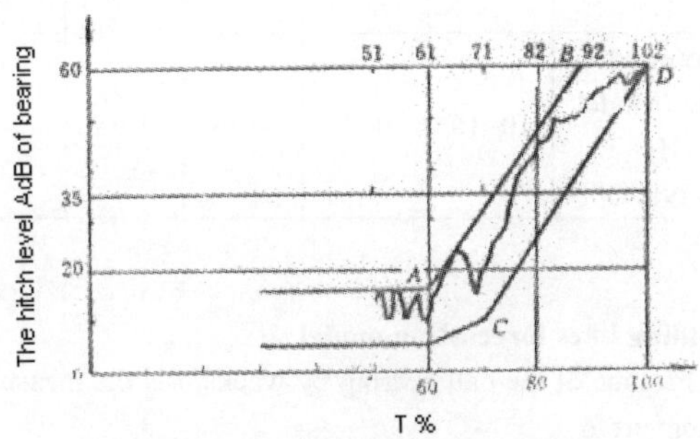

Fig.3 The variance function of the stochastic processes A_{dB} ($T\%$, ω)

deduction but can be expressed approximately with two folding lines after graphic analysis, as shown in Fig.3, where the two approximate variance functions are separately:

$$\overline{AB}:\ A_{dB} = \frac{11}{7}t\% - \frac{548}{7} \qquad (88\% \geq t\% \geq 60\%) \qquad (1)$$

$$\overline{CD}:\ A_{dB} = \frac{5}{3}t\% - \frac{320}{3} \qquad (100\% \geq t\% \geq 80\%) \qquad (2)$$

The mean function $m(t\%)$, as following, can be reached by applying vertical polynomial fitting:

$$m(T) = -5478.797 + 412.3429T - 12.54023T^2 + 0.196401T^3 \\ -1.658815\times10^{-3}T^4 + 7.10929\times10^{-6}T^5 - 1.191001\times10^{-8}T^6 \tag{3}$$

where $T = t\%$.

From equation (1), (2) and (3), we can deduce two kinds of forecasting models for life diagnosis by the forecast principles, which are approximate folding lines forecasting model and the trend extrapolation-forecasting model with the mean function.

Table 2a: The true value and forecasting value with two model of the remnant life of the ball bearing in motor

No. of the ball bearing		1	2	3	4	5	6	7
The remnant life (weeks)	True value	24	6	5	10	25	15	17
	Forecast value (Model I)	19~25	6~8	4~9	8~12	23~25	13~17	12~21
	Forecast value (Model II) $R(z)$=95.45%	20~28	4~12	2~10	7~15	19~27	13~20	13~21
No. of the ball bearing		15	16	17	18	19	20	21
The remnant life (weeks)	True value	14	14	10	13	12	4	14
	Forecast value (Model I)	9~16	11~16	6~11	9~16	8~14	5~6	10~17
	Forecast value (Model II) $R(z)$=95.45%	10~18	10~18	5~13	9~17	8~16	3~13	10~18

4.1 Approximate folding lines forecasting model

Suppose L is the lifetime of the ball bearing by weeks, t is the measurement time by weeks, then equation (1) and (2) convert to

$$L_{\overline{AB}} = \frac{1100}{7A_{dB} + 548}t, \quad L_{\overline{CD}} = \frac{500}{3A_{dB} + 320}t \tag{4,5}$$

where $L_{\overline{AB}}$ is the upper limit of lifetime and $L_{\overline{CD}}$ is the lower limit of lifetime.

If N pairs data (A_{dB}, t_i) from 16 dB $\le A_{dB} \le$40 dB have been obtained, how can we forecast the lifetime or the remnant life of the ball bearing by equation (4) and (5)?

In fact, substituting every pair of (A_{dB}, t_i) into equation (4) and (5) we may get the corresponding $L_{\overline{ABi}}$ and $L_{\overline{CDi}}$, so the upper and lower limit of the life, L_U and L_L, separately, are:

$$L_U = \min\{L_{\overline{AB}i}\}, \qquad L_L = \max\{L_{\overline{CD}i}\} \tag{6,7}$$

If T_N is the time how long the bearing has been in operation by weeks, the upper and lower limits of the remnant work life L_{RU} and L_{RL}, respectively, are:

$$L_{RU} = L_U - T_N, \qquad L_{RL} = L_U - T_N \tag{8,9}$$

Equation (8) and (9) are the approximate folding lines forecasting model presented by the writer. The credibility is controlled by how close it is between the approximate folding lines and the actual variance curve. If the approximate folding lines completely enclose the actual variance curve, the forecast credibility is 100%.

4.2 The trend extrapolation-forecasting model with the mean function

Basic ideas: It is reasonable to forecast the lifetime of other ball bearings that have the same data conditions as the above mentioned 171 bearings with the mean function and error property of obtained from the previous 171 group data. Here, the sample space is considered big enough.

Suppose we have N pairs of data of a certain bearing, $A_{dB} \leq 40$, express as $\{A_{dBi}, t_i\}, (t_i, weeks, i = 1,2,...,N)$, then how can we forecast the remnant life of this bearing? It contains the following two steps:

Step 1: Deduce the life and remnant life

Set T_j as the lifetime, and T_M (weeks) as the time when A_{dB} arrive the first value of $A_{dB} \geq 40$ dB. Then by analyzing above data of 171 bearings, we can find that the actual lifetime of the bearings will not exceed $T_M / 0.6$ (weeks). Thus we have $T_M + 1 \leq T_j \leq T_M / 0.6$. Now what we shall consider further is where the exact value of T_j lies in this domain.

Take step length as one week, and let T_j to increase from $T_M + 1$ to $T_M / 0.6$ (total M steps), then convert N groups data into the data with t%, namely

$$\{A_{dBi}, t_i\} \to \left\{A_{dBi}, \frac{100t_i}{T_j}\right\} = \{A_{dBi}, t_i\%\}(i = 1,2,...,N; j = 1,2,...,M).$$

Then compute remnant difference square sum:

$$SU_j = \sum_{i=1}^{N} [A_{dB} - m(t\%)]^2 .$$

By the principle of minimum two times, T_j should make SU_j to reach its minimum. So that the lifetime L_E and remnant life L_{ER} of the bearing are separately

$$L_E = T_j, \text{ when } \min(SU_j), \qquad L_{ER} = L_E - T_M \tag{10,11}$$

Step 2: Indicate the prediction credibility degree

Analyze the 171 groups of sample data with equation (10) and compute the errors δ_i by $\delta_i = L_{Ei} - L_i$, where L_i is the known lifespan, $i = 1,2, \ldots, 171$..

The statistics property of δ_i, which value domain is [-4,4] and step length is 1, is that it submits to the normal distribution with the mean $\mu = 0$ and standard deviation $\sigma = 2$.

Table 2b: The true value and forecasting value with two model of the remnant life of the ball bearing in motor (cont.)

No. of the ball bearing		8	9	10	11	12	13	14
The remnant life (weeks)	True value	15	20	15	11	9	8	21
	Forecast value (Model I)	11~18	19~26	8~15	8~12	8~14	7~10	20~23
	Forecast value (Model II) *R(z)*=95.45%	11~19	20~28	9~17	7~15	8~16	5~13	17~25
No. of the ball bearing		22	23	24	25	26	27	28
The remnant life (weeks)	True value	23	9	12	5	7	5	12
	Forecast value (Model I)	20~25	7~12	7~13	4~6	5~10	4~10	6~12
	Forecast value (Model II) *R(z)*=95.45%	21~29	7~15	8~16	2~10	5~13	4~12	5~13

5 The application of two predicting models

The writer got the data of other newly used 28 bearings. We start the forecasting job with the data of 16dB $\le A_{dB} \le$ 40dB, and when the first $A_{dB} \ge$40dB arises we have the results as shown in Table 2 of the two models. Table 2 shows:

1. Both models can tell the upper and the lower limit of the remnant life, and the second one can even indicate the credibility degree;
2. No. 1 bearing has 27 weeks (true value) of the remnant life after A_{dB} =40dB and No. 27 bearing has 5 weeks to go after reaching the same A_{dB}. The remnant lives of the two bearings differ in approximate 5 months. So it is apparently not reasonable to take the bearing state value $A_{dB} \ge$40dB as a criteria of replacement;
3. The two models presented by this paper are validated by the application.

This paper gives error limit as $[-\Delta,+\Delta]$, Δ may equal to 0, 1, 2, 3 or 4 (weeks). Then the corresponding credibility degrees are:

If Δ=4, the credibility degree $R(z) = 95.45\%$;
If Δ=3, the credibility degree $R(z) = 86.638\%$;
If Δ=2, the credibility degree $R(z) = 68.286\%$.

That is to say, the lower the value of error, the lower the credibility degree. In petroleum chemical plants we must take the high credibility degree for the sake of safety. Therefore we need take Δ=4, and $R(z) = 95.45\%$. Then we reach the following trend extrapolation-forecasting model with the mean function from step 1 and step 2

$$L_{RU} = L_E + 4 - T_M, \quad L_{RL} = L_E - 4 - T_M \qquad (12,13)$$

where L_{RU} and L_{RL} are separately the upper limit and lower limit of the remnant life of the ball bearing. The credibility degree of this forecast is $R(z) = 95.45\%$.

6. Concluding remarks

A random processes model for diagnosing the hitch level (A_{dB}) of induction motor ball bearings is established in the present research. A folding line method for the variance curve and the extrapolation method for the trend mean curve are also presented. Both methods can be employed to predict the remnant life of ball bearings for the motors. It is found in the research that the hitch level developing process of the bearings is non-stationary stochastic if the time is t%. Application of the model and the methods established in the research in predicting the remnant life of the motor ball bearings with the real samples collected in the fields shows the validation of the model and the methods. The findings of this research may bring significant benefits to the petroleum industries in diagnosing the defects of the ball bearings and predicting the remnant life of the bearings used in the industries.

Acknowledgement

The authors greatly acknowledge the financial support provide by Natural Sciences and Engineering Research Council of Canada (NSERC) and Canada Foundation for Innovation (FCI) to this research.

References

[1] I.N.Kovalenko, N.Yu.Kuznetsov, V.M.Shurenkov. *Models of Random Processes*. Boca Raton New York: CRC, 1996.

[2] Hodowanec, Mark M. Evaluation of Antifriction Bearing Lubrication Methods on Motor Life-Cycle Cost. *IEEE transactions on industry applications*. 35, No.6, (1999): 1247 (5 pages).

[3] Lonsdale, C P; Lutz, M L. Locomotive Traction Motor Armature Bearing Life Study. *Lubrication engineering*. 53, No. 8, (1997): 12 (8 pages).

[4] Motor-Gearhead Interface Impacts Motor Bearing Life, Motion System Size. *Power-conversion & intelligent motion*. 21, No. 11, (1995): 39 (9 pages).

Nonlinear Airway Smooth Muscle Stiffness Changes Using Artificial Neural Network

L. Chen, A.M. Al-Jumaily* and Y. Du
Diagnostics and Control Research Centre, Engineering Research Institute, AUT University, New Zealand

Abstract
The stiffness reduction due to longitudinal oscillations of the airway smooth muscles normally exhibits nonlinear behaviour. The latter is normally expressed in terms of nonlinear relationships between stiffness ratios and oscillation frequencies. In this paper these relationships are tackled using a feed-forward neural network (FNN) model. The structure of the FNN was selected through the training and validation using data collected from 11 experiments conducted on trachea smooth muscles dissected from pigs with different muscle lengths, muscle masses, oscillation frequencies and amplitudes. The robustness of the neural network model was improved to match the nonlinearity by using data pre-processing methods. The validation results show that the FNN model can predict the stiffness ratio changes with a mean square error of 0.0042.

Keywords :Nonlinear Airway, Muscle Stiffness, Artificial Neural Network

1. Introduction

It has been found that the stiffness of activated airway smooth muscle could be altered by cyclic stretching [1, 2, 3]. This behaviour has been ascribed to a dynamic interaction between the imposed stretch and the number of actin-myosin interactions in the muscle [4, 5, 6]. The quantitative relationship between stiffness ratio changes and oscillation frequencies is nonlinear and difficult to model due to a number of factors, such as muscle masses, which may affect the accuracy of the model.

Artificial neural networks (ANNs) are computational systems whose architecture and operation are inspired from our knowledge of biological neural cells (neurons) in the brain. They can be described either as mathematical and computational models for static and dynamic (time-varying) non-linear function approximation, data classification, clustering and non-parametric regression or as simulations of the behaviour of collections of model biological neurons. Neural modelling has shown incredible capability for emulation, analysis, prediction, and association of nonlinear behaviours. ANNs are able to solve difficult problems in a way that resembles human intelligence [7]. What is unique about neural networks is their ability to learn by examples. The main purpose of this work is to develop an application of existing ANN techniques to model the underlying nonlinear relationship between stiffness ratio changes and oscillation frequencies.

2. Neural Network Model

A simple FNN model was used in this study to estimate stiffness ratio. The general structure of the model is shown in Figure 1. The four inputs to the model are the oscillation frequency F, the oscillation amplitude A, the smooth muscle length L and the smooth muscle weight M. The output is the stiffness ratio SR. The stiffness of the smooth muscle was measured using the same method as described in [8], i.e. muscle stiffness $S = dF / dL$, where dF is the resulting force perturbation and

Corresponding author: Tel.: 64-9-921-9777; Fax: 64-9-921-9973
Email: ahmed.al-jumaily@aut.ac.nz

dL is the length perturbation amplitude. The stiffness ratio was computed using the method given in [9], i.e. $SR = S_{prior} / S_{post}$, where S_{prior} and S_{post} are the stiffness before and after oscillation, respectively.

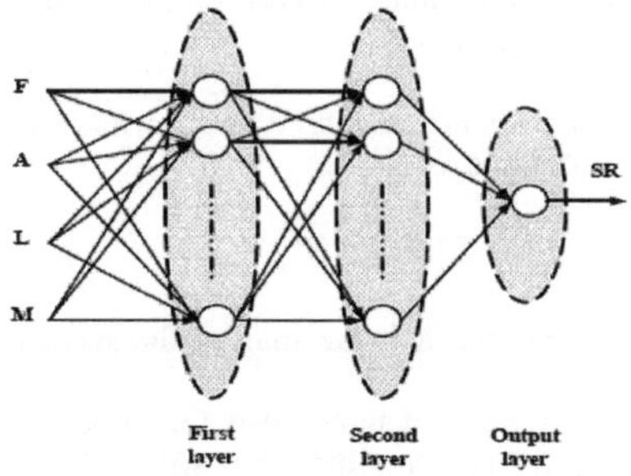

Figure 1: The structure of the FNN model used for predicting the stiffness ratio changes of the airway smooth muscle strips.

The model has three layers, the first layer, the second layer and the output layer. Sigmoid activation functions are used for the two hidden layers (the first layer and the second layer), and a pure linear function is used for the output layer.

3. Neutral Network training.

The general procedure for developing neural networks [10] consists of data pre-processing, appropriate training procedure, generalization and topology optimization.

3.1 Data Processing

Before training an RNN, the inputs and target data are pre-processed (scaled), such that they are within a specified range, [-1, 1]. This specified range is the most sensitive area of the sigmoidal functions, which are the hidden layer activation functions. In such case, the output of the trained network will also be in the range [-1, 1]. A post-processing procedure has to be performed in order to convert the output back to its original unit.

In the proposed neural networks, a sigmoid function is used in the activation layers:

$$y = \tanh(\beta x) = \frac{e^{\beta x} - e^{-\beta x}}{e^{\beta x} + e^{-\beta x}} \tag{1}$$

where, x is the input to the neuron, y is the output of the neuron and $\beta \in \mathbb{R}$

The most sensitive input area of the above function is in the range of [-1, 1]. The mathematical equation used for input data scaling is given as follows:

$$x_n = 2 \cdot (x - x_{min}) / (x_{max} - x_{min}) - 1 \tag{2}$$

where, x_n, x, x_{min}, x_{max} are processed value, original value, minimum and maximum value of the original data, respectively.

As the inputs have been transformed into the range of [-1, 1], the output of the trained network will also be in the range [-1, 1]. Thus the output data of the neural network have to be converted back to their original units using:

$$y = 0.5 \cdot (y_n) \cdot (y_{max} - y_{min}) + y_{min} \tag{3}$$

where y, y_n, y_{max}, y_{min} are the converted value, network output value, maximum value of target data and minimum value of target data, respectively.

Some aspects of data scaling problems were discussed by Koprinkova and Petrova [11]. A major issue was the scaling factor S, which is defined as:

$$SF = R / B \tag{4}$$

where, B is the highest value of the input, R is the span of the specific range to which the input data will be transformed.

In the current study, the highest value of inputs was the upper bound of frequencies, which was 75. The specific range was [-1, 1], so that the span R = 2. Thus the scaling factor was $SF = 2/75 = 0.021$, which was much higher than 0.009 [11]. Furthermore, the scaling factors that were used for all inputs by Koprinkova and Petrova were the same, whereas in this study, different inputs had different scaling factors, such that the scaled values were distributed uniformly through the whole specific range of [-1, 1]. This can further reduce the loss of information caused by the different input ranges.

3.2 Training Algorithm

During the training procedure, the goal is to minimize the mean square error (*MSE*) between the measured value and the output from the *FNN* by adjusting the weights and biases. The performance function that is used for training the neural networks is the mean square error function, which is defined as:

$$MSE = \frac{1}{N} \sum_{i=1}^{n} \left(X_t^m - \hat{X}_t \right)^2 \tag{5}$$

where, N is the number of sampling data pairs; X_t^m is measured (actual) value of biomass concentration; $\hat{X}_t$ is the corresponding estimated value predicted by the neural soft sensors.

The Levenberg-Marquardt back propagation training algorithm is adopted to train the neural networks due to its faster convergence and memory efficiency [7, 12]. The algorithm can be summarized as follows:

1. Present input sequence to the network. Compute the corresponding network outputs with respect to the parameters (i.e. weights and bias) X_k. Compute the error e and the overall *MSE*.
2. Calculate the Jacobian matrix J through the back propagating of Marquardt sensitivities from the final layer of the network to the first layer.
3. Calculate the step size for updating network parameters using:

$$\Delta X_k = -\left[J^T (X_k) J(X_k) + \mu_k I \right]^{-1} J^T (X_k) e \tag{6}$$

where, μ_k is initially chosen as a small positive value (e.g., $\mu_k = 0.01$).

4. Recompute the *MSE* using $X_k + \Delta X_k$. If this new *MSE* is smaller than that computed in step 1, then decrease μ_k, let $X_{k+1} = X_k + \Delta X_k$ and go back to step 1. If the new *MSE* is not reduced, then increase μ_k and go back to step 3.

The algorithm terminates when i) the norm of gradient is less than some predetermined value or, ii) *MSE* has been reduced to some error goal or, iii) μ_k is too large to be increased practically or, iv) a predefined maximum number of iterations has been reached.

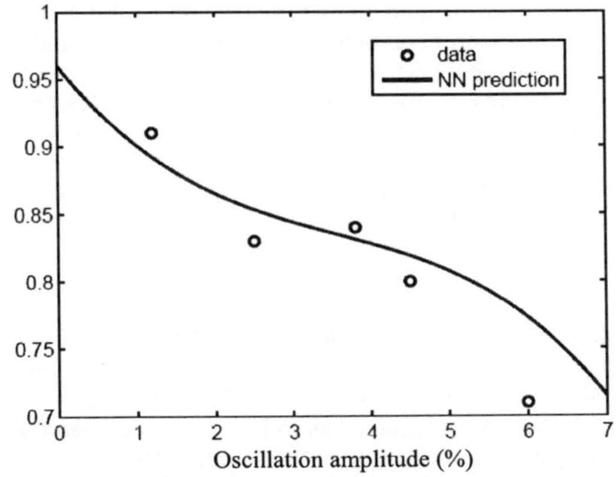

Figure 2: FNN model prediction of stiffness ratio versus oscillation amplitude. The circles are the experimental data. The solid line is the neural network model's output. *Test conditions*: Frequency = 55 Hz, L_{ref} = 16.5 mm, Mass = 0.0627 g

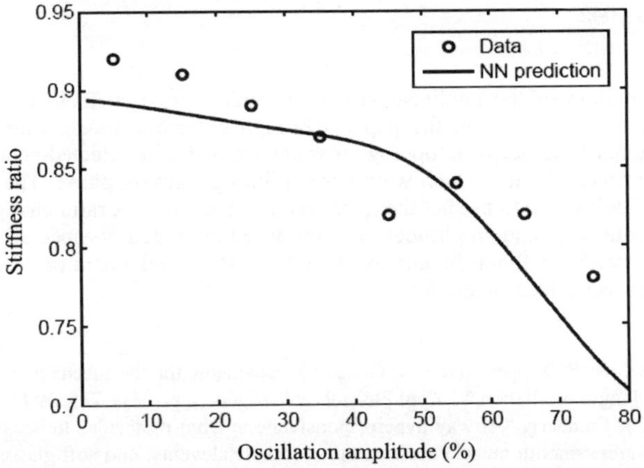

Figure 3: FNN model prediction of stiffness ratio versus oscillation frequency. The circles are the experimental data. The solid line is the neural network model's output. *Test conditions:* Oscillation Amplitude = 3.8 %, L_{ref} = 16.5 mm, Mass = 0.0627 g

3.3 Neural Network Model

The experimental data were divided into three groups: the training data set, validation data set and testing data set. Four sets of data were used as training data sets. Another four sets of data were used as validation data sets.

To prevent the neural network from being over-trained, an early stopping method was used. The error on the validation sets was monitored during the training process. The validation error would normally decrease during the initial phase of training. However, when the network began to over-fit the data, the error on the validation set would typically begin to rise. When the validation error increased for a specified number of iterations, the training was stopped, and the weights and biases at the minimum of the validation error were obtained. The rest of the data sets, which were not seen by the neural network during the training and validation period, were used in examining the trained network.

There are no general rules or guidelines for selection of optimal number of hidden neurons in recurrent neural network. The most commonly used method is trial and error. Fewer neurons result in inadequate learning by the network; too many neurons create over-training and result in poor generalization. One straightforward approach adopted by many researchers is to start with the smallest possible network and gradually increase the size until the performance begins to level off. From an engineering point of view, the smallest possible size of neural network for solving the problem is always of great interest. The approach used in this paper works in the opposite way to the aforementioned method. That is, to start with a reasonable big network and gradually "shrink" it until the error that appeared on the test data is beyond acceptance.

4. Results

The *MSE* between the network output and the target output of test data set was used to evaluate the "goodness" of the network. Extensive simulations were carried out. For each network structure, 500 networks were trained; the one that produced the smallest mean square error for the test data sets was retained. The final structure was determined as follows: 5 hidden neurons in the first layer; 35 hidden neurons in the second layer; and one neuron in the output layer.

Figure 2 shows a typical neural network prediction of stiffness ratio changes versus oscillation amplitudes (percentage of the reference muscle length) for the test data. It can be seen that the neural model predicts the main trend of the stiffness ratio changes with reasonable accuracies. Figure 3 shows a typical neural network predictions of stiffness ratio changes versus oscillation frequencies for the test data, which were unseen in the training and validation procedures. The lowest prediction *MSE* for the test data was 0.0042.

5 Conclusions

FNN models for estimation of the nonlinear airway smooth muscle stiffness ratio changes due to length oscillation were investigated in the paper. Through extensive model training and structure selection, an appropriate neural network topology was determined. The selected neural network model was tested using experimental data, which were unseen during training phase. The results show that the neural network model is able to predict the main trend of the stiffness ratio changes corresponding to both oscillation frequencies and amplitudes. The proposed approach is especially promising when limited knowledge is available about the airway smooth muscle mechanism or when it is costly and unfeasible to obtain a mechanistic model for it.

References

[1] X. Shen, M. F. Wu, R. S. Tepper, and S. J. Gunst, "Mechanisms for the mechanical response of airway smooth muscle to length oscillation," J Appl Physiol, vol. 83, no. 3, pp. 731-738, 1997.

[2] S. J. Gunst and J. J. Fredberg, "Airway hyperresponsiveness: from molecules to bedside, invited review: the first three minutes: smooth muscle contraction, cytoskeletalevents, and soft glasses," J Appl Physiol, vol. 95, pp. 413-425, 2003.

[3] L. Wang, P. D. Paré, and C. Y. Seow, "Changes in force-velocity properties of trachealis due to oscillatory strains," J Appl Physiol, vol. 92, pp. 1865-1872, 2002.

[4] J. J. Fredberg, D. Inouye, B. Miller, M. Nathan, S. Jafari, S. Helioui Aboudi, J. P. Butler, and S. A. Shore, "Airway Smooth Muscle, Tidal Stretches, and Dynamically Determined Contractile States" Am. J. Respir. Crit. Care Med., vol. 156, no. 6, pp. 1752-1759, 1997.

[5] R. A. Meiss and R. M. Pidaparti, "Active and passive components in the length-dependent stiffness of tracheal smooth muscle during isotonic shortening," J Appl Physiol, vol. 98, no. 1, pp. 234-241, 2005.

[6] R. A. Meiss, "An analysis of length-dependent active stiffness in smooth muscle strip," in Regulation of smooth muscle contraction (R. S. Moreland, ed.).

[7] M. Hagan, H. Demuth, and M. Beale, Neural network design. PWB publishing company, 1996.

[8] S. J. Gunst and M.-F. Wu, "Plasticity in skeletal, cardiac, and smooth muscle, selected contribution: plasticity of airway smooth muscle stiff-ness and extensibility: role of length-adaptive mechanisms," J App Physiol, vol. 90, pp. 741-749, 2001.

[9] A. Richard and M. P. Ramana, "Mechanical state of airway smooth muscle at very short lengths," J App Physiol, vol. 96, pp. 655-667, 2004.

[10] J. Almeida, \Predictive non-linear modelling of complex data by artificial neural network," Current Opinion in Biotechnology, vol. 13, pp. 72-76, 2002.
[11] P. Koprinkova and M. Petrova, "Data-scaling problems in neural-network training," Engineering Application of Artificial Intellingence, vol. 12, pp. 281-296, 1999.
[12] L. Scales, Introduction to non-linear optimization. New York: Springer-Verlag, 1985.

In-plane free vibrations of compound high speed rotating disks

Hamid R. Hamidzadeh
Department of Mechanical and Manufacturing Engineering, Tennessee State University,Nashville, TN 37209

Abstract
An analytical method is presented for determination of vibration characteristics of high speed Double-Segment Compound Rotating disks. More specifically, a systematic approach based on established solution for linear in-plane vibration of each segment satisfying the displacement and stresses compatibility is developed. Fixed and free boundary conditions for the compound spinning annular disks are considered, and natural frequencies and mode shapes of rotating the disks are computed. The medium for each segment is considered to be homogenous, isotropic, and elastic. The developed analytical solution was achieved by implementing two-dimensional plane stress theory. The modal displacements and stresses at both inner and outer boundaries are determined. The dimensionless natural frequencies for different modes, rotating speeds, and thickness ratios are computed. The effect of stiffness changes for each segment on the natural frequencies are also studied.

Keywords: Free vibrations, Transverse vibrations, High speed, In-plane, Rotating disks, and Critical speeds.

1. Introduction

In view of vast potential applications the flexible thin rotating disks received emphasis in recent years. The application of rotating disk includes many mechanical systems such as flywheels, disk drives, CD-ROMs, turbine rotors, DVDs, torsional disk dampers, etc. It is noted that for the last few years research works are done to increase the capacity of a hard disk drive, increase the rotating speed, and reduce size, as well as, access time. Therefore, vibration analysis for disks is emphasized on these research works. Disk vibration occurs in two directions, in-plane and transverse. The steady dynamic stresses induced by a concentrated load moving at a constant angular speed at the outer boundary were then evaluated through a Galilean transformation. Chonan and Hayase [2] analytically studied the in-plane stress distribution in a spinning annular disk. The disk was clamped at the inner boundary and subjected to a stationary distributed load along the outer boundary. The results obtained showed that the stress components induced from the rotation become predominant and the relative importance of the in-plane grinding force degenerates in the stress distribution as the rotation speed of the wheel increases. Burdess et al. [3] also studied the in-plane stress vibration in rotating disk and the general response of a rotating disk was considered. In their work, the equations of motion of a thin rotating disk were derived and solved for the general case by expressing the disk displacement in terms of scalar and vector potential. Then, properties of the free and forced solutions are determined and results relating to the stability and resonant behavior of the disk were derived. Chen and Jhu [4] investigated the free in-plane vibration of a thin spinning

The corresponding Autho, Tel: (615)963-5387, Fax: (615)963-5387
E-mail: hhamidzadeh@tnstate.edu

annular disk. The equations of motions were first derived with respect to a stationary coordinate system, and then Lame's potentials were used to simplify the coupled equations. It was observed that the critical speeds of the modes with nodal diameters approach an asymptotic value as the number of the nodal diameters increases. Their numerical simulations showed that this asymptotic critical speed was independent of the clamping ratio, while it was dependent on the Poisson's ratio of the disk. Chen and Jhu [5] also studied the in-plane response of a thin rotating annular disk under concentrated edge load with both the radial and tangential components. They presented numerical results for the natural frequencies and steady state response of a disk for a radius ratio for 0.3. Tzou et al. [6] studied the three dimensional vibration of an arbitrarily thick disk for two different boundary conditions: all surface traction free and all free except for the clamped inner radius. For a disk of significant thickness, vibration modes in which motion occurs within the disk's equilibrium plane can play a substantial role in setting its dynamic response. The equations for three-dimensional motions were described through the Ritz technique, yielding natural frequencies and mode shapes for coupled axial, radial, and circumferential deformations.

Chen and Jhu [7] studied the in-plane stress and displacement distributions in a spinning annular disk under stationary edge loads. By using Lame's potentials, solutions to the case of a spinning disk under stationary edge tractions were obtained. Hamidzadeh and Dehghani [8] studied linear in-plane vibration of an elastic rotating disk considering the stiffness and it was found that stiffness plays a major role on the natural frequency. Hamidzadeh and Wang [9] studied the effect of rotational speed and radius ratio on natural frequency and elastic stability of the free rotating disks. It was observed that mode of vibration, type of circumferential wave occurring, and boundary conditions have great effect on natural frequency. Critical speeds for different radius ratios were also obtained. Hamidzadeh [10] also developed an analytical solution for in-plane vibration of spinning rings. Main emphasis has always been given to linear and non-linear transverse vibration of rotating disk. This is primarily due to the fact that many systems exhibit a higher degree of transverse vibration than in-plane vibration. It has been found in recent years that in-plane vibration can play a significant role in rotating disks. Applications of rotating disks are numerous, some of which includes circular saws, grinding wheels, disk dampers, and disk brakes.

The research work presented in this paper will be on linear in-plane vibration of a thin rotating disk with two parts. The equation of motion of a rotating disk is derived and solved for the general case by the wave equations. The general solution is obtained using Bessel functions for the time dependent displacements and stresses. Computational analysis is then made to obtain the natural frequencies and mode shapes for the compound disks.

2. Nomenclature

Ψ	Elastic rotation factor
$\Omega_1^* = pc/c_1$	Non-dimensional natural frequency in rotating coordinate
$\Omega_2^* = \omega c/c_1$	Non-dimensional angular speed of rotation
Ω_F^*	Non-dimensional natural frequency in fixed coordinate
a	Inner radius of the disk
b	Outer radius of the inner disk
c	Outer radius of the disk
E	Modulus of elasticity
m	Number of nodal circles
n	Number of nodal diameter
p	Natural frequency experienced on rotating coordinate
r	Radial coordinate
(r,θ)	Rotating polar coordinate fixed to the disk
$U^* = u/c$	Non-dimensional radial displacement
u	radial displacement
$V^* = v/c$	Non-dimensional tangential displacement

v	Tangential displacement
ε_r	Radial strain
$\gamma_{r\theta}$	Shear strain
σ_r	Radial stress
$\tau_{r\theta}$	Shear stress
$\tau_{r\theta}^*$	Non-dimensional shear stress
ω	Angular speed of rotation
∇^2	Laplacian operator
ρ	Density of the medium
ν	Poisson's ratio
μ	Shear modulus
Δ	Volumetric strain

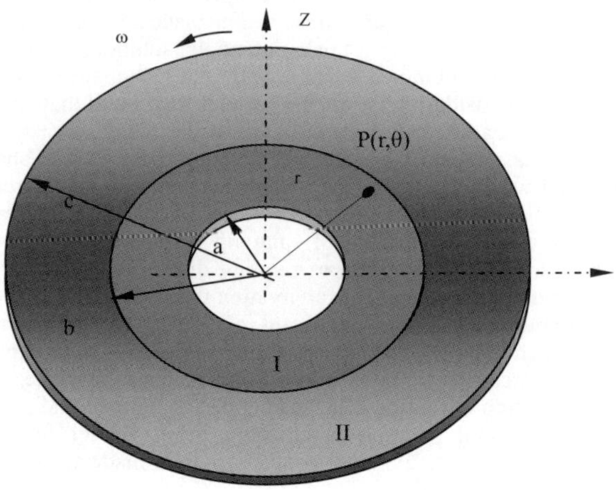

Figure 1: A typical two part compound disk

3. Governing equation of motion

A typical double compound disk rotating about its axis is shown in Figure 1. Each part is assumed to be homogeneous, isotropic and elastic. As it was presented by Hamidzadeh and Dehghani [8] equations of motion in terms of volumetric strains Δ and elastic rotation Ψ for each part is given by:

$$c_1^2\nabla^2\Delta-\ddot{\Delta}+\omega^2\Delta+2\omega\dot{\psi}=-2\omega^2 \tag{1a}$$

$$c_2^2\nabla^2\psi-\ddot{\psi}+\omega^2\psi+2\omega\Delta=0 \tag{1b}$$

where

$$\Delta=\frac{\partial u}{\partial r}+\frac{u}{r}+\frac{1}{r}\frac{\partial v}{\partial\theta},\qquad \psi=\frac{\partial v}{\partial r}+\frac{v}{r}-\frac{1}{r}\frac{\partial u}{\partial\theta} \tag{2.a,b}$$

$$c_1^{\;2}=\frac{E}{(1-\nu)\rho},\quad c_2^{\;2}=\frac{\mu}{\rho} \tag{3.a,b}$$

u and v are the radial and tangential displacement. E and μ are elastic and Shear moduli.

4. Modal displacements and stresses for each part

Assuming harmonic radial and tangential displacements, one can write

$$u(r,t) = u_n(r)e^{i(n+pt)}, \; v(r,t) = v_n(r)e^{i(n+pt)} \quad (4.a,b)$$

$$\sigma_r(r,\theta,t) = \sigma_{rn}(r)e^{i(n\theta+pt)} \quad \tau_{r\theta n}(r,\theta,t) = i\tau_{r\theta n}(r)e^{i(n\theta+pt)} \quad (4.c,d)$$

Introducing the non-dimensonalized variables

$$U_n^* = U_n/b, \; V_n^* = V_n/b, \; \sigma_n^* = \sigma_n/\mu, \; \tau_n^* = \tau_n/\mu \quad (5a\text{-}d)$$

then modal displacements and stresses at any radius for each segment according to Hamidzadeh and Dehghani [8] will be presented in the following equation.

$$\{U_n^*(r), U_n^*(r), \; \sigma_n^*(r), \; \tau_n^*(r)\}^T = [A(r)]\{B_n, C_n, D_n, E_n\}^T \quad (6)$$

Elements of $[A(r)]$ are in terms of material properties and Bessel functions of First and Second kinds. These elements are presented in the above mentioned paper.

5. Modal Displacements and Stresses For Compound Disks

For the first disk equation (6) can be written at r = a and r = b and after rearranging, results can be written as:

$$\begin{Bmatrix} U_n^*(b) \\ V_n^*(b) \\ \sigma_{rn}^*(b) \\ \tau_n^*(b) \end{Bmatrix} = [S_{nI}] \begin{Bmatrix} U_n^*(a) \\ V_n^*(a) \\ \sigma_r^*(a) \\ \tau_{r\theta n}^*(a) \end{Bmatrix} \quad (7.a)$$

where

$$S_{nI} = [A_I(r=b)][A_I(r=a)]^{-1} \quad (7.b)$$

Similarly for the second disk equation (6) for its inner and outer radii of r=b and r=c results

$$\begin{Bmatrix} U_n^*(c) \\ V_n^*(c) \\ \sigma_{rn}^*(c) \\ \tau_n^*(c) \end{Bmatrix} = [S_{nII}] \begin{Bmatrix} U_n^*(b) \\ V_n^*(b) \\ \sigma_r^*(b) \\ \tau_{r\theta n}^*(b) \end{Bmatrix} \quad (8.a)$$

Where

$$S_{nII} = [A_{II}(r=c)][A_{II}(r=b)]^{-1} \quad (8.b)$$

Equating dimensionalized $\{u,v,\sigma,\tau\}_{II(b)}$ and $\{u,v,\sigma,\tau\}_{I(b)}$ requires:

$$\begin{bmatrix} U(r)/c \\ V(r)/c \\ \sigma(r)/\mu_2 \\ \tau(r)/\mu_2 \end{bmatrix}_{II(r=b)} = [k] \begin{bmatrix} U(r)/b \\ V(r)/b \\ \sigma(r)/\mu_1 \\ \tau(r)/\mu_2 \end{bmatrix}_{I(r=b)} \tag{9.a}$$

where

$$[K] = \begin{bmatrix} b/c & 0 & 0 & 0 \\ 0 & b/c & 0 & 0 \\ 0 & 0 & \mu_1/\mu_2 & 0 \\ 0 & 0 & 0 & \mu_1/\mu_2 \end{bmatrix} \tag{9.b}$$

Then compatibility of stresses and displacements at r=b for disks I and II requires

$$\begin{bmatrix} U_n^*(r) \\ V_n^*(r) \\ \sigma_{rn}^*(r) \\ \tau_{r\theta n}^*(r) \end{bmatrix}_{r=c} = [Q] \begin{bmatrix} U_n^*(r) \\ V_n^*(r) \\ \sigma_{rn}^*(r) \\ \tau_{r\theta n}^*(r) \end{bmatrix}_{r=a} \tag{10.a}$$

Where

$$[Q] = [S_{nII}][K][S_{nI}] \tag{10.b}$$

Applying the boundary conditions at radius r = a and at radius r = c (clamped condition at inner side and free at outside) require: $u_{inne} = 0$, $v_{inner} = 0$, $\sigma_{outer} = 0$ and $\tau_{outer} = 0$. Adopting the above conditions equation (10) reduces:

$$\begin{bmatrix} U_n^*(r) \\ V_n^*(r) \\ 0 \\ 0 \end{bmatrix}_{outer} = [Q] \begin{bmatrix} 0 \\ 0 \\ \sigma_{rn}^*(r) \\ \tau_{r\theta n}^*(r) \end{bmatrix}_{inner} \tag{11.a}$$

Separating the above equation results in

$$\begin{bmatrix} Q_{33} & Q_{34} \\ Q_{43} & Q_{44} \end{bmatrix} \begin{bmatrix} \sigma \\ \tau \end{bmatrix}_{inner} = \begin{bmatrix} 0 \\ 0 \end{bmatrix} \tag{11.b}$$

where

$$\begin{vmatrix} Q_{33} & Q_{34} \\ Q_{43} & Q_{44} \end{vmatrix} = 0 \tag{11.c}$$

is the frequency equation for the multi segmented thin disk. Displacements and stresses mode shapes are given by equation (13.b) and (13.a).

Table 1: Comparison for dimensionless natural frequencies versus dimensionless speed for compound and equivalent single rotating disks, n = 2, m = 0, m = 1

	M=0		M=1	
Ω_2	Compound	Equivalent Single	Compound	Equivalent Single
0.1	1.522375	1.522375	2.779198	2.779198
0.2	1.427538	1.427538	2.824411	2.82441
0.3	1.331421	1.331421	2.863944	2.863944
0.4	1.234328	1.234328	2.896788	2.896788
0.5	1.136453	1.136453	2.922065	2.922065
0.6	1.037922	1.037922	2.939033	2.939033
0.7	0.938815	0.938815	2.947115	2.947115
0.8	0.839176	0.839176	2.945934	2.945934
0.9	0.739018	0.739018	2.935328	2.935328
1	0.63833	0.63833	2.915358	2.915358
1.1	0.537069	0.537069	2.886284	2.886284
1.2	0.435156	0.435156	2.848522	2.848522
1.3	0.332461	0.332461	2.802596	2.802596
1.4	0.22877	0.22877	2.749082	2.749082
1.5	0.123738	0.123738	2.688555	2.688555
1.6	0.016763	0.016763	2.621554	2.621554

6. Numerical Results

The results of this study are presented here for different cases where dimensionless natural frequency is defined as pc/c_1 and dimensionless rotational speed is defined as $\omega c/c_1$. To check the validity of present dimensionless natural frequencies for a compound rotating disks with the same material properties, they are compared with those of Hamidzadeh and Wang [9] for equivalent single disk in rotating coordinates. Comparison for n = 2, m = 0 and m = 1 are provided in Table 4.1. The comparison shows excellent agreement for dimensionless natural frequencies versus dimensionless rotating speed.

To determine the effect of stiffened inner or outer segment on the dimensionless natural frequency versus rotating speeds computation were performed for three different modulus of elastic ratio E1/E2. The results are presented in Figure 4.2 for n = 2 and for m = 0. The effects of variation of E1/E2 = 1 and E1/E2 =1.2 on dimensionless natural frequency versus dimensionless speed are provided in Figure 4.3 for n = 1, n = 2, n = 3, n = 4 and m = 0 in rotating coordinate. Similarly, the effects of variation of E1/E2 = 1 and E1/E2 = 1/1.2 on dimensionless natural frequency versus dimensionless speed are shown in Figure 4.4 for n = 1, n = 2, n = 3, n = 4 and m = 0 in rotating coordinate. These results indicate that for E1/E2 = 1/1.2 natural frequencies would be higher than if E1/E2 = 1. This is due to higher stiffness of the outer segment. For E1/E2 = 1.2 the outer segment is more flexible than inner one and consequently dimensionless natural frequencies would be lower than a uniform disk with elastic modulus of E1. Similar results are obtained for other values of n. In general dimensionless natural frequencies are higher for higher values of n.

7. Conclusion

An analytical solution for in-plane free vibrations of compound high speed rotating disks was developed. The solution can provide natural frequencies and mode shapes of fixed-free compound annular disks at different rotational speeds. The validity of the presented solution was established by good agreements between the results computed for a compound disk with the same materials and its respective equivalent single disk. The presented results indicated that the natural frequencies for different modes of compound disks can be significantly increased by stiffening the inner or the outer annular disks.

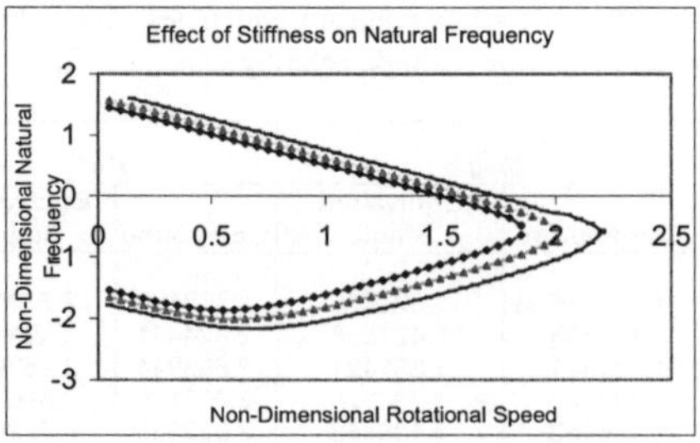

Case I: m = 0, n = 2, ♦ E1/E2 = 1.2, Δ E1/E2 = 1 and – E1/E2 = 1/1.2

Figure 2: Effect of stiffness on natural frequency for n=2, m=0 and for E1/E2=1.2, E1/E2=1.0 and E1/E2=1.0/1.2

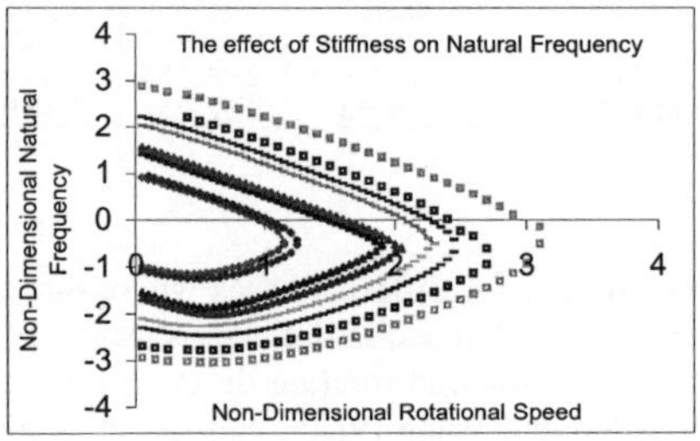

Case II: Inner E1/E2 = 1.2 and Outer E1/E2 = 1.0
♦ m = 0, n =1, Δ m = 0, n = 2, – m = 0, n = 3, □ m = 0, n = 4

Figure 3: Effect of stiffness on natural frequency for n = 1, n = 2, n = 3 and n = 4; for E1/E2 = 1.2 and E1/E2 = 1.0

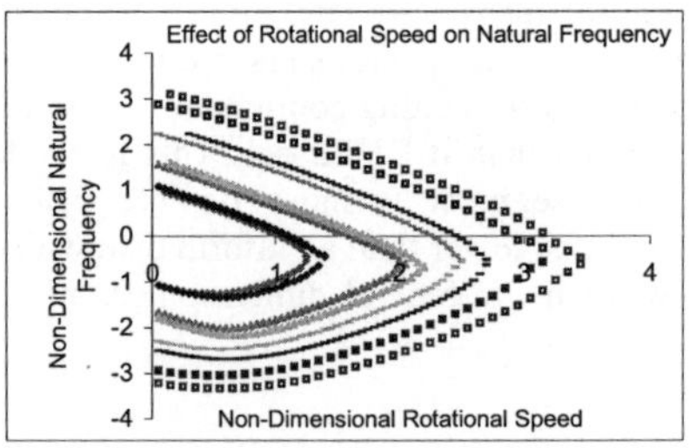

Case III: Inner E1/E2 = 1.0 and Outer E1/E2 = 1.0/1.2,
♦ m = 0, n = 1, Δ m = 0, n = 2, – m = 0, n = 3 , □ m = 0, n = 4

Figure 4: Effect of stiffness on natural frequency for n = 1, n = 2, n = 3 and n = 4; for E1/E2 = 1.0 and E1/E2 = 1.0/1.2

8. References

[1] Chonan, S. and Hayase, T. "Stress Analysis of a Spinning Annular Disk to a Stationary Distributed, In-Plane Edge Load," Journal of Vibration, Acoustics, Stress, and Reliability in Design, Vol. 109, July 1987, pages 277-282.

[2] Burdess, J. S., Wren, T. and Fawcett, J.N., "Plane Stress Vibrations in Rotating Disks," Proceedings of Institute of Mechanical Engineers 201, pp. 37-44, 1987.

[3] Chen, J S. and Jhu, J. L., " On the In-plane Vibration and Stability of a Spinning Annular Disk," Journal of Sound and Vibration 195(4), pp. 585-593,1996.

[4] Chen, J. S. and Jhu, J. L., "In-plane Response of a Rotating Annular Disk under Fixed Concentrated Edge Loads," International Journal of Mechanical Sciences 38(12), pp. 1285-1293, 1996.

[5] Tzou, K.I., Wickert, J.A., and Akay, A., " In-plane Vibration Modes of Arbitrarily Thick Disks," 1997 ASME Design Engineering Technical Conference, DETC97/VIB-4086, 1997.

[6] Chen, J. S. and Jhu, J. L., "In-plane Stress and Displacement Distributions in a Spinning Annular Disk Under Stationary Edge Loads," Response of a Rotating Annular Disk under Fixed Concentrated Edge Loads," Journal of Applied Mechanics 1997, Vol. 64, pages 897-904.

[7] Hamidzadeh, H. R. and Dehgani, M., "In-plane Free Vibration of a Rotating Disk," 1999 ASME Design Engineering Technical Conference, DETC/99/VIB-8146, pp. 592, 1999.

[8] Hamidzadeh, H. R. and Wang, H., "In-Plane Free Vibration of Spinning Rings An Analytical Approach," ASME International Mechanical Engineering Congress and Exposition, Vol. DE 108, pp. 9-16, 2000.

[9] Hamidzadeh, H. R., "Free Vibrations of Flexible Thin Rotating Disks," Professional Engineering Publishing, pp. 43-45, 2000.

Self-Excited vibration and stability analyses for axially traveling strings under steady wind loadings

Yuefang Wang* and Lefeng Lu

Department of Engineering Mechanics, Dalian University of Technology, Dalian 1160024, China

Abstract

The nonlinear, self-excited vibration is investigated for axially traveling strings loaded by steady wind actions. The equation of motion is obtained after modeling the steady wind loading as a deterministic, nonlinear function of mean wind speed. By using the Galerkin's approach, the traveling string is simplified as an approximate system with two degree of freedoms. The stability is analyzed for the equilibrium configuration after linearizing the discretized system. With the Routh-Hurwitz criterion, the stability region is identified and the condition for generation of stable limit cycles with multiple parameters via the Hopf bifurcation is pointed out. The Incremental Harmonic Balance method is adopted to determine the self-excited motion response of the string by taking the response frequency and expansion coefficients as unknowns that are solved by the Newton-Raphson method. The stability analysis is carried out by computation of the Floquet multipliers. Various stability conditions are presented considering the transport speed, wind speed and the viscous damping as operation parameters.

PACS: 46.70.Hg; 46.40.Ff

Keywords: Traveling string; Self-Excited vibration; Stability; Aerodynamic loading; Limit cycle

1. Introduction

Axially traveling strings can be found as prototyped structures of some engineering devices e.g. aerial string tramways and transmission belts that operate in open, windy environment. To these strings transverse vibration may be induced as wind flows across their cross sections. In general, this kind of vibration takes place when stability of equilibrium configuration of the string is lost. Consequently, a limit-cycle-typed is developed as the periodic motion with amplitude controlled by various operation parameters as well as aerodynamics of the airflow field. The safety of string operation will be jeopardized if amplitude of the periodic motion becomes excessively large. Although a variety of literatures have been published to dynamics and control of the traveling strings [1-3], no previous investigations are found devoted to the transverse vibration to the best knowledge of the authors.

* Corresponding author. Tel: +86-411-84708390; fax:+86-411-84708400.
E-mail address: yfwang@dlut.edu.cn (Y. Wang).

In this paper, the nonlinear vibration and motion stability are investigated for traveling strings with very small sag-to-span ratios under determinate, steady wind excitations. The governing equation is obtained and simplified by using the Galerkin's approach with a two-term displacement expansion approximation. The stability of the equilibrium configuration is provided through an eigenvalue analysis for loss of stability and generation of limit cycles via the Hopf bifurcation with application of the Routh-Hurwitz criterion. The periodic motion of the string as well as the response frequency is determined by means of the Incremental Harmonic Balance Method, with stability analyses carried out by eigenvalue computation for the Floquet multipliers.

2. Modeling and Approximation

Consider a traveling string with fixed-fixed boundary condition, very small sag-to-span ratio, uniform tension force and negligible static deformation from the gravitational force. Following the non-dimensionalization in [4] and discarding the geometrically nonlinear terms due to the coupling among displacements in different directions, the governing partial differential equation can be obtained, as

$$\ddot{y}(\eta,t)+2c\dot{y}'(\eta,t)+\delta(\dot{y}(\eta,t)+cy'(\eta,t))+(c^2-c_0^2)y''(\eta,t)=f(\eta,t), \tag{1}$$

where $y(\eta,t)$ is the transverse (the in-plane, or vertical) motion response, η being the curve coordinate and t being the temporal variable; c is the non-dimensional transport speed, c_0 is the non-dimensional axial rigidity related to the tension force; δ is the viscous damping factor per unit length of the string. The dot and prime denote the derivatives with respect to t and η, respectively. The wind loading is presumably applied on the string in the horizontal direction, which is parallel to the ground and perpendicular to the equilibrium configuration of the string. The wind loading is considered as a steady excitation which can be modeled as a nonlinear function of mean wind speed, albeit the airflows acting on the string in real operations is usually unsteady and turbulent. Mathematically, the distributed aerodynamic excitation is expressed, as in [5]

$$f(\eta,t)=f_1\dot{y}(\eta,t)+f_3\dot{y}^3(\eta,t), \tag{2}$$

where $f_1=0.5a\rho_0 hv_0\sqrt{gS}/\rho Ag$, $f_3=0.5b\rho_0 hv_0^{-1}\sqrt{gS}^3/\rho Ag$, a and b being coefficients drag and lift forces measured experimentally; v_0 is the mean wind speed, ρ_0, ρ, S, g, A and h the density of air, the density of the string material, the length of the string, the gravitational acceleration, the cross sectional area and the characteristic height of the cross-section [5]. Substitution of (2) into (1) yields

$$\ddot{y}+2c\dot{y}'+\delta cy'+(c^2-c_0^2)y''+(\delta-f_1)\dot{y}-f_3\dot{y}^3=0, \tag{3}$$

featuring a self-excited motion in the transverse direction. For an approximate solution the Galerkin's approach is used to discretize Eq. (3) with expansion of displacement: $y(\eta,t)=\xi_i(t)\varphi_i(\eta)$, (summation for $i, i=1,2,...,M$.) where $\varphi_i(\eta)=\sin i\pi\eta$ are trial functions and $\xi_i(t)$ the principal coordinates, respectively; M is the maximum truncation term in the displacement expansion. Substitution of the expansion into Eq. (3) leads to the following ordinary differential equation of the principal coordinates

$$\sum_{i=1}^{M}[\ddot{\xi}_i\varphi_i+2c\dot{\xi}_i\varphi_i'+\delta c\xi_i\varphi_i'+(c^2-c_0^2)\xi_i\varphi_i''+(\delta-f_1)\dot{\xi}_i\varphi_i]-\sum_{i=1}^{M}(f_3\dot{\xi}_i\varphi_i)^3=0. \tag{4}$$

Multiplying the foregoing with $\varphi_j(\eta)=\sin j\pi\eta$ and then integrating it over the total length of the string, one obtains the governing equations for each principal coordinate:

$$\ddot{\xi}_j+(c_0^2-c^2)j^2\pi^2\xi_j+(\delta-f_1)\dot{\xi}_j+2\sum_{i=1}^{M}d_{ij}(2c\dot{\xi}_i+\delta c\xi_i)=2f_3\int_0^1(\sum_{i=1}^{M}\dot{\xi}_i\varphi_i)^3\varphi_j d\eta, \quad (j=1,...,M) \tag{5}$$

where $d_{ij} = ij((-1)^{i+j} - 1)(i^2 - j^2)^{-1}(1 - \delta_{ij})$, δ_{ij} being the Kronecker delta.

3. Equilibrium Configuration and Its Stability

The string is sustained at its equilibrium configuration as long as the latter is stable. Based on Eq.(3) the equilibrium of the string is $y(\eta) = 0$ for any $c \neq 0$. Here the two-term expansion of displacement, i.e. M=2, is used in Eq.(5). As a result, equation (5) can be simplified for the principal coordinates, read as

$$\begin{cases} \ddot{\xi}_1 + (\delta - f_1)\dot{\xi}_1 - \frac{8}{3}(2c\dot{\xi}_2 + \delta c \xi_2) + \pi^2(c_0^2 - c^2)\xi_1 = \frac{3}{4} f_3 \dot{\xi}_1(\dot{\xi}_1^2 + 2\dot{\xi}_2^2), \\ \ddot{\xi}_2 + (\delta - f_1)\dot{\xi}_2 + \frac{8}{3}(2c\dot{\xi}_1 + \delta c \xi_1) + 4\pi^2(c_0^2 - c^2)\xi_2 = \frac{3}{4} f_3 \dot{\xi}_2(\dot{\xi}_2^2 + 2\dot{\xi}_1^2), \end{cases} \tag{6}$$

where the gyroscopic or Coriolis force expressed by the skew symmetric terms is included. Linearization of the forgoing in the state space leads to

$$\begin{Bmatrix} \dot{\xi}_1 \\ \ddot{\xi}_1 \\ \dot{\xi}_2 \\ \ddot{\xi}_2 \end{Bmatrix} = \begin{bmatrix} 0 & 1 & 0 & 0 \\ \pi^2(c^2 - c_0^2) & f_1 - \delta & \frac{8}{3}\delta c & \frac{16}{3}c \\ 0 & 0 & 0 & 1 \\ -\frac{8}{3}\delta c & -\frac{16}{3}c & 4\pi^2(c^2 - c_0^2) & f_1 - \delta \end{bmatrix} \begin{Bmatrix} \xi_1 \\ \dot{\xi}_1 \\ \xi_2 \\ \dot{\xi}_2 \end{Bmatrix}. \tag{7}$$

The stability property of the equilibrium is determined by the four eigenvalues of the coefficient matrix on the right hand side. To this end, the Routh-Hurwitz criterion is applied for explicit stability conditions expressed with operation parameters c, c_0, δ and the mean wind speed v_0, presented as

$$\begin{aligned} &\Delta_1 = \delta - f_1 \geq 0, \Delta_2 = 5\pi^2(c_0^2 - 1) + 256c^2/9 - 2f_1\delta + f_1^2 + \delta^2 \geq 0, \Delta_3 = 5\pi^2(\delta + f_1)(c_0^2 - c^2) + 256\delta c^2/9 \geq 0, \\ &\Delta_4 = 64\delta^2 c^2/9 - 8\pi^4 c_0^2 c^2 + 4\pi^4(c^4 + c_0^4) \geq 0, \Delta_5 = \Delta_3(\Delta_1\Delta_2 - \Delta_3) - \Delta_1^2\Delta_4 \geq 0. \end{aligned} \tag{8}$$

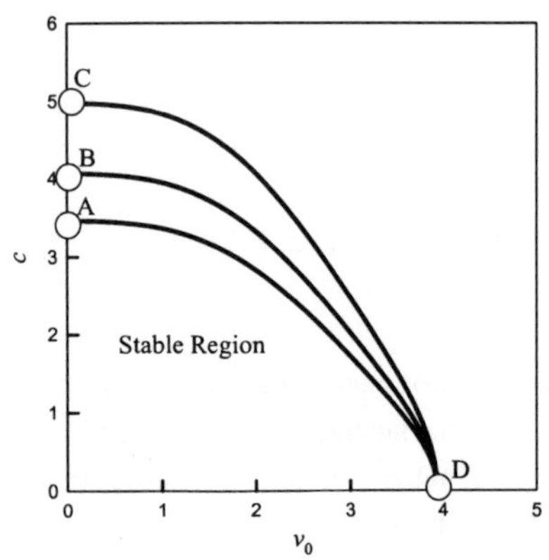

Fig. 1 Stability varying with c_0

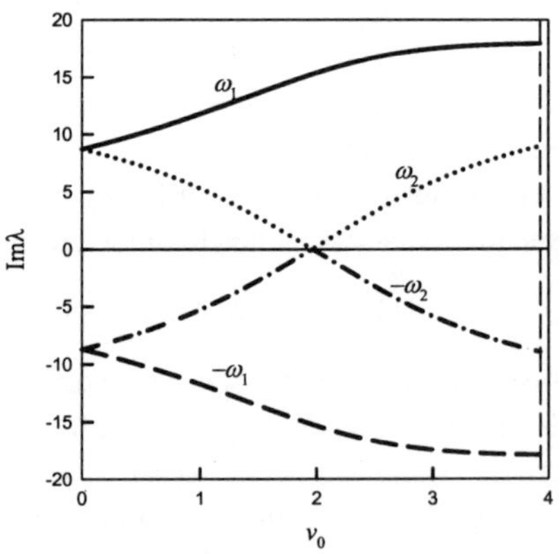

Fig.2 Imaginary parts of the eigenvalues

The stability region in the parametric space is geometrically partitioned by boundary surfaces on which the motion stays critically stable. Computationally, these boundary surfaces can be generated by making at least one of the five Δs becomes zero and the rest positive. On the other hand, at least one of the eigenvalues on the boundary possesses a zero real part, while others keep their real parts negative. To demonstrate the stability region and its boundary, a set of parameters are given: a=0.2992,

b=-0.2766, $\rho_0 = 1.293\text{kg/m}^3$, $\rho = 7800\text{kg/m}^3$, $S = 125.9\text{m}$, $g = 9.81\text{m/s}^2$, $A = 2.7745\times10^{-4}\text{m}^2$, h=0.0635m. Other fixed parameters are c_0=2.853 and $\delta = 0.08$. In the two-dimensional space of (v_0, c), the stability region in the parametric domain can be determined numerically by examination of criteria of Eq.(8) and then illustrated in Fig. 1.

The regions of stable motions and unstable motions are divided by the boundary curve AD in Fig.1, where point D ($v_0 = 3.932, c = 0$) is determined by Δ_1=0, the first equality condition of Eq. (8). The area cornered by the curve constitutes the region of asymptotic stability where all the four eigenvalues presenting in two conjugated pairs are with negative real parts. On the boundary curve the real part of one pair of eigenvalues becomes zero, while it remains negative for the other pair. As the boundary is surpassed by properly changing the parameters, the equilibrium becomes unstable due to the development of positive real part of at least one eigenvalue pair. This reveals the phenomenon that the equilibrium loses its stability via the Hopf bifurcation. Two additional boundaries are obtained and shown as curves BD and CD in Fig.1 by letting c_0=3.35 and 4.1, respectively. It is observed that the stability region is enlarged with increment of the tension force. Once the eigenvalues are computed the frequency of perturbed vibrations in the vicinity of the equilibrium can be determined. Figure 2 depicts the imaginary parts of the eigen-pairs varying with v_0 and c (not shown in that figure) on the boundary curve AD, where the vertical dash near the right margin represents the critical condition Δ_1=0. It is noticed that when $v_0 = 0, c = 3.4603$, the two imaginary parts collide with each other, which means the natural frequencies are identical for the linearized vibration response. With the increment of wind speed and decrement of transport speed, the first frequency, ω_1, goes up. By contrast, the second frequency, ω_2, drops between the interval $v_0 \in [0, 1.966)$ and goes up again between the interval $v_0 \in (1.966, 3.932]$. At $v_0 = 1.966, c = 3.3502$, the second frequency becomes zero, devoting to the motion an exponentially decaying since the real part of the second eigen-pair is still negative. Other parameters can be considered in a higher dimensional space of parameters for determination of stability region through the aforementioned analysis approach.

4. Periodic Motions and Stability Analysis

Like in other self-excited systems, a periodic motion is developed once the equilibrium becomes unstable. The Incremental Harmonic Balance Method (IHBM) [6,7] is adopted for investigation of the limit-cycle response of the traveling string under aerodynamic loadings. Introducing the rescaled temporal variable $\tau = \Omega t$, Ω being the response frequency, the governing equations of (8) are rewritten as

$$\Omega^2\mathbf{X}'' + \Omega\left(\bar{\mathbf{C}} + \bar{\mathbf{C}}_n(\mathbf{X},\mathbf{X}')\right)\mathbf{X}' + \bar{\mathbf{K}}\mathbf{X} = \mathbf{0}, \tag{9}$$

where $\mathbf{X}(\tau) = [\xi_1(\tau), \xi_2(\tau)]^T \triangleq [x_1(\tau), x_2(\tau)]^T$, the primes denote derivatives with respect to τ, and

$$\bar{\mathbf{C}} = \begin{bmatrix} \delta - f_1 & -\frac{16}{3}c \\ \frac{16}{3}c & \delta - f_1 \end{bmatrix}, \bar{\mathbf{C}}_n = \begin{bmatrix} -\frac{3}{4}f_3\Omega^2 x_1'^2 & -\frac{3}{2}f_3\Omega^2 x_1' x_2' \\ -\frac{3}{2}f_3\Omega^2 x_1' x_2' & -\frac{3}{4}f_3\Omega^2 x_2'^2 \end{bmatrix}, \bar{\mathbf{K}} = \begin{bmatrix} \pi^2(c_0^2 - c^2) & -\frac{8}{3}\delta c \\ \frac{8}{3}\delta c & 4\pi^2(c_0^2 - c^2) \end{bmatrix}$$

are linear and nonlinear damping matrices and stiffness matrix, respectively. The standard implementation of IHBM is carried out first by expressing the unknowns as linear combinations of 2N harmonic terms as

$$x_{1,2} = \overline{\mathbf{S}}\mathbf{A}_{1,2},$$

$\overline{\mathbf{S}} = \{1, \sin\tau, \cos\tau, \sin 2\tau, \cdots, \cos(N-1)\tau, \sin(N\tau)\}$ being the vector of harmonic terms and $\mathbf{A}_{1,2}$ the vectors of 4*N* coefficients for the two principal coordinates. The procedure of harmonic balance is then achieved by substituting $\mathbf{A} = \mathbf{A}_0 + \Delta\mathbf{A}, \Omega = \Omega_0 + \Delta\Omega$ into Eq.(9) followed by multiplying $\mathbf{S}^T = diag[\overline{\mathbf{S}}^T, \overline{\mathbf{S}}^T, ..., \overline{\mathbf{S}}^T]$ and integrating the product over the time interval 2π. Consequently, the equations of increments is obtained:

$$\mathbf{K}_{\mathbf{mc}}\Delta\mathbf{A} = \mathbf{R} - \mathbf{R}_{\mathbf{mc}}\Delta\Omega, \tag{10}$$

where

$$\left.\begin{aligned}
&\mathbf{K}_{\mathbf{mc}} = \Omega^2\mathbf{M} + \Omega(\mathbf{C}+\mathbf{C}_{\mathbf{nx}}) + \mathbf{K},\ \mathbf{R} = -\left(\Omega^2\mathbf{M} + \Omega(\mathbf{C}+\mathbf{C}_{\mathbf{n}}) + \mathbf{K}\right)\mathbf{A} + \mathbf{P},\ \mathbf{R}_{\mathbf{mc}} = \left(2\Omega\mathbf{M} + \mathbf{C} + 3\mathbf{C}_{\mathbf{n}}\right)\mathbf{A},\\
&\mathbf{M} = \int_0^{2\pi}\mathbf{S}^T\mathbf{S}''d\tau, \mathbf{K} = \int_0^{2\pi}\mathbf{S}^T\overline{\mathbf{K}}\mathbf{S}d\tau, \mathbf{C} = \int_0^{2\pi}\mathbf{S}^T\overline{\mathbf{C}}\mathbf{S}'d\tau, \mathbf{C}_{\mathbf{n}} = \int_0^{2\pi}\mathbf{S}^T\overline{\mathbf{C}}_{\mathbf{n}}\mathbf{S}'d\tau, \mathbf{C}_{\mathbf{nx}} = \int_0^{2\pi}\mathbf{S}^T\overline{\mathbf{C}}_{\mathbf{nx}}\mathbf{S}'d\tau,\\
&\overline{\mathbf{C}}_{\mathbf{nx}} = \begin{bmatrix} -\frac{9}{4}f_3\Omega^2x_1'^2 - \frac{3}{2}f_3\Omega^2x_2'^2 & -3f_3\Omega^2x_1'x_2' \\ -3f_3\Omega^2x_1'x_2' & -\frac{9}{4}f_3\Omega^2x_2'^2 - \frac{3}{2}f_3\Omega^2x_1'^2 \end{bmatrix}.
\end{aligned}\right\} \tag{11}$$

The Newton-Raphson method can be used to solve Eq. (10) in searching for solution of $\Delta\mathbf{A}$ and $\Delta\Omega$ to make $\mathbf{R}$ approach zero. Notice that there are 4*N*+1 unknowns, i.e. $\Delta\Omega$ and the 4*N* coefficients in $\Delta\mathbf{A}_{1,2}$, but only 4*N* equations in Eq. (10) are available for a solution. To make the computation through we fixed one of the 4*N* coefficients in $\Delta\mathbf{A}_{1,2}$ and exchange it with $\Delta\Omega$, yielding

$$\mathbf{K}'_{\mathbf{mc}}\Delta\mathbf{A}' = \mathbf{R}', \tag{12}$$

where

$$\mathbf{K}'_{\mathbf{mc}} = \left[\{\mathbf{K}_{\mathbf{mc},1}\}, \{\mathbf{K}_{\mathbf{mc},2}\}, ..., \mathbf{R}_{\mathbf{mc}}, ..., \{\mathbf{K}_{\mathbf{mc},4N}\}\right], \mathbf{R}' = \mathbf{R} - \{\mathbf{K}_{\mathbf{mc},i}\}, \Delta\mathbf{A} = \{\Delta\mathbf{A}_{,1}, ..., \Delta\Omega, ..., \Delta\mathbf{A}_{,4N}\}^T$$

with *i* representing the coefficient increment traded off for $\Delta\Omega$. By appropriately choosing the expansion of $\boldsymbol{x}$, the solution with some degree of accuracy may be achieved.

After the periodic solution of ξ_1 and ξ_2 is found, the stability of the limit-cycle motion solution can be analyzed by linearzing Eq. (6) and computing the Floquet multipliers of the perturbed, first-order system [7,8]:

$$(\Delta\mathbf{x})' = \mathbf{B}(\tau)\Delta\mathbf{x}, \tag{13}$$

where

$$\mathbf{B}(\tau) = \begin{bmatrix} 0 & 1 & 0 & 0 \\ \pi^2(c^2 - c_0^2) & f_1 - \delta + \frac{9}{4}f_3\dot{x}_1^2 + \frac{3}{2}f_3\dot{x}_2^2 & \frac{8}{3}\delta c & \frac{16}{3}c + 3f_3\dot{x}_1\dot{x}_2 \\ 0 & 0 & 0 & 1 \\ -\frac{8}{3}\delta c & -\frac{16}{3}c + 3f_3\dot{x}_1\dot{x}_2 & 4\pi^2(c^2 - c_0^2) & f_1 - \delta + \frac{9}{4}f_3\dot{x}_1^2 + \frac{3}{2}f_3\dot{x}_2^2 \end{bmatrix}$$

evaluated at the periodic solutions denoted by x_1 and x_2. The multipliers are defined as the eigenvalues of the matrix $\exp(\int_0^T \mathbf{B}(\tau)d\tau))$ periodic motion stays stable as long as all four modules of the multipliers are less than unity, where T is the period of the exponential matrix.

The steady limit cycle response of principal coordinates appears when the equilibrium loses its equilibrium, as discussed earlier. By using the technique of path following the coefficients of each harmonic terms in vector, **A,** varying with various parameters can be tracked and determined. Let c_0=2.853, c=0.2, $\delta = 0.08$, a=0.2992, b=-0.2766, $\rho_0 = 1.293\text{kg/m}^3$, $\rho = 7800\text{kg/m}^3$, $S = 125.9\text{m}$, $A = 2.7745\times10^{-4}\text{m}^2$, h=0.0635m. The IHBM gives the limit cycle solutions of the two principal coordinates expressed in the expansion:

$$x_1(\tau)=a_1\cos\tau,\ x_2(\tau)=a_2\cos\tau + b_2\sin 2\tau,$$

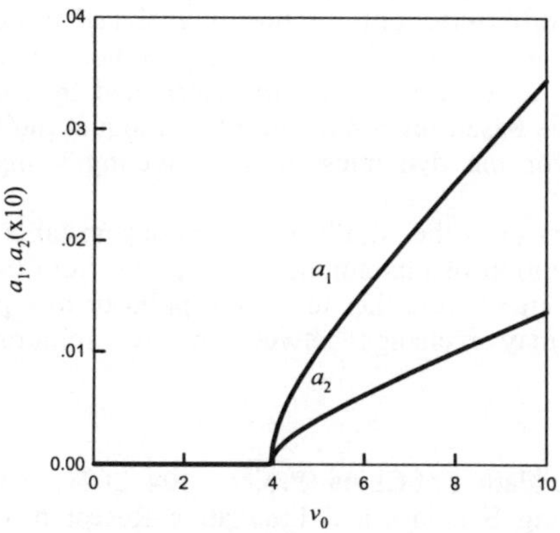

Fig. 3 Amplitudes of a_1 and a_2 against v_0.

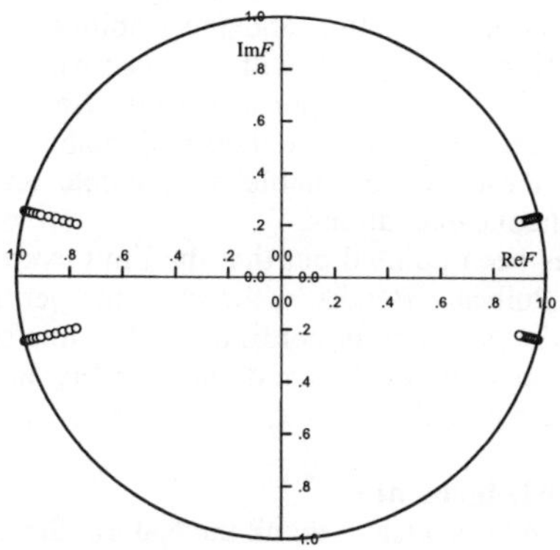

Fig. 4 The Floquet multipliers varying with v_0.

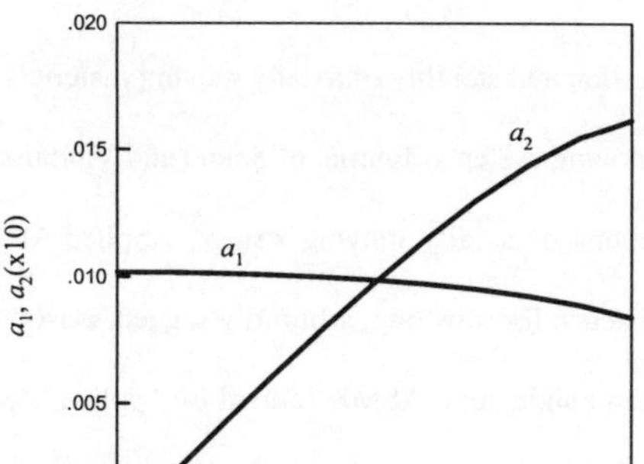

Fig. 5 Amplitudes of a_1 and a_2 against c.

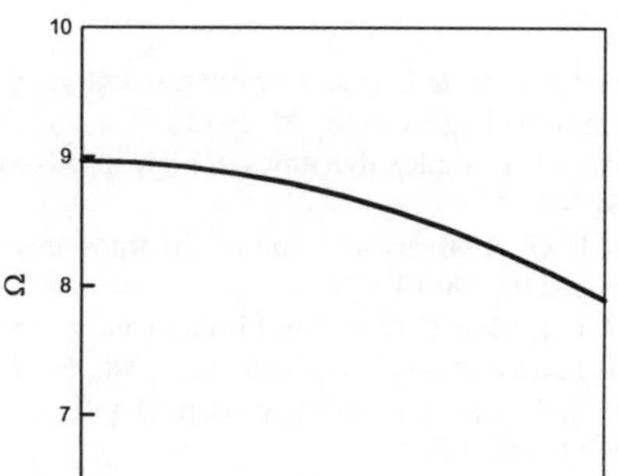

Fig. 6 The frequency of periodic motion against c.

where b_2 being very small compared to a_2. As shown in Fig. 3, the two first amplitudes varying with the wind speed, v_0, remain zero before the Hopf bifurcation takes place at $v_0 = 3.932$. After that there is a growth of the amplitudes with increment of v_0. In Fig. 4 the two conjugated Floquet multipliers are shown, where the limit cycle is found stable with all multipliers staying within the unit circle.

The influence of the transport speed on the amplitudes can be seen in Fig. 5 by computation with wind speed $v_0 = 5\text{m/s}$ and other parameters as used before. The a_1 and a_2, both coefficients of the harmonic term $\cos\tau$, evolve against each other with the transport speed. This means the dynamic configuration loses the geometrical symmetry with respect to the central point as the transport speed increases due to the participant of the second trial function. On the other hand, the natural frequency of the limit cycle decreases with the transport speed.

5. Conclusions

The self-excited vibration and stability of traveling strings with very small sag-to-span ratio under aerodynamic excitation are investigated in this paper. The Galerkin's approach is adopted to discretize the equation of motion and the stability of the equilibrium configuration is analyzed through the Routh-Hurwitz criterion with multiple parameters of wind and string operation. The periodic motion, i.e. the limit cycle response in this paper, as well as its frequency, are determined by using the Incremental Harmonic Method with stability analysis based on computations of Floquet multipliers. All of these efforts enable a complete analysis for the dynamics of the traveling strings with aerodynamic excitations.

It has been pointed out that the limit cycle appears once the equilibrium becomes unstable. Based on the Poincare-Bendixon theorem, the perturbed motion of this autonomous system from the fixed point (the equilibrium configuration in this case) is attractive either to a fixed point or to a periodic orbit (the limit cycle), and the stability property may exchange between the two solutions with multiple parameters.

Acknowledgement

The authors wish to thank the Natural Science Foundation of China (Projects 10472021, 10421002), the SRF for ROCS and the Program for Changjiang Scholars and Innovative Research Team in University of China (PCSIRT) for their financial supports.

References

[1] Wickert J A, Mote C D Jr. Current research on the vibration and stability of axially moving materials. Shock and Vibration Digest 1988; 20: 3–13.

[2] Pellicano F. Complex dynamics of high speed axially moving systems. Journal of Sound and Vibration 2002; 258: 31–44.

[3] Chen L Q. Analysis and control of transverse vibrations of axially moving strings. Applied Mechanics Reviews 2005; 58: 91–116.

[4] Luo A C J, Mote C D Jr. Equilibrium solutions and existence for traveling, arbitrarily sagged, elastic strings. ASME Journal of Applied Mechanics 2000; 67: 148–154.

[5] Yu P, Shah A H et al. Initially coupled galloping of iced conductors. ASME Journal of Applied Mechanics 1992; 59: 140–145.

[6] Lau S L, Cheung Y K. Amplitude incremental variational principle for nonlinear vibration of elastic system. ASME Journal of Applied Mechanics 1981; 48: 959–964.

[7] Leung A Y T, Chui S K. Nonlinear vibration of coupled Duffing oscillators by an improved incremental harmonic balance method. Journal of Sound and Vibration 1995; 181: 619–633.

[8] Raghothama A, Narayanan S. Periodic response and chaos in nonlinear systems with parametric excitation and time delay. Nonlinear Dynamics 2002; 27: 341–365.

Nonlinear dynamical analysis of a micro self-acting gas journal bearing

Hai Huang*, Guang Meng, Long Liu

State Key Laboratory of Vibration, Shock & Noise, Shanghai JiaoTong University, 800 DongChuan Road, Shanghai, 200240, People's Republic of China

Received 30 January 2006; received in revised form 25 March 2006; accepted 1 May 2006

Abstract

The motion equations have been established for a micro gas bearing, and the non-linear gas film force of the finite journal bearing is calculated. The Reynold's equation of the gas film is solved by the finite difference method and the motion of the journal has been simulated with fourth-rank Runge-Kutta method. The linear stiffness and damping coefficients of the micro gas bearing are gotten and the linear threshold speed of the system has been derived. While, the non-linear gas film force is adopted to simulate the dynamical response of the micro rotor, which is perturbed slightly from the static equilibrium position. The orbit of the rotor center, frequency spectrum and phase trajectory of the system are used to analyze the non-linear characters of the rotor center in the horizontal and vertical directions at different operating conditions. It is shown that the dynamic behavior of the micro gas bearing system varies with the rotational speed in the case of the eccentricity is unchanged.

Keywords: Micro gas bearing; Linear threshold speed; Nonlinear gas film force; Dynamical response

1. Introduction

The development of micro electro mechanical systems (MEMS), which take a characteristic length of micrometer, has achieved success in various applications in the last two decades. The most important target of the MEMS design is to achieve a mature micro-actuator, and a number of studies have reported the development of operational micro motors, which are fabricated through micro fabrication approaches such as poly-silicon surface micro machining and high aspect ratio LIGA processing. Some types of micro rotational devices are the top-drive[1] and side-drive[2] variable-capacitance micro-motors (VCM), the electroquasistatic induction micro-motors (IM)[3], and the ultrasonic micro-motors (USM)[4-7]. One feature common to these devices is that the rotation rates have been limited to much less than 10,000 revolutions per minute (RPM). The typical power levels of these micro motors range from picowatts to milliwatts. This limitation comes from bearing drag associated with driving or restraining of axial and lateral rotor movement. When the axial or later loads applied on these devices, the rotating members make contact with their adjoining surfaces and result in "dry rubbing".

As the demand for greatly improved compact power sources with the rapidly expanding capability of Micro-machining technology, a new class of MEMS, *power MEMS* are developed at MIT[8]. These machines include gas turbine engines that supported by gas bearing for mechanical-to-electrical energy conversion[9, 10], and electrostatic induction micro motor supported on gas lubricated bearing for electrical-to-mechanical energy conversion[11]. These power MEMS can achieve a rotation rate of more than a million RPM, and the rotor tip speed is on the order of 200 to 300 ms^{-1}, produce energy conversion on the order of watts which means the

* Corresponding author. Tel.: 086-021-54744990; fax: 086-021-.

E-mail address: huanghai2003@sjtu.edu.cn (H. Huang).

PhD candidate of School of Mechanical Engineering, Shanghai Jiao Tong University, Shanghai, 200240, People's Republic of China.

power-densities of such machines are comparable to those of their full scale counterparts. Under this condition, bearing elements are the major obstacles of successful development. Because of their small scale, it is difficult to use a typical bearing, such as rolling element or oil lubricated bearing. Naturally, a gas-lubricated bearing come to the best choice for a high-speed micro rotor. In face, the gas lubricated bearing has been used widely in conventional rotating systems. Compared with the other bearings, the gas bearing has distinct advantages. First, it can achieve an extremely high rotation speed with very low drag and almost no running wear, which means long service life. Second, there is no defilement due to leakage of lubricant such as oil. Finally, the fabrication of micro gas bearing is relatively simple for the micro fabrication techniques.

Operating at such high speed, an accurate prediction of the vibration characteristics of the micro gas bearing system must be made so as to design highly reliable rotary machines and avoid the occurrence of dangerous operations. Considering some unique challenges dictated by MEMS fabrication processes of micro fabricated gas bearing, the typical characters such as large length-to-diameter (*L/D*) and small clearance (*C/R*) are none possible in the micro gas bearing. The unusual geometry, combined with the very high rotational speed, lead to a complex interdependence between the rotor speed and eccentricity of the micro gas bearing. And the coupling between different fields makes the design and analysis difficult. To discover these relations, extensive research over the past two decades has been performed on it [12-16]. Huang discussed the advantages and scale-down of gas lubricated bearings[12]. Czolczynski investigated the stability of a new type of gas journal bearing, the so called "high stiffness bearing"[13]. Piekos developed a new computing method, called pseudo-spectral method to research on orbit simulation of journal bearings[14]. Savoulides constructed a small-scale gas bearing model by reference to Newton's second law for the rotor and employed stiffness and dumping coefficients[15]. Wang analyzed the bifurcation of a self-acting gas journal bearing[16]. The analysis focuses on the dynamic behavior of the gas bearing with changes in rotor parameters.

In this paper, the linear stiffness and damping coefficients are used to estimate the linear threshold speed. While, the nonlinear gas film force is adopted to acquire the dynamical response. The dynamical response of rotor center due to position perturbation is obtained by solving the gas governing equation and journal motion equation sequentially. Operating at differential speed, the rotor's response takes differential characters on whether convergent to a stable state.

2. Dynamic Model

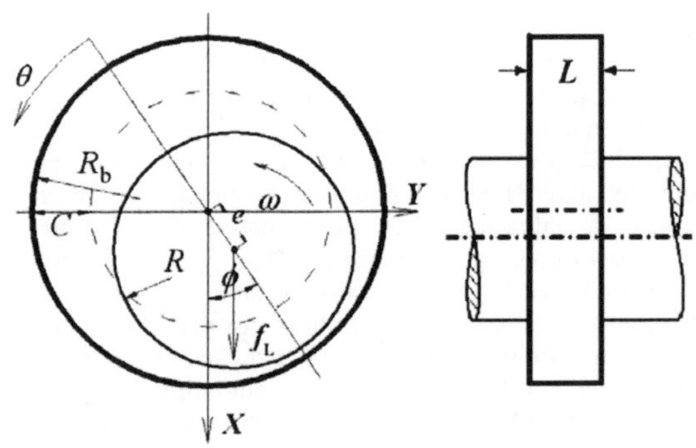

Figure 1 Model of a micro gas bearing

2.1 Bearing Geometry

The geometry and parameters of the micro gas bearing are schematically illustrated in Fig.1. The journal spins at an angular velocity, ω, inside a stationary. The bushing's radius is R_b, and the journal's radius is R. When the journal's center coincides with the bushing's center, the space between the outer edges of journal and bushing defines the average clearance, $C=R_b-R$. The rate of clearance to the radius of the journal is denoted by ψ ($\psi=C/R$). Shown here in cross section, the bearing has length L with the ends either exhausted to the ambient or treated with a uniform pressure distribution, denoted by p_a and the fluid viscosity by μ. The mass of the journal is denoted by m. The space between the center of mass and the center of geometry in the journal, known as unbalance eccentricity and denoted by e_m, is most often normalized by the average clearance and expressed as an "unbalance ratio", and is denoted by $\rho(\rho=e_m/C)$. The external lading force, denoted by f_L, applied to the journal and causes the journal center to move away from the bearing center. The distance between the two centers, known as the eccentricity and denoted by e, is most often normalized by the average clearance and expressed as an "eccentricity ratio", and is denoted by ε ($\varepsilon=e/C$). So $\varepsilon=0$ and $\varepsilon=1$ represent fully-centered and "crashed" journals, respectively. A coordinate system is defined as that the X-axis is parallel to the direction of the applied force, the Y-axis is perpendicular to the applied force in the journal cross section. The angle between the X-axis and the line connecting the journal center to the bearing center defines the attitude angle and is denoted by ϕ.

2.2 Motion Equation

The analysis presented in this paper considers a micro rotor-bearing system comprised of a rigid rotor and rigid mounted bushing. The journal is perfectly co-axial with the bushing. With these idealizations, it is possible to consider the rotating rotor has two degrees of translatory oscillation in the transverse plane. The idealized micro gas bearing model is shown in Fig.1. In the transient state, the equations of motion of the journal center in the Cartesian coordinates may be written as follows:

$$\left.\begin{aligned} m\ddot{x} &= -f_x + f_L + me_m\omega^2\cos(\theta+\phi) \\ m\ddot{y} &= -f_y + me_m\omega^2\sin(\theta+\phi) \end{aligned}\right\} \tag{1}$$

where the x and y , f_x and f_y are the journal center's displacements and gas-film force components in the directions of X and Y.

It is usually advantageous to convert the equations to the non-dimensional form. Denoting the following dimensionless variables:

$$X=\frac{x}{C},\quad Y=\frac{y}{C},\quad \rho=\frac{e_m}{C},\quad \tau=\omega t \tag{2}$$

$$M=\frac{m\cdot P_a}{72\mu^2 L}\psi^5,\quad F=\frac{f}{2RLP_a},\quad \Lambda=\frac{6\mu\omega}{P_a\cdot\psi^2} \tag{3}$$

Substituting the Eqs. (2) and Eqs.(3) into Eqs.(1), the non-dimensional motion equations are gotten as following:

$$\left.\begin{aligned} X'' &= \frac{-F_x+F_L}{M\Lambda^2}+\rho\sin(\theta+\phi) \\ Y'' &= \frac{-F_y}{M\Lambda^2}+\rho\cos(\theta+\phi) \end{aligned}\right\} \tag{4}$$

Rewritten Eqs. (4) in state space form as follows:

$$\left.\begin{aligned} Z_1' &= Z_3 \\ Z_2' &= Z_4 \\ Z_3' &= \frac{-F_x+F_L}{M\Lambda^2}+\rho\sin(\omega t+\phi) \\ Z_4' &= \frac{-F_y}{M\Lambda^2}+\rho\cos(\omega t+\phi) \end{aligned}\right\} \tag{5}$$

where $Z_1=X$, $Z_2=Y$, $Z_3=X'$, $Z_4=Y'$ are the state variables of the system.

It is obvious that the system would be operating at steady state when the imbalance eccentricity ratio equals to 0 ($\rho=0$) and the following relationship is satisfied: $Z_{30}=0$, $Z_{40}=0$, $F_{x0}=F_L$, $F_{y0}=0$.

2.3 Reynolds equation

The micro self-acting gas journal bearing is shown in Fig.1. The journal is supported by the pressure difference generated by it own rotation. A gas journal bearing incorporates the following design assumptions[16]:

1 The lubricating gas obeys the perfect gas law.

2 Gas lubricating films are very nearly isothermal because the ability of the bearing materials to conduct away heat is greater than the heat generating capacity of the gas-film. Thus, the flow may therefore be considered to be isothermal

3 As gas viscosity is somewhat insensitive to changes in pressure and the temperature is virtually constant, it may be assumed to be constant.

4 The flow of gas in and out of the sides of the bearing (side flow) is neglected.

Applying the above assumptions, the dimensionless gas pressure in the micro self-acting gas journal bearing should satisfy the dynamic Reynold's equation:

$$\frac{\partial}{\partial\theta}\left(PH^3\frac{\partial P}{\partial\theta}\right)+\frac{\partial}{\partial\xi}\left(PH^3\frac{\partial P}{\partial\xi}\right)=6\Lambda\left(\frac{\partial PH}{\partial\theta}+2\frac{\partial PH}{\partial\tau}\right) \tag{6}$$

where θ is the coordinate in circumferential direction, and ξ is the coordinate in axial direction.

$$P=\frac{p}{p_a},\ \theta=\frac{x}{R},\ \xi=\frac{z}{R},\ H=\frac{h}{C}=1+\varepsilon\cos\theta \tag{7}$$

where P is the dimensionless gas pressure distribution, H is the dimensionless gas-film thickness, Λ is the dimensionless bearing number.

The gas-film pressure distribution fulfill the following boundary conditions:

1 Gas pressure on both ends of the housing is equal to the atmospheric pressure p_a, $P(\theta,\pm\xi)=1$.

2 Gas pressure P is an even function for ξ, $P(\theta,\xi)=P(\theta,-\xi)$.

3 Gas pressure P is continuous at $\xi=0$, $\partial P/\partial\xi|_{\xi=0}=0$.

4 Gas pressure P is a periodic function for θ, $P(\theta,\xi)=P(\theta+2\pi,\xi)$, $\partial P/\partial\theta|_\theta=\partial P/\partial\theta|_{\theta+2\pi}$.

In the model of the bearing, the Reynolds' equation is a nonlinear partial differential equation, so a direct numerical method is used for its solution. Eq. (6) and the corresponding boundary conditions are discretized with the central-difference scheme in the θ and ξ directions, and the implicity-backward-difference scheme in time τ. For simplicity, a uniform mesh size is used. Eq.(6) can be transformed into the following form:

$$\begin{aligned}
&3\left(H_{i,j}^{n+1}\right)^2\left(\frac{H_{i+1,j}^{n+1}-H_{i-1,j}^{n+1}}{2\Delta\theta}\frac{Q_{i+1,j}^{n+1}-Q_{i-1,j}^{n+1}}{2\Delta\theta}\right)+\left(H_{i,j}^{n+1}\right)^3\left(\frac{Q_{i+1,j}^{n+1}-2Q_{i,j}^{n+1}+Q_{i-1,j}^{n+1}}{(\Delta\theta)^2}\right)+\\
&3\left(H_{i,j}^{n+1}\right)^2\left(\frac{H_{i,j+1}^{n+1}-H_{i,j-1}^{n+1}}{2\Delta\xi}\frac{Q_{i,j+1}^{n+1}-Q_{i,j-1}^{n+1}}{2\Delta\xi}\right)+\left(H_{i,j}^{n+1}\right)^3\left(\frac{Q_{i,j+1}^{n+1}-2Q_{i,j}^{n+1}+Q_{i,j-1}^{n+1}}{(\Delta\xi)^2}\right)\\
&=\Lambda P_{i,j}^{n+1}\left(\frac{H_{i+1,j}^{n+1}-H_{i-1,j}^{n+1}}{2\Delta\theta}\right)+\frac{\Lambda H_{i,j}^{n+1}}{2P_{i,j}^{n+1}}\left(\frac{Q_{i+1,j}^{n+1}-Q_{i-1,j}^{n+1}}{2\Delta\theta}\right)+\Lambda\frac{H_{i,j}^{n+1}}{P_{i,j}^{n+1}}\left(\frac{Q_{i,j}^{n+1}-Q_{i,j}^{n}}{\Delta\tau}\right)\\
&+2\Lambda P_{i,j}^{n+1}\left(\frac{H_{i,j}^{n+1}-H_{i,j}^{n}}{\Delta\tau}\right)
\end{aligned} \tag{8}$$

where $P^2=Q$.

Using the finite difference method (FDM), the Reynold's equation is solved and the dimensionless gas film pressure is acquired. Then, the dimensionless gas film force components, which act between the shaft and the bushing, are estimated as the integrals:

$$\left.\begin{aligned}
F_e&=\int_{-\xi}^{\xi}\int_0^{2\pi}P\cos\theta\cdot rd\theta d\xi\\
F_\theta&=\int_{-\xi}^{\xi}\int_0^{2\pi}P\sin\theta\cdot rd\theta d\xi
\end{aligned}\right\} \tag{9}$$

where F_e is the dimensionless bearing forces components at the radial direction, F_θ is the dimensionless bearing force component at the circumference direction.

2.4 The linear threshold speed of the system

The gas-film forces are nonlinear function, however, in most practical calculations, it is assumed that the perturbation of journal from its initial position is small enough to consider that the gas-film forces are linear. Thus the gas-film force increments are modeled by a first-order Taylor expansion [17, 18]:

$$\left.\begin{aligned}\Delta F_x &= K_{xx}\Delta Z_1 + K_{xy}\Delta Z_2 + D_{xx}\Delta Z_3 + D_{xy}\Delta Z_4 \\ \Delta F_y &= K_{yx}\Delta Z_1 + K_{yy}\Delta Z_2 + D_{yx}\Delta Z_3 + D_{yy}\Delta Z_4\end{aligned}\right\} \tag{4}$$

where ΔF_x and ΔF_y are the bearing reaction force increments, K_{ij} is the dimensionless stiffness coefficient and D_{ij} is the dimensionless damping coefficient.

Substituting Eqs.(10) into Eqs.(5), and take the assumption that the unbalance eccentricity ratio is equal to 0 (ρ=0), the system free vibration equations can be derived as follows:

$$\left.\begin{aligned}\Delta Z_1' &= \Delta Z_3 \\ \Delta Z_2' &= \Delta Z_4 \\ M\Lambda^2 \cdot \Delta Z_3' &= -\left(K_{xx}\Delta Z_1 + K_{xy}\Delta Z_2 + D_{xx}\Delta Z_3 + D_{xy}\Delta Z_4\right) \\ M\Lambda^2 \cdot \Delta Z_4' &= -\left(K_{yx}\Delta Z_1 + K_{yy}\Delta Z_2 + D_{yx}\Delta Z_3 + D_{yy}\Delta Z_4\right)\end{aligned}\right\} \tag{11}$$

Equation (11) denotes a linear system and it can be rewritten in matrix form:

$$\begin{pmatrix}\Delta Z_1' \\ \Delta Z_2' \\ \Delta Z_3' \\ \Delta Z_4'\end{pmatrix} = A \cdot \begin{pmatrix}\Delta Z_1 \\ \Delta Z_2 \\ \Delta Z_3 \\ \Delta Z_4\end{pmatrix} \tag{12}$$

where,

$$A = \frac{1}{M\Lambda^2}\begin{bmatrix} 0 & 0 & M\Lambda^2 & 0 \\ 0 & 0 & 0 & M\Lambda^2 \\ -K_{xx} & -K_{xy} & -D_{xx} & -D_{xy} \\ -K_{yx} & -K_{yy} & -D_{yx} & -D_{yy}\end{bmatrix}$$

The common approach adopted for investigating the stability of rotor bearing systems depends mainly upon solving the eigenvalue problem of Eq. (12). The real part of each eigenvalue δ is the damping rate and corresponds to how quickly a perturbation of the journal decays. $\delta<0$ implies a decay motion (stable), while $\delta>0$ implies a growth motion (unstable). There are four eigenvalues of matrix A in Eq. (12). We only consider the eigenvalue with the biggest positive real part, because it represents the most dangerous mode and can be observed in an experimental measurement. At the threshold of instability, the matrix A has an eigenvalue with real part equal to zero. The following equations can be derived:

$$K_{eq} = M\Lambda^2 \cdot \gamma^2 \tag{13}$$

$$\left.\begin{aligned}K_{eq} &= \frac{K_{xx}D_{yy} + K_{yy}D_{xx} - K_{xy}D_{yx} - K_{yx}D_{xy}}{D_{xx} + D_{yy}} \\ \gamma^2 &= \frac{\left(K_{eq} - K_{xx}\right)\left(K_{eq} - K_{yy}\right) - K_{xy}K_{yx}}{D_{xx}D_{yy} - D_{xy}D_{yx}}\end{aligned}\right\} \tag{14}$$

where K_{eq} denotes the equivalent stiffness and γ is the whirl ratio of the system respective.

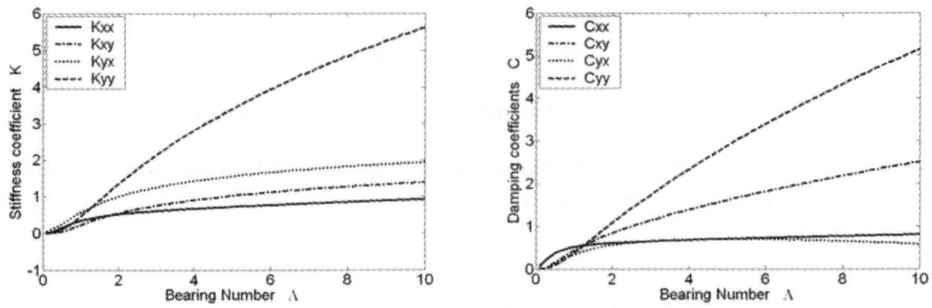

(a) Stiffness coefficient (b) Damping coefficient

Figure 2 The stiffness and damping coefficients versus the bearing number at ε=0.9

Figure 2 show the variation of stiffness and damping coefficients versus the bearing number at ε=0.9. It indicates that the dynamical coefficients increase with the increase of the bearing number except for two damping coefficients, C_{xx} and C_{yx}. The two damping coefficients decrease slightly after tending to a maximum at about Λ=1.5. The lower stiffness and damping coefficients imply that the system capacity of resistance to the disturbance is lower, and the system is likely to lose stability. The higher the stiffness and damping coefficients, which can enhance the resistant capacity, the better the system will keep operating stable after a slight disturbance. Thus, it can be predicted that the micro rotor-bearing system is more stable at higher rotating speed. Substituting the coefficients into Eq. (13) and Eqs. (14), it is found that the threshold speed is nearby the point Λ=0.3 when ε=0.9.

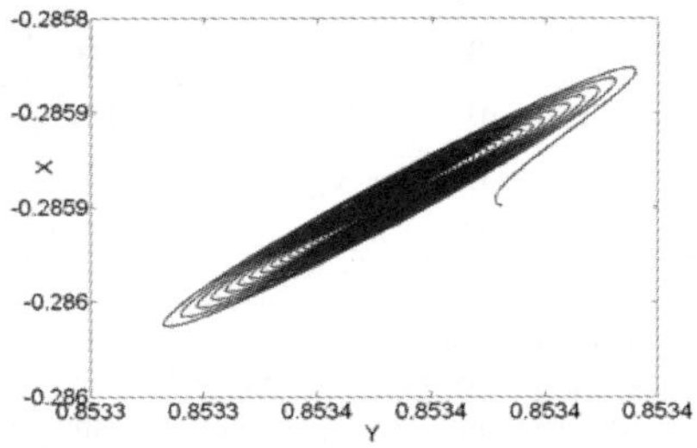

(a) The orbit of the journal center

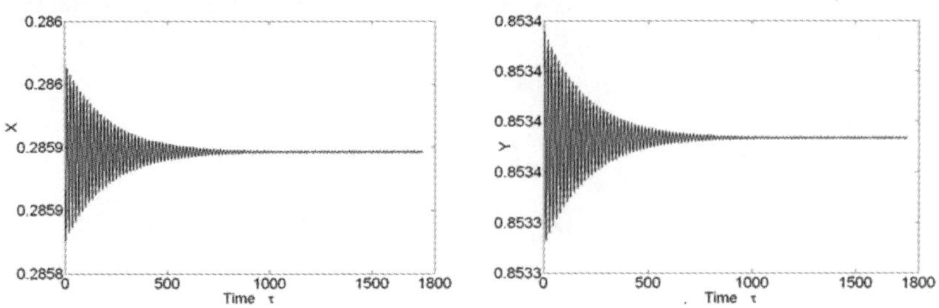

(b) Displacement at X direction (c) Displacement at Y direction

Figure 3 The orbit of journal center and the journal center displacement in X and Y direction at ε=0.9, Λ=0.35

3. Dynamical simulation and disturbance response

Numerical simulation for the micro gas journal bearing can be achieved by solving the Reynold's equation with initial values. For a balanced journal, the Eq. (5) for dynamical motion can be solved by fourth-order Rouge-Kutta method[19], and a series of data of displacements and velocities for journal can be obtained. Since the gas-film forces in Eq. (5) changes continuously along with system status parameters, they must be computed at each time step. Therefore, the essential work of numerical simulation is to solve the coupled equations (5) and (6) sequentially. In order to illustrate the difference dynamic behavior of the micro gas-bearing system at different rotational speed, the response of a rotor, which is disturbed away from it's static equilibrium position, is simulated for Λ=0.35, 0.3 and 0.28, respectively.

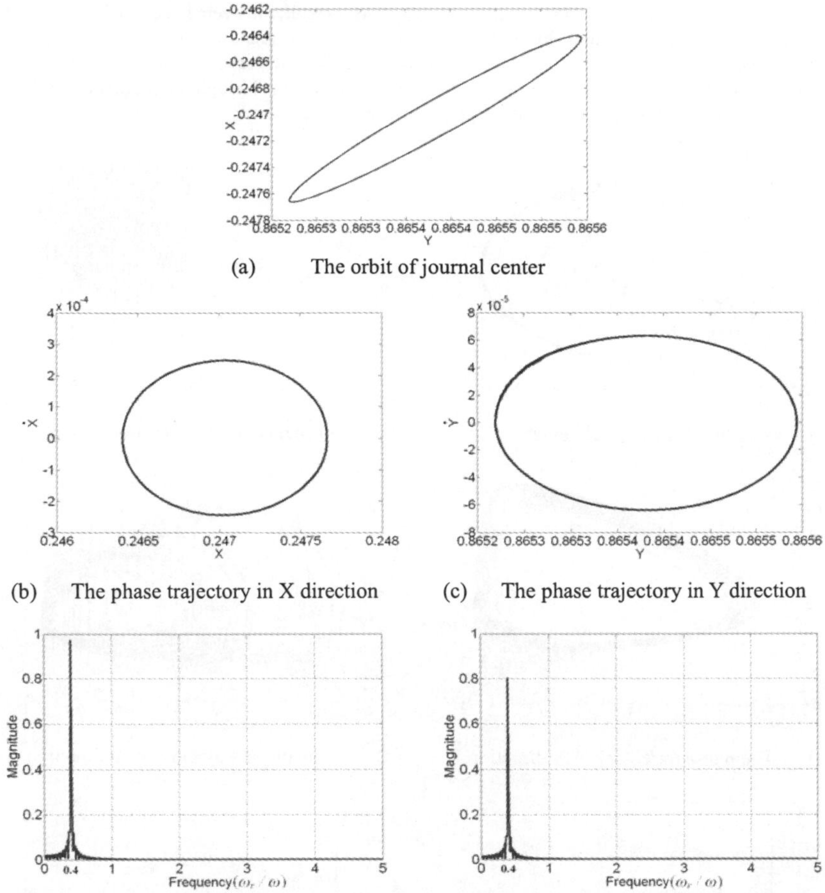

(a) The orbit of journal center

(b) The phase trajectory in X direction (c) The phase trajectory in Y direction

(d) Frequency spectrum in X direction (e) Frequency spectrum in Y direction

Figure 4 The orbit of journal center, phase trajectory and frequency spectrums of journal displacement in X and Y direction (at ε=0.9, Λ=0.3)

Figure 3 shows the orbit of the center and the displacements of the rotor in X and Y directions ($\varepsilon = 0.9$, $\Lambda = 0.35$). It indicates that the system with some given slight disturbance can still converge to its static equilibrium position if the rotational speed is above the threshold speed of the system. Figure 4 displays the response of the micro-rotor at ε=0.9 and Λ=0.3, including the orbit of the rotor center, displacements, phase portrait and the frequency spectrums. The oscillation amplitudes of the rotor center in X and Y directions keep constant after a short time. In this condition, the orbit of the rotor center is elliptical orbit, which the long axis is in circumference direction and the short axis is in radius direction. It implies that the stiffness in radius direction is higher than that in circumference direction. From the frequency responses, it is found that the whirl rate $\gamma^2(\omega_r/\omega)$ is 0.4, which is a little lower than half of the rotation frequency. The system would not converge to its equilibrium position if the rotational speed equals to the threshold speed of the system. However, the system will still keep stable in a small amplitude whirl motion around the static equilibrium position.

Figure 5 illustrates the response of the micro-rotor at $\varepsilon = 0.9$ and $\Lambda = 0.28$. Fig.5 (a) shows that the eccentricity ratio keeps increase for a large amplitude oscillation. As displayed in Fig.5(a, b, c, d), the vibrations of the journal take the characters of quasi-period motion. The rotor center whirls in a large range and the limit circle no longer exists. Since the stiffness coefficients increase with the increase of the eccentricity ratios, the resistance on the journal is larger with the increase of ε than that with the decrease of ε. As a result, the average eccentricity ratio of the whirl is lower than that of equilibrium position. In Fig.5 (e) and (f), it could be found that several frequency components appear and the lowest frequency equals to 0.41ω. It reveals that a slight disturbance can cause a large-radius whirl gradually when the rotating speed is under the threshold speed of the system. Though the journal doesn't "crash" into the bushing in the simulation, it is likely to occur for practically operating.

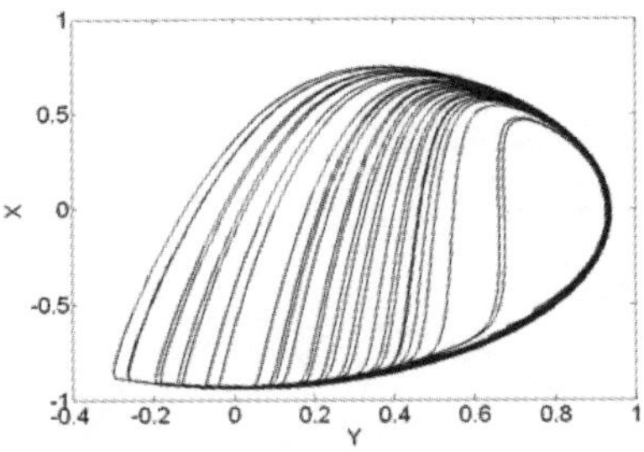

(a) Orbit of the journal center

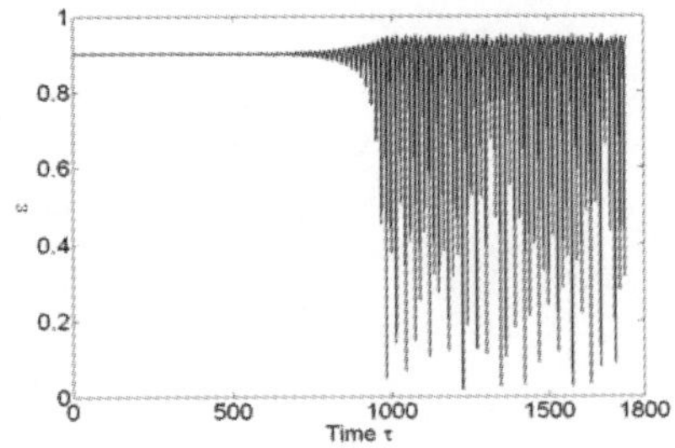

(b) Journal center's eccentricity ratio

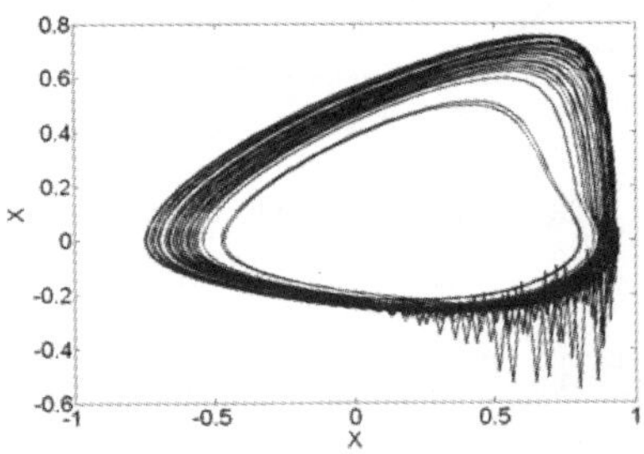

(c) The phase trajectory in X direction

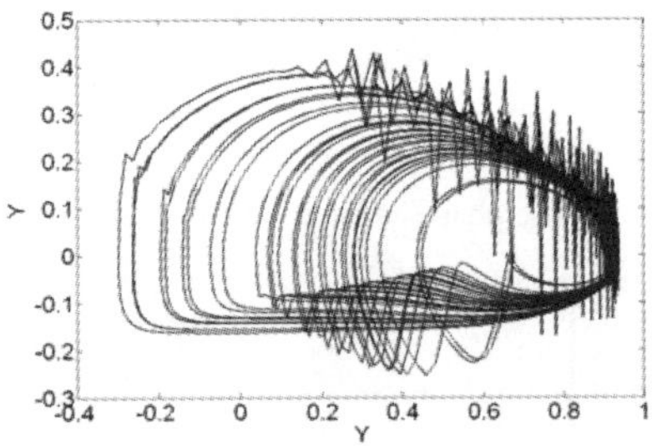

(d) The phase trajectory in Y direction

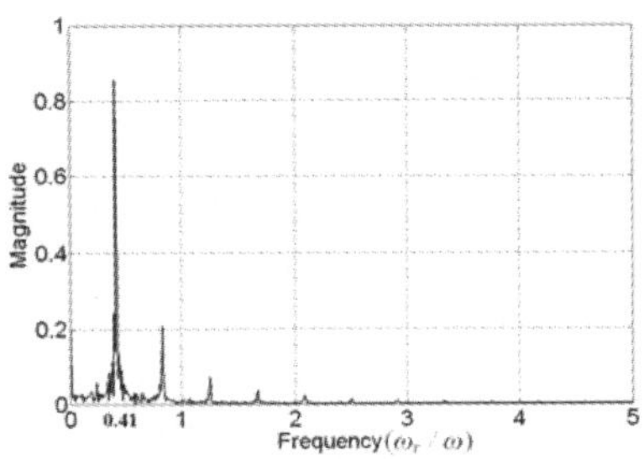

(e) Frequency spectrum in X direction

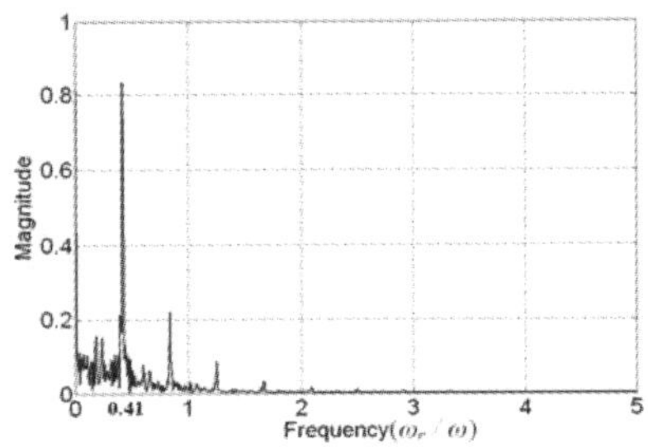

(f) Frequency spectrum in Y direction

Figure 5 The orbit of journal center, eccentricity ratio, phase trajectory and frequency spectrums of journal displacement in X and Y direction (at ε=0.9, Λ=0.28)

4. Conclusion

A rigid micro rotor supported by a micro self-acting gas film bearing is studied in this paper. The journal displacement, the system state trajectory, phase trajectory and frequency spectrum were used to analyze the dynamic behavior of this system. The analysis demonstrates that the dynamical behavior of the system changes with the rotational speed. The linear stiffness and damping coefficients are adopted to get the linear threshold speed of the system. While, the nonlinear gas-film force is adopted to get an accurate simulation of the journal's motion. For the stiffness and damping coefficients increase with the increase of the bearing number, the capacity of resisting the disturbance enhances with the increase of the bearing number. As a result, the micro gas bearing system is more stable at high rotating speed than at low rotating speed. Given a slight perturbation, the motion of the journal takes different characters at different operating speed. If the operating speed of the micro rotor is not less than the linear threshold speed, the journal will converges to the static equilibrium position or whirls around it with small amplitude. While, if the operating speed under the threshold speed, the system will lose linear stability and the disturbance will cause a large whirl, and it will even lead to "crash" into the bushing. For a practically micro gas bearing system, the latter operating is dangerous and is prohibited.

Acknowledgment

The supports from the National Natural Science Foundation of China (grant No. 10325209) are gratefully acknowledged.

References

[1] Fan LS, Tai YC, and Muller RS. IC-processed electrostatic micro-motors. Technical Digest - International Electron Devices Meeting. 1988: 666-669.
[2] Hameyer K and Belmans R. Design of very small electromagnetic and electrostatic micro motors. IEEE Transactions on Engrgy Conversion 1999; 14: 1241-1246.
[3] BART S and LANG J. An Analysis of Electroquasistatic Induction Micromotors. Sensors and Actuators 1989; 20: 97-106.
[4] Sun XQ, Masuzawa T, and Fujino M. Micro ultrasonic machining and its applications in MEMS. Sensors and Actuators a-Physical 1996; 57: 159-164.
[5] Morita T, Kurosawa M, and Higuchi T. An ultrasonic micromotor using a bending cylindrical transducer based on PZT thin film. Sensors and Actuators a-Physical 1995; 50: 75-80.
[6] Carotenuto,R, et al. Low voltage piezoelectric micromotor using a thin circular membrane. Proceedings of the 1997 IEEE Ultrasonics Symposium. 1997: 459-462.
[7] Dubois MA. and Muralt P. PZT thin film actuated elastic fin micromotor. IEEE Transactions on Ultrasonics, Ferroelectrics, and Frequency Control 1998; 45: 1169-1177.
[8] Epstein AH, et al. Power MEMS and microengines. International Conference on Solid-State Sensors and Actuators. 1997: 753.
[9] Fréchette L, et al. Demonstration of a Microfabricated High-Speed Turbine Supported on Gas Bearings. The 2000 Solid State Sensor and Actuator Workshop. 2000: 43-47.
[10] Epstein AH. Millimeter-scale, micro-electro-mechanical systems gas turbine engines. Journal of Engineering for Gas Turbines and Power-Transactions of the Asme 2004; 126: 205-226.
[11] Fréchette LG, et al. An electrostatic induction micromotor supported on gas-lubricated bearings. 14th IEEE International Conference on Micro Electro Mechanical Systems (MEMS 2001). 2001: 290-293.
[12] Huang JB, Tong QY, and Mao PS. Gas-lubricated micro-bearings for microactuators. 1991: 894-897.
[13] Czolczynski K Stability of high stiffness gas journal bearings. Wear 1994; 172: 175-183.
[14] Piekos ES. Numerical Simulation of Gas-Lubricated Journal Bearings for Microfabricated Machines. Cambridge, USA; Massachusetts Institute of Technology, 1999
[15] Savoulides N, Breuer NS, and Jacobson S. Low-Order Models for Very Short Hybrid Gas Bearings. Journal of Tribology 2001; 123: 368-375.
[16] Wang CC and Chen CK. Bifurcation analysis of self-acting gas journal bearings. Journal of Tribology 2001; 123: 755-767.
[17] Lund J. Review of the concept for dynamic coefficients for fluid film journal bearings. Journal of Tribology 1987; 109: 37-41.
[18] Zhao S, et al. Stability and response analysis of symmetrical single-disk flexible rotor-bearing system. Tribology International 2005; 38: 749-756.
[19] Wu C. Numerical mathematics and computing. Shanghai: Fudan Uniersity Press; 1991.

Parametric Modelling of Microresonators Dynamics with Considering Thermal Effects

M. A. Tadayon*†, H. Sayyadi*, G. N. Jazar**
*Department of Mechanical Engineering, Sharif University of Technology, Tehran, Iran.
**Department of Mechanical Engineering and Applied Mechanics North Dakota State University, Fargo , USA.

Abstract

Thermal dependency of material characteristics in MicroElectroMechanical Systems strongly affects their performance, design, and control. Hence, it is essential to understand and model that in MEMS devices to optimize their designs. Thermal dependency of material characteristics is one of the most important phenomena affecting the motion of Microresonator systems. A thermal phenomenon introduces two main effects: damping due to internal friction, and softening due to Young modulus-temperature relation. Based on some reported theoretical and experimental results, we qualitatively model the thermal phenomena and present two Lorentzian functions to describe the restoring and damping forces caused by thermal phenomena. In order to emphasize the thermal effects, a nonlinear model of the MEMS, by considering capacitor nonlinearity and midplane stretching, have been used. The response of the system is developed by employing multiple time scales perturbation method on nondimensionalized form of equations. Frequency response, resonant frequency and peak amplitude are examined for variation of dynamic parameters involved.

PACS: 43.40Ga, 43.58.Wc, 65.70.+y
Keywords: MEMS dynamics; Thermoelastic; Microcantilever; Microresonators; Nonlinear Modeling; Thermal damping; Thermal Relaxation; Capacitive Sensors.

---

1. Introduction

Designing MEMS devices are sometimes based on trial and error because most MEMS are modeled by simplified analytical tools, resulting in a relatively approximate prediction of performance behavior. Therefore, Microsystems design process requires several iterations before the desired performance are finally achieved [1-2]. The reduced-order models, on the other hand, need to be improved as a basis for prediction and optimization tool of the proposed behavior. Reduced-order models have shown their effectiveness in research and design, and are developed to capture the most significant characteristics of a MEMS behavior in a few variables [2-3].Difficulties have arisen in the process of creating high quality factor (Q) MEMS or NEMS where the Qs are found to be lower than expected from scaling

† *E-mail address:* tadayon@mech.sharif.edu

considerations of fundamental loss mechanism [4]. Some of the sources of quality factor decreasing are considered extrinsic. However the intrinsic sources of energy loss play significant role in attainable quality factor.

Most electric actuated microbeam-based resonators, sensors and actuators must work at resonance. Typical microresonator devices are made by a parallel capacitor, in which one electrode is fixed and the other is allowed to move using some flexibility. The movable electrode, fabricated in the form of microbeam, microplate, or microcantilever, serves as a mechanical resonator. It is actuated electrically and its motion can be detected by capacitive changes. This motion of the movable electrode can be converted to an electric signal in the capacitance, which is related to the physical quantity being measured [5].

The purpose of this paper is to investigate the mathematical modeling of thermal effects and nonlinearities achieved from capacitive and midplane stretching in dynamic behavior and sensitivity analysis of microresonators. Temperature dependent properties of the microbeam material play a significant role in affecting the design and application of Microsystems utilizing a microbeam or microcantilever resonator [4,6-8].The effects of thermal phenomena are modeled as an increase in damping [9] and decrease in stiffness rates [10], both as Lorentzian function of excitation frequency[11].

The steady state response frequency-amplitude dependency of system will be derived utilizing the multiple time scale perturbation method by considering the nonlinearity of the actuated force [12-14]. The developed analytic equation describing the frequency response of the system around resonance can be utilized to explain the dynamic of the system, as well as resonant frequency and peak amplitude.

2. Mathematical Modeling

The thermoelastic effects will be investigated in two parts, the first one is the thermal damping which is the energy dissipation mechanism [9,11] and the other one is thermal relaxation which affects the rigidity of material [10,11].

Thermoelastic damping is proportional to frequency; hence, when the principal natural frequency moves up while the size of devices decreases, the thermoelastic damping becomes more significant. Thermal energy dissipation is caused by irreversible heat flow across the thickness of the microcantilever as it oscillates.

For simulating the damping force corresponding to thermal damping Jazar et al. [9] introduce a frequency dependent force

$$f_{Td} = c_T L\left(\frac{\omega^2}{\omega_i^{\,2}}\right)\dot{z}\ ,L(x) = \frac{x}{1+x^2} \tag{1}$$

where c_T defines the thermal damping per unit length of the microbeam which depends on geometric and material properties of the microbeam and must be determined experimentally, ω_i is the natural frequency, ω is frequency of vibration, $\dot{z}$ is velocity in system with one degree of freedom, and L is Lorentzian function as defined above.

Since the warming of the microbeam material is Lorentzian frequency dependent, the effect of stiffness softening of the microbeam is also a frequency dependent characteristic. So, we present a negative softening function to define this behavior. More specifically, a negative restoring force with stiffness as a Lorentzian function of excitation frequency [10,11]

$$f_{Ts} = -k_T\,\frac{\omega/\omega_i}{1+(\omega/\omega_i)^2}w = -k_T L\left(\frac{\omega^2}{\omega_i^{\,2}}\right)w \tag{2}$$

determines the drop in linear rigidity stiffness force, $EI\left(\partial^4 w/\partial x^4\right)$. The breaking frequency of the thermal stiffness softening is also at the fundamental resonance frequency. The softening stiffness coefficient per unit length, k_T , which depends on geometric parameters and material properties of the microbeam must be determined experimentally.

The microresonator is composed of a beam resonator, a ground plane underneath in contact with the beam, and one (or more) capacitive transducer electrode(s).The one dimensional electrostatic force, f_e , between two electrodes is

$$f_e = \frac{\varepsilon_0 A (v - v_p)^2}{2(d - w)^2} \tag{3}$$

where, $\varepsilon_0 = 8.85 \times 10^{-12} As/Vm$ is permittivity in vacuum, A is the area of the microplate, $w=w(x,t)$ is the lateral displacement of the microbeam, and The electric load composed of a DC polarization voltage, v_p , and an AC actuating voltage, $v = v_i \sin(\omega t)$ [13] .

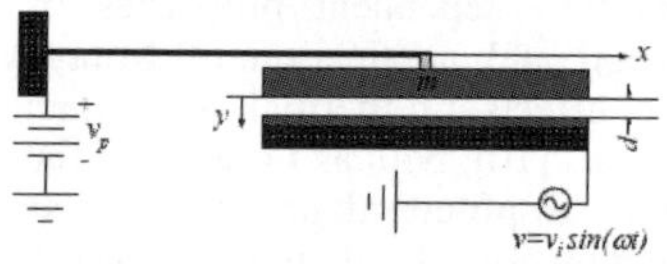

Fig. 1. A microcantilever model of microresonators

The equation describing lateral vibrations of the microbeam can be summarized and simplified to the following equation when the beam's geometry is uniform.

$$\frac{\partial^2 y}{\partial \tau^2} + a_2 \frac{\partial y}{\partial \tau} + \frac{\partial^4 y}{\partial z^4} + a_6 \frac{r}{1+r^2}\frac{\partial y}{\partial \tau} - a_7 \frac{r}{1+r^2} y = \frac{\varepsilon_0 A (v - v_p)^2}{2(1-y)^2} + \left(\hat{N} + \frac{EA_0}{2L}\int_0^L \left(\frac{\partial w}{\partial x}\right)^2 dx \right)\frac{\partial^2 w}{\partial x^2} \tag{4}$$

where the parameters are(n is a constant depending on mode shape of the microbeam.)

$$\tau = \omega_1 t \ , \omega_1 = \frac{n^2}{L^2}\sqrt{\frac{EI}{\rho}}, z = \frac{x}{L}, y = \frac{w}{d}, Y = \frac{w_0}{d}, r = \frac{\omega}{\omega_1}, a_1 = \frac{\varepsilon_0 A L^4}{2n^2 d^3 EI}, a_2 = \frac{cL^2}{n\sqrt{\rho EI}}, \tag{5a}$$

$$a_6 = \frac{c_T L^2}{n\sqrt{\rho EI}}, a_7 = \frac{k_T L^4}{n^2 EI}, \alpha_1 = 3\left(\frac{nd}{h}\right)^2 = \frac{A_0}{4I}(nd)^2, N = \frac{\hat{N} l^2}{EI}, r_x = \frac{\omega_x}{\omega_1} \tag{5b}$$

We consider the axial load is function of axial initial condition and the axial excitation

$$N = \frac{EA_0}{l}\left(x_0 + x_d \cos\left(\omega_x t\right)\right) \tag{6}$$

where x_0 and x_d are the initial stretch and the axial excitation amplitude respectively and ω_x is the axial excitation frequency then nondimensionalize above equation and achieve to following parameter

$$a_8 = \frac{nlAx_0}{I}, a_9 = \frac{nlA_0 x_d}{I} \tag{7}$$

We apply a separation solution $y = Y(\tau).\varphi(z)$ where the spatial function $\varphi(z) = \cos(\frac{\pi z}{2})$ is called mode shape function Then the required differential equation for the temporal function $Y(\tau)$ related to a microcantilever, would be

$$\ddot{Y}+\left(h+a_6\frac{r}{1+r^2}\right)\dot{Y}+\left(1-a_7\frac{r}{1+r^2}\right)Y=$$

$$\frac{1}{(1-Y)^2}\left[(\alpha+\beta)+2\sqrt{2\alpha\beta}\sin(r\tau)-\beta\cos(r\tau)\right]-a_8Y-a_9\cos(r_x\tau)-\alpha_1Y^3 \quad (8)$$

where, $h=a_2$, $\alpha=a_1v_p{}^2$, $\sqrt{2\alpha\beta}=a_1v_pv_i$, $\beta=\frac{a_1}{2}v_i{}^2$.

The third order expansion (Taylor series) will be used to model the electrostatic force on microresonator.

Applying the multiple time scales method produces $Y=a(\tau)\cos(r\tau+\gamma(\tau))$, and the following coupled equations yields(considering $r_x\approx 2$)

$$a'=\frac{1}{4(1+r^2)}\Big(-2(1+r^2)\sqrt{2\beta\alpha}\cos(\sigma\tau-\gamma)+a(-2ra_6-(1+r^2)$$

$$(a_9\sin\left(2\sigma^*\tau-2\gamma\right)+2(h+3a\sqrt{2\beta\alpha}\cos(\sigma\tau-\gamma)+\beta(1+4a^2)\sin(2\sigma\tau-2\gamma))))\Big) \quad (9.a)$$

$$\gamma'=\frac{1}{4(1+r^2)a}(-2\sqrt{2\beta\alpha}\sin(\sigma\tau-\gamma)+a(-ra_7+\left(1+r^2\right)(-18a\sqrt{2\beta\alpha}\sin(\sigma\tau-\gamma)+a_9\cos(2\sigma^*\tau-2\gamma)$$

$$+2(-2(\beta+\alpha)+\beta\cos\left(2\sigma\tau-2\gamma\right)a_8+(-12(\beta+\alpha)+8\beta\cos(2\tau\sigma-\gamma)+3\alpha_1)a^2)))) \quad (9.b)$$

where $\sigma=r-1, \sigma^*={(r_x-2)}/{2}$.

Assuming a' and $\gamma'-\sigma$ remain zero in steady state response and $\sigma^*=\sigma$. ($\gamma-\sigma\tau$ is argument of sinusoidal term and must invariant in time when $\tau\to\infty$).Eliminating $\gamma(\tau)$ and assuming $0\le a\le 1$ provide a relationship between the parameter of the system to have a periodic steady state response with frequency r give the equation that shows the a as a function of $h, a_6, a_7, \beta, \alpha, r$.

From Nonlinear modeling the amplitude coupled equation is:

$$\frac{1}{4(1+r^2)}\Big(-2(1+r^2)\sqrt{2\beta\alpha}\cos(\gamma)+a(-2ra_6-(1+r^2)$$
$$(-a_9\sin\left(2\gamma\right)+2(h+3a\sqrt{2\beta\alpha}\cos(\gamma)-\beta(1+4a^2)\sin(2\gamma))))\Big)=0 \quad (10.a)$$

$$\frac{1}{4(1+r^2)a}(2\sqrt{2\beta\alpha}\sin(\gamma)+a(-ra_7+\left(1+r^2\right)(18a\sqrt{2\beta\alpha}\sin(\gamma)+a_9\cos(2\gamma)$$

$$+2(-2(\beta+\alpha)+\beta\cos\left(2\gamma\right)a_8+(-12(\beta+\alpha)+8\beta\cos(2\gamma)+3\alpha_1)a^2))))-\sigma=0 \quad (10.b)$$

As illustrated in Fig.5(a,b) The resonance frequency is monotonically decreasing function of increasing both polarization and excitation voltages. The behavior of resonance shifting looks linear with variation of both voltages. Fig.5(c) shows the effects of the nonlinearity parameter on the resonance frequency. It can be seen this parameter linearly increasing the resonance frequency. The behavior of resonant frequency is not linear non-monotonic when damping is varied. As shown in Fig.5(d) the resonance frequency is approximately unchanged with variation of the system damping and as mentioned the damping effects on the resonance frequency is coupled with system other parameter same as nonlinearity.

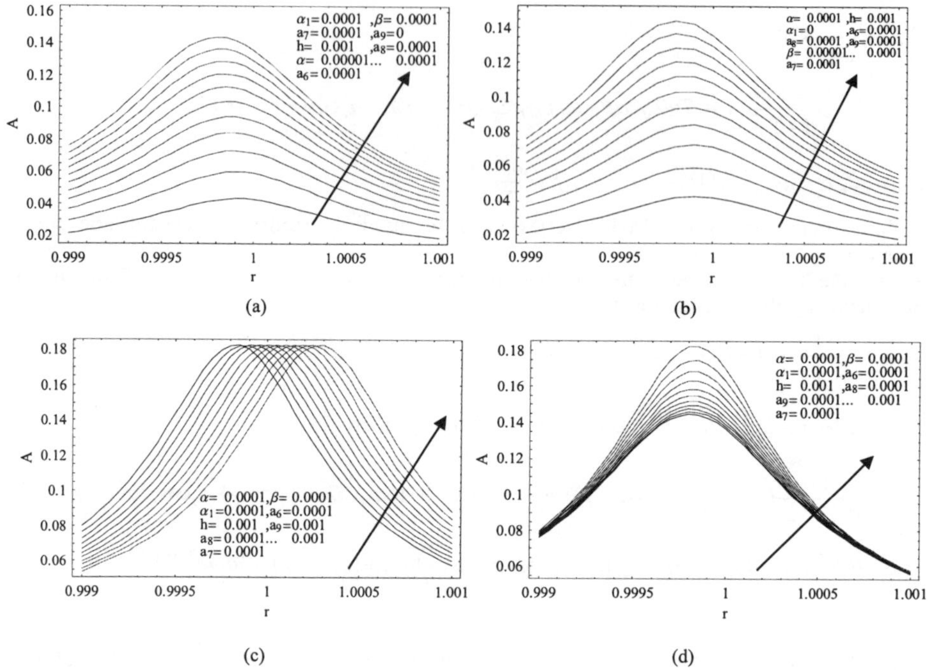

Fig. 2. Effect of variation of (a) DC voltage, (b) AC voltage, (c) axial load (d) axial vibration on frequency response. (Arrows show increasing of the parameter)

3. Conclusion

We have modeled the thermal phenomena by considering the nonlinearities from actuated force and midplane stretching in the nondimensionalized equation. The thermal phenomena have been translated to effective forces per unit length of the vibrating microbeam. Thermal properties of microbeam contribute into damping system due to warming and heat energy dissipation called "thermal damping", and into restoring system due to material heat softening called "temperature relaxation".

The achieved equation solved in primary resonance by utilizing the multiple time scales perturbation method and receives to two differential equations that are able to give amplitude respect time response. The differential equations solved in steady state then plotting the frequency response of the system by varying the effective dynamic parameters and the effects of them were discussed.

The nonlinearity from midplane stretching affect the system resonance frequency and its increasing reasoned that the variation of damping also affects the system resonance frequency while if it is not considered or it is small then damping doesn't affect on resonance frequency. Temperature relaxation reduces the peak amplitude little, while Thermal damping has a reduction effect on peak amplitude with a dominant effect. In addition, temperature relaxation shows a significant effect on shift of resonance frequency to lower values. Resonance shifting is a very important phenomenon especially in resonator-based sensors. It seems to be the most effective source of errors in resonant-sensors which are designed based on constant stiffness assumption.

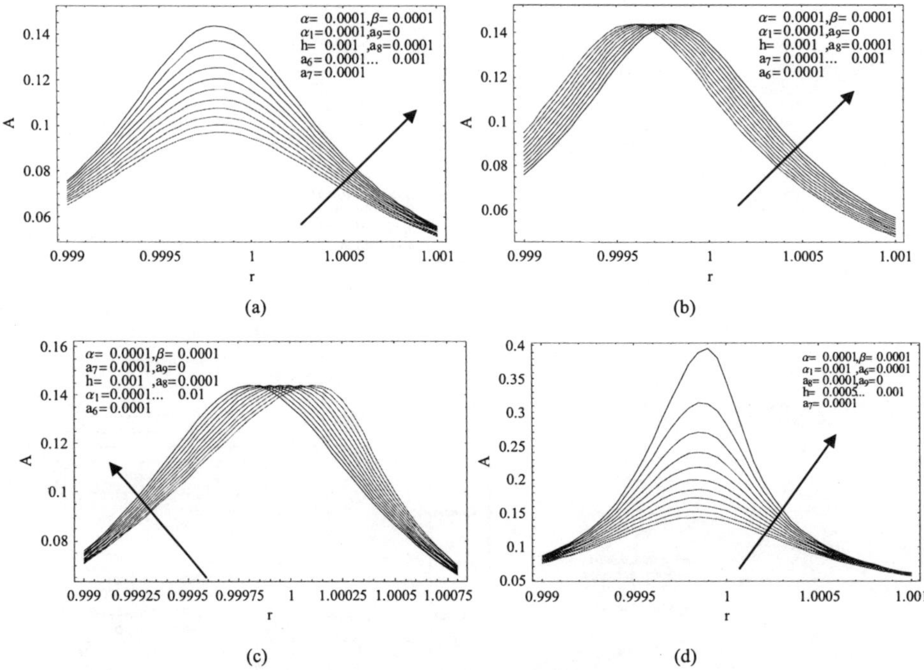

Fig. 3. Effect of variation of (a) thermal damping, (b) thermal relaxation, (c) nonlinearity (d) damping on frequency response. (Arrows show parameters reduction)

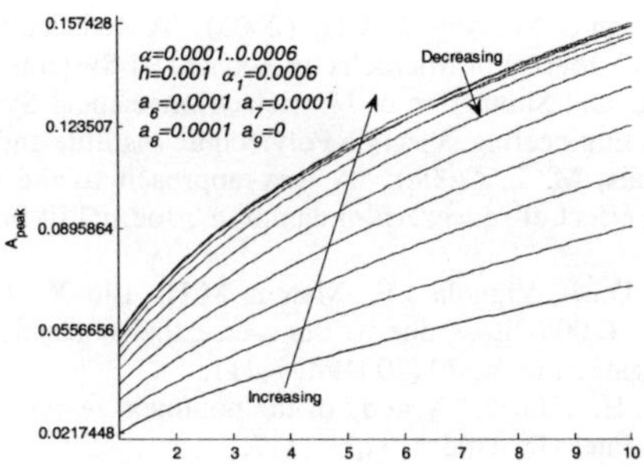

Fig. 4. Effect of variation of Ac Voltage on peak amplitude

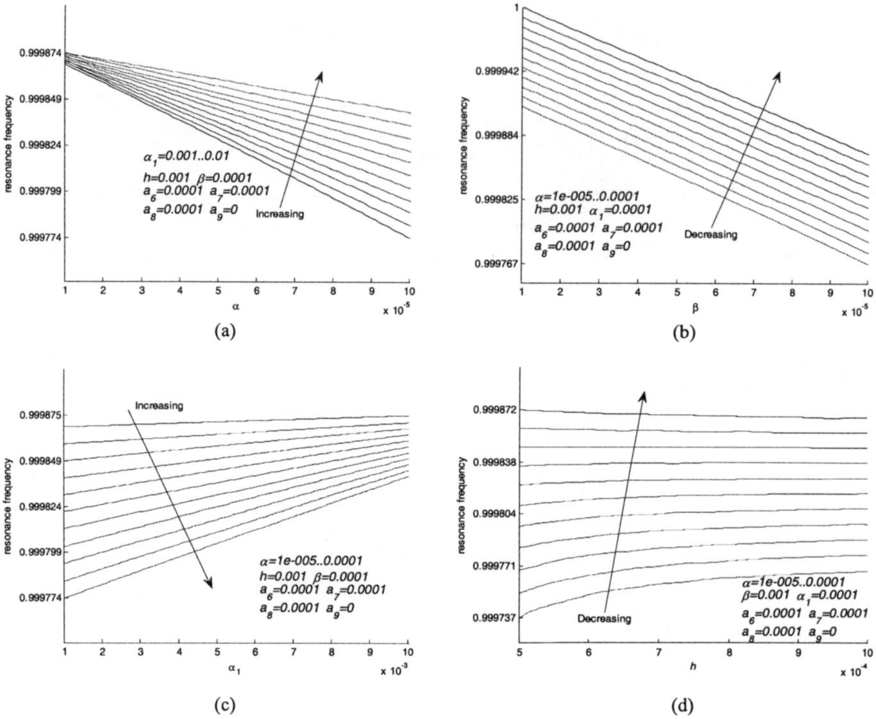

Fig. 5.Effects of variation of (a) DC voltage (b) AC Voltage (c) Nonlinearity (d) damping on resonance frequency

References

[1] Younis M. I., Abdel-Rahman E. M., Nayfeh A.H., (2003), "A Reduced-Order Model for Electrically Actuated Microbeam-Based MEMS", Journal of Microelectromechanical Systems, 12(5), 672-680.

[2] Younis M. I., , "Modeling and Simulation of Micrielectromecanical System in Multi-Physics Fields", Ph.D., thesis, (2004) Mechanicsl Engineering, Virginia Polytechnic Institute and State University.

[3] Nayfeh A. H., and Younis, M. I., (2004), "A new approach to the modeling and simulation of flexible microstructures under the effect of squeeze-film damping", Journal of Micromechanics and Microengineering, 14, 170-181.

[4] Houston B.H., Photiadis D.M., Vignola J.F., Marcus M.H., Liu X., Czaplewski D., Sekaric L., Butler J., Pehrsson P., Bucaro J.A., (2004)"Loss due to transverse thermoelastic currents in microscale Resonators", Materials Science and Engineering A 370 (2004) 407-411.

[5] Younis, M. I., Nayfeh, A. H., (2003), "A study of the nonlinear response of a resonant microbeam to electric actuation", Journal of Nonlinear Dynamics, 31, 91-117.

[6] Karami G., Garnich, M., "Micromechanical study of thermoelastic behavior of composites with periodic fiber wariness", Journal of Composites B, 36, 241-248, 2005.

[7] Duwel A., Gorman J., Weinstein M., Borenstein J., Ward P.,(2003), "Experimental study of thermoelastic damping in MEMS gyros", Sensors and Actuators A, 103, (2003), 70-75.

[8] Photiadis D.M., Houston B.H., Xiao L., Bucaro J.A., Marcus M.H.,(2002) "Thermoelastic loss observed in a high Q mechanical oscillator", Physica B,316-317(2002) 408-410.

[9] Jazar G.N., Mahinfalah M., Aagah M.R., Mahmoudian N., Khazaei A., Alimi M.H.,(2005), "Mathematical

modeling of thermal effects in steady state dynamics of microresonators using Lorentzian function: part 1 – thermal damping",2005 ASME International Mechanical Engineering Congress and R&D Expo, Orlando, Florida, USA, November 5-11.

[10] Aagah M.R., Mahmoudian N., Jazar G.N., Mahinfalah M., Khazaei A., Alimi M.H.,(2005), "Mathematical modeling of thermal effects in steady state dynamics of microresonators using Lorentzian function: part 2 – temperature relaxation",2005 ASME International Mechanical Engineering Congress and R&D Expo, Orlando, Florida, USA, November 5-11.

[11] Tadayon M.A., Sayyaadi H., Jazar G.N.," Nonlinear Modeling and Simulation of Thermal Effects in Microcantilever Resonators Dynamic", Journal of Physics: Conference Series at Institute of Physics(IOP),

[12] Tadayon M.A., Rajaeii A., Sayyaadi H., Jazar G.N., Alasty A., "Nonlinear Dynamic of MicroResonators", Journal of Physics: Conference Series at Institute of Physics(IOP),

[13] Zhang W., Meng G.,(2004), "Nonlinear dynamical system of micro-cantilever under combined parametric and forcing excitations in MEMS", Sensors and Actuators A,2004,Article In Press.

[14] Luo A.C.J., Wang F.Y.,(2002),"Choatic motion in a micro-electro mechanical system with non-linearity from capacitors ", Communications in Nonlinear Science and Numerical Simulation 7 (2002) 31–49.

Appendix

$$\alpha^2\beta^2(-512(1+r^2)^5\alpha^4\beta^4 a_9^{\,2}(-2\beta+a_9)^2(2\beta+3a_9)^2(-8a^2\beta^3-2\beta(\alpha+2a^2\beta)a_9+2a^2\beta a_9^{\,2}+a_9^{\,3}a^2)^2$$

$$(-2a^2ra_7(2\beta+a_9)+(1+r^2)(a^2a_9^{\,2}+2a^2a_8(2\beta+a_9)+2a^2a_9(-2(-1+r+\alpha)+3a^2\alpha_1)+$$

$$\beta(\alpha-4a^2(-2+2r+2\alpha+\beta)+12a^4\alpha_1)))^3+a^4(2\beta+a_9)^8(8a^4hr(1+r^2)a_6(2\beta+a_9)^4+$$

$$4a^4r^2a_6^{\,2}(2\beta+a_9)^4+(1+r^2)^2(32a^2\beta(a^2h^2-\alpha\beta)a_9^{\,3}+2a^2a_9^{\,4}(2a^2h^2-9\alpha\beta-\frac{1}{(1+r^2)}(-2a^2ra_7(2\beta+$$

$$a_9)+(1+r^2)(a^2a_9^{\,2}+2a^2a_8(2\beta+a_9)+2a^2a_9(-2(-1+r+\alpha)+3a^2\alpha_1)+\beta(\alpha-4a^2(-2+2r+2\alpha+\beta)$$

$$12a^4\alpha_1))))4\beta a_9(\beta^2(-32a^4h^2-3\alpha^2+32a^2\alpha\beta)+\frac{1}{(1+r^2)}(2\alpha\beta(-2a^2ra_7(2\beta+a_9)+(1+r^2)$$

$$a^2a_9^{\,2}+2a^2a_8(2\beta+a_9)+2a^2a_9(-2(-1+r+\alpha)+3a^2\alpha_1)+\beta(\alpha-4a^2(-2+2r+2\alpha+\beta)12a^4\alpha_1))))+$$

$$\frac{1}{(1+r^2)^2}((-2a^2ra_7(2\beta+a_9)+(1+r^2)(a^2a_9^{\,2}+2a^2a_8(2\beta+a_9)+2a^2a_9(-2(-1+r+\alpha)+3a^2\alpha_1)+$$

$$\beta(\alpha-4a^2(-2+2r+2\alpha+\beta)+12a^4\alpha_1)))^2))+4\beta^2(\beta^2(16a^4h^2+\alpha^2-8a^2\alpha\beta)-\frac{1}{(1+r^2)}$$

$$(2\beta(\alpha+4a^2\beta)(-2a^2ra_7(2\beta+a_9)+(1+r^2)(a^2a_9^{\,2}+2a^2a_8(2\beta+a_9)+2a^2a_9(-2(-1+r+\alpha)+3a^2\alpha_1)+$$

$$\beta(\alpha-4a^2(-2+2r+2\alpha+\beta)+12a^4\alpha_1))))+\frac{1}{(1+r^2)^2}((-2a^2ra_7(2\beta+a_9)+(1+r^2)$$

$$(a^2a_9^{\,2}+2a^2a_8(2\beta+a_9)+2a^2a_9(-2(-1+r+\alpha)+3a^2\alpha_1)+\beta(\alpha-4a^2(-2+2r+2\alpha+\beta)+12a^4\alpha_1)))^2))$$

$$+a_9^{\,2}(\beta^2(96a^4h^2+9\alpha^2-176a^2\alpha\beta)+\frac{1}{1+r^2}(2\beta(11\alpha+8a^2\beta)(-2a^2ra_7(2\beta+a_9)+(1+r^2)$$

$$(a^2a_9^{\,2}+2a^2a_8(2\beta+a_9)+2a^2a_9(-2(-1+r+\alpha)+3a^2\alpha_1)+\beta(\alpha-4a^2(-2+2r+2\alpha+\beta)+12a^4\alpha_1))))+$$

$$\frac{1}{(1+r^2)^2}((-2a^2ra_7(2\beta+a_9)+(1+r^2)(a^2a_9^{\,2}+2a^2a_8(2\beta+a_9)+2a^2a_9(-2(-1+r+\alpha)+3a^2\alpha_1)+$$

$$\beta(\alpha-4a^2(-2+2r+2\alpha+\beta)+12a^4\alpha_1)))^2)))^2)$$

Design and Analysis of General and Travelling Dielectrophoresis

T. Kinkeldei[1,2], and W.H. Li[1*]

[1]*School of Mechanical, Materials and Mechatronic Engineering, University of Wollongong, Wollongong, NSW 2522, Australia*

[2]*School of Mechanical, University of Karlsruhe, Forschungsuniversitaet Karlsruhe, Baden Wuerttemberg 76131, Germany*

Abstract:

This paper presents the simulation of general and travelling dielectrophoretic forces, as well as the movement of the particles in a sandwich structure micro-device. The electrode geometry of the micro device used for simulation is an interdigitated bar electrode. The simulation method used to solve the equations is based on the least square finite difference method (LSFD). The simulation first calculates all forces acting at any place in the chamber, with these forces the trajectory of a particle can now be proposed. All of the particles parameters like radius, voltage, initial height, etc can easily be changed and the simulation can be redone. With this continuous trial we receive different behavior of the particles and examine the relevancy of the different changes made. This detailed information about the influences of the parameters on the procedure in the micro-device can be used for the development of further micro-devices.

Keywords: Dielectrophoresis, least square finite difference, simulation, general DEP, travelling wave

1. Introduction

The development of micro-devices for handling particles in micrometer size and smaller, remains a challenge. Although a lot of work has been done in this research area there is still a need to focus on this topic, due to the big opportunities this research area unwraps. Different techniques have been used to handle these kinds of particles, from mechanical grippers to adhesive grippers to electrical handling devices. The latter have recently done a very good job in controlled manipulation and separation of particles especially in very small dimensions. One efficient effect used in electrical manipulation is called dielectrophoresis. DEP arises from the interaction of a non-uniform AC electrical field with the induced dipole in a micro-particle [1,2]. This interaction creates forces that can be used for its controlled manipulation. The utilization of DEP is steadily increasing. Latest research shows, that DEP Forces can be used to handle and separate carbon nanotubes [3] due to semi conducting or metallic. Other particles have also been dealt with successfully, particularly biological particles such as cells [4], viruses [5], and DNA [6] but also synthetic particles like latex spheres [7]. The different particles have different properties, which can be used to clearly identify and handle them and can be used to separate particle mixtures. Examples for these properties are different radius or different behaviour to the frequent of the applied

* Corresponding author. Phone: +61 2 4221 3490; Fax: +61 2 4221 3101.
E-mail address: weihuali@uow.edu.au

voltage in the device. Also, the electrical device, and especially the electrode geometry, can be used for special applications like mixer [8], filter [9], etc. The development of DEP simulation is based on the theory used to visualize the forces of the electrical field. Due to this, for different approaches different simulation methods are used. A common method to describe the DEP forces is based on FEM [10-11]. This method is also used by most commercial software, like Ansys or Maxwell, to calculate the potential distribution for different electrode geometries. Other methods used for simulation are distributed Lagrange multiplier (DLM) [12], method of moments (MoM) [13] or finite difference (FD) [14]. Electrical fields and DEP forces have been simulated for different electrode geometries like castellated electrodes [15], cage electrodes [16], interdigitated electrodes [11] and polynomial electrodes [17]. The simulated forces show close agreement with the experimental studies. Most simulation work is focused on just the calculation of the force and doesn't deal with the movement in the microdevice. An example [11] shows the behaviour of a micro particle in a device with interdigitated electrode geometry, undergoing general DEP and hydrodynamic forces.

2. Theoretical Background

Pohl introduced the theory of the dielectrophoretic force in 1951 [18]. It is derived from the dipole in a spherical particle induced by an AC non uniform electrical field. The derivation of both forces for gDEP and twDEP are given by the time averaged Force F(t)

$$F(t) = [m(t)\nabla]E(t) \tag{1}$$

where m(t) and E(t) are the time-dependant dipole moment and electrical field. Following the derivation from Huang the dipole moment $m_x(t)$ is taken as an example.

$$m_x(t) = 4\pi e_m r^3 E_{x0}\{\mathrm{Re}[f_{cm}]\cos(\omega t + \varphi_x) - \mathrm{Im}[f_{cm}]\sin(\omega t + \varphi_x)\} \tag{2}$$

Here φ_x is the phase and E_{x0} the field magnitude as spatial functions and ωt is the frequency. The other parameters are e_m the absolute permittivity of the suspending medium, r the radius and f_{cm} the Clausius-Mosetti factor, which is given by

$$f_{CM} = \frac{\varepsilon_p^* - \varepsilon_m^*}{\varepsilon_p^* + 2\varepsilon_m^*} \tag{3}$$

In this equation ε_p^* and ε_m^* are the particle and suspending medium complex permittivity, defined by $\varepsilon^* = \varepsilon - j(\sigma/\omega)$ with ε as permittivity, σ conductivity and j as square root from -1. This factor can be positive or negative depending on the frequency and is therefore responsible for positive or negative forces. Now solving equation (1) for the x component with equation (2) and the missing moments in y and z direction we receive the solution in x direction for the DEP force time dependant and time averaged by

$$\begin{aligned} F_x(t) &= m_x(t)\frac{\partial E_x(t)}{\partial x} + m_y(t)\frac{\partial E_x(t)}{\partial y} + m_z(t)\frac{\partial E_x(t)}{\partial z} \\ \langle F_x(t)\rangle &= 2\pi\varepsilon_m r^3\left[\mathrm{Re}(f_{cm})\frac{\partial\left(E_{x0}^2 + E_{y0}^2 + E_{z0}^2\right)}{2\partial x} + \mathrm{Im}(f_{cm})\left(E_{x0}^2\frac{\partial\varphi_x}{\partial x} + E_{y0}^2\frac{\partial\varphi_y}{\partial x} + E_{z0}^2\frac{\partial\varphi_z}{\partial x}\right)\right] \end{aligned} \tag{4}$$

With equation (4) for every direction x,y,z we receive the total DEP force acting on a particle from non uniform electrical fields, which is the base for our simulation work.

3. Numerical Integration

3.1 Calculation of DEP forces

The numerical equations of the simulation are based on the least square finite difference method, introduced by [1]. It can be viewed as a further development of the general finite difference method GFD, an established method to solve numerical problems. The potential distribution is satisfied with the Lapace equation

$$\frac{\partial\phi^2}{\partial x^2}+\frac{\partial\phi^2}{\partial y^2}=0 \tag{5}$$

where φ is potentional distribution. The electrical field can be obtained by differentiating the potential. The general and travelling wave DEP forces are numerically represented as

$$\begin{aligned} F_{cDEP} &= 2\pi\varepsilon_m a^3 \mathrm{Re}[f_{CM}]\nabla(|\mathrm{Re}[\widetilde{\mathrm{E}}]|^2+|\mathrm{Im}[\widetilde{\mathrm{E}}]|^2) \\ &= 2\pi\varepsilon_m a^3 \mathrm{Re}[f_{CM}]\nabla(|\nabla\phi_R|^2+|\nabla\phi_I|^2), \end{aligned} \tag{6}$$

$$\begin{aligned} F_{twDEP} &= -2\pi\varepsilon_m a^3 \mathrm{Im}[f_{CM}]\nabla\times(E_{Rx}E_{Iy}-E_{Ry}E_{Ix})\hat{z} \\ &= -2\pi\varepsilon_m a^3 \mathrm{Im}[f_{CM}][-(E_{Rx,x}E_{Iy}+E_{Rx}E_{Iy,x}-E_{Ry,x}E_{Ix}-E_{Ry}E_{Ix,x})\hat{x} \\ &+(E_{Rx,y}E_{Iy}+E_{Rx}E_{Iy,y}-E_{Ry,y}E_{Ix}-E_{Ry}E_{Ix,y})\hat{y}] \end{aligned} \tag{7}$$

3.2 Particle motion

The forces acting on the particle are the gravitational force, the drag force of the fluid and finally the DEP force. The two missing forces are given by

$$\begin{aligned} F_{Flow} &= 6\cdot\pi\cdot k\cdot r\cdot\eta(\vec{v}_m-\vec{v}_p) \\ F_{Grav} &= \frac{4}{3}\pi\cdot r^3(\rho_p-\rho_m)\cdot g \end{aligned} \tag{8}$$

The force due to the flow is constant along the x direction and just varies in y direction with a parabolic profile; the gravitational force is a constant. The basic for the movement is equation of the momentum, it is given by

$$\begin{aligned} m\ddot{x}(t) &= F_{DEP,x}+6\cdot\pi\cdot\eta\cdot k\cdot r\cdot(v_m-\dot{x}(t)) \\ m\ddot{y}(t) &= F_{DEP,y}-(\rho_p-\rho_m)v\cdot g-6\cdot\pi\cdot\eta\cdot k\cdot\dot{y}(t) \end{aligned} \tag{9}$$

4. Results and Discussion

4.1 General DEP

The parameters used are $\mathrm{Re}(f_{C-M})=\pm 0.5$, medium flow rate of 487.76 μm/s and conductivity σ_m of 10^{-4} S/m, a dielectric constant ε_m of 80 and mass density of 1 g/cm^3. The particle used for simulation is a micro latex sphere with conductivity σ_p of 10^{-18} S/m, a dielectric constant ε_p of 2.5 and a mass density of 1.05 g/cm^3. A relatively low gravitational force for the particle is observed, which is just relevant over long distance travel. The other preferences of the micro device like chamber height, applied voltage, particle size and others were varied and the results were used to find out more about the particle behavior in the device.

The simulation results for both negative and positive DEP with the same particle sizes are shown in Fig. 1, where the particle size is 1 μm. The left part of Fig.1 shows negative DEP for a release Voltage of 2 Vpp, the electrodes are marked by the grid and the bold lines at the bottom of the plot. The maximum DEP forces act at the electrode edges on the particle so that the biggest gradient of the movement occurs at the electrode edges. The trajectory of the particles of various initial heights move closer together and after a certain time and distance, much longer than this device, all the particles levitate at the same height. This is due to the balance between the gravitational force and the DEP force. At a certain height above the electrode, the gravitational force gets bigger than the DEP force and the trajectory is levelling off to an even. In the right part of Fig. 1, the trapping of the particle can be seen, with a release Voltage of 3Vpp, the electrodes are again marked with bold lines and the grid. Depending on the initial height of the particle the stopping distance varies. The maximum force in x direction, which is responsible for trapping has to overcome the force by the flow that trapping can occur.

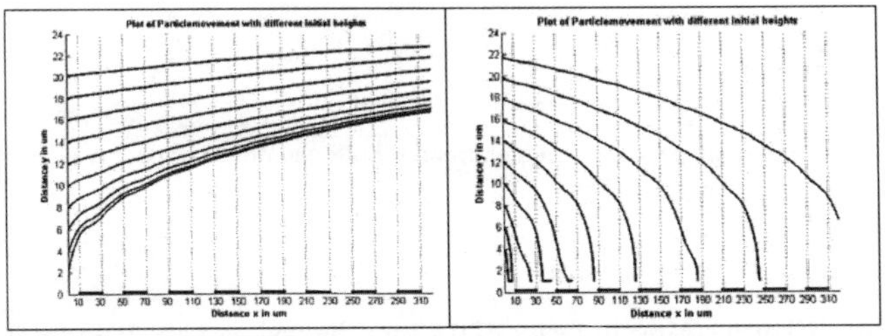

Fig. 1. Plot of particle movement for left negative DEP and right positive DEP. The chamber height is 24 um. The viewed Device length is 320 um. The electrode length and the gap between are both 20 um

Fig. 2 shows both negative and positive DEP acting on particles with different radius. The particles start at the same initial height. For negative DEP, in the left part of Fig.2, for a Voltage of 1 Vpp and the chamber height of 20 um, it is received a higher levitation height for bigger radius. To see the trajectory migrating into an even the domain of the device has to be expanded. For negative DEP in the right part of Fig.2 with 3 Vpp and a chamber height of about 26 um the maximum stopping distance gets smaller for bigger radius.

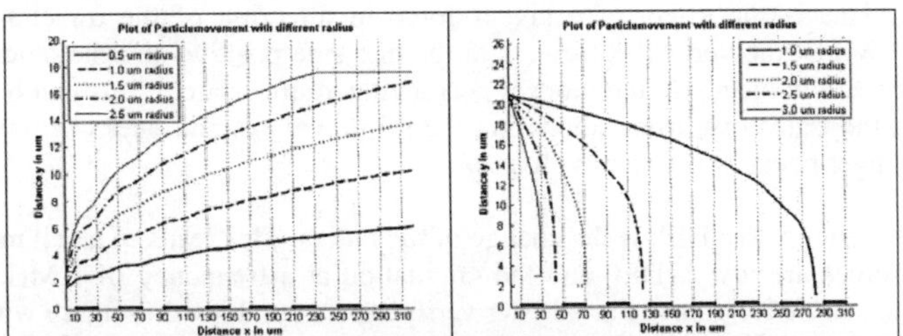

Fig. 2. Negative (left) and positive DEP for different particles sizes.

The effect of electrode geometry on both negative and positive DEPs are shown in Fig. 3. As can be seen from this figure, for an electrode width of 25 um we receive the highest DEP forces and for the electrode of 15 um the lowest forces. The reason for this is due to the potential distribution. Although there is no change at the electrode surface in x direction, there is a big change in y direction. This leads to bigger average forces in the device above the electrode and therefore to a higher levitation or earlier stopping with increasing width of the electrode. With electrode geometries of bigger width particles can be levitated higher or trapped earlier in the device.

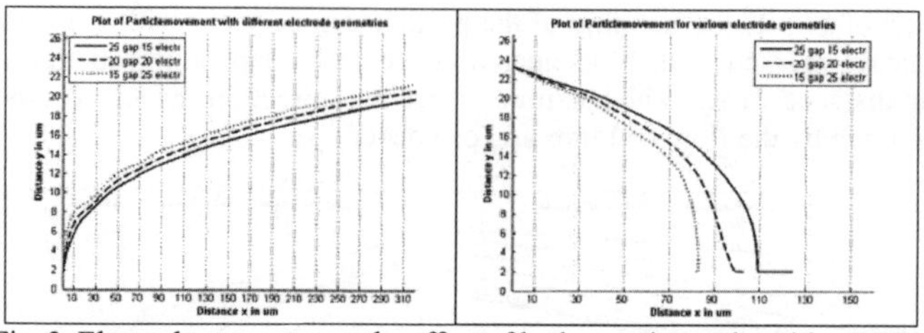

Fig. 3. Electrode geometry on the effect of both negative and positive DEPs.

4.2 Travelling Wave DEP

The DS 19 cells, with a Clausius-Mosotti factor of 0.17 for the real part and 0.45 for the imaginary part at a frequency of 1 MHz, is employed to investigate the travelling wave DEP. The density of the cells is 1.07 g/cm^3. The cell movement is shown in Fig. 4. As can be seen from this figure, the cells move in circles to a centre above the electrodes due to just traveling wave DEP at a frequency of 1 MHz and applied Voltage of 2 Vpp and a mass density of the cell changed to 1.4 g/cm^3. As reported by Li ea al [1], the twDEP force in x direction acts also in higher regions of the chamber, so that even the particle at the top are moved into traveling wave direction and travel further then at lower initial heights till they get finally caught by the forces in y direction. The effect of voltages on the travelling wave is shown in Fig. 5. The left part shows the movement of the cells as similar already seen in Fig.4. The right part shows the traveling distance that the cells cover. This curve has first a slightly parabolic increase but then converts into a straight linear increase. Unlike to general DEP it can be seen that for higher Voltage the stopping distance increases. This is the effect of a higher force in direction of the traveling wave due to the increasing voltage. Whilst the vertical forces are at the one side positive and the other side negative, the closer the particle is to the electrode the more it gets levitated afterwards. This can be figured out in the increasing wave of the trajectory, from comparing the single valleys and peeks of one trajectory and the comparing of the bigger peeks at different voltages.

A special case in traveling wave DEP is the change of the radius. This leads to a cell movement as seen in Fig.6. The acting forces are just twDEP and the gravitation at a frequency of 1 MHz and 2 Vpp and a mass density for the cell of 1.4 g/cm^3. For a better verifying we combine this figure with a plot of tw- and gDEP forces acting on particle with different radius, here we have a mass density of 1.07 g/cm^3.

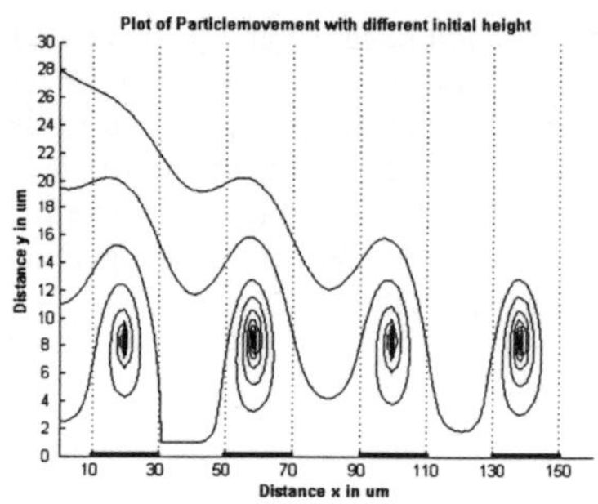

Fig. 4. The effect of the initial heights on the travelling wave particle motion.

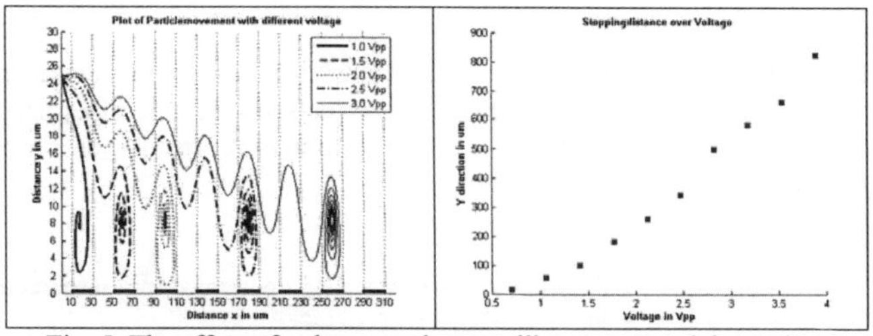

Fig. 5. The effect of voltage on the travelling wave particle motion.

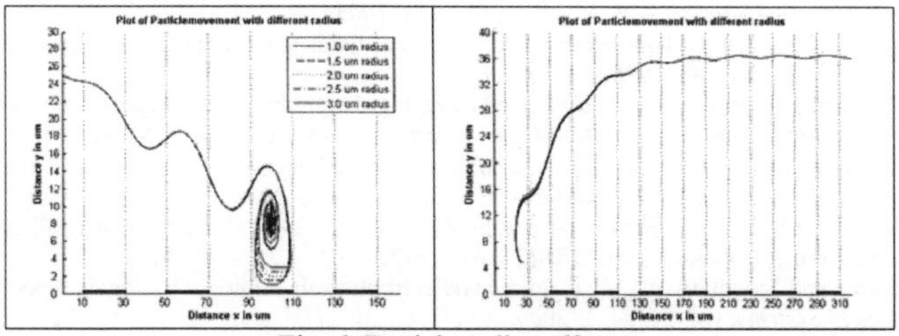

Fig. 6. Particle radius effect.

As can be seen from Fig. 6, the particle radius has no effect on the movement of the cells. The only clear thing to see is that there are different radiuses is in the left plot, where the cells collide with the bottom of the chamber. In the right plot we can see a slightly thicker trajectory at the beginning over the first electrode. This little difference is due to the gDEP force, which is of more importance the closer a cell is to the electrode.

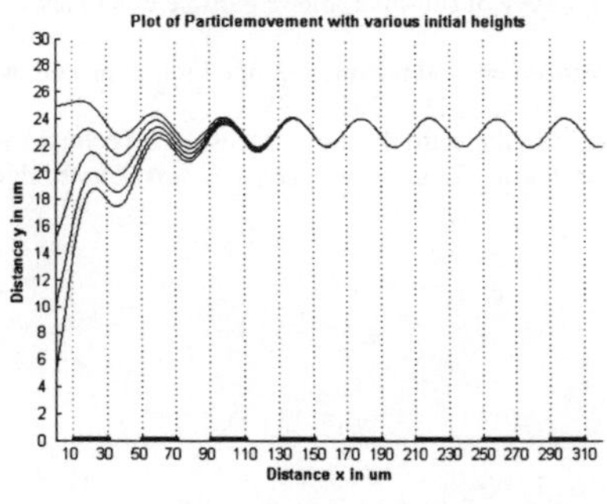

Fig. 7. The initial height effect.

Similar to the radius in a combination of tw- and gDEP the initial height has no effect on the final trajectory of the cells, as shown in Fig. 7. The difference to single twDEP is the levitation of the particle against the former trapping at the centers above the electrode. The reason for the levitation is clearly the gDEP force. As seen for negative DEP with just general DEP in the figures before, the particle are levitated to a certain height and then follow a horizontal even. It is shown that the increasing of levitation stops at a certain value and the trajectory continues instead of an even in a simple sinus oscillation due to the twDEP forces. The frequency in this plot is at 1 MHz and the Voltage is 2 Vpp, the cell size is 1 um.

4. Conclusion

The numerical simulation methodology was developed to study both general and traveling wave DEP based an interdigitated microelectrode array. The effects of particle size, applied voltage, initial heights on the particle motion were analysed. These simulation results are expected to optimize interdigitated microelectrodes based DEP devices.

References

[1] W.H. Li, H.Du, D.F.Chen, and C.Shu, "Analysis of dielectrophoretic electrode arrays for nanoparticle manipulation", *Computational Materials Science,* Vol. 30, pp. 320-325, 2004.

[2] T. B. Jones, *Electromechanics of particles.* Cambridge; New York: Cambridge University Press, 1995.

[3] D.F. Chen, H. Du, W.H. Li, and C. Shu, "Numerical modeling of dielectrophoresis using a meshless approach," *Journal of Micromechanics and Microengineering, Vol. 15, pp. 1040-1048, 2005.*

[4] R. Pethig, "Dielectrophoresis: Using inhomogeneous AC electrical fields to separate and manipulate cells," *Critical Reviews in Biotechnology*, vol. 16, pp. 331-348, 1996.

[5] M. P. Hughes, H. Morgan, F. J. Rixon, J. P. H. Burt, and R. Pethig, "Manipulation of herpes simplex virus type 1 by dielectrophoresis," *Biochimica Et Biophysica Acta-General Subjects*, vol. 1425, pp. 119-126, 1998.

[6] M. Washizu, O. Kurosawa, I. Arai, S. Suzuki and N. Shimamoto, "Applications of Electrostatic Stretch-and-Positioning of DNA," *IEEE Transactions on Industry Applications*, vol. 31, pp. 447-456, 1995.

[7] N. S. Green, "Dielectrophoretic separation of nano-particles," *Journal of Applied Physics*; vol. 30, 1997, pp. 41-44

[8] J. Deval, "A Dielectrophoretic Chaotic Mixer," *IEEE*; vol. 2, 2002

[9] H. Li, "Characterization and Modeling of a Microfluidic Dielectrophoresis Filter for Biological Species," *Journal of Microelectromechanical Systems*, vol. 14, no. 1, 2005

[10] H. Morgan, "Numerical solution of the dielectrophoretic and travelling wave forces for interdigitated electrode arrays using the finite element method," *Journal of Electrostatics*, vol. 56, pp. 235-254, 2002.

[11] H. Li, "On the Design and Optimization of Micro-Fluidic Dielectrophoretic Devices: A Dynamic Simulation Study," *Biomedical Microdevices*, vol. 6, pp.289-295, 2004.

[12] A. T. J. Kadaksham, "Dielectrophoresis of nanoparticle", *Electrophoresis*, vol. 24, pp. 3625-2632, 2004.

[13] M. P. Hughes, "Simulation of Travelling Electric Field Manipulation of Particle", pp. 48-52

[14] B. Malnar, "Separation of latex spheres using dielectrophoresis and fluid flow", IEEE, vol. 150, pp. 66-69, 2003.

[15] N. G. Green, "Ac electrokinetics: a survey of sub-micrometre particle dynamics", *J. Phys. D Appl. Phys*, vol. 33, pp. 632-641, 2000.

[16] J. Suehiro, "The dielectrophoretic movement and positioning of a biological cell using a three-dimensional grid electrode system", *J. Phys. D Appl. Phys.*, vol. 31, pp. 3298-3305, 1998.

[17] Y Huang, "Electrode design for negative dielectrophoresis", Meas. Sci. Technol., vol. 2, pp. 1142-1146, 1991.

[18] H. Pohl, *The Behavior of Neutral Matter in Nonuniform Electric Fields*; Cambridge University Press: Cambridge, 1978.

Self-Propulsion of a Capsule on a Lubricated Surface Driven by Piezoelectric Actuators

Z.C. Feng
Department of Mechanical and Aerospace Engineering, University of Missouri-Columbia
Columbia, MO 65211, USA

Abstract

This paper studies the self-propulsion of a capsule. The self-propulsion is achieved with piezoelectric actuators using the impact mechanism. The capsule is moving on a lubricated surface with an interaction force similar to that of a viscous liquid. It is shown that nonlinearity in the interaction force results in net displacement of the capsule when piezoelectric electric actuator is periodically excited.

Keywords: self-propulsion, impact drive, piezoelectric actuator

1. Introduction

Invasiveness in medical diagnosis is often imposed by the need to deploy and manipulate the sensor inside the human body. As a specific example, there have been extensive research activities in developing self-propelled endoscopic robots for colonoscopy [1-4]. Traditional colonoscopes are pushed inside the human colon by external maneuvers, a process which is very uncomfortable and even painful to the patients since the human intestinal track has several acute bends. The introduction of a self-propelled device, which generates just "internal" forces and which does not need any external pushing actions and special manual practice of endoscopists, would improve the colonoscopy procedure in terms of patient pain reduction and ease of advancement.

The mechanism for propulsion that has been investigated extensively in the literature is the "inchworm locomotion" [1-5]. Mimicking the motion of an inchworm, the propulsion device consists of two clampers at its ends and one extensor at its mid-section. The propulsion follows the sequenced actions of clamping and extension. Because of the adhesion process, clamping to the soft tissues is not easily accomplished. The device is rather complicated and the efficiency still needs to be improved.

A promising alternative propulsion mechanism is the "Impact Drive Mechanism"[6]. Consider the schematic diagram in Figure 1. The two masses are connected by a linear actuator. If the linear actuator expands slowly, the friction force between the mass M and the surface prevents any movement of the mass M. Consequently, the smaller mass m is displaced to the right, Figure 1(b). When the linear

The corresponding author, Tel: Phone: (573) 884-4624, Fax: (573) 884-5090
E-mail: FengF@missouri.edu

actuator is driven to contract suddenly, it generates tensions to pull the two masses together. If the tension is large enough to overcome the friction force, the mass M will have a small displacement, Figure 1(c). This net displacement is driven by the un-even expansion and contraction of the actuator. This cycle can be repeated to obtain large cumulative displacement. Moreover, by changing the rate in expansion and contraction, the direction of motion can be reversed.

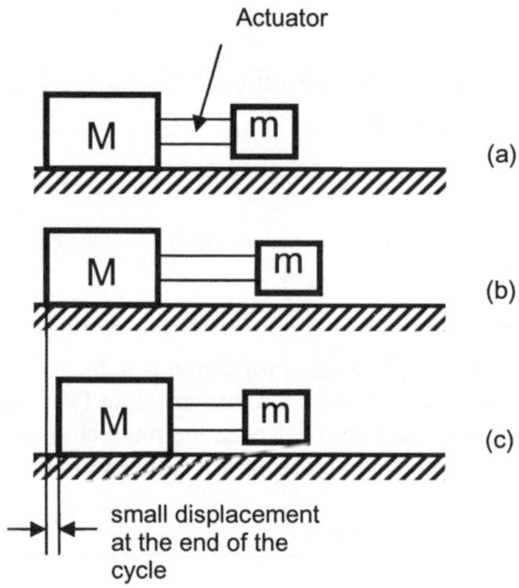

Figure 1. Schematic diagrams of the "Impact Drive Mechanism". (a) The beginning of the cycle. (b) Slow expansion of the actuator is accompanied by the displacement of the smaller mass. The larger mass sticks to the surface because of the friction. (c) Sudden contraction of the actuator pulls the two masses together. Tension is large enough to overcome the friction. The larger mass slides to the right resulting in a net displacement. The cycle can be repeated.

The impact drive mechanism can be easily implemented. Piezoelectric materials can be used as the linear actuator. The whole system is very simple. In fact the system can be packaged in a capsule with the actuator and the smaller mass all within the capsule. In that case, the Impact Drive Mechanism becomes a self-propelled capsule as shown in Figure 2.

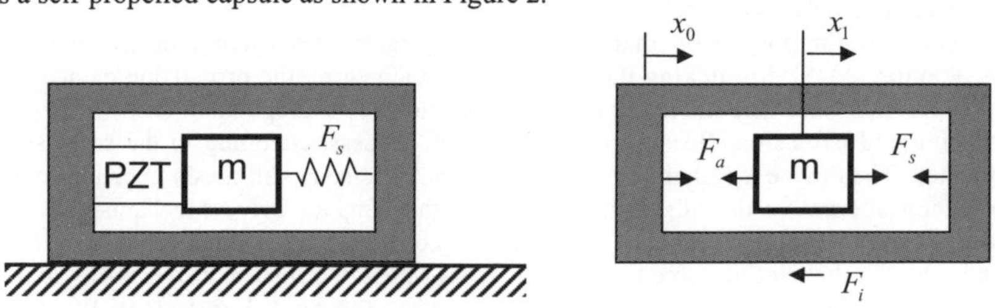

Figure 2 The mechanical model.

Before the above concept can be applied to self-propelled capsules as miniature sensor carriers in medical diagnosis, we must note that the nonlinear characteristic of the friction is essential for the propulsion. Here the nonlinear characteristic is the non-proportional displacement corresponding to the forcing applied on the larger mass. This characteristic is also known as the stick-slip phenomenon. When the capsule moves on a lubricated surface, stick-slip may no longer be present. Is self-propulsion still possible? In this paper, we assume that the interaction force between the capsule and the surface is similar to that of a viscous liquid. In the following section, we present the mechanical model. The self-propulsion is then studied using analytical and numerical methods. We first consider the case when the forcing function is a single frequency sinusoidal function. We then study multiple-frequency forcing.

2. The Mechanical Model

The mechanical model of the impact drive mechanism is shown in Figure 2. The mass M lies on a surface. Mass m is attached to mass M through the piezoelectric actuator. Gravity provides the normal force to generate friction between the surfaces. Since the motion in the vertical direction can be ignored, the normal force is a constant. We can thus ignore the other effects of gravity to simplify the model. The piezoelectric actuator elongates along the horizontal direction when subjected to an electric field. Depending on the specific design of the actuator, the electric field may be applied along the longitudinal or the transversal directions of the actuator. The constitutive law of piezoelectric material under one-dimensional stress is

$$\varepsilon = d_{33}E_3 + \frac{\sigma}{E}. \tag{1}$$

Here, we assume without the loss of generality that the electric field is applied along the longitudinal direction of the actuator. The coefficient d_{33} represents the electromechanical coupling effect; it is a material constant as the Young's modulus E. Recall that the strain is determined by the displacements of the masses and the actuator initial length, and the stress is the actuator force divided by the cross-sectional area of the actuator, i.e.

$$\varepsilon = \frac{x_1 - x_0}{L_0}, \sigma = \frac{F_a}{A}. \tag{2}$$

where x_0 and x_1 are the displacements of the capsule and the inner moving mass. Therefore:

$$F_a = -EAd_{33}E_3 + \frac{EA}{L_0}(x_1 - x_0) = -F_a^{'}(t) - k^{'}(x_0 - x_1). \tag{3}$$

Consider the mechanical model shown in Figure 2. The equations of motion can be written as

$$M\ddot{x}_0 = -F_a^{'}(t) - k^{'}(x_0 - x_1) - F_s(x_0 - x_1) - F_i(\dot{x}_0) \tag{4}$$

$$m\ddot{x}_1 = F_a^{'}(t) + k^{'}(x_0 - x_1) + F_s(x_0 - x_1), \tag{5}$$

where $F_a^{'}(t)$ represents the forcing from the applied electric field; $F_i(\dot{x}_0)$ represents the force acting at the interface between the large mass and the lubricated surface. $F_s(x_0 - x_1)$ represents the nonlinear spring force. Introduce the following change of variables:

$$v = \dot{x}_0,\ y = x_0 - x_1. \tag{6}$$

The equations of motion can be written as

$$\dot{v} = \frac{1}{M}[-F_a^{'}(t) - k^{'}y - F_s(y) - F_i(v)] \tag{7a}$$

$$\ddot{y} = -\frac{1}{m_e}F_a^{'}(t) - \frac{1}{m_e}[k^{'}y + F_s(y)] - \frac{1}{M}F_i(v), \tag{7b}$$

where

$$m_e = \frac{mM}{m+M}. \tag{8}$$

3. Self-Propulsion with a Single Frequency Sinusoidal Excitation

In the absence of nonlinearity, the steady-state solution of the above equations can be found easily. Corresponding to sinusoidal excitations, the solutions will be sinusoidal too. Self-propulsion is not possible. If the excitation to the actuator is given by

$$F_a'(t) = a\cos\omega t \tag{9}$$

Self-propulsion becomes possible when the interfacial force is nonlinear. For example, if

$$F_i(v) = cv + bv^2,\ F_s = k''y, \tag{10}$$

that is, the interfacial friction force is different whether the mass M is moving forward or backward as shown in Figure 3. The quadratic term acts as a rectifier. Net translation occurs. This mechanism is often observed in nature. Grains of certain plants are seen to move on a surface when they are caused to vibrate relative to the surface. The direction dependent interfacial force is similar to that of a fish hook. The direction of the propulsion is determined by the nature of the interfacial forces.

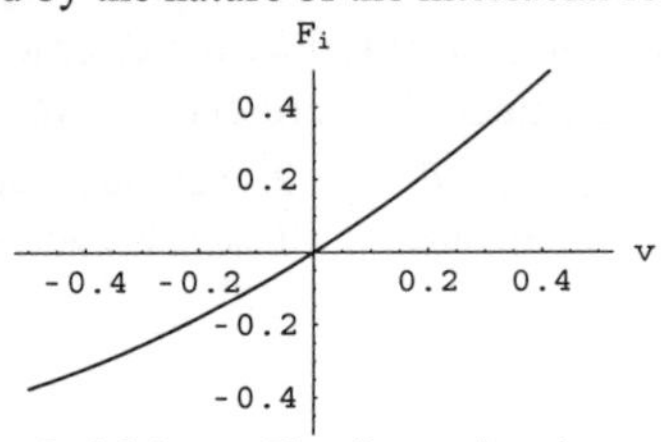

Figure 3 Direction dependent interfacial force. The forward resistance is larger than backward resistance. $F_i(v) = v + 0.5v^2$.

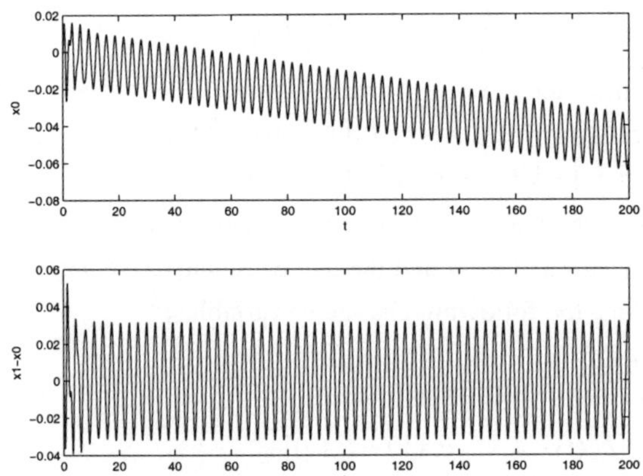

Figure 4. Translational motion resulting from direction dependent interfacial force. Parameters are: $M = m = 1$, $k = 10$, $a = 0.25$, $\omega = 2$. $F_i = v + 0.5v^2$.

The propulsion resulting from directional dependence of the interfacial force has limited usefulness in practice since the translational direction cannot be controlled without changing the interfacial properties. In the absence of the direction dependence of the interfacial forces, self-propulsion is still possible. Consider a specific nonlinear spring and a nonlinear interface force as the following:

$$F_s(y) = k^{"} y + \varepsilon h y^2 \,,\; F_i(v) = cv + \varepsilon d v^3 \tag{11}$$

where $k^{"}$ is the linearized spring constat; c is the linearized damping coefficient. The spring is assumed to have quadratic nonlinearity for reasons that will become clear shortly. The small parameter ε is introduced to make explicit the smallness of nonlinearity in our system. Let the forcing term be sinusoidal:

$$F_a^{'}(t) = a\cos(\omega t)\,. \tag{12}$$

The equations of motion can be written as follows:

$$M\dot{v} + cv + ky = -a\cos(\omega t) - \varepsilon(h\,y^2 + d\,v^3) \tag{13a}$$

$$m_e \ddot{y} + \frac{m_e}{M} cv + ky = -a\cos(\omega t) - \varepsilon(h\,y^2 + \frac{m_e}{M} d\,v^3) \tag{13b}$$

where $k = k^{'} + k^{"}$.

Let

$$v = v_0 + \varepsilon v_1 + \varepsilon^2 v_2\,, \tag{14a}$$

$$y = y_0 + \varepsilon y_1 + \varepsilon^2 y_2\,. \tag{14b}$$

Equations at the leading order are:

$$M\dot{v}_0 + cv_0 + ky_0 = -a\cos(\omega t) \tag{15a}$$

$$m_e \ddot{y}_0 + \frac{m_e}{M} cv_0 + ky_0 = -a\cos(\omega t)\,. \tag{15b}$$

Equations at $O(\varepsilon)$ are:

$$M\dot{v}_1 + cv_1 + ky_1 = -(h\,y_0^2 + d\,v_0^3) \tag{16a}$$

$$m_e \ddot{y}_1 + \frac{m_e}{M} cv_1 + ky_1 = -(h\,y_0^2 + \frac{m_e}{M} d\,v_0^3) \tag{16b}$$

Equations at $O(\varepsilon^2)$ are:

$$M\dot{v}_2 + cv_2 + ky_2 = -(2h\,y_0 y_1 + 3d\,v_0^2 v_1) \tag{17a}$$

$$m_e \ddot{y}_2 + \frac{m_e}{M} cv_2 + ky_2 = -(2h\,y_0 y_1 + 3\frac{m_e}{M} d\,v_0^2 v_1) \tag{17b}$$

The steady-state solution of the leading order is given by

$$v_0 = A_0 \cos(\omega t) + B_0 \sin(\omega t) \tag{18a}$$

$$y_0 = C_0 \cos(\omega t) + D_0 \sin(\omega t) \tag{18b}$$

Substituting the above into (15a) and (15b) and solving the resulting algebraic equations, we get:

$$A_0 = -\frac{a\,c\,m_e M^3[k(m_e - M) + m_e M\omega^2]\omega^2}{D}\,, \tag{19a}$$

$$B_0 = -\frac{a\;m_e M^3[-k + m_e\omega^2]\omega^3}{D}\,, \tag{19b}$$

$$C_0 = -\frac{a\{M^4\omega^2(k - m_e\omega^2) + c^2(m_e - M)[k(m_e - M) + m_e M\omega^2]\omega^2\}}{D} \tag{19c}$$

$$D_0 = -\frac{a\,c m_e^2 M^2 \omega^3}{D} \tag{19d}$$

where

$$D = M^4\omega^2(k - m_e\omega^2)^2 + c^2[k(m_e - M) + m_e M\omega^2]^2 . \tag{20}$$

Solving (16a) and (16b), we get

$$v_1 = A_1 + B_1\cos(\omega t) + C_1\sin(\omega t) + D_1\cos(2\omega t) + E_1\sin(2\omega t) + F_1\cos(3\omega t) + G_1\sin(3\omega t)$$

$$y_1 = H_1 + I_1\cos(\omega t) + J_1\sin(\omega t) + K_1\cos(2\omega t) + L_1\sin(2\omega t) + M_1\cos(3\omega t) + N_1\sin(3\omega t)$$

where the coefficients are cumbersome expressions, most of which are not needed for the following analysis. We note in particular, however, that

$$A_1 = 0 , \tag{21a}$$

$$D_1 = -\frac{2hm_e M\omega^2\{4C_0 D_0 M^2\omega(k - 4m_e\omega^2) + c(C_0^2 - D_0^2)[k(m_e - M) + 4m_e M\omega^2]\}}{4M^4\omega^2(k - 4m_e\omega^2)^2 + c^2[k(m_e - M) + 4m_e M\omega^2]^2} , \tag{21b}$$

$$E_1 = \frac{4hm_e M\omega^2\{(C_0^2 - D_0^2)M^2\omega(k - 4m_e\omega^2) - cC_0 D_0[k(m_e - M) + 4m_e M\omega^2]\}}{4M^4\omega^2(k - 4m_e\omega^2)^2 + c^2[k(m_e - M) + 4m_e M\omega^2]^2} , \tag{21c}$$

In other words, at this order of approximation, there is no translation.

Equations (17a) and (17b) can be written as

$$M\dot{v}_2 + cv_2 + ky_2 = -(2h\,R + 3d\,S) + \text{harmonic terms} \tag{22a}$$

$$m_e\ddot{y}_2 + \frac{m_e}{M}cv_2 + ky_2 = -(2h\,R + 3\frac{m_e}{M}d\,S) + \text{harmonic terms} \tag{22b}$$

where

$$R = \frac{1}{2}(C_0 I_1 + D_0 J_1) \tag{23}$$

$$S = \frac{1}{2}[2A_1(A_0^2 + B_0^2) + D_1(A_0^2 - B_0^2) + 2A_0 B_0 E_1] \tag{24}$$

We obtain the constant part of the solution for v_2, i.e.

$$\bar{v}_2 = -3\frac{d}{c}S \tag{25}$$

Substituting (19) and (21) into (24), we obtain:

$$\bar{v}_2 = \frac{3d\,a^4 hm^3\omega^6\{-c^2(k - 4m\omega^2) - (m + M)\omega^2[3k(m + M) + 4m(m - 3M)\omega^2]\}}{2\{c^2(k - 4m\omega^2)^2 + 4\omega^2[k(m + M) - 4mM\omega^2]^2\}\{c^2(k - m\omega^2)^2 + \omega^2[k(m + M) - mM\omega^2]^2\}^2} \tag{26}$$

We call $\bar{v}_2$ the drift velocity. Note that for $c = 0$, the denominator in the above is zero if $\omega = \omega_n$ or $\omega = \omega_n/2$, where

$$\omega_n = \sqrt{\frac{k}{m_e}} . \tag{27}$$

In Figure 5, the drift velocity is plotted as the forcing frequency. It is interesting to note that the drift may reserve direction upon frequency changes. The dynamics of the masses at two different frequencies are shown in Figure 6 and Figure 7.

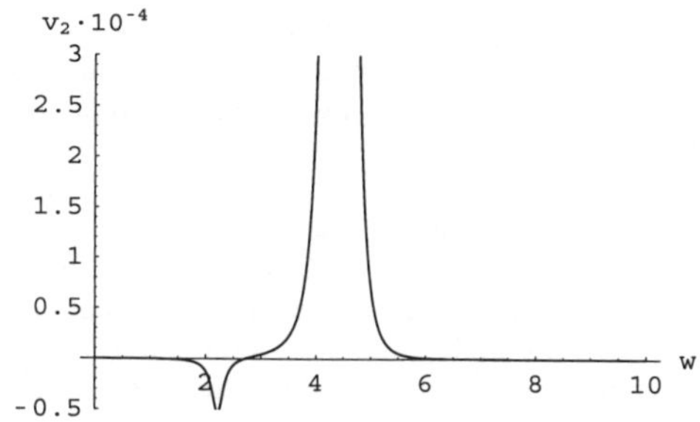

Figure 5 Translational velocity as a function of the forcing frequency. Parameters: $M = m = 1$, $k = 10$, $a = 0.25$. $F_i = v + 0.9v^3$, $c = 1$, $h = 50$.

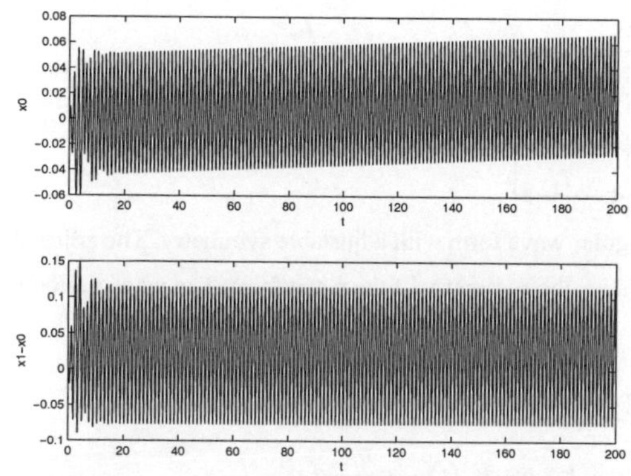

Figure 6 The dynamics when forced at the primary resonance frequency. Parameters: $M = m = 1$, $k = 10$, $a = 0.25$, $\omega = 4.47$. $F_i = v + 0.9v^3$, $c = 1$, $h = 50$.

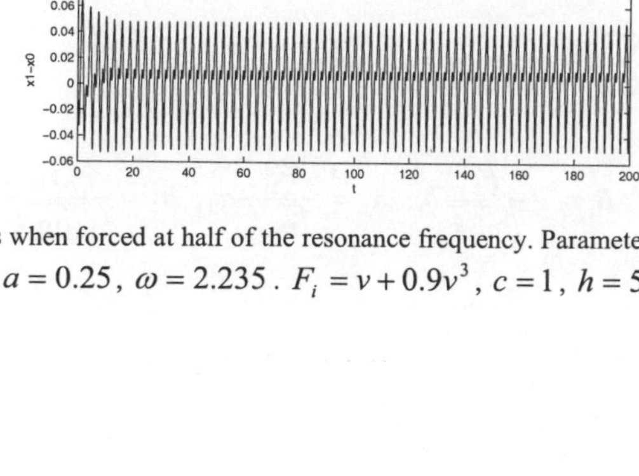

Figure 7 The dynamics when forced at half of the resonance frequency. Parameters: $M = m = 1$, $k = 10$, $a = 0.25$, $\omega = 2.235$. $F_i = v + 0.9v^3$, $c = 1$, $h = 50$.

The above analysis shows that $\bar{v}_2 = 0$ if either d or h is zero. Therefore, for symmetric nonlinear interfacial relationship, the spring must have a quadratic nonlinear term to produce propulsion under sinusoidal excitation. It is also possible to reverse the direction of the propulsion by changing the frequency of the sinusoidal excitation.

4. Self-Propulsion with Multiple Frequency Excitation

The self-propulsion is achieved through the combined effect of the nonlinear interfacial force and the nonlinear spring. A more convenient design of a self-propelled capsule is to use select wave forms as excitation to the actuators. For example, Figure 8 shows a triangular wave form. Because of the asymmetric nature of the forcing wave form, the nonlinearity in the spring is not needed.

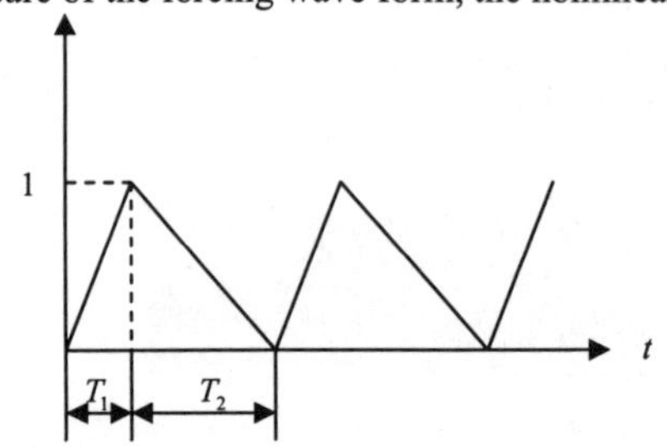

Figure 8 A triangular wave form with adjustable symmetry. The adjustable parameter $r = T_2 / T_1$ determines the symmetry of the wave form. The wave period is $T = T_1 + T_2$.

With the above assumptions, equations in (13) become:

$$M\dot{v} + cv + ky = -F_a^{'}(t) - \varepsilon d\, v^3 \tag{28a}$$

$$m_e\ddot{y} + \frac{m_e}{M}cv + ky = -F_a^{'}(t) - \varepsilon \frac{m_e}{M} d\, v^3 . \tag{28b}$$

where

$$F_a^{'}(t) = F_0 + \sum_{n=1}^{\infty} (F_n \sin n\omega t) . \tag{29}$$

Follow the same procedures in the above section. The solution at the leading order can be written as

$$v_0 = \sum_{n=1}^{\infty} (A_n \cos n\omega t + B_n \sin n\omega t) \tag{30a}$$

$$y_0 = C_0 + \sum_{n=1}^{\infty} (C_n \cos n\omega t + D_n \sin n\omega t) \tag{30b}$$

where

$$A_1 = \frac{F_1}{\sqrt{m_e k}} a_1,\ B_1 = -\frac{F_1}{\sqrt{m_e k}} b_1,\ A_2 = \frac{F_2}{\sqrt{m_e k}} a_2,\ B_2 = -\frac{F_2}{\sqrt{m_e k}} b_2$$

$$a_1 = \frac{p^3(p^2-1)\rho^3}{p^2(1-p^2)^2\rho^4 + 4[1+(p^2-1)\rho]^2\varsigma^2}$$
$$b_1 = \frac{2p^2\rho[1+(p^2-1)\rho]\varsigma}{p^2(1-p^2)^2\rho^4 + 4[1+(p^2-1)\rho]^2\varsigma^2}$$
$$a_2 = \frac{2p^3(4p^2-1)\rho^3}{p^2(1-4p^2)^2\rho^4 + [1+(4p^2-1)\rho]^2\varsigma^2}$$
$$b_2 = \frac{2p^2\rho[1+(4p^2-1)\rho]\varsigma}{p^2(1-4p^2)^2\rho^4 + [1+(4p^2-1)\rho]^2\varsigma^2}.$$

The dimensionless parameters are defined in the following:

$$\omega_n = \sqrt{k/m_e},\ p = \frac{\omega}{\omega_n},\ \rho = \frac{M}{m_e} = 1 + \frac{M}{m},\ \varsigma = \frac{c}{2\sqrt{m_e k}}.$$

The solution for v_1 contains oscillating terms and a constant. The constant is the drift velocity. When only the first two harmonic terms are included in the actuator force, the drift velocity is

$$\bar{v}_1 = -\frac{d}{c}[3(A_1^2 - B_1^2)A_2 + 6A_1B_1B_2] = -\frac{3dF_1^2F_2}{c\sqrt{(m_e k)^3}}[(a_1^2 - b_1^2)a_2 + 2a_1b_1b_2]. \quad (31)$$

The above formula can be much simplified if c is assumed to be small. Ignoring quadratic or higher order terms of c, we obtain

$$\bar{v}_1 = -\frac{3dF_1^2F_2m_e^3\omega^3}{2cM^3(-k+m_e\omega^2)^2(-k+m_e\omega^2)}.$$

In dimensionless parameters:

$$\bar{v}_1 = -\frac{3dF_1^2F_2}{2c[k(M+m)]^{3/2}}\left(\frac{1}{\rho-1}\right)^{3/2}\frac{1}{(p^2-1)^2(4p^2-1)}. \quad (32)$$

The more general result in (31) is too complex to write down explicitly. The result in (32) clearly indicates the dependence of the drift velocity upon the various physical parameters. However, it contains singularities close to $p = 1$, and $p = 1/2$ due to the omission of the damping terms.

At $p = 1$, the result in (31) also takes a simpler form:

$$\bar{v}_1 = \frac{18dF_1^2F_2M^5\omega_n}{c^3m_e[c^2m_e(m_e+3M)^2 + 36kM^4]} \quad (33)$$

Note also that this result is obtained by keeping only the first two terms in the Fourier series. If $p < 1/2$, those neglected terms may have a frequency close to the resonance frequency. The result above will again be inaccurate. But the simple result is useful in choosing parameters for maximum drift. It is not surprising that the nonlinear coefficient, d, in the interfacial force must be maximum. In the meantime, we want to minimize c, k, and M. The coefficient c limits the oscillation amplitude. Therefore the smaller the parameter c, the larger the propulsion effect. The Fourier series corresponding to the wave form in Figure 8 is

$$F_a'(t) = \frac{a}{2} + \frac{a}{\pi}\sum_{n=1}^{\infty}\left(\frac{1}{n}\sin n\omega t\right) \quad (34)$$

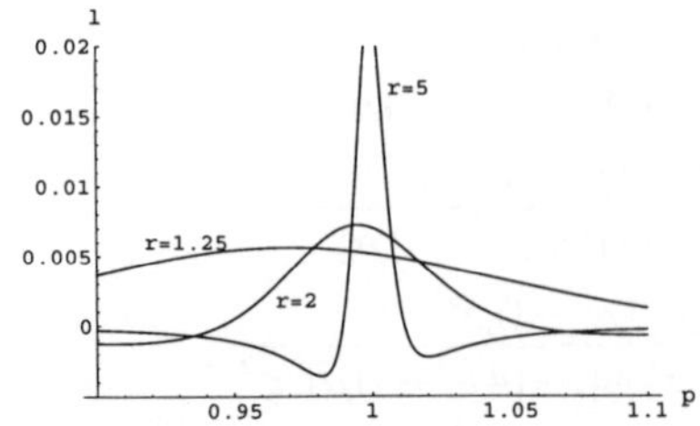

Figure 9. Drift velocity as a function of the forcing frequency. Parameters: $M+m=2$, $k=10$, $F_1=1/\pi$, $F_1=1/(2\pi)$ $F_i=v+v^3$, $c=1$. Note that the forcing frequency can change the drift direction.

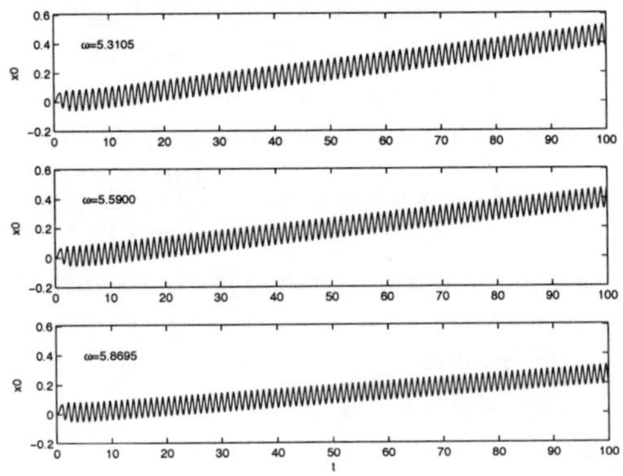

Figure 10. Displacement of the capsule at different forcing frequencies for mass ratio $\rho=1.25$. Parameters: $M+m=2$, $k=10$, $F_1=1/\pi$, $F_1=1/(2\pi)$ $F_i=v+v^3$.

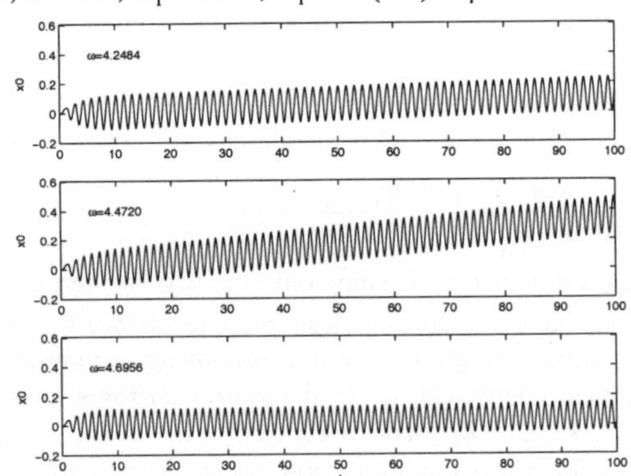

Figure 11. Displacement of the capsule at different forcing frequencies for mass ratio $\rho=2$. Parameters: $M+m=2$, $k=10$, $F_1=1/\pi$, $F_1=1/(2\pi)$ $F_i=v+v^3$.

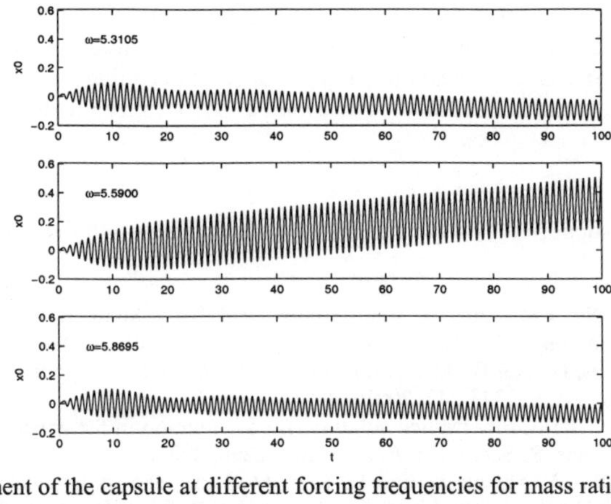

Figure 12. Displacement of the capsule at different forcing frequencies for mass ratio $\rho = 5$. Parameters: $M + m = 2$, $k = 10$, $F_1 = 1/\pi$, $F_1 = 1/(2\pi)$ $F_i = v + v^3$. Note that the drift direction changes as the forcing frequency changes.

for $T_1 = 0$. The results below are obtained for this forcing wave form. If all sinusoidal terms change algebraic signs, the symmetry of the wave form changes. The drift velocity is plotted against the forcing frequency for different mass ratios in Figure 9. It is important to note that our model includes damping only on the lubricated surface. For small mass ratios, the mass of the capsule is relatively small. The actuator will cause the capsule to move more than the inner mass. More effective damping occurs. At large mass ratio, the capsule hardly moves, the effective damping from the lubricated surface is very small. The system exhibits a sharp resonance. Although larger drift velocity appears to be possible at the resonance, the sensitivity to the forcing frequency is less desirable. The displacements of the capsule for three different mass ratios excited at frequencies are shown in Figures 10-12.

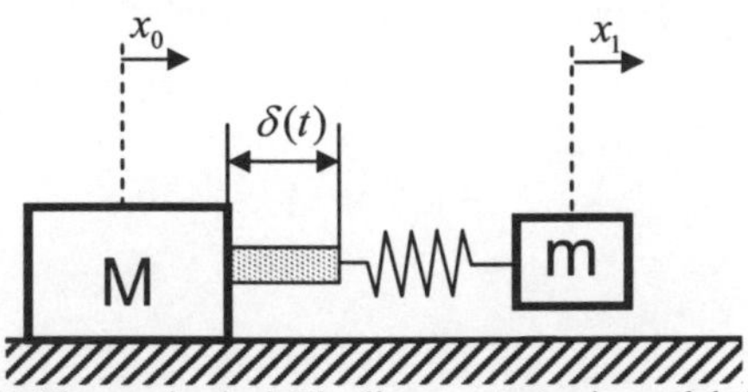

Figure 13 The mechanical. The linear actuator is modeled by a prescribed displacement in series with a spring.

5. Discussions and Conclusions

The piezoelectric actuator in this paper is assumed to be a force actuator. When the stiffness connecting the two masses is small, the actuator may be regarded as a displacement actuator as shown in Figure 13. For a linear spring, we found that the equations of motion identical to those in (4) and (5) are obtained. However, the actuator force is replaced by $F_\delta(t) = k\delta(t)$.

Assuming that the interaction force between the capsule and the surface be that of a viscous liquid, we have shown that self-propulsion is possible if the interaction force is nonlinear. The interaction force we assumed is that of a shear thickening liquid. Shear thickening is common in many liquids. Our analysis also shows how the mass should be distributed to maximize the drift velocity.

Acknowledgement

This work is supported by a grant from the Research Board of the University of Missouri.

References

[1] S.J. Phee, W.S. Ng, I.M. Chen, F. Seow-Choen, and B.L. Davies, Locomotion and steering aspects in automation of colonoscopy, IEEE Engineering in Medicine and Biology Magazine, 16, 85-96, 1997.
[2] J. Peirs, D. Reynaerts, H. Van Brussel, A miniature manipulator for integration in a self-propelling endoscope, Sensors and Actuators A, 92, 343-349, 2001.
[3] A. Menciassi and P. Dario, Bio-inspired solutions for locomotion in the gastrointestinal tract: background and perspectives, Phil. Trans. R. Soc. Lond. A, 361, 2287-2298, 2003.
[4] P. Dario, P. Ciarletta, A. Menciassi, and B. Kim, Modeling and experimental validation of locomotion of endoscopic robots in the colon, The International Journal of Robotics Research, 23, 549-556, 2004.
[5] N. Lobontiu, M. Goldfarb, and E. Garcia, A piezoelectric-driven inchworm locomotion device, Mechanism and Machine Theory, 36, 425-443, 2001.
[6] T. Higuchi, M. Watanabe, Apparatus for effecting fine movement by impact force produced by piezoelectric or electrostrictive element. US Patent Number 4894579. 1990.

Decomposition-Modeling Smart Struts for Vehicle Suspension Development

Xubin Song
Eaton Corporation, 26201 Northwestern Highway, Southfield, MI 48076, USA

Abstract

Model and simulation study is the starting point for engineering design and development, especially for developing vehicle control systems. This paper presents a methodology to develop models for application of smart struts for vehicle suspension control development. The modeling approach is based on decomposition of the testing data. Per the strut functions, the data is dissected according to both control and physical variables. Then the data sets are characterized to represent different aspects of the strut working behaviors. Next different mathematical equations can be built and optimized to best fit the corresponding data sets, respectively. In this way, the model optimization can be facilitated in comparison to a traditional approach to find out a global optimum set of model parameters for a complicated nonlinear model from a series of testing data. Finally, two struts are introduced as examples for this modeling study: magneto-rheological (MR) dampers and compressible fluid (CF) based struts. The model validation shows that this methodology can truly capture macro-behaviors of these struts.

PACS: 07.10Fq; 07.05.Tp
Keywords: Model; Decomposition; Magneto-rheological (MR); Compressible fluid; Suspension

1. Introduction

Smart materials exhibit physical features of energy transformation among electrical, magnetic, thermal, chemical and mechanical energy. Typical studied materials are piezoelectric and electrostrictive elements, electro-rheological (ER) and magneto-rheological (MR) fluid, shape memory alloy, and fiber optics. The structure embedded with smart materials as intelligent/smart structures are widely studied for the applications into structural acoustics, fluid/structure interactions, vibration absorbers in machines, helicopter rotor design, and flexible structures [1-3]. One of the significant commercial success stories is the application of MR dampers for vehicle semiactive suspensionsystem from the automotive industry. MagneRide has been jointly developed by Delphi and Lord Corp and is now becoming standard equipment for GM luxury Cadillac cars and being pushed into Buick too. For more details about MagneRide, please refer to the websites of both Lord and Delphi Corporation.

For these application studies, rigorous mathematical representations of these materials and systems are critical for researchers to best utilize their properties and potential. Quantifying

Corresponding author. Tel:+1-248-226-1738; fax:+1-248-226-6818.
Email: xubinsong@eaton.com

coupled field responses and interactions with the underlying sub-systems can be physics based formulations [4], finite element method [3], and (partial) differential equations [1,2]. In [3], a nonlinear finite element method is presented to characterize flexible structures with application of piezo-ceramic sensors/actuators. These methods look into the material properties and try to characterize the micro-natures for representing macro-behaviors of such systems. Thus, even though these methods are accurate and effective, intense computations are typically required even for modern computers to achieve numerical solutions. Considering the control system design, the sampling rate should be fast enough for better performance. However, if computational burden is too heavy, then model based control algorithms cannot be implemented for real-world applications at a competitive cost, not to say commercial success.

In the following of this paper, the focus is on generalizing the modeling methodology from the author's previous work on modeling MR dampers and compressible fluid (CF) based struts. The proposed approach should be useful for engineering study and system design of other smart materials.

2. Background

In modern vehicle suspension developments, control plays an important role to further improve vehicle safety, ride and handling. In comparison to a passive suspension, a controllable suspension can provide more properly tuned vertical compliance so the wheels can follow the uneven road, isolating the chassis from roughness in the road; can better maintain the wheels in the proper steer and camber attitudes to the road surface; can be controlled to react more properly to the dynamic forces produced by the tires – longitudinal (acceleration and braking) forces, lateral (cornering) forces, and braking and driving torques; can more reduce roll of the chassis; can keep the tires in tighter contact with the road with less load variations. Two noted technologies are MR dampers in Figure 1, and compressible fluid based struts shown in Figure 2. Like other control system development, modeling these two devices is an important aspect of engineering activities.

Magneto-rheological fluids exhibit rheological properties that are controllable by a magnetic field. This property is used in MR dampers, such as the one shown in Figure 1, to provide different damping forces according to the magnetic field that is created within the damper. The magnetic field is controlled by the electrical current supplied to the coil of the MR valve, which is commonly used to restrict the fluid flow as the damper piston moves relative to the damper body.

There exist two general methods for modeling devices, such as MR and ER dampers. One is the parametric modeling technique that characterizes the device as a collection of (linear and/or nonlinear) springs, dampers, and other physical elements. The second method, called non-parametric modeling, employs analytical expressions to describe the characteristics of the modeled devices based on both testing data analysis and device working principles. Ehrgott and Marsi [5] used the Tchebycheff function to build a non-parametric model for ER fluid. Their model is quite complicated, as it requires a large number of higher terms in order to maintain sufficient accuracy. The other way is to apply physics to understand the device mechanism and based on such analysis parametric models can be built in which the characteristics of the MR dampers are represented by a series of linear and nonlinear elements with defined parameters, such as springs and dampers. The Bingham viscoelastic-plastic model, described by Shames and Cozzarelli [4], was used in the modeling studies on ER fluid and devices such as Kamath and Wereley [6]. In another study, Spencer et al. [7] provide a parametric model for MR dampers, based on the extension of the Bouc-wen model [8]. The Bouc-Wen model is derived from a Markov-vector formulation to model nonlinear hysteresis systems. CF struts can replace coil spring and shock absorber as one device. The silicone-fluid based struts can be connected to hydraulic pumps so that CF struts can work as active actuators as ABC struts (developed by Mercedes-Benz) but cost less. Inside the cylinder the strut is filled with such as compressible silicone fluid under a certain pressure. The rod can be moved around as suspension travel so that the strut can change its spring and damping forces. On the piston, there are orifices similar to shock absorbers so that the fluid can be exchanged between two chambers to create damping forces. In [9], a parametric model for CF struts is presented. But the model does not clearly characterize the strut functions per the strut capability. Furthermore, its complexity may limit the choices of control design.

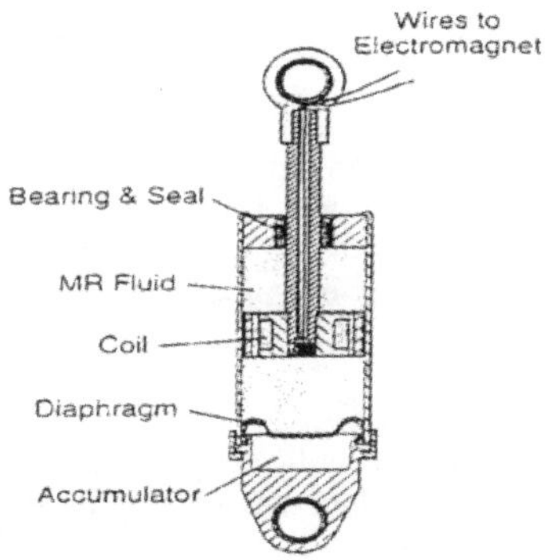

Figure 1. Magneto-Rheological (MR) Damper (In Courtesy of Lord Corporation)

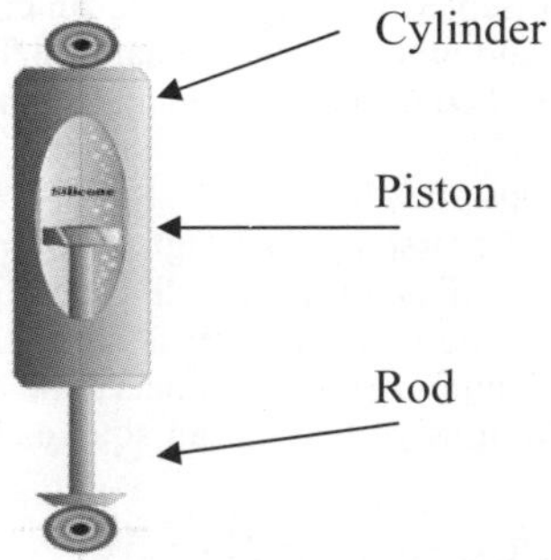

Figure 2. Compressible Fluid Struts

3. Modeling Methodology

For the automotive industry, the cost is always a priority consideration as well as reliability. One of the cost reduction efforts is to shorten development cycle for new products of both vehicle and components. That means automotive engineers have to find out approaches to quickly understand new developments and efficiently achieve new designs. There is no wonder that engineers use testing data to set up lookup-table models when how to model a new component or subsystem is not clear or too cumbersome. In this paper, a generalized methodology is proposed to analyze the available data and build a more mathematical model for unclear plants. One advantage of the mathematical formulations is that they can more clearly disclose some internal relationships between physical and controllable variables and be more computation-efficient than lookup-tables. Furthermore, since such a model can be continuous and differential as well as tractable computationally, it can be included into those advanced model-based control algorithms.

In this paper, the suspension devices are used to explain this modeling procedure. The strut behavior model can be described by three aspects: strut functionality (elastic and damping force), controls (current or hydraulic pressure), and internal variables (velocity and displacement). It is worth noting that controls and internal variables are independent dimensions for testing, or they are orthogonal in the modeling dimensions. Thus this kind of data dissection can facilitate the model validation and optimization.

Step-1 Testing Range Setup

For a vehicle suspension, physical constraints of such as suspension travel and maximum strut forces can be pre-determined for device design. When the suspension involves control functions, the working frequency range and min and max corner forces can be further used to specify the limits of control variables, for example, current for MR dampers and precharged pressure for CF struts. Then testing can be designed to cover the desirable working ranges.

Step-2 Data Decomposition

Form the above testing setup, the collected data can be used for modeling purposes. Usually the data can be presented in the data sets of amplitude vs control signal, and force vs velocity (or displacement). These data sets can disclose the fundamental relationships between strut behaviors (force) and control (current), and between strut behavior and internal variables (suspension velocity and displacement).

Step-3 Formulation Selection

A data set can be usually represented by one or a group of mathematic equation. Some examples are listed here. One simple selection is the polynomial function. Bilinear and sharp transition phenomena can be represented by a hyperbolic function $tanh(x)$. State space equations can be used to capture hysteresis and other internal dynamics. The principle is that such selection must be appropriate for model based control algorithm and simulation study.

Step-4 Rough Machining: Local Optimization

In mechanical workshop, the first step is rough machining, to machine a coarse part first and make the piece approximately satisfactorily with the specifications. The optimization of each selected formulation with the corresponding data set is similar to rough machining. This can be quickly achieved, because the decomposing process makes the optimization straightforward. That may just need to use a standard routine of least mean square (LMS) method, gradient search, or simplex method in MATLAB.

Step-5 Precision Machining: Global Optimization

Global optimization of the built model from Step 1 to 4 is to use the original testing data to continue to improve the model accuracy from Step 4. This is just like precision machining to get the final product. Since every formulation is optimized per the corresponding data set, the final optimization is to micro-tune the parameters so that the overall model accuracy can be improved to better represent the strut.

The methodology presented above is based on testing data analysis. Then differential and continuous formulations are created and/or chosen to represent data sets that are extracted from the testing data. This decomposition-modeling approach is more oriented to non-parametric but does not preclude the parametric modeling methods. In the following sections, two modeling examples are used to further elaborate this modeling methodology.

4. Example One: Modeling MR Dampers

The first example with application of the above modeling methodology is for modeling a product-oriented MR damper. The testing data is provided by the Lord Corp. The data is graphically presented in Figures 3 and 4. In Figure 3, the damper force is subjected to the control signal, the current to the MR damper from 0 to 2A. The larger the current is, the higher force peak the damper can achieve. Figure 5 shows one decomposed data set to describe the relationship between current (A) and the peak force (N). Damping force vs velocity in Figure 4 shows the natures of the MR damper: bilinear behavior and hysteresis loop near the zero velocity region. In this section, appropriate mathematical functions are proposed to capture these damper features as following.

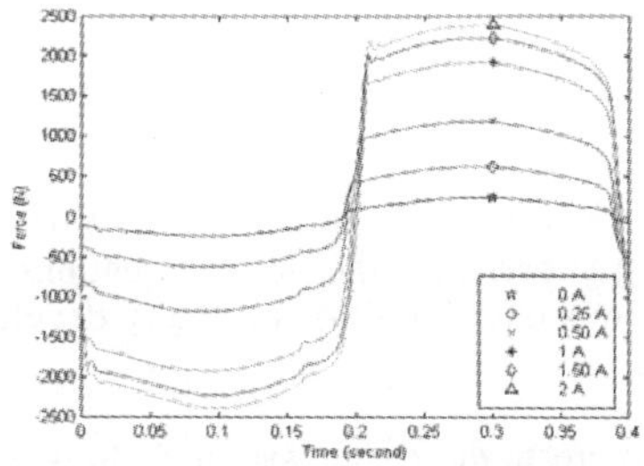

Figure 3. Force Time Trace for a MR damper

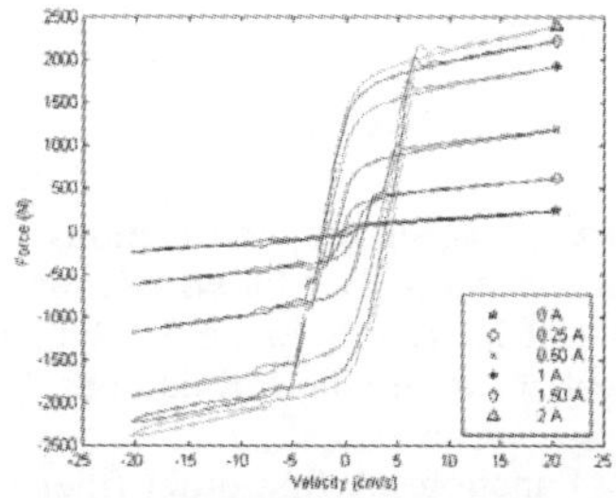

Figure 4 orce-Velocity for a MR Damper

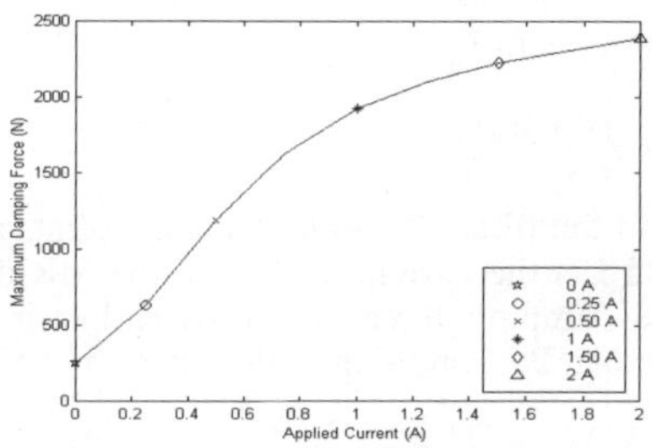

Figure 5. Maximum Damping Force vs. Applied Current for a MR Damper

1. A Polynomial Function: A function such as

$$A_{mr}(I) = \sum_{i=0}^{n} a_i I^i \tag{1}$$

is used to describe the maximum damping force as a function of the applied current. In Eq. (1), A_{mr} is the maximum damping force, a_i is the polynomial coefficients with appropriate units, n is the order of the polynomial, and I is the current applied to MR dampers. However, if such as differentiation is not required for application, this function can be replaced by spline functions, linear (or nonlinear) interpolation, and even lookup tables, though not further discussed in this paper.

2. A Shape Function: Several functions proposed in Eq. (2) are used to preserve the frequency correlation between the damper force and relative velocity across the damper, and also represent the bilinear behavior of the force-velocity curve.

$$S_b(V) = \frac{(b_0 + b_1 | V - V_0 |)^{b_2(V-V_0)} - (b_0 + b_1 | V - V_0 |)^{-b_2(V-V_0)}}{b_0^{\;b_2(V-V_0)} + b_0^{\;-b_2(V-V_0)}} \tag{2a}$$

In Eq. (2), b_i, i=0 to 2, are constants, V is the velocity across the MR damper, and V_0 is a constant. Other candidate functions can be

$$S_b(V) = \tanh[(b_1 I + b_0)V] \tag{2b}$$

$$S_b(V) = \mathrm{sgn}(V)[1-\exp(-b_0|V|/V_0)] \tag{2c}$$

Combining Eqs. (1) and (2) yields the damper force as a function of damper current and relative velocity, i.e.,

$$F_s = A_{mr}(I)S_b(V) \tag{3}$$

However, other proper functions can also be used to capture the transition change of bilinear characteristics. But based on our modeling experience, the above equations can work well even out of the testing range for dynamic system study. Eq. (2a) is actually developed from Eq. (2b) to better capture the MR damper dynamics.

3. A Delay Function: A first-order filter is used to create the hysteresis loop. In its state space form, this filter is formulated as

$$\dot{x} = -(h_0 + h_1 I + h_2 I^2)x + h_3 F_s \tag{4a}$$

$$F_h = (h_0 + h_1 I + h_2 I^2)x + h_4 F_s \tag{4b}$$

where x is a state variable of the filter, h_i, i=0 to 2, are constants and h_i, i=3 to 4 are constants or functions of current, and I is the current applied to the MR damper, as defined earlier in Eq. (1). F_h combines the damping force F_s shown earlier in Eq. (3) and the hysteresis function. It is worth noting that the condition of the filtering coefficients

$$h_0 + h_1 I + h_2 I^2 > 0$$

must be satisfied in order for Eq. (4) to be stable (i.e., have a decaying solution).

4. Offset Function: In some cases, the damping force is not centered at zero because of the effect of the gas-charged accumulator in the damper. Therefore, it is necessary to include a force bias in the model, such as

$$F_{mr} = F_h + F_{bias} \tag{5}$$

where F_{bias} represents the non-zero centered damping force, and F_{mr} is the MR damping force.

The combination of the four functions mentioned above provides a complete MR damper model, as presented in Eq. (5). For selecting model parameters in Eqs. (1) to (5), the local optimization is applied first. For example, the data set in Figure 5 is used to determine the coefficients of Eq. (3). Given a specific force vs velocity data set, the parameters in Eq. (2) can be roughly estimated. For the coefficients h_i of the delay function, actually the first step is to determine the filtering coefficient for each current, and then use LMS method to find out h_i. Finally, the testing data is applied for global optimization of the MR damper model that consists

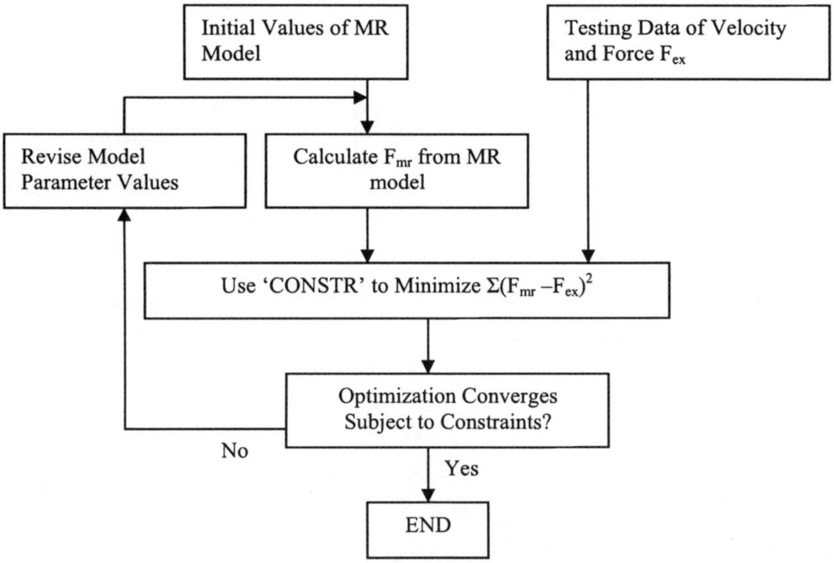

Figure 6. Flow Chart for Globally Optimizing MR Damper Model

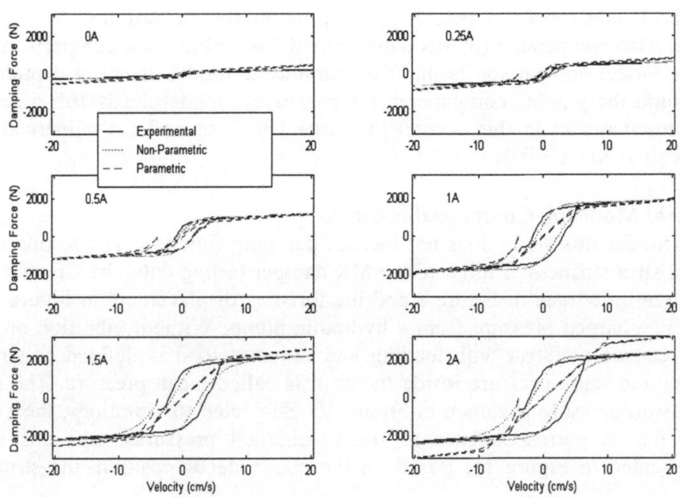

Figure 7. MR Damper Model Comparison and Validation

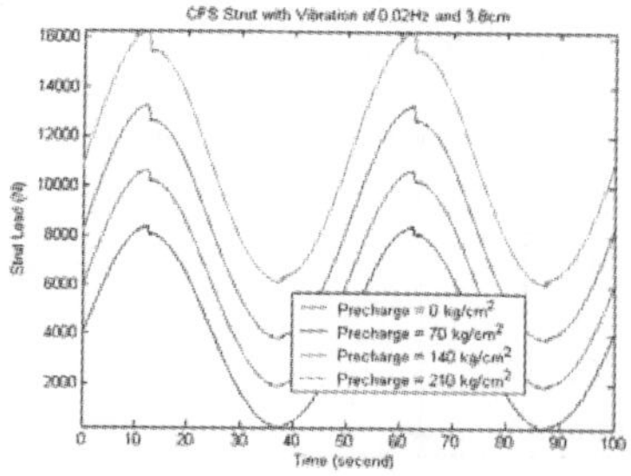

Figure 8. Strut Forces vs. Time at 0.02Hz Vibration.

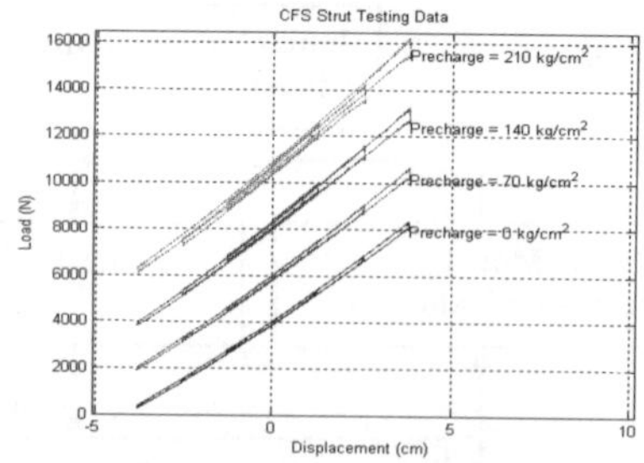

Figure 9. Strut Forces vs. Displacement withPrecharges,

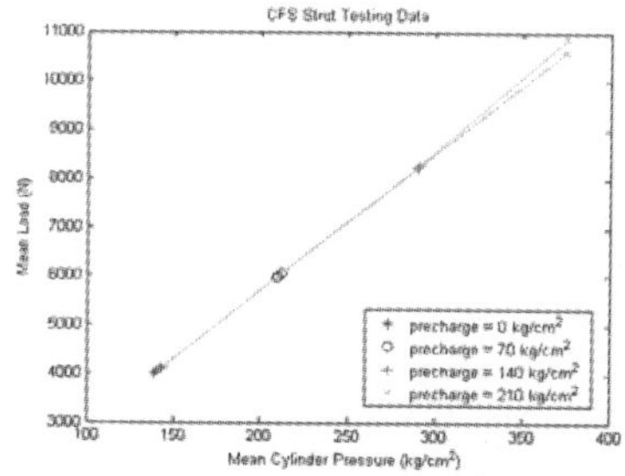

Figure 10. Mean Strut Load vs. Mean Strut Pressure.

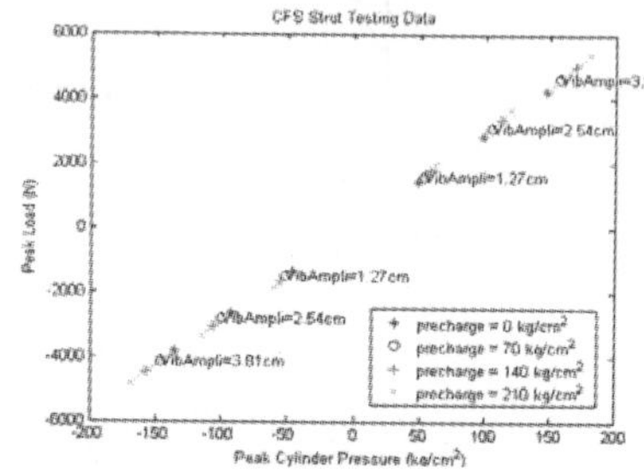

Figure 11. Peak Strut Load vs. Peak Strut Pressure

of Eqs. (1) to (5). The global optimization procedure for this case is shown in Figure 6, which uses the MATLAB constraint function. Figure 7 presents the validation of the built model from the testing data. Also one parametric MR damper model is included for comparison purposes. The reason to show model comparison is that that parametric model does not capture the damping hysteresis. Beyond the graphic comparison, the parametric model needs 10E-5 second step size while the presented model in this section just uses 10E-2 second for numerical simulation in MATLAB as well as SIMULINK.

5. Example Two: Modeling Compressible Struts

The CF strut for this study does not include damping function. The testing data is used to characterize the strut stiffness. Similar to the MR damper testing data, the CF strut testing data is presented as force time trace in Figure 8 and the force vs displacement in Figure 9. The control variable is the precharged pressure from a hydraulic pump. Without vibration on the struts, the static pressure caused the strut with loading and the strut load is defined as strut mean load. Correspondingly the static pressure inside the strut is called mean pressure. The relationship of mean load vs mean pressure is shown in Figure 10. Subjected to vibrations, the peak strut loads can be found out in correspondence to the precharged pressures, and this relationship is graphically presented in Figure 11. Based on the above decomposition, the strut force can be represented by

$$F_{CFS} = F_{mean} + F_{vib} + F_{friction} \tag{6}$$

where F_{CFS} is the total force from the CF strut, F_{mean} the mean load dependent on the prechagre levels, F_{vib} the strut load due to the vibration on the strut, and $F_{friction}$ the friction force from the relative motions between cylinder and rod. The mean strut load is assumed in this paper to have a linear relationship with respect to the mean cylinder pressure (P_{mean}) that is from the known initial cylinder pressure (P_{ic}) and the tunable external prechagre pressure (P_{pp}) as following

$$F_{mean} = \alpha P_{mean} + F_{bias} \text{ and } P_{mean} = P_{ic} + P_{pp} \qquad (7,8)$$

Finally the friction force is described by a function (or a two-dimension lookup table) that employs the F_{vib} and the P_{pp}.

Vibration-dependent load for the CF strut can be derived from Figure 10. In the figure, F_{vib} calculation is laid out, which comes from strut vibration under a certain cylinder pressure. K_{dp} and K_{pf} are used to implicitly characterize the fluid compressibility. It is worth noting that K_{dp} has two values dependent on the displacement signals, because the strut has bilinear behavior as shown in the testing data presented above. So does K_{pf}. The friction force can also be decomposed from the original testing data, and should be properly modeled.

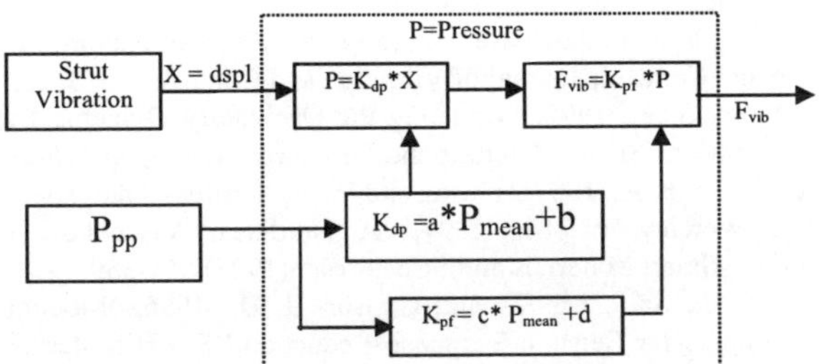

Figure 12. Model of Vibration-Dependent Strut Load

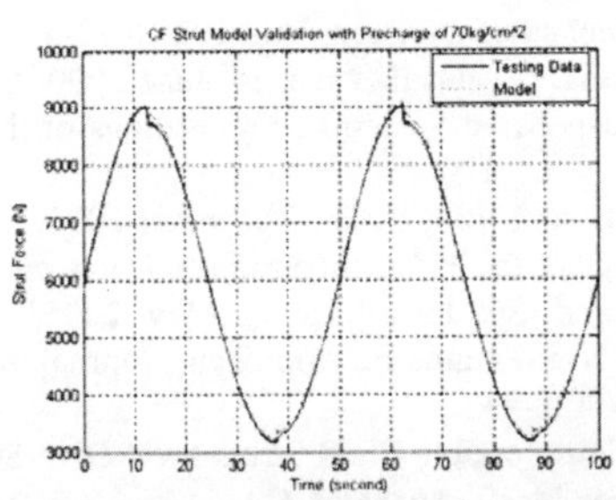

Figure 13. Validation of CF Strut Model at 70kg/cm^2 Precharge

According to the modeling procedure mentioned above, the data sets are used to rough machining of friction, and those coefficients in Equation (9) and Figure 12. Then precision machining is applied to optimizing the CF strut model. For brevity, the model is compared with the testing data at the precharge pressure 70kg/cm^2 as shown in Figure 13.

6. Summary

In this paper, the existing modeling approaches for smart materials are briefly reviewed. Then a new methodology is proposed for engineering applications, especially for automotive research and development. The paper elaborates this modeling method with MR damper and CF struts for vehicle suspensions, referring to [11, 12] to learn more about model-optimizing results. For suspension system design, usually it is possible to pre-determine the working ranges for suspension devices such as dynamic and static loads, and maximum suspension travels. In addition, the controllable devices have amplitude and bandwidth limits for control signals. Thus the testing need be properly set up to reflect these requirements in order to make the built model accurate for applications. Based on the testing data, the next critical step for this methodology is to decompose the data so that sub-models could be found out and optimized with the corresponding data sets. This facilitates the global optimization of the complete model that is integrated from the sub-models. Two examples are used to further demonstrate this modeling methodology. The model validation shows that this approach is effective. The models have shown their feasibility for model based control algorithm development, though not presented here. For those interested readers, please refer to [13, 14].

References

[1] Banks, H. T., Smith, R. C., and Wang, Y., October 1996, Smart Material Structures: Modeling, Estimation and Control (Wiley-Masson Series Research in Applied Mathematics), John Wiley & Sons

[2] Fabrizio, M., Lazzari, B., and Morror, A., December 2002, Mathematical Models and Methods for Smart Materials, World Scientific Publishing Co., Inc

[3] Song, X., Schulz, M.J., Pai, F. and Abdelnaser, A., January 1999, 'An Automated Design Technique for Nonlinear Structure and Control', Journal of Vibration and Control, Vol. 5, No. 1, pp. 123-150

[4] Ehrgott, R. C., and Marsri, S. F., 1992, 'Modeling the Oscillatory Dynamic Behavior of Electro-Rheological Materials in Shear,' Smart Materials and Structures, Vol. 4, pp.275-285.

[5] Shames, I. H., and Cozzarelli, F. A., 1997, 'Elastic and Inelasitc Stress Analysis,' Taylor & Francis

[6] Kamath, G. M., and Wereley, N. M., 1997, 'A Nonlinear Viscoelastic-plastic Model for Electrorheological Fluids,' Smart Materials and Structures, 6(1997) 351-359.

[7] Spencer, B. F., Dyke, S. J., Sain, M. K., and Carlson, J. D., 1996, 'Modeling and Control of Magnetorheological Dampers for Seismic Response Reduction,' Smart Materials and Structures, 5 (1996) 565-575

[8] Wen, Y. K., and Asce, M., April 1976, 'Method for Random Vibration of Hysteretic Systems,' Journal of the Engineering Mechanics Division, Vol. 102 No .1-3

[9] Campos, J., Lewis, F. L., Davis, L., and Ikenaga, S., June 2000, 'Backstepping Based Fuzzy Logic Control of Active Vehicle Suspension Systems', Proceedings of the American Control Conference, Chicago, IL

[10] Mattsson, S. E., Astrom, K. J., and Bell, R. D., December 1996, 'A Natural Approach to Modeling Physical Systems', Proceedings of the 35th Conference on Decision and Control, Kobe, Japan

[11] Song, X., Ahmadian, M., and Southward, S.C., May 2005, 'Modeling Magneto-Rheological Dampers with Application of Non-Parametric Approach', Journal of Intelligent Material Systems and Structures, Vol. 16, No. 5, pp. 421-432

[12] Song, X., 2004, 'Modeling Compressible Fluid Struts for Active Suspension Development', ASME IMECE2004-59590, November 14-19, Anaheim, CA

[13] Song, X., December 1999, 'Design of Adaptive Vibration Control Systems with Application of Magneto-Rheological Dampers', Dissertation, Virginia Tech

[14] Song, X., Ahmadian, M., Southward, S.C., and Miller, L., 2005, 'Superharmonic-Free Adaptive Semiactive Magneto-Rheological Suspension', ASME IMECE2005-79355, November 5-11, Orlando

A direct model reference adaptive control system design and simulation for the vibration suppression of a piezoelectric smart structure

Tamara Nestorović Trajkov *, Heinz Köppe, Ulrich Gabbert
Institute of Mechanics, Otto-von-Guericke University of Magdeburg, Universitätsplatz 2, D-39106 Magdeburg, Germany

Abstract

The paper presents control system design based on a non-linear model reference adaptive control law (MRAC) used for the vibration suppression of a smart piezoelectric mechanical structure. Numerical simulation of the proposed control system is performed based on the finite element (FE) model of the structure, modally reduced in order to meet the requirements of the control system design. The MRAC problem is defined and a direct control algorithm described in the paper is suggested as a solution to the control problem. The basic MRAC algorithm is modified by augmenting the integral term of the control law in order to provide the robustness of the control system with respect to the stability. This approach provides preserving the boundness of the system states and adaptive gains, with small tracking errors over large ranges of non-ideal conditions and uncertainties. The efficiency of the suggested control for the vibration suppression is tested and shown through a numerical simulation of the funnel-shaped piezoelectric structure.

PACS: 45.80.+r; 02.30.Yy; 46.40.–f; 07.10.Fq
Key Words: Direct MRAC system; Vibration suppression; Simulation of a piezoelectric smart structure

1. Introduction

In the development of smart structures, controller design plays a significant role. It is therefore meaningful to investigate the possibilities for the controller design and its implementation, as a part of the overall design procedure of smart structures. In this paper the model reference adaptive control (MRAC) is proposed as a method for the controller design in smart structures. Unlike in the model-based approaches, which assume the knowledge about the model (parameters) of a structure, in MRAC techniques the model parameters are not explicitly defined. The idea of the MRAC is based upon the existence of the reference model, specified by the designer, which reflects the desired behaviour of the controlled structure. The controller is designed in such a manner that the output of the controlled structure should track the output of the reference model [1], [2].

The motivation for development and application of direct MRAC to smart structures comes from the need to design a robust and stable control system using minimal prior knowledge about the controlled plant. Fixed feedback gain controllers require either a full state vector to be measurable or the use of appropriate observers [3]–[6]. On the other hand they also require the knowledge about the system parameters, which can be determined e.g. through identification procedures [5]–[7]. Development and implementation of direct MRAC techniques represent an attempt to design a simple

* Corresponding author. Tel.: +49-391-67 11724; fax: +49-391-67 12439
E-mail address: tamara.nestorovic@mb.uni-magdeburg.de (T. Nestorović)

control system [8], [9] in the sense that it will not require identifiers or observers in the control loop, maintaining at the same time and even improving the performance and robustness of the control system in comparison with some fixed gain controllers or providing the solution to control problems when the fixed gain controller cannot be applied. In the authors' opinion a direct MRAC algorithm can perform well in cases of the insufficient prior knowledge or the unknown changes of the system parameters. In direct methods the control gains are computed directly without an explicit identification of the system parameters. Therefore, due to fewer computations that have to be performed, one advantage of the direct over indirect control methods can be viewed through the reduced computational effort.

In this paper a direct MRAC algorithm is suggested for the control of piezoelectric smart structures. The model reference adaptive control problem is stated and the control algorithm described in the paper is suggested as a solution to the defined control problem. The basic MRAC algorithm is modified by augmenting the integral term of the control law [10], [11] in order to provide the robustness of the control system with respect to the stability. This approach provides preserving the boundness of the system states and adaptive gains, with small tracking errors over large ranges of non-ideal conditions and uncertainties.

This adaptive control algorithm is proposed for the application with smart structures. Testing and verification of the controlled system is performed using the example of the piezoelectrically controlled funnel-shaped shell structure, which was modelled using the finite element method (FEM) approach [12], [13].

2. Direct robust MRAC algorithm

Direct robust MRAC algorithm is derived from the general model reference adaptive tracking problem. MRAC system is based on the reference model, specified by the designer, which reflects the desired behaviour of the controlled structure (Fig. 1).

For a controlled plant described by the state space model in a general form:

$$\dot{\mathbf{x}}(t) = \mathbf{A}\mathbf{x}(t) + \mathbf{B}\mathbf{u}(t) + f_{\mathrm{x}}(t), \tag{1}$$

$$\mathbf{y}(t) = \mathbf{C}\mathbf{x}(t) + f_{\mathrm{y}}(t), \tag{2}$$

the control objective is to find without an explicit knowledge of the matrices **A**, **B** which contain the system parameters, such control law **u**(*t*) that the plant output **y**(*t*) follows the output $\mathbf{y}_m$ of a specified reference model, Eqs. (3), (4), with the least possible error.

$$\dot{\mathbf{x}}_m(t) = \mathbf{A}_m\mathbf{x}_m(t) + \mathbf{B}_m\mathbf{u}_m(t) \tag{3}$$

$$\mathbf{y}_m(t) = \mathbf{C}_m\mathbf{x}_m(t). \tag{4}$$

The notations used, have the following meanings: $\mathbf{x}(t) \in R^{n\times 1}$ represents the state vector, $\mathbf{u}(t) \in R^{m\times 1}$ is the control input vector, $\mathbf{y}(t) \in R^{m\times 1}$ is the controlled output vector, **A** and **B** are the state and the control matrices respectively, $\mathbf{x}_m(t) \in R^{n_m\times 1}$ is the state vector of the reference model, $\mathbf{u}_m(t) \in R^{m\times 1}$ is the command vector and $\mathbf{y}_m(t) \in R^{m\times 1}$ the output of the reference model. General bounded, unknown and unmeasurable plant and output disturbances are denoted with $f_{\mathrm{x}}(t)$ and $f_{\mathrm{y}}(t)$, respectively. The output tracking error is defined as:

$$\mathbf{e}_{\mathrm{y}}(t) = \mathbf{y}_m(t) - \mathbf{y}(t). \tag{5}$$

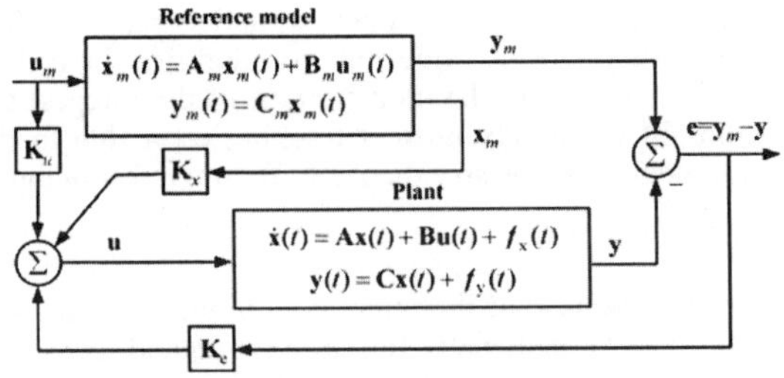

Fig. 1. General form of the MRAC system

The reference model is designed to meet some desired performance properties and it has the same number of outputs as the plant. Otherwise it is independent of the controlled plant. Further it is required that the reference model is asymptotically stable. The model is assumed to be bounded-input/bounded-state stable. Since the reference model only represents desired behavior of the controlled structure, the dimension n_m of the model state vector may be much less than the dimension n of the plant state, which is practically the case with large flexible smart structures. However, since $\mathbf{y}(t)$ should track $\mathbf{y}_m(t)$, the number of the reference model outputs has to be equal to the number of the plant outputs. Regarding these requirements, the reference model can be designed by selecting the parameters which provide asymptotic stability. Desired responses of the reference model can be obtained by an appropriate parameter selection and confirmed through an iterative simulation and through tuning procedures. The development of the adaptive control algorithm is based on the command generator tracker concept [2], which represents a control law for linear systems with known coefficients. Based on this concept the control law is expressed in the following form.

$$\begin{aligned}\mathbf{u}(t) &= \mathbf{K}_u(t)\mathbf{u}_m(t) + \mathbf{K}_x(t)\mathbf{x}_m(t) + \mathbf{K}_e(t)\mathbf{e}_y \\ &= \mathbf{K}_u(t)\mathbf{u}_m(t) + \mathbf{K}_x(t)\mathbf{x}_m(t) + \mathbf{K}_e(t)[\mathbf{y}_m(t) - \mathbf{y}(t)]\end{aligned} \tag{6}$$

For the convenience, the adaptive gains are concatenated within the $m \times n_r$ matrix of the adaptive gains $\mathbf{K}_r(t)$, while the measured values are concatenated in the $n_r \times 1$ vector $\mathbf{r}(t)$.

$$\mathbf{K}_r(t) = \begin{bmatrix}\mathbf{K}_e(t) & \mathbf{K}_x(t) & \mathbf{K}_u(t)\end{bmatrix}, \qquad \mathbf{r}(t) = \begin{bmatrix}\mathbf{e}_y(t) \\ \mathbf{x}_m(t) \\ \mathbf{u}_m(t)\end{bmatrix} \tag{7}$$

Then the control law can be written as:

$$\mathbf{u}(t) = \mathbf{K}_r(t)\mathbf{r}(t) \tag{8}$$

The adaptive gain $\mathbf{K}_r(t)$ is represented as a sum of a proportional and integral part:

$$\mathbf{K}_r(t) = \mathbf{K}_p(t) + \mathbf{K}_I(t) \tag{9}$$

According to the basic MRAC algorithm the proportional and integral gains are adapted in the following way:

$$\mathbf{K}_p(t) = \mathbf{e}_y\mathbf{r}^{\mathrm{T}}(t)\overline{\mathbf{T}}, \qquad \dot{\mathbf{K}}_I(t) = \mathbf{e}_y\mathbf{r}^{\mathrm{T}}(t)\mathbf{T}, \qquad \mathbf{K}_I(0) = \mathbf{K}_{I0} \tag{10}$$

where $\mathbf{T}$ and $\overline{\mathbf{T}}$ are $n_r \times n_r$ time-invariant weighting matrices and $\mathbf{K}_{I0}$ is the initial integral gain. Selection of the weighting matrices $\mathbf{T}$ and $\overline{\mathbf{T}}$ and the plant output matrix $\mathbf{C}$ is limited by the sufficient conditions of stability, which can be expressed in terms of the condition for the almost-strictly-positive-realness of the plant (see [2], [5]).

Since the control system is assumed to operate under real conditions in the presence of disturbances and noise, the basic MRAC algorithm is modified in order to account for nonideal realistic conditions. The idea is based on the modification of the integral term in order to prevent its divergence under conditions of constantly present tracking error due to disturbances. According to [11] the integral term is augmented by the so called σ-term in the following way.

$$\dot{\mathbf{K}}_I(t) = \mathbf{e}_{\mathrm{y}}(t)\mathbf{r}^{\mathrm{T}}(t)\mathbf{T} - \sigma\mathbf{K}_I(t) \tag{11}$$

In the robust MRAC algorithm the role of the integral gain with σ-term is to guarantee convergence, whereas the proportional term adds the penalty for large errors and leads the system to small tracking errors. Without the σ-term, the adaptive gain $K_I(t)$ is calculated using an integrator and its value can steadily increase whenever perfect tracking (with zero output error) is not possible, which is the case in a real environment in the presence of disturbances and noises. Thus the integral term can reach unnecessarily large values or can even diverge. With the σ-term the integral gain $K_I(t)$ is obtained from a first order filtering of $\mathbf{e}_{\mathrm{y}}(t)\mathbf{r}^{\mathrm{T}}(t)\mathbf{T}$ and therefore cannot diverge, unless $\mathbf{e}_{\mathrm{y}}(t)$ diverges.

3. Simulation results for the robust MRAC algorithm applied to a funnel shaped piezoelectric structure

Described direct robust MRAC algorithm was implemented for the vibration suppression of a funnel-shaped shell structure attached with piezoelectric actuators and sensors. The simulation results are presented below. The funnel represents the inlet part of a magnetic resonance tomograph used in medical diagnostic (for more details see [5]).

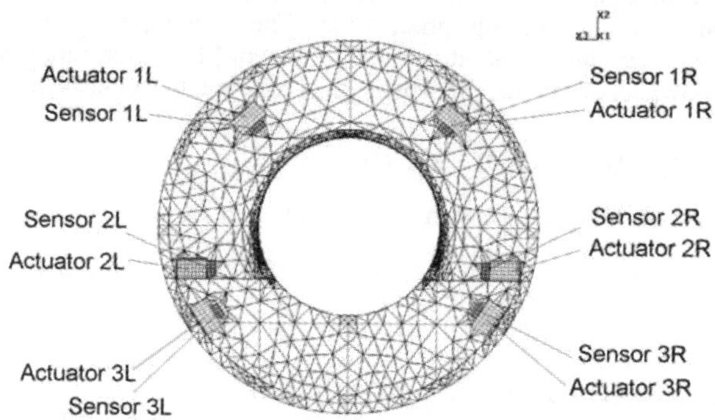

Fig. 2. The FEM mesh of the funnel-shaped structure and placement of the attached piezoelectric actuators and sensors

In order to implement and test the controller, the simulation was performed with the model of the funnel obtained using the FEM approach [12], [13]. Based on the geometry of the funnel with the attached piezoelectric actuators and sensors, the FEM model was obtained using the FEM software COSAR [14]. The resulting FEM mesh together with the actuator and sensor placement is represented in Fig. 2. The funnel and the piezoelectric patches were modeled using the Semiloof shell elements. For the modeling of the active piezoelectric material, the finite elements included also the electric degrees of freedom. After an appropriate procedure for the modal reduction, a state-space model of the funnel was obtained, which was used for the MRAC system design and the controller testing through a numerical experiment with Matlab/Simulink. In the notation used in the text for actuators and sensors A and S stand for *actuator* and *sensor*, respectively. Further, the appropriate patches are denoted as 1L, 2L, 3L for the left-hand side actuators and sensors and 1R, 2R, 3R for the right-hand side ones.

The robust algorithm with the modified integral gain is used in order to provide the integral gain convergence and the overall stability of the control system with respect to boundness of the states, adaptive gains and of the output error. The motivation for the use of the MRAC comes from the fact that a relatively simple control algorithm can provide the desired behavior of the controlled system specified by the reference model without estimation of the system parameters. Since the control algorithm is not based on the state variables measurement, but rather on the output measurement i.e. the output error, the estimation of the state variables is also not needed. The excitations in this case are sinusoidal with frequencies corresponding to the selected eigenfrequencies of the funnel (e.g f_1=9.573 Hz, f_2=23.333 Hz) obtained through the procedure of modal analysis [4], [5]. Excitations corresponding to the eigenfrequencies are assumed to be the worst case, due to the possibility of resonance.

The selected value σ=0.1 for the coefficient in the integral gain of the control algorithm provides the convergence of the adaptive gains and the boundness of the output responses and of the control inputs. For a single-input single-output case the coefficients of the reference model, Eqs. (3), (4), are chosen in the following way: $A_m = -3$, $B_m = 1$, $C_m = 3$. The actuator/sensor pair A2R–S1R is considered. In the simulation examples a zero reference input is considered. Sensor responses in the following figures represent simulated outputs of the structural model (sensor voltage signals) obtained on the basis of the finite element approach.

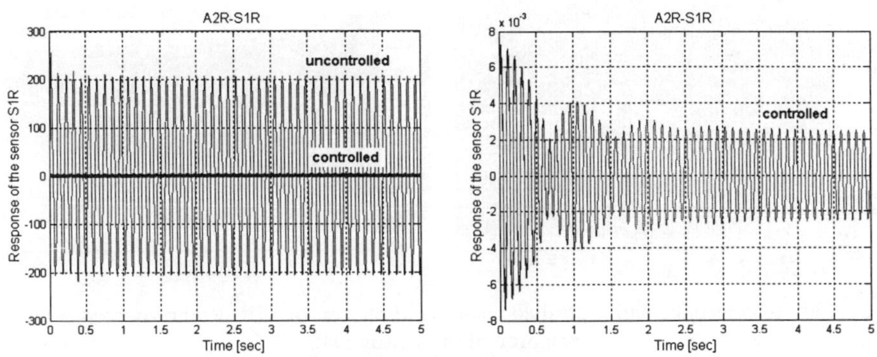

Fig. 3. Left: uncontrolled and controlled responses of the sensor S1R; Right: zoomed controlled response of the sensor S1R

First the excitation $\sin(2\pi f_1 t)$ is considered. Appropriate elements of the matrices **T** and $\overline{\mathbf{T}}$ corresponding to integral and proportional gains in Eq. (10) respectively, are chosen to be 1000. The uncontrolled and the controlled output are represented in Fig. 3. The left portion of the figure represents the uncontrolled and controlled responses in the same diagram. The control signal and the adaptive gain K_{e_y} are represented in Fig. 4.

The behavior of the controlled outputs and adaptive gains was investigated through many numeric experiments for different values of the elements in the matrices **T** and $\overline{\mathbf{T}}$. Through the simulation it was observed that for the higher values of the elements in **T** and $\overline{\mathbf{T}}$, the time response of the output is reduced and the adaptive gain is increased. With approximately the same control effort, increasing the **T** and $\overline{\mathbf{T}}$ elements 10 times, results in a reduction of the controlled output and in increasing the adaptive gain approximately 2 times.

With the excitation $\sin(2\pi f_2 t)$ corresponding to the second selected eigenfrequency, the simulation results for the sensor and actuator signals are represented in Fig. 5 (σ=0.1 and the elements of **T** and $\overline{\mathbf{T}}$ are 1000). Similarly like in the case of the excitation with the first selected eigenfrequency, for greater values of the elements in **T** and $\overline{\mathbf{T}}$ with approximately the same control effort, smaller output responses are obtained with increased adaptive gains.

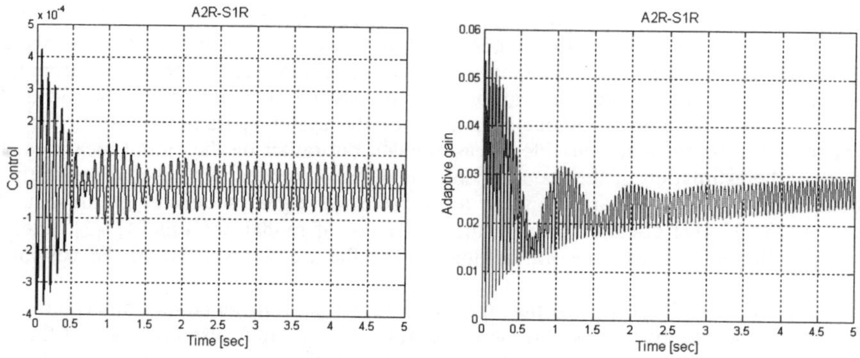

Fig. 4. Control and adaptive gain (elements of the matrices $\mathbf{T}$ and $\overline{\mathbf{T}}$ are 1000)

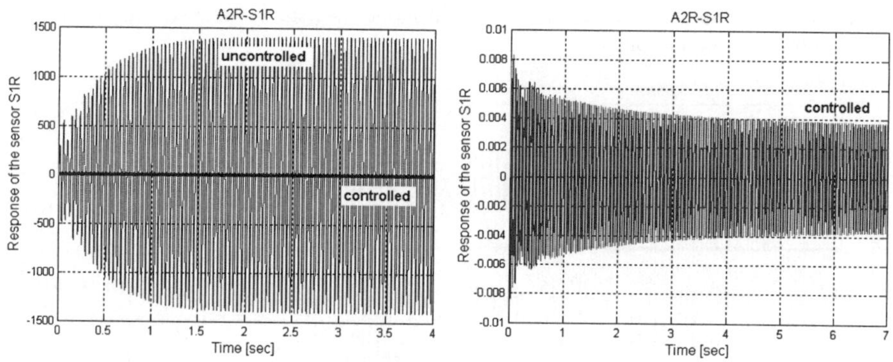

Fig. 5. Left: uncontrolled and controlled response of the sensor S1R; Right: zoomed controlled response of the sensor S1R

Obtained results show a considerable vibration suppression of the controlled system. An appropriate choice of the matrices $\mathbf{T}$ and $\overline{\mathbf{T}}$ provides suppressed vibration magnitudes in comparison with the uncontrolled system. On the other hand special attention has to be paid to the required control effort and calculation time for the control algorithm to be executed during the sampling intervals. In order to control higher frequencies, higher sampling frequencies are required which in turn imposes limits to real time applications.

4. Conclusion

Direct MRAC was suggested as a control technique for the dynamic vibration control of piezoelectric structures. The technique does not require explicit identification of the model parameters. The control law is based on states and inputs of the reference model and on the output error. Other advantages of the method are: stability in the presence of disturbances and unmodelled dynamics; specification of the desired performance through the reference model; the order of the controlled system can be much larger than the order of the reference model. The suggested controller was tested in the numeric experiment based on the state space model obtained using the FEM approach. The results show vibration suppression and robustness in the sense of the boundness of the output error, states and adaptive gains. In order to guarantee the robust stability, perfect tracking is not obtained in general, but the adaptive controller maintains a small tracking error over large ranges of nonideal conditions and uncertainties. Due to fast calculations which are required for the real time implementation of the control law and the objective technical limitations of the computer system, the effect of the control was tested and shown through the simulation results using the FEM model of a funnel shaped piezoelectric structure.

References

[1] P.N. Paraskevopulos, Digital Control Systems, Prentice Hall, New Jersey, 1996.
[2] H. Kaufman, I. Barkana, K. Sobel, Direct adaptive control algorithms: theory and applications, Second Edition, Springer-Verlag, New York, 1998.
[3] T. Nestorović Trajkov, U. Gabbert, H. Köppe, Vibration Control of a Plate Structure Using Optimal Tracking Based on LQ Controller and Additional Dynamics, in (ed.) Fabio Casciati, Proceedings of the 3rd World Conference on Structural Control, 7-12 April 2002, Como, Italy, Vol. 3, John Wiley & Sons Ltd., 2003, pp. 85-90.
[4] T. Nestorović Trajkov, U. Gabbert, H. Köppe, Controller design for a funnel-shaped smart shell structure, in (ed.) Katica (Stevanović) Hedrih, Facta Universitatis, Series mechanics, automatic control and robotics, Special issue: Nonlinear Mechanic, Nonlinear Sciences and Applications II, 3(15), 2003, pp. 1033-1038.
[5] T. Nestorović, Controller Design for the Vibration Suppression of Smart Structures, Fortschritt-Berichte VDI Reihe 8, Nr. 1071, VDI Verlag, Düsseldorf, 2005.
[6] T. Nestorović Trajkov, U. Gabbert, Active control of a piezoelectiric funnel-shaped structure based on the subspace identification, Structural Control and Health Monitoring, John Wiley & Sons, Published online 3 Oct 2005, DOI 10.1002/stc.94.
[7] P. Van Overschee, B. De Moor, Subspace Identification for Linear Systems: Theory, Implementation, Applications, Kluwer Academic Publishers, Boston, 1996.
[8] K. Sobel, H. Kaufman, L. Mabius, Model reference output adaptive control systems without parameter identification, Proceedings of the 18th IEEE Conference on Decision and Control, 1979, pp. 347-351.
[9] I. Bar-Kana, Adaptive control – a simplified approach, in (ed.) C. Leondes, Control and Dynamic Systems – Advances in Theory and Applications, XXV, 1987, 187-235.
[10] P. Iannou, J. Sun, Robust Adaptive Control, Prentice-Hall, Inc., 1996.
[11] P. Iannou, P. Kokotovic, Adaptive Systems with Reduced Models, Springer-Verlag, Berlin, 1983.
[12] H. Berger, H. Köppe, U. Gabbert, F. Seeger, On Finite Element Analysis of Piezoelectric Controlled Smart Structures, in (eds.) U. Gabbert, H.S. Tzou, Smart Structures and Structronic Systems, Kluwer Academic Publishers, 2001, pp. 189–196.
[13] U. Gabbert, H. Köppe, T. Nestorović Trajkov, Controller Design for Engineering Smart Structures Based on Finite Element Models, SPIE's 9th Annual International Symposium on Smart Structures and Materials, San Diego, Proceedings of SPIE, 4693, 2002, pp. 430-439.
[14] COSAR – General Purpose Finite Element Package: Manual, FEMCOS mbH Magdeburg (see also http://www.femcos.de)

A New Methodology of Modeling a Novel Large-scale Magnetorheological Impact Damper

Yancheng Li, Jiong Wang* and Linfang Qian
School of Mechanical Engineering, Nanjing University of Science & Technology, Nanjing 210094, China

Abstract

Because of its significant traits, magnetorheological (MR) damper becomes to be one of the most promising devices for vibration reduction. Many investigations have been done in the fields as automobiles, civil engineering and medical treatment. However, the applications of vibration-reduction under impulsive loads, which are essential for practical uses such as rocket launcher, weapon recoil system and many other applications are not well explored. A lot of dynamic models have been developed to describe the dynamic characteristics of MR damper for its employment when the load is random and smooth. While, when the loads are impulsive, little dynamic model can be used to describe the dynamic behaviour of the MR damper. In this paper, a novel MR impact damper for impulsive load with two damping passages in series, four long-thin flow passages in the piston head was exhibited and the model of this impact damper was developed for its use under impulsive load. Inertia damping force, which is caused by abrupt acceleration of the damper, was introduced. It is indicated that damping force of this novel MR damper generated is quite large, while the dynamic range of this impact damper is relative small.

Keywords: MR Impact Damper; Nonlinear Model; Large-scale; Impulsive Load;

1. Introduction

The significant characteristic of MR fluids, which is the ability to reversibly change in milliseconds from Newtonian liquids to semi-solids having controllable yield strength when exposed to a magnetic field [1], gives the MR impact dampers a broad field of application. Many models, such as Bingham plasticity model[2], Bouc-wen[3] model and Herschel-Bulkley[4-5] model, are used to describe the dynamic behaviour of the MR damper in all conditions except that the loads are impulsive. The impulsive load, which has a large amplitude and short period of time, can be easily found in almost every cases of modern engineering. Ahmadian[6]and Wang[7] studied the dynamic performance of the damper under impulsive and impact load and found that there is a uncontrollable region that the performance of the damper can't be controlled by changing the applied currents. This phenomenon can't be well explained by above models.

In this paper, a novel MR impact damper with two damping passages in series, four long-thin flow passages in the piston head was exhibited and the model of this impact damper was developed. Inertia damping force, which is caused by abrupt acceleration of the damper and can not be neglected in the cases of impulsive loads, was introduced. It is indicated that damping force of this novel MR damper generated is quite large, while the dynamic range of this impact damper is relative small.

2.Innovative Structure of MR Impact Damper

A MR impact damper is an intelligent device whose damping force can be changed by altering the

Corresponding author. Tel: +86-25-84315327; Fax: +86-25-84315327.
E-mail adress: wjiongz@mail.njust.edu.cn (Jiong Wang)

magnetic density in the magnetic coils, leading to the continuous change of viscosity in the piston gap. Typically, the structures of MR impact damper are divided by the differences as double-ended and single-ended or twin tube and monotube. The accumulator is the difference between the double-ended MR impact damper and single-ended ones and it provides a barrier between the MR fluid and a compressed gas (usually nitrogen) that is used to accommodate the necessary volume changes. The work principle of the accumulator is researched by James A Norris et al.[8], it is revealed that the accumulator works only before the accumulator piston reaches the stops and the contribution of the accumulator force to the "damping force" will be constant until the pressure exerted by the MR fluid on the accumulator is reduced to the below a maximum pressure. However, in the case of high flow velocity, the accumulator cannot accommodate the necessary volume changes within few milliseconds. A long-stroke single-ended MR impact damper (without the accumulator) is brought forward firstly in this paper, as shown in Fig1.

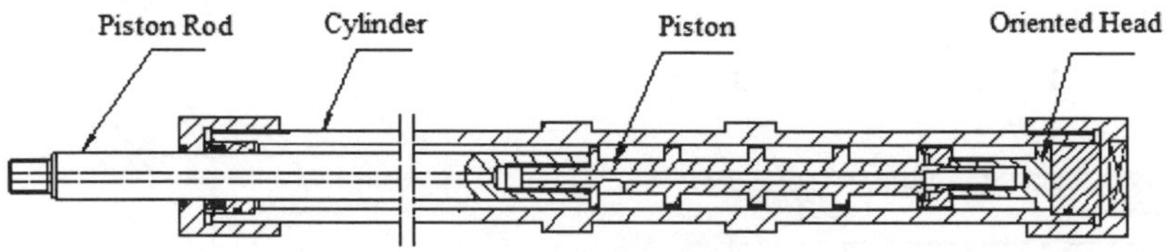

Figure 1. Sketch of the MR Impact Damper

The damper uses a particularly simple geometry in which the outer cylindrical housing is part of the magnetic circuit. The effective fluid orifice is the annular space between the outside of the piston and the inside of the damper cylinder housing. Movement of the piston causes fluid to flow through this entire annular region. In order to improve strength and uniformity of magnetic field, more than one magnetic coil is used in large-scale MRF damper. In the end of the piston rod, an oriented head was used to lead the way of piston. Four flow orifices were arranged centrosymmetrically, which also arise the damping force when the fluid flows through these orifices.

3. Modeling of MR Impact Damper

During the motion of the MR impact damper piston, fluid flows through the annular gap from high pressure chamber to low pressure chamber. It should be noted that the flow in the gap is a non-steady flow because of the jerk of the impulsive force. During this analysis, a quasi-static flow model is proposed to describe the flow of MR fluid in the damper. For the quasi-static analysis of MR fluid dampers, assume that: (1) MR fluid flow is fully developed and cannot be compressed; (2) volume change of the damper is neglected because of the volume of piston rod is so small compared with total volume; and (3) a simple Bingham plasticity model may be employed to describe the MR fluid behavior.

A simple Bingham plasticity model is effective in describing the essential field dependent fluid characteristic [9]. In this model, the total shear stress is given by

$$\left.\begin{array}{ll} \tau = \tau_0(H)\,\mathrm{sgn}(\dot{\gamma}) + \eta\dot{\gamma} & |\tau| > |\tau_0| \\ \dot{\gamma} = 0 & |\tau| < |\tau_0| \end{array}\right\} \tag{1}$$

where τ_0 is the yield stress caused by the applied field; $\dot{\gamma}$ is the shear strain rate; H is the amplitude of the applied magnetic field; and η is field-independent post-yield plastic viscosity, which is defined as the slope of the measured shear stress versus the shear strain rate. Due to its simplicity, the Bingham model is very effective, especially in the damper design phase.

Velocity Profile

The damper structure is shown in Fig 2 and the structure of oriented head is shown in Fig 3. During the motion of the piston, volume change is neglected because of the volume of piston rod is so small compared with total volume. Since MR fluid is uncompressed, a volume balance exists in the damper. Supposed that the velocity of the piston is v, according to volume balance

$$A_P v = A_g v_g \tag{2}$$

Wherein, v_g is the average velocity of MR fluids in the gap; A_P is active area of piston; and A_g is the area of the gap.

Noted that

$$A_P = \pi(d_1^2 - d_2^2)/4 \text{ and } A_g = \pi(D^2 - d_1^2)/4 \tag{3}$$

Where, D is the diameter of the cylinder; d_1 is the diameter of piston; d_2 is the diameter of piston rod.

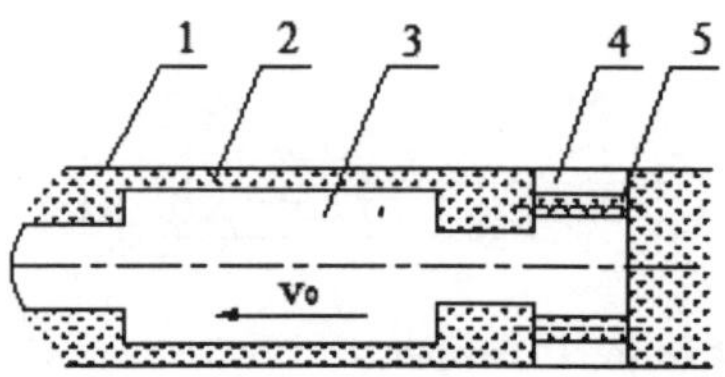

1 Cylinder 2 MR Fluid 3 Piston 4 Oriented Head 5 Flow Orifice

Figure 2 Velocity profile of the MR impact damper

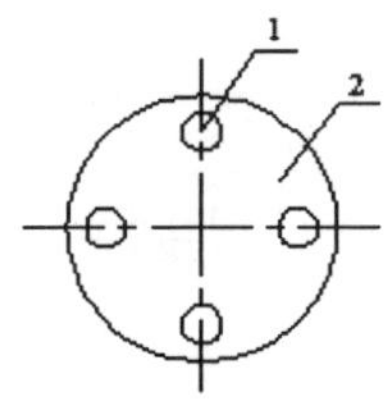

1 Flow orifice 2 Oriented head

Figure 3. Sketch map of oriented head

Average velocity of the MR fluids in the orifice can also be obtained by the volume balance

$$A_h v = m A_o v_o \tag{4}$$

Where, m is the number of the orifice, here, m=4; v_o is the average velocity in the orifice; A_h is active area of oriented head; and A_o is the area of the orifice.

Noted that

$$\left.\begin{aligned} A_h &= \pi(D^2 - d_h^2)/4 \\ A_o &= \pi d_o^2/4 \end{aligned}\right\} \tag{5}$$

Wherein, d_h is the diameter of oriented head rod; d_o is the diameter of orifice.

Fluid Inertia

Fluid inertia is the force caused by flow acceleration. If the flow velocity is relative low, the fluid inertia can be neglected. However, in impact damper, the impulsive force is very large which lead to an extremely large acceleration. Because of the effect of a large acceleration, fluid inertia cannot be ignored. In fluid mechanics, the fluid inertia[10] can be expressed as:

$$F_i = C_i \dot{v}_x(r) \quad \text{and} \quad C_i = l\rho \frac{A_v^2}{A_g} \tag{6,7}$$

where l is the length of the flow passage; ρ is the density of MR fluid; A_v is the area of the cylinder section;.It should be noticed that the fluid inertia is opposite to the direction of flow acceleration.

In this impact damper, there are two flow passages which are the gap and the flow orifice, so equation (3) is made up of two different parts

$$F_I = F_{IP} + F_{IO} = L_g \rho \frac{A_v^2}{A_g} \dot{v}_g(r) + mL_O \rho \frac{A_v^2}{A_o} \dot{v}_o(r) \tag{8}$$

Substitution of equation (2) (3) (4) (5) into equation (8) yields

$$F_I = \rho L_g A_p \frac{A_v^2}{A_g^2} \dot{v} + \rho L_O A_h \frac{A_v^2}{A_o^2} \dot{v} = C_1 \dot{v} \tag{9}$$

$$C_1 = \rho A_v^2 (L_g \frac{A_p}{A_g^2} + L_O \frac{A_h}{A_o^2}) \tag{10}$$

Where L_g and L_O are the length of the gap and the orifice; A_g and A_O are the section area of the gap and the orifice; $v_g(r)$ and $v_o(r)$ are the flow velocity in the gap and the orifice; m is the number of the orifices.

Viscosity damping force

Viscosity damping force is the force comes from the pressure drop through two flow passages. There are three viscosity damping force, which are the forces exist in the front of the flow passages and within the flow passages. It is noted that MR fluid in the gap is under the action of a magnetic field, so the force yields will be taken account of in the section 4.4.

$$F_\eta = F_{\eta g} + F_{\eta o} + F_{\eta l} \tag{11}$$

Where, $F_{\eta g}$ is throttle damping force just in the front of the gap, which arises from the volume change between the cavity and the gap. $F_{\eta o}$ is throttle damping force just in the front of the orifice, which arises from the volume change between the cavity and the orifice. $F_{\eta o}$ is damping force in the orifice.

In front of the gap, flow flux of the fluid can be expressed as:

$$Q_g = A_v \cdot v = C_d \gamma_1 A_v \sqrt{\frac{2|\Delta P_v|}{\rho}} \operatorname{sgn}(\Delta P_v) \tag{12}$$

So

$$F_{\eta g} = A_v \cdot \Delta P_v = \frac{\rho A_v^3 v^2}{2C_d^2 \gamma_1^2 A_g^2} \operatorname{sgn}(v) \tag{13}$$

Supposed that

$$C_2 = \frac{\rho A_v^3}{2C_d^2 \gamma_1^2 A_g^2} \operatorname{sgn}(v) \tag{14}$$

Then

$$F_{\eta g} = C_2 v^2 \tag{15}$$

Where, C_d is dynamic load coefficient, γ_1 is the equivalent coefficient of gap section.

In front of the orifice, flow flux of the fluid can be expressed as:

$$Q_h = A_h \cdot v_0 = C_d \gamma_1 A_o \sqrt{\frac{2|\Delta P_o|}{\rho}} \operatorname{sgn}(\Delta P_o) \tag{16}$$

So

$$F_{\eta o} = A_h \cdot \Delta P_o = \frac{\rho A_h^3 v_0^2}{2C_d^2 \gamma_2^2 A_o^2} \operatorname{sgn}(v) = \frac{\rho A_h^5 v^2}{2m^2 C_d^2 \gamma_2^2 A_o^4} \operatorname{sgn}(v) \tag{17}$$

Supposed that

$$C_3 = \frac{\rho A_h^5}{2m^2 C_d^2 \gamma_2^2 A_o^4} \operatorname{sgn}(v) \tag{18}$$

Where, γ_2 is the equivalent coefficient of orifice area. Then

$$F_{\eta o} = C_3 v^2 \tag{19}$$

When MR fluid flows through the orifices, it behaves as a Newton liquid because that there is no magnetic field. So

$$F_{\eta l} = m A_o \eta \dot{\gamma} \tag{20}$$

Coulomb damping force

Coulomb damping force is the damping force due to changing viscosity of MR fluids exposed to magnetic field. As mentioned before, the flow of MR fluid in the gap can be considered as flow between two parallel plates. So

$$F_{MR} = \pi dL \cdot \tau = \pi dL(\tau_0 \operatorname{sgn}(\dot{\gamma}) + \eta \dot{\gamma}) = \frac{\pi dL \eta}{h} v_g + \pi dL \tau_0 \operatorname{sgn}(\dot{\gamma}) \tag{21}$$

So

$$F_{MR} = \frac{\pi dL \eta A_p}{h A_g} v + \pi dL \tau_0 \operatorname{sgn}(\dot{\gamma}) \tag{22}$$

Where, d is the average girth of the gap, L is the total length of the gap filled with the magnetic field vertical with flow direction.

Friction

Friction between piston rod and MR fluid should be noticed due to a very high velocity during the employment of the impact damper. This friction can be write as

$$F_f = F_{fe} \operatorname{sgn}(\dot{v}) \tag{23}$$

Wherein, F_{fe} is the estimated value of friction. It is noted that the direction of friction is aligned with the velocity of piston.

4. Controllable force and dynamic range

Controllable force and dynamic range are the two most important variables in evaluating the overall performance of the MR damper. Controllable force F_τ is the force due to the field induced yield stress τ_0. Coulomb damping force is the only force which can be controlled by changing the current in the coil. The dynamic range is defined as the ratio between the total damper output force F and the uncontrollable force F_{uc}. In general, the uncontrollable force F_{uc} includes the fluid viscosity force F_η and friction force F_f. Therefore the dynamic range D_r can be calculated by

$$D_r = \frac{F}{F_{uc}} = \frac{F_I + F_\eta + F_{MR} + F_f}{F_I + F_\eta + F_f} = 1 + \frac{F_{MR}}{F_I + F_\eta + F_f} \tag{24}$$

While dynamic range of a normal MR damper is

$$D_r = \frac{F}{F_{uc}} = \frac{F_{MR} + F_\eta + F_f}{F_\eta + F_f} = 1 + \frac{F_{MR}}{F_\eta + F_f} \tag{25}$$

It is obvious that dynamic range of this novel large-scale MR damper is smaller than the normal MR damper because of the effect of fluid inertia and additional structure force.

The total damping force is

$$F_d = C_1\dot{v} + (C_2 + C_3)v^2 + mA_o\eta\dot{\gamma} + \frac{\pi dL\eta A_p}{A_g h}v + \pi dL\tau_0 \operatorname{sgn}(\dot{v}) + F_{fe}\operatorname{sgn}(\dot{v}) \tag{26}$$

From equation (26) we can see that the uncontrollable damping force is connected with the velocity and the acceleration of piston, which can help us to understand the uncontrollable region of MR damper under impact loads.

5. Conclusions

In this paper, a novel MR impact damper with two damping passages in series, four long-thin flow passages in the piston head was exhibited and the model of this impact damper was developed. Inertia damping force, which is induced by abrupt acceleration of the damper and can not be neglected in the cases of impulsive loads, was introduced. It is indicated that damping force of this novel MR damper generated is quite large, while the dynamic range of this impact damper is relative small.

References

[1] G. Yang, B.F. Spencer, J.D. Carlson, and M.K. Sain, Large-scale MR fluid dampers: modeling and dynamic performance considerations. Engineering Structures. 24 (2002) 309-323.

[2] G. A. Dimock, Jason E. Lindler, Norman M. Wereley, Bingham Biplastic Analysis of Shear Thinning and Thickening in Magnetorheological Dampers, In: Norman M. Wereley (Eds), Proceedings of SPIE, 3985 (2000) 444-455.

[3] B.F. Spencer Jr,S.J.Dyke,M.K.Sain and J.D.Carlson, Phenomenological Model of a Magnetorheological Damper, J. Engineering Mechanics, 10(1996)1-23.

[4] Lee DY, Wereley NM. Analysis of electro- and magneto-rheological flow mode dampers using Herschel–Bulkley model. In: In: T.Tupper Hyde (Eds.), Proceedings of SPIE Smart Structure and Materials Conference, Newport Beach, CA, 2000, pp. 244–52.

[5] Wang X, Gordaninejad F. Study of field-controllable, electro- and magneto-rheological fluid dampers in flow mode using Herschel–Bulkley theory. In: T.Tupper Hyde (Eds.), Proceedings of SPIE Smart Structure and Materials Conference, Newport Beach, CA, 2000, pp.232–43.

[6] Mehdi Ahmadian, James A Norris. Rheological controllability of double-ended MR dampers subjected to impact loading. Proc. of SPIE, 5386(2004) 185-94.

[7] Jiong Wang and Yancheng Li.Dynamic Simulation and Test Verification of MR Shock Absorber under Impact Load. Journal of Intelligent Material Systems and Structures, 17(2006) 309-14.

[8] Norris, J.A. and Ahmadian, M. Behavior of Magnetorheological Fluids Subject to Impact and Shock Loading, In: Proceedings of IMECE'03-2003 ASME International Mechanical Engineering Congress, Washington, DC, USA, 2003, pp. 1–5.

[9] Phillips RW. Engineering applications of fluids with a variable yield stress. PhD thesis, University of California, Berkeley,CA, 1969.

[10] Ping Zhang, Nonlinear impulsive vibration-reduction dynamics and design, first ed., National Defenses Publishing Company, Beijing, 2003.

Nonlinear Characteristics of Magnetorheological Damper under Base Excitation

Yancheng Li, Jiong Wang, Linfang Qian
School of Mechanical Engineering, Nanjing University of Science and Technology, Nanjing 210094, P.R.China

Abstract

As technology advances, semi-active suspensions continue to gain considerable attention. MR damper has been proved to be a promising new device employed in vehicle primary suspension. However, Magnetorheological (MR) damper is a typical nonlinear device and has the characters such as inherent nonlinearity, time-delaying and uncertainty. Nonlinearity of MR damper is of great influence on its behaviors. The primary purpose of this study is to analyze the nonlinear dynamic behavior of MR damper under base excitation in order to design a more effective MR damper for vehicle use. The response of a harmonic base-excited isolation system using MR damper, representing a quarter car suspensions is studied use the method of revised harmonic solution. Dynamic characteristics of MR base-excited isolation system in different conditions are analyzed and compared. Some useful suggestions are brought forward for designing a more effective MR damper to avoid the nonlinearity during the process of design.
Keywords: Magnetorheological Damper; Nonlinear Performance; Base Excitation;

1. Introduction

As technology advances, semi-active suspensions continue to gain considerable attention. MR damper has been proved to be a promising new device employed in vehicle primary suspension [1-2]. A lot of analytical models ranging from a single suspension to a complete vehicle model has often been used to access the performance and befits of such systems [3-4]. The majority of these studies have examined how well semi-active dampers can compromise between ride comfort and vehicle handling, as compared to passive and full active dampers [5]. Recently, nonlinear behaviors of vehicle suspensions incorporating with MR damper gain noticeable attentions [6-8] since it can highly affect the performance of vehicle suspension. A variety of researches focus its issues on the influence of nonlinear behavior on both ride comfort and vehicle handling [9]. However, little research has focused on how this compromise can be approached on the process of designing a MR damper.

The primary purpose of this study is to analyze the nonlinear dynamic behavior of MR damper under base excitation in order to design a more effective MR damper for vehicle use. The response of a harmonic base-excited isolation system using MR damper, representing a quarter car suspensions is studied use the method of revised harmonic solution [10]. Dynamic characteristics of MR base-excited isolation system in different conditions are analyzed and compared. Some useful suggestions are brought forward for designing a more effective MR damper to avoid the nonlinearity during the process of design.

* Corresponding author. Tel: +86-25-84315327; Fax: +86-25-84315327.
E-mail adress: wjiongz@mail.njust.edu.cn (Jiong Wang)

2. Bingham Model

To analyze the dynamic performance of MR damper, Bingham model is the most widely used one to describe the behaviors of the MR fluid. In this model, the relationship [11] between shear stress and strain is:

$$\tau = \tau_y \operatorname{sgn}(\frac{du}{dt}) + \eta \frac{du}{dt} \tag{1}$$

Wherein, τ is the shear stress of MR fluid; τ_y is the yield stress of MR fluid and is the function of magnetic density; η is dynamic viscosity of the MR fluid; *du/dt* is the velocity of the MR fluids in damping passage.

Damping force of MR damper is associated with the structure and the size of the damper. Based on Bingham model, damping force of MR damper can be simply expressed as:

$$F = f_c \operatorname{sgn}(\dot{x}) + C_0 \dot{x} \tag{2}$$

Wherein, f_c is the Coulomb force correlative to magnetic density; C_0 is the viscosity damping coefficient of the MR damper, which is correlative to the damper structure; $\dot{x}$ is the relative velocity of the piston rod.

Typically, for a two-ended MR damper, coefficients of equation (2) are:

$$f_c = \frac{3K_0 L \tau_y A_p}{h} \text{ and } C_0 = \frac{12\eta L A_p^2}{\pi D h^3} \tag{3,4}$$

Wherein, L is effective length of the piston rod, D is the diameter of inner cylinder, Ap is the effective area of the piston, and Ap $=\pi (D^2-d^2)/4$, d is the diameter of piston rod, h is the gap between the piston and cylinder, K_0 is a revised coefficient between 0.8 and 1.

It is shown that, damping force of the MR damper can be divided into two parts, which are viscosity damping force[11] depend on the viscosity of the MR fluid and coulomb damping force[2] correlative to magnetic density in damping passages. If the structures and the size of the MR damper are given, viscosity damping force depends on the piston velocity and is uncontrollable. However, the coulomb damping force is connected with the applied magnetic field in electromagnetic coil and is controllable. So this isolation system using a MR damper, as a whole, is a typical nonlinear system, whose dynamic character should be paid more attentions.

3. MR Isolation System under Base Excitation

To investigate the performance of vehicle primary suspension, a base-excited isolation system is used to present the quarter car suspension. A typical magnetorheological vibration-reduction system under base excitation is shown in Figure 1, there are a spring K and a damping force F_{MR} connected between the base and the mass parallelly. Its mathematic model can be expressed as:

$$m\ddot{y} + F_{MR} + k(y - x) = 0 \tag{5}$$

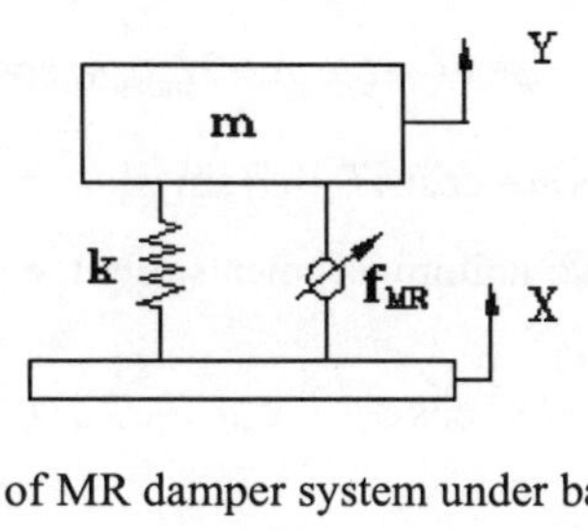

Fig1 Scheme of MR damper system under base excitation

Substitution of the equation (2) into equation (5) yields

$$m\ddot{y} + f_c \operatorname{sgn}(\dot{y} - \dot{x}) + C_0(\dot{y} - \dot{x}) + k(y - x) = 0 \tag{6}$$

Supposed that *z=y-x*, then

$$m\ddot{z} + f_c \operatorname{sgn}(\dot{z}) + C_0\dot{z} + kz = m\ddot{x} \tag{7}$$

Wherein, z is the displacement of the mass relative to the displacement of the base; *x* is the base excitation; z and x is the function of time. In this study, a sine excitation is adopted to investigate the performance of this isolation system, $x=A\sin(\omega t)$, so equation (7) turns to be:

$$m\ddot{z} + f_c \operatorname{sgn}(\dot{z}) + C_0\dot{z} + kz = -mA\omega^2 \sin(\omega t) \tag{8}$$

Where A is the displacement amplitude of the base excitation, ω is the frequency of the base excitation.

4. Harmonic Solutions

To solve equation (8), some assumptions are made to simplify this solving process. Assumed that

$$\omega_0 = \sqrt{k/m}\,, T = \omega_0 t\,, n = \omega/\omega_0, q = z/A\,, \xi_c = f_c/(2mA\omega_0^2)\,, \xi_0 = C_0/(2m\omega_0)$$

Then equation (8) turns to be:

$$\ddot{q} + 2\xi_c \operatorname{sgn}(\dot{q}) + 2\xi_0\dot{q} + q = n^2 \sin(nT) \tag{9}$$

where $\dot{q}$ and $\ddot{q}$ are derivative and duplicate derivative of q. Equation (9) has a solution of the form

$$q = q_0 \sin(nT + \varphi) \tag{10}$$

$$\dot{q} = q_0 n \cos(nT + \varphi) \text{ and } \ddot{q} = -q_0 n^2 \sin(nT + \varphi) \tag{11}$$

For the sign function, it can be transformed into

$$\operatorname{sgn}[\cos(\omega t + \phi)] \approx \frac{4}{\pi}\cos(\omega t + \phi) \tag{12}$$

$$n^2 \sin(nT) = n^2[\sin(nT + \varphi)\cos\varphi - \cos(nT + \varphi)\sin\varphi] \tag{13}$$

Substitution of the equation (11) (12) (13) into (9) yields:

$$\begin{aligned}&(-q_0 n^2 + q_0)\sin(nT + \varphi) + (\frac{8}{\pi}\xi_c q_0 n + 2\xi_0 q_0 n)\cos(nT + \varphi)\\ &= n^2[\sin(nT + \varphi)\cos\varphi - \cos(nT + \varphi)\sin\varphi]\end{aligned} \tag{14}$$

Both sides of equation (14) have two uniform segments. Let the coefficients of each segment in both sides equals

$$-q_0 n^2 + q_0 = n^2 \cos\varphi, \quad \frac{8}{\pi}\xi_c q_0 n + 2\xi_0 q_0 n = n^2 \sin\varphi \tag{15}$$

Solving equation (15) with respect of φ yields

$$q_0 = \left[\frac{n^4\pi^2}{((1-n^2)\pi)^2 + (2\xi_0 n\pi + 8\xi_c n)^2} \right], \text{ and } \varphi = -\arctan\frac{2\xi_0 n\pi + 8\xi_c n}{\pi(1-n^2)} \tag{16}$$

The acceleration transmissibility is defined as the ratio of the magnitude of the acceleration transmitted through the spring and damper to the fixed base to the sinusoidal force applied by the machine. Acceleration transmissibility of MR damper can be figured as:

$$T_a = \left|\frac{\ddot{y}_{\max}}{\ddot{x}_{\max}}\right| = \left|\frac{(\ddot{z}+\ddot{x})_{\max}}{\ddot{x}_{\max}}\right| = \left|\frac{(\ddot{z} - A\omega^2 \sin\omega t)_{\max}}{(-A\omega^2 \sin\omega t)_{\max}}\right| \tag{17}$$

Noted that $q = z/A$ and substitution of equation (10) and (11) into equation (17) yield

$$T_a = \left|(-q_0 \sin(\omega t + \varphi) - \sin\omega t)_{\max}\right| = \sqrt{1 + q_0^2 + 2q_0 \cos\varphi} \tag{18}$$

The displacement transmissibility describes how a steady-state displacement (Y) of the base of a device mounted on an isolator is transmitted into a motion of the device (X). Displacement transmissibility can be expressed as:

$$T_d = \left|\frac{y_{\max}}{x_{\max}}\right| = \left|\frac{(z+x)_{\max}}{x_{\max}}\right| = \left|\frac{(Aq + A\sin\omega t)_{\max}}{(A\sin\omega t)_{\max}}\right| \tag{19}$$

Noted that $q = z/A$ and substitution of equation (10) and (11) into equation (19) yield

$$T_d = \left|(q_0 \sin(\omega t + \varphi) - \sin\omega t)_{\max}\right| = \sqrt{1 + q_0^2 - 2q_0 \cos\varphi} \tag{20}$$

5. Results of Numerical Simulations

From equation (3), it is shown that the coulomb damping coefficient f_c is correlated with applied currents and in direct ration with it. Simulations are done to analyze the dynamics of acceleration transmissibility and displacement transmissibility on 5 different conditions, which are the conditions when the coulomb damping coefficients are 0.2, 0.8, 2.0, 4.0 and 8.0. Results are shown in Figure 2-4. In order to describe the dynamic of this transmissibility, three areas are marked according to the different viscosity damping ratio ξ_0. Under-damped area, critical-damping area and over-damping area are corresponding with the cases when$0<\xi_0<1$, $\xi_0=1$, $\xi_0>1$ separately. Multi-value performance appears in the condition when the viscosity damping ration is below1 ($n<1$), shown in Fig2, 3 and 4. That is to say, Multi-value performance appears in the condition when the viscosity damping coefficient is relative small. Because that the viscosity damping is determined by the structure parameters of the MR damper, it is recommended that some optimal decisions must be taken to choose the relationship between active length L, effective area A_p, diameter of the cylinder D and the gap h in the progress of design. MR damper should be arranged to work in the area of vibration-reduction, that is in the area when damping ratio is great than 2, shown in Figure2-4.

The formulas for transmissibility of acceleration and displacement are very useful in the design of system to provide protection from unwanted vibration. It is known that the coulomb damping coefficients increase with the increase of the applied current, however acceleration transmissibility goes down in the vibration-reduction area on the contrary and the displacement transmissibility go arise on the same conditions. Acceleration transmissibility and displacement transmissibility are all correlative with ride quality and vehicle handling and are incompatible with each other because that

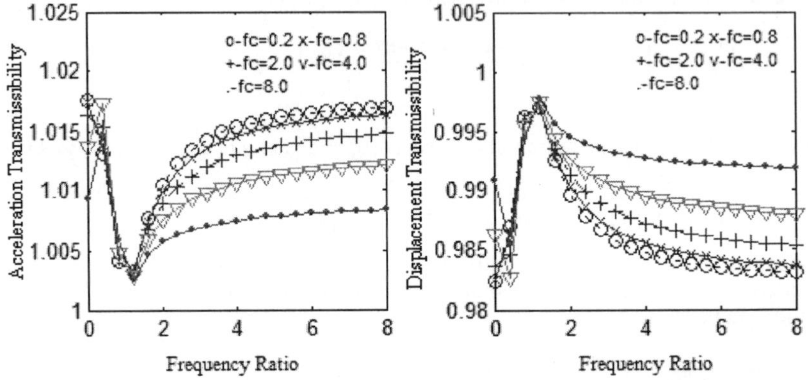

Fig.2 Acceleration transmissibility and displacement transmissibility (under damping, $\xi_0=0.05$)

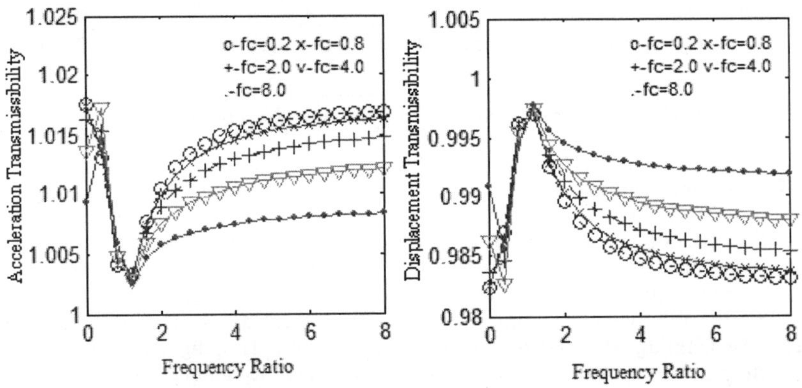

Fig.3 Acceleration transmissibility and displacement transmissibility (critical damping, $\xi_0=1$)

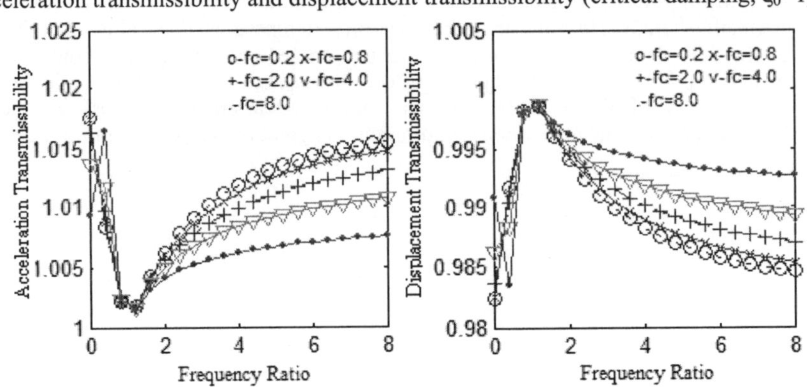

Fig.4 Acceleration transmissibility and displacement transmissibility (over damping, $\xi_0=2$)

there is a trade off between ride quality and vehicle handling. It is very important that a proper coulomb damping coefficient should be carefully chosen in practical work.

6. Conclusions

In this paper, the response of a harmonic base-excitation isolation system using MR damper is exhibited with the method of revised harmonic solution. Dynamic characteristics of MR isolation system in different conditions are analyzed. Some useful suggestions are brought forward for designing a more effective MR damper.

1. A multi-value performance appears in transmissibility when the damping ration is below1. Because that the viscosity damping is determined by the structure parameters of the MR damper, it is recommended that some proper measures must be taken to choose the relationship between active length, effective area, diameter of the cylinder and the gap in the progress of design.
2. Coulomb damping coefficients increase with the increase of the applied current, however acceleration transmissibility goes down in the vibration-reduction area on the contrary and the displacement transmissibility go arise on the same conditions. Structure parameters should be optimized together with the property of magnetic field

References:

[1] Karakas E S and Gordaninejad F. Control of a quarter HMMWV suspension system using a magneto-rheological fluid damper. Proc of SPIE 2004; 5386:204-13.

[2] Yao G Z, Yap F F, Chen G, Li W H, and Yeo S H. MR damper and its application for semi-active control of vehicle suspension system . Mechatronics2002; 12:963-73.

[3] Mehdi Ahmadian and John Gravatt. A comparative analysis of passive twin tube and skyhook MRF dampers for motorcycle front suspension. Proc of SPIE 2004; 5386:238-49.

[4] Mehdi Ahmadian and Nader Vahdati. Effect of hybrid semi-active control on steady state and transient dynamics of vehicle suspensions. Proc of SPIE 2004; 5390:34-45.

[5] Michael J. Craft; Gregory D. Buckner, and Richard D. Anderson. Fuzzy logic control algorithms for magneshock semi-active vehicle shock absorbers: design and experimental evaluations. Proc of SPIE2003; 5049:577-88.

[6] Amit Shukla, Michael Bailey Van Kuren. Nonlinear dynamics of a magnetorheological fluid based active suspension system for a neonatal transport. Proc of SPIE 204; 5386:83-92.

[7] Smitht C B and Wereley N M. Nonlinear damping identification from transient data. Proc of SPIE2000; 3672:2-19.

[8] Snyder R A; Kamath G M, and Wereley N M. Characterization and analysis of magnetorheological damper behavior due to sinusoidal loading. Proc of 2000; 3989:213-29.

[9] Ying Z G, Ni Y Q, and Ko J M. Non-clipping optimal control of randomly excited nonlinear systems using semi-active ER/MR dampers. Proc of SPIE 2002; 4696:209-18.

[10] Ping Zhang, Nonlinear impulsive vibration-reduction dynamics and design, first ed., National Defenses Publishing Company, Beijing, 2003.

[11] Changrong Liao, Weimin Chen, Miao Yu, Shanglian Huang, Design consideration of Magnetorheological Damper for Vehicle use, J. of Chinese Mechanical Engineering2002, 13:723-26.

Analysis of Distributed Micro-Control Actions on Free Paraboloidal Membrane Shells

H. H. Yue[a], Z. Q. Deng [a] and H. S. Tzou[b]

[a]*School of Mechatronic Engineering, Harbin Institute of Technology,Harbin, 150001, China*

[b]*Department of Mechanical Engineering, StrucTronics Lab, University of Kentucky, Lexington, KY 40506-0503, USA*

Received 15 March 2006; received in revised form 10 May 2006; accepted 15 May 2006

Abstract

Flexible paraboloidal shells, as key components, are increasingly utilized in antennas, reflectors, optical systems, aerospace structures, etc. To explore precise shape and vibration control of the paraboloidal membrane shells, this study focuses on analysis of microscopic control actions of segmented actuator patches laminated on the surface of a free paraboloidal membrane shell. Governing equations of the membrane shell system and modal control forces of distributed actuator patches are presented first, and followed by the analysis of dominating micro-control actions based on various natural modes, actuator locations and geometrical parameters. Finally, according to the parametric analysis, simulation data reveal main factors significantly influencing active control behavior on smart free-floating paraboloidal membrane shell systems, thus providing design guidelines to achieve optimal control of paraboloidal shell systems.

PACS: 07.10.Eq; 77.65.-J; 46.70.De

Keywords: Micro-control action; Distributed actuator; Segmentation; Free paraboloidal membrane shells; Parametric analysis

1. Introduction

Free-floating paraboloidal membrane shells are fully utilized in the field of advanced aerospace and telecommunication, such as focusing viewfinders, antennas, etc. [1]. Because of its flexibility, this structure can vibrate and reshape easily, and if without effective control, the vibration can prolong to degrade its precision and accuracy, or even become unstable leading to finally self-destruction [2,3]. Along with development of "intelligent" structural systems, various control techniques and smart materials are used to adjust static shapes or to counteract the undesirable vibration of the structronic system [4]. Among various smart materials and structures, the distributed control technique using piezoelectric actuator patches is widely investigated and applied to engineering systems [5-9].

Distributed control of beam and plate has been widely studied over years [10]. Modal control forces of arbitrarily located quarterly segmented piezoelectric actuators laminated on cylindrical shell panels were studied and their modal actuation factor, modal feedback factor and controlled damping ratios were derived and evaluated [11,12]. Spherical shells with piezoelectric actuators were also studied [13-15]. Micro-control actions and distributed control effectiveness of segmented actuator patches laminated on hemispheric shells was evaluated [16]. Distributed modal voltages and their

*Corresponding author. Tel.: +86-0451-8641-3802; fax: +86-0451-8641-3857.

E-mail address: block@hit.edu.cn (H.H. Yue).

spatial strain characteristics of toroidal shells and conical shells were recently investigated [17,18]. Micro-control characteristics of a deep paraboloidal shell were evaluated [19]. The precision distributed control effectiveness of adaptive paraboloidal shells laminated with segmented actuator patches was investigated [20]. However, distributed control actions of free-floating paraboloidal membrane shells still need further investigation in order to achieve precision vibration and shape control.

This study mainly focuses on micro-control actions and control effectiveness of distributed piezoelectric actuator patches laminated on free-floating paraboloidal shells. Control behavior and laminated paraboloidal shell control system equation are discussed first, and modal control force expression is deduced. In order to evaluate modal micro-control actions of paraboloidal membrane shells with distributed actuators, a series of mode shape functions based on the membrane approximation are used in the formulation of modal control forces. Because the transverse control action is considered, microscopic control actions of segmented actuators in transverse directions are evaluated. Finally, the contributing control force components, control effects and normalized control actions on various modes, actuator locations, and geometrical parameters are analyzed in case studies.

2. Distributed modal control of thin paraboloidal shell

2-1 Control behavior and laminated shell control system equation

It is assumed that piezoelectric actuator is laminated on the surface of shell, and it is free from external in plane normal forces, as Figure 1. Because of the converse piezoelectric effect, the induced strain, which is due to imposed control voltages, can counteract the shell vibration. It is assumed that the applied voltage is more significant than the self-induced voltage from the direct piezoelectric effect. Thus, the self-generated voltage is often neglected in the active vibration control system. For an open-loop control, the applied voltage can be an imposed reference voltage, and for a closed-loop control, the control signal is set up as a function of the distributed sensor signal.

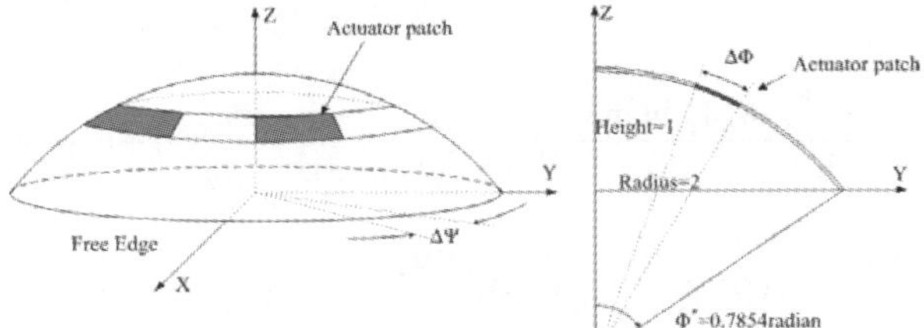

Fig.1. Shallow paraboloidal shell with distributed actuator patches.

For a piece of bi-axially polarized piezoelectric material used as an actuator, a control voltage can lead to two normal strains in the plane of actuator patch. The electromechancial coupling process is called "the converse piezoelectric effect", and such strains can induce control forces and moments, which are used to actuate static shape or to counteract the shell oscillations. Since the control forces N_{ij}^a and moments M_{ij}^a are induced by the distributed actuators, the detail membrane force and moment expressions can be defined as:

$$N_{\phi\phi}^a = d_{3\phi} Y_p \phi^a, \quad N_{\psi\psi}^a = d_{3\psi} Y_p \phi^a, \quad M_{\phi\phi}^a = r_\phi^a d_{3\phi} Y_p \phi^a, \quad M_{\psi\psi}^a = r_\psi^a d_{3\psi} Y_p \phi^a \qquad (1,2,3,4)$$

Note that d_{3i} is the piezoelectric strain constant; h^a is the actuator thickness; Y_p is Young's modulus of the piezoelectric actuator; r_i^a is defined the moment arm, namely the distance from shell's neutral surface to the actuator's mid plane; and ϕ^a is the control voltage signal. According to thin shell theory, the in-plane shear effects of piezoelectric patch are not considered, namely $N^a_{ij} \approx 0$ and $M^a_{ij} \approx 0$. Thus, the resultant forces N_{ii} and moments M_{ii} can be defined:

$$N_{\phi\phi} = N^e_{\phi\phi} - N^a_{\phi\phi} = N^e_{\phi\phi} - d_{3\phi} Y_p \phi^a \tag{5}$$

$$N_{\psi\psi} = N^e_{\psi\psi} - N^a_{\psi\psi} = N^e_{\psi\psi} - d_{3\psi} Y_p \phi^a \tag{6}$$

$$M_{\phi\phi} = M^e_{\phi\phi} - M^a_{\phi\phi} = M^e_{\phi\phi} - r^a_\phi d_{3\phi} Y_p \phi^a \tag{7}$$

$$M_{\psi\psi} = M^e_{\psi\psi} - M^a_{\psi\psi} = M^e_{\psi\psi} - r^a_\psi d_{3\psi} Y_p \phi^a \tag{8}$$

Where N^e_{ii} and M^e_{ii} are elastic membrane force and moment in the i-th direction respectively. Based on the membrane approximation, the transverse shear forces and moments are neglected. Substituting those parameters into the system equations, the control equations of laminated paraboloidal membrane shell are carried out:

$$b\frac{\partial[(N^e_{\phi\phi} - N^a_{\phi\phi})\tan\phi]}{\partial\phi} + \frac{b}{\cos^3\phi}\frac{\partial N^e_{\psi\phi}}{\partial\psi} - \frac{b}{\cos^2\phi}(N^e_{\psi\psi} - N^a_{\psi\psi}) - \cos^3\phi\frac{\partial(M^a_{\phi\phi}\tan\phi)}{\partial\phi}$$
$$+ M^a_{\psi\psi}\cos\phi + \frac{b^2\sin\phi}{\cos^4\phi}\cdot F_1 = \frac{b^2\sin\phi}{\cos^4\phi}\rho h\ddot{u}_\phi \tag{9}$$

$$b\frac{\partial(N^e_{\phi\psi}\tan\phi)}{\partial\phi} + \frac{b}{\mathrm{con}^3\phi}\frac{\partial(N^e_{\psi\psi} - N^a_{\psi\psi})}{\partial\psi} - \frac{b}{\mathrm{con}^2\phi}N^e_{\psi\phi} - \frac{1}{\cos^2\phi}\frac{\partial M^a_{\psi\psi}}{\partial\psi} + \frac{b^2\sin\phi}{\mathrm{con}^4\phi}F_2$$
$$= \frac{b^2\sin\phi}{\mathrm{con}^4\phi}\rho h\ddot{u}_\psi \tag{10}$$

$$\frac{\partial}{\partial\phi}\left[\frac{b}{\cos^2\phi}M^a_{\psi\psi} - \frac{\cos^3\phi}{b}\frac{\partial(M^a_{\phi\phi}b\tan\phi)}{\partial\phi}\right] - \frac{1}{\cos^2\phi\sin\phi}\frac{\partial}{\partial\psi}\left(\frac{\partial M^a_{\psi\psi}}{\partial\psi}\right) - b\tan\phi(N^e_{\phi\phi} - N^a_{\phi\phi})$$
$$- \frac{b\sin\phi}{\cos^3\phi}(N^e_{\psi\psi} - N^a_{\psi\psi}) + \frac{b^2\sin\phi}{\cos^4\phi}F_3 = \frac{b^2\sin\phi}{\cos^4\phi}\rho h\ddot{u}_3 \tag{11}$$

In order to completely evaluate the distributed micro-control membrane force and moment components of actuator patches on free-floating paraboloidal shell, the moment control effects are still kept in above system equations, although the elastic moment of thin paraboloidal shell are neglected in the membrane approximation. The modal control force of paraboloidal membrane shell is defined next.

2-2 Distributed control force of paraboloidal shell

It is assumed that the dynamic response of the paraboloidal shell is composed of all participating modes and its modal expansion is:

$$u_i(\phi,\psi,t) = \sum_{k=1}^{\infty}\eta_k(t)U_{ik}(\phi,\psi), i = \phi,\psi,3 \tag{12}$$

Using the orthogonal condition and substituting the modal expansion into the simplified piezoelectric shell equation, one can derive the modal control equation of the paraboloidal shell as [21]

$$\ddot{\eta}_k + \frac{c}{\rho h}\dot{\eta}_k + \omega_k^2\eta_k = \hat{F}_i + \hat{F}^a_k = \hat{F}_k \tag{13}$$

Where k denotes the circumferential wave number; ω_k is the k-th natural frequency; η_k is a modal coordinate of the k-th mode; c is the damping constant; $\hat{F}_i$ is mechanical excitation in the i-th direction; and $\hat{F}_k^a$ is electrical control excitation. $L_i^c\{\phi_\phi, \phi_\psi, \phi_3\}$ is an operator derived from the converse piezoelectric effect with only the transverse control voltage ϕ_3. It is assumed that the actuator is made of a hexagonal piezoelectric material, i.e., $d_{3\phi} = d_{3\psi}$, and both the actuator and shell have uniform thickness, i.e., $r_\phi^a = r_\psi^a$, thus $N_{\phi\phi}^a = N_{\psi\psi}^a$, $M_{\phi\phi}^a = M_{\psi\psi}^a$. Substituting all parameters into the operator expressions gives

$$L_\phi^c\{\phi_3\} = -\frac{\cos^3\phi}{b}\frac{\partial N_{\phi\phi}^a}{\partial\phi} - \frac{\cos^6\phi}{b^2}\frac{\partial M_{\phi\phi}^a}{\partial\phi} \tag{14}$$

$$L_\psi^c\{\phi_3\} = -\frac{\cos\phi}{b\sin\phi}\frac{\partial N_{\psi\psi}^a}{\partial\psi} - \frac{\cos^2\phi}{b^2\sin\phi}\frac{\partial M_{\psi\psi}^a}{\partial\psi} \tag{15}$$

$$\begin{aligned} L_3^c\{\phi_3\} = &-\frac{\cos^6\phi}{b^2}\frac{\partial^2 M_{\phi\phi}^a}{\partial\phi^2} + \frac{3\cos^5\phi\sin^2\phi - 2\cos^5\phi}{b^2\sin\phi}\frac{\partial M_{\phi\phi}^a}{\partial\phi} - \frac{\cos^2\phi}{b^2\sin^2\phi}\frac{\partial^2 M_{\psi\psi}^a}{\partial\psi^2} \\ &+ \frac{\cos^5\phi}{b^2\sin\phi}\frac{\partial M_{\psi\psi}^a}{\partial\phi} + \frac{\cos^3\phi}{b}N_{\phi\phi}^a + \frac{\cos\phi}{b}N_{\psi\psi}^a \end{aligned} \tag{16}$$

Recall that the objective of this study is to investigate spatially distributed micro-control actions of actuator patches at various locations, not the paraboloidal shell dynamics, so the mechanical excitation is neglected. Hence the modal control force can be defined as:

$$\begin{aligned} \hat{F}_k &= \frac{1}{\rho h N_k}\int_\psi\int_\phi [\sum_i^3 (F_i + L_i^c\{\phi_3\})\cdot U_{ik}(\alpha_1,\alpha_2)] A_1 A_2 d\alpha_1 d\alpha_2 \\ &= \frac{1}{\rho h N_k}\int_\psi\int_\phi [L_\phi^c\{\phi_3\}U_{\phi k} + L_\psi^c\{\phi_3\}U_{\psi k} + L_3^c\{\phi_3\}U_{3k}]\frac{b^2\sin\phi}{\cos^4\phi} d\phi d\psi \end{aligned} \tag{17}$$

Where

$$N_k = \int_{\alpha_1}\int_{\alpha_2} [\sum_i^3 U_{ik}^2(\alpha_1,\alpha_2)] A_1 A_2 d\alpha_1 d\alpha_2$$

This generic modal force denotes a control operator derived from the converse piezoelectric effect with a transverse control voltage ϕ_3 [22]. Detailed microscopic membrane and bending control components induced by actuator patches are evaluated next.

3. Micro-control actions of segmented patches with free B.C.s

Based on the membrane approximation, three suitable mode shape functions of paraboloidal membrane shells with free boundary conditions are selected as:

$$U_{\phi k} = A_k \cos\frac{(2k+1)\pi}{\phi^*}\phi(\sin\phi)^{k+1}\cos k\psi \tag{18}$$

$$U_{\psi k} = -A_k\cos\phi(\sin\phi)^{k+1}\sin k\psi \tag{19}$$

$$U_{3k} = A_k (k+1)\cos\phi(\sin\phi)^k \cos k\psi \tag{20}$$

Where $U_{\phi k}$ is the meridional mode shape function; $U_{\psi k}$ is the circumferential mode shape function; U_{3k} is the transverse mode shape function; ϕ is the meridional angle measured from the pole; ψ is the circumferential angle; and A_k is the modal amplitudes. Substituting the operator $L_i^c\{\phi_3\}$ into the modal force equation, one can carry out the detail modal force expression.

$$\begin{aligned}\hat{F}_k &= \frac{1}{\rho h N_k}\int_\psi\int_\phi [L_\phi^c\{\phi_3\}U_{\phi k} + L_\psi^c\{\phi_3\}U_{\psi k} + L_3^c\{\phi_3\}U_{3k}]\frac{b^2\sin\phi}{\cos^4\phi}d\phi d\psi \\ &= \frac{A_k Y_p d_{3i}\phi^a}{\rho}T_k = \frac{A_k Y_p d_{3i}\phi^a}{\rho}(T_{k\_meri} + T_{k\_cir} + T_{k\_trans})\end{aligned} \tag{21}$$

Where T_k denotes total control action by actuator, and T_{k_meri}, T_{k_cir} and T_{k_trans} denote the control actions respectively in the meridional, circumferential and transverse directions. Furthermore, if the electrode resistance is neglected, the potential is constant over the effective electrode surface for piezoelectric patch. Thus, the actuator control signal applied to a segmented actuator patch can be defined as [22]:

$$\phi^a(\phi,\psi,t) = \phi^a(\phi,\psi,t)[u_s(\phi-\phi_0) - u_s(\phi-\phi_1)][u_s(\psi-\psi_0) - u_s(\psi-\psi_1)] \tag{22}$$

Note $u_s(*)$ is a unit step function: $u_s(\phi-\phi_1)=1$, when $\phi \ge \phi_1$; $u_s(\phi-\phi_1)=0$, when $\phi < \phi_1$. Thus its spatial derivatives are:

$$\frac{\partial\phi^a(\phi,\psi,t)}{\partial\phi} = \phi^a(\phi,\psi,t)[\delta(\phi-\phi_0) - \delta(\phi-\phi_1)][u_s(\psi-\psi_0) - u_s(-\psi_1)] \tag{23}$$

$$\frac{\partial\phi^a(\phi,\psi,t)}{\partial\psi} = \phi^a(\phi,\psi,t)[u_s(\phi-\phi_0) - u_s(\phi-\phi_1)][\delta(\psi-\psi_0) - \delta(\psi-\psi_1)] \tag{24}$$

Where $\delta(*)$ is a Dirac delta function, $\int \delta(\phi-\phi_i)d\phi = 1$ when $\phi=\phi_i$ and $\int \delta(\phi-\phi_i)d\phi = 0$ when $\phi \neq \phi_i$. Furthermore, the meridional and circumferential oscillations are usually small as compared with the transverse oscillation. Thus, this study primarily focuses on the control effects and micro-control actions that can counteract the transverse oscillation. As a result, control forces in both meridional and circumferential directions are neglected. Assume 1) both piezoelectric strain constants are equal, i.e., $d_{31} = d_{32}$, 2) the moment arm $r_\phi^a = r_\psi^a = r^a$, and 3) the actuator boundary is defined from $\phi_0 \rightarrow \phi_1$ and $\psi_0 \rightarrow \psi_1$. Thus, the modal control force and its microscopic contributing control actions becomes

$$\hat{F}_{k\_trans} = \frac{1}{\rho h N_k}\int_{\psi}\int_{\phi}\{[-\frac{\cos^6\phi}{b^2}\frac{\partial^2 M^a_{\phi\phi}}{\partial\phi^2} + \frac{3\cos^5\phi\sin^2\phi - 2\cos^5\phi}{b^2\sin\phi}\frac{\partial M^a_{\phi\phi}}{\partial\phi}$$
$$-\frac{\cos^2\phi}{b^2\sin^2\phi}\frac{\partial^2 M^a_{\psi\psi}}{\partial\psi^2} + \frac{\cos^5\phi}{b^2\sin\phi}\frac{\partial M^a_{\psi\psi}}{\partial\phi} + \frac{\cos^3\phi}{b}N^a_{\phi\phi} + \frac{\cos\phi}{b}N^a_{\psi\psi}]$$
$$A_k(k+1)\cos\phi\cdot\sin^k\phi\cdot\cos k\psi\cdot\frac{b^2\sin\phi}{\cos^4\phi}\}d\phi d\psi \tag{25}$$
$$= \frac{d_{3i}A_k Y_p \phi^a}{\rho}[T^{\phi\_mem}_{k\_trans} + T^{\phi\_bend}_{k\_trans} + T^{\psi\_mem}_{k\_trans} + T^{\psi\_bend}_{k\_trans}]$$

Note that $T^{\phi\_mem}_{k\_trans}$, $T^{\phi\_bend}_{k\_trans}$, $T^{\psi\_mem}_{k\_trans}$, $T^{\psi\_bend}_{k\_trans}$ respectively denote contributing microscopic meridional and circumferential membrane/bending components of the total transverse control action. Substituting the mode shape functions into Eq.(25) yields detailed microscopic contributing control actions of a free-floating membrane shell.

$$T^{\phi\_mem}_{k\_trans} = \frac{b}{khN_k}(k+1)(\sin k\psi_1 - \sin k\psi_2)\int_{\phi_0}^{\phi_1}\sin^{(k+1)}\phi d\phi \tag{26}$$

$$T^{\phi\_bend}_{k\_trans} = -\frac{(k+1)}{khN_k}r^a(\sin k\psi_1 - \sin k\psi_2)\{(\cos^3\phi_0\sin^{(k+1)}\phi_0 - \cos^3\phi_1\sin^{(k+1)}\phi_1)$$
$$-[\sin^{(k-1)}\phi_0\cos^2\phi_0(k\cos^2\phi_0 - 3 + 3\cos^2\phi_0)$$
$$-\sin^{(k-1)}\phi_1\cos^2\phi_1(k\cos^2\phi_1 - 3 + 3\cos^2\phi_1)] \tag{27}$$
$$+[(3\cos^2\phi_0\sin^2\phi_0 - 2\cos^2\phi_0)\sin^k\phi_0$$
$$-(3\cos^2\phi_1\sin^2\phi_1 - 2\cos^2\phi_1)\sin^k\phi_1]\}$$

$$T^{\psi\_mem}_{k\_trans} = \frac{b}{khN_k}(k+1)(\sin k\psi_1 - \sin k\psi_2)\int_{\phi_0}^{\phi_1}\frac{\sin^{(k+1)}\phi}{\cos^2\phi}d\phi \tag{28}$$

$$T^{\psi\_bend}_{k\_trans} = -\frac{(k+1)}{hN_k}r^a\{[(\cos k\psi_1 - \cos k\psi_0) + \frac{(\sin k\psi_0 - \sin k\psi_1)}{k}]\int_{\phi_0}^{\phi_1}\frac{\sin^{(k-1)}\phi}{\cos\phi}d\phi \tag{29}$$
$$+[(\sin k\psi_1 - \sin k\psi_0)(\cos^2\phi_0\sin^k\phi_0 - \cos^2\phi_1\sin^k\phi_1)]\}$$

Furthermore, for a closed shell, the circumferential angle is 0 to 2π and the curvature angle is from 0 to ϕ^* in the meridian direction. Thus, the modal force operator can be derived as:

$$N_k = \int_{\phi}\int_{\psi}\sum_i^3 U^2_{ik}(\phi,\psi)\frac{b^2\sin\phi}{\cos^4\phi}d\psi d\phi$$
$$= \pi A_k^2 b^2\int_0^{\phi^*}\{\frac{\sin^{(2k+1)}\phi}{\cos^2\phi}[\cos^2\frac{(2k+1)\pi}{2\phi^*}\phi\cdot\tan^2\phi + (k+1)^2 + \sin^2\phi]\}d\phi \tag{30}$$

Note that although four microscopic control actions of actuator patches are defined, two bending control actions should be minimal due to intrinsic membrane-dominated shell dynamics. This characteristic should be revealed in parametric studies presented later. Also, to keep generality, actuator material constants and control voltage, i.e., $d_{3i}Y_p\phi^a A_k/\rho$, are assumed constants. Once actuator materials and input voltages are specified, detailed control characteristics of paraboloidal membrane shells can be defined. Accordingly, parametric studies presented next can be inferred to account for various actuator materials and shell structures.

4. Case studies

In order to evaluate four micro-control actions and the overall control effect, a shallow paraboloidal shell model is selected, as shown in Figure 1. Geometric parameters are: major radius of revolution $a = 2m$, height $c = 1m$, shell thickness $h = 0.001m$, and the piezoelectric patch thickness $h^a = 40\mu m$. Base on dynamic characteristics and natural modal oscillations of a free-floating paraboloidal shell, actuator patches are symmetrically laminated on the shell surface. Piezoelectric patches are laminated on the shallow shell as $\phi_0 \rightarrow \phi_1$ (i.e., 0~0.1, 0.1~0.2, 0.2~0.3, 0.3~0.4, 0.4~0.5, 0.5~0.6, 0.6~0.7 radians) in the meridian direction, and are divided at every $(2n-1)\pi/(2k)$ radians (i.e., $\psi_0 \rightarrow \psi_1$) in the circumferential direction, where $n = 1,2,3...,2k$ and k is the circumferential wave number. Recall that the transverse oscillation dominates. Thus, the transverse modal control force T_{k_trans} and its four contributing components $T_{k\_trans}^{\phi\_mem}$, $T_{k\_trans}^{\phi\_bend}$, $T_{k\_trans}^{\psi\_mem}$, $T_{k\_trans}^{\psi\_bend}$ are evaluated in this section. Accordingly, transverse control actions of actuator patches at various shell locations for the first four modes (k=1-4) are respectively calculated and plotted in Figures 2 to 5.

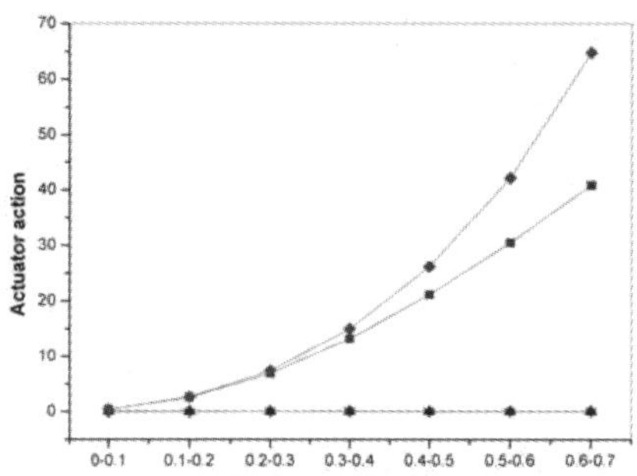

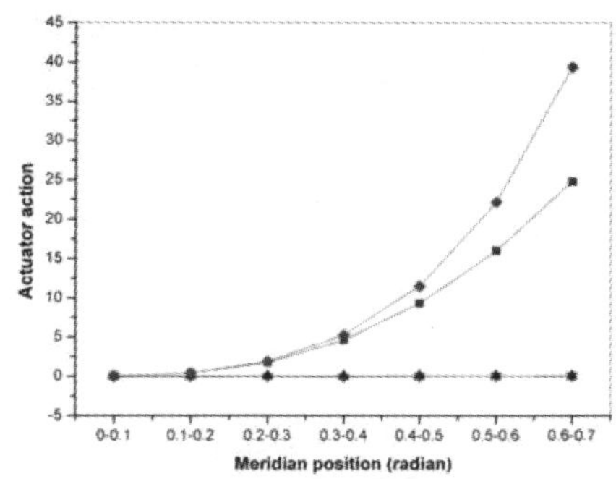

(k=1) (k=2)

Fig.2. Transverse actuator control action (k=1). Fig.3. Transverse actuator control action (k=2)

(◆: $T_{k\_trans}^{\psi\_mem}$; ■: $T_{k\_trans}^{\phi\_mem}$; ▼: $T_{k\_trans}^{\phi\_bend}$; ▲: $T_{k\_trans}^{\psi\_bend}$)

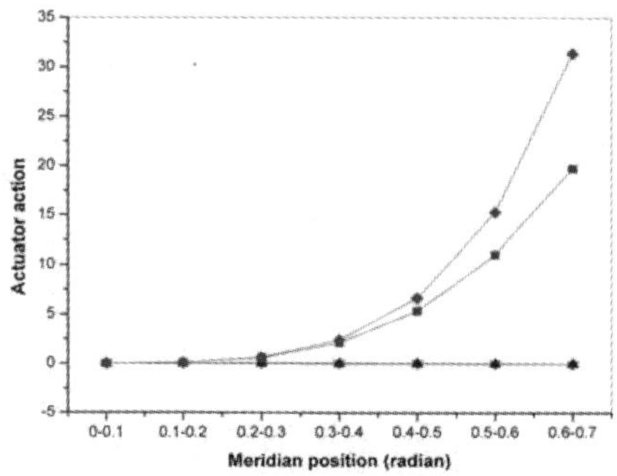

(k=3)

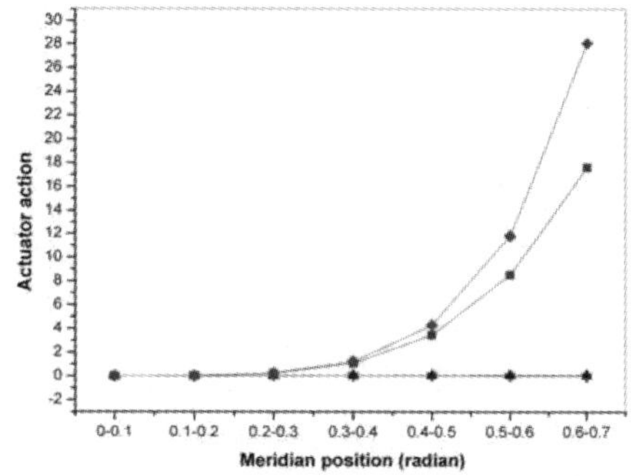

(k=4)

Fig.4. Transverse actuator control action (k=3) Fig.5. Transverse actuator control action (k=4).
(◆: $T_{k\_trans}^{\psi\_mem}$; ■: $T_{k\_trans}^{\phi\_mem}$; ▼: $T_{k\_trans}^{\phi\_bend}$; ▲: $T_{k\_trans}^{\psi\_bend}$)

These four figures reveal that the circumferential and meridional membrane control forces $T_{k\_trans}^{\psi\_mem}$ and $T_{k\_trans}^{\phi\_mem}$ dominate in all control actions, and the bending control actions $T_{k\_trans}^{\phi\_bend}$ and $T_{k\_trans}^{\psi\_bend}$ are near to zero. Furthermore, the control action obviously increases when the actuator patch is close to the free boundary, and the circumferential membrane control action $T_{k\_trans}^{\psi\_mem}$ is larger than the meridional control action $T_{k\_trans}^{\phi\_mem}$ .Via comparing the magnitudes of actuator action of various natural modes, Figure 6 illustrates that the actuator action gradually decreases at higher mode, since the inherent membrane effect diminishes as the mode increases.

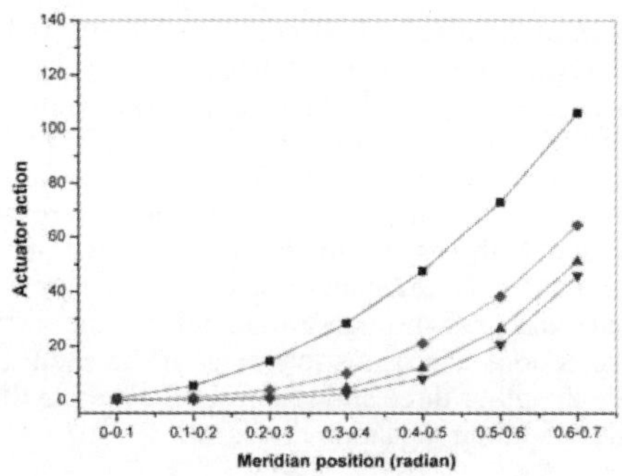

Fig.6. Total modal transverse control action T_{k_trans} at various locations.
(■: k=1; ◆: k=2; ▲: k=3; ▼: k=4)

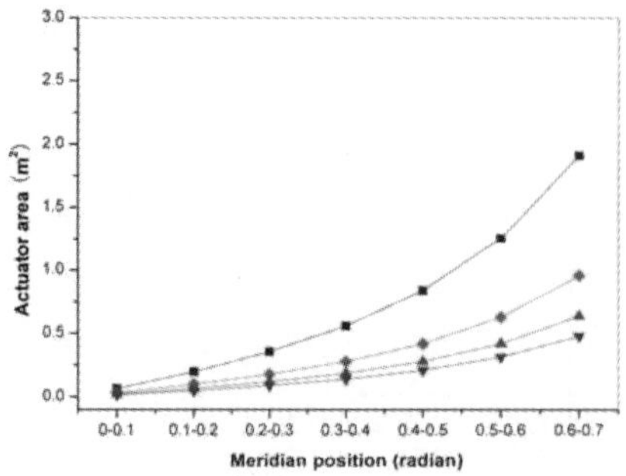

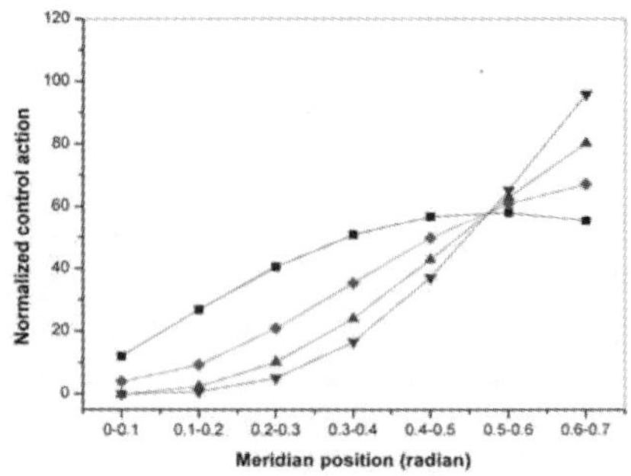

Fig.7. Modal dependent actuator area . Fig.8. Normalized control action

(■: k=1; ◆: k=2; ▲: k=3; ▼: k=4)

However, observing the size of actuator patches at various locations from the pole to the free rim indicates that the size actually increases and becomes the largest neat the free boundary when a constant circumferential division is imposed. Conceptually, a larger actuator patch certainly induces a significant control force, as well as individual micro-control actions. Accordingly, one needs to normalize the control force with respect to its actuator size. The normalized modal control force is calculated so as to evaluate the effective control force per unit area. The effective area of actuator patches can be calculated by

$$S^a = \int_{\phi}\int_{\psi} A_1 A_2 d\psi d\phi = b^2(\psi_1 - \psi_0)\frac{\cos^3\phi_0 - \cos^3\phi_1}{3\cos^3\phi_0\cos^3\phi_1} \tag{31}$$

Recall that piezoelectric patches are respectively 0~0.1, 0.1~0.2, 0.2~0.3, 0.3~0.4, 0.4~0.5, 0.5~0.6, 0.6~0.7 radians in the meridian direction, and are divided at every $(2n-1)\pi/(2k)$ radians in the circumferential direction, where $n = 1,2,3...,2k$ and k is the circumferential wave number (Note that the Lamé parameter of the shell are $A_1 = R_\phi$ and $A_2 = R_\psi \sin\phi$). Figure 7 shows the effective area of actuator patches for first four modes, in which the patch area for the first mode is larger than the other three modal divisions due to its modal dependent circumferential division. Thus, the total control action can be normalized with respect to the actuator area. The modal-dependent control action of the first four shell modes is calculated at various locations and plotted in Figure 8. Interestingly, all four modal actions pass through a common transition point, i.e., ϕ~0.53 radians. Among the four modal control actions at various locations of the shell, the 1st modal control action drops after the transition, while the other three still rise from the pole to the free boundary. This could be induced by the natural modes and their actuation divisions.

5. Conclusions

Aiming at oscillation and shape control of free-floating paraboloidal shell systems, this study focuses on mathematical modeling, analysis and evaluation of microscopic control forces and micro-control actions of distributed actuator patches laminated on shallow membrane shells. Mathematical model of laminated paraboloidal shell control system was presented and distributed modal control forces of actuator patches was defined. The dominating micro-control actions of segment patches were evaluated based on the membrane approximation, and in addition, the detailed microscopic contributing control force components, modal control effects, and actuator locations were evaluated in case studies. Finally the effective actuator area and normalized control actions at various locations were investigated. On a basis of parametric analysis of a shallow paraboloidal membrane shell laminated distributed actuators, the following conclusions can be drawn.

(1) The distributed membrane control actions (include meridional and circumferential membrane micro-control components) dominate in the total control force for a given mode.

(2) Micro-control action of actuator patches varies at various locations on the paraboloidal shell, and it increases as the patch moves from the pole to the free edge.

(3) The magnitude of control action decreases while the natural mode number increases.

(4) Based on the shell curvature variation and specified segmentation method of actuators, the effective actuator area increases when the segment patch location approaches the free rim of shallow paraboloidal shell.

(5) The normalized control action increases along the meridian direction for k=2, 3, 4 modes, and the magnitude of normalized control action is maximal at the free boundary. Whereas, the 1st modal control action increases first and then decreases for k=1 after the transition location.

These conclusions serve as general design guidelines for actuator selection and placement on paraboloidal membrane shells to achieve high-precision and accuracy in practical applications.

6. Acknowledgement

This research is supported, in part, by a grant from the Spaceflight Technical Innovation Foundation of China (HTCX2005-01). Prof. Tzou also likes to thank the Visiting Professorship program at the Harbin Institute of Technology.

References

[1] Crawley. E.F., Intelligent structures for Aerospace: A Technology Overview and Assessment, AIAA Journal. Vol. 32, No. 8, 1994, pp. 1689-1699.

[2] Nurre G S, Ryan R S, Scofield H N, etal., Dynamics and control of large space structures. Journal of Guidance, 1984, 7(5), pp. 514-5262

[3] Hyland D C, Junkins J L, Longman R W., Active control technology for large space structures. Journal of Guidance, 1993, 16(5), pp. 801～821

[4] H.S. Tzou, Guran A. (Editors), Gabbert, U., Tani, J. and Breitbach, A. (Associate Editors), 1998, Structronic System-Smart Structure, Devices, and Systems, Volume-2: System and Control, World Scientific Publishing Co., New Jersey/Singapore.

[5] Dongi F, Dinkler D, Kroplin B. Active panel flutter suppression using self-sensing piezoactuators. AIAA Journal, 1996, 34 (6), pp. 1224-1230

[6] Anderson, E.H., Hagood, N.W., and Goodliffe, J.M., 1992, "Self-sensing Piezoelectric Actuation: Analysis and Application to Controlled Structures," AIAA Paper: AIAA-92-2465-CP at 33 SDM Conference.

[7] Baz A. and Ro, J., 1994, "The Concept and Performance of Active Constrained Layer Damping Treatments," Sound and Vibration, Vol. 28, No.3, pp. 18-21.

[8] Hanagud S. and Obal, M.W., 1988, "Identification of Dynamic Coupling Coefficients in a Structure with Piezoelectric Sensors and Actuators," AIAA paper No.88-2418.

[9] H.S. Tzou, 1987, "Active Vibration Control of Flexible Structures via Converse Piezoelectricity," Development in Mechanics, Vol. 14(b), Proceedings of the 20th Midwest Mechanical Conference, pp. 1201-1206.

[10] H.S. Tzou and G.L. Anderson (eds), Intelligent Structural System, Dordrecht Boston London: Kluwer Academic Publishers, 1992.

[11] Qiu. J., Tani, J., Vibration Control of a Cylindrical Shell Using Distributed Piezoelectric Sensors and Actuators, Journal of Intelligent Material Sys & Structures, Vol.6, No.4, 1995, pp.474-481.

[12] H.S. Tzou, Y. Bao and V. B. Venkayya, 1996, Parametric Study of Segmented Transducers Laminated on Cylindrical Shells, Part 2 Actuator Patches, Journal of Sound and Vibration. 197(2), Oct., pp. 225-249.

[13] Jayachandran. V., and Sun. J. Q., Modeling Shallow-spherical-shell Piezoceramic Actuators as Acoustic Boundary Control Elements, Smart Materials and Structures, 1998, Vol.7, No.1, pp.72-84.

[14] Birman. V., Griffin. S., and Knowles. G., 2000, Axisymmetric Dynamics of Composite Spherical Shells with Active Piezoelectric/Composite Stiffeners, Acta Mechanica, 141(1), pp. 71-83.

[15] Ghaedi, S. K., and Misra, A. K., 1999, Active Control of Shallow Spherical Shells Using Piezoceramic Sheets, Proceedings of SPIE-The International Society for Optical Engineering, v 3668 n II, Mar 1-4, 1999, pp. 890-912.

[16] P. Smithmaitrie, H.S. Tzou, Micro-control Action of Actuator Patches Laminated on Hemispherical Shells, Journal of Sound and Vibration, Vol.277, 2004, pp. 691-710.

[17] H.S. Tzou, D.W. Wang, and W.K. Chai, Dynamics and Distributed Control of Conical Shells Laminated with Full and Diagonal Actuators, Journal of Sound and Vibration, Vol. 256(1), 2002, pp.65-79.

[18] H.S. Tzou and D.W. Wang, Micro-sensing Characteristics and Modal Voltages of Piezoelectric Laminated Linear and Nonlinear Toroidal Shells, Journal of Sound and Vibration Vol. 254(2), 2002, pp.203-218.
[19] H.S. Tzou, J.H. Ding, et al., Micro-control Actions of Distributed Actuators Laminated on Deep Paraboloidal Shells, JSME Int. J., Series C, 2002, Vol.45, No1, pp.8-15.
[20] H.S. Tzou and J.H. Ding, Actuator Placement and Micro-Actuation Efficiency of Adaptive Paraboloidal Shells, Journal of Dynamic Systems, Measurement, and Control, Dec, 2003, Vol. 125, pp. 577-584.
[21] H.S. Tzou, J.P. Zhong, and J.J. Hollkamp, Spatially Distributed Orthogonal Piezoelectric Shell Actuators (Theory and Applications), Journal of Sound and Vibration, Vol.177, No.3, 1994, pp. 363-378.
[22] H.S. Tzou, Piezoelectric Shells, Kluwer Academic Publishers, Boston, Dordrecht, 1993.

MATLAB simulation of semi-active skyhook control of a quarter car incorporating an MR damper and a fuzzy logic controller

W.H. Li[1*], R.S. Ujszaszi[1], B. Liu[1], X.Z. Zhang[1], P.B. Kosasih[1], and X.L. Gong[2]
[1]*School of Mechanical, Materials and Mechatronic Engineering, University of Wollongong, Wollongong, NSW 2522, Australia*
[2]*CAS Key Laboratory of Mechanical Behavior and Design of Materials, University of Science & Technology of China, Hefei, 230027, China*

Abstract:

This paper investigates the performance of skyhook control designed for use with a magnetorheological damper through MATLAB simulation. The fuzzy logic simulation was observed for increased performance over conventional passive damper systems, and for behaviour in keeping with the skyhook control theory. Once it was established that the controller was accurately simulating skyhook control, the overall performance and system response was analysed. It was found that the semi-active system improved the response of the vehicle sprung mass at the expense of unsprung mass control, proving that greater occupant comfort can be achieved via semi-active suspension control.

Keywords: MR damper, skyhook control, MATLAB simulation, fuzzy-logic controller

1. Introduction

Characterised as a smart material, Magnetorheological fluid (MR) can change its rheological behaviour from an oily substance to a semi-solid when subjected to a magnetic field. This change takes place within the presence of a magnetic field. As the magnetic field strength is increased, the fluid becomes firmer and as such the yield strength is also increased. MR fluid can be manufactured with the ability to elongate or to keep a constant volume and is therefore suitable for many different applications. Systems based on this fluid are particularly attractive because they are simple, quiet and have a response time typically less than 25ms [1-4].

From the early 1990's the number of MR fluid applications has been increasing dramatically. MR fluid devices require constant feedback to appropriately vary density and damping characteristics and up until this point in time computers did not have adequate processing capabilities and were much too large.

* Corresponding author. Phone: +61 2 4221 3490; Fax: +61 2 4221 3101.
E-mail address: weihuali@uow.edu.au

With the high processing rates available in today's computers and micro processors MR fluid applications are becoming much more commercially available. Some of the MR applications that have proven to be useful are semi-active suspension truck seats and semi-active passenger car suspensions. The Lord Corporation had successfully implemented an MR damper to improve driver comfort in large trucks [5]. These MR damper equipped seats can vary damping according to road conditions, but also driver characteristics such as weight and posture. General motors have also utilised an MR equipped damper on some of their passenger cars. The Delphi sourced MagnaRide damper can now be found on car models such as the Cadillac Seville STS and the SRX SUV (Sports Utility Vehicle) [6]. The automotive MR damper has been found to improve the tyre contact patch by reducing roll and pitch while maintaining a much more compliant ride. Conventional passive damping systems are essentially a compromise between occupant ride comfort and the ability of the suspension to keep the tyre in contact with the road surface. This allows for only one damping coefficient and therefore a compromise must be made [7]; however, if the Sky-Hook strategy is implemented, two damping coefficients may be dealt with. In practice, it is necessary to monitor this absolute speed, e.g., by placing an accelerometer on the chassis and integrating the signal itself [7].

Fuzzy logic has greatly increased in popularity over the years, and as such can now be found in applications from household appliances to industrial process control [8]. It is a very convenient way to implement a control system such as a skyhook controller. After binding the input vales to their respective sets, the fuzzy logic controller applies a set of rules and determines the corresponding output. For the case of the skyhook controller the inputs would be the vehicle sprung and unsprung mass velocities, and the output would be the new damping force to adjust the damper. Using the predetermined rules for skyhook control the fuzzy logic controller would use the input and output conditions to appropriately control the MR damper. This objective of this project is to investigate dynamic performances of a quarter car incorporating a MR damper based on a fuzzy logic control. The simulation analysis was carried out using one of the toolboxes found in the software package MATLAB, known as Simulink, which is very effective in analysing non-linear systems.

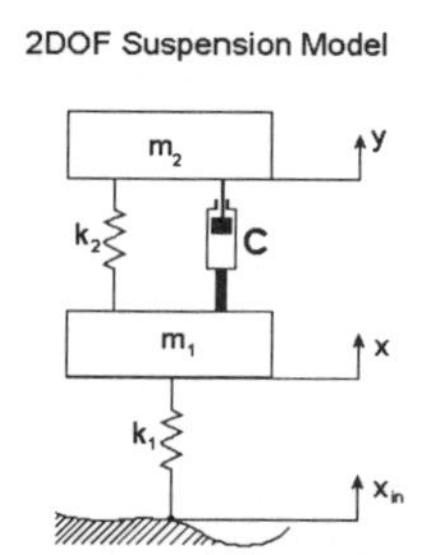

Fig.1 2-DOF suspension system.

2. Suspension mathematical model

2.1 2-DOF passive suspension model

A commonly used 2-DOF quarter car model is shown in Fig. 1, which incorporates an unsprung mass connected via a spring to the vehicle sprung mass. The unsprung mass is described as the combined mass of vehicle components that are not isolated from input disturbances by a spring. The sprung mass consists of the combined weight of the components that are isolated from road disturbances by the spring, and is therefore the gross vehicle mass minus the unsprung mass. The component required to complete the 2-DOF quarter car model is the suspension damper. While the spring is specified to control road input, the role of the damper is to control the spring. Once the spring is excited by a bump in the road, the damper dissipates the energy stored in spring to provide high occupant comfort levels and improve road holding.

The motion of the quarter model is given by

$$\begin{aligned} m_1\ddot{x} + k_1(x - x_{in}) + c(\dot{x} - \dot{y}) + k_2(x - y) = 0 \\ m_2\ddot{y} + k_2(y - x) + c(\dot{y} - \dot{x}) = 0 \end{aligned} \tag{1}$$

where m_1 and m_2 are unsprung mass and sprung mass; k_1 and k_2 are tire stiffness and spring stiffness; c is damping coefficient; x and y are unsprung, sprung mass displacement; and x_{in} is the system input displacement.

The transfer functions for the sprung mass and unsprung mass are given by

$$\frac{Y(s)}{X_{in}(s)} = \frac{k_1(Cs + k_2)}{m_1 m_2 s^4 + (m_1 + m_2)Cs^3 + [k_1 m_2 + (m_1 + m_2)k_2]s^2 + k_1 Cs + k_1 k_2} \tag{2}$$

$$\frac{X(s)}{X_{in}(s)} = \frac{m_2 k_1 s^2 + s(k_1 + k_1 k_2)}{m_1 m_2 s^4 + (m_1 + m_2)Cs^3 + [k_1 m_2 + (m_1 + m_2)k_2]s^2 + k_1 Cs + k_1 k_2} \tag{3}$$

2.2 Skyhook on/off controller

If the damper, as shown in Fig. 1, is replaced with a MR damper whose damping coefficient, C_{MR} can be controlled, the passive suspension system will be changed to a semi-active suspension system. As shown in Fig. 1 again, the sprung mass and unsprung masses are assigned velocity constants v_1 and v_2 respectively, and the relative velocity of the sprung mass to the unsprung mass was designated v_{12}. The skyhook control strategy is introduced to control the system, which is represented by

$$\begin{aligned} v_1 v_{12} > 0 \qquad C_{MR} \neq 0 \\ v_1 v_{12} < 0 \qquad C_{MR} = 0 \end{aligned} \tag{4}$$

Therefore, the strategy is based on the reasoning that the unsprung mass motion should affect the sprung mass as little as possible.

2.3 Skyhook fuzzy logic controller

A fuzzy logic controller is employed to implement the skyhook control principle. The Fuzzy Logic Toolbox, integrated into the MATLAB and Simulink environments, allows the creation of fuzzy inference systems through its FIS (Fuzzy Inference System) editor, membership editor and rule editor. The FIS editor provides an overall view of the fuzzy system showing all inputs, outputs, and parameters covering the method of fuzzy control. Similar to the on off controller, the fuzzy controller has inputs v_1, v_2, and outputs the damping force, C.

Following the FIS editor, the membership editor creates conditions for the fuzzy logic controller to categorise the inputs and outputs. The level of precision required dictated the number of conditions that inputs and outputs were bound to; and for the purposes of skyhook control, 16 to 18 conditions on each input, and two conditions on the output were selected.

With the input and output conditions defined, the rule editor was employed to create relations governing the system output. Based on the inputs v_1 and v_2, 281 rules were generated to completely define the fuzzy logic skyhook controller and describe overall system behaviour.

3. Simulation results and analysis

In this paper, the simulation parameters are listed in Table 1. Using these parameters, the overall suspension system performance in terms of displacement, velocity, and acceleration for both sprung mass and unspung mass will be addressed.

Table 1. Simulation parameters.

Sprung Mass m_2	400kg
Unsprung Mass m_1	40kg
Spring Stiffness k_2	20000N/m
Tyre Stiffness k_1	150000N/m
Reference Damping Force C	1000Ns/m
Skyhook Damping Force C_{MR}	0Ns/m when off 2000Ns/m when on

3.1 Skyhook on/off controller performance

Using MATLAB simulink, the graphs for displacement, velocity and acceleration are produced. Perhaps the most significant of these graphs for all the simulations was the sprung mass displacement. The displacement plot provided a map of the vehicle body motion and could directly give an indication of harmonic isolation and thus, occupant comfort. In every displacement plot, the skyhook controlled damper was plotted against road input conditions, and the passive suspension displacement, used as a reference wave. Fig. 2 illustrates the response of the sprung mass when subject to multiple step inputs. As can be seen from this figure, the overall skyhook performance is better than the passive performance. However, the responses at several points need improvement. Further examination points at $t = 3$ and $t = 4$, the waveform experienced some noise or distortion, assumed to have been caused by the rapid switching of the damping force. The waveform also did not reach steady state between points $t = 3$ and $t = 7$ as the steady state error was compounded by the controller feedback. After time $t = 7$, the sprung mass did reach steady state earlier than the passive damper. This was probably due to the skyhook system having a higher average damping force.

Distortion was also observed in the unsprung mass waveform, as shown in Fig. 3. As shown in this figure, the waveform was of almost identical shape to the passive damper displacement, and thus, a conclusion on whether the controller was implementing the skyhook principle could not be established. Possible reasons for the waveform not exhibiting these certain characteristics could have also been attributed to the choice of input and suspension parameters.

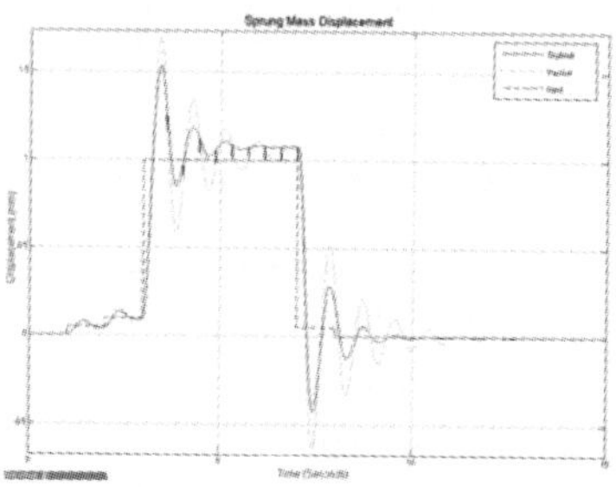

Fig. 2. Sprung mass displacement for on/off controller

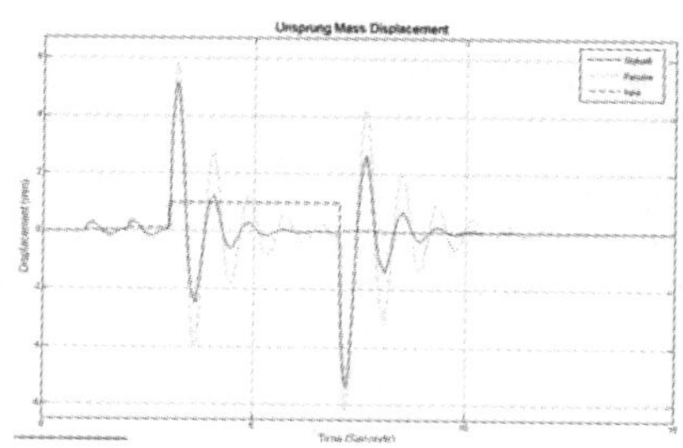

Fig. 3. Unsprung mass displacement for on/off controller

3.2 Fuzzy logic controller performance

To compare the fuzzy logic controller system response with that of on/off controller, the identical system input was used. The sprung mass displacement at fuzzy logic controller is shown in Fig. 4. When compared to the sprung mass motion from the on/off controller in Fig. 2, the benefits on the fuzzy logic controller become apparent. This fuzzy logic controller could be seen to have reduced the rise time of the system oscillations, dramatically reduced settling time; and produced a clean waveform with very little noise. The unsprung mass displacement, as shown in Fig. 5, is reaffirmed that the fuzzy logic controller was emulating an ideal skyhook damper. What could be seen is that the reference wave is far more controlled than the skyhook waveform. This was in contrast to the sprung mass motion of the same system showing a more controlled nature than the reference wave; thus, the skyhook theory of adding damping to the sprung mass whilst reducing damping on the unsprung mass has been demonstrated.

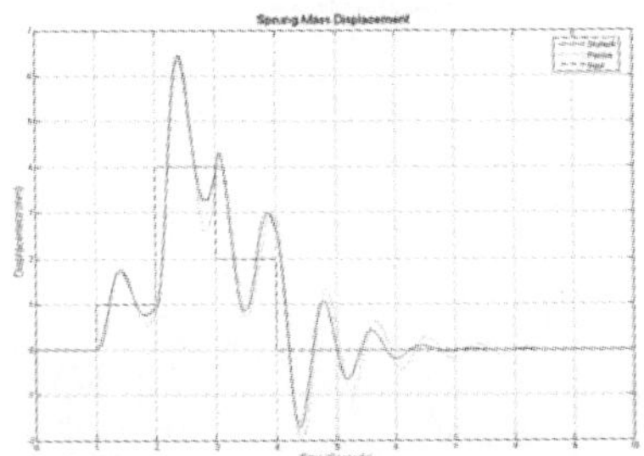

Fig. 4. Sprung mass displacement for fuzzy logical controller at stair inputs

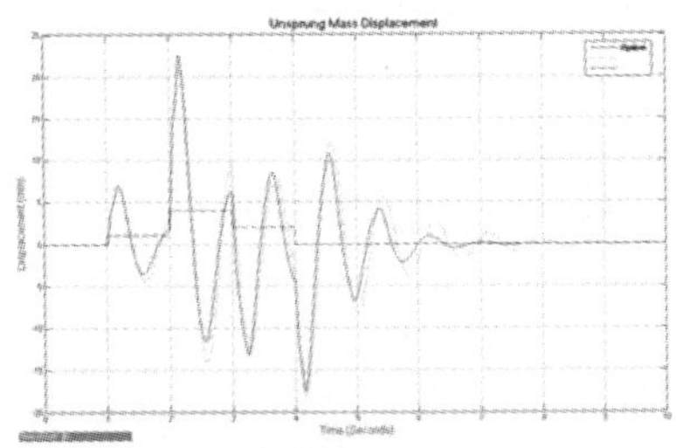

Fig. 5. Unsprung mass displacement for fuzzy logical controller at stair inputs

To further test the system response, a random stair sequence was input into the Simulink model and the sprung mass was displacement was observed, as shown in Fig. 6. Again, the controller appeared to be working effectively. The skyhook waveform did not appear to have any noise, nor were there any unexpected results. Of particular significance were the regions between time $t = 4$ and $t = 5$ seconds and times $t = 8$ and $t = 8.5$ where the waveform began to settle rapidly. The unsprung mass displacement was also examined; however, there did not seem to be any significant factors pertaining to this response.

The final input condition sequenced was a 0.05Hz sine wave. As shown in Fig. 7, the sprung mass displacement for the skyhook controller showed a speedier initial settling time, however, had a larger overshoot than the reference waveform. The same was also observed in the unsprung mass response.

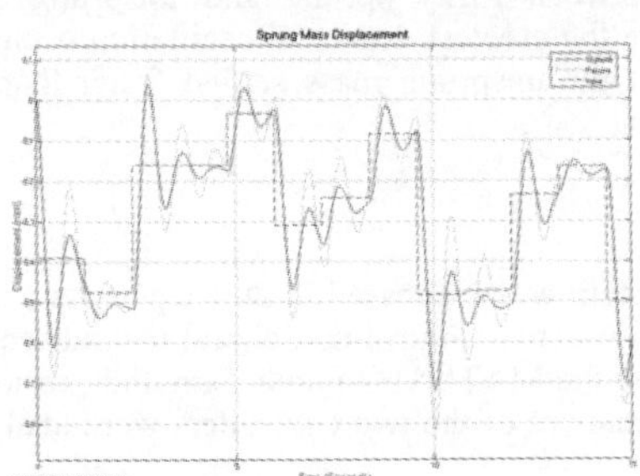

Fig. 6. Sprung mass displacement for fuzzy logic controller at random inputs.

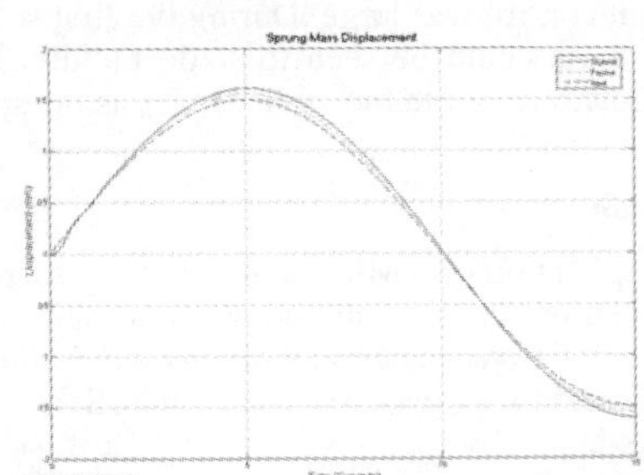

Fig. 7. Sprung mass displacement for fuzzy logic controller (0.05Hz sine input).

3.3 System response comparison

To examine the suitability of the two controllers for automotive applications, both controllers are assumed to be subjected to an input step function and compared against a reference wave to determine overall response and performance. The system performance was compared from the following aspects: overshoot, settling time, steady state error and noise. Fig. 8 shows the sprung mass displacements at on/off and fuzzy logic controllers as well as the reference wave. Immediately identifiable in this figure is the steady state error and the settling time differences between the two systems. The steady state error of the fuzzy logic waveform was first investigated and found not to have been a side affect of the controller. The settling time of the fuzzy logic controller was the shortest, and settled almost three seconds quicker than the reference. Consequently, the on/off controller did not settle at all, resulting in noise. This was due to controller limitations and subsequent magnification of the steady state error.

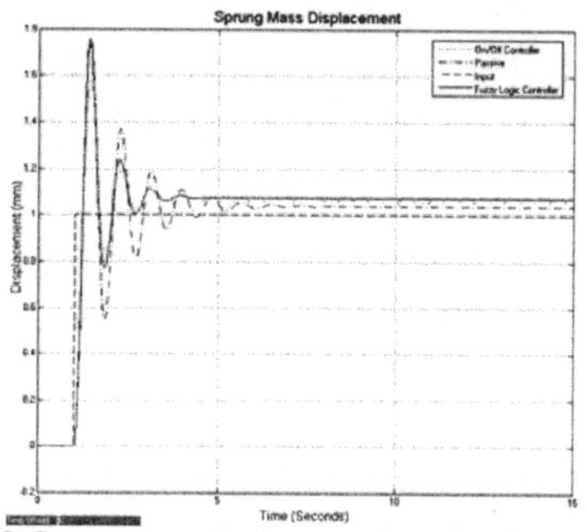

Fig. 8. Sprung mass displacement (step input)

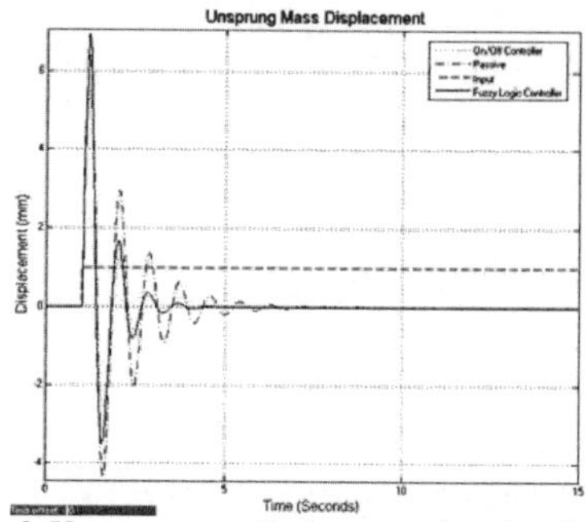

Fig. 9. Unsprung mass displacements (step input)

Generally characterised by the initial damping force, the overshoot was observed to be largest from the fuzzy logic controller, followed by the reference, and then the on/off controller. This is an important characteristic that demonstrated the superiority of the skyhook fuzzy logic controller. What was observed was that the fuzzy logic simulation settled the quickest whilst having the largest initial displacement, thus, proving the theoretical benefits of the skyhook model.

Illustrated in Fig. 9 is the unsprung mass displacement for the simulated system. Once again, the unsprung mass was observed for behaviour mimicking the ideal skyhook damper. The fuzzy logic controller exhibited this effectively, and could be observed when the sprung and unsprung responses were compared. What was found was that the discrepancy between fuzzy sprung and unsprung mass displacement control was large. During the first second where the greatest degree of oscillation occurred, the sprung mass could be seen to settle rapidly. In contrast the unsprung mass settled faster than the reference, however, not to the same degree as the sprung mass.

4. Conclusion

This application of a MR damper in a car suspension system was addressed in this paper. As a basis for the simulation, the skyhook model was translated to a semi-active control model and the subsequent equations were derived. These conditions were simulated using the MATLAB toolbox Simulink, resulting in the on/off and fuzzy logic skyhook controllers. The performances of the two controllers were analysed and compared.

The fuzzy logic controller resulted in a much smoother system output and a marked improvement in overall response. Sprung and unsprung mass accelerations were reduced, and the system settling time was greatly improved. The controller did also reinforce the theory of adding damping to the sprung mass at the expense of unsprung mass control, and an improvement in the sprung mass motion was observed. After examining sprung and unsprung mass behaviour, it was concluded that the controller and MR damper would have a beneficial affect on vehicle passenger comfort levels.

The on/off controllers simulated also exhibited characteristic skyhook behaviour, but did not perform to the same standard as the fuzzy based system. More noise was found in the on/off waveform, and the overall system response was found to be inferior though it is still better than the passive system.

References

[1] M.R. Jolly, J.W. Bender, J.D. Carlson, Properties and applications of commercial MR fluids, J. Intell. Mater. Syst. Struct., 10 (1999) 5-13.

[2] X.J. Wang, F. Gordaninejad, Lyapunov-based control of a bridge using magneto-rheological fluid dampers, J. Intell. Mater. Syst. Struct., 13 (2002) 415-419.

[3] Y.T. Choi, N.M. Wereley, Y.S. Jeon, Semi-Active Vibration Isolation Using Magnetorheological Isolators, J. Aircraft, 42 (2005) 1244-1251.

[4] W.H. Li, G.Z. Yao, G. Chen, S.H. Yeo, F.F. Yap, Testing and Steady State Modeling of a Linear MR Damper under Sinusoidal Loading, Smart Mater. Struct., 9 (2000) 95-102.

[5] J.D. Carlson, M.R. Jolly, MR fluid, foam and elastomer devices, Mechatronics, 10 (2000), 555-569.

[6] J. DeGaspari, Hot Stuff, in: Mechanical Engineering Magazine: ASME, 2002.

[7] M. Ahmadian, C. A. Paré, A Quarter-Car Experimental Analysis of Alternative Semi-Active Control Methods, J. Intell. Mater. Syst. Struct., 11(2000) 604-612.

[8] D. Ruan, Fuzzy Logic Foundations and Industrial Applications, in: Boston, Kluwer Academic Publishers, 1996.

An effective permeability model to predict field-dependent modulus of Magnetorheological Elastomers

X.Z. Zhang, W.H. Li*, B. Liu, P.B. Kosasih and X.C. Zhang
School of Mechanical, Materials and Mechatronic Engineering, University of Wollongong, Wollongong, NSW 2522, Australia

X.L. Gong and P.Q. Zhang
CAS Key Laboratory of Mechanical Behavior and Design of Materials, Department of Mechanics and Mechanical Engineering, University of Science and Technology of China, Hefei 230027, China

Abstract

Magnetorheological Elastomers (MREs) are composites where magnetic particles are suspended in a non-magnetic solid or gel-like matrix. Solid MREs are shown to have a controllable, field-dependent shear modulus. Up to now, most of conventional MREs models are based on the magnetic dipole interactions between two adjacent particles of the chain. These models can express the field-dependent properties of MREs with simple chain-like structure. In this paper, a effective permeability model is proposed to predict the field-induced modulus of MREs. This model is based on effective permeability rule instead of dipole interaction rule, which take into account the particle's saturation and can predict the mechanical performances of MREs with complex structure and components. A novel MREs is also designed to improve the magnetic energy density and field-dependent performance by using the iron particles with magnetizable soft shell.

Keywords: magnetorheological elastomers, mechanical properties, magnetic field, field-dependent shear modulus

1. Introduction:

Magnetorheological (MR) material is a class of smart materials whose rheological properties can be controlled rapidly and reversibly by the application of an external magnetic field. Traditionally, it is composed of MR fluids and MR foams, while MR elastomers became a new branch of them. MR materials typically consist of micron-sized magnetic particles suspended in a non-magnetic matrix. The magnetic interactions between particles in these composites depend on the magnetization orientation of each particle and on their spatial relationship, coupling the magnetic and strain fields in these materials and giving rise to a number of interesting magnetomechanical phenomena [1-6].

MR elastomers (MREs) are composites where magnetic particles are suspended in a non-magnetic solid or gel-like matrix. The particles inside the elastomer can be homogeneously distributed or they can be grouped (e.g. into chain-like columnar structures). To produce an aligned particle structure, the magnetic field is applied to the polymer composite during crosslinking so that the columnar structures can form and become locked in place upon the final cure. This kind of processing imparts special anisotropic properties to the viscoelastic materials. Only recently has the field responsiveness of the viscoelastic properties of these elastomers been explored [7-11].

MREs have a controllable, field-dependent modulus while MR fluids and MR foam have a field-dependent yield stress. This makes the two groups of materials complementary rather than competitive to

* Corresponding author. Phone: +61 2 4221 3490; Fax: +61 2 4221 3101.
E-mail address: weihuali@uow.edu.au

each other. In other words, the strength of MR fluids is characterized by their field dependent yield stress while the strength of MREs is typically characterized by their field dependent modulus. Other obvious advantages of MREs are that the particles are not able to settle with time and that there is no need for containers to keep the MR material in place. Because the chain-like or columnar structures have been locked in the rubber-like matrix during curing, the particles need no time to arrange again while MREs are applied an external magnetic field, thus the response time of MREs is much less than that of MR fluids (several ms).

MR fluids' field-dependent yield stress makes them to be widely used in various smart devices, such as dampers, clutches, and brakes. There is little doubt that there are numerous applications that can make use of controllable stiffness and others unique characteristics of MREs, such as adaptive tuned vibration absorbers (TVAs), tuneable stiffness mounts and suspensions, and variable impedance surfaces. [2, 4].

There are many models of MREs have been developed. The mechanical properties of MREs can be divided into two distinctive regimes: the composite properties without applied magnetic field and the composite properties with applied magnetic field. Usually the host composite in MREs is a rubber-like material with a nonlinear stress-strain relationship. Ogden's model has been widely used to model rubber-like materials[8]. MREs modulus is also a function of filler (iron particles) volume fraction and their zero-field modulus can be given by Guth model [12]. Most models of MR material field-dependent behavior are based on the magnetic dipole interactions between two adjacent particles of the chain. Ginder et al. and Davis [5, 12] have used finite element analysis method (FEM) to determine the values of the modulus under a varied magnetic field. For elastomer composites containing magnetically soft particles dispersed in natural rubber, a 40% maximum change in modulus was observed upon the application of a saturating magnetic field. The theoretical approaches show that the maximum shear modulus increment of conventional MREs is about 50% [12]. These dipole models can explain the simple ball-chain structure but did not take into account the complex structure or mixed components. The intergradation between unsaturation and saturation of MREs is very hard to be expressed by these models too. In order to fabricate high quality MREs, the complex structure and components ought to be presented, and a new model of MREs must be used to explain the mechanical properties of them.

In this paper, a new model of MREs is presented to explain field-induced modulus of MREs with complex structure and components and the particle's saturation is taken into account. A novel structure MREs is also be introduced for example.

2. Model of magnetic-field induced increase in shear modulus

In this chapter, the conventional MREs with simple chain-like structure are presented to introduce a new effective permeability model at first. Then, the field-induced modulus of MREs with complex structure can be predicted by this model because the conventional magnetic dipole and correlative theory can not explain the field-induced modulus of such complex structure. Finally, the comparison with conventional and novel MREs' field-dependent modulus is given.

For structural constructions as shown in Figure 1 (b), by using the Maxwell Garnett mixing rule [13], the effective permeability of chains can be predicted as:

$$\mu_{ceff} = \mu_m + 2\phi\mu_m \frac{\mu_p - \mu_m}{\mu_p + \mu_m - \phi(\mu_p - \mu_m)} \tag{1}$$

Here, particles have permeability μ_p and that of matrix is μ_m, radius of particle is R, distance between adjacent particles is d, volume fraction of particles is ϕ_p, volume fraction of particles in column is $\phi = 4R/3d$, volume fraction of column structure in MREs is $\phi_c = \phi_p/\phi = 3\phi_p d/4R$. (The column is composed of the particles chain and the rubber between the particles.)

At the cross-section normal to the columns, the MREs are composed by columns domain and matrix domain. The effective permeability along the direction of columns is calculated by parallel-connection rule:

$$\mu_{eff} = \sum_{i=1}^{n} \phi_i \mu_i \quad \text{and} \quad \sum_{i=1}^{n} \phi_i = 1 \tag{2}$$

where the ϕ_i and μ_i are the volume fraction and relative permeability of component i.
So the effective relative permeability of conventional MREs is:

$$\mu_{eff} = \mu_{ceff}\phi_c + \mu_m(1-\phi_c) = \mu_m + 2\phi_p\mu_m \frac{\mu_p - \mu_m}{\mu_p + \mu_m - \frac{4R}{3d}(\mu_p - \mu_m)} \tag{3}$$

According to the equation: $\tau = -\frac{1}{2}\mu_0 H_0^2 \partial\mu_{eff}(\varepsilon)/\partial\varepsilon$ [14] and $\varepsilon = x/d$ for shear mode (As shown in Fig.2), the shear stress of MREs can be expressed as:

$$\tau = 12\phi_p\mu_0\mu_m(\frac{R}{d})H_0^2 \frac{(\mu_p - \mu_m)^2\varepsilon}{\sqrt{1+\varepsilon^2}[3\sqrt{1+\varepsilon^2}(\mu_p + \mu_m) - 4(\frac{R}{d})(\mu_p - \mu_m)]^2} \tag{4}$$

And shear modulus is:

$$G = 12\phi_p\mu_0\mu_m(\frac{R}{d})H_0^2 \frac{(\mu_p - \mu_m)^2}{\sqrt{1+\varepsilon^2}[3\sqrt{1+\varepsilon^2}(\mu_p + \mu_m) - 4(\frac{R}{d})(\mu_p - \mu_m)]^2} \tag{5}$$

Because $\varepsilon << 1$, $\mu_p >> \mu_m$, and $R/d \approx 1/2$, the equation can be simplified as:

$$G \approx 6\phi_p\mu_0\mu_m H_0^2 \tag{6}$$

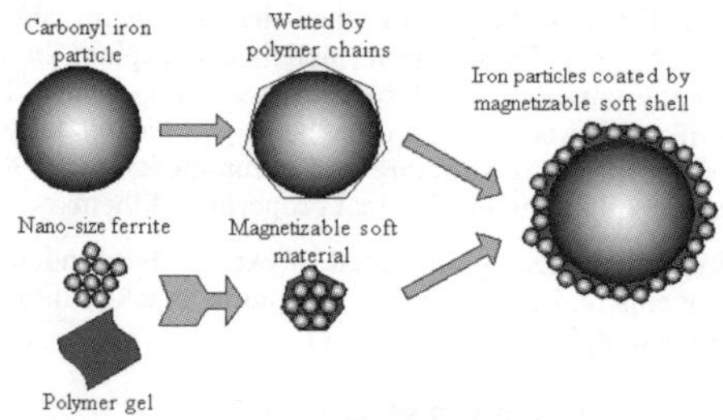

(a) Fabricate process

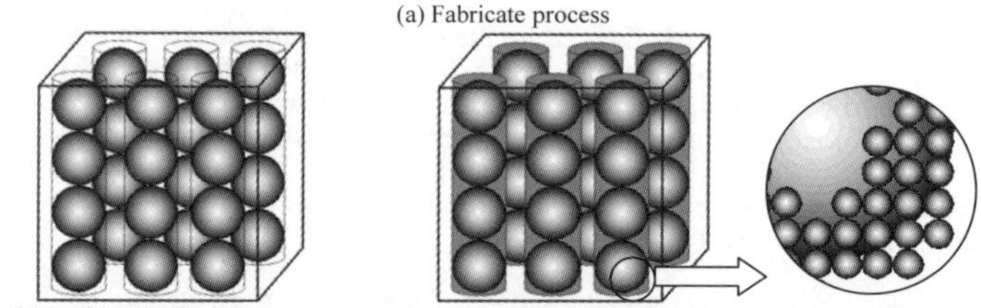

(b)Traditional MR elastomers construction (c)MR elastomers with nano-size particles additive

Fig.1 New construction MREs

Conventional MREs is general composed of magnetizable particles with average diameter about several microns and polymer matrix such as rubber [2]. In this paper, a new material design is used to improve the performance of MREs. Different from conventional methods, the iron particles are coated with magnetizable soft shell composed of nano-size ferrite powder and polymer gel. As shown in Fig.1(a), in order to fabricate this kind of magentizble soft shell, nano-size ferrite and polymer gel are pre-requisite. Firstly, it is needed for forming a continuous composite structure to wet the particles by polymer chains. Then mix the nano-size particles with polymer to produce the soft magnetism material. After that, coat the micro-size particles with the soft magnetism material and add this kind of coated particles into liquid rubber matrix to fabricate MREs. At last, the mixture is poured into a mould and a strong external magnetic field is applied to the mixture of particles and liquid rubber matrix to form chain like structure. With this process, magnetizable soft shell will deform and fill in the void space existing in micro-size particles chains. Thus, this method will increase the pack/energy density the effective permeability of MREs in special place and direction, and consequently will improve the MRE performances because the shear modulus depends on the energy density. The structural comparison between conventional MREs and the proposed new MREs is shown in Fig.1(b) and (c), where the nano-particles additives around micron-particles are zoomed out.

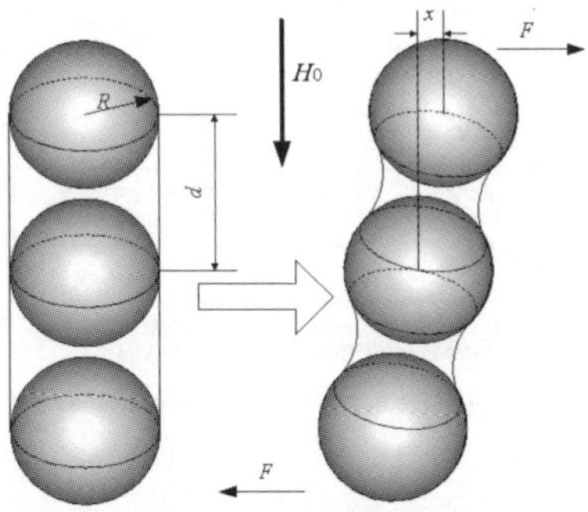

Fig.2 The shear model of particles chain

For novel MREs, the column structure without deformation is composed of micro-particles and soft shell. When MREs are deformed by force, the distance between particles is increased, some matrix around the column intrudes the column to refill the gap and the effective permeability is changed too. The changes result in the increase of magnetic energy and the field-dependent modulus of MREs (Fig.2).

The magnetizable soft shell is composed of nano-particles and rubber and the volume fraction of nano-particles in soft shell is ϕ_n. At the shear mode, when the shear strain is $\varepsilon = x/d$, the volume fraction of particles (including nano-size and micro-size particles) in the column can be expressed as:

$$\phi_{pc} = \frac{4R + \phi_n(3d - 4R)}{3d\sqrt{1+\varepsilon^2}} \tag{7}$$

By using the same deduction as above-mentioned, the shear modulus of nano-additive MREs is:

$$G_n = \frac{3}{4}\phi_p \mu_0 \mu_m (\frac{d}{R}) H_0^2 \frac{(\mu_p - \mu_m)^2 [\frac{4R}{d}(\phi_n - 1) - 3\phi_n]^2}{\sqrt{1+\varepsilon^2}\{3\sqrt{1+\varepsilon^2}(\mu_p + \mu_m) + [\frac{4R}{d}(\phi_n - 1) - 3\phi_n](\mu_p - \mu_m)\}^2} \tag{8}$$

Because $\varepsilon << 1$, $\mu_p >> \mu_m$, and $R/d \approx 1/2$, the equation can be simplified as:

$$G_n \approx \frac{3}{2}\phi_p \mu_0 \mu_m H_0^2 \left(\frac{2+\phi_n}{1-\phi_n}\right)^2 \tag{9}$$

When $\phi_n = 0$, equation 9 has the same form as equation 6.

Actually, the relative permeability of particles is the function of magnetic field intensity and if the saturation ought to be taken into account, it is just need replace μ_p by $\mu_p(H)$ in equation 9, where $\mu_p(H)$ can be obtained by experiment.

Here an empirical equation about $\mu_p(H)$ is given as[15]:

$$\mu_p(H) = \frac{H(\mu_p - 1) + \mu_p M_s}{H(\mu_p - 1) + M_s} \tag{10}$$

where μ_p is the largest relative permeability of particles, and M_s is the saturation magnetization and $\mu_0 M_s$=2.1T for Fe.

Assuming the relative permeability of particles is 1000 original and that of matrix is 1, the volume fraction of micro-paritcles in MREs is 27% and the volume fraction of nano-particles in magnetizable soft shell is 27% too, the field-dependent shear modulus of conventional MREs and that of the novel MREs are calculated and shown in Fig.3. Obviously, if the iron particles are covered by magnetizable soft shell, the shear modulus is increased significantly.

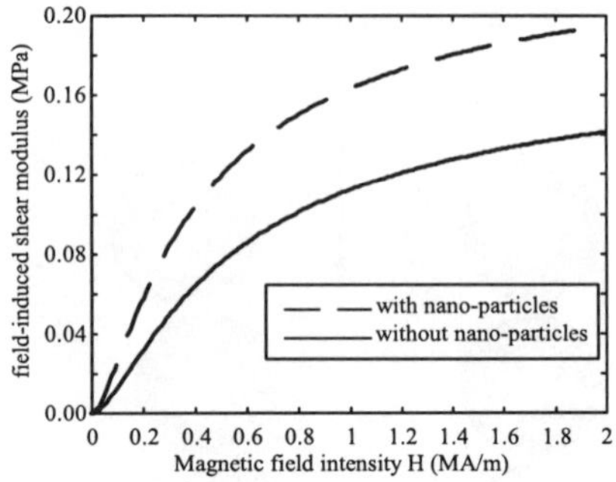

Fig.3 The predicted field dependent shear modulus

3. Zero-field shear modulus

The mechanical properties of MREs without applied magnetic field can use the conventional model to predict. The material can be seen as the composite properties with different filler. The approximated shear modulus G_{ran} of elastomer filled with randomly distributed, spherical rigid particles is simply given by the equation [12]:

$$G_{ran} = G_0(1 + 2.5\phi_t + 14.1\phi_t^2) \tag{11}$$

where G_0 is the shear modulus of the unfilled elastomer and ϕ_t is the volume fraction of filler. The modulus calculated by this equation is similar to the value of anisotropic MREs [12].

According to the Chapter 3, the volume fraction of filler in novel MREs is $\phi_t = \phi_p[4R + \phi_n(3d - 4R)]/4R$. Fixing the volume fraction of nano-size particles in film is 27% and R/d=1/2, the zero-field shear modulus can be calculated and shown in Fig.4.

According to the Fig.4, the added nano-size magnetism film can hardly improve the zero-field shear modulus of MREs.

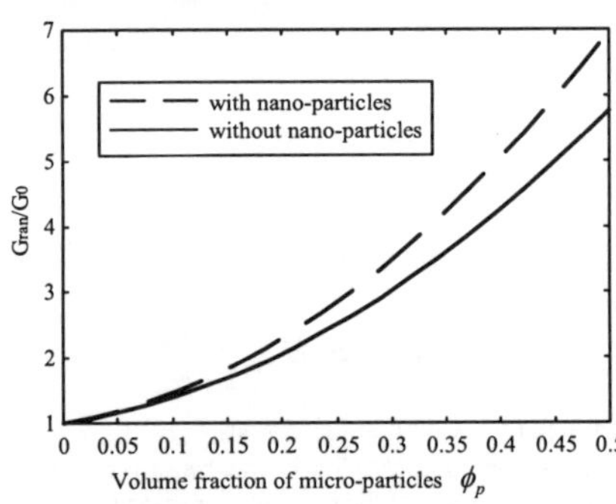

Fig.4 The predicted zero-field shear modulus

4. Conclusion

In this paper, a new equivalent permeability model is presented to explain field-induced modulus of MREs with complex structure and components. This model is based on efficiency permeability rule and takes into account the particle's saturation.

A novel structure MREs is also be introduced as an example. It is designed to improve the magnetic energy density and field-dependent performance. The new method will use the iron particles which are coated with magnetizable soft shell composed of nano-size ferrite powder and polymer gel. Mechanical performances of the newly proposed MREs are expected to be improved.

According to the simulated results, the novel MREs have the much larger field-dependent modulus than that of conventional ones. At the same time, the zero-field shear modulus of the MREs is not improved obviously by using the magnetizable soft shell.

References

[1] Shiga T, Okada A, Kurauchi T. Magnetroviscoelastic behavior of composite gels, J. Applied Polymer Science 58 (1995): 787–792.

[2] Ginder JM, Nichols ME, Elie LD, Tardiff JL. Magnetorheological elastomers: properties and applications, SPIE – The International Society for Optical Engineering. Proceedings of the 1999 Smart Structures and Materials on Smart Materials Technologies 3675 (1999):131–138.

[3] Carlson JD, Jolly MR. MR fluid, foam and elastomer devices, Mechatronics 10 (2000):555–569.

[4] Li WH, Yao GZ, Chen G, Yeo SH, Yap FF, Testing and Steady State Modeling of a Linear MR Damper under Sinusoidal Loading, Smart Mater. Struct., 9(2000): 95-102.

[5] Zhang XZ, Zhang PQ, Gong XL, Wang QM Study on the mechanism of the squeeze-strengthen effect in magnetorheological fluids, J. Appl. Phys., 96(2004): 2359-2364.

[6] Demchuk SA, Kuzmin VA. Viscoelastic properties of magnetorheological elastomers in the regime of dynamic deformation, J. Engineering Physics and Thermophysics 75(2002): 396–400.

[7] Lokander M, Stenberg B. Improving the magnetorheological effect in isotropic magnetorheological rubber materials, Polymer Testing 22 (2003): 677–680.

[8] Shen Y, Golnaraghi MF, Heppler GR. Experimental research and modeling of magnetorheological elastomers, J. Intelligent Material Systems and Structures 15 (2004): 27–35.

[9] Yalcintas M, Dai H. Vibration suppression capabilities of magnetorheological materials based adaptive structures, Smart Mater. Struct. 13 (2004): 1–11.

[10] Dorfmann A, Ogden RW. Magnetoelastic modeling of elastomers, European J. Mechanics A/ Solids 22 (2003):497–507.

[11] Farshad M, Benine A. Magnetoactive elastomer composites, Polymer Testing 23 (2004):347–353.

[12] Davis LC. Model of magnetorheological elastomers, J. Applied Phys, 85(1999):3348–3351.

[13] Kärkkäinen KK, Sihvola AH, Nikoskinen KI. Effective Permittivity of Mixtures: Numerical Validation by the FDTD Method, Ieee Transactions on Geoscience And Remote Sensing, 38(2000):1303-1308.

[14] Davis LC. Polarization Forces and Conductivi-ty Effects in Electrorheological Fluids, J.Appl.Phys, 72 (1992):1334-1340.

[15] Ginder JM, Davis LC. Shear Stresses in Magnetorheological Fluids: Role of Magnetic Saturation, Appl.Phys.Lett. 65(1994): 3410-3412.

The Simulation of Magnetorheological Elastomers Adaptive Tuned Dynamic Vibration Absorber for Automobile Engine Vibration Control

X.C. Zhang, X.Z. Zhang, W.H. Li* and B. Liu
School of Mechanical, Materials and Mechatronic Engineering, University of Wollongong, Wollongong, NSW 2522, Australia

X.L. Gong and P.Q.Zhang
CAS Key Laboratory of Mechanical Behavior and Design of Materials, Department of Mechanics and Mechanical Engineering, University of Science and Technology of China, Hefei 230027, China

Abstract:

The aim of this article is to investigate the use of a Dynamic Vibration Absorber to control vibration of engine by using simulation. Traditional means of vibration control have involved the use of passive and more recently, active methods. This study is different in that it involves an adaptive component in the design of vibration absorber using magnetorheological elastomers (MREs) as the adaptive spring.

MREs are kind of novel smart material whose shear modulus can be controlled by applied magnetic field. In this paper, the vibration mode of a simple model of automobile engine is simulated by Finite Element Method (FEM) analysis. Based on the analysis, the MREs Adaptive Tuned Dynamic Vibration Absorber (ATDVA) is presented to reduce the vibration of the engine. Simulation result indicate that the control frequency of ATDVA can be changed by modifing the shear modulus of MREs and the vibraion reduction efficiency of ATDVA are also evaluated by FEM analysis.

Keywords: magnetorheological elastomers, field-dependent shear modulus, Dynamic vibration absorber, Finite Element Method

1. Introduction

Most of mechanical, civil and construction systems are suffered from undesirable vibrations, which may damage the systems or even make the systems fail. In addition, vehicle manufacturers are continuously searching for ways in which vehicle refinement can be improved and its cost reduced. This involves the study of a wide range of noise sources from engine structure-borne noise to airborne type noise. Consumers now expect high levels of comfort and refinement from all passenger cars. In response, car manufacturers are placing vehicle refinement at the heart of their product development strategies. It is, therefore, of great importance to develop vibration control devices to reduce or suppress system vibrations [1-2].

Among all vibration control devices, research on Tuned Dynamic Vibration Absorbers (TDVAs) is, perhaps, the richest in terms of total number of investigations and the time devoted to these investigations. Since their invention in 1900s [1], TDVAs have effectively suppressed vibrations of machines and structures. However, the effectiveness of conventional TDVA is always limited due to the narrow frequency ranges. In many practical applications, off-tuning of a TDVA occurs because of varying usage patterns and loading conditions. To overcome these shortcomings, Adaptive Tuned Dynamic Vibration Absorbers (ATDVAs) are extensively studied. The ATDVA is similar to conventional TDVA but with adaptive elements that can be used to change the tuned condition.

* Corresponding author. Email: weihuali@uow.edu.au

Generally, ATDVA designs adopt two major groups: one is to use variable geometries and the other is to use smart or intelligent materials [3]. From the important view of system reliability and maintainability of ATDVA designs, Sun et al. [4] suggested the use of intelligent materials as alternatives. Davis et al. [5] suggested the use of a piezoelectric ceramic elements as part of the device stiffness, and reported the use capacitive shunts to vary the device natural frequency from 290 Hz to 350 Hz. Flatau et al. [6] used the magnetostrictive material Terfenol-D to develop an ATDVA that achieved a natural frequency variation from 1375 Hz to 2010 Hz. Williams et al.[3] used the thermal actuation of shape memory alloys (SMA) as tuning materials to develop an ATDVA. Testing showed that the SMA ATVA natural frequency could be varied by approximately 15% from 38.5 Hz to 46.5 Hz. Though many attempts have been made, there are still some shortcomings, such as low responses, complex structures, prune to working conditions, etc.

This paper proposes to design a new ATDVA working with a new intelligent material: MR elastomers. With the magnetic field increases, the shear modulus of MR elastomers increases steadily. Removal of the magnetic field, MR elastomers immediately reverse to their initial status. Based on these unique characteristics that their mechanical properties can be magnetically controlled, MR elastomers have found a lot of applications, such as dampers, engine mounts, and shock absorbers [8]. The use of MREs to develop ATDVAs are expected to have many advantages: very fast response (less than a few milli seconds), simple structure, easy implementation, good maintenance, high stability, and effective control. Ginder et al. [8] did a pioneer work that utilized MR elastomers as variable-spring-rate elements to develop an ATDVA. Their results indicated that a natural frequency range from 580 Hz to 710 Hz at the magnetic field 0.56 Tesla. Gong et al. [9] used MR elastomers to develop an ATDVA and reported the natural frequency varying 50% from its centre frequency.

The crankshaft of automobile engine is subject to dynamic torsional deformation as well as dynamic axial displacement, both causing vibration. Vibration control is of great importance because excessive vibration levels may damage the engine and create an annoying human environment. In this paper, the purpose of the new ADTVA is to suppress the vibration of an engine when the external stimulant forces with variant frequencies are added on this engine. Hence, before developing the new ATDVA, it is proposed to examine the validation of this ATDVA. FEM can be used over the complete frequency range of interest. It provides a valuable tool for the rapid assessment of the noise implications of design changes, thereby reducing the number of options requiring detailed analysis. Following this, on the basis of the principle of the MREs and the vibration suppression, the finite element method is applied to build an engine model according to the properties of the engine 1.3L VVT2. The transmissibility of the engine model corresponding to stimulant forces with variant frequencies is calculated and recorded by the FEM too. Based on the simulation results, MATLAB is used to analyze and conclude the relationship between the properties of the MREs and the effect of the vibration suppression.

This paper is structured as four sections. In Section 1, the intelligent material MR Elastomer is introduced, and its' changeable range of the shear modulus is discussed. In Section 2, the principle of the TDVA is explained. Section 3 is to demonstrate the Finite Element Modeling and the simulation results. Finally, the conclusion is given.

2. The MR Elastomer in the ATDVA

MR elastomers are suspensions of magnetized particles dispersed in a polymer medium, such as rubbers [7]. When individual particles are exposed to an applied magnetic field, magnetic dipole moments pointing along the magnetic field are induced in the particles. Pairs of particles then form head-to-tail chains. After the matrix is cured, the particles are locked into place and the chains are firmly embedded in the matrix[11]. For particle concentrations of interests, the shear modulus G of the elastomer filled with dispersed rigid particles can be calculated with acceptable accuracy by following function [10],

$$G = G_0\ (1 + 2.5\phi + 14.1\phi^2) \qquad (1)$$

Here G_0 is the shear modulus of the unfilled elastomer, while ϕ is volume fraction of particles.

According to [8], the field induced modulus of MREs is in direct proportion to the magnetic intensity controlled by the electrical current intensity in the coil enlaced on the absorber. By assuming MRE is as same as the one in [8], corresponding to the elastomer with variant density of rigid parcels, the changeable range of the shear modulus ΔG can be represented by

$$\frac{\Delta G}{G(0)}(\max) = \frac{0.1911\,MPa}{G_0} \qquad \text{when } \phi = 0.266. \qquad (2)$$

Here typically, G_0 is close to 1.1 MPa. Therefore, $\Delta G/G(0)\,(\max)$ could be roughly equal to 50%. Following this, the changeable range of the shear modulus of the MRE is located between 1.1MPa and 1.5MPa.

3. ATDVA

Suppose a mass-spring system or a primary system is excited by a harmonic driving force $P\sin\omega t$. When the driving frequency equals to the natural frequency of the system, the response is infinite. The TDVA is essentially a secondary mass attached to the primary system via a spring and a damper. When the natural frequency of the TDVA is tuned such that it coincides with the frequency of unwanted vibration in the primary system, the vibration of the primary system is greatly reduced. Thus, the energy of the primary system is apparently "absorbed" by the TDVA (Fig.1).

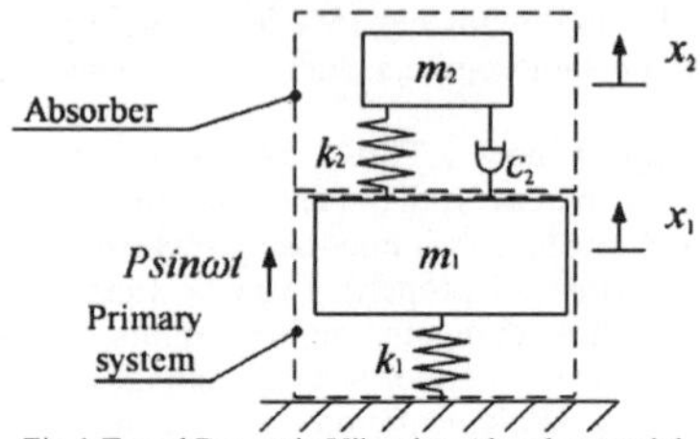

Fig.1 Tuned Dynamic Vibration Absorber model

The vibration reduction efficiency can be preparatory predicted by using a simplest model based on classical vibration theory. According to [12], a series of parameters are defined and shown in table 1.

Table.1 Parameters for building the mathematical model

$\omega_{n1} = \sqrt{\frac{k_1}{m_1}}$	Natural frequency of structure	$\omega_{n2} = \sqrt{\frac{k_2}{m_2}}$	Natural frequency of TDVA
$\varsigma_1 = \frac{c_1}{2\omega_{n1} m_1}$	Damping ratio of structure	$\varsigma_2 = \frac{c_2}{2\omega_{n2} m_2}$	Damping ratio of TDVA
$\mu = \frac{m_2}{m_1}$	Mass ratio	$r = \frac{\omega}{\omega_{n1}}$	Force frequency ratio
$g = \frac{\omega_{n2}}{\omega_{n1}}$	Natural frequency ratio	δ_{st}	Static deflection

Assume the stimulant force as $F = k_1 \delta_{st} e^{st}$, the transmissibility equation for the force-exited system can be expressed as following:

$$\frac{X_1}{\delta_{st}} = \frac{(g^2 - r^2 + 2\varsigma_2 grj)}{(-r^2 + 2\varsigma_1 rj + 2\varsigma_2 \mu grj + 1 + \mu g^2)(-r^2 + 2\varsigma_2 rgj + g^2) - \mu(2\varsigma_2 grj + g^2)^2} \tag{3}$$

As expressed in this equation, without changing the mass of the structure and the TDVA, the only method to change the displacement of the system is to adjust the spring and damper of the TDVA.

In this example, by assuming the engine as the primary system, and the TDVA replaced by the ATDVA, the vibration suppression can be achieved successfully by adjusting k_2 and c_2. Because the values of the k_2 is related to the values of the shear modulus of the MREs, changing the shear modulus of the elastomer could be an efficient method to adapt the requirements of the vibration suppression. As discussed in Section 2, since the shear modulus of the MREs can be adjusted from 1.1MPa to 1.5MPa by changing the flux, it is proposed that using MREs can directly develop a relationship between the flux and the natural frequency of the system. In the following section, based on the parameters of the MR Elastomer, a FEM is applied to examine the validation of the new ATDVA.

4. Finite Element Modeling and Analyzing

In order to examine the validation of the MRE semiactive vibration absorber, the engine 1.3L VVT2 is introduced in this paper. By respectively setting the passive tuned vibration adapter and the MRE semiactive vibration absorber on this engine in ANSYS simulation environment, the structural vibrations of the engine system are tracked and analyzed so as to demonstrate the effects of the proposed MRE vibration absorber.

4.1 Engine 1.3L VVT2 and Finite Element Modeling

Based on the properties of engine 1.3L VVT2, this simulation model is simplified as a block with overall dimension of 607×634×605 and mass of 138kg. Furthermore, because treating the source of the vibration is the most effective and often the most economical solution to vibration problems, it is proposed to isolate the engine from the car base by the engine mounts with rubber part. Normally, there are several pairs of engine mounts to fix the engine on the car base. Therefore, in the simulation model, the block representing the engine is attached by four rubber pads which are fixed in the simulation environment. Another function of the engine mounts is to isolate the car from the engine vibration in working status. The vibration transmissibility depends on disturbance frequency. Because a passive system is only effective for disturbance with frequencies much higher than its natural frequency, the first compulsory work is to set the natural frequency of the system much less than the rotation speed of the engine which is from 63Hz to 100Hz so as to eliminate the influence from the engine. In this example, by predefining the properties of the four rubber pads, the natural frequency of the system is located around 43Hz.

Table.2 properties of 1.3L VVT2

Model	Unit	Value
Number of Cylinders		4
Number of Valves		16
Displacement	cm3	1297.5
Rated Power/Speed	kW/rpm	65/6000
Max. Torque/Speed	Nm/rpm	118/3800
Fuel Quality	RON	93
Emission	Euro IV	
Overall Dimension(L×W×H)	mm	607×634×605
Net Weight	kg	138

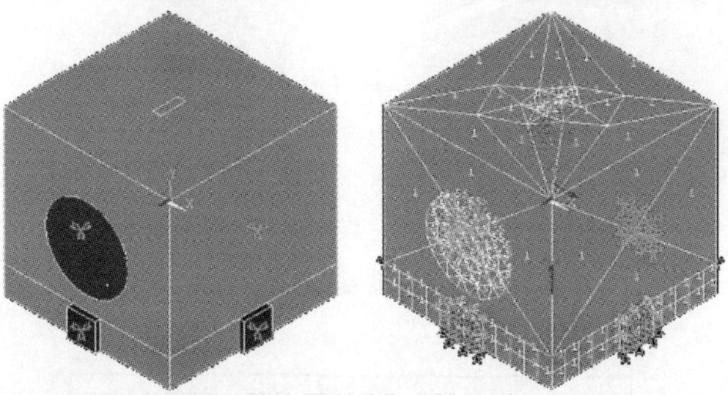

Fig.2 FE Modeling of the engine

As the basis for numerical parametric studies for further evaluation of the dynamic performance of the proposed new ATDVA, the baseline model is built as shown in Fig.2. In order to reduce the calculations without influencing the accuracy of the simulation, the element used to build this model is solid brick with 20 nodes; the meshing coefficient is set as '9' which is rough but acceptable.

On the purpose of observing the transmissibility of the engine model, a series of stimulant forces are added on this engine. By considering the working status of the engine, the forces from the engine rotation provide the vibration in the vertical (Z) and horizontal (X and Y) directions. In this simulation process, the frequency domain of the stimulant forces provided by the internal forces from the engine rotation are both set from 20Hz to 100Hz, while the phase difference between the two directions (X and Z) of the forces from the engine rotation is set as $\pi/2$ to simulate the rotary eccentric force.

By setting the frequency independent damping of the system as 0.05, the transmissibility of the engine block in the three orthogonal directions can be illustrated in Fig.3. In this figure, it is obvious that the most prominent vibration is in the vertical (Z) direction, while the others' amplitudes of the vibrations are much smaller. Hence, the focus of this FEM should be put on suppressing the vibration in the vertical direction. Because the amplitude of the vibration is directly related to the stimulant force, the transmissibility of the system could be described as $Amplitude/Force$ (m/N).

4.2 Tuned Vibration Absorber

In order to suppress the vibration in a continuous frequency domain, an ATDVA model is added on the engine model as shown in Fig.4. The materials of the ATDVA are referred to the requirements of the design. Most part of the ATDVA is meshed roughly, while the MR Elastomer is meshed in detail so as to guarantee the accuracy of the simulation.

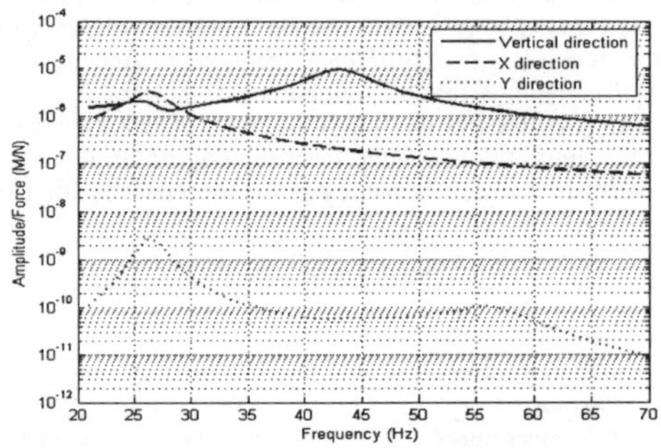

Fig.3 Transmissibility response of the engine

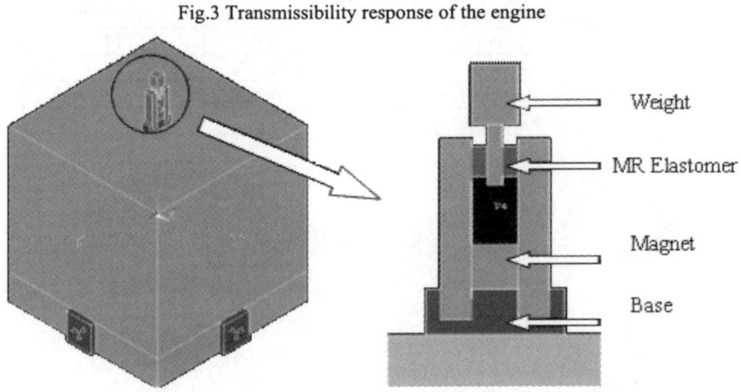

Fig.4 FE Modeling of the engine equipped with ATDVA

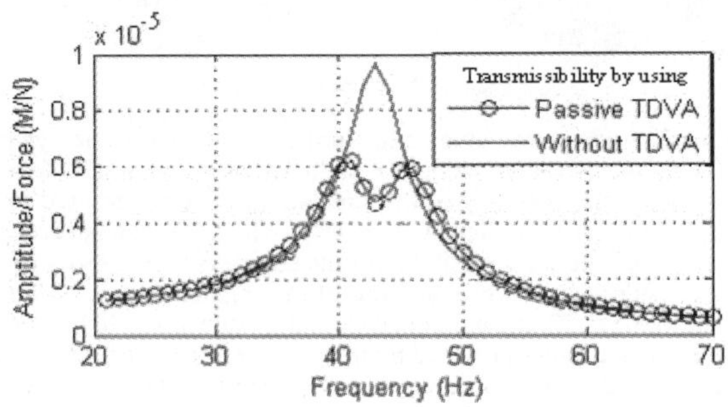

Fig.5 Effect of TDVA

Based on the discussion in Section 2 and 3, in this paper, the way to reduce the disturbance from the car base is to change the shear modulus of the MR Elastomer so as to keep the natural frequency of the ATDVA as same as the frequency of the stimulant forces. In this example, the mass of the ATDVA is set as 1% of the engine, while the shear modulus is set around 1.3MPa. The simulation results are illustrated in Fig.5.

As shown in this figure, there are two peaks around the Resonance frequency, and this passive TVA is only capable of reducing the amplitudes between the two peaks. In order to extend the capability of the TVA, it has been proposed that an Adaptive Tuned Dynamic Vibration Absorber (ATDVA) employing MREs must be designed and manufactured. The basic magnetic field is generated by a permanent magnet and the accessional magnetic intensity is controlled by the electrical current intensity in the coil. The induced magnetic field is imposed in the direction of particles' chains in MREs and it works at shear modulus. Hence, this absorber is capable of instantaneously changing its stiffness, thus it can switch between resonance frequencies, increasing its effective bandwidth as compared to classical TDVA for vibration control. Because a specific MR Elastomer with changeable shear modulus from 1.1MPa to 1.5MPa is applied in this simulation, the validation of the new ATDVA must be examined based on this restriction. Therefore, this simulation could be considered as achieving a function to study the correlation between the shear modulus of the MREs and the vibration amplitude. Following this, by increasing the values of the shear modulus of the MREs, the amplitudes of the system stimulated by a series of multi-frequency forces are derived as shown in Fig.6. By adjusting the values of the MREs' modulus according to disturbing force frequency, the effect of the new ATDVA is highlighted by comparing to the passive TVA in Fig.7.

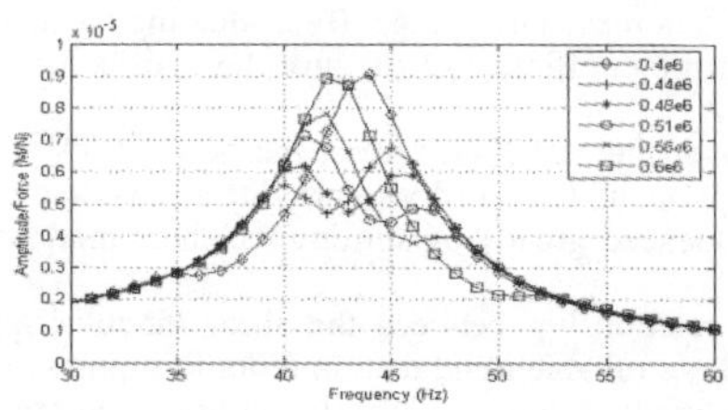

Fig.6 Vibration amplitudes corresponding to variant shear modulus of MREs

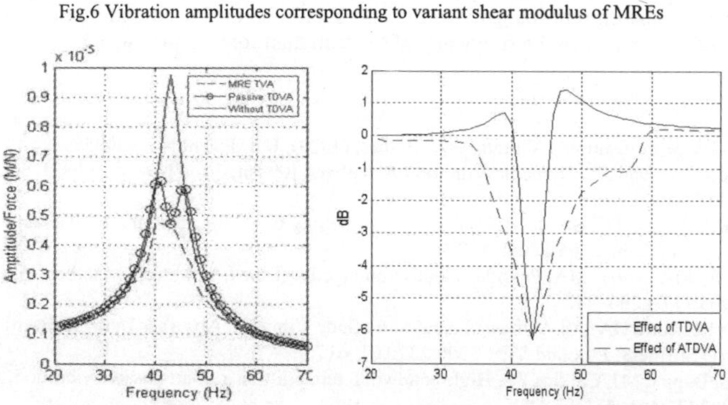

Fig.7 The transmissibility of the system with passive and Semiactive TDVA

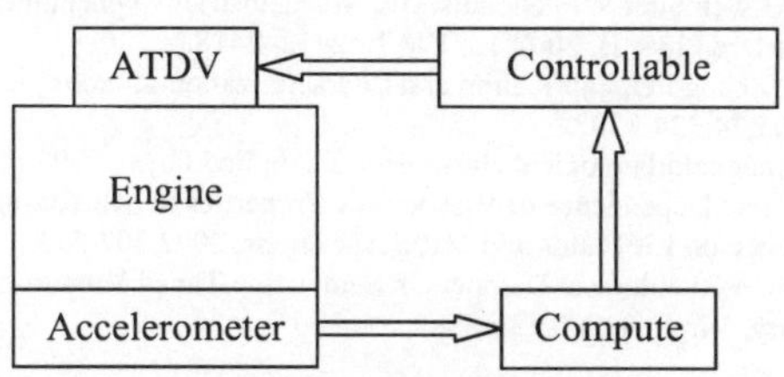

Fig.8 ATDVA System

As demonstrated in above figure, it is obvious that the new ATDVA performs much better than the passive TDVA whose shear modulus cannot be adjusted. In the frequency domain of interest which is from 35Hz to 50Hz, the vibration of the system cannot only be suppressed in the natural frequency of the system, but also can be suppressed in other frequencies. Therefore, it is valid to design and manufacture this new ATDVA and equip it on the engine for vibration suppression.

Based on the validation provided by simulation, the principle of the new ATDVA can be illustrated as in Fig.8. Detecting the vibration by the accelerometers mounted on the engine, a system will analyse the vibration information and provide control command to the controllable power so as to change the current intensity in coils which adjust the magnetic field intensity around the MREs. As discussed in Section 2 and 3, the ATDVA will vary the natural frequency of the system in variant vibration situations, so that the new system will not only achieve the vibration isolation at different frequencies but also fulfil the overall performance requirements.

5. Conclusion

In context of this paper, a new ATDVA employing MR Elastomer has been introduced to the engine vibration problem. By analyzing the principle of the MRE ATDVA, FEM is applied to evaluate the validation of this new vibration absorber. From the simulation results, there are several conclusions can be made as following.

Changing the magnetic intensity controlled by the electrical current intensity in the coil, the shear modulus of the MREs can shift in a reasonable range. By evaluating the analysis results of the simulation, it can be concluded that the elastic coefficient of the absorber can be continuously changed to shift the natural frequency of the ATDVA.

By tracking the frequency of the stimulant forces, the new ADTVA is capable of suppressing the vibration in a wider frequency domain. The FEM also provide a reliable evidence that the transmissibility of the whole system can be influenced greatly by shifting the shear modulus of the MRE in the practical range.

Finally, investigating the relationship between the shear modulus and the transmissibility of the engine system, the FEM provides a reliable trend that, in order to suppress the vibration of the engine, the higher the frequency of the stimulant force is, the larger the shear modulus of MREs is required. Thereupon, the electrical current intensity in the coil should be increased continuously so as to suppress the vibration of the system when the frequency of the stimulant force is increasing.

Reference

[1] Frahm H, Device for Damping Vibrations of Bodies, (1909), U.S. Patent No. 989958.

[2] Liu K, Liu J, The Damped Dynamic Vibration Absorbers: Revisited and New Results, J. Sound Vib., 2005;284:1181-1189.

[3] Williams KA, Chiu GTC, Bernhard RJ, Dynamic modelling of a shape memory alloy adaptive tuned vibration absorber, J. Sound Vib., 2005;280:211–234.

[4] Sun JQ, Jolly MR, Norris MA, Passive, adaptive and Active Tuned. Vibration Absorbers-A Survey, J. Mech. Design, 1995;117B:234–242.

[5] Davis CL, Lesieutre GA, An Actively Tuned Solid State Vibration Absorber Using Capacitive Shunting of Piezoelectric Stiffness, J. Sound Vib., 2000;232:601–617.

[6] Flatau AB, Dapino MJ, Calkins FT, High-bandwidth tunability in a smart passive vibration. Absorber, Proc. SPIE, 1998;3327:463–473.

[7] Carlson JD, Jolly MR, MR fluid, foam and elastomer devices, Mechatronics, 2000;10:555–569.

[8] Ginder JM, Clark SM, Schlotter WF, Nichols ME, Magnetostrictive phenomena in magnetorheological elastomers, Int. J. Modern Phys. B, 2002;16, 17&18:2412–2418.

[9] Gong XL, Zhang XZ, Zhang PQ, Fabrication and Characterization of Isotropic Magnetorheological Elastomers, Polymer Testing, 2005;24:324-329.

[10] Davis LC, Model of magnetorheological elastomers, J. Applied Phys., 1999;85(6):3348–3351.

[11] Bellan C, Bossis G, Field Dependence of Viscoelastic Properties of MR Elastomers, Proceedings of the 8th International Conference on ER Fluids and MR Suspensions, 2002:507-513.

[12] Koo JH, Using Magneto-Rheological Dampers in Semiactive Tuned Vibration Absorbers to Control Structural Vibrations, Blacksburg, Virginia, 2003:31-35.

Spatial Signal Characteristics of Shallow Paraboloidal Shell Structronic Systems

H.H. Yue[a], Z.Q. Deng[a], H.S. Tzou[b]

[a]*School of Mechatronic Engineering, Harbin Institute of Technology,Harbin, 150001, China*

[b]*Department of Mechanical Engineering, StrucTronics Lab, University of Kentucky, Lexington, KY 40506-0503, USA*

Received 15 March 2006; received in revised form 10 May 2006; accepted 15 May 2006

Abstract

Based on the smart material and structronics technology, distributed sensor and control of shell structures have been rapidly developed for the last twenty years. This emerging technology has been utilized in aerospace, telecommunication, micro-electromechanical systems and other engineering applications. However, distributed monitoring technique and its resulting global spatially distributed sensing signals of thin flexible membrane shells are not clearly understood. In this paper, modeling of free thin paraboloidal shell with spatially distributed sensor, micro-sensing signal characteristics, and location of distributed piezoelectric sensor patches are investigated based on a new set of assumed mode shape functions. Parametric analysis indicates that the signal generation depends on modal membrane strains in the meridional and circumferential directions in which the latter is more significant than the former, when all bending strains vanish in membrane shells. This study provides a modeling and analysis technique for distributed sensors laminated on lightweight paraboloidal flexible structures and identifies critical components and regions that generate significant signals.

PACS: 07.10.Eq; 46.70.De; 73.43.cd; 77.65.-J;

Keywords: Modeling; Distributed piezoelectric sensor; Free membrane shell; Mode shape function

---

1. Introduction

Smart structures and structronic system are widely used in the field of active shape and vibration control of high performance structures. Detailed reviews of the state of smart structure technologies are provided in a number of papers [1-3]. Because piezoelectric material exhibits both the direct and converse piezoelectric effects, it can serve as distributed sensor and actuator for structural health monitoring and precision control [4-7]. Crawley has realized active control of structure with piezoelectric actuators [8]. Clark proposed dynamic modeling of piezoelectric beam and panel [9]. NASA has investigated actuation of smart girder grillage structure with initiative PZT in their CSI (Control-Structures Integration) program [10]. In the development of aerospace technology, there exists the higher demand for precision control of structural shape and vibration. Large-scale, flexibility and low stiffness are often associated with space structures, e.g., solar panel, satellite aerial, optic system and their support structures, while still requiring high-precision and accuracy in their operations [11, 12]. Paraboloidal shells of revolution are often used as key components in many advanced mirror and reflector structures, so active control of multifarious thin

*Corresponding author. Tel.: +86-0451-8641-3802; fax: +86-0451-8641-3857.

E-mail address: block@hit.edu.cn (H. H. YUE).

paraboloidal shell structures is a challenge due to their non-constant radii induced nonlinear local behavior.

To explore techniques for exact control and sense of paraboloidal shells, a large number of studies have been carried out for years. Independent modal control of flexible rings using orthogonal convolving piezoelectric sensors and actuators was studied [13]. Distributed excitation and control of cylindrical shells with fully distributed actuator, partially distributed actuators, segmented actuator patches, line actuators, etc. were also investigated [14,15]. Distributed sensing and control of shallow spherical shells have also been investigated in the past few years [16]. Micro-signals of conical shells and shell panels were evaluated [17]. Distributed modal voltages and their spatial strain characteristics of toroidal shells and spherical shells were recently investigated [18,19]. Spatially distributed sensing of paraboloidal shell based on the bending approximation theory was studied recently [20]. However, distributed sensing and control of flexible paraboloidal shells and their applications to micro and large precision optical and antenna membrane structures are still lacking and need to be fully explored. Based on a *new* set of mode shape function, this study is to investigate spatially distributed sensing signals of lightweight membrane shells of revolution, and to evaluate the factors that influence the effects of sensing signals, e.g., shell curvature, sensor segment location and major signal components. So that precision control strategies can be developed later accordingly.

2. Modeling and Mode Shape Functions of Free Membrane Paraboloidal Shells

Fig.1. shows a generic paraboloidal shell of revolution that is placed in a tri-orthogonal global coordinate system (X, Y, Z) and the shell itself is defined in a tri-orthogonal curvilinear coordinate system (ϕ, ψ, α_3). Two radii of the double curvatures respectively are R_ϕ and R_ψ; ϕ denote angular change in the meridian direction and ψ denote angular change in the circumferential direction. The Lamé parameter of the shell are $A_1 = R_\phi$ and $A_2 = R_\psi \sin\phi$.

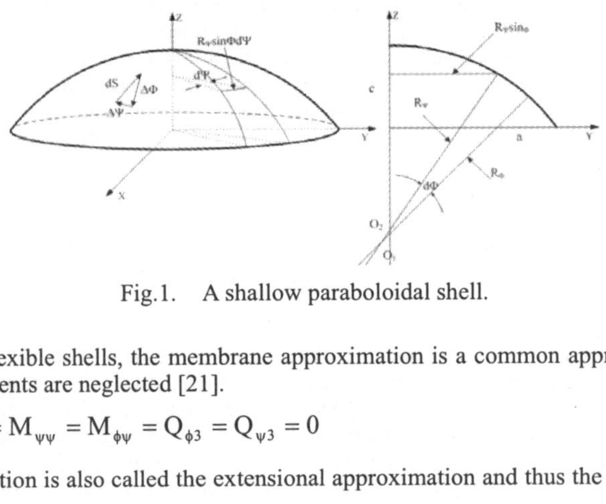

Fig.1. A shallow paraboloidal shell.

For thin and flexible shells, the membrane approximation is a common approximation in which all bending components are neglected [21].

$$M_{\phi\phi} = M_{\psi\psi} = M_{\phi\psi} = Q_{\phi 3} = Q_{\psi 3} = 0 \tag{1}$$

This approximation is also called the extensional approximation and thus the system equation can be derived:

$$\frac{\partial(R_\psi N_{\phi\phi}\sin\phi)}{\partial\phi} + R_\phi \frac{\partial N_{\psi\phi}}{\partial\psi} - N_{\psi\psi} R_\phi \cos\phi + R_\phi R_\psi \sin\phi F_1 = R_\phi R_\psi \sin\phi \rho h \ddot{u}_\phi \tag{2}$$

$$\frac{\partial(R_\psi N_{\phi\psi}\sin\phi)}{\partial\phi} + R_\phi \frac{\partial N_{\psi\psi}}{\partial\psi} - N_{\psi\phi} R_\phi \cos\phi + R_\phi R_\psi \sin\phi F_2 = R_\phi R_\psi \sin\phi \rho h \ddot{u}_\psi \tag{3}$$

$$-R_\phi R_\psi \sin\phi(\frac{N_{\phi\phi}}{R_\phi}+\frac{N_{\psi\psi}}{R_\psi})+R_\phi R_\psi \sin\phi F_3 = R_\phi R_\psi \sin\phi \rho h \ddot{u}_3 \tag{4}$$

Here F_1, F_2 and F_3 respectively are excitation forces in meridional, circumferential and transverse directions. With the membrane approximation, the membrane forces become:

$$N_{\phi\phi} = K\{\frac{\cos^3\phi}{b}(\frac{\partial u_\phi}{\partial\phi}+u_3)+\mu\frac{\cos\phi}{b\sin\phi}(\frac{\partial u_\psi}{\partial\psi}+u_\phi\cos\phi+u_3\sin\phi)\} \tag{5}$$

$$N_{\psi\psi} = K\{\frac{\cos\phi}{b\sin\phi}(\frac{\partial u_\psi}{\partial\psi}+u_\phi\cos\phi+u_3\sin\phi)+\mu\frac{\cos^3\phi}{b}(\frac{\partial u_\phi}{\partial\phi}+u_3)\} \tag{6}$$

$$N_{\phi\psi} = N_{\psi\phi} = \frac{K(1-\mu)}{2}\frac{\cos\phi}{b\sin\phi}(\frac{\partial u_\phi}{\partial\psi}+\cos^2\phi\sin\phi\frac{\partial u_\psi}{\partial\phi}-u_\psi\cos\phi) \tag{7}$$

Note the membrane stiffness $K = Yh/(1-\mu^2)$ and the bending stiffness $D = Yh^3/12(1-\mu^2)$, here Y is Yong's Modulus and μ is Poisson's ratio, and $b = a^2/2c$. For evaluating distributed sensor signals and free oscillation behavior of paraboloidal shells, fundamental structure dynamics of shells need to be investigated. **Fig.2.** shows a laminated shell with distributed piezoelectric layer and its resultant force /moment distribution of a shell element near the shell edge are also illustrated.

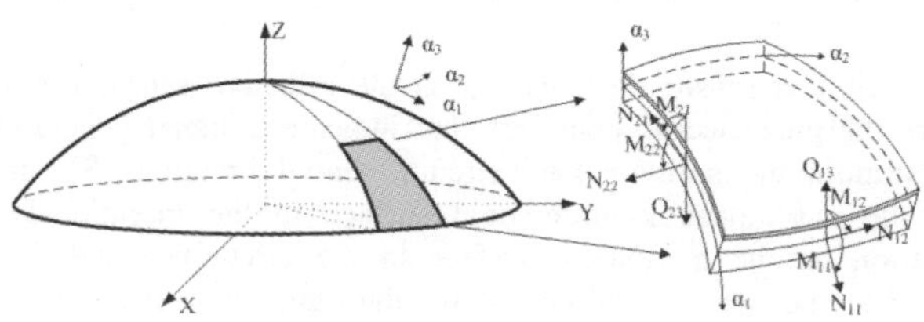

Fig.2. Stresses of an element of paraboloidal shell with distributed sensor patch.

All external mechanical and electric excitations are zero in free vibration which exhibits shell's intrinsic dynamic characteristics. For a free-floating paraboloidal membrane shell, all force and moment on the edge are zero, i.e., the displacement and rotation angle are not zero. These boundary conditions (B.C.s) are defined as follows:

$$N_{\phi\phi} = 0\,(\phi = \pm\phi^*);\; N_{\psi\psi} = 0;\; Q_{\phi 3} = 0;\; M_{\phi\phi} = 0 \tag{8}$$

Note that ϕ^* is the shell's meridional angle, ϕ is a meridional variable between $0\sim\phi^*$. Recall that the membrane approximation requires all moment terms zero, so the mode shape function needs to satisfy the membrane force B.C.s on the boundary. Thus, three *new* mode shape functions satisfying the B.C.s are selected as:

$$U_{1k} = A_k \cos\frac{(2k+1)\pi}{\phi^*}\phi(\sin\phi)^{k+1}\cos k\psi \tag{9-a}$$

$$U_{2k} = -A_k \cos\phi(\sin\phi)^{k+1}\sin k\psi \tag{9-b}$$

$$U_{3k} = A_k(k+1)\cos\phi(\sin\phi)^k \cos k\psi \tag{9-c}$$

Here the mode number k=1, 2, 3…, and A_k is the k-th modal amplitude. The total dynamic response can be represented by the summation of all participating natural modes and their respective modal participation factor:

$$u_i(\phi,\psi,t) = \sum_{k=1}^{\infty} \eta_k(t) U_{ik}(\phi,\psi), i = 1,2,3 \tag{10}$$

Note that $\eta_k(t)$ is the modal participation factor; $U_{ik}(\phi,\psi)$ is the mode shape function; k denotes the k-th mode. Micro-signal characteristics of distributed sensor patches on paraboloidal shells are discussed next.

3. Micro-Sensing Signal of Distributed Piezoelectric Patches

Piezoelectric patches spatially distributed on shell surface provide distributed global dynamic signals of elastic paraboloidal shells. For paraboloidal shell, two dimensional distributed sensors are considered and output signals from these distributed sensors are evaluated.

It is assumed that the distributed piezoelectric sensor is uniformly thin, as comparing with shell thickness. Thus, the piezoelectric sensor strains are constant and it is equal to the outer surface strains of the paraboloidal shell. It is worth noting that as the distributed sensor, only the direct piezoelectric effect is considered. So one can define an open-circuit voltage ϕ^s in the transverse direction as

$$\phi^s = \frac{h^s}{S^e} \int_{\alpha_1}\int_{\alpha_2} (h_{31}S_{11}^s + h_{32}S_{22}^s + h_{36}S_{12}^s) A_1 A_2 d\alpha_1 d\alpha_2 \tag{11}$$

Where h^s is the distributed sensor thickness; S^e is the effective sensor electrode area; h_{31} and h_{32} respectively are the piezoelectric constants that indicate a signal generation in the transverse direction due to the strain in the meridional and circumferential direction; S_{11}^s and S_{22}^s respectively are the strains on the meridional/circumferential surface in the meridional and circumferential direction; S_{12}^s is strain on the meridional surface in the circumferential direction. Since many piezoelectric materials are not sensitive to shear strain, the signal expression can be simplified to

$$\phi^s = \frac{h^s}{S^e} \int_{\alpha_1}\int_{\alpha_2} (h_{31}S_{11}^s + h_{32}S_{22}^s) A_1 A_2 d\alpha_1 d\alpha_2 \tag{12}$$

For paraboloidal shells, sensor sensitivities can be defined in three principal motions, namely along the meridian, circumferential and transverse directions. For thin membrane shells with free boundary, vibration in the transverse direction dominates and thus it is only considered in this study. Assuming $h_{31} = h_{32}$, substituting mode shape function into the strains, and using shell Lamé parameters yields the sensor signal:

$$\begin{aligned}\phi^s &= \frac{h^s}{S^e} \int_{\alpha_1}\int_{\alpha_2} (h_{3\phi}S_{\phi\phi}^s + h_{3\psi}S_{\psi\psi}^s) A_1 A_2 d\alpha_1 d\alpha_2 \\ &= \frac{h^s h_{3i}}{S^e} \int_{\phi}\int_{\psi} b[(k+1)\sin^{k+1}\phi\cos k\psi + (k+1)\sin^{k+1}\phi\sec^2\phi\cos k\psi] d\phi d\psi \\ &= h^s h_{3i}[(\phi_{\phi\phi})_{mem} + (\phi_{\psi\psi})_{mem}]\end{aligned} \tag{13}$$

Where $b = a^2/2c$ (as shown in Fig.1); $(\phi_{\phi\phi})_{mem}$ and $(\phi_{\psi\psi})_{mem}$ are respectively the meridional and circumferential membrane signal components. Assume the distributed sensor is defined by $\phi_1 \sim \phi_2$ and $\psi_1 \sim \psi_2$, the two signal components $(\phi_{\phi\phi})_{mem}$ and $(\phi_{\psi\psi})_{mem}$ are

$$(\phi_{\phi\phi})_{mem} = \frac{b(k+1)}{S^e}(\sin k\psi_2 - \sin k\psi_1)\int_{\phi_1}^{\phi_2}\sin^{k+1}\phi d\phi \quad (14)$$

$$(\phi_{\psi\psi})_{mem} = \frac{b(k+1)}{S^e}(\sin k\psi_2 - \sin k\psi_1)\int_{\phi_1}^{\phi_2}\sin^{k+1}\phi \sec^2\phi d\phi \quad (15)$$

Where

$$S^e = \int_{\phi_1}^{\phi_2}\int_{\psi_1}^{\psi_2} A_1 A_2 d\psi d\phi = \int_{\phi_1}^{\phi_2}\int_{\psi_1}^{\psi_2} \frac{b^2 \sin\phi}{\cos^4\phi} d\psi d\phi = b^2(\psi_2 - \psi_1)\frac{\cos^3\phi_1 - \cos^3\phi_2}{3\cos^3\phi_1 \cos^3\phi_2} \quad (16)$$

Consequently, the microscopic sensing signal components of distributed sensor patches defined by ($\phi_1 \rightarrow \phi_2$) and ($\psi_1 \rightarrow \psi_2$) can be evaluated at different locations. And detail parameter analysis of distributed sensors on free thin paraboloidal shell can be investigated.

4. Numerical experiments

In order to evaluate the micro-sensing signal and sensing effect, a laminated shallow paraboloidal shell model is selected in this study, which is a shallow paraboloidal shell with meridional angle from 0 to 0.7854 radians ($\phi^* = 0.7854$) in the meridional direction and a circumferential angle from 0 to 2π radians ($\psi = 0 \sim 2\pi$) in the circumferential direction, as shown in Fig.3. Based on thin shell theory, the parameters of shell are defined: maximal radius of revolution $a = 2m$, height $c = 1m$, and shell thickness $h = 0.002m$.

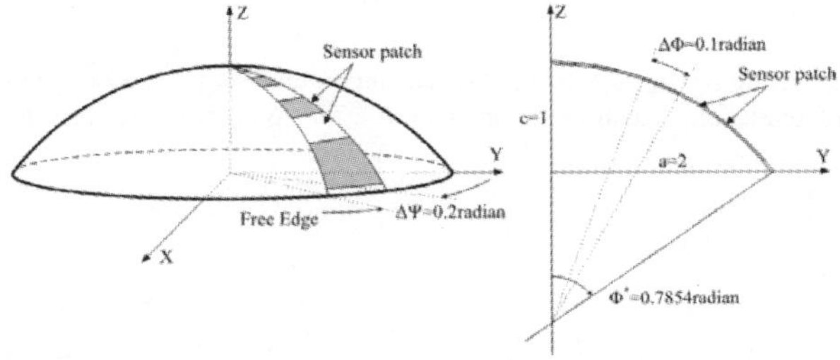

Fig.3. Sensor patches laminated on the shallow paraboloidal membrane shell

To evaluate the spatial micro-sensing signal characteristic, the size of distributed sensor patches is defined as: $\Delta\phi = \phi_2 - \phi_1 = 0.1$ rad, $\Delta\psi = \psi_2 - \psi_1 = 0.2$ rad, except patches close to the free edge with $\Delta\phi = \phi_2 - \phi_1 = 0.08$ rad. Thus, there are 248 sensor patches (i.e., 8 patches in the meridional direction and 31 patches in the circumferential direction) laminated on the shallow paraboloidal membrane shell and their sensor areas can be calculated. Using the sensing signal component expressions defined previously, one can calculate the detail signal components of sensor patches at various locations for different natural modes and those signals, i.e., $(\phi_{\phi\phi})_{mem}$, $(\phi_{\psi\psi})_{mem}$ and $(\phi)_{mem} = (\phi_{\phi\phi})_{mem} + (\phi_{\psi\psi})_{mem}$, are respectively plotted. Figs.4-9 illustrate these spatially distributed signal components of the first six modes (k=1 to 6) on shallow paraboloidal membrane shells. The micro-sensing signal is always negligible at the centre pole and varies with respective to different wave numbers at the free edge.

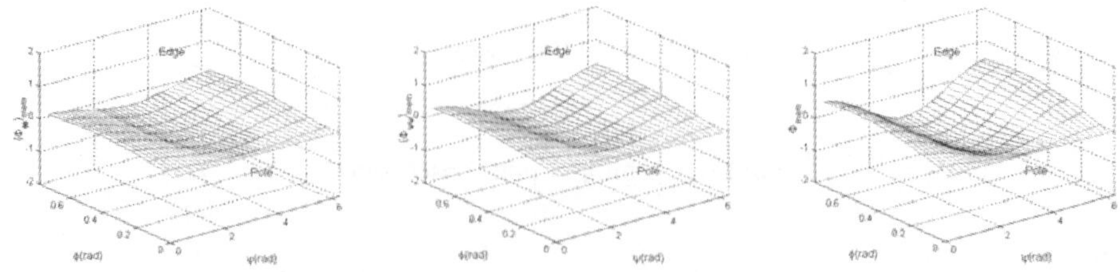

(a) (b) (c)

Fig.4. Micro-sensing signals, when k=1. (a): The meridional sensing signal component; (b): The circumferential sensing signal component; (c): The total distributed sensing signal.

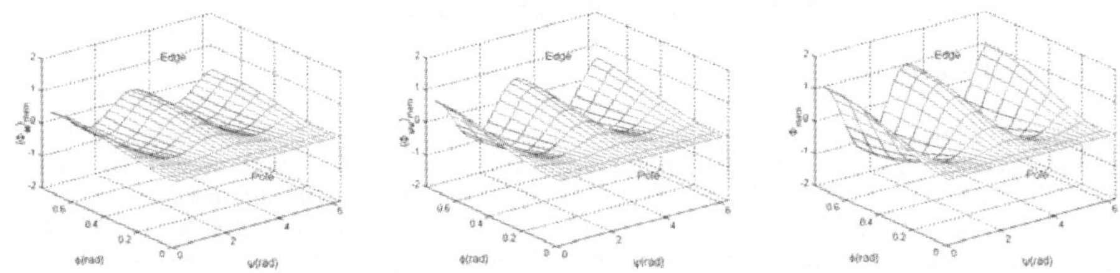

(a) (b) (c)

Fig.5. Micro-sensing signals, when k=2. (a): The meridional sensing signal component; (b): The circumferential sensing signal component; (c): The total distributed sensing signal.

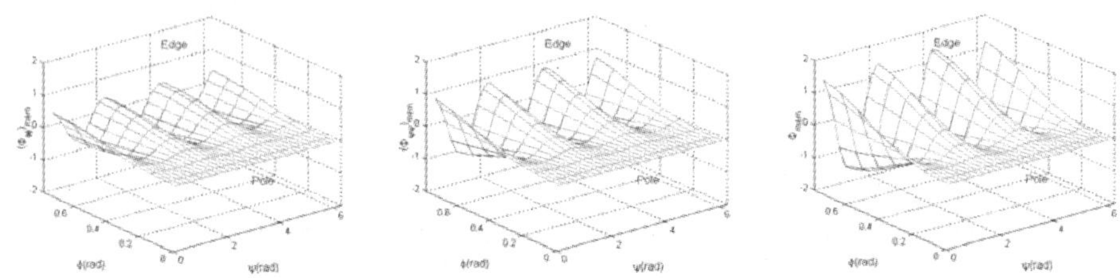

(a) (b) (c)

Fig.6. Micro-sensing signals, when k=3. (a): The meridional sensing signal component; (b): The circumferential sensing signal component; (c): The total distributed sensing signal.

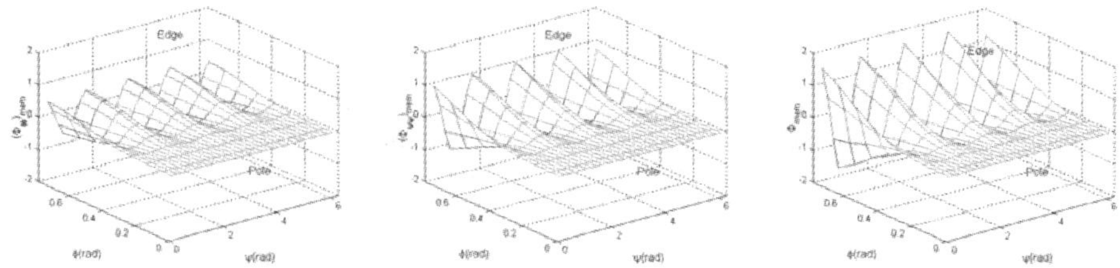

(a) (b) (c)

Fig.7. Micro-sensing signals, when k=4. (a): The meridional sensing signal component; (b): The circumferential sensing signal component; (c): The total distributed sensing signal.

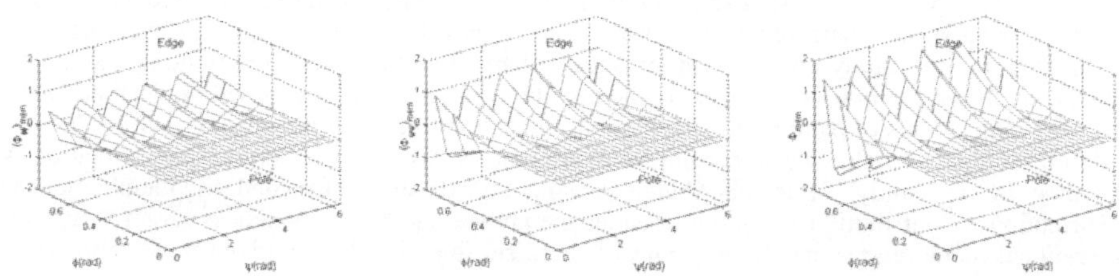

(a) (b) (c)

Fig.8. Micro-sensing signals, when k=5. (a): The meridional sensing signal component; (b): The circumferential sensing signal component; (c): The total distributed sensing signal.

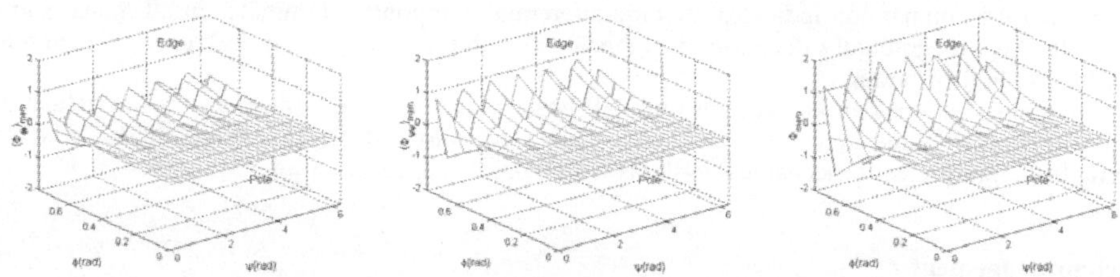

(a) (b) (c)

Fig.9. Micro-sensing signals, when k=6. (a): The meridional sensing signal component; (b): The circumferential sensing signal component; (c): The total distributed sensing signal.

With the mode number increasing, the signal wave pattern grows at the free edge corresponding to modal strain variations, and the signal magnitude decreases due to diminishing membrane strains at higher modes, though the modal amplitude A_k assumed unity. In practice, the modal amplitude usually decreases at higher modes, too. Furthermore, the circumferential membrane modal signal component $(\phi_{\psi\psi})_{mem}$ is larger than the meridional component $(\phi_{\phi\phi})_{mem}$ in overall membrane modal signal, and the signals gradually decrease at higher modes, due to diminishing influence of membrane effects to higher modes, as follows Fig.10.

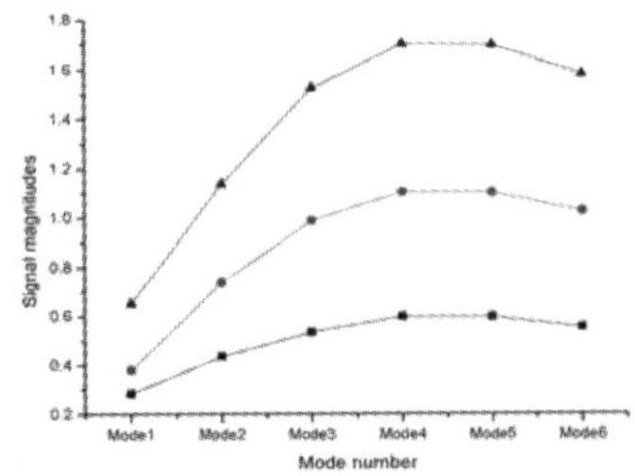

Fig.10. Maximal value of various micro-sensing signal components of shallow shells, mode number from 1 to 6. (■ : $(\phi_{\phi\phi})_{mem}$, ● : $(\phi_{\psi\psi})_{mem}$, ▲ : $(\phi)_{mem} = (\phi_{\phi\phi})_{mem} + (\phi_{\psi\psi})_{mem}$)

5. Conclusions

In order to achieve precision sensing and vibration control of smart free thin paraboloidal shell system, the micro-sensing signal characteristics of distributed patches laminated on shallow membrane shell are evaluated in this study. A *new* set of mode shape functions for free-floating paraboloidal membrane shells were proposed and used in the distributed signal analysis. Based on detailed analysis of micro-sensing signals of paraboloidal membrane shells, the following conclusions can be drawn.

1) Aiming at thin and flexible paraboloidal membrane shells, the membrane approximation was adopted in this study. Thus, distributed sensing signals are dominated by membrane strain distributions. Contribution of circumferential membrane signal component is larger than that of meridional membrane signal component in overall sensing signal for a given mode, although the difference between the two components is not significant in this study.
2) Micro-sensing signal of the sensor patches differs at various locations and modes on paraboloidal shells. Signal is zero at the shell pole and it fluctuates with the change of wave numbers at the free boundary in accordance with modal dynamic behaviors. Natural modal strain variations of thin paraboloidal shells generate distributed sensor signals in sensor patches, which are distinct among shell natural modes.
3) Quantitative comparison indicates the circumferential component dominates in all modal signals and the signals gradually decrease at higher modes, due to diminishing influence of membrane effects to higher modes.

Above observations can be used for optimal sensor placements, and these results are valuable to develop dynamic vibration control and static shape actuation of flexible paraboloidal structures ranging from micro-electromechanical systems to large-scale space structure.

6. Acknowledgement

This research is supported, in part, by a grant from the Spaceflight Technical Innovation Foundation of China (HTCX2005-01). Prof. Tzou also likes to thank the Visiting Professorship program at the Harbin Institute of Technology.

References

[1] E. F. Crawley, Intelligent structures for Aerospace: A Technology Overview and Assessment, AIAA Journal. Vol. 32, No. 8, 1994, pp. 1689-1699.
[2] Rogers. C. A., Intelligent Material System-The Dawn of a New Materials Age, Journal of Intelligent Material System and Structures. Vol. 4, 1993, pp. 4-12.
[3] U. Gabbert, H.S. Tzou, Smart Structures and Structronic System, Kluwer Academic Publishers, Dordrecht, Boston, London, 2001.
[4] H.S. Tzou and G.L. Anderson (eds), Intelligent Structural System, Dordrecht Boston London: Kluwer Academic Publishers, 1992.
[5] J. Callahan and H. Baruh, Modal Sensing of Circular Cylinder Shells Using Segmented Piezoelectric Elements, Smart Material and Structures, Vol. 8, 1999, pp. 125-135.

[6] Hui-Ru Shih and Roderick Smith, Photonic Control of Cylindrical Shells with Electro-Optic Photostrictive Actuators, AIAA Journal. Vol. 42, No. 2, 2004, pp. 341-347.

[7] H.S. Tzou, Piezoelectric Shells (Distributed Sensing and Control of Continua), Kluwer Academic Publishers, Boston, Dordrecht, 1993.

[8] E.F. Crawley, Intelligent structures for aerospace: a technology overview and assessment [J]. AIAA Journal, 1994, 32(8), pp. 1689-1699.

[9] R.L. Clark, William R. Saunders and G.P. Gibbs, Adaptive structures: dynamics and control [M]. New York: John Wiley & Sons, Ltd, 1998.

[10] A Salama, R Ramaker, M Milman. Recent experience in simultaneous control-structure optimisation. 3rd Annual NASA/DoD CSI Conference, 1989, SanDiego, USA, pp. 271-283.

[11] A. Das, S.V. Rao, B.K. Wada, Adaptive structure: challenges, issues and opportunities [A]. Proceedings of the 30th conference on decision and control[C].Piscataway, NJ, USA: IEEE, 1991, pp. 2538-2542.

[12] P.N. Swanson, J.B. Breckinridge, A. Diner, et al,, System concept for a moderate cost large deployable reflector[J], Optical engineering, 1986, 25(9), pp. 1045-1054.

[13] H.S. Tzou, J.P. Zhong and J.J. Hollkamp, Spatially Distributed Orthogonal Piezoelectric Shell Actuators (Theory and Applications), Journal of Sound and Vibration, Vol.177, No.3, 1994, pp. 363-378.

[14] J.Qiu., J.Tani, Vibration Control of a Cylindrical Shell Using Distributed Piezoelectric Sensors and Actuators, Journal of Intelligent Material Sys & Structures, Vol.6, No.4, 1995, pp. 474-481.

[15] A.R.Faria and S.F. M Almeida, Axisymmetric Actuation of Composite Cylindrical Thin Shells with Piezoelectric Rings, Smart Materials & Structures, Vol.7, No.6, 1998, pp. 843-850.

[16] V. Birmax, G.J. Knowles and J.J. Murray, Application of Piezoelectric Actuators to Active Control of Composite Spherical Caps, Smart Materials and Structures Vol. 8, 1999, pp. 128-222.

[17] H.S. Tzou, D.W. Wang, and W.K. Chai, Dynamics and Distributed Control of Conical Shells Laminated with Full and Diagonal Actuators, Journal of Sound and Vibration, Vol. 256(1), 2002, pp. 65-79.

[18] H.S. Tzou and D.W. Wang, Micro-sensing Characteristics and Modal Voltages of Piezoelectric Laminated Linear and Nonlinear Toroidal Shells, Journal of Sound and Vibration Vol. 254(2), 2002, pp. 203-218.

[19] H.S. Tzou, P. Smithmaitrie and J.H. Ding, Sensor Electromechanics and Distributed Signal Analysis of Piezoelectric-elastic Spherical Shells, Mechanical System and Signal Processing, Vol.16 (2/3), 2002, pp. 185-199.

[20] J.H. Ding and H.S. Tzou, Micro-electromechanics of Sensor Patches on Free Paraboloidal Shell Structronic Systems, Mechanical System and Signal Processing, Vol.18, 2004, pp. 367-380.

[21] W. Soedel, Vibrations of Shell and Plates, New York: Material Dekker Inc, 1981.

Investigation of Mechanical Characteristics of EAPap Actuator under Ambient Effects

Lijie Zhao [1,2], Yuanxie Li[1], Heung Soo Kim[1,*], Jaehwan Kim[1], Chulho Yang[3]

[1] *Department of Mechanical Engineering, Inha University, 253 Yonghyun-Dong, Nam-Ku, Incheon, 402-751, South Korea*

[2] *Department of Mechanical Engineering, Shenyang Institute of Aeronautical and Engineering, 52 North Huanghe Street, Huanggu District, Shenyang, 110034, China*

[3] *School of Mechanical Engineering, Andong National University, 388 Songchun-Dong, Andong, Kyungbuk, 760-749, South Korea*

Abstract

Cellulose paper is used as a basis of Electro-active paper (EAPap) actuator. EAPap actuator is a new and promising biomimetic actuator due to its characteristics of lightweight, biodegradable, dryness, large displacement output, low actuation voltage and low power consumption. EAPap is a complex anisotropic material, which has not been extensively characterized and additional basic testing is required before developing application devices. Therefore, mechanical properties of EAPap are investigated in this work under different environmental condition such as humidity and temperature. The pulling test results provide that the humidity and temperature heavily impact the mechanical properties of EAPap. Electro-mechanical coupling effects are also investigated by applying electric field during the pulling test. Strong shear electro-mechanical coupling effects are observed when comparing relative increment of mechanical properties under electric excitation. It is concluded that strong shear piezoelectricity of EAPap actuator are existed.

Keywords: Electro-Active polymer, Electro-active paper actuator, Mechanical property, Electro-mechanical coupling, Temperature, Humidity, Electric field

1. Introduction

Electro-active polymers (EAP) have been received much attention due to the development of new EAP materials that exhibit a large displacement output. This characteristic is valuable attribute that has enabled a myriad of potential applications, and it has evolved to offer operational similarity to biological muscles. Some of the current available materials are ionic polymer metal composites (IPMC), gel-polymers, conductive polymers, electron irradiated P(VDF TrFE), electrostrictive polymer artificial muscle (EPAM), electrorheological fluids (ER), EAPap, and so on [1-6]. Much of the ongoing research efforts on Electro-Active Polymer (EAP) materials have been conducted in the development and understanding of new polymer materials.

Electro-Active paper (EAPap) has been studied as a new EAP material that was first demonstrated by Kim *et al* [1, 7]. EAPap actuator has been made with cellulose paper by depositing very thin electrodes on both sides of the paper. When an electric field is applied across the thickness direction of the paper, it produces a large bending deformation: the maximum tip displacement of 4.3 mm out of 40 mm long

[*] Corresponding Author, E-mail: heungsookim@inha.ac.kr

EAPap sample was obtained under the 0.25 V/m electric field at the resonant frequency of the actuator [8]. The tip displacement tends to be linearly increased along with the excitation voltage and saturated after the maximum displacement. The exciting electric field of EAPap actuator is quite low compared to that of other electronic EAP materials. When the blocked tip force was measured, 1.1 mN of the maximum force was observed from the EAPap actuator. The electrical power consumption was 10 mW/cm^2. Since the power requirement of EAPap actuator is less than 15 mW/cm^2, which is below the safety limit of microwave driven power, the EAPap actuators can be driven by remote microwave power. This idea is useful for applications that require ultra-lightweight multifunctional capabilities such as smart skin, micro insect robots, flapping wing for insect-like flying objects, smart wall paper, MEMS, and so on. One drawback of this actuator is that environmental factors impact its performance including temperature, humidity, and electric fields. Therefore, the characteristics of EAPap actuator under different environmental conditions are investigated by pulling test. The electro-mechanical coupling of EAPap actuator is also investigated to understand the driving mechanism of EAPap.

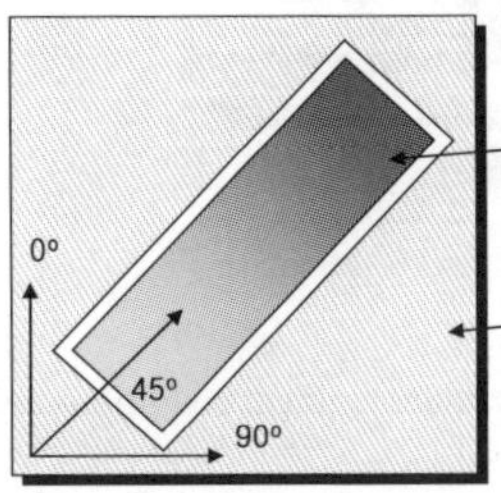

Figure 1. Orientation of cellulose film and schematic of EAPap

2. Experiments

2.1 Sample preparation

Cellulose film is a basis of EAPap actuator. To make Electro-Active Paper, very thin gold electrodes are deposited on both side of film to apply electric field into the cellulose as shown in Fig. 1. Commercially available cellulose film is known to have mechanical direction generated in the manufacturing process. Since the effect of material orientation is important to characterize the performance of EAPap actuator, three different material orientations classified as 0°, 45° and 90°are studied in the present study as shown in Fig. 1. To understand basic mechanical behavior of EAPap under ambient conditions, cellulose film is investigated first and the samples are classified as C9 to be discriminated from the previously investigated EAPap actuators [7, 8]. The size of C9 sample is 125 mm in length, 5 mm in width and 20 μ m in thickness.

Two driving mechanisms of EAPap actuators are ion migration and piezoelectric effects [8]. The study of electro-mechanical coupling of EAPap actuator is required for better understanding of the actuating behavior of EAPap. In the present paper, electro-mechanical coupling of EAPap actuator is investigated by applying electric field on the EAPap during pulling tests. The prepared samples for electro-mechanical coupling test are classified as C10 and the size of sample is 50 mm in length, 12 mm in width and 20 μ m in thickness. The gold electrodes are deposited on both side of the sample and the area of electrode is 40 mm X 10 mm on both sides of the sample. The thickness of electrode is less than 0.2 μ m. The samples are also prepared according to three different orientations of material direction.

2.2 Test setups

The experimental setups comprised of two systems. One is pulling test machine and the other is constant temperature and humidity chamber as shown in Fig. 2. Two systems are combined together to verify the ambient factors how to impact the mechanical properties of cellulose-based material. Data acquisition is coded by Labview commercial software.

Pulling test machine: In the system, two sensors are utilized besides the pulling machine (manufactured by Daeil System Korea, Model No.: DVIO-B-4545M-100t). Load cell (made by Daecell Korea, Model No. UU-K010) is used to measure the applied load. Linear scaler (made by Sony Japan, Model No.: GB-BA/SR128-015) is used to measure displacement while the load is applied. **Constant temperature and humidity chamber:** The environmental chamber is manufactured by Labcamp Co. Ltd. Korea, Model No.: CTHC-500P. The chamber provides ambient conditions in order to control the humidity and the temperature.

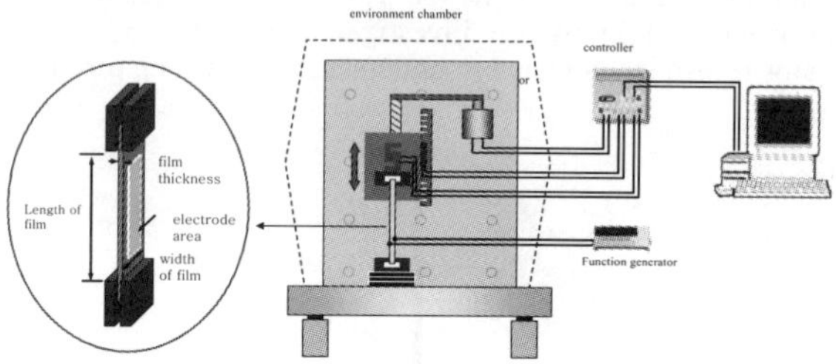

Figure 2. Schematic of pulling test under ambient conditions and EAPap samples

3. Experimental procedure

3.1 Tests for the effects of temperature and humidity

The test method for C9 samples is performed in accordance with ASTM D 882-97, which is a standard test method for tensile elastic properties of thin plastic sheeting. Environmental conditions are designated in terms of humidity and temperature separately; (i) Room condition is determined first (24°C, 20% humidity). (ii) The temperature is fixed at 24°C, the humidity are designated as 75% and 90%. (iii) The humidity is fixed at 20%, the temperature is designated as 50°C.

3.2 Tests for electro-mechanical coupling of EAPap

To investigate the electro-mechanical coupling of EAPap, the pulling test under electric excitation is conducted. The gold electrodes are deposited on the both sides of cellophane film using Vacuum Evaporation System (manufactured by Sam-Han Vacuum Development Co. Ltd., Model No.: SHE-6D-350T). The test condition is in room condition, that is at 24°C and 20% relative humidity. Three different electric excitations such as 150 V/mm, 300 V/mm and 450 V/mm are applied to the EAPap during the pulling tests.

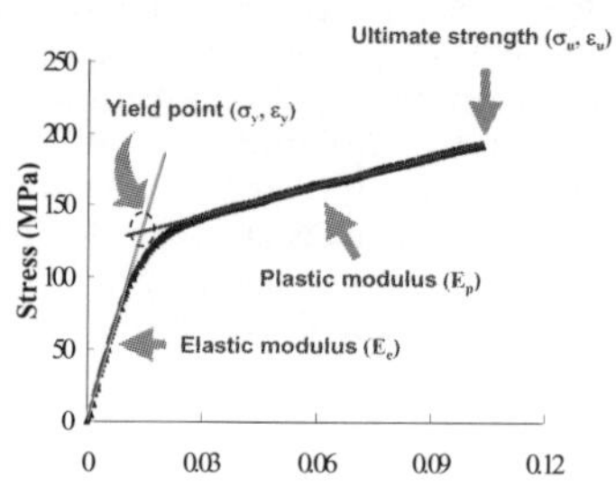

Figure 3. Typical stress-strain curve of cellulose film

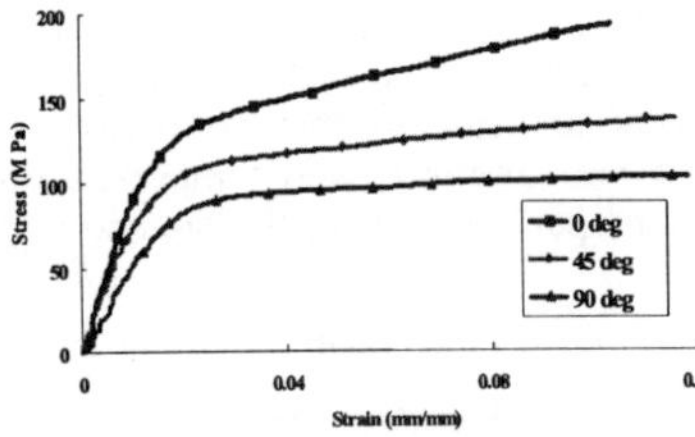

Figure 4. Stress-strain curve of C9 sample with three different material orientations (0°,45° and 90°) under room condition

Table 1. Material properties of C9 sample under room condition(24℃, 20% humidity)

Sample No.		E_e (MPa)	E_p (MPa)	σ_y (MPa)	ε_y (mm/mm)	σ_u (MPa)	ε_u (mm/mm)
C9-0	1	9536	705	131	0.023	193	0.103
	2	9400	821	125	0.022	154	0.06
	3	8430	699	122	0.023	181	0.106
C9-45	1	7262	282	110	0.024	137	0.116
	2	6841	302	107	0.036	132	0.119
	3	6990	261	113	0.025	142	0.140
C9-90	1	4996	114	91	0.038	102	0.124
	2	5133	97	93	0.026	101	0.123
	3	5130	113	92	0.019	103	0.120

4. Experiments results

4.1 Effects of temperature and humidity

Mechanical pulling tests are performed to investigate the tensile mechanical properties. C9 samples are tested for tensile elastic properties of thin plastic film. Elastic modulus, plastic modulus, ultimate stress, ultimate strain, and yielding point are determined in different environmental conditions. Figure 3 shows typical stress-strain curve of cellophane film. Stress-strain curve of typical plastic film provides the bi-modal trend as shown in Fig. 3. The initial elastic modulus is determined for the high, initial stress-strain curve. A second quite pronounced quasi-constant slope is found beginning at about 20% of the failure strain or about 70% of the failure stress. This modulus is dubbed the plastic modulus, E_p. The yielding point (σ_y, ε_y) is determined by passing a best fit straight line along the initial portion of the σ-ε curve and a second straight line along the secondary portion of the σ-ε curve and finding the intersection point as shown in the figure. Stress-strain curves of C9 with different material orientations are presented in Fig. 4. It is clearly observed that mechanical properties such as elastic and plastic modulus, yielding and ultimate stresses are decreased as increasing material angle, which is well known properties of orthotropic material. It can be inferred that EAPap has orthotropic properties. Table 1 provides experimental results of C9 cellulose film under room conditions (24°C and 20% humidity). The pulling tests under different humidity conditions are conducted and the effect of humidity is investigated next. Tables 2 , 3 provide material properties of C9 samples under 75 % and 90% humidity conditions. Representative mechanical properties such as elastic modulus and ultimate strength under different humidity conditions are compared and presented in Fig. 5. As increasing relative humidity, elastic modulus and ultimate strength are gradually decreased and this trend is independent of material direction. The effect of temperature is also investigated. Table 4 presents material properties of C9 sample under high temperature condition (50°C and 20% humidity). Elastic modulus and ultimate strength of C9 sample are compared in Fig. 6. It is also observed that mechanical properties are gradually decreased as increasing temperature. Material orientation does not affect this trend. It is well known that the performance of EAPap actuator is sensitive to environmental conditions [8]. From the above observations, it is concluded that humidity and temperature seriously impact the mechanical properties of cellulose film and also the performance of EAPap actuator.

Table 2. Material properties of C9 sample under 20℃, 75% humidity

Sample No.		E_e (MPa)	E_p (MPa)	σ_y (MPa)	ε_y (mm/mm)	σ_u (MPa)	ε_u (mm/mm)
C9-0	1	6897	817	94	0.014	163	0.113
	3	6827	958	88	0.014	122	0.050
	4	7295	879	92	0.014	169	0.116
C9-45	1	4247	230	76	0.018	110	0.189
	2	4216	213	76	0.018	120	0.234
	4	4436	290	73	0.017	94	0.187
C9-90	3	3659	166	66	0.018	75	0.159
	4	3725	156	62	0.016	72	0.169
	5	3852	168	67	0.018	83	0.148
	5	910	36	14	0.018	40	0.775
	6	1194	41	19	0.015	36	0.486

Table 3. Material properties of C9 sample under 20℃, 90% humidity

Sample No.		E_e (MPa)	E_p (MPa)	σ_y (MPa)	ε_y (mm/mm)	σ_u (MPa)	ε_u (mm/mm)
C9-0	2	5832	958	64	0.012	130	0.087
	3	6256	926	76	0.018	153	0.113
	4	6186	895	94	0.014	156	0.093
C9-45	2	3382	359	72	0.018	114	0.245
	3	3251	392	74	0.018	116	0.211
	4	3987	417	77	0.022	117	0.251
C9-90	1	815	32	17	0.019	49	0.744
	5	910	36	14	0.018	40	0.775
	6	1194	41	19	0.015	36	0.486

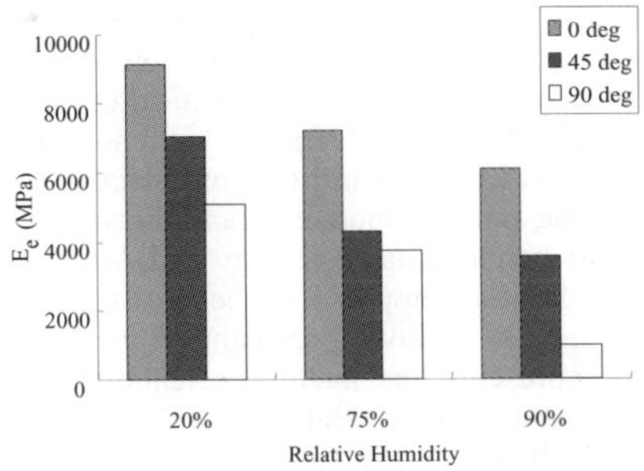

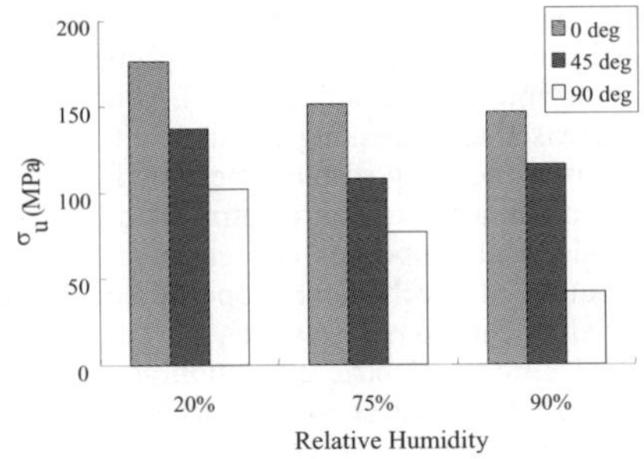

(a) Elastic modulus (b) Ultimate strength

Figure 5. Variation of mechanical properties with humidity

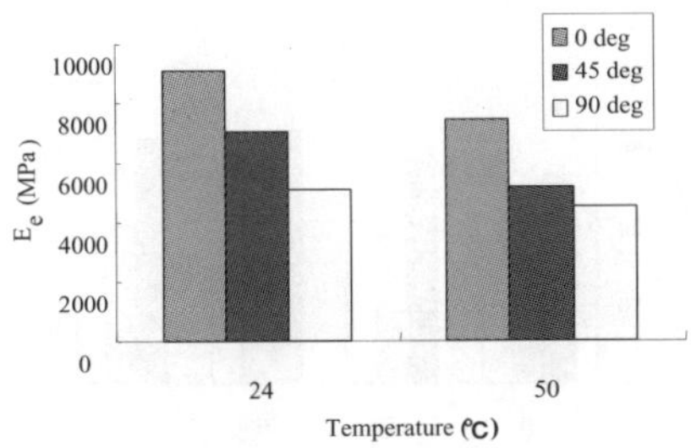

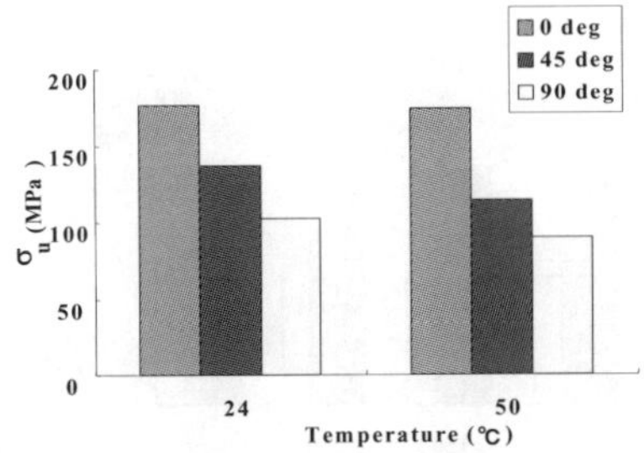

(a) Elastic modulus (b) Ultimate strength

Figure 6. Variation of mechanical properties with temperature

Table 4. Material properties of C9 sample under 50℃, 20% humidity

Sample No.		E_e (MPa)	E_p (MPa)	σ_y (MPa)	ε_y (mm/mm)	σ_u (MPa)	ε_u (mm/mm)
C9-0	1	7201	660	99	0.014	183	0.172
	2	6613	721	105	0.017	192	0.172
	3	8433	727	104	0.012	149	0.09
C9-45	1	5092	202	83	0.018	110	0.176
	2	5055	231	82	0.019	116	0.192
	3	5384	254	84	0.019	116	0.177
C9-90	1	4756	214	71	0.017	92	0.142
	2	4538	214	75	0.019	91	0.117
	3	4180	136	72	0.018	87	0.154

4.2 Electro-mechanical coupling

In this section, the electro-mechanical coupling of EAPap actuator is investigated by applying electric field during the pulling test. C10 sample is coated with gold electrodes on the cellulose film. However, the thickness of gold electrode is less than 1 percent of cellulose film. Therefore, the mechanical properties of EAPap do not differ from those of cellulose film. Three different electric fields (150 V/mm, 300V/mm and 450V/mm) are applied and room condition (24 ℃ and 20% relative humidity) is preserved during the pulling tests. Mechanical properties are extracted by the pulling tests and elastic modulus and ultimate strength are compared to investigate electro-mechanical coupling of EAPap actuator. Figure 7 provides the variation of elastic modulus and ultimate strength of EAPap under different electric excitations. The mechanical properties are increased as increasing electric field until 300 V/mm, but decreased at 450 V/mm. It is reported that the induced bending displacement of EAPap is linearly increased as increasing exciting electric field, but saturated after certain electric field [8]. Figure 7 provides the same phenomenon and the saturation of mechanical properties can cause the saturation of induced bending displacement of EAPap. The magnitudes of mechanical properties of 0 degree sample are largest compared to those of other two material orientations. However, 45 degree sample provides largest relative increment of mechanical properties under electric excitation as shown in Fig 8. This represents that material orientation of 45 degree provides largest electro-mechanical coupling effect of EAPap. This is related to shear piezoelectricity of EAPap actuator. It is well known that cellulose has larger shear piezoelectricity compared to in-plane normal piezoelectricity [8] and relative increment of mechanical properties under electric excitation provides the clue of old observations.

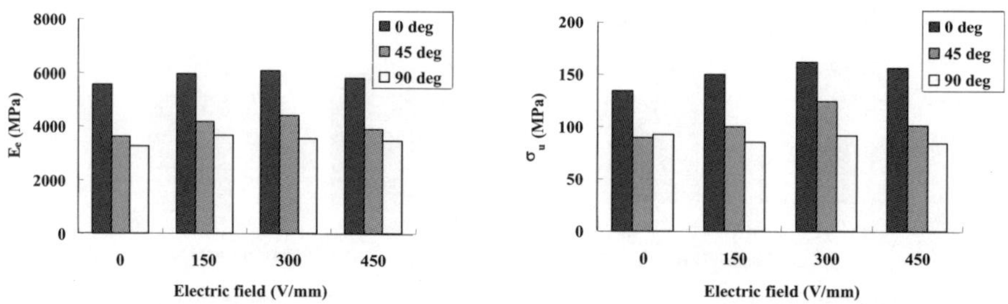

(a) Elastic modulus (b) Ultimate strength

Figure 7. Variation of elastic modulus and ultimate stresses with electric excitation

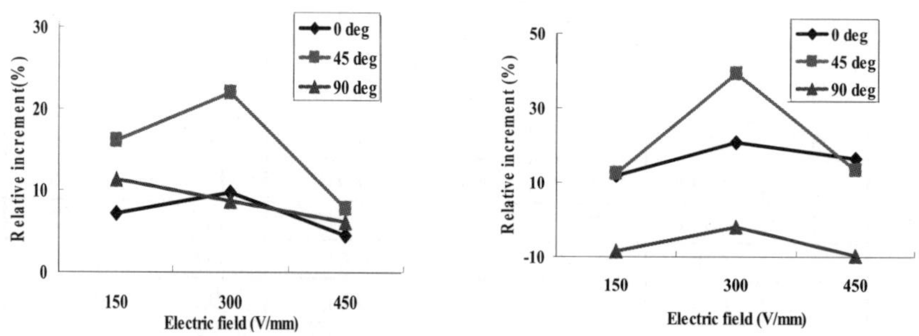

(a) Elastic modulus (b) Ultimate strength

Figure 8. Relative increment of elastic modulus and ultimate stress with electric excitation

5. Conclusion

In this paper, mechanical behaviors and electro-mechanical coupling of EAPap actuators are investigated under different temperature and humidity conditions. Two sets of samples classified as C9 and C10 are tested in the constant temperature and humidity chamber. The pulling test results provide that humidity and temperature heavily impact the mechanical properties of EAPap. The elastic strength and stiffness are generally decreased when the humidity and temperature are increased. However, mechanical properties are increased as increasing electric field. Saturation of mechanical properties is observed as increasing electric field. It is considered that this causes the saturation of the performance of EAPap actuator. Finally, the strong shear electro-mechanical coupling effects of EAPap are observed when comparing relative increment of mechanical properties under electric excitation. This provides the clue of strong shear piezoelectricity of EAPap actuator.

Acknowledgement

This work was supported by the Korea Research Foundation Grant funded by the Korean Government (MOEHRD) (KRF-2005-042-D00029)

References

[1] Bar-Cohen Ed Y. Electroactive Polymer (EAP) Actuators as Artificial Muscles: Reality, Potential, and Challenges. SPIE Press; 2001.

[2] Kim J, Ryu YH, Choi SB. New shunting parameter tuning method for piezoelectric damping based on measured electrical impedance. Smart Mater. Struct 2000;9; 868-877.

[3] Shahinpoor M, Bar-Cohen Y, Simpson JO and Smith J. Ionic polymer-metal composites (IPMCs) as biomimetic sensors, actuators and artificial muscles—a review. Smart Mater. Struct 1998; 7; 15-30.

[4] Dresselhaus MS, Dresselhaus G, Avouris Ph. Carbon Nanotubes: Synthesis, Structure, Properties and Applications. Topics in Applied Physics;80; Springer, Berlin, Germany, 2001.

[5] Pelrine R, Kornbluh R, and Joseph J. Electrostriction of Polymer Dielectrics with Compliant Electrodes as a Means of Actuation. Sensor and Actuators A: Physical 1998;64 ;77-85.

[6] Zhang QM, Bharti V and Zhao X. Giant Electrostriction and Relaxor Ferroelectric Behavior in Electron-Irradiated Poly(vinylidene fluoride -trifluoroethylene) Copolymer. Science 1998;280; 2101-4.

[7] Kim J, Seo YB. Electro-active paper actuators, Smart Mater. Struct 2002;11; 355-360.

[8] Kim J, Song CS and Yun SR. Cellulose based electro-active papers: performance and environmental effects; Smart Mater. Struct. 2006;15;719-723.

[9] Bazhenov VA. Piezoelectric Properties of Wood. Consultants Bureau; New York; 1961.

On the Modeling of Ferromagnetic Shape Memory Alloy Actuators

H. Tan, M. H. Elahinia

Dynamic and Smart Systems Laboratory
Department of Mechanical, Industrial, and Manufacturing Engineering
The University of Toledo
Toledo, Ohio 43606, USA

Abstract

Ferromagnetic Shape Memory Alloys (FSMAs) such as Ni-Mn-Ga have attracted significant attention over the last few years. As actuators, these materials offer high energy density, large stroke, and high bandwidth. These properties make FSMAs potential candidates for a new generation of actuators. The preliminary dynamic characterization of Ni-Mn-Ga illustrates evident nonlinear behaviors including hysteresis, saturation, first cycle effects, and dead zone. In this paper, in order to precisely control the position of an FSMA actuator a dynamic model is developed. The Ni-Mn-Ga actuator model consists of the dynamics model of the actuator, the kinematics of the actuator, the constitutive model of the FSMA material, reorientation kinetics of the FSMA material, and the electromagnetic model of the actuator. Furthermore, a constitutive model is proposed to take into account the elastic deformation as well as the reorientation. Simulations results are presented to demonstrate the dynamic behavior of the actuator.

Keywords: Ferromagnetic shape memory alloys; Actuator; Modeling; Phenomenology; Control

1 Introduction

Lightweight electro-hydraulic-actuators, such as pumps or linear actuators, are needed in the aerospace industry. To date, smart materials such as piezoelectric and magnetostrictive materials have been successfully applied in these actuators. The important properties of smart materials that make them attractive for such applications are high energy density and high actuation frequency. Although the strokes of such transducers are small, their ability to work at higher frequencies makes it possible to achieve the required performance. However, a smart material with larger stroke is potentially a better candidate for these applications. Large stroke becomes more important, when considering the effects of fluid viscosity and compressibility, which are limiting factors at higher working frequencies.

Compared with piezoelectric and magnetostrictive materials, FSMAs, such as Ni-Mn-Ga, exhibit significantly larger strain of up to 9.5%, which is induced by external magnetic fields of less than 400kA/m[1, 2]. The strain is the same order as the thermally-activated strain of shape memory alloys. However, FSMAs' strain is the result of field-induced martensite twin boundary motions in response to the applied magnetic fields. As a result, it is possible to develop a large strain transducer with higher frequency bandwidths[3, 4]. The actuation frequency of these materials is between the frequency of

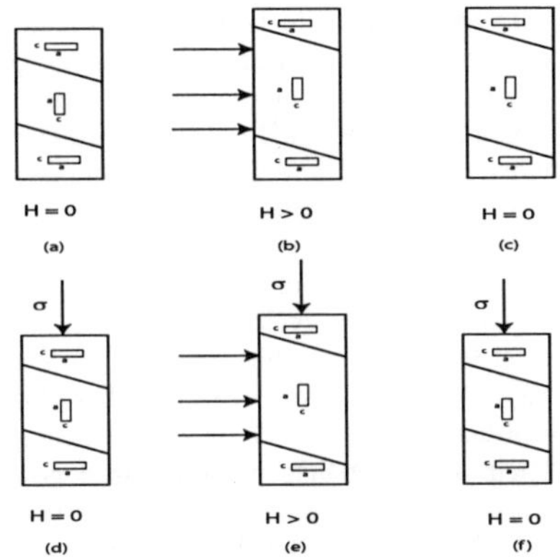

Figure 1: Strain mechanism of Ni-Mn-Ga with external field and mechanical stress. (a-c): actuation without mechanical loading; (d-f): actuation under mechanical loading

conventional shape memory alloys (10 Hz) and that of piezoelectric materials (10 kHz). Furthermore, the energy density of Ni-Ga-Mn materials is higher than that of other smart materials with electromagnetic-mechanical energy transfer features when operating in specific frequency range, such as piezoceramic and magnetostrictive materials. Because of high energy density and broad bandwidth, FSMAs, such as Ni-Mn-Ga, have recently received extensive attention as a new class of actuator material[5, 6, 7, 8].

The details of the strain mechanism have been well established in the literature[9, 10, 11]. A two-dimensional schematic representation of the Ni-Mn-Ga material is shown in Fig. 1. When the magnetic field is zero, at low temperatures, we assume that the Ni-Mn-Ga consists of two martensitic variants separated by twin boundaries as illustrated in panel (a). Actually, there are many kinds of martensite variants. This assumption is valid because the strain along the measurement direction is determined essentially by those two variants. Each variant has an easy magnetization direction (parallel to the c axis) and a hard magnetization direction (parallel to the a axis). The martensitic variants can be categorized into two types. If the easy magnetization direction of the variant is parallel to the field, the variant is an axial variant (a-variant). Otherwise, it is a transversal variant (t-variant). As the field strength increases, the transverse field will cause the increase of volume fraction of axial variants at the expense of transversal variants' volume fraction due to twin boundary motions, as illustrated in panel (b). The total vertical length of the Ni-Mn-Ga specimen increases with the field increment until the material is saturated. This variants reorientation procedure is also called forward process. When the magnetic field is strong enough that all potential variants' rotation occur, it is defined as the saturation stage. As shown in panel (c), when the magnetic field is removed, the strain does not change because of the lack of a driving, or restoring, force. In addition to the magnetic field, the external mechanical force also induces twin boundary motion, as depicted in panels (d-f). In panel (f), rotation of axial variants induced by the compressive mechanical stress brings about the decrease of the macroscopic strain. This procedure is defined as reverse reorientation procedure.

Over the past decade, several research groups have focused on developing models to describe the magnetic field-induced strain of Ni-Mn-Ga due to twin boundary motions. In 1998, James and Wuttig proposed a micromagnetic model to analyze the magnetic field induced microstructure change of martensitic variants and the field-induced strain in FSMA through the "constrained theory of magnetostriction"[12]. The model illustrates the linear strain-field characteristics and predicts the detailed magnetization distributions in the twins. O'Handley et al. presented the two dimensional field-induced strain model through free energy analysis[13, 9]. In addition to illustrating that the large magnetization anisotropy and easier twin-boundary movement are necessary material properties for generating the large field-induced strain, their model exhibits good agreement with experiments. Hirsinger and Lexcellent introduced an internal variable into the free energy expression[14, 15]. Consequently, a phenomenological model with internal variables was developed. This model exhibits good strain predication capability on the experiments of field-induced strain under various external mechanical loading. With a similar approach Kiefer and Lagoudas developed a thermodynamic phenomenological model based on the internal variables[16]. By considering the magnetization limitation induced by variants' rotation, they further presented a thermodynamic model exhibiting first cycle effect and the internal magnetization[17]. Faidley et al. presented a constitutive model for the reversible strain in Ni-Mn-Ga solenoid actuators in which the actuation direction and the magnetic field is parallel[18]. Likhachev and Ullakko presented a general thermodynamic model of the field-induced strain[19, 20], which could be applied to the multi-dimensional cases involving multi-variants. Their approach has demonstrated good agreement with experimental results. Based on Likhachev's approach, Tan and Elahinia presented an enhanced phenomenological model describing the magnetostrain behavior under both changing mechanical stress and changing magnetic field[21, 22].

Recently, Ni-Mn-Ga based transducers have been applied in various linear actuators[3, 23, 6, 7, 8]. One common feature of all these actuators is a mechanism for achieving reversible strain. This is due to the fact that the deformation of Ni-Mn-Ga elements does not change without a restoring mechanism after the magnetic driving force is removed. Henry et al. utilized a spring along the measured strain direction of the FSMA element to provide the restoring compressive stress[3]. Plessis et al. designed latching valves with antagonistic FSMA actuators[23]. In this arrangement, the restoring force for one actuator is provided by the opposing FSMA actuator. Tickle has constructed a rotating magnet around the FSMA element to study its dynamic response properties[24]. This way, by changing the magnetic field direction, the field-induced strain can be restored. Aside the mechanism of external compressive stress, the internal restoring force can also induce the reversible strain. Faidley and et al. observed the reversible compressive strain of about 0.41% in an $Ni_{50}Mn_{28.7}Ga_{21.3}$ actuator without an externally applied restoring force when the magnetic field was applied along the same direction as the measured strain[18]. In their work, the restoring force is believed to be the result of pinning sites existing in the martensite. Actually, the internal restoring force is due to pinning sites distributing in the bulk.

However, all the present models for Ni-Mn-Ga materials are quasi-static model and the dynamics of materials are not explored so far. To the best of authors' knowledge, there is no published report on the integrated mechanical-electromagnetic modeling of these actuators. In other words, the affect of mechanic and electric dynamics of actuators on materials behaviors have not been investigated. In this work, we proposed a complete model of an Ni-Mn-Ga based actuator including the dynamics characteristics of materials, the mechanical subsystem, and the electromagnetic subsystem. In the next section, the proposed structure of an Ni-Mn-Ga based transducer is explained. An enhanced phenomenological model is presented to address the interaction between compressive stresses and the field-induced strain. The actuator's model consists of the field-induced strain model, the material constitutive model, the mechanical dynamic model, the kinematic model, and the electromagnetic model. In the following section, simulations are performed to demonstrate that the model is able to capture the dynamic feature. Also, the simulation model is utilized to investigate the effect of different parameters on the performance of actuators.

2 Modeling of the Ni-Mn-Ga Actuator

The structure of the actuator studied in this work is illustrated in Fig. 2. When the control voltage is applied to the Helmholtz coils, the coils generate a magnetic field in the y-direction. Due to the field-induced strain, the Ni-Mn-Ga material deforms along the x-direction recovering the strain that was

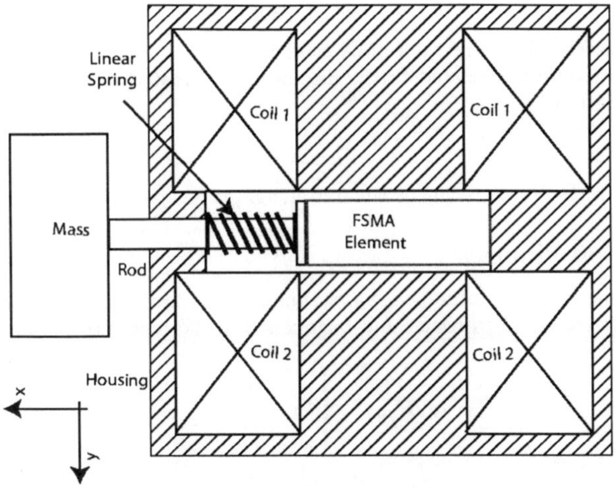

Figure 2: Ni-Mn-Ga based actuator structure with restoring spring

previously induced by a compressive stress along the x-direction. The rod which is attached to the top of the FSMA moves along the x-direction transmitting the displacement of the FSMA material. The spring between the upper housing and the rod provides the restoring compressive stress along the measured strain direction. The restoring mechanism enables large force/stroke and therefore enhances the energy density. It is worth noting that different restoring mechanisms affect the driving force, which is an important characteristic for FSMA based transducers. Furthermore, the stroke is dependent on the restoring mechanism. For example, the internal restoring stress could only produce strain in the order of 0.4%, while the compressive spring enables significantly larger reversible strain[4, 3].

The model for an FSMA-based actuator consists of the following sub-models: the constitutive model, the reorientation model, the dynamics/kinematics of the transducer, and the electromagnetic model. In this section, the reorientation model is developed based on the previous work of Kiefer and Lagoudas. Furthermore, a constitutive model is presented that takes into account the magnetic saturation stage of the transducer as well as the dynamics of the actuator. Finally, the electromagnetic model is derived from the magnetic circuit analysis by considering the effect of strain on the magnetic properties.

2.1 Kinematics and dynamics of the actuator

The dynamic equation for the FSMA transducer, which is shown in Fig. 2 can be written as:

$$m\ddot{x} + c\dot{x} + k(x0 + x) - F_{out} = 0 \tag{1}$$

where m, c, and k are the mass that the transducer moves, the stiffness of the spring, and the damping coefficient of the material, respectively. It is worth noting that the stiffness of material is account for in the constitutive model specifically. $x0$ is the initial displacement of the spring causing bias compressive stress. F_{out} is the force from the Ni-Mn-Ga element. This force is transmitted through the rod to the moving load as shown in Fig. 2. Defining σ as the reaction stress applied by the FSMA element to the moving load, F_{out} is expressed by σ as:

$$F_{out} = \sigma A \tag{2}$$

where A is the cross-sectional area of the FSMA element.

The strain of the actuator is related to the displacement created in the actuation. This kinematic relationship can be written as:

$$\Delta x = \epsilon l \tag{3}$$

where l is the length of the FSMA element when it elongates completely with no mechanical loading.

2.2 Reorientation kinetics

The reorientation functions are obtained by assuming the maximum dissipation in the variants' reorientation process. According to Kiefer and Lagoudas[17], the reorientation function is defined as:

$$\Phi^{\xi}(\sigma, \bar{H}, \xi) = \begin{cases} \pi^{\xi} - Y^{\xi}, \dot{\xi} > 0 \\ \\ \pi^{\xi} - Y^{\xi}, \dot{\xi} < 0 \end{cases} \tag{4}$$

Here Φ^{ξ} is the reorientation function defining the threshold values for the activation of the reorientation process that depends on the independent state variables, where: σ is the external mechanical stress; $\bar{H}$ is the magnetic strength vector; and $0 \leq \xi \leq 1$ is the volume fraction of axial variants to transversal variants. π^{ξ} is the driving force for variant reorientation defined as:

$$\pi^{\xi} = \sigma \epsilon^{r,max} + \frac{1}{2}\sigma^2 \Delta S - \mu_0 M^{sat}[\sin(\theta_3) - 1]H + \rho K_1(\sin(\theta_3))^2 - \frac{\partial f^{\xi,p}}{\partial \xi} \tag{5}$$

where

$$\sin\theta_3 = \frac{\mu_0 M^{sat}}{2\rho K_1} H \tag{6}$$

$$\frac{\partial f^{\xi,p}}{\partial \xi} = \begin{cases} A^p \xi + B_1^p + B_2^p, \dot{\xi} > 0 \\ C^p \xi + B_1^p - B_2^p, \dot{\xi} < 0 \end{cases} \tag{7}$$

$\triangle S$ is the difference in elastic compliance between the long and short axis directions of the tetragonal martensitic crystal. In this case, $\triangle S = 0$ [17]. $\epsilon^{r,max}$ is the maximum strain; μ_0 is the permeability of free space and ρ is the mass density; H is the magnitude of field strength; $\frac{\partial f^{\xi,p}}{\partial \xi}$ is the partial derivative of hardening function f with respect to ξ. Saturated magnetization M^{sat} and magnetic anisotropy energy ρK_1 are both determined by experimental measurements. The hysteresis parameter Y^{ξ} in Equation (4) and hardening coefficients A^p, B_1^p, B_2^p, and C^p in Equation (7) are dependent on experimental measurements.

The reorientation function must satisfy the Clausius-Duhem inequality, which can be written as following in this case:

$$\dot{\xi} > 0,\ \pi^{\xi} - Y^{\xi} = 0 \tag{8}$$

$$\dot{\xi} < 0,\ \pi^{\xi} - Y^{\xi} = 0 \tag{9}$$

$$\Phi^{\xi} < 0,\ \dot{\xi} = 0 \tag{10}$$

The volume fraction can be used to derive the strain due to variants reorientation by $\dot{\epsilon^r} = \Lambda^r \dot{\xi}$. If the reorientation strain tensor Λ^r is a constant, integration of the previous equation leads to:

$$\epsilon^r = \epsilon^{r,max} \xi \tag{11}$$

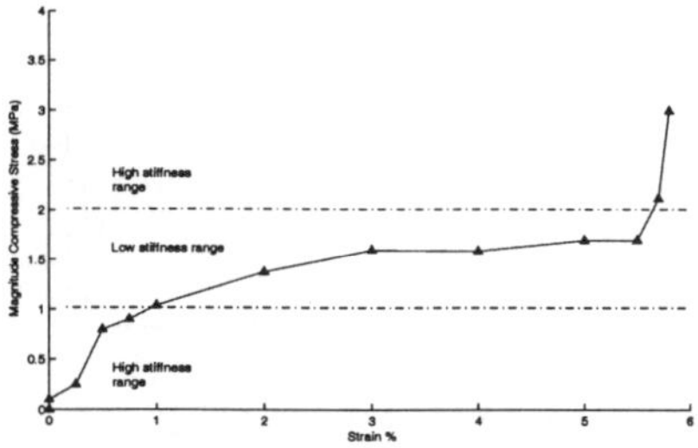

Figure 3: Fitting curve of experimental results of axial compression strain in $Ni_{40}Mn_{28}Ga_{12}$ martensite[11]

2.3 Constitutive model

In order to analyze the dynamics of the actuator, both elastic strain and field-induced strain need to be taken into account. Particularly, in the saturation stage with $u0H > \frac{2\rho K1}{M^{sat}}$ and $sin\theta_3 = 1$, the Clausius-Duhem inequality (Equation (10)) shows $\dot{\xi} = 0$. However, at the end of the actuation phase and the FSMA saturates, the mass will continue to movement because of its inertia. This extra motion will continuously modify the mechanical loading on FSMA element through the mechanical connection between the FSMA element and mass. As both the displacement of the mass and the reaction force on the FSMA are varying, the FSMA element strain is changing accordingly. It is well known that both the transverse field and axial mechanical stress are capable of rotating the twin variants[21]. In the saturation stage, the magnetic field is not able to further rotate the variant. Therefore the additional strain will take place through the reorientation induced by the reduced external mechanical loadings on the FSMA element. The variation of strain is also due to the elastic deformation in the twinning.

It is assumed that the total strain consists of elastic strain, field-induced strain, and strain due to phase transformation. In this study, it is assumed that the temperature is constant and there is no phase transformation and no thermal expansion in the actuation. Therefore, the total strain can be written as:

$$\epsilon = \epsilon^e + \epsilon^r \tag{12}$$

where ϵ^e is the elastic strain. The variation of ϵ^r was defined in Section 2.2.

The stress-strain behavior of an FSMA element is highly nonlinear under a typical compression test. This behavior can be categorized into two ranges: low stiffness E_l and high stiffness E_h, as shown in Fig. 3. Before the compressive stress surpasses the threshold value, the strain changes slightly. When the compressive stress reaches approximately 1.04MPa, the axial variants begin to rotate in favor of the transverse variants. This is similar to applying the magnetic field along the axial direction (x-direction). It is worth noting that the strain in the low stiffness range is essentially the result of variants reorientation.

It is assumed that the total strain is decoupled into conventional elastic strain and twinning strain when the FSMA element is under compression. In both the forward reorientation process and reverse reorientation process, the twinning strain has been taken into account in the driving force function (Equation(5)) by introducing the elastic energy in Gibbs free energy. Therefore, ϵ^e is defined as:

$$\epsilon^e = \frac{\sigma}{E_h} \tag{13}$$

where σ is the external compressive stress.

By submitting Eqs.(11),(13) into Equation (12), the constitutive equation can be rewritten as:

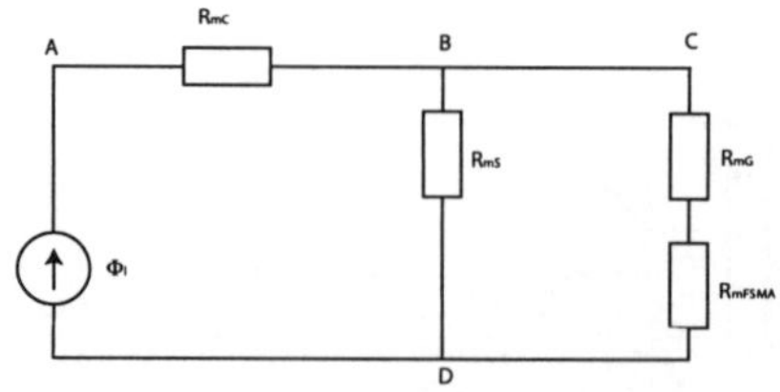

Figure 4: Magnetic circuit of the Ni-Mn-Ga based transducer

$$\epsilon = \frac{\sigma}{E_h} + \xi \epsilon^{r,max} \tag{14}$$

2.4 Electromagnetic Model

The magnetic circuit corresponding to the actuator is shown in Fig. 4. The top section of the FSMA element is made of ferromagnetic material so that the majority of the flux will go through it before entering the upper soft iron. By applying Kirchhoff law at point B of the magnetic circuit, the magnetic field flux density in the Ni-Mn-Ga element is:

$$\phi_{FSMA} = \frac{\phi_i Rm_S}{Rm_S + Rm_G + Rm_{FSMA}} \tag{15}$$

where $\phi_i = Li$ is the flux generated by the coils, L and i are the inductance and current of the coils, respectively; Rm_{FSMA} is the reluctance of FSMA; Rm_S is the reluctance of the stray field; Rm_G is the reluctance of the air gap between the FSMA and coils, and Rm_C in Fig. 4 represents the reluctance of the coils. It is worth noting that Rm_G and Rm_{FSMA} are not constant. Instead, they are also affected significantly by the strain of the FSMA element. Rm_G and Rm_{FSMA} are defined by Suorsa *et al.* as [8]:

$$Rm_G = Rm_{G0}(1 + 4\epsilon) \tag{16}$$

$$Rm_{FSMA} = Rm_{FSMA}(\epsilon = 0)(1 - \epsilon) \tag{17}$$

where Rm_{G0} is the initial reluctance dependent on the geometric parameters of the FSMA element when $\epsilon = 0$; $Rm_{FSMA}(\epsilon = 0)$ is a constant[25].

The magnetic field through the FSMA material can be written as:

$$h_{FSMA} = \frac{\phi_{FSMA} Rm_{FSMA}}{b} \tag{18}$$

where b is the width of the Ni-Mn-Ga element measured along the field direction. The magnetic field strength of the Ni-Mn-Ga element can be written as a function of the applied current to the coils. Substituting for ϕ_{FSMA} in Equation (18):

$$h_{FSMA} = \frac{Rm_{FSMA}}{b} \frac{Rm_S Li}{Rm_S + Rm_G + Rm_{FSMA}} \tag{19}$$

It can be seen that h_{FSMA} is dependent on the current through the coils. As the parameters in Equation (15), the flux density B_{FSMA} is the produce of h_{FSMA} and permeability coefficient in vacuum. The current and voltage of the coils are related through Farady's law as following:

$$\frac{U}{R} - \frac{L}{R}\frac{di}{dt} - i = 0 \tag{20}$$

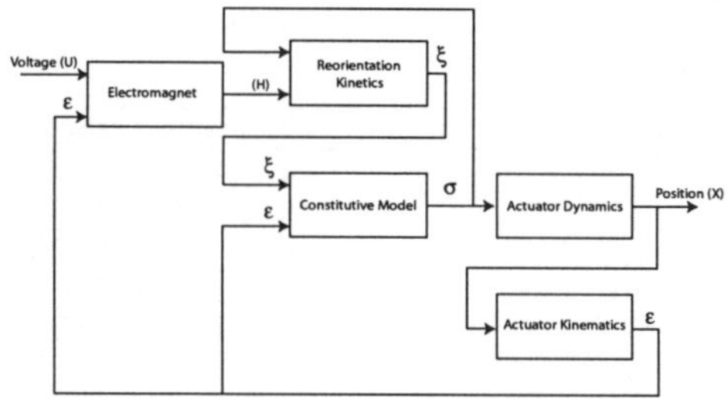

Figure 5: System diagram of simulation model

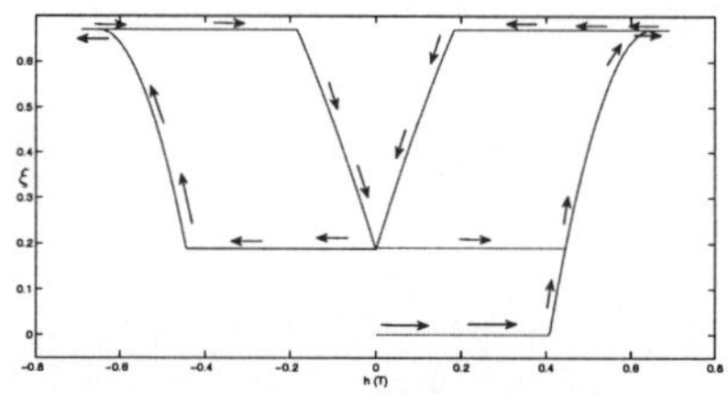

Figure 6: Simulations result with constant compressive stress 1.2MPa as input

where U and R are the applied voltage and the resistance of the coils, respectively.

An important dynamics feature in the electromagnet is the eddy current which can cause extra power losses and deterioration of the field distribution in the air gap. Two components are involved in the eddy current analysis: iron core and housing of the coil and the Ni-Mn-Ga specimen. Regarding the first component, the core is made of laminated sheets with the thickness less than 0.35mm so that the eddy current effect can be neglected. For the Ni-Mn-Ga specimen, the thickness of the specimen is 2mm and it has been shown that eddy current effect is not significant in the actuators with small tickness[26]. Therefore, in this paper, the eddy current effect is not included in the model.

3 Results

The computer simulation model is developed in Simulink, as shown in Fig. 5. It consists of the reorientation, constitutive, kinematic, and dynamic submodels. The construction of subsystems is according to their physical properties. The interface variables between constitutive submodel and actuator dynamics submodel is selected as σ so that the integrator blockages of Simulink are utilized in the dynamics submodel. Otherwise, if ξ is chosen the interface variable, the derivative blockages are adopted consequently in the submodel resulting in much higher possibility of simulation failures. It is worth noting that the reorientation model is base on a piecewise function and the differentiability of variable σ and ϵ is not guaranteed. In order to avoid singularity, the stress is derived form the constitutive model.

At the beginning of the actuation, the FSMA element consists of only transverse variants, that is $\xi = 0$. This status is obtained through the appropriate annealing treatment in the phase transformation from austenite to martensite. The material parameters used for the simulations are taken from experimental results published by Heczko et. al.[27]. The simulation parameters are listed in Table 1.

Table 1: Simulation parameters of the FSMA actuator

Material Parameters	Hardening Parameters
$\rho = 8000\ Kg/m^3$	$A_p = 0.034568$MPa
$\rho K_1 = 1.67\times10^5$	$B_1^p = -0.018100$MPa
$M^{sat} = 5.14\times10^3$	$B_2^p = 0.033266$MPa
$\epsilon^{r,max} = 0.062$	$C^p = 0.167632$MPa
$\Delta S = 0$	$Y^{\xi} = 0.054285$MPa
Parameters of Dynamics Model	Coil properties
$m = 0.5$ Kg	Coil inductance=0.0698H
$k = 3300$ N/m	Number of turn=400Turn
$c = 100$ N-s/m	Resistance=16Ω
$E_h = 200$MPa	

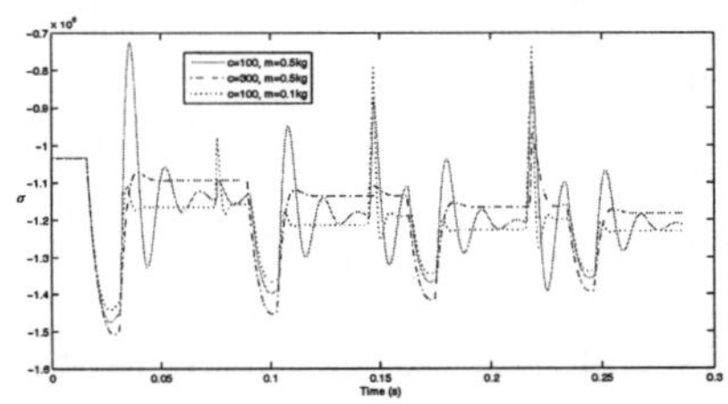

Figure 7: σ dependence on mass and damping coefficient

The validation of the model has been done through taking a constant compressive stress (instead of the changing spring stress) as an input. Simulations show the relationship between volume fraction ξ and the magnetic field h as depicted in Fig. 6. The result reproduces the model calculation result given by Kiefer[17].

The dynamic simulations illustrate the significant impact of parameters k, c, and m on the actuation behavior as shown in Fig. 10. The first cycle of actuation is shown in panel (a). It is evident that the simulation captures the movement due to the mass inertia after the FSMA element is saturated ($\sin\theta_3 = 1$). The dramatic increase of σ in the forward process eliminates the potential space for reverse process. In order to explain the dramatic increase of σ, it is worth noting that the σ variation induced by the spring is not the main contributing factor. As shown in panel (b), after a quarter period (of input signal), ξ (Xi in the figure) reaches about 0.2. Therefore, the displacement change in the forward process is $\Delta x = \xi\epsilon^{rmax}l = 1.2 \times 10^{-4}$. This change causes the stress variation $\Delta\sigma = \frac{k\Delta x}{Area} = 0.1MPa$. The total σ variation in the forward process is about 0.4MPa as illustrated in Panel(c). According to Eqs. (1),(2), the variation of σ also depends on the acceleration, velocity, damping coefficient, and mass.

It is interesting to investigate the amplitude of stress σ applied on the Ni-Mn-Ga element because the blocking stress of Ni-Mn-Ga is small (about 3MPa). To this end, the simulations are performed to study the stress augment with respect to the change of mass, damping coefficient, and input frequency. It is worth noting that input frequency is chosen as the control variable. The reason is that both the velocity and acceleration is highly dependent on the frequency. Furthermore, the frequency is a system state variable under control. The simulations illustrate evidently that the increase of frequency actually contributes significantly to the ascending amplitude of stress, as shown in Fig. 8. On the other hand,

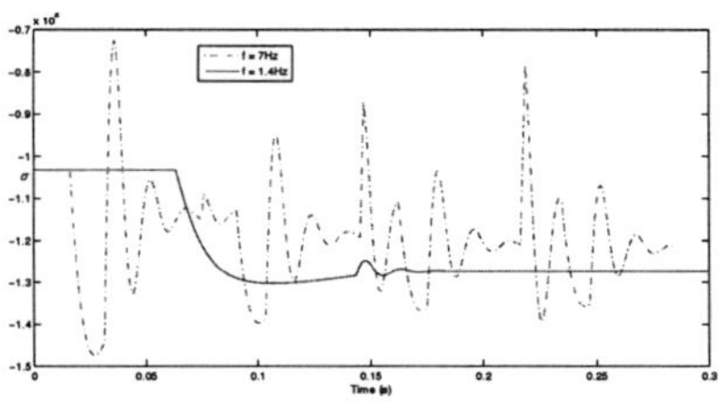

Figure 8: Effect of change of the input signal frequency on σ

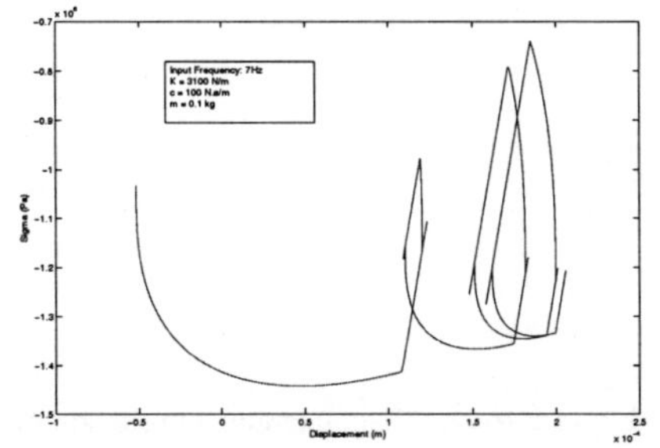

Figure 9: σ-*displacement* relationship of the actuator

compared with change of frequency, the equivalent increase of mass or damping coefficient increase does not induce the same extent of amplitude augmentation of the stress, as shown in Fig. 7. This can be explained by the fact that impact of the frequency augmentation are amplified by both the inertia force item $m\ddot{x}$ and damping force $c\dot{x}$ at the same time. Another observation is that the increase of mass or damping coefficient also contributes to the augmentation of the stress as shown in Fig. 7. In order to evaluate the performance of the actuator, the stress(traction)-displacement relationship is important. The simulation model is able to provide this relationship with corresponding design parameters, as shown in Fig. 9.

After several cycles, the actuation cycle will converge to a constant loop. With the constant compressive stress, the actuation will always achieve the cyclic period in the second cycle as suggested in Fig. 6. However, with the dynamic components, the number of periods necessary before achieving the repeated cyclic actuation (marked by the arrows in panel (d)) will dependent on k, c, and m. The displacement amplitude of the cyclic actuation is inversely proportional to the spring stiffness k as shown in Fig. 11. Also, the damping coefficient determines the amplitude and interval of the overshot shown in Fig. 11.

4 Conclusion

The paper develops a simulation model to investigate the dynamic characteristics of the FSMA based actuators for future control studies. A constitutive model is proposed to take elastic deformation into account as well as the reorientation strain. The constitutive model, the reorientation kinetics, the kinematic model, and the dynamic model of the actuator are integrated to study the dynamic behavior

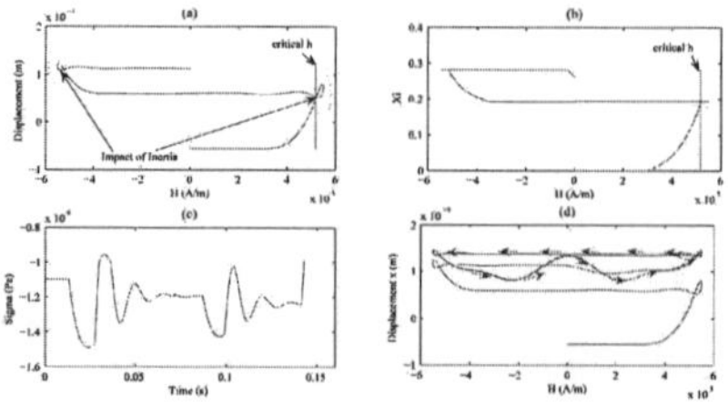

Figure 10: Simulation results illustrating the dynamics in the actuation

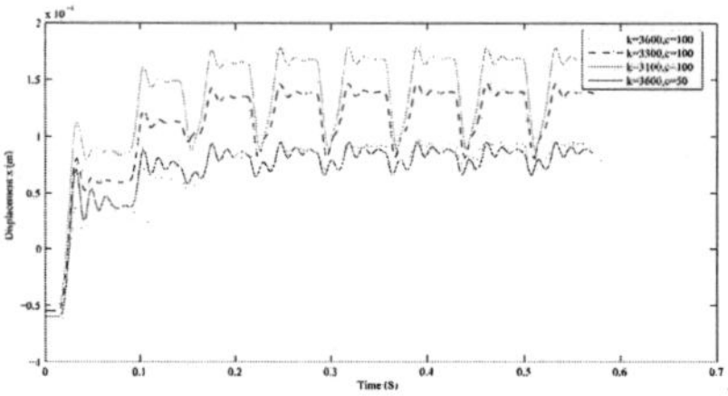

Figure 11: Effects of k and c to the displacement amplitude of the cyclic actuation

of the actuator. The simulation results show that the model can capture the dynamics features in the actuation, including the oscillation and the mass movement in the saturation region. Furthermore, it is found that the spring stiffness k and damping coefficient c have significant impacts on the stress applied on the FSMA element and the number of the oscillation cycle before achieving repeatable actuation. Finally, the increase of input frequency dominates the augmentation of the stress amplitude in the actuation.

Acknowledgements

This research was supported through the start up funds provided by the University of Toledo. The authors would like to express their appreciation for the support.

References

[1] S. J. Murray, M. Marioni, P. G. Tello, S. M. Allen, and R. C. O'Handley. Giant magnetic-field-induced strain in Ni-Mn-Ga crystals: experimental results and modeling. *Journal of Magnetism and Magnetic Materials*, 226-230:945–947, 2001.

[2] A. Sozinov, A. Likhachev, N. Lanska, and K Ullakko. Giant magnetic-field-induced strain in NiMnGa seven-layered martensitic phase. *Applied Physics Letters*, 80:1746–1748, 2002.

[3] C. P. Henry, D. Bono, J. Feuchtwanger, Allen S. M., and O'Handley R. C. AC field-induced actuation of single crystal NiMnGa. *Journal of Applied Physics*, 91(10):7810–7811, May 2002.

[4] L. E. Faidley, M. J. Dapino, G. N. Washington, T. A. Lograsso, and R. C. Smith. Analytical and experimental issues in Ni-Mn-Ga transducers. *Smart Structures and Materials 2003: Active Materials: Behavior and Mechanics Proceeding of SPIE*, 5049.

[5] AdaptMat. Msm properties. *http://www.adaptamat.com/technology/.*

[6] N. Lanska, O. Soderberg, A. Sozinov, Y. Ge, K. Ullakko, and V. K. Lindroos. Composition and temperature dependence of the crystal structure of Ni-Mn-Ga alloys. *Journal of Applied Physics*, 95(12):8074–8078, June 2002.

[7] J. Tellinen, I. Suorsa, A. Jaakelainen, I. Aaltio, and K. Ullakko. Basic properties of magnetic shape memory actuators. *8th international conference ACTUATOR 2002, Bremen, Germany*, June 2002.

[8] I. Suorsa, E. Pagounis, and K. Ullakko. Magnetic shape memory actuator performance. *Journal of Magnetism and Magnetic Materials*, 272-276:2029–2030, 2004.

[9] R. C. O'Handley, S. J. Murray, M. Marioni, H. Nembach, and S. M. Allen. Phenomenology of giant magnetic-field-induced strain in ferromagnetic shape-memory materials. *Journal of Applied Physics*, 87(9):4712–4717, May 2000.

[10] R. Tickle and R. D. James. Magnetic and magnetomechanical properties of Ni2MnGa. *Journal of Magnetism and Magnetic Materials*, 195(3):627–638, June 1999.

[11] A. A. Likhachev and K. Ullakko. Magnetic-field-controlled twin boundaries motion and giant magneto-mechanical effects in NiMnGa shape memory alloy. *Physics Letters A*, 275:142151, October 2000.

[12] R. D. James and M. Wuttig. Magnetostriction of martensite. *Philosophical Maganize A*, 77(5):1273–1299, 1998.

[13] R. C. O'Handley. Model for strain and magnetization in magnetic shape-memory alloys. *Applied Physics Letters*, 83(6):3263–3270, September 1998.

[14] L. Hirsinger and C Lexcellent. Modelling detwinning of martensite platelets under magnetic and (or) stress actions on Ni2MnGa alloys. *Journal of Magnetism and Magnetic Materials*, 254-255:275–277, 2003.

[15] L. Hirsinger. Ni2MnGa shape memory aoolys: Modelling of magneto-mechanical behavior. *International Journal of Applied Electromagnetics and Mechanics*, 19:473–477, 2004.

[16] B. Kiefer and D. C. Lagoudas. Phenomenological modeling of ferromagnetic shape memory alloys. *Smart Structures and Materials 2004: Active Materials: Behavior and Mechanics Proceeding of SPIE, Bellingham, WA*, 5387:164–176.

[17] B. Kiefer and D. C. Lagoudas. Modeling of the magnetic field-induced martensitic variant reorientation and associated magnetic shape memory effect in MSMAs. *Smart Structures and Materials 2005: Active Materials: Behavior and Mechanics Proceeding of SPIE, Bellingham, WA*, 5761:454–465.

[18] L. E. Faidley, M. J. Dapino, G. N. Washington, and T. A. Lograsso. Reversible strain in Ni-Mn-Ga with collinear field and stress. *Smart Structures and Materials 2004: Active Materials: Behavior and Mechanics Proceeding of SPIE, Bellingham, WA*, 5387.

[19] A. A. Likhachev and K. Ullakko. Quantitative model of large magnetostrain effect in ferromagnetic shape memory alloys. *EPJdirect B2*, pages 1–9, 1999.

[20] A. A. Likhachev and K. Ullakko. Quantitative model of large magnetostrain effect in ferromagnetic shape memory alloys. *The European Physical Journal B*, 14:263–267, 2000.

[21] H. Tan and H.M. Elahinia. Modeling of Ferromagnetic Shape Memory Alloy Based Transducers for Electro-Hydraulic Actuators. *International Mechanical Engineering Congress and Exposition, Orlando, FL, USA*, Nov 2005.

[22] H. Tan and H.M. Elahinia. Dynamics Modeling of Ferromagnetic Shape Memory Alloys (FSMA) Actuator. *SPIE Conference: Smart Structures and Integrated Systems, San Diego, CA, USA*, 6173, Mar 2006.

[23] A. J. du Plessis, A. W. Jessiman, C. J. Muller, and M. C. van Schoor. Latching valve control using ferromagnetic shape memory alloy. *Smart Structures and Materials 2003: Industrial and Commercial Applications of Smart Structures Technologies. Proceedings of SPIE*, 5054:320–331, 2003.

[24] R. Tickle. Rotating Magnet. *http://www.aem.umn.edu.*

[25] I. Suorsa, J. Tellinen, K. Ullakko, and E. Pagounis. Voltage generation induced by mechanical straining in magnetic shape memory materials. *Journal of Applied Physics*, 95(12):8054–8058, 2004.

[26] M. A. Marioni. Pulsed magnetic field-induced twin boundary motion in Nickl-Manganese-Gallium. *MIT PhD dissertation*, May 2003.

[27] O. Heczko, L. Straka, and K Ullakko. Relation between structure, magnetization process and magnetic shape memory effect of various martensites occuring in Ni-Mn-Ga alloys. *Journal de Physique IV*, 112:959–962, 2003.

Dynamic performance analysis of nonlinear tuned vibration absorbers

Jeong-Hoi Koo[1], Amit Shukla[1], and Mehdi Ahmadian[2]

[1]*Department of Mechanical and Manufacturing Engineering, Miami University,Oxford, OH 45056 USA*
[2]*Department of Mechanical Engineering, Virginia Tech, Blacksburg, VA 24061 USA*

Abstract

The main purpose of this study is to evaluate the dynamic characteristics of a damping controlled semi-active tuned vibration absorber (TVA). A base-excited, single-degree-of-freedom structure coupled with a TVA model is adopted as the baseline model for our analysis. A non-model based groundhook control (displacement-based on-off control or "On-off DBG") was used to control the damping of the TVA. After developing numerical models of a nonlinear TVA along with the on-off DBG control and a passive TVA, the optimal tuning parameters of both TVA models were obtained using an optimization routine. Using the two optimally tuned models, parametric studies were preformed by varying the values of both on and off-state damping to evaluate the dynamic performances of the TVAs using the peak transmissibility criteria. The results showed that the peak transmissibility of the semi-active TVA is nearly 21% lower than that of the passive, indicating that the semi-active TVA is more effective in reducing vibration levels. The results further showed that increases in off-state damping decreases the effectiveness of semi-active TVA in reducing maximum vibration levels.

Keywords: Nonlinear Absorber, Tuned vibration absorber

1. Introduction

Since their invention in the early 1900s, Tuned Vibration Absorbers (TVAs) have been effective engineering devices that suppress vibrations of machines and structures. A TVA is a vibratory subsystem, normally consisting of a mass, a spring, and a damper, that is mounted on a primary system. The TVA counteracts the motions of the primary system, "absorbing" the primary structure's vibrations. However, a conventional passive TVA is only effective when it is tuned properly; otherwise, it can magnify the vibration levels of the primary system, hence, the name "tuned" vibration absorber. In many practical applications, the off-tuning of a TVA often occurs because of system parameter changes over time. To cope with these problems, extensive studies over the past two decades have been performed to develop new designs and concepts for passive TVAs. These designs include adaptive, semi-active, and active TVAs. A comprehensive review of configurations, developments, and applications of TVAs can be found in the papers by Sun et

[1] Corresponding author, tel.: +1-513-529-7587; fax: +1-513-529-1454
E-mail address: koo@muohio.edu

al. [1], Housner et al. [2], Symans and Constaninou [3], and Soong and Spencer [4]. Fig. 1 shows a conventional TVA model. In this model, the primary structure is coupled with a damped TVA, and the mass of the structure and absorber are defined by m_1 and m_2, with their corresponding displacements as x_1 and x_2, respectively. The absorber's spring (k_2) and damper (c_2) are mounted on the structure. The stiffness and damping of the structure are represented by k_1 and c_1, respectively.

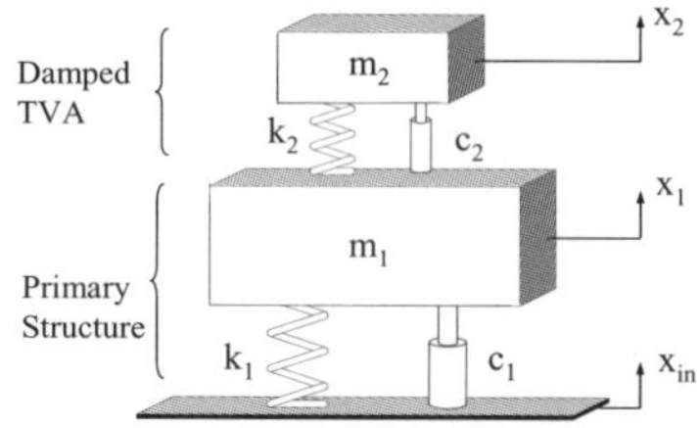

Fig. 1. Conventional Passive TVA Model (Base-Excited Model)

In an effort to enhance the performance of the traditional TVAs, this study considers a semi-active TVA by controlling its damping element. The next section provides an overview of a semi-active TVA, including a non-model based control method. After introducing system parameters, an optimal tuning technique is explained. Following the optimal tuning, parametric studies of the semi-active and equivalent passive TVA models will be discussed. Finally, a detailed analysis and comparison of the simulation results conclude the work.

2. Semi-Active Tuned Vibration Absorbers

The semi-active TVA model (Fig. 2b) replaces a passive damping element with a controllable damping element, which distinguishes it from the conventional passive system (Fig. 2a). A controllable damping element, such as a magnetorheologoical damper, is the key element for the semi-active TVA. Fig. 2b is used to derive the dynamic equations of motion for the semi-active model. The equations of motion that describe this system are given in Eq. (1).

$$\begin{bmatrix} m_1 & 0 \\ 0 & m_2 \end{bmatrix}\begin{Bmatrix} \ddot{x}_1 \\ \ddot{x}_2 \end{Bmatrix}+\begin{bmatrix} c_1 + c_{controllable} & -c_{controllable} \\ -c_{controllable} & c_{controllable} \end{bmatrix}\begin{Bmatrix} \dot{x}_1 \\ \dot{x}_2 \end{Bmatrix}+\begin{bmatrix} k_1 + k_2 & -k_2 \\ -k_2 & k_2 \end{bmatrix}\begin{Bmatrix} x_1 \\ x_2 \end{Bmatrix}=\begin{bmatrix} c_1\dot{x}_{in} + k_1 x_{in} \\ 0 \end{bmatrix} \quad (1)$$

This mathematical model will be used in the development of the numerical model of the semi-active TVA in a later section. Fig. 3 illustrates how a controllable damper can provide a wide range of damping force. At a given velocity, V_g, the corresponding damper force for the passive damper is a constant force, F_{pa}. On the other hand, the controllable damper offers a damper force ranging from F_{off} to F_{on}. This controllable damper, which provides a wide dynamic force range, can significantly improve the dynamic performance of the semi-active TVA with proper control methods. To effectively control this semi-active damping, this study uses the displacement-based, on-off groundhook (On-off DBG). This is because it was identified as a suitable control policy in the previous simulation work by the authors [5]. The On-off DBG control adjusts the damping level of the semi-active damping element based on the product of the primary systems' displacement and the relative velocity across the damper. The on-off DBG groundhook control policy can be written as

$$x_1 v_{12} \geq 0 \qquad c_{controllable} = c_{on}$$
$$x_1 v_{12} < 0 \qquad c_{controllable} = c_{off} \tag{2}$$

where x_1 is the displacement of the structure and v_{12} represents the relative velocity across the damper.

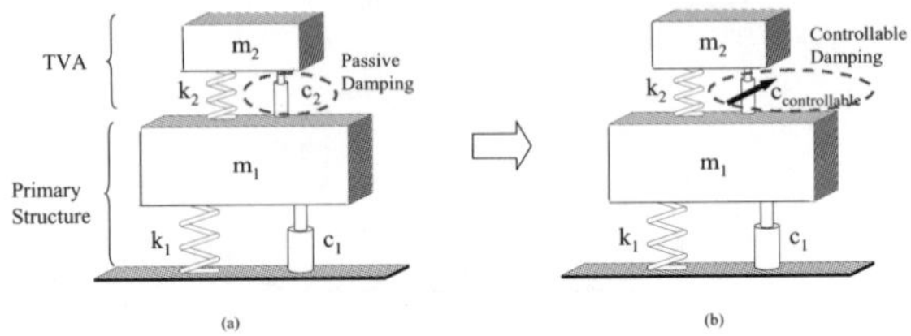

Fig. 2. Passive TVA versus Semi-Active TVA; (a) Passive Model and (b) Semi-Active Model

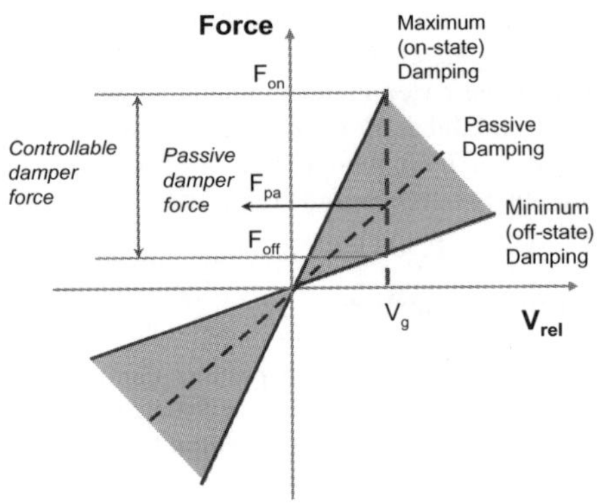

Fig. 3. Semi-active Damping

3. Optimal Tuning of the TVAs

This section presents the simulation parameters and the optimal tuning of simulation models. The tuned systems provide for an equal comparison of system performance. The simulation models are tuned by using a numerical optimization technique, which is necessary for the models because of the presence of damping in the primary system and the non-linearity of semi-active control. The goal of the optimization is to find the TVA parameters, such as on/off-damping ratios (ζ_{on} and ζ_{off}), and stiffness (k_2), which generate the "best" performance of each model. The optimization routine uses a minimization of the maximum value of transmissibility in a frequency range of interest as an objective function.

3.1 System Parameters

The simulation models are developed by coupling the controller and the mathematical models that were discussed earlier. The actual realization of these simulation models is done with Matlab. The Simulink toolbox within Matlab is used to build block diagrams for each of these simulation models. Table 1 shows the system parameters of simulations, and Table 2 shows a summary of the simulation parameters used in the Simulink.

Table 1. System Parameters

Parameter	Value
Structure Mass (m_1)	575 lb
TVA Mass (m_2)	32 lb
Structure Stiffness (k_1)	1909 lb/in
Structure Damping Ratio (ζ_1)	3 %

Table 2. Numerical Parameters Used in Simulink Models

Simulation Input	Initial Frequency	Target Frequency	Simulation Time	Equation Solver	Fixed-step Size
Chirp	0.5 Hz	10 Hz	150 sec	Runge-Kutta	1/100

3.2 Optimization Routine

Fig. 4 shows the flow chart for the optimization technique. The execution of the optimization routine involves three steps. In the main program, the system parameters are defined, along with the initial values and ranges of the simulation parameters. The main program calls the Simulink numerical models, which are responsible for generating the peak transmissibilities. These peak transmissibilities are then sent to the optimizer, "fmincon.m", where the minimum values of the peak transmissibilities are returned, along with the corresponding simulation parameters. Table 3 summarizes the initial values and parameter ranges for this optimization.

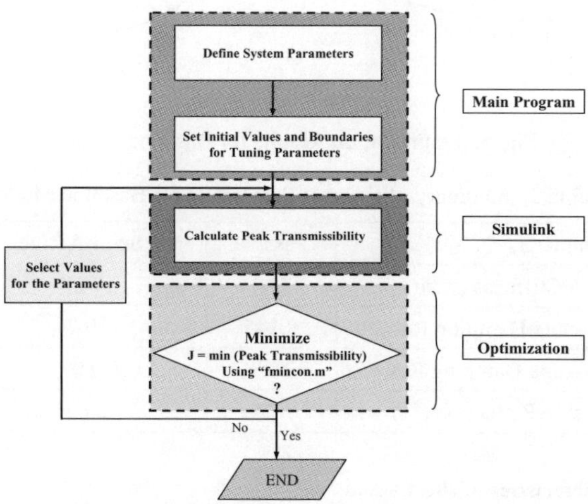

Fig. 4. Flow Chart for Optimization Routine

Table 3. Summary of initial values and parameter ranges

Model	Parameter	Initial Value	Parameter Range
Passive	Damping Ratio	0.1	$0.1 \le \varsigma_{pa} \le 0.7$
	TVA Stiffness (lb/in)	75.3	$10000 \le k_{a\_pa} \le 100000$
On-off DBG	On-state Damping Ratio	0.7	$0.1 \le \varsigma_{on\_od} \le 0.7$
	Off-state Damping Ratio	0.07	$0.01 \le \varsigma_{off\_od} \le 0.07$
	TVA Stiffness (lb/in)	71.1	$10000 \le k_{a\_od} \le 100000$

Fig. 5 shows the transmissibility plots of the optimally tuned TVAs. The on-off DBG controlled semi-active TVA outperforms the passive system in reducing the peak transmissibility. The semi-active TVA achieved nearly 21 % of more reduction of the maximum vibrations levels over the passive TVA. Table 4 shows a summary of the values of baseline parameters. These results will serve as the baseline values for parametric studies in the subsequent sections.

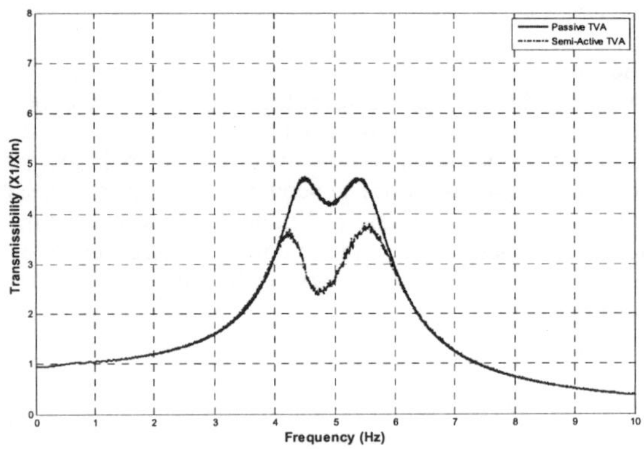

Fig. 5. Optimization Results for Baseline Models

Table 3. Summary of Optimal Parameters for Baseline Models

Parameters	Passive	Semi-Active
TVA Stiffness (lb/in)	75.3	70.8
On-state Damping Ratio	0.145	0.7
Off-state Damping Ratio	N/A	0.07
Passive Reduction (%)	N/A	20.96

4. Dynamic Characteristics of the TVAs

This section contains the results of the parametric studies performed on the baseline numerical models. These parametric studies provide an understanding of the dynamics of the TVAs as their parameters change. It uses transmissibility and phase plots to evaluate each TVA's performance. The phase angle analysis adds valuable explanations in analyzing the results.

4.1 Performance Analysis of the Passive TVA

Fig. 6a shows the transmissibility, or the ratio between the output and the input displacement of the primary structure, for a passive TVA as the damping ratio (ζ_2) changes from 0.0 to 0.9. When the damping ratio is 0.0, there are two large peaks, and a complete isolation occurs at the valley. Increasing the damping ratio to its baseline value lowers these resonant peaks and widens the valley between the two peaks. However, this has a negative effect; it raises the valley floor. Further increasing the damping ratio above its baseline value gradually causes the two peaks become a single peak. This means that the structure and the TVA become strongly coupled, and function nearly as a single mass, effectively negating any benefits of the TVA.

Fig. 6b shows the phase angles between the TVA mass and the structure mass as the damping ratio increases. The phase plots support the above discussion. When the damping ratio is at its baseline value, the phase angle around the tuned frequency is relatively close to 90 degrees; the TVA effectively counteracts the motions of the structure mass, achieving the minimum transmissibility over the entire frequency range of interest. However, as the damping ratio increases, the phase angle decreases. When the damping ratio is 0.9, the phase angle drops to about 30 degrees, indicating that the two masses have become strongly coupled. Thus, a single resonant peak forms in the transmissibility plot at this high damping ratio. For the passive system, excessively increasing the damping ratio results in a coupling of the TVA, and effectively renders it useless in reducing the vibrations of the structure mass.

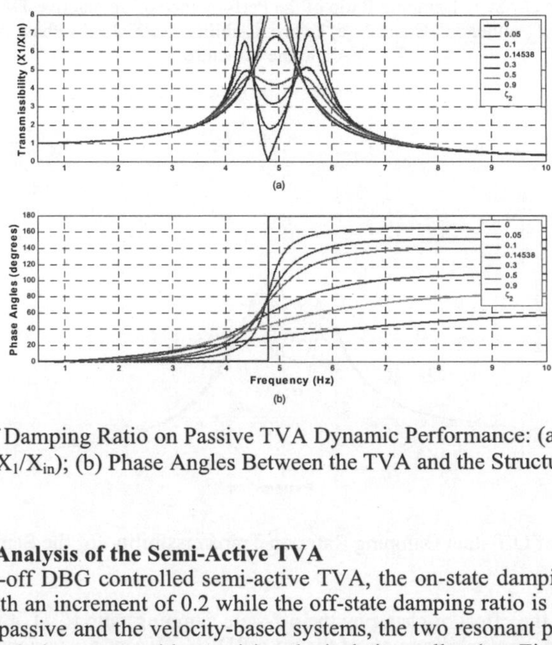

Fig. 6. Effect of Damping Ratio on Passive TVA Dynamic Performance: (a) Transmissibility (X_1/X_{in}); (b) Phase Angles Between the TVA and the Structure

4.2 Performance Analysis of the Semi-Active TVA

For the case of on-off DBG controlled semi-active TVA, the on-state damping ratio (ζ_{on}) varies from 0.1 to 0.9, with an increment of 0.2 while the off-state damping ratio is fixed at the baseline value. Unlike the passive and the velocity-based systems, the two resonant peaks decrease as the on-state damping ratio increases, without raising the isolation valley (see Fig. 7a). This indicates that the on-off DBG control keeps the TVA and the structure masses decoupled at the tuned frequency, enabling the TVA to effectively counteract the motions of the structure at a high on-state damping ratio. This is because the on-off DBG control policy ensures the minimum (off-state) damping ratio at the valley, independent of the on-state damping ratio, preventing lock-up (coupling of the two bodies). This result is one of the key benefits of this semi-active system.

Fig. 7b shows the phase angle changes of the on-off DBG system as the on-state damping ratio increases. The phase angle of at the tuned frequency (valley) stayed close to 90 degrees, regardless of the on-state damping ratio. Thus, the on-off DBG TVA achieved the minimum transmissibility. Moreover, the phase angles at the two resonant frequencies approach 90 degrees as the on-state damping ratio increases, reducing the resonant peaks, as shown in Fig. 7a.

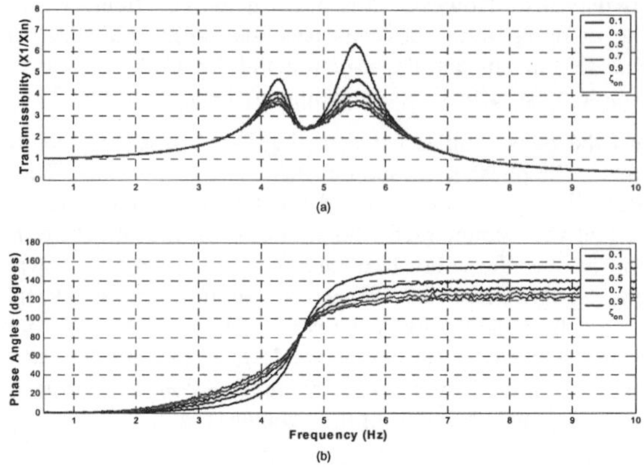

Fig. 7. Effect of On-State Damping Ratio of the Performance of Semiactive TVA with On-Off Displacement-Based Groundhook Control: (a) Transmissibility (X_1/X_{in}); (b) Phase Angles Between the TVA and the Structure

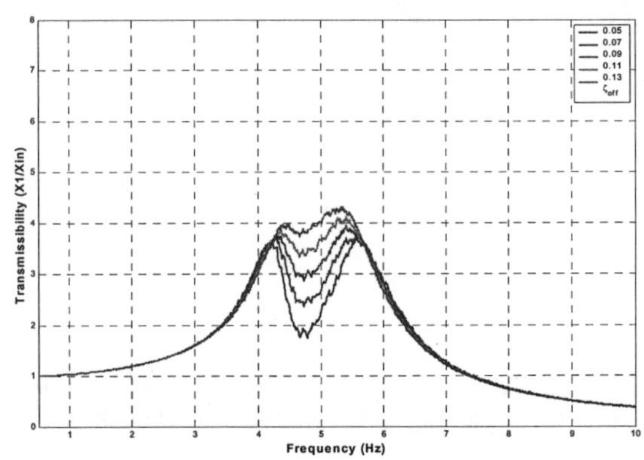

Fig. 8. Effect of Off-state Damping Ratio on Transmissibility for the Semi-active TVA

To evaluate the effect of changing the off-state damping ratio (ζ_{off}), it varied from 0.05 to 0.13. Fig. 8 shows transmissibility plots of the on-off DBG TVA as ζ_{off} varied. When the off-state damping ratio increases, the valley floor also increases. Moreover, the amplitudes of the two peaks grow, and they tend to become one. This observation indicates that high off-state damping ratios negate the performance of the on-off DBG TVA. This analysis suggests that the off-state damping ratio should be tuned at its optimal value in order to offer its maximum performance gains.

5. Conclusions

In this study, the dynamic performance of a damping controlled nonlinear TVA and its equivalent passive TVA were evaluated. A non-model based groundhook control called "On-off DBG" control policy was adopted for damping control of the semi-active TVA. Using an optimization routine, the semi-active TVA and its equivalent passive TVA were optimally tuned for equal evaluation. The performances of each of the optimized cases are then compared using the peak transmissibility criteria. The results indicate that the semi-active TVA outperforms the passive in reducing the peak transmissibility. The semi-active reduced more than 20% of the maximum vibrations levels as compared its passive counterpart. Furthermore, the results suggest that the off-state damping ratio of the semi-active TVA should be tuned at its optimal value in order to offer its maximum performance gains.

References

[1] Sun, J. Q., Jolly, M. R., and Norris, M. A., " Passive, Adaptive and Active Tuned Vibration Absorbers-A survey," *Transactions of the ASME, 50th Anniversary of the Design Engineering Division*, Vol. 117, pp. 234-242, 1995.

[2] Housner, G. W., Bergman, L. A., Caughey, T. K., Chassiakos, A. G., Claus, R. O., Masri, S. F., Skelton, S. E., Soong, T. T., Spencer, B. F., and Yao, J. T. P., "Structural Control: Past, Present, and Future," *Journal of Engineering Mechanics*, 123(9), pp. 897-971, 1997.

[3] Symans, M. D. and Constantinou, M. C., "Semi-Active Control Systems for Seismic Protection of Structures: A State-of-Art Review," Journal of Engineering Structures, 21, pp. 469-487, 1999.

[4] Soong, T. T. and Spencer, B. F. Jr., "Supplemental Energy Dissipation: state-of-the-art and state-of-the-practice," *Journal of Engineering Structures*, 24, pp. 243-259, 2002.

[5] Koo, J-H., Ahmadian, M., Setareh, M., and Murray, T., "In Search of Suitable Control Methods for Semi-Active Tuned Vibration Absorbers," *Journal of Vibration and Control,* Vol. 10, No. 2, pp 163-174 February, 2004

Application of Magnetorheological Elastomer to vibration control

Hua-xia Deng, Xing-long Gong*

CAS Key Laboratory of Mechanical Behavior and Design of Materials, Department of Mechanics and Mechanical Engineering, University of Science and Technology of China, Hefei 230027, China

Abstract

Traditional dynamic vibration absorber (DVA) is widely used in industries as vibration absorption equipment. However, it is only effective at narrow working frequency range. This shortcoming has limited its stability and application. This paper develops an adaptive tuned vibration absorber (ATVA) based on unique characteristics of magnetorheological elastomers (MREs), whose modulus can be controlled by an applied magnetic field. This ATVA works in shear mode and consists of dynamic mass, static mass and smart spring elements with MREs. Based on the double pole model of MR effects, the shift-frequency capability of the ATVA has been theoretically and experimentally evaluated. The experimental results demonstrated that the natural frequency of the ATVA can be tuned from 27.5 Hz to 40 Hz. To study its vibration absorption capacity, a beam structure with two ends supported has been employed. To analyze the vibration absorption capacity, a dynamic model of coupling beam and absorber has been established. Both the calculation and experimental results show that the absorption capacity of the developed ATVA is better than the traditional TVA and can achieve as high as 25dB which was justified by the experiment.

Keywords: Magnetorheological Elastomer (MRE); Adaptive Tuned Vibration Absorber (ATVA); Vibration Control

1. Introduction

Magnetorheological (MR) materials are smart materials which have MR effects and many unique properties with magnetic field. MR effect is that the rheological properties will be changed under applied magnetic field. Since its first discovery by Rabinow at 1948 [1], MR materials have developed into a family with MR fluids, MR foams and MR elastomers [2]. The most common MR material is MR fluids (MRFs) which are magnetically polarizable particles suspended in viscous fluids [3]. The general criterion to estimate the MR effect of MRFs is the variation capability of dynamic yield stress within a post-yield regime under applied magnetic field. Numerous applications based on MRFs benefit from the properties that the dynamic yield stress can be continuously, rapidly and reversibly controlled by the applied magnetic field. Such applications have been applied on varies of fields such as automotive industry [4-6], earthquake-resistance [7] and vibration control [8] and have been commercialization and industrialization. For instance, LORD Corporation of American have been professionally research, manufacture, distribute MRFs and series of applications based on these materials such as brakes, clutches

* Corresponding author. Tel.&Fax: +86-551-3600419

Email address: gongxl@ustc.edu.cn (X.L.Gong)

and variable-friction dampers. But MRFs also exhibit some shortcomings which hinders in the applications such as deposition, environmental contamination [9] and sealing problems.

MR elastomers (MREs), the structural solid analogues of MRFs may be a good solution to overcome these disadvantages. MREs are composed of polarizable particles dispersed in a polymer medium. Typically, magnetic fields are applied to the polymer composite during crosslinking so that particles form chainlike or columnar structures, which are fixed in the matrix after curing [10]. The comparison of MREs and MRFs is that the MR effect of MREs is the field-dependent modulus with in pre-yield regime [11]. Such properties induced that MREs have promising in many applications such as adaptive tuned vibration absorbers (TVAs), stiffness tunable mounts and suspensions, and variable impedance surfaces. Relative to MRFs, the application of MREs is still at exploring stage. Watson [12] applied an patent using MREs, method and apparatus for varying the stiffness of a suspension bushing; Ginder et al [13, 14] constructed and tested tunable automotive mounts and bushings based on MR elastomers which can be applied to minimize the effect of suspension resonances excited by torque variation due to worn brake rotors by shifting the resonance away from the excitation frequency. Ginder and coworkers [15] did pioneer work on the development of an adaptive tunable vibration absorber (ATVA) using MREs. Their initial experimental results indicated that the ATVA had the capability to shift frequency from 500 Hz to 610 Hz.

Other smart materials could also be applied to ATVA. Traditional dynamic vibration absorber (DVA) theoretically brings the object base to rest at a single excitation frequency, the resonance frequency of the DVA. It means that the absorber is usually used to suppress a single harmonic excitation of the vibrating systems. For many practical systems, which have complex vibration sources or wide vibration frequency bandwidth, the passive absorber will lose its effect and potentially aggravate the base vibration. Williams with his group [16, 17] did a lot of work on ATVA with SMA, which can vary its frequency by changing environmental temperature. Davis et al.[18,19] applied a piezoceramic inertial actuator(PIA) to ATVA which vary nature frequency from 243 Hz to 257 Hz. Magnetostrictive material was used by Flatau et al.[20] to develop an ATVA achieving a natural frequency variation from 1400Hz to over 2000 Hz through application of DC magnetic fields.

The objective of this project is to investigate the application of MREs to vibration control. After the prototype has been succeeded by our lab, a shear mode vibration absorber which is more compact and more utilizable was developed in this paper. Both theoretical analysis and experimental testing were carried out to investigate its frequency-shift property and vibration absorption capacity.

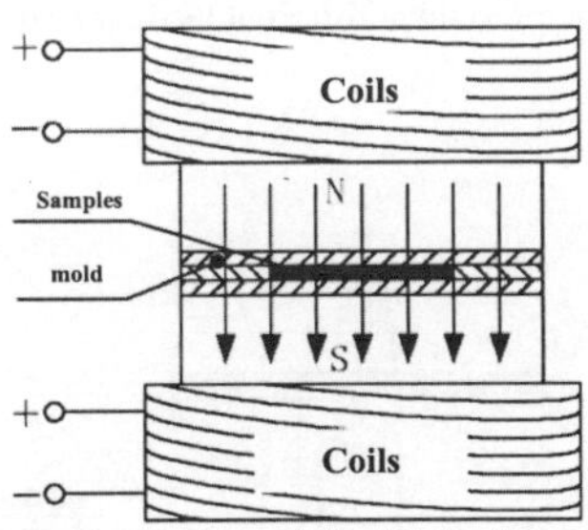

Fig. 1 The schematic diagram of the fabrication method

Fig. 2 Photograph of the modified DMA

2. ATVA Design

2.1 Material Preparation

The MRE materials consist of 704 silicone rubbers as a matrix, carbonyl iron particles with size of 3 to 5 microns, and a small amount of silicone oil. The schematic diagram of the fabrication method is shown in Fig.1. At the first step of the fabricating process, all ingredients are thoroughly blended with an agitator. Then the mixture is packed into an aluminum mold and placed in the vacuum to take out the air in the mixture. After that, seal the mold and place it in a magnetic field of 1 Tesla for 24 hours. Particles will be arranged in chain formation resulted from the anisotropic magnetic forces among the particles. When the elastomer is cured, such ordered structure is locked in the matrix. The experimental results reveal that MREs with chains of iron particles have good MR effect.

2.2 MRE Characterization

Figure 2 is a photograph of the MRE testing system, which is modified by our group on the base of the Dynamic Mechanical Analyzer(DMA) system from the Triton Co. (model: tritec2000). To measure MRE properties under various magnetic fields, an electromagnet was developed, which can provide magnetic field intensity up to 1100 mT. With this system, both the shear modulus and the loss factor against magnetic field at various excitation frequencies were measured. The results are shown in Fig. 3. It can be seen from Fig. 3(a) that the shear modulus shows an increasing trend with magnetic field intensity. However, the increasing slope decreases with the increment of magnetic fields, which is due to the magnetic saturation. In addition, the modulus also shows an increasing trend with loading frequency, which justifies that MRE performs viscoelastic behavior. Fig. 3(b) shows the loss factor is independent of magnetic field and frequency, which was demonstrated before [21].

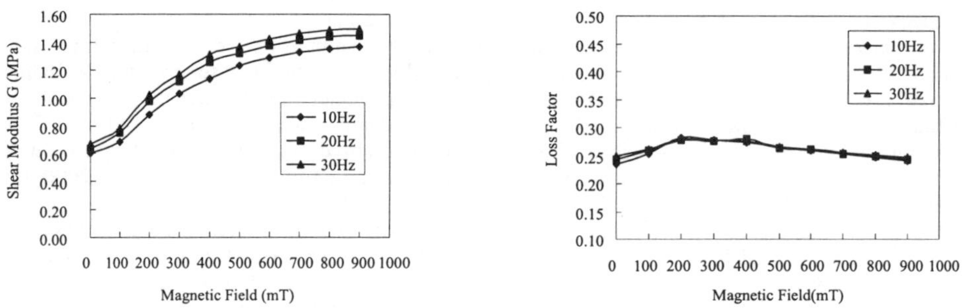

Fig. 3 Variation of shear modulus and loss factor with magnetic field at different loading frequency

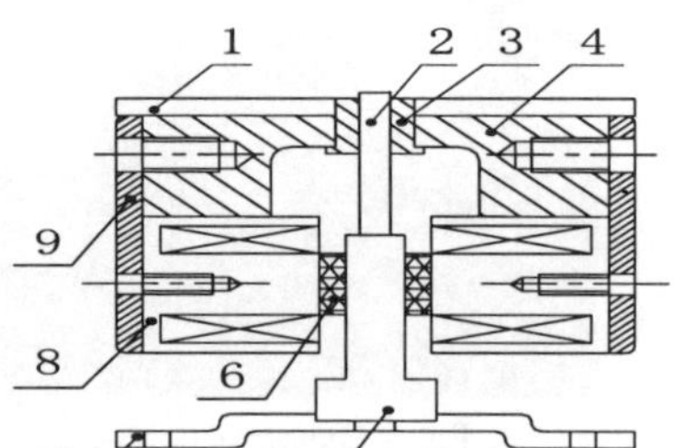

Fig. 4 The schematic diagram of the developed ATVA

1.Cover; 2.Guiderod; 3.Linear bearing; 4.Magnetic conductor; 5.Shear plate; 6.MREs; 7.Base; 8.Electromagnet; 9.Mounting Shell

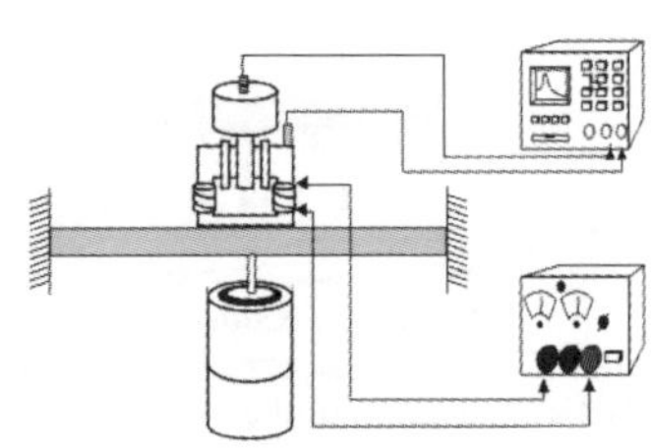

Fig.5 The evaluation system of frequency-shift property

2.3 ATVA Structure

The core of this design is to utilize the limit space efficiently as possible. The efficient component for ATVA to vibration absorption is its oscillator, dynamic mass, in another word. The schematic diagram of the developed ATVA is shown in Fig. 4. As shown in this figure, the ATVA consists of three main parts: dynamic mass, static mass and smart spring element with MREs. The electromagnets and magnetic conductors form a closed C-shape magnetic circulate, which are assembled at mounting shell to be the dynamic mass. This development makes the ATVA more compact and more efficient because no additional oscillator and nearly most components were moved to be dynamic mass. The static mass consists of the shear plate and the base. Through the shear plate, MREs, the smart spring elements connect the dynamic mass and the static mass. To ensure the ATVA works in the shear mode, a guiderod and a linear bearing are employed.

The working principle of the system is as below. The magnetic field is created by two coils in the electromagnets and the field strength is controlled by the coil current, provided by an external DC power. As MRE's shear modulus depends on the field strength, the equivalent stiffness of the ATVA changes with the field strength as well as the coil current. Consequently, the natural frequency of the ATVA can be controlled by the coil current. Thus, the ATVA natural frequency can be changed by tuning the coil current to trace the external excitation frequency. When the tuned ATVA frequency matches the excitation frequency, the vibration can be attenuated significantly. This point will be theoretically addressed in the following section.

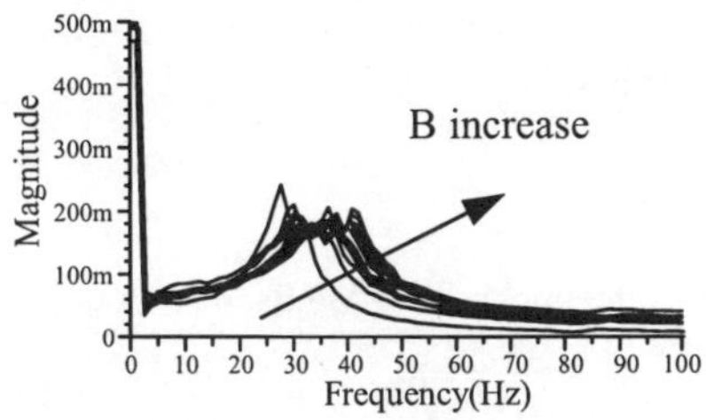

Fig.6 The magnitude curve versus frequency at various magnetic fields

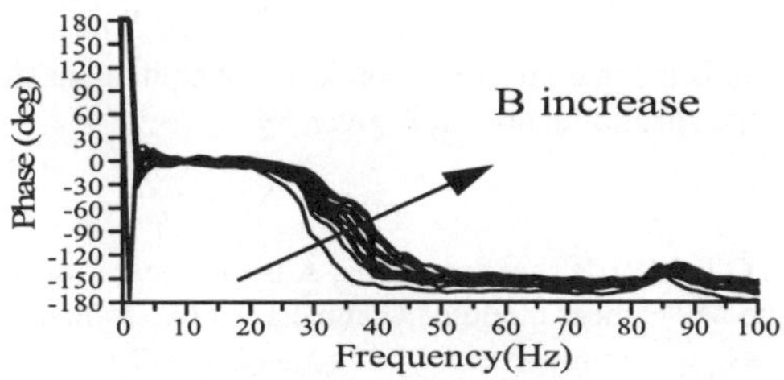

Fig.7 The phase curve versus frequency at various magnetic fields

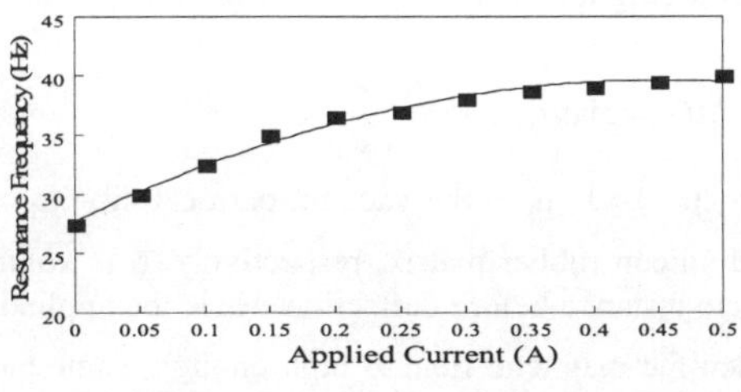

Fig.8 The relationship between applied current and resonance frequency

3 ATVA Testing

3.1 Experimental Study of Frequency-shift Property

For simplicity, a beam with both ends supported is employed to investigate the frequency-shift capability of the MREs. The experimental setup is shown in Fig. 5. The beam is made up of low carbon steel, with the size of 930mm×100mm×11mm. The material parameters are: Young's modulus E=200

GPa, density ρ=7800kg/m^3, Poisson's ratio ν=0.3, and the loss factor η=0.05. Using the modal analysis, its first and second mode natural frequencies are 60 Hz, and 250 Hz, respectively. As shown in Fig.5, the experimental procedure is as below. The developed ATVA is placed at the centre of the beam. White noise excitation applied to the beam is generated by an exciter (model: JZK-10, manufactured by Sinocera Piezotronics INC., China). Two accelerometers (model: PCB 3510A) are placed on the oscillator and the base beam to measure their responses, respectively. The measured signals are sent to the Dynamic Signal Analyzer (model: Signal Calc ACE DP240, Data Physics Corp.). The transfer function can be obtained by using FFT analysis. The frequency-shift capability at various magnetic fields of the ATVA based on MREs is shown in Fig. 6 and Fig.7. The two figures show that the magnitude curve and the phase curve of transfer function moves rightward with the increase of the magnetic field. The resonance frequency can be obtained by reading the peak-to-peak value or the frequency at 90 degree at the phase curve. By tuning the coil current, the relationship of the resonance frequency of the absorber versus the coil current can be obtained and it is shown in Fig.8. The resonance frequency increases from 27.5 Hz at 0 A to 40 Hz at 0.5 A. Its relative frequency change is as high as 145%.

3.2 Theoretic Analysis of Frequency-shift Property

The resonance frequency of the TVA is represented as

$$f = \frac{1}{2\pi}\sqrt{\frac{k_\tau}{m}} \tag{1}$$

where m is the mass of oscillator, k_τ is the equivalent shear stiffness.

In the shear direction, k_τ is given by

$$K_\tau = \frac{GA}{h} \tag{2}$$

where G is MRE's shear modulus, A is the shear area, and h is the thickness of MREs.

The MRE shear modulus G consists of two terms, as below

$$G = G_0 + \Delta G_d \tag{3}$$

where G_0 is the initial shear modulus without any magnetic field, ΔG_d is the shear modulus increment under the applied magnetic field. The double pole model is widely used to predict the magnetic induced shear modulus increment [22]. Using this model, the shear modulus increment under the magnetic field is given by [23]

$$\Delta G_d = 36\phi\mu_f\mu_0\beta^2 H_0^2\left(\frac{R}{d}\right)^3\zeta \tag{4}$$

where $\beta = (\mu_p - \mu_f)/(\mu_p + 2\mu_f) \approx 1$, μ_0 is the vacuum permeability, $\mu_p \approx 1000$ and $\mu_f \approx 1$ are relative permeability of particles and silicon rubber matrix, respectively. ϕ is volume fraction, R is the average particle radius, d is the particle distance before deflection, H_0 is the applied magnetic field intensity, and $\zeta = \sum_{k=1}^{\infty} 1/k^3 \approx 1.202$. When the magnetic field is high enough, some particles will saturate, equation (4) is not long valid. When all particle saturate, equation (4) will be replaced as

$$\Delta G_d = 4\phi\mu_f\mu_0 M_s^2\left(\frac{R}{d}\right)^3 \tag{5}$$

where M_s is particle's saturation intensity.

Substituting Equations (2) and (3) into Equation (1), the natural frequency is rewritten as

$$f = f_0 + \Delta f \tag{6}$$

where f_0 is the resonant frequency without magnetic field, Δf is the shift of resonance frequency due to the applied magnetic field.

$$f_0 = \frac{1}{2\pi}\sqrt{\frac{G_0 A}{mh}} \tag{7}$$

$$\Delta f = \frac{1}{2\pi}\sqrt{\frac{G_0 \times A}{m \times h}} \bullet (\sqrt{1+\frac{\Delta G_d}{G_0}} - 1) \tag{8}$$

From the Eqn. (7), the initial resonant frequency of the ATVA can be designed to match the vibration system, the focus of the vibration attenuation. Equation 8 reveals that the frequency-shift capacity is not only proportional with the MR effect but also relative to the initial shear modulus. The larger initial modulus with the same MR effect will cause the wider frequency-shift bandwidth. When $G_d << G_0$, Eqn. (8) can be approximately substituted by Eqn. (9)

$$\Delta f = \frac{\Delta G_d}{4\pi}\sqrt{\frac{A}{G_0 mh}} \tag{9}$$

The frequency-shift is proportional to the shear modulus increases, and it can be seen from the beginning of the curve shown in Fig. 8.

When G_d is as large as G_0, Eqn. (8) is not valid. Substituting Eqn. (4) into Eqn. （7）, the frequency-shift can be rewritten as below

$$\Delta f = \frac{1}{2\pi}\sqrt{\frac{G_0 \times A}{m \times h}} \bullet (\sqrt{1+\frac{36\phi\mu_f\mu_o\beta^2 H_o^2 (\frac{R}{d})^3 \zeta}{G_0}} - 1) \tag{10}$$

Similarly, substituting Eqn. (5) into Eqn. (7), the frequency-shift at saturation status is a constant as given by

$$\Delta f = \frac{1}{2\pi}\sqrt{\frac{G_0 \times A}{m \times h}} \bullet (\sqrt{1+\frac{4\phi\mu_f\mu_o M_s^2 (\frac{R}{d})^3}{G_0}} - 1) \tag{11}$$

The theoretical trend presented by Eqs. (9), (10), (11) agrees well with the results shown in Fig. 8 qualitatively.

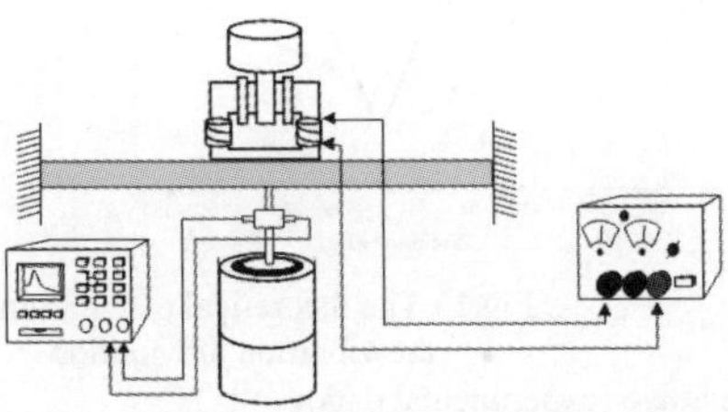

Fig. 9 The evaluation system for vibration attenuation

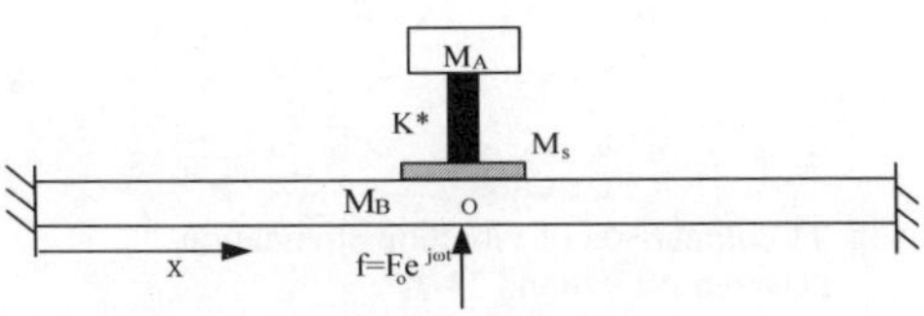

Fig.10 The simplification model of evaluation system

3.3 The Study on Vibration Attenuation

The effect of vibration absorption can be evaluated by the system shown in Fig.9. Compared with the frequency-shift experiment, the major difference is that an impedance head is induced. As shown in this figure, the impedance head connects with the beam and the exciter. The exciter provides sinusoidal excitation to the beam with a linear frequency scan from low to high. The natural frequency will be tuned by changing coil current to trace the excitation frequency. The base point impedance with various magnetic fields can be measured by the dynamic signal analyzer. If the magnetic field is fixed, the absorber can be looked as a classical passive TVA with fixed resonance frequency. To evaluate the vibration absorption capacity, a ratio γ of base point admittance with TVA and without TVA is employed.

The system in Fig. 9 can be simplified as shown in Fig. 10. Suppose the force and velocity applied to the base point O is F_0, and V_0, respectively. The resultant forces and velocities at the lower and upper points of the static mass M_s are F_s^b, V_s^b and F_s^t, V_s^t, respectively. The resultant force and velocity of the absorber mass M_A are F_A and V_A. The admittance at point O is defined as H_o^b and given by

$$H_o^b = \frac{V_o}{F_o} = \frac{j\omega}{M_B}\sum_{k=1}^{n}\frac{\Phi_k^2(x_o)}{\omega_k^2(1+j\eta)-\omega^2} \tag{12}$$

where ω_k and ϕ_k are the kth natural frequency and mode function, x_o is the coordinator of the point O, n is the mode number, η is the loss factor, ω is the excitation frequency, $j = \sqrt{1}$. Eqn. (12) can be simplified as

$$V_o = \frac{\Delta H}{H_o^b + \Delta H} H_o^b F_o, \quad \Delta H = 1/(j\omega M_s + \frac{j\omega M_A K^*}{K^* - \omega^2 M_A}) \tag{13}$$

where ΔH is the additional admittance at the point O due to the absorber and the mass, K* is the complex stiffness of absorber, The complex stiffness can be expressed by conventional spring's stiffness K, damping factor C and damping ratio ξ:

$$K^* = K(1 + j\omega C/K) = K(1 + 2j\omega\omega_0\xi M_A/K) \tag{14}$$

The admittance at the point O with absorber is given by

$$H_A = \frac{\Delta H}{H_o^b + \Delta H} H_o^b \tag{15}$$

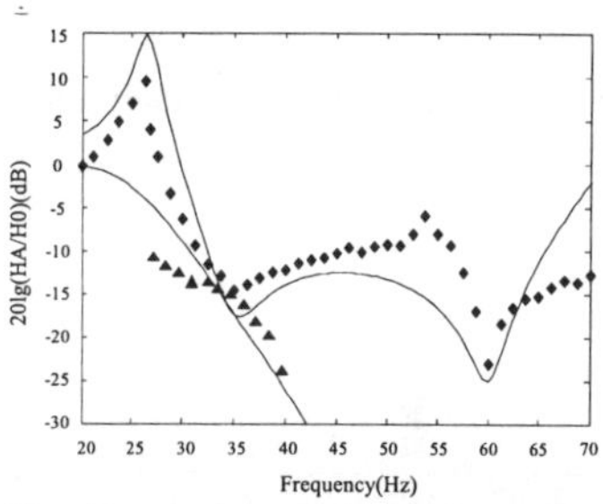

Fig. 11 comparison of vibration attenuation between ATVA and TVA

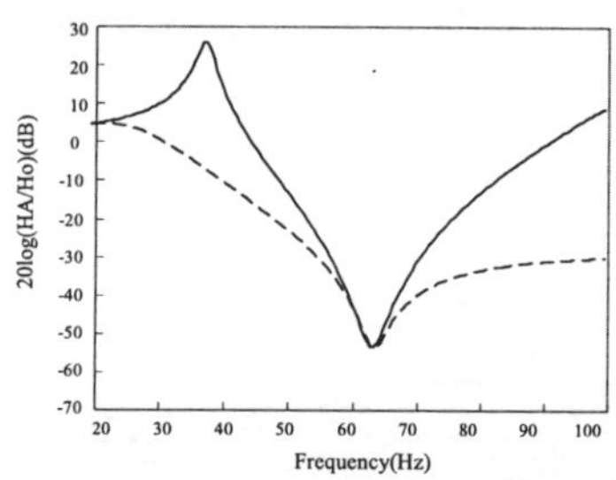

Fig.12 The theoretical calculation of the vibration attenuation

▲adaptive(experimental data); ◆ passive (experimental data);
- - adaptive(numerical data); — passive(numerical data)

When there is no active mass, the admittance at the point O is given by

$$H_o = \frac{\Delta H^s}{H_o^b + \Delta H^s} H_o^b, \quad \Delta H^s = 1/(j\omega M_s) \tag{16}$$

The ratio γ to reflect the effect of vibration absorption capacity can be defined as

$$\gamma=20(|H_A / H_O|) \tag{17}$$

The comparison of vibration absorption capacity which is apart from the natural frequency of the primary system between the adaptive TVA and the passive one is shown in Fig. 11, where the upper data are the results of passive TVA whose resonance frequency is fixed at 35 Hz and the lower data are the results of ATVA whose frequency is tuned to trace the excitation frequency. As shown in Fig. 11, experimental data and numerical curve are very close. It indicates that the dynamic model is reasonable and the experimental data is reliable. For the passive TVA, the best vibration attenuation efficiency occurs at the natural frequency of the primary system. The effect become worse sharply while the excitation frequency is apart from beam natural frequency and a new resonance hump occurs at 27Hz. For ATVA, limited to the MR effect of spring element, ATVA can not be experimentally evaluated at the entire frequency band. In tunable frequency band, the comparison between adaptive TVA and passive one indicates that ATVA has better absorbing effects than passive TVA. For example, at 40Hz, the ATVA's effect of vibration absorption is -22dB while it is -8dB for the passive TVA. For even wider bandwidth, theoretical analysis which is shown at Fig.12 predicts that the ATVA also has better effects and there is no resonance hump at the formant frequency of the passive TVA.

4 Conclusion

A shear mode ATVA with MRE was developed in this paper. It consists of dynamic mass, static mass and smart spring elements with MREs. Both theoretical and experimental results indicate that the resonance frequency of the developed ATVA can be controlled by electrical currents. The resonance frequency varies from 27.5 Hz at 0 A to 40 Hz at 0.5 A. Its relative frequency change is as high as 145%. To evaluate the vibration absorption capability of TVA, a beam with two ends supported is used as an object base and the ratio γ of base point admittance with TVA and without TVA is employed. The experimental results agree well with theoretical calculation and they all demonstrate that the developed ATVA has better performance than conventional passive absorber in terms of frequency-shift property and vibration absorption capacity.

5 Acknowledgement

The authors want to thank Prof. P.Q. Zhang, H.L. Sun, Y.S. Zhu, J.F. Li, Z.B. Xu for their useful discussions. This work is supported by BRJH Project of Chinese Academy of Sciences and SRFDP of China (No. 20050358010).

References

[1] J.Rabinow. The Magnetic Fluid Clutch. AIEE Transactions 1948; 67:1308-1315.
[2] J.David Carlson, Mark R. Jolly. MR fluid, Foam and elastomer devices. Mechatronics 2000; 10:555-569.
[3] Xianzhou Zhang, Xinglong Gong, Peiqiang Zhang, Qiming Wang. Research on mechanism of squeeze-strengthen effect in magnetorheological fluids. Journal of Applied Physics 2004; 96:2359-2364.
[4] P.P. Phule, J.M.Ginder. Synthesis and Properties of Novel Magnetorheological Fluids Having Improved Stability and Redispersibility. International Journal of Modern Physics B 1999; 13(14-16):2019-2027.
[5] J.M.Ginder. Behavior of Magnetorheological Fluids. MRS BULLETIN 1998;23-8:26-29.
[6] J.M.Ginder, L.C.Davis, L.D. Elie. Rheology of Magnetorheological Fluids: Models and Measurements. International Journal of Modern Physics B 1996; 10 (23-24): 3293-3303.
[7] S.J.Dyke, B.F.Spencer, M.K.Sain, J.D.Carlson. An Experimental Study of MR Dampers for Seismic Protection. Smart Materials & Structures1998;7-5:693-703.
[8] H.J.Jung, B.F.Spencer, I.W.Lee. Control of Seismically Excited Cable-stayed Bridge Employing Magnetorheological Fluid Dampers. Journal of Structural Engineering-ASCE2003; 129-7:873-883.
[9] Y.Shen, M.F.Golnarghi, and G.R.Heppler. Experimental Research and Modeling of Magnetorheological Elastomers. J.Intell.Mater.Syst.Struct 2004;15:27-35.
[10] X.L.Gong X.Z.Zhang and P.Q.Zhang. Fabrication and Characterization of Isotropic Magnetorheological Elastomers. Polymer Testing 2005; 24-5: 669-676.

[11] G.Y.Zhou. Shear properties of a magnetorheological elastomer. Smart Mater. Struct 2003;12: 139-146.
[12] J.R.Watson,canton,Mich.Method and applaratus for varying the stiffness of a suspension bushing. U.S.Patent 5,609,353,1997.
[13] W. M. Stewart, J. M. Ginder, L. D. Elie, and M. E. Nichols.Method and apparatus for reducing brake shudder. U. S.Patent 5,816,587, 1998.
[14] J.M.Ginder, M.E.Nichols, et al. Controllable-Stiffness Components Based on Magnetorheological Elastomers. Proceedings of SPIE 2000;3985 :418-425.
[15] J.M.Ginder, W.F.Schlotter, and M.E.Nichols. Magnetorheological Elastomers in Tunable Vibration Absorbers. Proc.SPIE 2001; 4331:103-110.
[16] K.A. Williams, G.T.-C. Chiu, R.J. Bernhard. Passive–adaptive vibration absorbers using shape memory alloys. Proc.SPIE 1999; 3668:630–64.
[17] K.A.Williams, G.T. -C. Chiu, R.J. Bernhard. Dynamic modeling of a shape memory alloy adaptive tuned vibration absorber. J. Sound & Vibration 2005;280:211-234.
[18] C.L. Davis, G.A. Lesieutre, J. Dosch.A tunable electrically shunted piezoceramic vibration absorber. Proc.SPIE 1995; 3045:51–59.
[19] C.L. Davis, G.A. Lesieutre. An actively tuned solid-state vibration absorber using capacitive shunting of piezoelectric stiffness. J. Sound & Vibration 2000; 3:601–617.
[20] A.B. Flatau, M.J. Dapino, F.T. Calkins. Highbandwidthtenability in a smart vibration absorber. Proc.SPIE 1998; 3327: 463–473.
[21] X.L.Gong, X.Z.Zhang, P.Q.Zhang. Fabrication and Characterization of Isotropic magnetorheological Elastomers. Polymer Testing 2005; 24-5:669-676.
[22] T.Shiga, A.Okada, and T.Kurauchi. Magnet-roviscoelastic behavior of composite gels. J. Appl. Polym. Sci. 1995;58: 787-792.
[23] L.C.Davis. Model of magnetorheological elastomers. J. Appl. Phys. 1999; 85:3348- 3351.

Emulation and analysis of the response time of the magnetrheological fluid damper

Na Shen[1] Jiong Wang[*]

School of Mechanical Engineering, Nanjing University Science and Technology, Nanjing, 210094, China

Received 14 March 2006; received in revised form 10 May 2006; accepted 31 May 2006

Abstract

The paper mainly discusses the response time of the magnetrheological fluid damper. Emulate and analyze the response time of the magnetrheological fluid and the response time of the apparatus. And build the model of Smith Precompensation of the MRFD hysteresis system to compensate the time lag.

PACS: 83.85.Ns

***Keywords*:** The magnetrheological fluid damper, the response time, emulation

1 Instruction

The magnetrheological fluid damper (MRFD) is promising appliance in vibration control, with the advantages of rapid response, small bulk, being easy to be integrated to computer, having continuously adjusted damp force, etc. Because of the nonlinearity of the kinetic performance of the MRFD, it is hard to build a precise model for control. The response time is one of the important components of the MRFD performance, which decides the control period, appliance area, and control effect of the MRFD.The response time of the MRFD is different from the response time of magnetrheological fluid. The former is the response time of the apparatus, which refers to the period between the moment that the MRFD receives an external control signal and the moment that it reaches to the stabile working stage. While the later means the time that the magnetrheological fluid needs to become Bingham stabile flow in external magnetic field. Apparently, the response time of magnetrheological fluid is a part of the response time of the MRFD. Here we analyze them separately.

2 The response time of magnetrheological fluid

The academic model of MRFD based on Bingham model is as follow:

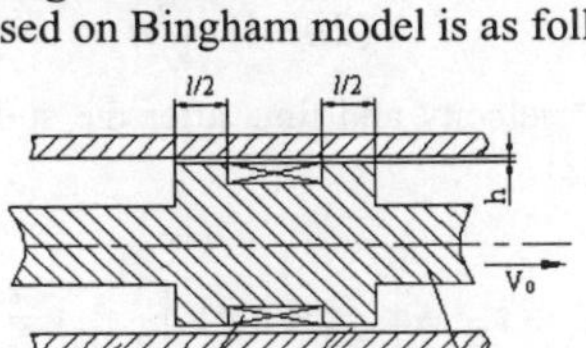

Fig1 the magnetrheological fluid damper

l is the valid length of the damp passage; h is the width of the damp passage; V_0 is the moving velocity of the piston

* Corresponding author. Tel.: 86-025-84315327; fax: 86-025-84315327

E-mail address: **wjiongz@mail.njust.edu.cn**

The damp force provided by the damp passage is

$$F_R = -A_p \Delta p = \frac{12\eta A_p{}^2 lu}{wh^3} P = (A_p u + \frac{wh}{2} u)\frac{12\eta l A_p}{wh^3} + c\frac{\tau_y l A_p}{h}\operatorname{sgn}(u) \tag{1}$$

The c is compensating coefficient.

$$c = 2.07 + \frac{12A_p u\eta}{12A_p u\eta + 0.4wh^2\tau_y} \tag{2}$$

F_η is viscosity damp force, which is in direct ratio to velocity, and is independent of magnetic field intensity.

$$F_\eta = (A_p u + \frac{wh}{2}u)\frac{12\eta l A_p}{wh^3} \tag{3}$$

F_τ is coulomb damp force, which is in direct proportion to yield stress, and is independent of

velocity. $F_\tau = c\frac{\tau_y l A_p}{h}\operatorname{sgn}(u)$ (4)

u is the flow velocity of magnetrheological fluid in the damp passage. τ_y is the critical yield shear stress, and has relation with magnetic field intensity H. η is plastic viscidity of the MRFD. Ap is the working area of the piston. w is the average perimeter of the damp passage. h is the width of the damp passage; l is the valid length of the damp passage. (See fig1)
The relationship between the flow velocity and time after the sudden change of the pressure of the stabile Newton liquid is [2]:

$$\begin{aligned} u(y,t) = {} & \sum_{k=0}^{\infty}\frac{(-1)^k 32u_0}{(2k+1)^3\pi^3} e^{-\lambda\left[\frac{2k+1}{h}\right]^2 t}\cos\frac{(2k+1)\pi}{h}y \\ & + \frac{(\Delta P_0 + \Delta P_1)h^2}{8\eta l}\left[(1-ay^2) + \frac{32}{\pi^3}\sum_{(2n-1)^3}^{(-1)^n} e^{-\frac{t}{\tau_n}}\cos\frac{(2n-1)\pi y}{h}\right] \end{aligned} \tag{5}$$

$$\tau_n = \frac{h^2\rho}{(2n-1)^2\pi^2\eta} \tag{6}$$

The relationship between the flow velocity and time after the sudden change of the flow of the stabile Newton liquid in same pressure is [2]

$$u(y,t) = u_0(1-ay^2) - \frac{\tau_{y0}}{\eta}\frac{h}{\pi}\left[\sum_{n=1}^{\infty}\frac{(-1)^{n+1}}{n}\left[1 - e^{(-\frac{t}{\tau_n})}\right]\sin\left(\frac{2n\pi\Delta y}{h}\right)\right] \tag{7}$$

$$\tau_n = \frac{h^2\rho}{4n^2\pi^2\eta} \tag{8}$$

3 The response time of the apparatus

The damp force is adjusted by changing the input current of the excitation loop that enlaced around the piston axis, which causes transform of the magnetic field and thus changes the viscidity of the magnetrheological fluid.

According to the parameter of the excitation loop, the relationship between the magnetic field intensity H and current I is:

$$H = \frac{NI}{2d} \tag{9}$$

N is the circle numbers of the loop, *d* is the diameter of the loop.

Nowadays, the control current source of the MRFD usually works in pulse with modulation (PWM) method. Thereby, the response time of the excitation loop caused by the PWM wave of control signal is an emphasis of studying. The following instances of the response of the MRFD's excitation loop are emulated in different parameters by PSpice programme. Fig 2 is a kind of PWM current source of MRFD. The excitation loop of MRFD takes the parameters of RD-1005-3 as reference, whose default resistance is 4.7Ω, and the default inductance is 4.3 mH. The input PWM pulse is 3.3V in peak. The default pulse width is 30us. And the default period is 60us. The emulating time is 10 ms.

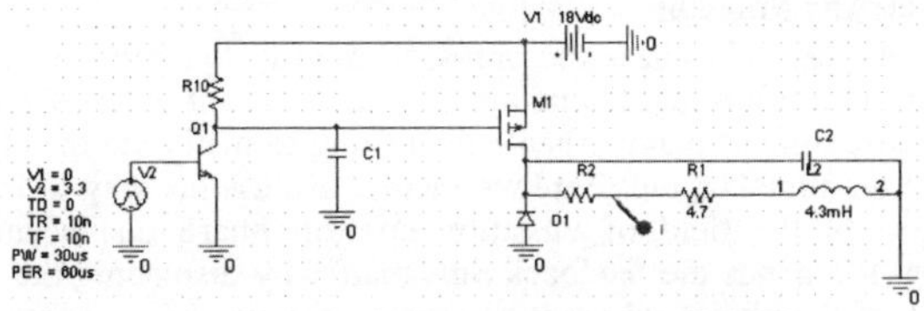

Fig 2 the emulated circuit of PWM current source of MRFD

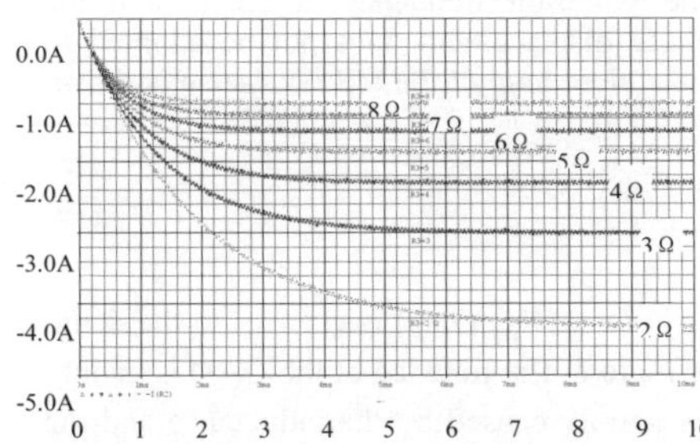

Fig3 current –time fig. in different resistance

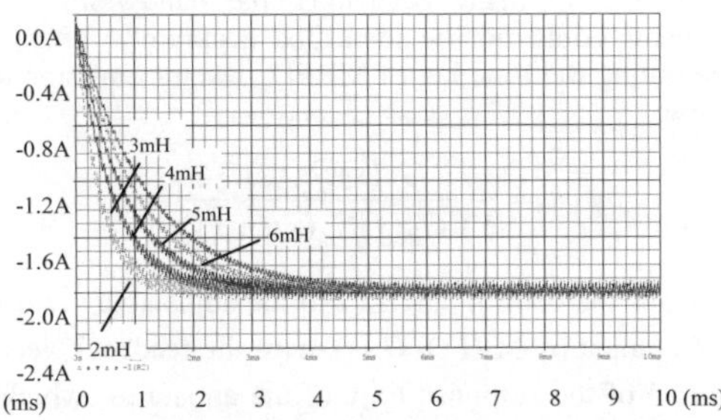

Fig4 current –time fig. in different inductance

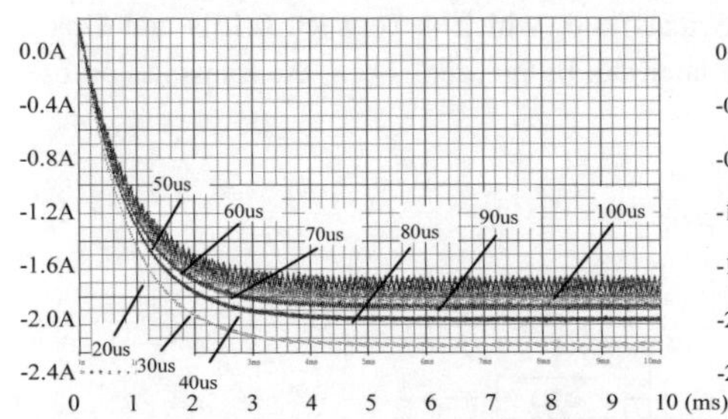

Fig5 current –time fig. in different PWM Cyc.

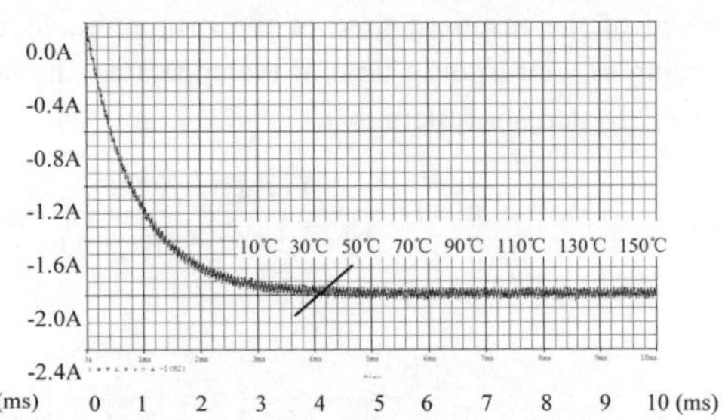

Fig6 current –time fig. in different temperatures

Fig3 shows the loop current changing with time, when the loop resistance changes from 2Ω to 8Ω. When the loop resistance is 2Ω, the found time of the loop current is about 9ms. While when the loop resistance is 8Ω, the found time of the loop current is about 3ms. With the accretion of the equivalent resistance of the loop, the response time of the current decrement gradually. Fig4 shows the loop current changing with time, when the loop inductance changes from 2mH to 6mH. When the loop inductance is 2mH, the found time of the loop current is about 2ms. While when the loop inductance is 6mH, the found time of the loop current is about 6ms. With the increase of the equivalent inductance of the loop, the response time of the current also increases. In addition, the different connect methods of the loop have great influence to the response time. The response time in parallel-wound method is shorter than in series-wound method. Fig5 is the emulation of the response time of the loop current in different frequency of the PWM wave. When the duty circle of PWM is 50% and the working cycle changes from 20us to 100us, the influence to the response time of the loop current is week. Fig6 is the emulation of the response time of the loop current in different temperature. When the temperature changes from 10℃ to 150℃, all the curves almost superposes. Thus, the influence to the response time of the loop current can be ignored. With these analysis and emulation, we know that the response time of the loop increases in direct proportion to the equivalent inductance, and in inverse proportion to the equivalent resistance. The influences of the working frequency of the PWM wave and temperature are inconspicuous to the response time of the loop current.

4 The way of compensate the time lag

To reduce the time lag, we can take these measures: to modify the structure of MR damper and its current source to shorten response time of apparatus; to input instant inverted current to the excitation loop to shorten the response time of magnetrheological fluid; to meliorate the algorithm of the control strategies to compensate the time lag, etc. Here we mainly discuss the way of meliorate the algorithm of the control strategies. In the field of vibration control, phase displacement is often used to compensate the time lag. It amends the feedback plus matrix by displacing the phase of the feedback speed and displacement of the object vibrating in certain frequency. Therefore, the control force is adjusted. But in many occasions, the bandwidth of the vibration frequency is broad, and the components could not be identified accurately. Thus, phase displacement does not be adopted in compensating the time lag of MRFD. The transmission of the close loop of the MRFD system is shown as follow:

$$\phi(s) = \frac{G_C(s)G_M(s)}{1+G_B(s)G_C(s)G_M(s)} \tag{10}$$

$G_C(s)$ maps the control current I to the adjustable yield shear stress τ_y, $G_M(s)$ maps τ_y to the damp force F_R output by MRFD. $G_B(s)$ maps the feedback vector F_R to I. The time lag of the $G_C(s)$ is mainly composed of the response time of the apparatus, which is mainly caused by the inductance and the resistance of the excitation loop. For a certain MRFD, the parameters of the loop are fixed. So the time lag is certain. The hysteresis function is $G_C(S)e^{-\tau s}$. The time lag of the $G_M(s)$ is mainly composed of the response time of the magnetrheological fluid, which is between 0.1ms and 1ms. Comparing to the response time of the apparatus, the later can be ignored. Then, the transmission of the MRFD hysteresis system is:

$$\phi(s) = \frac{G_C(s)G_M(s)e^{-\tau s}}{1+G_B(s)G_C(s)G_M(s)e^{-\tau s}} \tag{11}$$

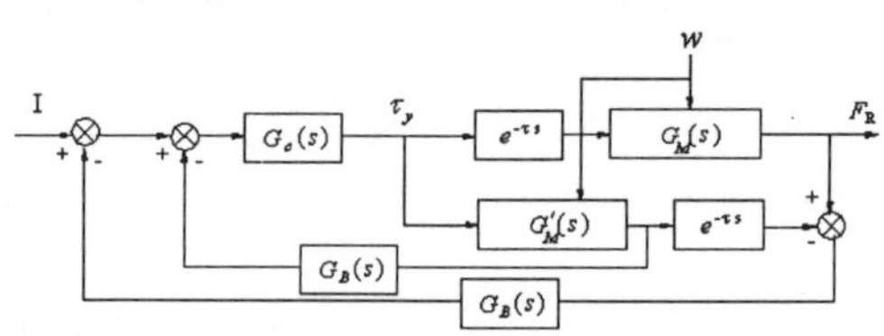

Fig 7 the model of Smith Precompensation of the MRFD hysteresis system

From the function 11, we can see that hysteresis term $e^{-\tau s}$ makes the root of the eigenfunction transfer right in the s-plan, and reduces the systemic stability. Here we can use Smith Precompensation to remove the hysteresis term from the eigenfunction. In Fig.7, $G_M'(s)$ is the estimated static model of $G_M(s)$. w is disturbance vector. In the model, the error caused by w is:

$$e(s) = H(s)(1-e^{-\tau s})F_I(s) \tag{12}$$

H(s) maps w to F. In this function, the error will increase with w. On the other hand, the precision of $G_M'(s)$ will also cause error. To reduce these error, the difference of the damp force F_R and the hysteresis function of $G_M'(s)$ is feedback to the input port to correct the control current I.

5 Conclusions

In real-time control of the MRFD, the Smith Precompensation of the MRFD hysteresis system can fulfill compensation of the time lag. But this method is sensitive to the precision of the estimation model. To improve the compensatory effect, we can use experimental data correct the dynamic model, or choose intelligent control strategies such as sliding mode control.

References:

[1] 关新春, 欧进萍. 磁流变减振驱动器的响应时间试验与分析. 地震工程与工程振动. 2002; 6: 96-102

[2] Abdulazim, H.F., William W.C. and Pradeep P. Phule, Modeling of Magnetorheologicial fluid damper with Parallel Plate Behavior, SPIE, 1998; 3327: 276-283.

[3] 徐赵东,沈亚鹏. 磁流变阻尼结构的双态控制和三态控制弹塑性分析. 西安交通大学学报. 2003; 7: 754-758

[4] 汪建晓, 孟光. 磁流变液阻尼器用于振动控制的理论及实验研究. 振动与冲击, 2001; 20

On Dynamic Transmitting Property of Circular Plate MR Clutch

Chongzhi Guo[1], Jiangchuan Guo[1], Yu Guo[2], Ziyang Ma[1]
[1].*College of Industrial Equipment and Control Engineering,* ,
[2]. *College of Computer Science , South China University of Technology* , *uangzhou* 510640, *China*

Abstract:

This study focuses on the analysis of relationship between the current density and the torque of a circular plate Magneto-Rheological (MR) clutch. In order to get the expression of magnetic induction intensity, the Finite Element Method(FEM) is used for the magnetic analysis on the given geometry of circular plate MR clutch under different current density. With some reasonable assumptions, the discrete values of the magnetic induction intensity along some defined paths are obtained. The fitted expression of magnetic induction intensity is derived from discrete points and the analysis of these discrete data. Based on the expression and the Bingham model which is used to describe the constitutive characteristics of the MR fluids flow between two circular plates subject to an applied magnetic field induced by current density, the mathematical model to transmit the torque is established. From the model and the fit expression, the relationship of the torque and the current density is deduced. The numerical results show that the torque transferred under magnetic induction density by control current density is increased smoothly as the current density is increased except a very short time after initial start. Results also indicate that the torque can be controlled continuously by changing the current density. The analysis provides the theoretical foundation for the design of the MR clutch, and the equation of the torque provides the information by which the torque transmitted by the clutch can be manipulated accurately through adjusting the current density.

Keywords: MRF; Clutch; FEM; Torque;

1. Introduction

The concept of Magneto-rheological (MR) was initially proposed by Rabinow. MR fluids consist of stable suspensions of micro-sized, magnetizable particles dispersed in a carrier medium such as silicon oil or water. In the absence of an applied magnetic field, MR fluids exhibit Newtonian-like behavior. Upon application of a magnetic field, the suspended particles in the MR fluids become magnetized and align themselves, like chains, with the direction of the magnetic field. The formulation of these particle chains restricts the movement of the MR fluids, thereby increasing the yield stress of the fluids and at this time the behavior of the controllable fluids is often represented as a Bingham fluid. The MR effect directly influences the mechanical properties of the MR fluids.The change is rapid, reversible and controllable with the magnetic induction intensity[1,2].

The yield strength of the resulting MR fluids is entirely adequate for most applications. Additionally,

The corresponding author, *Email*: guochongz@263.net

the stability of these suspensions is remarkably good. It is certainly adequate for most common types of MR fluid application. MR fluids can be used in the construction of magnetically controlled devices such as the MR fluid rotary brake, or clutch. An MR clutch is a device to transmit torque, by the shear stress of MR fluids. According to the shape of the MR fluids in clutch, they can be classified as circular plate clutch in which the MR fluid between two circular plates is like a circular plate and cylindrical clutch in which the MR fluid between the inner and outer cylinders is a cylinder [3-5]. Each has the merits and demerits, in this paper, we only put attention on circular plate clutch. The torque and the output speed is controlled by the magnetic induction intensity which is influenced by current density. The FEM(finite element method) model is established to analysis the magnetic induction intensity on the given geometry of circular plate MR clutch under different current density. The expressions of the magnetic induction intensity and current density are fitted. And base on this, the equations of the torque and the current density are derived to provide the theoretical foundation in the analysis of the clutch.

2. Operational principle of circular plate MR clutch

The MR clutch transmits torque by the shear stress of MR fluids that can be continuously adjusted by the magnetic induction intensity. By this means, the magnitude of the torque can be manipulated smoothly. The operational principle of the circular plate MR clutch is shown in Fig.1. The driving-plate and the driven-plate are connected to the driving-shaft and the driven-shaft, respectively. The shafts are supported by bearings. The MR fluid fills the working gap between the driving-plate and driven-plate. The driving-shaft is driven by out power and it circumrotate at a speed of ω_1, so do the driving-plate. In the absence of an applied magnetic field, the suspended particles of the MR fluid dispersed randomly, exhibiting Newtonian-like behavior, so that cannot restrict the relative motion between the driving-shaft and the driven-shaft [6]. However, in the course of operation, a magnetic flux path is formed when electric current is put through the coil. As a result, The suspended particles in the MR fluids become magnetized and align themselves to form chain-like structures, in the direction of the magnetic flux path. These chain-like structures restrict the motion of the MR fluid, thereby increasing the shear stress of the fluid[7].

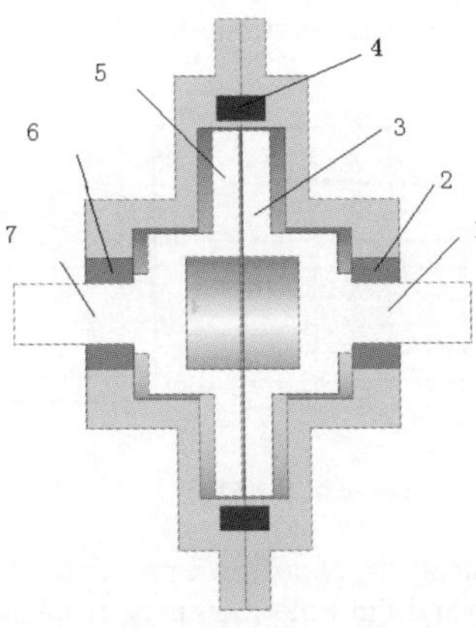

1-output shaft, 2-bearing, 3-driven plate, 4-ccoil, 5-driving plate, 6-bearing, 7-input shaft
Fig.1 A three-dimensional view of circular plate MR clutch

The power is transmitted by the shear stress of the fluid from the driving-shaft to the driven-shaft. The torque developed by the MR clutch goes up rapidly and continuously as the applied magnetic field increases. There are many factors who can affect the performance of the clutch, such as the properties of the MRF, the distribution of the magnetic induction intensity.

3. Electromagnetic finite element modeling and solution

Ansys is a popular FEM software with engineer in following area Multiphysics, Mechanics, Fluid Dynamics, Electromagnetics, etc. This software can be used in the Industries of Aerospace, Automotive/Ground, Chemical, Processing, Civil Engineering, Consumer Products, Educational, Electronics, Environmental, Government/Defense, HVAC and refrigeration, Industrial Equipment, Marine/Offshore, Medical/Biomed, Microsystems Power Generation, etc. Magnetic analyses, available in the ANSYS , can calculate the magnetic field in devices such as: Power generators, Magnetic tape/disk drives, Transformers, Waveguides, Solenoid actuators, Resonant cavities, Electric motors, Connectors, Magnetic imaging systems, Antenna radiation, Video display device sensors, Filters, Cyclotrons[8]. Consider the tstricting equations for the analysis of electromagnetic field, the ANSYS program uses Maxwell's equations as the basis for electromagnetic field analysis [9].

$$\begin{cases} \oint_l H \bullet dl = \int_\Gamma \left(J_S + J + \dfrac{\partial D}{\partial t} \right) d\Gamma \\ \oint_l E \bullet dl = -\int_\Gamma \dfrac{\partial B}{\partial t} \bullet d\Gamma \\ \oint_S B \bullet dS = 0 \\ \oint_S D \bullet dS = \int_V \rho dV \end{cases} \quad (1)$$

where l is the boundary of the surface Γ. S is the close surface of the space V. H is magnetic strength. J_S is outer current density while J is the inner current density put through the conductor. D is the electric displacement. E is electric strength. B is the magnetic induction intensity and t is the time.

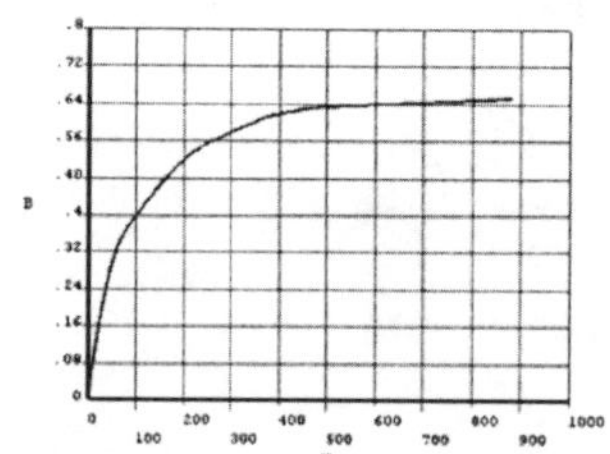

Fig.2 The B-H curve of the MRF

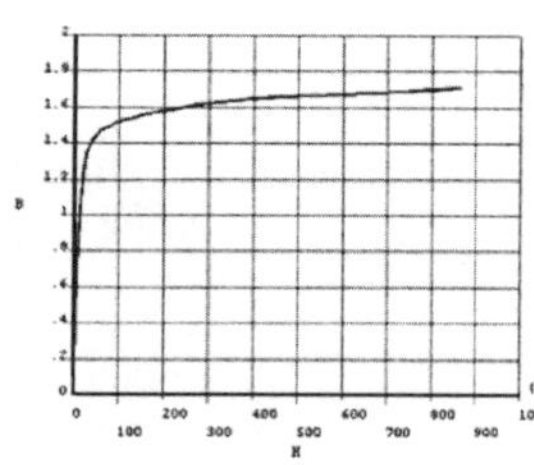

Fig.3 The B-H curve of the shell

During the due course of the analysis, choosing a proper model is very important. The reason is that the choice will determine the cost and the time the analysis taking and the accuracy of the result. 3-D model maybe more precise, but it need your computer have an advanced hardware configuration. Sometimes 2-D model could be more proper, especially when the model is axial symmetry. The current put through the coil is direct current. For these reasons above, 2-D static model is enough for this analysis. Plane53 specified as element type have 8 nodes and its shape is quad. The MRF type is SG-MRF2035 provided by a company in NingBo city. Both the MRF and the shell are nonlinear materials [11]. Their B-H curves are shown in Fig.2 and Fig.3.

The discrete model is built for the Ansys analysis is denoted in Fig.4, area A1 defined as driving-plate, area A5 defined as driven-plate, area A6 defined as MRF, area A8 defined as coil, area A4 defined as shell. There are various boundary conditions condition, Neumann condition and the one synthesis of the two. Dirichlet condition means the magnetic flux flow paralleled to the boundary while Neumann condition implies the magnetic flux flow normal to the boundary [11]. In this analysis, Dirichlet condition is adequate.

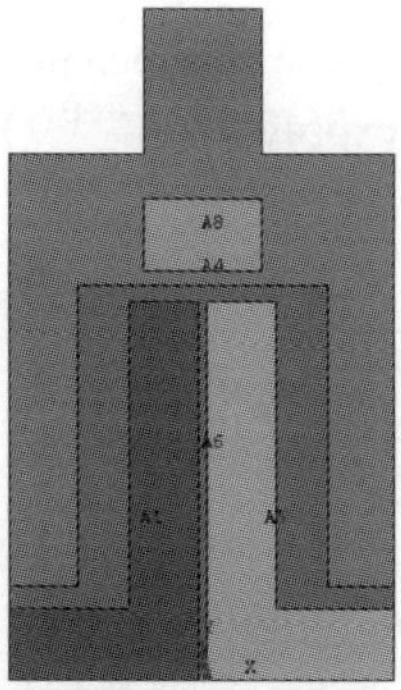

Fig.4 The FEM model of the clutch

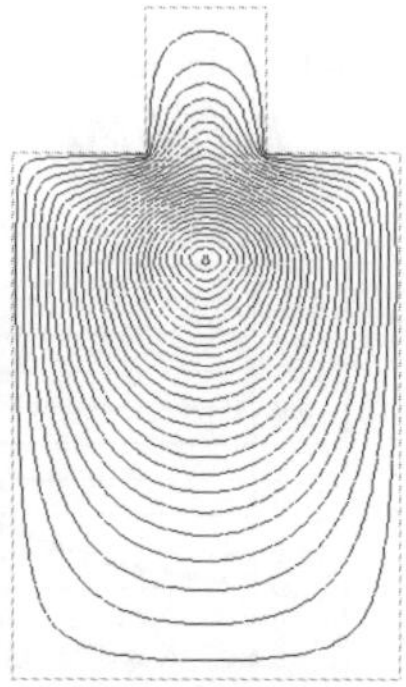

Fig.5 The distribution of the flux lines

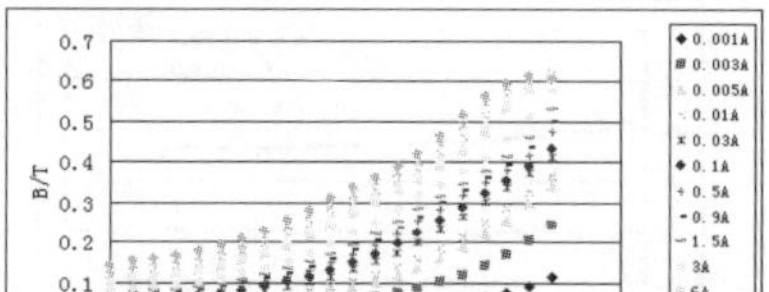

Fig.6 Radial magnetic induction intensity distribution versus current density

Comparing with the dimension of radius direction, the distance between the driving-plate and the driven-plate is tiny. So the assumption that the magnetic induction intensity along z axis is uniformly is taken. It just changes along radius direction. Fig.5 show the flux line obtained from FEM model established in Fig.4. As Fig.5 shown, the flux lines are paralleled in the zone near the symmetry axis. This proves the assumptions are reasonable.

With some assumptions, the discrete values of the magnetic induction intensity at some key points along a defined path are obtained. The distributions of magnetic induction intensity under various current densities are shown in Fig.6. From Fig.6, we can learn even the current density vary from a small one to a big one, the variation trend of magnetic induction intensity is very similar. In order to analyze the magnetic field quantificationally, the fitted expressions of magnetic induction intensity are derived from experience and the analysis of these discrete data. With the fitted expressions, the magnitude of the magnetic induction intensity will be exactly known when the current density is given. The fitted expression obey the following exponential Eq.(2)

$$B = a\exp(br) \quad (2)$$

Where a and b are fitted modulus. Fig.7 show a fitted curve when the magnitude of the current is 0.5A.

Through Fig.7, we can see these discrete data distribute near the curve. Taking polyfit also have a favorable result, but when repeat the fitting, the fitted modulus a and b become oscillating. Comparing with the two methods, the exponential one is more suitable.

Table 1 show the values of the fitted modulus under various current density. After observing these data, power fitting is selected. The curve for the modulus b and the current density is shown in Fig.8. Eq.(3) is the fitted result.

$$\left.\begin{aligned} a &= 0.024I^{0.2917} \\ b &= 40.076I^{-0.0431} \end{aligned}\right\} \tag{3}$$

Integrating Eq.(2), the expression to calculate the magnetic induction intensity is obtained.

$$B = 0.024I^{0.2917}\exp(40.076I^{-0.0431}r) \tag{4}$$

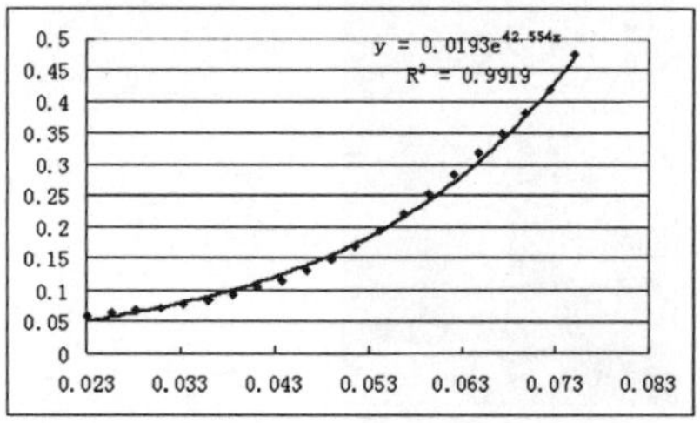

Fig.7 The fitted curve when the magnitude of the current is 0.5A

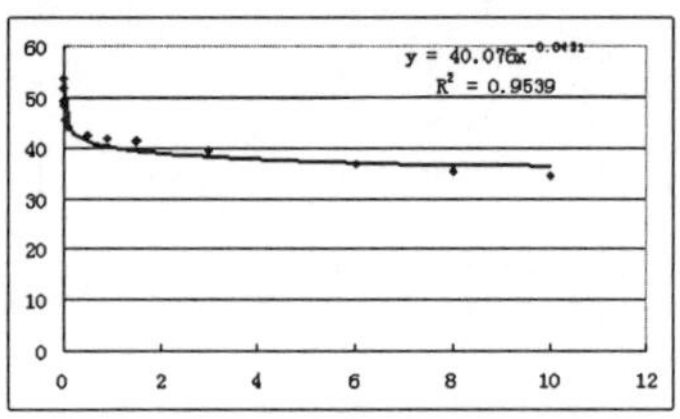

Fig.8 The curve of the modulus b versus the current density

Table 1 The fitted modulus

Current (A)	Coefficient a	Coefficient b
0.001	0.0016	53.809
0.003	0.0041	51.943
0.005	0.0075	49.434
0.010	0.0091	48.314
0.030	0.0130	45.706
0.100	0.0158	44.065
0.500	0.0193	42.554
0.900	0.0211	41.896
1.500	0.0234	41.220
3.000	0.0291	39.529
6.000	0.0396	36.775
8.000	0.0450	35.417
10.000	0.0504	34.203

4. Analysis model for the torque and current

When electric current is put through the coil, magnetic field is induced. The behavior of the controllable fluid filling in the circular clutch is often represented as a Bingham fluid having a variable yield strength. It can be described as follow equations [12-14]:

$$\tau = \tau_y(B) + \eta\dot{\gamma} \qquad \tau \geq \tau_y(B) \tag{5}$$

$$\dot{\gamma} = 0 \qquad \tau < \tau_y(B) \tag{6}$$

Where where $\tau_y(B)$ is the yield stress developed in response to the applied magnetic field. Its value is dependent upon the magnetic induction intensity B. η is the viscosity of MR fluids with no applied magnetic field , $\dot{\gamma}$ is the shear rate.

In order to elicit the torque equation, analyzing the flowing condition of MRF filling in the gap between driving-plate and the driven-plate is our first step. For the object of that, the following assumptions are given: (1) The flow of the fluid is incompressible. (2) The flow of the fluid is steady. (3)There is no flow in axial and radial direction. (4) The gravity of the MRF is neglected. (5) Pressure gradient along z axis is zero[15]. According to the above assumptions, as shown in Fig.9, in the cylindrical coordinate system (r,θ,z), velocity distribution is as following:

$$V_r = 0, \quad V_\theta = r\omega(z), \quad V_z = 0 \tag{7}$$

Where V_r, V_θ and V_z are the velocities along *r* axis, θ axis and *z* axis, respectively. $\omega(z)$ is the rotational speed of the MR fluid and the function of *z*. The boundary conditions of the clutch are: $z = 0$, $\omega(z) = \omega_1$; $z = h$, $\omega(z) = \omega_2$

Where ω_1 and ω_2 are the rotational speed of driving-plate and driven-plate respectively, *h* is the

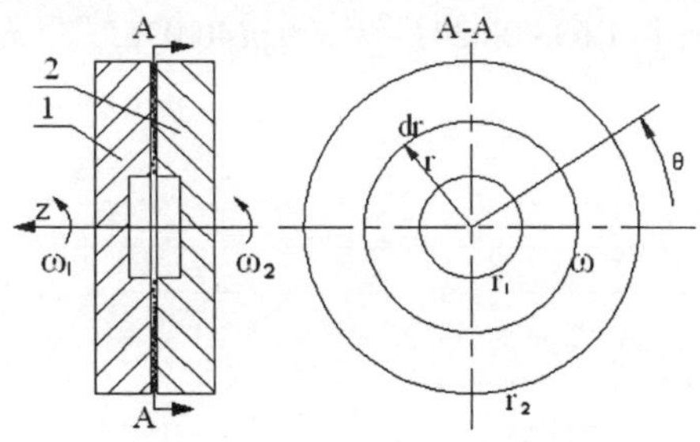

Fig.9 Analysis model of the clutch

width of the gap between driving-plate and driven-plate. The shear rate $\dot{\gamma}$ in above Eq.(5) can be calculated by[14]:

$$\dot{\gamma} = r\frac{d\omega(z)}{dz} \tag{8}$$

Where $d\omega(z)/dz$ is the rotational angel speed gradient in *z* axis. From Eq.(5) and Eq.(8) we can obtain:

$$d\omega(z) = \frac{1}{\eta\, r}\left[\tau - \tau_y(B)\right]dz \tag{9}$$

Integrating Eq. (9) and applying the boundary conditions of the MR fluid clutch

$$\int_{\omega_1}^{\omega_2} d\omega(z) = \int_0^h \frac{1}{\eta r}\left[\tau - \tau_y(B)\right]dz \tag{10}$$

When the clutch work steadily, the magnetic induction intensity distribute uniformly in the gap along

z axis. From Eq.(10) the shear stress produced by MRF can be brought forward as follows:

$$\tau = \tau_y(B) + \frac{\eta\Delta\omega}{h} r \tag{11}$$

Where $\Delta\omega$ is the difference of rotational angel speed between driving-plate and driven-plate. The tiny shear force transmitted by this circular plate fluid can be calculated by:

$$dF_\tau = \tau dA \tag{12}$$

dA is tiny area of circular palate clutch when the value of radius is r. It can be expression as: $dA = 2\pi\ rdr$. Equation (12) can be mathematically manipulated to yield the torque as follow:

$$T = \int dT = \int r dF_\tau = \iint \tau r dA = 2\pi \int_{r_1}^{r_2} \left[\tau_y(B) + \frac{\eta\Delta\omega}{h} r \right] r^2 dr \tag{13}$$

Where r_1 is the inner radius of circular palate clutch and r_2 is the outer radius of circular palate clutch[15]. The MRF here we use is SG-MRF2035 and its essential equation described as following

$$\tau_y(B) = 64720 \times \left(1 - \exp\left(-1.63B\right)\right) \tag{14}$$

Base on equations above, the torque T developed by the MR fluid can be mathematically manipulated to yield as follow:

$$T = 2\pi \int_{r_1}^{r_2} \left(64720 \times \left(1 - \exp\left(-1.63 \times 0.024 I^{0.2917} \times \exp(40.076 I^{-0.0431} r)\right)\right) + \frac{\eta\Delta\omega}{h} r \right) r^2 dr \tag{15}$$

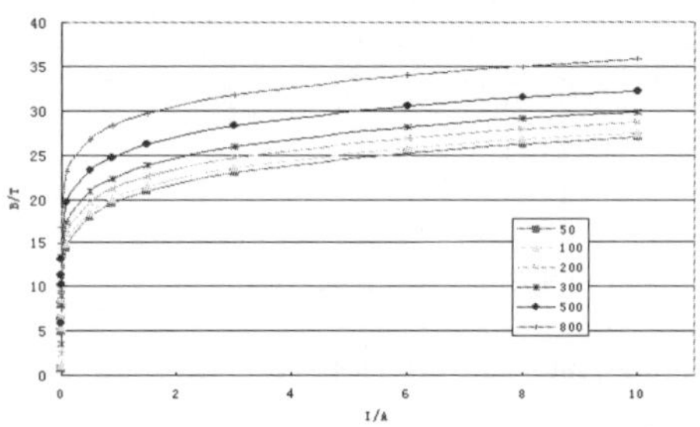

Fig.10 the curves of output torque with the current density at various rotational angel speed differences

5. Result and discussion

In this calculation, assume the following parameters are given: based fluid: silicone oil; density: 3.09g/cc; weight percent solids: 81.24%; viscosity: when the temperature is η=0.24Pa.s. The geometric dimensions of the clutch are given following: inner radius of the driving-plate r_1=0.023m, outer radius of the driving-plate r_2=0.075m, width of the gap $h = 0.001m$. Fig.10 shows the changes of output torque with the current density at five various rotational angel speed differences. The lowermost curve in Fig.7 shows the diversification of the torque when rotational angel speed difference is 50rad/s. From the curve we can see when electric current is not put through the coil, the torque transmitted by the viscosity of the MRF is only 0.696N.m. When the current density goes up smoothly from zero, the torque rise rapidly. Especially in the course of the current going up from zero smoothly, the slope of the curve is very big. But after the current density exceeding 0.5A, the curve become gently. The main reason for this is the saturation of the magnetic induction intensity. When the current density arrive at 3A, the torque transmitted by the clutch up to 23N.m. It means the torque goes up more than 32times. This implies the torque transmitted by the clutch can be manipulated accurately through adjusting the current density.

Reference

[1] Carlson D J, Weiss K D. A growing attraction to magnetic fluids. Machine Design,1994, 8: 61~64

[2] Lee U, Kim D, Hur N. Design analysis and experimental evaluation of an MR fluid clutch. Proceeding of the 6th International Conference on Electrorheological Fluids. Magnetorheological Suspensions and their Application. July 22~25 1997: 674~681. Yonezawa Japan (Copyright: World Scientific Publishing, Co Pte, Ltd)

[3] Tang X, Zhang X, Tao R. Enhace the yield stress of magnetorheological fluids.Proceeding of the 6th International Conference on Electrorheological Fluids. Magnetorheological Suspensions and their Application. July 22~25 1997: 3~8. Yonezawa Japan (Copyright: World Scientific Publishing, Co Pte, Ltd)

[4] Carlson, J.David. and Michael J. Chrzan. (1994) Magnetorheological Fluid Dampers United States Patent Number 5,277,281. Jan.

[5] Shen R,Flores G A,Liu J,In vitro investigation of anovel cancer therapeutic method using embolizing properties of magnetorheological fluids. J Magnetism & Mater,1999,194:167

[6] R. Bolter and H. Janocha, Design rules for MR fluid actuators in different working modes, Proceedings of the SPIE's 1997 Symposium on Smart Structures and Materials 1997: 148-159.

[7] J. Huang, GH. Deng.Y.Q. Wei and J.Q. Zhang, Application of magnetorheological fluids to variable speed transmission. Proceedings of the International Conference on Mechanical Transmissions (ICMT'2001), Chongqing, Peoples R China, April 5-9, 2001: 296-298.

[8] ANSYS is a trademark of SAS Inc. www.ansys.com

[9] L.Q. Liao, C.R. Liao, J.G. Cao and L.J. Fu: A design methodology of magnetorheological fluid damper using Herschel-Bulkley model, SPIE Vol. 5253(2003) 710-717.

[10] Yu A, Andreev, High-power ultrawideband electromagnetic pulse source[C]. 14th IEEE International Pulsed Power Conference, PPC-2003.

[11] Nakata T, etc. Practical Analysis of 3D Dynamic Nonliner Magnetic Field Using Time Periodic Finite Element Method. IEEE Transactions on Magnetic,1995,31(3)

[12] Felt W D, Hagenbuchle M, Liu J. Rheology of magnetorheological fluid. Proceeding of the 5th International Conference on Electrorheological Fluids. Magnetorheological Suspensions and Associated Technology, July 10~14 1995: 738~746. Sheffield. UK. (Copyright: World Scientific Publishing, Co Pte, Ltd)

[13] Kordonsky W I. Magnetorheological fluids and their Application. Materials Technology, 1993,8(11): 240~242

[14] J. Huang, J.Q. Zhang, Y. Yang, Y. Q. Wei: Analysis and design of a cylindrical magneto-rheological fluid brake, Journal of Materials Processing Technology, 2002,129: 559-562.

[15] J.D. Carlson, D.M. Catanzarite and K.A.St. Clair, Commercial magnetorheological fluid devices, Proceeding of the 5th International Conference on Electrorheological Fluids, Magnetorheological Suspensions and Associated Technology, Sheffield. UK, July 10~14 1995: 20-28.

Web-based Traffic Noise Control Support System for Sustainable Transportation

Lisa Fan[a,*], Liming Dai[b], Anson Li[a]
[a]Department of Computer Science, University of Regina, Regina, S4S 0A2
[b]Department of Industrial Engineering, University of Regina, Regina, S4S 0A2

Abstract

Traffic noise is considered as one of the major pollutions that will affect our communities in the future. This paper presents a framework of web-based traffic noise control support system (WTNCSS) for a sustainable transportation. WTNCSS is to provide the decision makers, engineers and publics a platform to efficiently access the information, and effectively making decisions related to traffic control. The system is based on a Service Oriented Architecture (SOA) which takes the advantages of the convenience of World Wide Web system with the data format of XML. The whole system is divided into different modules such as the prediction module, ontology-based expert module and dynamic online survey module. Each module of the system provides a distinct information service to the decision support center through the HTTP protocol.

Keywords: Traffic Noise Control, Web-based Support System, Ontology, Sustainable Transportation.

1. Introduction

Traffic noise reduces our ability to concentrate, causes communication disturbances and disturbances to sleep and recreation; it also produces negative influences on human vegetative nervous systems and even damages to hearing [1]. Enormous research has been carried out for understanding the traffic noise and its effects to human being; and tremendous efforts have been put into controlling the traffic noise in the residential areas near the highways. However, it is still a great challenge to the engineers, road designers and planners to reduce or control the traffic noise in an efficient and cost-effective manner. As a huge amount of factors involves in the traffic noise, and noise level is subjectively valued, the control of traffic noise is very complicated in practice.

* Corresponding author: Dr. Lisa Fan, Computer Science, University of Regina, Regina, SK, Canada S4S 0A2, Phone: (306) 585-4110, Fax: (306) 585-4745, E-mail: fan@ca.uregina.ca

In the City of Regina, traffic noise provides negative impact to our communities. Based on the acoustic experiments [2], the noise level near the roads of heavy traffic can be very high. During a regular rush hour with a traffic flow of less than 50 vehicles per minute, the equivalent noise intensity level L_{Aeq} can be as high as 75.4 dBA in comparing with 67 dBA and 72 dBA for residential and commercial lands respectively, as the exterior criteria set forth by the Federal Highway Administration (U.S.) for different land uses close to highways. The maximum peak noise $L_{Cpk,}$ usually generated by heavy trucks, can reach 100 dB or higher. To develop a sustainable community for the future, reducing traffic noise is obviously a necessary consideration for the transportation development. A pilot project of using rubber asphalt to build quiet pavement is constructed to study the traffic noise reduction. The history of rubber asphalt pavement can be traced back to 1940s, the capability of this technology of reducing traffic noise level was not noticed until early of 1980s. With the more public attention was paid on environmental issues, especially the problem of tire recycle, the rubber asphalt pavement, a sustainable use of tire rubber, was widely spread around world [3]. In order to have a better understanding of acoustical performance of rubber asphalt pavement, a thorough and systematic experimental investigation have been carried out on the mechanism of the tire-pavement noise and the key factors affecting the control of the traffic noise on the AR road with considerations of various traffic volume/speed, vehicle types, road structure and road pavement conditions by Dr.Dai and his acoustic research groups. In this paper, we presents a framework of web-based noise control support system to assist design engineers effectively and efficiently access the data and be able to analyze the data, predict the traffic noise level; to help the decision makers to make appropriate decisions; and provide the publics a platform to discuss and exchange ideas and give feedbacks to decisions makers.

2. Web-based Decision Support System

Decision Support System was started in the early 1970s. It can be defined as computer technology solutions that can be used to support complex decision making and problem solving [Shim4]. To account for complexity and uncertainty associated with decision problems, the DSS is a set of computer-based tools that provide decision maker with interactive capabilities to enhance his understanding and information basis about considered decision problem through usage of models and data processing, which in turn allows reaching decisions by combining personal judgement with information provided by these tools. The advances in computer technologies have affected everyone in the use of computerized support in various activities. Traditional decision support systems (DSS) focuses on computerized support for making decision with respect to managerial problems [5]. With the introduction of Web technology, one has to reconsider the existing methods and re-design or modify the existing systems to meet the challenges, as well as take the advantages of Web technology. The Web is used as a universal interface and the underlying infrastructure for Intelligent Web Information Systems. There is an emerging and fast growing interest in web-based support systems in many other domains, such as medical, business, water treatment, research, knowledge management etc. The benefits of Web technology are: it provides a distributed infrastructure for information processing; the Web can be used as a channel to discuss one of the most popular support systems; it can deliver timely, secure information and tools with user friendly interface; Users can access the system at any time, any place; Users can control and retrieve results remotely and instantly.

3. A Framework of the Web-based Traffic Noise Control Support System

This section presents the modeling and architecture of the traffic noise control support system.

3.1 System Architecture

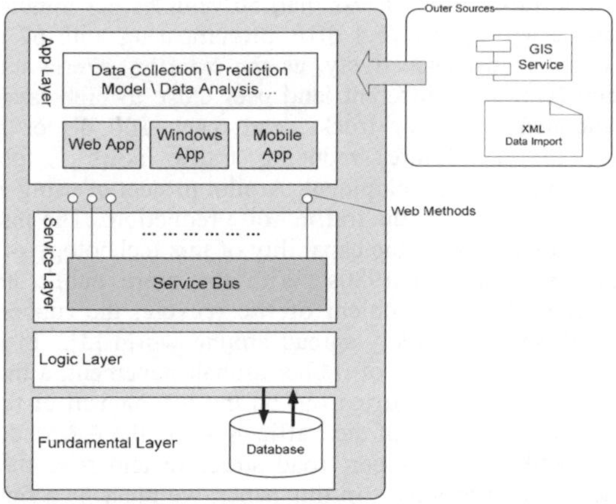

Figure 1 Architecture of the Web-based Support System for Traffic Noise Control

The system architecture takes the advantages of SOA, providing the different services through XML technology over various clients/servers platforms, such as Mac OS, Linux, and Windows. Database is located on the fundamental level of the system architecture. It stores all the data and the data relationships for the whole system, and it is in charge of all the data transactions and processing all the data queries requested from the logic layer.

3.2 Mobility

The time factor plays an increasingly important role in the economic efficiency of transport industry business processes. In order to work more efficiently, a company's operations have to be working as quickly and smoothly as possible. The mobile client application is designed for the engineers who need to work out of the office regularly. For example, the road pavement designer will go to the construction spot for collecting data to evaluate the noise level prediction. Therefore, a mobile application helps the engineers to finish their work efficiently and smoothly. Because of the functionality limitation of the mobile devices, the mobile client application is mainly used to input real time data and data query through wireless connection. There are two kinds of situations will happen to the mobile client application: The application is connected to service center through the wireless connections, such as GPRS and Wi-Fi. The users can query the database very easily. The mobile device is not connected or the wireless connection is not available, and then the user cannot query data through Internet, but can input the data which will be temporally stored in the device. When the user goes back to the office, the connection is available again. The data can be synchronized to the data server automatically.

3.3 GIS Service integration

Presenting the traffic data on a digital map undoubtedly will enhance the user experience dramatically. In another word, it will make the traffic data easily to be understood by the users. Nowadays, there are lots of geographic information system services available on internet, such as Google map and MSN virtual earth. Through using these public map APIs (Application Programming Interface), the client web application of the traffic noise control system is able to integrate the map service seamlessly.

3.4 Functional Modules

3.4.1 Traffic Noise Prediction Module

The module is aim to build up a model for traffic noise prediction based on the data of traffic flow, pavement types of the road surfaces and the environment factors. This prediction module is playing a key role for developing a sustainable transportation system. It provides the prediction result of the noise impact in the future for the road designers to adjust their design so that it can prevent re-paving and wasting the cost on the noise barriers. Mathematical models for prediction of traffic noise usually extract the functional relationship between the parameter of noise emission, Leq, and measurable parameters of traffic and roads. The classical functional relationships available in literature have based on data measured through semi-empirical models, typically regression analysis. The following equations are some typical examples.

$$Leq=55.5+10.2\log Q+0.3p-19.3\log(L/2) \quad (1)$$

$$Leq=38.8+15\log Q-10\log L \quad (2)$$

$$Leq=10\log(N_c+N_m+8N_{hv}+88N_b)+33.5 \quad (3)$$

$$Leq=10\log(N_c)+p \quad (4)$$

Here p is the percentage of heavy vehicles, L is the road width, Q is the total number of vehicles per hour, N_c is the number of light vehicles per hour, N_m is the number of motorcycles per hour, N_{hv} is the number of heavy vehicles per hour, N_b is the number of buses per hour. The total number of vehicle per hour, Q, is expressed as the equivalent number of cars and obtained under hypothesis that one heavy vehicles is equivalent to 6 light vehicles and one motorcycle to 3 light vehicles. These models do not provide accurate approximation because any models includes the flow and composition of the road traffic that may be different than the examined areas. The prediction module should provide user tools to aid in determining the appropriate models and allow user to play with different scenarios. Simulation software is a necessary component of this module.

3.4.2 Expert Support Module

In the Expert Support Module, we propose a domain ontology knowledge support system to help engineers in the model construction and problem solving. Domain knowledge will be represented into the ontology which is a new research area in artificial intelligence of computer science. An ontology is an explicit specification of a conceptualization. An ontology refers to an engineering artifact, constituted by a specific vocabulary used to describe a certain reality, plus a set of explicit assumptions regarding the intended meaning of the vocabulary [8]. Basically, ontology is a hierarchical conceptual graph, consisting of concepts and relations. It captures the existing domain knowledge and describes it into machine-understandable structure in order to help computer to do inferences without human's involvement [10].

The reasons we choose the ontology technology to build our expert support system are: First, the knowledge that is represented inside the ontology serves as specification of common conceptualization in the road pavement engineering domain [11]. Second, the crossing-boundary and explicit web ontologies can be re-used and updated by different organizations. Third, the ontology guided resources search is able to largely enhance material searching for the engineers. For example, one engineer encountered a complex problem which he can not solve based on his current knowledge background. An ontology base information retrieval can provide more accurate and useful searching results in the specific engineering domain than the current popular search engines, such as Yahoo and Google.

From Figure 2, we can see that the domain ontology will be stored in the relational database. The Ontology guided structure will directly enhance the quality and efficiency of resources searching.

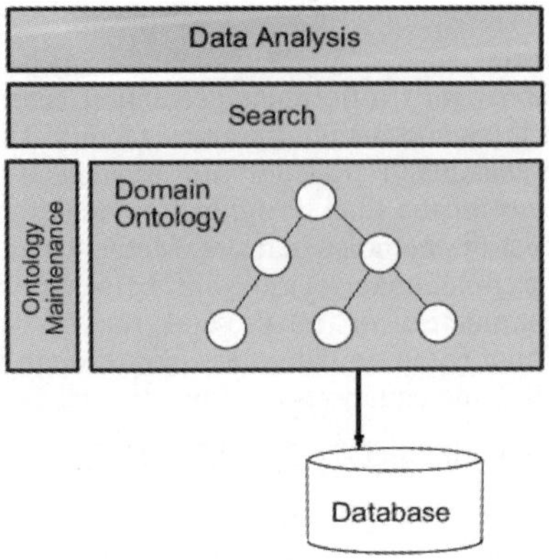

Figure 2. The Structure of the Ontology-based Expert Support System

For example, an engineer wants to search some articles and documents about road construction in the company's document library. The articles about asphalt rubber pavement will be discovered even though the keyword road construction is not mentioned in those articles. The reason why the system can achieve this kind of functions is because the computer understands the word "asphalt rubber pavement" semantically which is one type of road constructions. The current keyword based searching technique will be extremely hard to have this kind of functionality as the ontology guided system do. As we presented earlier, the ontologies are also constructed above the SOA architecture. Semantic Web Services are responsible for providing an explicit Web Ontology for different platform to access. The Web Ontology language, RDF (Resources Description Framework) and OWL (Ontology Web language) recommended by WWW Consortium are chosen [13]. The advantages of building Semantic Web Services are very obvious. For example, in our system, there are several different client modes that have been proposed, such as mobile client application, web application, and windows client application. Those client applications would possibly be written in various different programming languages and platforms, such as Java, .Net and PHP. To access the ontologies through a unified interface, Semantic Web Services which relies on XML document exchange and HTTP protocol is the best choice.

Figure 3. An Example of Web Ontology for Rubber Asphalt Pavement

3.4.3 Dynamic Online Survey Module

One of the useful tools that WTNCSS provides is the dynamic online survey. Surveys are a widely used tool to measure individuals' opinions, preferences and behavior. The web-based survey is very effective way to reach large numbers of people from all over the world compared to paper, in-person, or telephone surveys [12]. Figure 4 shows a screen shot of the dynamic survey of public opinions about traffic noise. The system can analyze the survey results and present to the decision makers. Therefore it provides useful feedbacks from the public to assist the decision makers to make appropriate decisions.

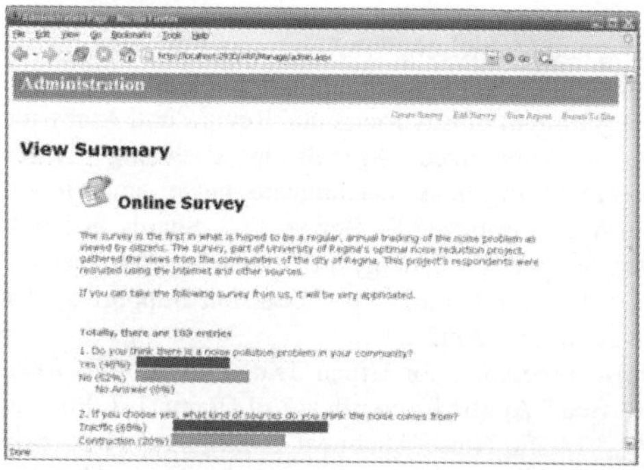

Figure 4 A Screen shot of Dynamic Online Survey

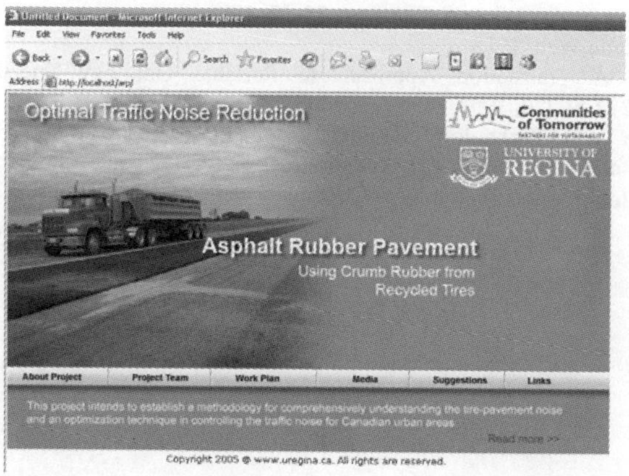

Figure 5 Screen shot of the Website of Traffic Noise Reduction by Using Asphalt Rubber Pavement

4. Conclusive Remarks

Tire-pavement noise is a complex phenomenon and affected by many factors, such as road conditions, pavement materials, traffic volume and speed, road environment and road constructions. In the research of this paper, a preliminary study on the establishment of a web-based support system for traffic noise control is presented. The system, namely WTNCSS is to provide a platform for engineers, decision makers and publics to effectively and efficiently access the information for their design, decision making and discussion processes. The prediction module, ontology-based expert module and the dynamic online survey module are the key components of the WTNCSS system. This paper demonstrates the initial design and implementation of the Web based support system to the traffic noise control application. Further implementations of the system with different modules can be performed on the basis of the framework established in the present research.

References

[1] Cvetkovic, D., Prascevic, M., and Stojanovic, V., "NAISS – Model for Traffic Noise Prediction", The Scientific Journal of Working and Living Environmental Protection, Vol.1, No.2, pp.73-81, 1997.

[2] Dai, L., Cao, J., Fan, L., and Mobed, N., "Traffic Noise Evaluation and Analysis in Residential Areas of Regina." Journal of Environmental Informatics. Volume 5, 17-25, 2005.

[3] Sacramento County Department of Environmental Review and Assessment and Bollard and Brennan, Inc. "Effectiveness of Rubberized Asphalt in Reducing Traffic Noise", available at : http://www.rubberpavements.org/library/sacramento_noise _study/index.html

[4] Shim, J.P., Warkentin M., Courtney, J.F., Power, D.J., Shards, R., and Carlsson, C., "Past, Present and Future of Decision Support Technology", Decision Support Systems, pp. 111-126, 2002.

[5] Turban, E., Aronson, J.E., and Liang, T.P., "Decision Support Systems and Intelligent System", Pearson Education, New Jersey, 2005.

[6] Burgess, M. A., "Noise Prediction for Urban Traffic Conditions-Related to Measurements in the Sydney Metropolitan Area." Applied Acoustics, Vol.10, pp. 1-7, England, 1977.

[7] Fagotti, C., Poggi, A., "Traffic Noise Abatment Strategies – The Analysis of Real Case not Really Effective." Proceeding of 18th International Congress for Noise Abatment, pp. 223-233, Italy, 1995.

[8] Guarino, N., Giaretta, P., "Ontologies and Knowledge Bases: Towards a Terminological Clarification." In N. Mars (ed.), Towards Very Large Knowledge Bases: Knowledge Building and Knowledge Sharing. IOS Press, Amsterdam: 25-32, 1995.

[9] Lee, T.B., Semantic Web, 1998

[10] Gruber, T.R., "A Translation Approach to Portable Ontology Specifications." Knowledge Acquisition, 5(2):199-220, 1993.

[11] "Semantic Web" W3C Technology and Society Domain. Available at: http://www.w3.org/2001/sw/

[12] Pargas, R.P. Witte, J.C. "OnQ: An Authoring Tool for Dynamic Online Surveys" International Conference on Information Technology: Computers and Communications, p.717, 2003.

Visual Tracking and Recognition Using Adaptive Probabilistic Appearance Manifold in Particle Filter

Yanxia Jiang[a], Hongren Zhou[b], Zhongliang Jing[c]
[a]*School of Electronics, Information and Electrical Engineering ,Shanghai Jiaotong University, Shanghai 200030, P.R. China*
[b]Advisory Committee for State Informatization, Beijing 100089, P. R. China
[c] Institute of Aerospace Science and Technology, Shanghai Jiaotong University, Shanghai 200030, P.R. China

Abstract
Unlike traditional face tracking and recognition systems, which have usually been embodied with two independent components: the tracking and recognition modules, a framework for simultaneous visual tracking and recognition is proposed in this paper. In our framework, tracking and recognition share the same appearance manifold. During training, K-means is applied to partition the normalized training face images into pose subsets. Appearance manifold and dynamics are learned in this period. During testing, Appearance manifold is updated after tracking and recognition are conducted. The updated appearance manifold is used for future frames. Experimental results show that our proposed framework is effective for tracking and recognition.

PACS: 05.45.-a; 47.27.eb
Keywords: Face Tracking; Face Recognition; K-means; Appearance Manifold; Particle Filter

1.Introduction

Face recognition has long been an active area of research, and numerous algorithms have been proposed over the past few years. While current face recognition systems perform well under relatively controlled environments, the performance of face recognition is affected by different kinds of variations such as expression, illumination and pose. Thus researchers start to look at the video-to-video (or video-based) face recognition [1][2][2][3][4][5], where both training and test set are video sequences containing face.

Many works on video-based face recognition have been proposed in recent years. Edwards et al. [1] proposed an adaptive framework on learning the human identity by using the motion information along the video sequence, which improves both face tracking and recognition. In [2], Liu et al. proposed an updating-during-recognition scheme, where the current and past frames in a video sequence are used to update the subject models to improve recognition results for future frames. X. Liu and T. Chen [3] applied adaptive Hidden Markov Model temporally to perform video-based face recognition. In [4], K. C. Lee et al. built an appearance manifold, which was approximated by linear subspaces and the dynamics among them embodied in a transition matrix learned from an image video sequence. In the works cited above, tracking and recognition are performed as two separated modules. Thus in recent years, some works on

The corresponding author, *Email*: zljing@sjtu.edu.cn

simultaneously tracking and recognition are proposed [5][6]. S. Zhou et al. [5] proposed a generic framework to track and recognize human faces simultaneously. This probabilistic approach is effective in capturing small 2-D motion but may not work well with large 3-D pose variation. In [6], K. C. Lee et al. presented an algorithm for modeling, tracking, and recognizing human faces in video sequences within one integrated framework. This was accomplished through a novel appearance model, which was utilized simultaneously by tracking and recognizing modules. Motivated by the work proposed in [6], a framework for video-based face tacking and recognition using adaptive appearance manifold is proposed in this paper.

The rest of this paper is organized as follows. In section 2, we briefly introduce the experimental data. Our online tracking and recognition framework is described in detail in section 3. Experimental results are presented in section 4, with conclusion in section 5.

2. Experimental Data

In this section, we briefly describe the data used in our experiments. We consider MoBo database [7] and Honda/UCSD [8]. MoBo database contains 96 face sequences of 24 different subjects walking on a treadmill. 4 different walking situations are considered: slow walking, fast walking, incline walking and carrying a ball. Each sequence consists of 300 frames. The resolution of each video sequence is 486×640. For each object, we choose one sequence for training and the other three for testing. During training, the cropped face images are obtained by a tracker proposed in [9] and then manually corrected. The size of the cropped face images is set to be 20×20.

Honda/UCSD database has been collected and used by K. C. Lee. The database contains video sequences of 20 different individuals. All the video sequences contain significant 2-D and 3-D head rotations and facial expression changes. We selected 20 video sequences, one for each individual, for training, and the other 25 sequences for testing. The resolution of each video sequence is 640×480.

3. Integrated Framework for Face Tracking and Recognition

In video-based face recognition, one of the foremost challenges is image variation, which include pose variation, expression variation and illumination variation. The integrated framework for face tracking and recognition is to simultaneously locate face in the input image and determine its identity. In this paper, probabilistic appearance manifold is both utilized in tracking and recognition. Suppose there are N persons we wish to track and recognize. Let k denote the person index. If the appearance manifold M_k for each individual is known, tracking and recognition become straightforward. Let I_t denote an input image, Z_t represent the face image in frame t, and θ_t denote the state vector. Z_t can be describes as:

$$Z_t = f(\theta_t, I_t) \tag{1}$$

Particle filter is applied in the paper to perform tracking. The likelihood function of observation $p(Z_t|\theta_t)$ can be represent as:

$$p(Z_t|\theta_t) = \frac{1}{\Lambda_1}\exp\left(-\frac{d_H^2(Z_t, M_{k_{t-1}^*})}{2\sigma^2}\right) \tag{2}$$

Where $d_H(Z_t, M_{k_{t-1}^*})$ denotes the $L^2 - Hausdorff$ distance between image Z_t and $M_{k_{t-1}^*}$. Λ_1 is a normalization term. Let Z_t^* denote the cropped face image by the tracker, recognition is then performed by:

$$k_t^* = \arg\max_k p(k|Z_t^*) = \frac{1}{\Lambda_2}(\exp(-\frac{d_H^2(Z_t^*, M_k)}{2\sigma^2})) \quad (3)$$

where Λ_2 is a normalization term. After tracking and recognition in current frame are performed, pose estimation and pose manifold updating are conducted.

3.1 Face Tracking

Face tracking is to locate the face region in a video image. In this paper, particle filter is used for face tracking. Non-parameterized Monte Carlo method is used to realize iterative Bayesian deduction in particle filter. It does not restrict the models to be linearity or Gauss. We briefly introduce particle filter below.

Assuming that $p(\theta_t|\theta_{t-1})$ denote the state transition probability of the motion model. $p(Z_t|\theta_t)$ is the likelihood function of observation model. A set of weighted samples $\{\theta_t^{(j)}, \omega_t^{(j)}\}_{j=1}^J$ is used to approximately represent the posterior probability $\pi_t(\theta_t) = p(\theta_t|Z_{0:t})$, where J is the number of samples. We set the importance function as $q(\theta)$, the weight of the samples can be described as:

$$\omega_t^{(j)} = \pi(\theta_t^{(j)}) / q(\theta_t^{(j)}) \quad (4)$$

This technique is called Importance Sampling (IS). To accommodate a video, importance sampling is used in a sequential fashion, which leads to Sequential Importance Sampling (SIS). SIS propagates S_{t-1} according to the sequential importance function $q(\theta_t|\theta_{t-1})$, and calculates the weight using

$$\omega_t = \omega_{t-1} p(Z_t|\theta_t) p(\theta_t|\theta_{t-1}) / q(\theta_t|\theta_{t-1}) \quad (5)$$

In the condensation algorithm, the importance function is set as $p(\theta_t|\theta_{t-1})$. Eq. (5) can be substituted as:

$$\omega_t = \omega_{t-1} p(Z_t|\theta_t) \quad (6)$$

Condensation algorithm is applied in this paper.

In particle filter, two important models are needed. They are state transition model and observation model. Affine transition model is adopted in this paper. The motion is characterized by $\theta = (S, \alpha, T_x, T_y)$, where $\{S, \alpha\}$ are deformation parameters, $\{T_x, T_y\}$ denote the 2-D translation parameters. The initial region of object is set as $[X_0, Y_0]^T$. The candidate region can be represented as:

$$\begin{bmatrix} X_t \\ Y_t \end{bmatrix} = S \times \begin{bmatrix} \cos\alpha & \sin\alpha \\ -\sin\alpha & \cos\alpha \end{bmatrix} \times \begin{bmatrix} X_0 \\ Y_0 \end{bmatrix} + \begin{bmatrix} T_x \\ T_y \end{bmatrix} \quad (7)$$

The motion model can be described as:

$$\theta_t = F(\theta_{t-1}, U_t) = \theta_{t-1} + U_t \quad (8)$$

where U_t denotes the state noise. The observation model can be described as:

$$Z_t = G(\theta_t, V_t) \tag{9}$$

where V_t is the observation noise.

3.2 Face Recognition

Face recognition is to approximate the identity of the cropped face image. In this paper, appearance manifold learned during training is used for recognition. Probabilistic Face Recognition is then conducted by Bayesian rule.

During training, face images are cropped out by an online appearance-adaptive face tracker proposed in [9] and then manually corrected. There are three steps in learning manifolds and dynamics for each person [4]. In the first step, K-means is applied to classify faces into subsets. Suppose one video of person k containing l consecutive images $G_k = \{g_{k1}, g_{k2}, \ldots, g_{kl}\}$. Samples are partitioned into m disjoint subsets $\{G_{k1}, G_{k2}, \ldots, G_{km}\}$ according to their poses. Secondly, for each subset G_{ki}, a linear approximation of the underlying subset $C^{ki} \subset M_k$ is obtained by computing PCA plane L_{ki} in every subset. For each subset G_{ki}, we obtain the mean face μ_{ki}, the number of face images P_{ki} in each subset, the M largest eigenvalues $\{\lambda_{ki1}, \lambda_{ki2}, \ldots, \lambda_{kiM}\}$ and the corresponding eigenvectors $\{\Phi_{ki1}, \Phi_{ki2}, \ldots, \Phi_{kiM}\}$. Finally, dynamics are learned. The transition probability $p(C^{ki} | C^{kj})$ is defined by counting the actual transitions between different G_{ki} observed in the image sequences:

$$p(C^{ki} | C^{kj}) = \frac{1}{\Lambda_{ki}} \sum_{q=2}^{l} \delta(I_{q-1} \in G_{ki}) \delta(I_q \in G_{kj}) \tag{10}$$

where Λ_{ki} is a normalization term. It ensures that $\sum_{j=1}^{m} p(C^{ki} | C^{kj}) = 1$.

We also set $p(C^{ki} | C^{ki})$ to a small constant at the same time. The transition probability presents the transition probability between pose subspaces, and it captures the temporal dynamics of the face motion in the training video sequence. With G_{ki} and its linear approximation L_{ki} computed, we can obtain $p(Z_t | C^{ki})$ in the probability subspace as:

$$p(Z_t | C^{ki}) = \left[\frac{\exp(-\frac{1}{2} \sum_{r=1}^{M} \frac{y_r^2}{\lambda_{kir}})}{(2\pi \prod_{r=1}^{M} \lambda_{kir}^{1/2})} \right] \left[\frac{\exp(-\frac{1}{2\rho} \varepsilon^2)}{(2\pi\rho)^{\frac{(N-M)}{2}}} \right] \tag{11}$$

where N is the dimension of face image, $\rho = \frac{1}{N-M} \sum_{r=M+1}^{N} \lambda_{kir}$, $y = \Phi_{ki}^T (Z_t - \mu_{ki})$, $\varepsilon^2 = \|Z_t - \mu_{ki}\|^2 - \|y\|^2$.

During testing, let Z_t be a face image, conditional probability $p(k | Z_t)$ be the likelihood that originated from person k. $p(k | Z_t)$ can be computed according to Eq. (3). We can see that the distance between tracked face image Z_t and appearance manifold M_k determine the identification. However, such distance can be computed accurately only if M_k is fully known. In our case, M_k is usually unknown and can only be approximated with samples. We provide the similar probabilistic framework proposed in [6] to estimate x^* and $d(Z_t, x^*)$, where x^* is the point on M_k at minimal L^2 distance to Z_t. Readers can refer to [6] for details. Then $d_H(Z_t, M_k)$ can be computed as:

$$d_H(Z_t, M_k) = \sum_{i=1}^{m} p(C^{ki} \big| Z_t) d_H(Z_t, C^{ki}) \tag{12}$$

To incorporate temporal information, the conditional probability $p(C^{ki} \big| Z_t)$ should be taken as the joint conditional probability $p(C_t^{ki} \big| Z_t, Z_{0:t-1})$, where $Z_{0:t-1}$ denotes the frames from the beginning to time $t-1$. $p(C_t^{ki} \big| Z_t, Z_{0:t-1})$ can be obtained as:

$$p(C_t^{ki} \big| Z_t, Z_{0:t-1}) = \frac{1}{\Lambda_3} p(Z_t \big| C_t^{ki}) \sum_{j=1}^{m} p(C^{ki} \big| C^{kj}) p(C_t^{ki} \big| Z_{t-1}, Z_{0:t-2}) \tag{13}$$

Where Λ_3 is the normalization term. $p(C^{ki} \big| C^{kj})$ can be obtained by Eq. (10). $p(Z_t \big| C_t^{ki})$ can be computed by Eq. (11).

3.3 Appearance Manifold Updating

In order to improve the robustness of tracking and recognition, online adaptive appearance manifold is proposed in this paper. In [10], K. C. Lee et al. proposed an online learning of appearance manifold in the training video. In their method, a prior generic appearance manifold is constructed from multiple pre-training video sequences of different instances of the class. All the pose manifolds are updated during training period. Then the updated pose manifolds are utilized in the testing period. However in this paper, the distance between the tracked face image and the appearance manifolds of all the individuals are computed during recognition. If the ratio between the shortest distance and the second shortest is smaller than a predefined threshold, appearance manifold updating is then conducted. In order to update the manifold, pose estimation should be conducted first. We can estimate the pose according to:

$$C_t^{k*i*} = \arg\max_i p(Z_t^* \big| C_t^{k*i}) \tag{14}$$

where Λ_4 is the normalization term. $p(Z_t^* \big| C_t^{k*i})$ can be obtained by Eq. (11).

After pose estimation, pose manifold updating is performed. As mentioned before, linear approximation of the underlying subset of appearance manifold are obtained by computing a PCA plane of fixed dimension for the images in subset. When we wish to update the manifold, we need to adapt the eigenspace model of C_t^{k*i*}. Some algorithms have been proposed to update eigenbasis when new observations are available. The method proposed in [11] is applied in our framework to update the subspace parameters in the pose manifold.

In order to make it understood easily, assuming that pose manifold C^i is approximated by an affine subspace $L = \{\mu, \Phi, \Lambda, P\}$ constructed by PCA, where μ is the center of the subset, Φ is the eigenvector matrix, Λ is the corresponding diagonal matrix of eigenvalues, P is the number of image samples used to construct pose subspace. The size of Φ is $d \times M$. Λ is a $M \times M$ diagonal matrix. A new subspace needs to be created when the new observation becomes available. Let $L' = \{\mu', \Phi', \Lambda', P'\}$ denote the new affine subspace, x represent the new observation. L' can be obtained according to Eq. (16) to (19).

$$\mu' = \frac{1}{P+1}(P\mu + x) \tag{15}$$

$$P' = P+1 \tag{16}$$

$$\left(\frac{P}{P+1}\begin{bmatrix}\Lambda & 0\\ 0 & 0\end{bmatrix} + \frac{P}{(P+1)^2}\begin{bmatrix} gg^T & \gamma g \\ \gamma g^T & \gamma^2\end{bmatrix}\right)R = R\Lambda' \tag{17}$$

$$\Phi' = [\Phi, \hat{h}]R \tag{18}$$

where $x' = x - \mu$, $g = \Phi^T x'$. $h = x' - \Phi g$. $\gamma = \hat{h}^T x'$. $\hat{h}$ is unit residue vector, it can be represented as:

$$\hat{h} = \begin{cases} \frac{h}{\|h\|_2} & \text{if} \quad \|h\|_2 \neq 0 \\ 0 & \text{otherwise} \end{cases} \tag{19}$$

4. Experimental Results

In order to test our proposed method, MoBo database and Honda/UCSD database are used in this paper. The cropped images in the training are resized to 20×20. We choose the number of linear PCA planes as 10. The number of pose subset of every object is set to be 10. Fig. 1 shows an example of tracking and recognition result in MoBo database. Another example of tracking and recognition result is shown in Fig. 2. In Mobo database, the average recognition is 97.65%. While in Honda/UCSD database, the average recognition is 96.76%. We also do experiment using the fixed appearance manifold. The average recognition rate by this method can only reach 96.42% and 93.64% respectively.

Fig. 1. An example of tracking results for five key frames in Mobo database

Fig. 2. An example of tracking results for five key frames in Honda/UCSD database

5. Conclusion

A framework for face recognition and tracking in video sequences using online adaptive appearance manifold is proposed in this paper. During training, face images are cropped out by face tracker and manually corrected. K-means is applied to partition the training images into clusters. In every cluster, PCA is applied to get approximation to the pose manifold. During testing, tracking and recognition share the same appearance manifold. Pose estimation and pose manifold updating are conducted according to the recognition results. The updated manifold is then used for next tracking and recognition. Experimental results show that our proposed framework is effective for tracking and recognition.

Acknowledgments

We would like to thank Dr. R. Gross for providing us the MoBo database and Dr. K-C Lee for the Honda/UCSD database. This research is jointly supported by National Natural Science Foundation of China (60375008) and Shanghai Key Technologies Pre-research Project (035115009).

References

[1] G.J. Edwards, C.J. Taylor and T.F. Cootes. Improving Identification Performance by Integrating Evidence from Sequences, Proceedings of IEEE Conference on Computer Vision and Pattern Recognition, pp.486-491, June 23-25, 1999.

[2] X. Liu, T. Chen and S. M. Thornton. Eigenspace Updating for Non-Stationary Process and Its Application to Face Recognition, Pattern Recognition, Special issue on Kernel and Subspace Methods for Computer Vision, September 2002.

[3] X. Liu and T. Chen. Video-based face recognition using adaptive hidden markov models, Proceedings of IEEE Conference on Computer Vision and Pattern Recognition, pp. 340 - 345,2003.

[4] K. C. Lee, J. Ho, M. H. Yang and D. Kriegman. Video-Based Face Recognition Using Probabilistic Appearance Manifolds, Proceedings of IEEE Computer Conference on Computer Vision and Pattern Recognition, pp. 313-320,2003.

[5] S. Zhou and R. Chellappa. Probabilistic human recognition from video, Proceeding of European Conference Computer Vision, volume 3, pp. 681–697, 2002.

[6] K .C. Lee, J. Ho and M. H. Yang. Visual Tracking and Recognition Using Probabilistic Appearance Manifolds, Computer Vision and Image Understanding, pp. 303-331,2005.

[7] R. Gross and J. Shi. The CMU Motion of Body (MoBo) database, Technical Report CMU-RI-TR-01-18, Robotics Institute, Carnegie Mellon University, 2001.

[8] http://vision.ucsd.edu/~leekc/HondaUCSDVideoDatabase/HondaUCSD.html, CMU-RI-TR-01-18, Robotics Institute, Carnegie Mellon University, June, 2001.

[9] S. Zhou, R. Chllappa and B. Moghaddam. Visual tracking and recognition using appearance-adaptive models in particle filters, IEEE Transactions on Image Processing, vol. 13, No. 11, pp.1-34. 2004.

[10] K. C. Lee and D. Kriegman. Online Learning of Probabilistic Appearance Manifolds for Video-based Recognition and Tracking, Proceedings of IEEE Conference on Computer Vision and Pattern Recognition, pp. 852-859,2005.

[11] P. M Hall, D. R. Marshall and R. R. Martin. Incremental Eigenanalysis for Classification, British Machine Vision Conference, pp. 286-295,1998.

Region-Based Infrared and Visible Dynamic Image Fusion

Bo Yang[*], Gang Xiao, Zhongliang Jing
Institute of Aerospace Science and Technology, Shanghai Jiaotong University, 1954, Huashan Road, Shanghai, 200030, PR China

Abstract

A region-based infrared and visible dynamic image fusion scheme is proposed in this paper. Target detection technique is employed to segment the source images into target and background regions. Different fusion rules are adopted respectively in target and background regions. A limitedly redundant discrete wavelet transform (LR DWT) method is introduced to achieve shift invariant multi-resolution representation of each source images. Fusion experiments on real world image sequences indicate that the proposed method is effective and efficient, which achieves better performance than the generic fusion method.

PACS: 07.57.-c; 07.60.-j
Keyword: Infrared; Image fusion; Discrete wavelet transform; Region-based

1. Introduction

Image fusion is a specialization of the more general topic of data fusion, dealing with image and video data [1]. It is the process by which multi-modality sensor imageries from same scene are intelligently combined into single view of the scene with extended information content. Image fusion has the important applications in the military, medical imaging, remote sensing, and security and surveillance fields. The benefits of image fusion include improved spatial awareness, increased accuracy in target detection and recognition, reduced operator workload and increased system reliability [2].

Image fusion processing must satisfy the following requirements, as described in [3]: Preserve (as far as possible) all salient information in the source images; do not introduce any artifacts or inconsistencies; be shift invariant; be temporal stable and consistent. The last two points are especially important in dynamic image fusion (or image sequences fusion) as human visual system is highly sensitive to moving artifacts introduced by the

[*] Corresponding author, Tel. /fax: +86 21 6293 3108. *E-mail address:* yangbo_sd@sjtu.edu.cn (B. Yang)

shift dependent fusion process [3].

Fusion process can be performed at different levels of information representation, sorted in ascending order of abstraction: signal, pixel, feature and symbol levels [4]. From the simplest weighted pixel averaging to more complicated multi-resolution (MR) method (including pyramidal schemes and wavelet schemes), pixel-based fusion methods were well researched [3] [5]-[10]. Recently, feature level fusion with region-based fusion scheme has been reported both qualitative and quantitative improvements over the pixel-based method as more intelligent semantic fusion rules can be considered based on actual features [11]-[16].

2. Generic Pixel-Based Image Fusion Scheme

The generic pixel-based fusion scheme is briefly reviewed, more details can be found in [3] [5]-[16]. Fig.1 illustrates generic wavelet fusion scheme, which can be divided into three steps as following: At first, all source images are decomposed by using multi-resolution method, which can be the pyramid transform (PT) [5] [6], discrete wavelet transform (DWT) [7]-[9], discrete wavelet frames (DWF) [3] or dual-tree complex wavelet transform [10] etc. Then the decomposition coefficients are fusion by applying a fusion rule, which can be a point-based maximum selection (MS) rule or more sophisticated area-based rules [6] [7]. Finally, the fused image is reconstructed by using the corresponding inverse transform on the fused coefficients.

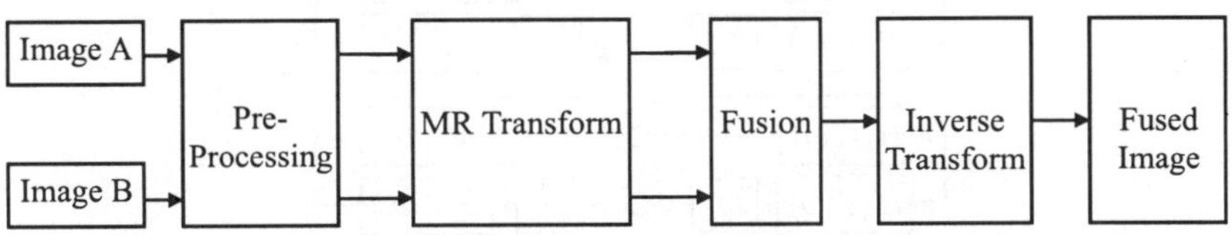

Fig. 1 Generic pixel-based image fusion scheme

3. The Region-based Dynamic Image Fusion Scheme

For pixel-based approaches, the MR decomposition coefficient is treated independently (MS rule) or filtered by a small fixed window (area-based rule). However, the most applications of a fusion scheme are interested in features within the image, not in the actual pixels. Therefore, it seems reasonable to incorporate feature information into the fusion process [14]. A number of region-based fusion schemes have been proposed [11]-[16]. However, most of region-based schemes are designed for still image fusion, and every frame of each source sequence is processed individually in image sequences case. These methods do not take full advantage of the wealth of inter-frame-information within source sequences.

The novel region-based fusion scheme proposed for fusion of visible and infrared (IR) image sequences is shown in Fig. 2, where the target detection (TD) techniques are introduced to segment target regions intelligently. For convenience, we assume both source sequences are registered well before fusion. Firstly, both the visible and IR sequences are enhanced by using pre-processing operator. Then each frame of the source sequences transformed by using a MR method (where the LR DWT is adopted in this paper, see the Sec. 3.1). Simultaneously, the frames are segmented into object and background regions by using a TD method. Different fusion rules are adopted in target and background regions. Finally, the fused coefficients belong to each region are combined, and fused frames are reconstructed by using the corresponding inverse transform.

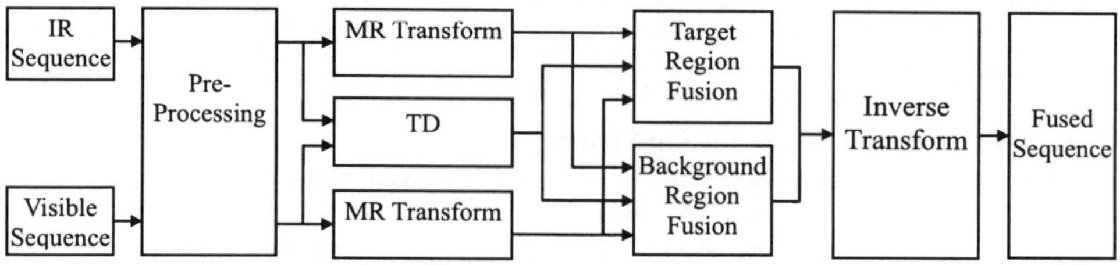

Fig. 2 Region-based IR and visible dynamic image fusion scheme

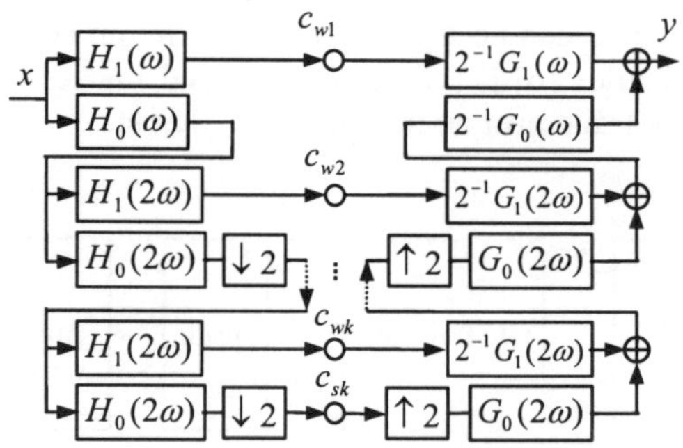

Fig. 3 The LR DWT and inverse transform.

3.1 The Limited Redundancy Discrete Wavelet Transform

It is well know that the standard DWT produces a shift dependent [17] [18] signal representation due to downsampling operations in every subband, which results in a shift dependent fusion scheme, as described by Rockinger [3]. To overcome the problem, Rockinger present a perfectly shift invariant wavelet fusion scheme by using DWF. However, this method is much computationally expensive due to high redundancy ($2^m \times n:1$ for *m*-D and *n*-level decomposition) of the representation. Bull et al [10] further develop the wavelets fusion method by introducing DT CWT, which provide approximately shift invariance by introducing limited redundancy ($2^m:1$ for *m*-D and any levels decomposition). However, the DT CWT employs two filters banks, which must be designed rigorously to achieve appropriate delays while satisfying perfect reconstruction (PR) conditions. Moreover, the decomposition coefficients from every tree should be regarded as the real or imaginary part of complex, increasing difficulty of subsequent processing in the fusion rule.

A new implementation of the DWT is introduced in this paper, which provides approximate shift invariance and nearly perfect reconstruction while preserving the properties of DWT: computational efficiency and easy implementation. Fig.3 shows decomposition and reconstruction scheme for the new transform in cascading form, which can be extended easily to 2-D by separable filtering along rows and then columns. For the limited redundancy (not more than $3:1$ for 1-D) of the new transform, we use LR DWT (the limitedly redundant discrete wavelet transform) here to distinguish it from DWT and DWF.

3.2 Region Segmentation Algorithm

The target detection (TD) operator aims for segmenting both source frames into target regions, in which the significant information is included such as moving human and vehicle, and background regions. A novel target detection method is proposed in this paper based on the characteristics of IR imaging. At first, a region merging method [19] is adopted to segment the initial IR frame. It is easy to find the target regions, which have high contrast with the neighboring background, in the segmented IR frame. A confidence measure [20] for each candidate region is computed. It is very inefficient to compute the confidence measure for each candidate within every frame. Therefore, a model matching method is adopted to find the target regions in the subsequent frames. A target model is obtained by using intensity information of the target region in pre-frame. Not the whole but a small region in post-frame which correspond with (and is little larger than) the target region in pre-frame is matched. The initial detection operator based on segmentation and confidence measure will be repeated in case no target being detected in certain successive frames. The target detection in the visible sequence is similar to the IR sequence.

3.3 Fusion Rules in the Target Region

To preserve the full information as far as possible in the target region, a special fusion rule should be employed in object region. Assume that target detection gives M target maps: $T_{IR} = \{t_{IR}^1, t_{IR}^2, \cdots, t_{IR}^M\}$ in IR frame and N target region maps: $T_V = \{t_V^1, t_V^2, \cdots, t_V^N\}$ in the corresponding visible frame. The target map is down sampled by 2^m (in accord with the resolution of decomposition coefficients) to give a decimated target map at each level. The target maps in both source frames are analyzed jointly $T_J = T_{IR} \bigcup T_V$. The frame is segmented into three sets: single, overlapped target region sets and background region set. Overlapped target regions $T_O = T_{IR} \bigcap T_V$. Single target regions are all the target regions where no overlap $T_S = T_J \bigcup \overline{T}_O$. Clearly, $T_J = T_S \bigcup T_O$. Background regions $B = \overline{T}_J$.

In the single target regions, fusion rule can be written as:

$$c_f(x,y) = \begin{cases} c_{ir}(x,y), \text{ if } (x,y) \in T_{IR} \\ c_v(x,y), \text{ if } (x,y) \in T_V \end{cases} \tag{1}$$

In a connected overlapped target region $t \in T_O$, a similarity measure between two source is define as:

$$M(t) = \frac{2 \cdot \sum_{(x,y)\in t} I_{ir}(x,y) \cdot I_v(x,y)}{\sum_{(x,y)\in t} [I_{ir}(x,y)]^2 + \sum_{(x,y)\in t} [I_v(x,y)]^2} \tag{2}$$

where I_{ir} and I_v denote IR and visible frames respectively. Then an energy index of the coefficients within the overlapped region is computed respectively in IR and visible frame,

$$S_i(t) = \sum_{(x,y)\in t} c_i(x,y)^2 \tag{3}$$

where $t \in T_O$ and $i = ir, v$ means the IR and visible frame respectively. A threshold of similarity α is introduced where $\alpha \in [0,1]$ and normally $\alpha = 0.85$ is appropriate.

In case $M(t) < \alpha$ the fusion rule in overlapped target region $t \in T_O$ can be written as:

$$c_f(x,y) = \begin{cases} c_{ir}(x,y), & \text{if } S_{ir}(t) \geq S_v(t) \\ c_v(x,y), & \text{otherwise} \end{cases} \tag{4}$$

In case $M(t) \geq \alpha$, a weight average method is adopted:

$$c_f(x,y) = \begin{cases} \varpi_{\max}(t) \cdot c_{ir}(x,y) + \varpi_{\min}(t) \cdot c_v(x,y), & \text{if } S_{ir}(t) \geq S_v(t) \\ \varpi_{\min}(t) \cdot c_{ir}(x,y) + \varpi_{\max}(t) \cdot c_v(x,y), & \text{if } S_{ir}(t) < S_v(t) \end{cases} \tag{5}$$

where the weights $\varpi_{\min}(t)$ and $\varpi_{\max}(t)$ can be obtained

$$\begin{cases} \varpi_{\min}(t) = \frac{1}{2}(1 - \frac{1 - M(t)}{1 - \alpha}) \\ \varpi_{\max}(t) = 1 - \varpi_{\min}(t) \end{cases} \tag{6}$$

Finally, in the background regions, the simplest MS rule is adopted.

4. Experimental Results and Analysis

The proposed fusion scheme was examined on real world dynamic images (image sequences). Fig.4 shows some results of fusion of spatially registered visual (visible light) and infrared (IR) (thermal 3-5 micro m) image sequences consisting of 32 subsequent frames of each sensor. Visual inspection of the results from each method shows that the proposed dynamic scheme is superior to the generic scheme in case of same transform method.

To quantify temporal stability and consistency, we adopt the quantitative measure $I((S_v, S_{ir}), F)$ proposed by Rockinger, more details can be found in [3]. We computed the average mutual information (AMI) over the 31 set of inter-frame-differences (IFDs) using each scheme with each wavelet fusion method. In Table 1, DB4 means the Daubechies wavelets of order 4, BIOR4.4 denotes the biorthogonal wavelets, where the two "4" are orders of synthesis and analysis filters respectively, Q-shift9 designed by Kingsbury [21], "Generic" and "Proposed" means using the generic wavelet fusion scheme (fusion frames one by one) and the proposed region-based dynamic fusion method. Clearly, the measure results in Table 1 are consistent with the conclusion of visual inspection. In addition, the LR DWT method performs better than the DWT and DT CWT methods in both generic and dynamic schemes, and it is comparable with the DWF method in respects of temporal stability and consistency while using lower computational cost because of lower redundancy of the LR DWT.

Table 1 The AMI for the IFDs of visual and IR image sequences.

Fusion Scheme	DWT DB4	DWT BIOR4.4	DT CWT Q-shift9	LR DWT DB4	LR DWT BIOR4.4	DWF DB4	DWF BIOR4.4
Generic	0.5233	0.5453	0.6362	0.8534	0.8624	0.9066	0.9173
Proposed	0.6021	0.6144	0.6809	0.9062	0.9081	0.9245	0.9201

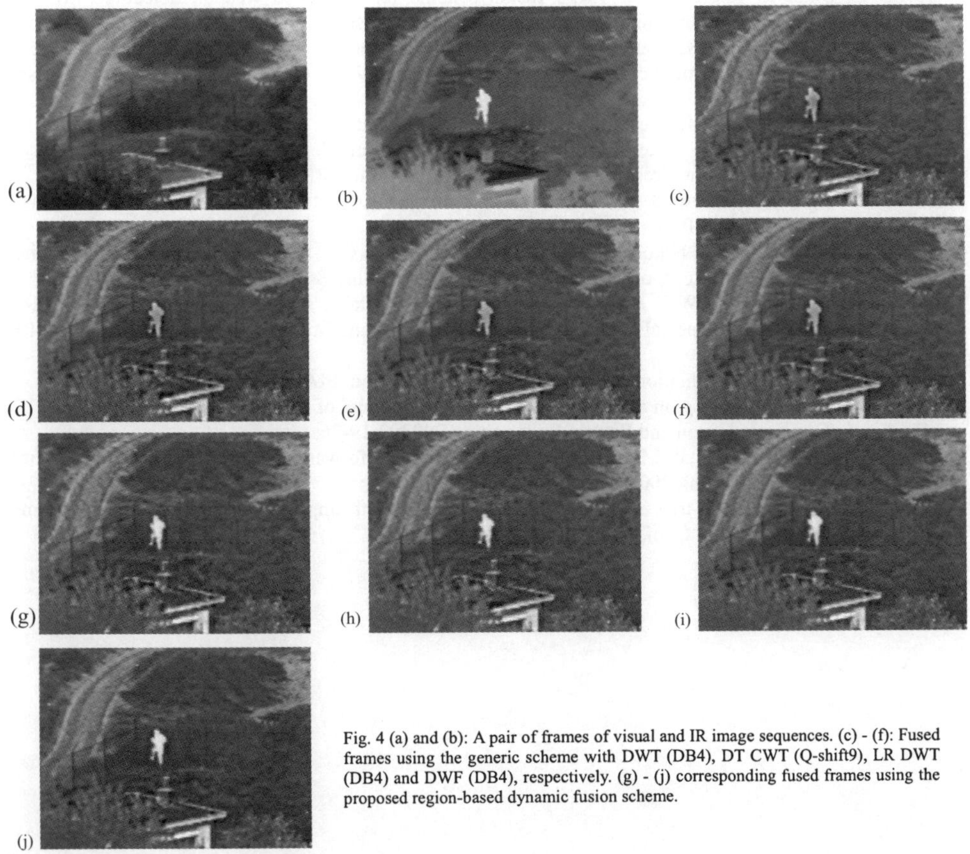

Fig. 4 (a) and (b): A pair of frames of visual and IR image sequences. (c) - (f): Fused frames using the generic scheme with DWT (DB4), DT CWT (Q-shift9), LR DWT (DB4) and DWF (DB4), respectively. (g) - (j) corresponding fused frames using the proposed region-based dynamic fusion scheme.

References

[1] Maitre H, Bloch I. Image fusion. Vistas in Astronomy 1997; 41(3):329–335.
[2] Smith MI, Heather J.P. A review of image fusion technology in 2005. Proc. SPIE 2005; 5782:29-45.
[3] Rockinger O. Image sequence fusion using a shift invariant wavelet transform. IEEE Trans. Image Processing 1997; 3:288-291.
[4] Abidi M, Gonzalez R. Data Fusion in Robotics and Machine Intelligence. Academic Press, USA 1992.
[5] Toet A. Hierarchical image fusion. Machine Vision and Applications 1990; 3:1–11.
[6] Burt PJ, Kolczynski RJ. Enhancement with application to image fusion. Proc. 4th Int. Conf. on

Computer Vision 1993; 173-182.
[7] Li H, Manjunath BS, Mitra SK. Mulitsensor image fusion using the wavelet transform. In Proc. IEEE Int. Conf. on Image Processing, Austin, Texas 1994; 1:51-55.
[8] Chipman LJ, Orr TM, Lewis LN. Wavelets and image fusion. IEEE Trans. Image Processing 1995; 3:248-251.
[9] Koren I, Laine A, Taylor F. Image fusion using steerable dyadic wavelet transforms. Proc. IEEE Int. Conf. on Image Processing, Washinton D.C. 1995; 232-235.
[10] Hill P, Canagarajah N, Bull D. Image Fusion using Complex Wavelets. Complex Proc. 13th British Machine Vision Conference, University of Cardiff 2002.
[11] Zhang Z, Blum R. Region-based image fusion scheme for concealed weapon detection. Proc. 31st Annual Conference on Information Sciences and Systems, March 1997.
[12] Matuszewski B, Shark LK, Varley M. Region-based wavelet fusion of ultrasonic, radiographic and shearographyc non-destructive testing images. Proc. 15th World Conf. on Non-Destructive Testing, Rome, October 2000.
[13] Piella G, Heijmans H. Multiresolution image fusion guided by a multimodal segmentation. Proc. Advanced Concepts of Intelligent Systems, Ghent, Belgium, September 2002.
[14] Piella G. A region-based multiresolution image fusion algorthim. ISIF, Annapolis, July 2002.
[15] Piella G. A general framework for multiresolution image fusion: from pixels to regions' Information Fusion 2003; 4:259–280.
[16] Lewis JJ, O'Callaghan RJ, Nikolov SG, Bull DR, Canagarajah CN. Region based fusion using complex wavelets. Proc. 7th Int. Conf. on Information Fusion, Stockholm, Sweden 2004; 555-562.
[17] Simoncelli E.P, Freeman W.T, Adelson EH, Heeger DJ. Shiftable multiscale transforms. IEEE Trans. Information Theory, Special Issue on Wavelet Transforms and Multiresolution Signal Analysis 1992; 38:587—607.
[18] Strang GT. Wavelets and dilation equations: A brief introduction. SIAM Rev. 1989; 31(4):614-627.
[19] Frank N, Richard N. On region merging: the statistical soundness of fast sorting with applications. IEEE Conf. on Computer Vision and Pattern Recognition 2003; 2:19-26.
[20] Yilmaz A, Shafique K, Shah M. Target tracking in airborne forward looking infrared imagery. Image and Vision Computing 2003; 21:623-635.
[21] Kingsbury N.G. The dual-tree complex wavelet transform with improved orthogonality and symmetry properties. IEEE Int. Conf. Image Proc. 2000; 375–378.

Modeling of Complex Systems for Diagnosis

Xudong W. Yu
Department of Computer Science, Southern Illinois University Edwardsville, Edwardsville, IL 62026-1656, USA

Abstract
This paper presents a bond-graph based modeling scheme that focuses on efficiently and effectively diagnosis of component failures. Our diagnostic task requires the generation of behavioral constraints that directly map output measurements to individual components of diagnostic interest. To achieve this, we adopted the bond graph framework which helps express models of a system in normal working condition by a set of output equations that represent measurable parameters as a function of parameters that represent individual components of the system. Once a bond graph model of the system is created, output equations are derived automatically from equations that are associated with basic bond-graph elements. These output equations are then used to generate conflict sets (i.e., suspected faulty components). A method for conflict resolution has been developed that identify an optimal sequence of tests to isolate the actual faults.

Keywords: bond graph, modeling for diagnosis, conflict set and resolution

1. Introduction

Our research focuses on developing a robust and efficient method for diagnosis of complex systems. We define the diagnostic task as: given a faulty system, identify a component or a set of components that are the primary causes for observed discrepancies between normal (predicted) behavior and observed behavior of the system. There have been many efforts at building automated diagnostic systems using Artificial Intelligence. Early approaches include *knowledge-based systems* (e.g., MYCIN [1], INTERNIST [2], PIP [3], CENTAUR [4], and MIDST [5] and [6]), *causal models* which use cause-effect relations between parameters and states as the organizing principle for diagnostic knowledge (e.g., CASNET [7], HF Program [8]), and *set covering techniques* which allow for simultaneous diagnosis of multiple diseases [9]. More recently, diagnosis systems have been developed that focus on using fundamental knowledge of physical system for efficient and effective diagnosis (e.g., [10] and [11]). The fundamental knowledge is typically constructed into a model of the correctly functioning system that can be analyzed to generate the expected behavior of the system. This behavior is compared with the observed behaviors of the system, and discrepancies are analyzed to derive possible faults. Typically, system models that derive behavior from structure are easier to construct for human-engineered systems, where design documents in the form of schematics and system specifications are often readily available. This is in contrast to systems, such as physiological processes, where the mapping from structure to function is not well defined, and the same component can play different roles in different contexts. The ability of a model-based diagnosis system to correctly identify different faults is very dependent on the existence of a good *model* of the system [12].

The corresponding author. Tel: (618)650-2321 Fax: (618)650-2555
Email: xyu@siue.edu

We have adopted an integrated approach which combines the rule-based, associational knowledge obtained from human experts with a model-based approach. The idea is to employ the more efficient and tailored associational knowledge for routine faults. When the associational module fails, the model-based diagnosis component is invoked to derive a correct solution. Thus the system maintains a good tradeoff between efficiency and robustness. In addition, our system employs knowledge refinement techniques to transfer new knowledge to the associational module from the model-based component to improve problem solving efficiency. In a way, this leads to transformations of more generic, fundamental knowledge into a form more useful for diagnosis. These ideas are incorporated into Multi-level Diagnostic System (MDS), a system that assists mechanics [1] in the diagnosis of complex aircraft subsystems [13] and [14]. These subsystems, such as the pneumatic system of an airplane, cover multiple domains (hydraulic, mechanic, thermodynamic, and electric), with individual components that can assume multiple behavior modes. MDS contains two diagnostic modules: (i) an associational module that uses heuristics extracted from troubleshooting manuals and human experts, (ii) an *MBD* (*Model-Based Diagnostic*) module that relies on behavior and functional knowledge of the system to be diagnosed. The activities of the two modules and the knowledge/data transfer between the two are handled by a third module, the *diagnostic controller.*

An important goal of our research is to develop efficient and effective schemes for component-based diagnosis of complex systems. A key issue in achieving this goal is the development of an adequate modeling scheme. In order to achieve this, it is important that: (i) individual components and component behaviors be explicitly represented, and (ii) relations between measurable parameters and individual components should be readily derivable from the models. We accomplish this by adopting the bond graph modeling language [15] from system dynamics. Our modeling scheme first constructs a bond graph model of the system. Equations that express the relations between measurable variables and component behaviors are derived from this model. In addition, considerable amount of our effort in building a successful modeling methodology focuses on the development of a bond-graph library of bond-graph fragments for commonly used primitive elements, and the method for selecting the appropriate fragments from the model for elements identified through system decomposition.

In rest of the paper, we first review existing modeling approaches in section 2. We discuss in detail our modeling scheme and illustrate this process by examining how the model is created for the pneumatic system in sections 4. A brief introduction of the pneumatic system is given in section 3. Section 5 shows how the generated model is used for efficient diagnosis. Section 6 contains implementation and discussion.

2. Background

In model-based reasoning, building the right model for the right task has always been the key to the success of a methodology. A powerful methodology for developing system models is compositional modeling, which involves first constructing model fragments for primitive elements of a system, and then combining them to build the model of a system. There are several different approaches in compositional modeling: (i) the component-oriented approach ([16], [17], [18], and [19]), and (ii) the process-oriented approach ([20], [21], and [22]).

In the component-oriented modeling paradigm, a model of the system is constructed from a set of physical components and connections among their terminals. For example, in the *component connection* (CC) approach, components are defined by their terminals, associated variables, and QSIM [23] qualitative constraint equations that are used to express behavior of the component. A system model is defined by introducing explicit constraints among terminal variables that are connected. Behavior of the system can then be predicted using the QSIM simulator. In the process-oriented approach [16], the model of a system is defined in terms of a set of processes that govern the dynamic behavior of the system, and a set of individual views that model the objects that make up the system and their properties. For example, both Falkenhainer and Forbus uses a *compositional modeling* technique that builds a composed model of a system or situation (referred to as *scenario*) which is sufficient for the task at hand while minimizing extraneous details [20] and [23]. To achieve this, domain knowledge is decomposed into semi-independent *model fragments*, each describing some fundamental piece of the domain's physics, such as process mechanisms, devices, and objects. Each model fragment imposes a set of relations that are linked to the individuals it represents and their associated parameters. The assumption is that the given scenario can be decomposed into a strict, structural part-of hierarchy or re-assembled from model fragments.

Although these modeling schemes are effective methods for generating possible behaviors of a system, their focus is not on diagnosis. As discussed earlier, our goal in this modeling work is to build a model of a system that can be used efficiently and effectively for component-based diagnosis. This requires generation of behavioral constraints that directly map output measurements to individual components of diagnostic interest. To achieve this, we adopted the bond graph framework which helps express models of a system in normal working condition by a set of equations that represent measurable parameters as a function of parameters that represent individual components of the system. These equations are derived from equations that are associated with basic bond-graph elements. The bond graph methodology provides a formal and systematic language for modeling dynamic systems. The use of bond graphs also helps to make a number of assumptions and issues about system functionality explicit, and provides the link between assumption classes and bond graph model fragments. Because of their very definition, bond graph model fragments are grounded in a physical reality, and are in 1-1 correspondence with the components and mechanisms of the physical system being modeled. Moreover, choice of bond graph elements to describe components and mechanisms makes explicit the assumptions that are in effect in describing component and system behavior.

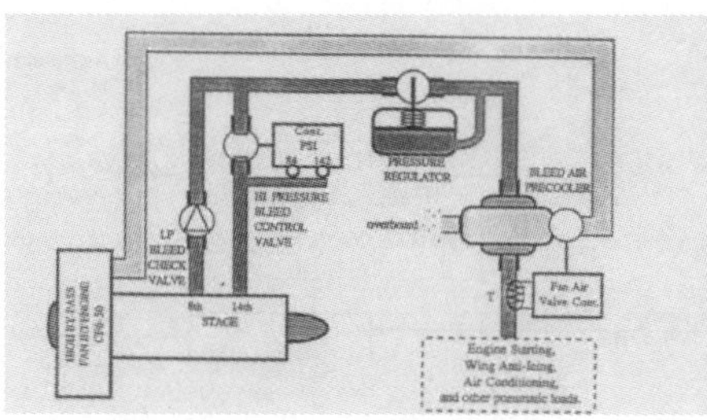

Figure 1: The Pneumatic System

3. The Pneumatic System

We will illustrate our modeling scheme by examining how models are generated for the pneumatic system of DC-10 aircraft. The pneumatic system (Figure 1) regulates air pressure and temperature drawn from one of three engines before it is delivered through the manifold system to different subsystems of the aircraft that constitute loads (e.g., the wing de-icing system). Specifically, the pneumatic pressure is regulated by a *pressure regulator subsystem*. The pressure regulator valve is modeled as a first-order system, where the opening of the regulating valve is determined by the changes in pressure at the regulator output. The temperature is controlled by a *pre-cooler subsystem*, whose primary component, a heat exchanger, draws cool air from a second source to cool the bleed air from the engine. Feedback mechanisms sense the temperature at the pre-cooler output. This information is fed back to the valve controller that fixes the opening of the valve to control the amount of cold air input to the heat exchanger, using the power obtained from the hot air transmitted through the sense line. For the diagnosis model, both the pressure regulator and pre-cooler subsystem are modeled in more detail in terms of primitive components. For example, the pre-cooler subsystem is modeled in terms of six primitive components (Figure 2): (i) the heat exchanger, (ii) the feedback controller, (iii) the valve, (iv) valve controller, (v) the temperature control sensor, and (vi) the sense line.

4. The Modeling Scheme

Effective problem solving using model-based approaches requires the ability to dynamically construct models of the system under consideration. The generated models should be both parsimonious and adequate for the specific task such as diagnosis of complex systems. For our modeling language, we have adopted bond graph, a popular language in system dynamics community which organizes domain independent structure based on the fundamental principle of *energy conservation* [15], i.e., all physical mechanisms in the real world are modeled in terms of energy transfers that occur between instances of primitive elements. Our modeling scheme can be summarized into two major steps:

1. Build a bond-graph model of the system that is both adequate and parsimonious, based on (a) a *schematic description S* and *functional description* F of the system, (b) A *domain theory D*, and (c) a *task description T* that defines the task to be accomplished
2. Generate *output equations* that relate observations (i.e., measurements on output parameters) to individual components of the system which will facilitate our diagnostic task.

Note that the bond graph model of a system may be made up of a number of individual bond graphs. In the modeling philosophy that we adopt, individual bond graphs usually correspond to different domains, such as thermodynamics and fluid mechanisms, that the system behavior covers. However, once the bond graph model of system is built, our modeling scheme automatically generates output equations that relate observable parameter to individual components under diagnostic scrutiny. The 5 primary steps in and creating bond graph model and generating output equations are discussed in detail in the rest of the sections.

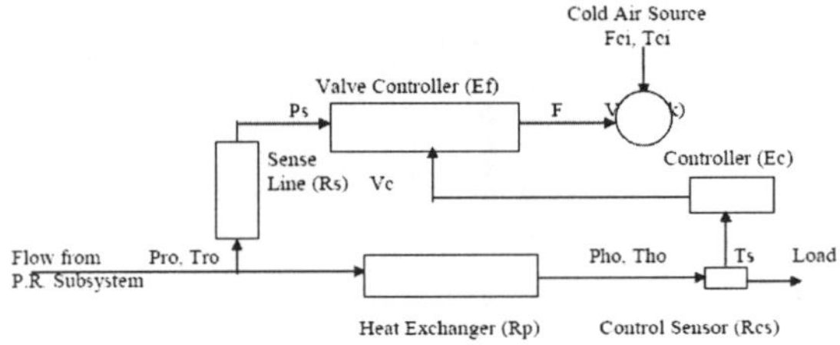

Figure 2: The Pre-cooler Subsystem

4.1 Parameter Recognition and Definition

To extend bond graph modeling for *component-oriented diagnosis*, individual components and their relations with measurable parameters (*OBS*) need to be represented explicitly. In the bond graph framework, primitive elements such as resistors and capacitors represent mechanisms [24] which may or may not be in 1-1 correspondence with individual system components. To deal with this problem, we extended parameter definitions used in the bond graph framework.

Parameters associated with primitive elements (e.g., R, C, etc) are divided into two sets: *component parameters* and *co-component parameters*. Component parameters directly relate to the functionality of components under diagnostic scrutiny. For example, the parameter R_p resistance directly relates to a primary functionality of the heat exchanger component. It models the junction at which heat transfer occurs between two substances because of temperature differences. Note that a component definition may include multiple component parameters, where each parameter represents an aspect of the functionality of the component. As part of a component parameter definition, its Possible Direction of Change (PDC) is also recorded as a list of ((+, P_{z+}), (-, P_z)) where P_{z+} and P_{z-}) represent the probabilities that the component parameter is likely to increase or decrease, respectively, in fault situations. Oftentimes, an individual component is more likely to change in one direction than the other. For example, with regard to the fluid resistance R of the pipe, PDC(R) = ((+, 0.2) (-, 0.001)) indicating that *R* almost always increases as it ages. Co-component parameters are not directly associated with primitive component functionality, but they represent bond graph elements that are introduced to complete system functionality description. For example, the thermal capacitances of the heat exchanger (Figure 3) in the pre-cooler system (Figure 2) do not represent individual components of the system but model the air masses that exchange heat, which in turn is related to flow rates of the incoming air streams.

Effort and flow variables are also characterized as *input*, *output*, and *state* variables. Input variables are associated with source elements of the bond graph, and, therefore, are exogenous to the system being analyzed. Output variables represent values which can be measured as part of the observation set, and state variables represent the minimum set of energy-related variables (e.g., heat flow rate and temperature in thermodynamics, velocity and force in mechanics) that uniquely describe the state of a dynamic system[2]. Characterization of parameters often depends on our viewpoint when we analyze a system. Parameters that are known to be insignificant or unchanged for a specific diagnostic task can be considered as constant. Parameters that express interactions between the system and other subsystems that are not modeled are also considered constant (because their effects are considered to be exogenous to the diagnostic situation). For the pneumatic system, jet engine pressure and temperature represent input variables, because the engine is not included as part of the diagnosis task. Possible output or measurable parameters are P_{ho} and T_{ho}, the output pressure and temperature of the bleed air at the load, and P_{ro} the pressure at the pressure regulator output. Resistance at the heat exchanger junction R_p and the resistance of the sense line to liquid flow R_s, are examples of component parameters for the precooler subsystem.

4.2 System Decomposition

In general, system decomposition is task and viewpoint dependent and difficult to automate. For example, consider the *jet engine* as part of the pneumatic system. For the diagnosis task, if it is sufficient to determine that the cause of a problem is engine failure, the engine can be modeled as an effort source. However, if the diagnosis task requires that the cause of the problem within the engine be determined, then it is important to model the *pistons* and *valves* within the engine explicitly, and the mechanisms that determine engine functionality need to be represented in more detail. We make the assumption that the modeler performs the system decomposition task. As discussed earlier, this involves decomposing the systems functionality by domain, and selecting a set of *primary mechanisms* that define the system functionality in that domain. In the bond graph framework, primary mechanisms specify how a subsystem affects system behavior by controlling energy transfers between components. More formally, the primary mechanisms in any domain can be classified as: (i) *energy sources*, (ii) *energy flow* and *storage mechanisms* (those that transfer energy from one location to another or store energy at a location), and (iii) *energy transformation mechanisms* (those that convert energy from one form to another).

In addition, we define a special class of mechanisms called *feedback mechanisms* to facilitate the modeling of engineering systems. Most engineering systems are used for automated control which involves coupling a low-order sensing and information-processing component with a high-power system. Generally, in this type of system, a parameter value is sensed in the main part of the system (the high-powered component), to generate the feedback signal, which is sent to an actuator that adjusts the behavior of the main system as necessary, to control and optimize its performance. The feedback mechanism can generally be modeled by a Modulated TransFormer (MTF) (shown in Figure 3), where

---

[2] State variables may also be measurable

the actuator obtains an effort E_1 from a source, and applies a certain portion of it to E_2 to the main system, based on the signal *S* it receives.

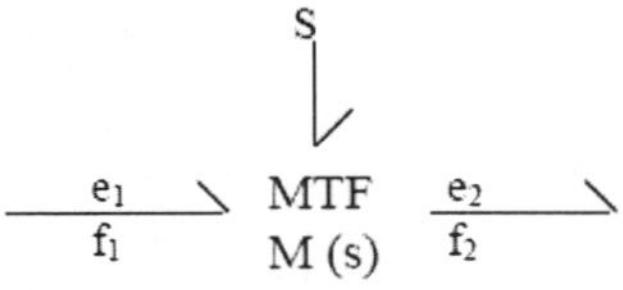

Figure 3.The MTF Module

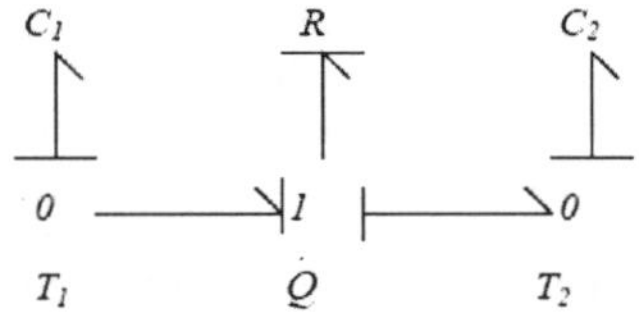

Figure 4.The Heat Exchange Module

Note that elements that transmit e_1 and receive e_2 can correspond to any of the primary mechanisms: *source*, *energy flow*, *storage*, and *energy transformation.* In some cases, the signal is preprocessed by an analyzer before it is sent to the actuator. To explicitly model the analyzer (i.e., signal processor), another MTF can be added to the model, which receives the original signal and formulates a new one to send to the actuator. Since feedback mechanisms usually involve a transformation of energy from one form to another (e.g., heat to electricity), it can be included within *energy transformation* mechanisms.

From a procedural viewpoint, the modeler begins the decomposition process by first identifying the different domains that describe system behaviors of interest. For example, in the pneumatic system, the domains of interest are: (i) thermal, (ii) fluid, and (iii) mechanical. Simultaneously, the modeler studies the specification task and the system schematics that describe the set of components of diagnostic interest. This leads to the selection of one or more component parameters that govern behaviors of interest for each component. The next step involves selection of the primary mechanisms in the bond graph framework. In our system, the jet engine is not of diagnostic interest, and therefore, it is modeled as an ideal effort source. Similarly, based on the schematics of the precooler system and its functional description, it is represented as a composition of three major subsystems: (i) heat-exchange subsystem, (ii) the valve subsystem, and (iii) the feedback mechanism. Each subsystem consists of one or more primary mechanisms. For example, the feedback subsystem consists of two primary mechanisms: (i) fluid source, (ii) resistive fluid flow through the sense line that provides pneumatic power (in the form of pressure) to the *valve controller*, and (iii) feedback mechanism for temperature control of the system.

4.3 Model Fragment Selection

Once primary mechanisms have been identified, model construction takes on a compositional modeling flavor. In our framework, this requires the modeler to first select bond graph fragments based on primary mechanisms and domain information. To facilitate this task, we have developed a *bond graph library* that represents collections of mechanisms in various domains. An example of an element in the bond graph library is the bond graph fragment for the heat exchange mechanism (Figure 4). Note that the system decomposition step produces primary mechanisms and their list of associated components. For each primary mechanism, the modeler's task is to index into the library and pick the appropriate bond graph fragment(s) that correspond to this mechanism, and then to map physical system components into this generic structure. For example, in the heat exchanger, if heat transfer occurs uniformly across a thin slab of material, the thermal resistance *R* is a function of the thermal conductivity, the cross-sectional area, and the thickness of the material. If heat exchange occurs between two blocks of metal, the capacitance *C* of each block is a function of its mass and specific heat. On the other hand, if heat exchange occurs between two fluids flowing through pipes, the capacitance value computations are more complex.

Given a description of a primary mechanism and its components, our modeler index into the bond graph library and pick the appropriate fragment by using *assumption classes.* Each assumption class represents a *consistent* combination of the physical setting of the system, its operating conditions, and the conditions that influence the behavior of its components. Collections of characteristic and component assumptions represent alternative ways to model the same aspect or phenomenon. These are then organized into mutually exclusive assumption classes, which form the basis for indexing into the bond graph library and retrieving appropriate model fragments. A model fragment is a bond graph segment that contains one or more primitive bond graph elements and a set of equations that define relations between effort and flow variables for individual bond-graph elements [15], e.g., $Q \bullet R = T_1 - T_2$ for a resistive heat junction.

4.4 Composing a Bond Graph Model

Once bond-graph fragments are selected for individual mechanisms, they need to be composed to form the bond graph model of the system. In previous work, automation of model composition has been considered to be a difficult task ([18], [20], and [22]). However, our use of the bond graph modeling language makes the task much easier. As discussed earlier, interactions between bond graph components are expressed in a domain independent way: as energy transfers which are represented as *directed bonds*, and links between segments in the same domain are established by *junctions*, i.e., *common flow* or 1-junctions, and *common effort* or 0-junctions. Connections between bond graph segments in different domains are established if there is energy transfer between the subsystems modeled by these segments. In this case, the connections are established through energy transform mechanisms: *transformers* and *gyrators*. To illustrate this modeling method, we now discuss the construction of the bond graph model of the precooler subsystem. As we discussed earlier, the temperature regulating part of the pneumatic system is composed of three subsystems: (i) the heat exchange subsystem, (ii) the pneumatic flow valve subsystem, and (iii) the feedback control subsystem. For each subsystem, one or more primary mechanisms are identified. For example, three primary mechanisms are identified for the heat exchange subsystem: the heat exchange mechanism, and two source mechanisms. The bond graph fragment corresponding to the particular heat exchange mechanism (1-junction in the center of bond graph 1 of figure 5) is selected from the library based on the assumption that the heat exchange between the two heat masses (represented as two capacitors) occurs uniformly through a resistive junction R_p. Bond graph fragments for the two sources (S_h and S_c) are selected based on the assumption that each source provides heat at constant temperature (T_h and T_c) along resistive pathways, i.e., there is heat loss during transportation of the air mass. The bond graph fragments are then connected using bonds 1 and 2 to form the model for the heat exchanger subsystem (bond graph 1 of figure 5). Note that the arrows on the bonds 1 and 2 indicate that the direction of energy transfer is from the hot source S_h to the hot capacitor C_h and from the cold capacitor C_c to the cold source S_c. The bond graph model for the valve subsystem (bond graph 2 of figure 5) is built by connecting the bond graph fragment for resistive flow (model of the valve) with the fragment for an ideal source (model of the source of air).

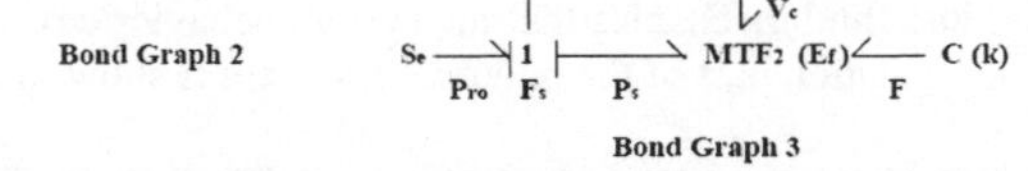

Figure 5: Bond Graph Model of the pre-cooler subsystem

This bond graph indicates that the amount of pneumatic flow is determined by the resistance in the system (i.e., the valve). The resistance, in turn, is determined by the *opening* of the valve, which is controlled by the feedback system. The feedback subsystem contains three primary mechanisms: (i) the pneumatic source (modeled by S_e), (ii) resistive flow that represents the function of the *sense line* (modeled by left segment of bond graph 3 of figure 5), (iii) the feedback mechanism that controls the temperature of the main pneumatic flow based on the a voltage signal V_c sent by the signal processor (the controller). This bond graph models the physical situation where the valve controller (MTF_2) transfers a fraction of the pneumatic power P_s obtained through a *sense line* (a pipe) into a mechanical force F that acts against the valve spring to determine the opening of the valve. The amount of power transferred is determined by a voltage signal V_c from the controller (MTF_1).

4.5. Equation Generation

Output equation generation is a three step process: (i) assign causal strokes to bonds in the bond graph, (ii) generate state equations, and (iii) generate output equations and manipulate them algebraically to convert them to the desired form. The first step is discussed in detail in [13] and [26] and is not repeated here. The algorithms we have developed for the second and third steps are here.

The bond graph framework adopts the state space approach to modeling dynamic systems. An nth order system is modeled as a set of *n* first order differential equations. In our methodology, a state equation has the following form:

$$\dot{x}_i = g_i(z_1,\ldots,z_n,x_1,\ldots,x_l,u_1,\ldots,u_r,c_1,\ldots,c_m)$$

Where x_i's are state parameters, $\dot{x}_i$'s are their derivatives, z_i's are component parameters, u_j's are input parameters, g_i's are algebraic functions, and c_k's are co-component parameters. The method for generating state equations from a bond graph summarized below:

Identifying key parameters of the system

Key parameters of the pneumatic system are identified and categorized into five different classes: (i) input, (ii) state, (iii) component, (iv), co-component, and (v) output. Here, component parameters represent the functions of individual components in the systems that are under diagnostic scrutiny. For the pneumatic system, jet engine temperature *Thi* and *Tci* represent input parameters, because the engine is not included as part of the diagnosis task. Possible output or measurable parameters are *Pho* and *Tho,* the output pressure and temperature of the bleed air at the load. Resistance at the heat exchanger junction *Rp* and the resistance of the sense line to liquid flow *Rs*, are examples of component parameters for the pneumatic system. *Qh* and *Qc,* which represent the flow of hot and cold air, are example of state parameters. Cc is an example of a co-component parameter, which models a subsystem as opposed to an individual component.

Build a bond graph model of the system from which behavior of the system can be derived

Since our focus is on component-based diagnosis, individual components are explicitly represented in our model as component parameters, each representing one aspect of the component's behavior. Note that the bond graph model of a system may be made up of a number of individual bond graphs. In the modeling philosophy that we adopt, individual bond graphs usually correspond to different domains, such as thermodynamics and fluid mechanics that the system behavior covers. For example, the bond graph model for the heat regulation part of the pneumatic system is show in Figure 5.

Formulate initial equations for basic bond graph elements such as I, R, and C

For the pneumatic system, five initial equations (1-5) are generated from bond graph 1[3]:

[3] The two air masses are modeled as lumped systems. More detailed piecewise models can also be created.

$$\text{Ch:} T_h = T_{hi} + \frac{Q_h}{C_h}, \quad \text{Rh:} \dot{Q}_1 = \frac{T_{hi} - T_h}{R_h}, \quad \text{Cc:} T_c = T_{ci} + \frac{Q_c}{C_c}, \quad \text{Rc:} \dot{Q}_3 = \frac{T_c - T_{ci}}{R_c}, \quad \text{Rp:} \dot{Q}_2 = \frac{T_h - T_c}{R_p}$$

Formulate first order differential equations for each state variable

Each state variable will be express in terms of other variables which linked to it through the same 0- or 1- junction. In our example, by making substitutions using equations (1)-(4), we get:

$$\dot{Q}_h = \dot{Q}_1 - \dot{Q}_2 = -\frac{Q_h}{C_h R_h} - \frac{T_{hi} - T_{ci}}{R_p} - \frac{Q_h}{C_h R_p} + \frac{Q_c}{C_c R_p}$$

Following the same process, the equation for $\dot{Q}_c$ is derived as:

$$\dot{Q}_c = -\frac{Q_c}{C_c R_c} - \frac{T_{hi} - T_{ci}}{R_p} - \frac{Q_c}{C_c R_p} + \frac{Q_h}{C_h R_p}$$

Generates output equations

Each output parameter is linked to various component parameters through a set of equations. The process involves symbolic manipulations, and is based on the following assumptions: (i) the physical systems we deal with are linear or are modeled by linear approximations, therefore, the algebraic functions derived for the state equations are also linear, and (ii) the systems was operating normally in a steady state, and diagnosis is initiated when system parameters deviate from their steady state values (i.e., we perform steady state diagnosis). Here, we illustrate how a set of output equations are generated for Tho. In steady state, we approximate $\Delta t = l/v$, where l is the length of the path the air masses traverse in the preschooler, and v is the velocity of the air flow, approximated as $(v_h + v_c)/2$[4].

Therefore, $\dot{Q} = Qv/l$ for both $\dot{Q}_c$ and $\dot{Q}_h$. Substituting for $\dot{Q}_c$ and $\dot{Q}_h$, and solving these equations produce:

$$Q_h = -C_h^2(T_{hi} - T_{ci})(v/l + l/C_c R_c D) \tag{6}$$

where,

$$D = C_c C_h R_p^2 \frac{l^2}{v^2} + \frac{cR_p}{R_h R_c l}(C_c R_c(R_h + R_p) + C_h R_h(R_c + R_p)) + \frac{R_p^2}{R_c Rh}(R_c + R_h + R_p) \tag{7}$$

Since the output parameter Tho is equal to Th after time period Δt. Using equation (1) & (6) we get:

$$T_{ho} = T_{hi} - C_h(T_{hi} - T_{ci})(\frac{v}{l} + \frac{l}{C_c R_c D}) \tag{8}$$

Next, we generate equations for co-component parameters using domain knowledge and first principles. For the pneumatic system,

$$C_c = Vc\rho = \Delta t F_c c\rho = F_c c\rho l/v \tag{9}$$

where V is the volume of the cold air in the heat exchanger unit, ρ is the density, c is the unit thermal capacitor of the air, and Fc is the flow rate of the cold air. In this case, both ρ and c are constant.

Using similar methods, the following equations are generated from bond graph 2 and 3:

$$F_c = \sqrt{P_3/C_3}(X_{\max} - X), \; X = \frac{A}{k}(P_{ro} - R_s F_s)E, \; E = E_f + E_c(T_{set} - R_{cs} Q_r) \tag{10,11,12}$$

where Fs is the flow rate through the sense line, $Tset$ is the desired temperature, $P3$ is the pressure difference over the valve, $C3$ is the valve constant, and $Xmax$ is the max length of valve opening. *Pro*

---

[4] A more exact solution would assign different velocities to the two air masses

is the pressure of the flow from the pressure regulator subsystem, *A* is the area of the opening of the sense line. A, *P3*, and *C3* are constants,

5. Diagnostic Results

We illustrated how a set of equations (7 through 12) for output parameter Tho of the pneumatic system (figures 1 and 2) were generated. Similarly, another set of output equation are generated for parameter Pho, the output pressure of the bleed air at the load:

$$P_{ho} = P_{in}(1 - R_p / (R_p + C_3/x_h), \quad X_h = X_{set} - A_2(P_{in} - R_{pt} f_{pt})\frac{E_p}{K_p} \qquad (13,14)$$

For diagnosis purposes, the parameters categorized as follows:

- Output parameters: *Tho* and *Pho.*
- Input parameters: *Pin, Thi,* and *Tci.*
- Component parameters: *Rp, Rs, Ef, k, Rs, Ec, Rcs, kp,* and *Rpt.* The set of component parameters defines the set of possible single fault hypotheses that are considered for diagnostic analysis.
- Co-component parameters: *Cc* (the thermal capacitance of the heat exchanger), *D, Fc, X, E, P,* and X_h.
- Constants: ρ (the density of the air) and *C3* (the cold air valve constant).

Given the set of measurements made on the system and the system description which include parameter definitions and the current set of output equations, the diagnostic algorithm is summarized as follows:

Repeat until one candidate is confirmed or all candidates are eliminated

{

1. Generate partial explanations, one for each new measurement.
2. Generate/update candidate set using best first search.
3. Perform test(s) selected using a decision theoretic scheme.

}

The diagnosis algorithm is based on the assumptions: (a) the system was operating in steady-state, (b) the measurement sampling rate is high enough to catch deviations soon after they manifest, and (c) the faults do not cause catastrophic changes in the system. Diagnosis is initiated when observed system behavior deviates from a steady state. Again, the pneumatic system is used to illustrate the process. Details of step 1 and 2 can be found in [25].

1. Generate partial explanations by performing qualitative causal analysis on the set of output equations. For each output parameter and a new measurement, first determine the sign of PDC(Pk,Wi(+|-)) (Possible Direction of Change of Pk caused by change in Wi) for all output parameter Pk and component parameter Wi pairs by computing the partial derivative $\frac{\partial p_k}{\partial w_i} = \frac{\partial p_k}{\partial s_1}\frac{\partial s_1}{\partial s_2}\cdots\frac{\partial s_{n-1}}{\partial s_n}\frac{\partial s_n}{\partial w_i}$

   Where s1, s2,…,si are co-component parameters that link Pk to Wi. For example, the relation between Rs, and Tho can be determined using equations (7) - (12):

$$\frac{\partial T_{ho}}{\partial R_s} = \frac{\partial T_{ho}}{\partial C_c}\frac{\partial C_c}{\partial F_c}\frac{\partial F_c}{\partial X}\frac{\partial X}{\partial R_s} = (-C_h(T_{hi} - C_{hi})\frac{l}{C_c^2 R_c D})\bullet(\frac{l}{v}c\rho)\bullet(-\sqrt{\tfrac{P_3}{C_3}})\bullet(-\frac{A}{k}F_s E)$$

   Since the qualitative values of variables A, k, E, Fs, Cc, Rc, c, ρ, D, l, Ch, C3, P3, and v are all known to be + and our steady-state assumption implies that Thi-Tci > 0 (i.e., the initial temperature difference between hot and cold air may fluctuate but is always positive), the partial derivative evaluates to -, therefore, PDC(Tho, Rs+) is -, and PDC(Tho, Rs-) is +. Next, partial explanations for each measurable parameter is generated based on whether it is normal (within 2% of the normal value), above-normal (+), or below-normal (-), as described below:

- For each deviant output parameter Y, form a proposition formula: X1(+|-) ∨ X2(+|-) ∨...∨ Xn(+|-), where Xi(+|-)'s are changes in the component parameters Xi's that are consistent with the observed deviation of Y, i.e., PDC(Y,Xi(+|-)) is equal to the deviation in Y. In our example, the following partial explanation for Tho_+ would be generated: *F(Tho+) = Rp+ ∨ Ef+ ∨ K- ∨ Rs- ∨ Ec ∨ Rcs-*. Notice that, based on previously established value of PDC(Tho,Rs-) (+) and PDC(Tho,Rs+) (-) Rs- is included in the formula while Rs+ is not. In other words, only a decrease in the resistance of the sense line is consistent with the observed above-normal temperature value. Further more, although Rs- is included in the explanation, the low probability associated with the event (decreased resistance) will result in it being dropped from the candidate list for further analysis.
- For each measurement Y that is reported to be normal, form a proposition formula:

$$(\neg X1 \wedge \neg X2 \wedge \ldots \wedge \neg Xn) \vee (X1(+|-) \wedge X2(+|-)) \vee \ldots \vee (Xn\text{-}1(+|-) \wedge Xn(+|-))$$

  Each pair of (Xi, Xj) is included if their influence on Y are complementary (i.e., if PDC(Y,Xi(+|-)) is +, then PDC(Y,Xj(+|-)) is -, and vice versa.). This formula suggests that the Xi's are either all normal or at least two of them are deviant and their combined effect on Y is null. In our example, assuming the output parameter Pho is normal, the following formula will be generated: *Pho = (¬Kp ∧ ¬Rp ∧ ¬Ep ∧ ¬Rpt) ∨ (Kp+ ∧ Ep+) ∨ (Kp- ∧ Ep-) ∨ ...* Note that each *¬Xi* implies both *¬Xi+* and *¬Xi-*.

2. Generate and update the candidate set based on the current set of partial explanations. Our system uses a best-first search algorithm that generates candidates in order of their prior probabilities until either (i) the number of candidates reaches a preset threshold k_1, or (ii) the prior probability of the best candidate is k_2 times greater than the prior probability of the last candidate. (Both k_1 and k_2 are predetermined constants and are set at 25 in the current system.) The following table shows the prior probabilities of failure of the components in the pneumatic system:

Ef+	Ef-	Ec+	Ec-	Kp+	Kp-	K+	K-	Rp+
0.2	0.1	0.18	0.01	0.15	0.18	0.15	0.15	0.2
Ep+	Ep-	Rpt+	Rpt-	Rcs+	Rcs-	Rs+	Rs-	Rp-
0.1	0.2	0.2	0.01	0.1	0.05	0.2	0.001	0.001

   For each component parameter, two probability values are used, one indicating the likelihood of its increasing, the other indicating the possibility of its decreasing. This use of prior probabilities allows the system to take into account that components are more likely to fail in one direction than the other. For example, the resistance of a pipe used for fluid transfer may increase over time but almost never decreases. In our example, given that *Tho* is above normal and all other measurable parameters are normal, the following candidates are generated:

   Cand = ((Ef+), (Ec+), (K-), (Rcs-), (Rp+, Rpt+), (Rp-, Ep-), (Kp+, Rp+), (Kp-, Rpt+), (Kp-, Ep-), (Ep+, Rpt+), (Ep+, Kp+))

   Notice that *Rs-* (the resistance of the sense line used for fluid transfer is decreasing) is not included in the initial set even though it is consistent with the current observation because of its extremely low probability.

3. Perform measurement selection based on the established relationship between measurable parameters and individual components (i.e., the sets of output equations) using an information-theoretic method similar to the one used in GDE ([10]). For each possible measurement *Oi*, we calculate ΔHe(*Oi*) to

evaluate the expected changes in the entropy of the system if the measurement *Oi* is made, using the formula:

$$\Delta\,\mathrm{He(Oi)} = \sum_{k=1}^{m} p(O_i = V_{ik})\log p(O_i = V_{ik}) + p(U_i)\log p(U_i) - \frac{np(U_i)}{m}\log\frac{p(U_i)}{m}.$$

In the above formula, each Vik (1≤k≤m) is a possible outcome of measurement Oi, Ui is the set of candidates that do not predict a value for Oi. In the qualitative framework, a measurement can take one of three values: above-normal (+), normal (0), and below-normal (-). Give the current set of candidates C, the probability P(Oi=Vik) can be calculated using the formula: $\mathrm{P(Oi = Vik)} = \sum_{C_l \in C} p(O_i = V_{ik} \mid C_l)p(C_l)$, Where p(Cl) is the current probability for the candidate Cl . P(Oi=Vik|Cl) (the probability that parameter Oi has value Vik given that the candidate is Cl) can be calculated using qualitative analysis on the equations for Oi. When Vik is either + or -, P(Oi=Vik|Cl) has value 1 if Cl contains only components that are in the corresponding partial explanation of Ol. It has value 1/3 when Cl contains components in partial explanations for both Oi+ and Oi-. It has value 0 when Cl does not contain any components in either of the partial explanations. When Vik is 0, P(Oi=Vik|Cl) has value 1 if Cl does not contain any components in either of the two lists. It has value 0 when Cl only contains components in one of the partial explanations. It has value 1/3 when Cl contains components in both partial explanations. For example, given the two possible partial explanations for Pro+ and Pro-:

F(Pro+) =Kp+ ∨ Ep- ∨ Rpt+ ∨ Rp-   *F(Pro-) =Kp- ∨ Ep+ ∨ Rpt-*

Since Ep- is only in F(Pro+) and Kp- is only in F(Pro-), (Ep-,Kp-) supports, with equal probabily, Pro+ (with Ep-), Pro- (with Kp-), P_{ro} normal (using both Ep- and Kp-). Therefore, we obtain *p(Pro=+ | (Ep-,Kp-)) = 1/3*. Similarly, *p(Pro=+ | Ecs-) = 0* because Ecs- is not included in the partial explanations for Pro.

After ΔHe(Oi) is calculated for all the remaining possible measurements Oi, the one with the smallest ΔHe(Oi) value is selected. In our example, the following ΔHe's are computed:

Test	Pvc	Fc	Vc	Pro	Phv	Pprc	Tcs	Op
ΔHe	-0.909	-0.890	-0.727	-0.676	-0.457	-0.457	-0.362	-0.175

Based on this information, *Pvc* (The output power at the *valve controller*) was chosen as the next measurement. The initial probability of a candidate *CL* is calculated from the prior probability of its components using the formula: $p(C_l) = \prod_{c \in C_l} p(c) \prod_{c \notin C_l} (1 - p(c))$. After each measurement, the probability of a candidate is updated using Bayes rule:

$$p(C_l \mid O_i = V_{ik}) = \frac{p(O_i = V_{ik} \mid C_l)p(C_l)}{p(O_i = V_{ik})}.$$

We do not compute p(Oi=Vik) since it is the same for all candidates and does not affect their relative order. Continuing with our example, given the new measurement, Pvc was recorded as being above normal indicating that the fault is in the pre-cooler subsystem. As a result, the new candidate set is *((Ef+), (Ec+), (Rcs-))*. By calculating ΔHe, the system identified that the measurement *Vc* (the voltage from the controller) would provide the most information. *Vc* was reported to be normal, and that left *Ef+* as the only single candidate. However, the system also noticed that a double fault involving *Rcs+* and *Ec+* could also explain the symptoms observed so far. As a result, (*Rcs+*, *Ec+*)

was also generated as a candidate. A measurement at *Tcs* was then suggested, and since it was normal, the system concluded that the actual faulty component is *Ef* (the valve controller).

6. Conclusion

In this paper, we presented a bond-graph based modeling scheme that focuses on the diagnosis task by first constructing bond graph model of a system and then generating output equations that can be used for effective diagnosis. Our methodology was successfully used in number of diagnostic systems including MDS [25] and DOC [13]. The bond graph model builder has been implemented in X window and C++ with graphic and menu-based interface. The menu enables the user to build a bond graph model of the system in a number of ways. We developed adequate bond graph libraries for the thermodynamic and fluid domains. The library of components in the thermodynamics domain contains basic bond graph elements, such as the resistive heat exchange junction, different models of heat capacitance, the temperature source S_e and the flow source S_f, and 0- and 1-junctions for building composed systems. In addition, we include descriptions of standard components, such as the heat exchanger and models of basic feedback mechanisms that generate signals with strengths proportional to temperature differences. Fragments in the library can be retrieved using a top-down, hierarchical selection process. The *Diagnostic Module* is implemented in C++ and uses subroutines from Mathematica to perform symbolic manipulation and calculate partial derivatives in order to generate partial explanations.

Our overall modeling philosophy is similar to some of the existing compositional modeling approaches ([18], [19], [20], [22], etc). However, the primary difference is that our modeling framework is based on the more formal bond graph language, and, therefore, we are better able to characterize and formalize the system decomposition and model composition tasks. There is also a difference in the way constraints are generated. In our framework, the equations are combined from basic equations associated with bond graph elements which are well grounded in physical reality. Starting with an analytic equation-oriented model also provides the opportunity to introduce successively more precise information (such as orders of magnitude information, and quantitative values for parameters) if available, and derive more accurate diagnostic results without altering our framework or modeling methodology.

References

[1] E. H. Shortliffe, *Computer-based medical consultations: MYCIN*, New York, 1976.

[2] Y. Peng and J. A. Reggia. "Plausibility of diagnostic hypothesis: the nature of simplicity", *AAAI-86*, pp. 140-145, 1986.

[3] S. Pauker, G. A. Gorry, J. Kassirer, and W. Schwartz, "Towards the simulation of clinical cognition – taking a present illness by computer", *American Journal of Medicine*, 60, pp. 981-996, 1976. Cambridge, MA, 1987.

[4] J. Aikins. "Prototypical knowledge for expert systems", *Artificial Intelligence*, 20, pp. 163-210, 1983.

[5] G. Biswas and X. Yu. "A rule network for efficient implementation of a mixed-initiative reasoning scheme", *proceedings of ACM*, pp. 123-130, 1989.

[6] G. Biswas, X. Yu, and et. al., "PLAYMAKER: A Knowledge-based approach to characterizing hydrocarbon plays", *International Journal of Pattern and Artificial intelligence*, vol. 4, pp. 315-339, 1990.

[7] S. Weiss, C. Kulikowsi, and A. Safir. "A model-based consultation system for the long-term management of glaucoma", *Proceedings of the 5th IJCAI*, pp. 826-832, 1977.

[8] W.J. Long, S. Naimi, M.G. Criscitiello, and R. Jayes. "The development and use of a causal model for reasoning about heart failure", *unpublished manuscript*, 1988.

[9] J. B. Weinberg and G. Biswas, "The Functional Modularity of Diagnosis Domain Structure," *Sixth Intl. Workshop on Principles of Diagnosis*, W. Nejdl, ed., Goslar, Germany, pp. 123-130, Oct. 1995.

[10] J. deKleer, "Focusing on Probable Diagnoses", *Proceedings of the 9th AAAI*, pp. 842-848, 1991.

[11] M.R. Genesereth. "The use of design descriptions in automated diagnosis", *Artificial Intelligence*, pp. 411-436, 1984.

[12] R. Davis and W.C. Hamscher, "Model-Based Reasoning: Troubleshooting", in *Exploring Artificial Intelligence: Survey Talks from the National Conferences on AI*, H.E. Shrobe, ed., pp. 297-346, Morgan Kaufmann, San Mateo, CA, 1988.

[13] G. Biswas, R. Kapadia, X. W. Yu, "Combined Qualitative- Quantitative Steady State Diagnosis of Continuous-valued Systems", *IEEE Transactions on Systems, Man, and Cybernetics*, July 1997.

[14] X. W. Yu, S. Manganaris, and G. Biswas, "A Component Connection Modeling Case Study: Result, and Future Directions", *Working papers of the 5th International Workshop on Qualitative Reasoning about Physical Systems*, Austin, TX, pp. 326 - 340, May, 1991.

[15] R.C. Rosenberg, and D.C. Karnopp, *Introduction to Physical System Dynamics*, McGraw-Hill, 1983.

[16] J. de Kleer, "How Circuit Work", *Qualitative Reasoning about Physical Systems*, pp. 205-280, MIT Press, 1985.

[17] D.W. Franke and D. L. Dvorak, "Component Connection Models", *Workshop on Model-Based Reasoning*, IJCAI-89, Detroit, MI, pp.~97-101, 1989.

[18] P. Nayak, L. Joskowicz, S. Addanki, "Automated Model Selection Using Context-Dependent Behaviors". *Proceedings of the Tenth National Conference on Artificial Intelligence*, pp. 710-716, 1992

[19] P. Nayak, L. Joskowicz, "Efficient Compositional Modeling for Generating Causal Explanations", Artificial Intelligence, vol 82(3), pp. 193-227, 1996.

[20] B. Falkenhainer and K. Forbus, "Compositional Modeling: Finding the right model for the job", *AI Journal*, vol. 51, pp. 95-143, 1991.

[21] B. Falkenhainer, "Ideal Physical Systems", *Proceedings of the Eleventh National Conference on Artificial Intelligence*, pp. 600-605, 1993.

[22] K. Forbus, "Qualitative process theory", *Artificial Intelligence*, pp. 24:85–168, 1984.

[23] B. Kuipers, "Qualitative Simulation", *Artificial Intelligence*, vol.29, pp. 289-388, 1986.

[24] J. Top and H. Akkermans, "Computational and Physical Causality", *Proceedings 12th IJCAI*, pp. 1171-1176, 1991.

[25] X. W. Yu, G. Biswas, and J. Weinberg, "MDS: An Integrated Architecture for Associational and Model-based Diagnosis", *Journal of Applied Intelligence*, Vol. 14, pp. 179-195, 2001.

[26] S. Manganaris and G. Biswas, "Modeling and Qualitative Reasoning about Perturbations in Mechanics Problem Solving", *Tech. Report CS-90-16*, Vanderbilt Univ., Nashville, TN, 1990.

Ada Programming for Solving Nonlinear Equations

Trong Wu

Department of Computer Science Southern Illinois University, Edwardsville, IL 62026-1656, USA

Abstract

This paper introduces the Ada programming for solving non-linear equations over a new class of *real numbers* which are based on the concepts of *model numbers* and *rough numbers* for a given computer system. We will study structures of Ada interval computation over *model numbers* and *rough numbers*. To do these, we must revise *commonly interval computation* from compact intervals to *closed-open intervals* for their initial intervals. This way, we can promise that the final resulting interval will always be a shorter than the result from the ordinary interval computation. Two examples are presented, one is use Newton method and the other is apply iterative method for solving non-line equations. The Ada programs and their the approximated solutions are given in both decimal and binary values.

Keywords: Ada language, Interval computation, Rough numbers, Model numbers, and Algebriac srtrctures.

1. Introduction

Most of computer users are not aware that the computation of *real numbers* and *model numbers* are not the same. Even more confusion, some programming languages such as FORTRAN, COBOL, Pascal, etc. allow programmers to declare variables with ***type*** *real* in their programs for computations. To make it clear to all computer users, we must study the algebraic structures for the sets of all *real numbers*, *model numbers*, and *rough numbers* over computer systems. This not only provides the theory foundation for computer arithmetic but also interprets rounding-errors from an algebraic viewpoint.

The Ada programming language [3, 6, and 16] calls a real number that can be stored exactly in a computer system with a radix of power 2 a *model number*. In 1982, Pawlak proposed a concept called *rough sets* that are used in the theory of knowledge for classification of features of objects [13]. He considered X to be a set and R to be a relation over X. The pair (X, R) is called *rough space*, and R is called the *rough relation* if xRy, then one could say that x is too close to y, x and y are indiscernible, and x and y belong to the same elementary set. The concept of rough sets is usually used in knowledge representation. In 1994 the author proposed a new class of real numbers that is based on *rough set theory* and is called *rough numbers* [18]. We will show that the set of all rough numbers over the set of all computer systems has the dyadic number structure.

Corresponding author: Tel.: 1-618-650-2393, Fax: 1-618-650-2555
E-mail: twu@siue.edu

In addition, we will revise the ordinary interval computation given by Moore [11] from a short compact interval $[a, b]$ with real numbers a and b to a closed-open interval $[a, b)$ with computer exactly represented numbers a and b for all initial intervals in computation. The ordinary interval could have infinite many points, at the upper bound of the interval, more than the revised closed-open interval. To implement this, we consider each real number x within the computer range, take a to be the downward rounding value of x, s to be the smallest positive number that the computer can provide, and take $b = a + s$. Thus $[a, b)$ is the shortest closed-open interval for x on the given computer system. In this way, we guarantee that this new method will provide the shortest resulting interval. To study the computation of this large class of numbers, it is important to study the structure of *rough numbers* with respect to a given computer system.

2. The structure of real numbers

The Ada programming language calls a real number that can be stored exactly in a computer system with a radix of power 2 a *model number*. Therefore, the set of all *model numbers* is a subset of *dyadic numbers* (the definition of a *dyadic number* is given in the Section 3), and we call it the set of *limited dyadic numbers* with respect to a given computer system, *M*. The word "*limited*" is subject to the given computer system's word size for the *floating-point numbers*. The concept of a *rough number* is that for a *real number* that is not a *model number* and that cannot be stored exactly in a given computer system with respect to the given computer system, *M*. Therefore, *a* could be a *rough number* for machine M_1, but not necessarily a rough number for machine M_2. For simplicity, most computer systems use a *downward rounding* policy to approximate a *rough number* by a *model number*. Thus, there are infinitely many *rough numbers* approximated by the same *model number*. Because of this fact, we will use a *downward rounding* policy to handle the approximation of *rough numbers* throughout this paper. The formal definition of *rough numbers* with respect to a given machine *M* is quote here:

Definition 1 Let E_M be the set of all real numbers within the range of a given computer system *M*, and we will define a binary relation *R* on E_M. For $x, y \in E_M$, we define $(x, y) \in R$, if *x* and *y* belong to the same smallest closed-open *model interval,* say $[a, b)$, where *a* and *b* are model numbers. Then, we say that *x* and *y* are *rough numbers* in the $[a, b)$. The pair (E_M, R) is called *rough space*, and *R* is called the rough relation over E_M [18].

This definition states that *x* and *y* are *rough numbers*, and they belong to the same smallest closed-open interval $[a, b)$ that the machine *M* can provide. Both *x* and *y* are approximated to the same real number *a* that can be exactly represented by the machine *M* for computation. It is known that $|a - x|$ is called the *rounding-error* of *x* on *M*. It is then easy to verify that *R* is a binary relation over E_M, and that *R* partitions E_M into a collection of smallest closed-open *model intervals*. These smallest closed-open *model intervals* are equivalence classes of the rough relation *R* with respect to given machine *M* [16]. The family of all these intervals of (E_M, R) is the quotient set E_M/R. The shortest closed-open *model intervals,* each containing infinitely many rough numbers and one model number--the lower bound, are one dimensional rough sets given by Pawlak [13], and the interval numbers given by Yao [19].

In fact for any rough number *x* whose value is within the range of a given computer system *M*, there exists a smallest closed-open interval

$$\left[i+\frac{k-1}{2^n}, i+\frac{k}{2^n}\right),$$

with respect to a given machine *M*, such that

$$x \in \left[i + \frac{k-1}{2^n}, i + \frac{k}{2^n} \right), \tag{1}$$

for some positive integer n, where the lower bound is the approximation of x, i is the integer part of x, k is an integer with $0 \le k \le 2^n - 1$, and $n = n(b_e, b_m)$, a function of the number of bits in the exponent b_e, and the number of bits in the mantissa b_m for a floating-point number format.

The difference between the rough number x and its approximation given in Eq.(1) is

$$x - \left(i + \frac{k-1}{2^n} \right) \tag{2}$$

which is called the *rounding-error* of x. Often, a programmer is either unaware of the existence of rounding-errors or unable to reduce or control them in the course of a computation.

It is known that the set of all real numbers, denoted E, together with arithmetic addition and multiplication, + and ∗, forms a *field* (see appendix for definition). Human arithmetic, + and ∗, over the set E always returns exact results. However, computer systems can store only certain real numbers exactly in memory. For example 100.5, 70.725, 12.375, 0.5, 87.125, etc. are such numbers. On the other hand, real numbers such as 0.1, 0.2, 0.3, 0.4, 0.6, etc. cannot be stored in computer system exactly. Actually, they are *rough numbers*.

The operations used, within a program, for usual addition, +, and multiplication, ∗, are not the same operations for real numbers. Even if we allow a computer system to have as many bits as required to store a floating-point number, not all non-zero numbers in the computer system will have a multiplicative inverse. Therefore, these numbers do not constitute a *field.*

The structures of the set of all real numbers and the set of all the numbers in the computer system are quite different. They are not *isomorphic*. So the computations of + and ∗ over the set of all numbers represented by a computer system could induce some errors. Thus, we need to find out: (1) what is the actual structure of computer represented numbers and (2) what kind of arithmetic can computer systems do? To answer these two questions, we will study the set of *dyadic* numbers.

3. The structure of dyadic numbers

We will begin with a *b-adic* expansion for a non-negative real number x, where b is a prime. Generally speaking, we want to write a real number x as the sum of multiples of powers of b, where the multiples are non-negative integers less than b. Clearly, the *b-adic* expansion of the number may fail to be unique in a decimal expansion. Decimals .9999. . . (all nines) and 1.000 . . . (all zeros) are expansions of the same real number. When $b = 2$, the *b-adic* expansions are then called *dyadic* expansions [8], and numbers which can be written as *dyadic* expansions are called *dyadic* numbers. Today, most computer systems use the floating-point number format to store real numbers. Each of these numbers is represented by a dyadic expansion. The following Theorem is given by Kelley [8].

Theorem 1 For each real number x, we have its *dyadic* expansion over the finite field $(B; +, *)$:

$$x = sign \sum_{i=-\infty}^{\infty} a_i 2^i , \tag{3}$$

where $a_i \in B = \{0, 1\}$ and $sign \in \{+,-\}$.

A computer system is a finite state machine; therefore, it is capable of representing only a finite set numbers of internally. Thus, any attempt to use a digital computer to do arithmetic in the set of all real numbers is doomed to failure. The set of all real numbers is an infinite set and most of the elements in the set cannot be represented in a computer. For theoretical reasons, we may assume that a computer system can have any finite number of bits to store its numbers, integers and floating-point numbers. For any non-negative integer n with respect to a given computer system, we consider the set of all numbers with the representation:

$$y = sign \sum_{i=-n}^{n} a_i 2^i , \tag{4}$$

where $a_i \in B$ and $sign \in \{+,-\}$.

This representation contains two parts. The first part is with positive indices $0 \leq i \leq n$, and the second part is with negative indices $-1 \leq i \leq -n$. Actually, these two parts are the integer part of y and fraction part of y, respectively. In this paper, we will define a structure for the set of all numbers provided by the set of all computer systems, including possible virtual computers. The representation is denoted by D and called the set of all *finite* (*term*) *dyadic numbers*:

$$D = \{\ y \mid y = sign \sum_{i=-N}^{N} a_i 2^i ,\ a_i \in B,\ sign \in \{+,-\},\ \text{and } N \text{ is a non-negative integer}\ \}. \tag{5}$$

The addition, subtraction, multiplication, and division of dyadic numbers are defined in [10]. The actual implementation of the arithmetic over computer systems is slightly different from one computer to another.

In fact, each computer system can have only a predetermined number of bits of memory space to store a number of its *type* such as *integer*, *float*, *double*, . . . , etc. If the set of numbers described in the Theorem 1 is not the set of numbers in one computer, they are the numbers that can be represented by some other computer. An *integer* within the computer predefined range often can be represented exactly. However, a real number within the range usually cannot represented correctly. Today, most computer systems implement the *IEEE floating-point number format* [7] for storing real numbers. For example, a 32 bit *floating-point number* format is divided into three areas: a sign bit, an 8 bit exponent, and a 23 bit mantissa:

$$sign\ (1.M) \times 2^{E\text{-}127}, \tag{6}$$

where $sign = 0$ indicates a positive value, $sign = 1$ indicates a negative value, 1 is a hidden bit, M is a 23 bit mantissa, and exponent E is with range $0 \leq E \leq 255$. It is clear that a floating-point number represented in Eq.(6) can be written as a *finite* (*term*) *dyadic number* given in the set D in Eq.(5). Most computer systems use a limited number of bits to represent a floating-number. Therefore, ring computation, + and $*$, can cause an overflow or underflow. Many computer systems use floating-point number arithmetic for numerical computation and many computer users are not aware that floating-point number arithmetic often can create rounding-errors during the course of a computation. These errors are unavoidable. In fact, the output of a numerical computation program executed on one machine can be different from the output of the same program from another machine.

4. Hierarchical structures and computations

In mathematics, we have learned properties of several classes of numbers such as *integers*, *rational numbers*, *irrational numbers*, and *real numbers*. However, a computer system can have only a limited word size; therefore, only a part of integers and a part of rational numbers can be computed correctly in a computer system. Other numbers cannot even be represented exactly in a computer system. Computer systems handle real numbers in a binary fashion. For each real number x, we have its dyadic expansion over the finite field $(B; +, *)$. Among all the dyadic numbers, only a limited number of dyadic numbers can be computed in a computer system. The relationship of the set of real numbers and the hierarchical structures of *dyadic numbers* is shown in *Figure 1* below:

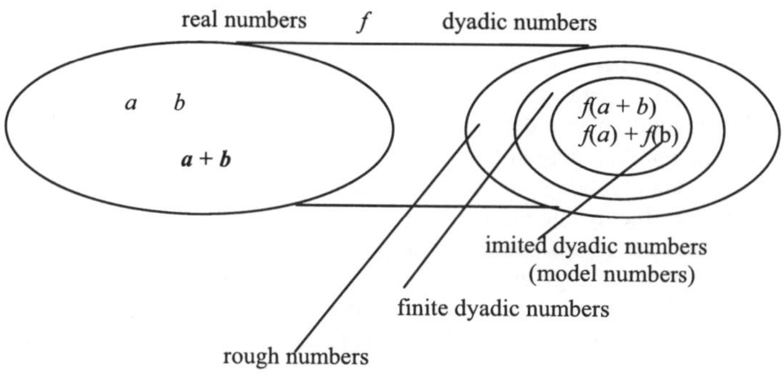

Figure 1 Hierarchical structures of dyadic numbers

On the set of real numbers, we perform a *field* computation which is an exact computation. However, when a real number within the given range is stored or read into a computer system, it is converted into a *dyadic* number. The computation over the set of *limited dyadic numbers* is the *dyadic number* computation. Let f be a mapping that takes each real number into its *dyadic representation*. Consider a and b to be two real numbers within a given range mapped into their *dyadic* representations $f(a)$ and $f(b)$ respectively. In general, we should have

$$f(a+b) \neq f(a) + f(b). \tag{7}$$

The set of real numbers has a *field* structure, while the set of *finite dyadic numbers* is not a *field*; it is a subset of *ring*. f does not preserve the structure, and f cannot even be a local homomorphism between the set of real numbers within a given range of machine M and the set of *limited dyadic numbers*. Hence the The addition '+' (or multiplication '∗') on the left hand side of Eq.(7) is a *field* addition and the addition '+' on the right hand side is a *ring* addition. The addition '+' is overloaded from *field* addition to the *dyadic number* addition. To avoid the overloading and possible confusion, we will introduce a new addition '⊕' for the *dyadic number* addition. To adjust the inequality in Eq.(7) we will add in an *error* term, *Err*:

$$f(a+b) = f(a) \oplus f(b) + Err. \tag{8}$$

This *error* term, *Err*, is called *rounding-error*. Unfortunately, most programming textbooks do not teach students the addition '+' in program code is not the usual real number addition, but a *dyadic number* addition. Many computer users are misled and disappointed that the computer system cannot provide accurate results for computations.

In the next section, we will suggest a new computational method for the set of all rough numbers called model interval computation that can provide verification of the magnitude of the absolute error.

5. The model interval computation

In numerical computation, computer calculations are often based on input values from the user. Usually, the input data are either integers or real numbers or combinations of both. The user expects the result to be accurate. Unfortunately in many cases, the computer system can only return approximate values as output, though strictly integer computation always provides an exact answer. What is the problem? We have found that the basic problem is that computer systems do not perform real number computations in the same way as a human does.

In recent years, researchers introduced a new idea for performing computer arithmetic with compact intervals of real numbers instead of the direct use of scalars [1, 2, 4, 9, 13, 14, and 15]. We revise ordinary interval computation to a dyadic interval computation which replaces the original interval real endpoints with dyadic number endpoints and the closed interval with a closed-open interval. In ordinary interval computation, the real endpoints usually cannot be represented by a computer system exactly. In the closed-open interval, the dyadic number endpoints are always exact numbers represented by computer systems. Furthermore, the closed-open interval with dyadic number endpoints always provides the user with the shortest interval in a given computer system. Also, the closed interval with real number endpoints will have infinitely many more points than the closed-open interval. Therefore, the closed-open initial interval will provide the user final computation results with the smallest maximum possible error in any given machine.

In addition, the Ada language, for example, permits the user to declare new floating-point number types [16], with various precision, as follows:

type *typename* **is digit** *dd;*

This new type "*typename*" is a *floating-point number type* with *dd* decimal digits (up to the system limitation) of precision. Moreover, the Ada language permits users to declare his/her own fixed-point number types [16], with various error limits, as follows:

type *typename* **is delta** *maxerror*;

The new type is a *fixed-point number type* with maximum error *maxerror*. The *maxerror* determines the number of bits used in the mantissa for computation. Therefore, users can define "much rough" rough numbers or "less rough" rough numbers for their computation needs. This way, we are able to guard against the accumulation of computation error due to rounding-errors which are created by all floating-point number operations.

For any real numbers x and y, there exist basic model intervals $X = [a_1, b_1]$ and $Y= [a_2, b_2]$ such that $x \in [a_1, b_1)$ and $y \in [a_2, b_2)$ where a_1, b_1, a_2, and b_2 are model numbers. For a real number z that is a model number, we consider a special case $Z = [z, z_1]$, where z_1 is the smallest model number greater than z. Then, the usual computer scalar arithmetic operations, addition +, subtraction −, multiplication ∗, and division /, are defined for interval computation. Therefore, the sum, difference, product, and quotient of two real numbers are all within a given interval respectively. We use revised closed-open intervals to exclude the upper endpoint for interval computation. The results are shown here:

Addition

$$x + y \in [[a_1, b_1] + [a_2, b_2]\,) = [a_1 + a_{2,},\ b_1 + b_2) \tag{9}$$

Subtraction

$$x - y \in [[a_1, b_1] - [a_2, b_2]\,) = [a_1 - b_{2,},\ b_1 - a_2) \tag{10}$$

Multiplication

$$x * y \in [\ min(a_1a_2,\ a_1b_2,\ b_1a_2,\ b_1b_2),\ max(a_1a_2,\ a_1b_2,\ b_1a_2,\ b_1b_2)) \tag{11}$$

Division

$1/y \in [1/b_2, 1/a_2)$, if $0 \notin$ B, then we define

$$x/y \in [min(a_1/b_2, a_1/a_2, b_1/b_2, b_1/a_2), max(a_1/b_2, a_1/a_2, b_1/b_2, b_1/a_2)) \quad (12)$$

The deficiency of interval arithmetic is the use of real numbers for the lower bound and upper bound for the interval. In general, a real number cannot be represented exactly by the floating-point number format. A better way to perform interval computation and reduce rounding-error is (1) to create the smallest closed interval, containing the given real number, with model numbers for the lower and upper bounds, and (2) to develop bit-to-bit computation over the mantissa, the internal representation of model numbers.

Since the product of two model numbers or the inverse of a model number might not be a model number, we round the result outward to the nearest model numbers. Let $m(x)$ be the greatest model number less than x for the lower bound and $m(y)$ be the smallest model number greater than y for the upper bound. Therefore, we redefine multiplication and division as follows:

Multiplication

$$x * y \in [m(min(a_1a_2, a_1b_2, b_1a_2, b_1b_2)), m(max(a_1a_2, a_1b_2, b_1a_2, b_1b_2))) \quad (13)$$

Division

$1/y \in [1/b_2, 1/a_2)$, if $0 \notin B$, then we define

$$x/y \in [m(min(a_1/b_2, a_1/a_2, b_1/b_2, b_1/a_2)), m(max(a_1/b_2, a_1/a_2, b_1/b_2, b_1/a_2))) \quad (14)$$

Some algebraic properties of interval arithmetic are given by Moore [12] and Yao [19]. A mathematical theorem that supports interval computation was given by Moore [11]. In 1966, Moore proved an important theorem that supports interval computation, and since then interval computation has become a new and growing branch of applied mathematics. We will call this theorem the fundamental theorem of interval computation. It is necessary to state the theorem here to support our work.

Theorem 2 The Fundamental Theorem of Interval Computation: Let $f(x_1, x_2, \ldots, x_n)$ be a rational function of n variables. Consider any sequence of arithmetic steps which serve to evaluate f with given arguments $x_1, x_2, \ldots, x_n$. Suppose we replace the arguments x_i by the corresponding closed interval X_i ($i = 1, 2, \ldots, n$) and replace the arithmetic steps in the sequence used to evaluate f by the corresponding interval arithmetic steps. The result will be an interval $f(X_1, X_2, \ldots, X_n)$. This interval contains the value of $f(x_1, x_2, \ldots, x_n)$ for all $x_i \in X_i$ ($i= 1, 2, \ldots, n$).

6. Examples

In this section, we will present two examples in solving non-linear equations by using interval conputation. One is using Newton method and the orther is using iterative method. The computation is done with an Ada Interval Mathematics Package [17].

(1) Newton Method

```
--| This program uses interval method (Interval_Math package) and
--| Newton method to solve equation:
--|     x**3 + 2(x**2) - 7 = 0

--| Let  f(x)  =  x**3 + 2(x**2) - 7 = 0
--|
--|       x   |      -3        -2        -1        0        1        2
--|     ______|_______________________________________________________
--|      f(x) |      -16       -7        -6        -7       -4       9
--|
```

```
--| From the above table: f(1) = -4
--|                       f(2) = 9
--| For f(x) = 0,    1 < x < 2 ,   x lies in (1,2)

--| To obtain c that lies in (a, b)
--|   Since,     f(2) > 0
--|              f"(x) = 6x + 4 > 0
--|   therefore,
--|              b = 2
--|              c = b - f(b)/f'(b)

with system, Text_Io, Interval_math;
use Text_Io;

procedure newton_1 is

  type My_Real is digits System.Max_Digits;
  package Real_Io is new Text_Io.float_Io (My_real);
  package Int_Io is new Text_Io.integer_Io (integer);
  package My_Real_Math is new Interval_Math(Real => My_Real);

  use Real_Io, Int_Io, My_Real_Math;

  a    : model_interval := interval(1.0);
  b    : model_interval := interval(2.0);
  x, f_b, f_d_b    : model_interval;

begin
    --put("a = ");    put_line(a);
    loop
        --put("b = ");    put_line(b);
        f_b := Poly_3(b, 1.0, 2.0, 0.0, -7.0);
        f_d_b := Deriv_3(b, 1.0, 2.0, 0.0, -7.0);
        x := b - f_b / f_d_b;
        put("x = ");    put_line(x);
        exit when abs(average_bound(x-b)) <= Float'Epsilon;
        b := x;
    end loop;

end newton_1;
```

This equation has one real root; the approsimated solutions for the root are given in the following decimal and binary values:

```
x = Lower_Bound : 1.54999999999999980460074766659724E+00
                 (1000110011001100110011001100110011001100110011001100010)
    Upper_Bound : 1.55000000000000020428103653102888E+00
                 (1000110011001100110011001100110011001100110011001101011)
x = Lower_Bound : 1.43596867424948260172357095143 5E+00
                 (0110111110011011101001001001110111111101101100100101)
    Upper_Bound : 1.43596867424949170555237287771 9E+00
                 (0110111110011011101001001001110111111101101100111010)
x = Lower_Bound : 1.42884470939842689318766133510 5E+00
                 (0110110111001000110001000101000111101101110010100 10)
    Upper_Bound : 1.42884470939844598902368488779 8E+00
                 (0110110111001000110001000101000111101101110011111 01)
x = Lower_Bound : 1.42881770216864167899473159195 6E+00
                 (0110110111000110111111110011011011000010100110110 11)
    Upper_Bound : 1.42881770216868053680059347243 5E+00
                 (0110110111000110111111110011011011000010101001100 11)
x = Lower_Bound : 1.42881770178134304138950483320 5E+00
                 (0110110111000110111111110011010100011000110001001 00)
    Upper_Bound : 1.42881770178142142313504336925 7E+00
                 (0110110111000110111111110011010100011000110110101 01)
```

(2) Iterative Method

```
--| This program uses interval method (Interval Math package) and iterative
--| method to solve equation     x**7 - x - 0.2 = 0;

--| Let x**7 = x + 0.2 or x = x**7 - 0.2
--| If graph functions y = x**7 and y = x + 0.2, three intersections will be
--| found. They are closed to points 1, 0, and -1.

--| In this program, the method is used to get value of x in x**(1/7) is:
--|    Let z = x**(1/7) ==> ln(z) = 1/7 * ln(x) ==> z = e**(1/7*ln(x))
--|

--| To compute the points closed to 1, -1
--|    let f1 = x(n)**7  and f2 = x(n) + 0.2  (n = 0, 1, 2, ...)
--|    therefore, x(n+1) = [ x(n) + 0.2 ] ** (1/7)
--| To compute the point closed to 0
--|    let f1 = x(n) and f2 = x(n)**7 - 0.2  (n = 1, 2, 3,...)
--|    therefore, x(n+1) = x(n)**7 - 0.2


with system, Text_Io, Interval_math;
use Text_Io;

procedure iter_1 is

  type My_Real is digits System.Max_Digits;
  package Real_Io is new Text_Io.float_Io (My_real);
  package Int_Io is new Text_Io.integer_Io (integer);
  package My_Real_Math is new Interval_Math(Real => My_Real);

  use Real_Io, Int_Io, My_Real_Math;

  x1, x2, y : model_interval;
  i : integer := 1;


begin
        new_line;put_line ("First solution ********** close to 1.0");new_line;
        x1 := interval(1.0);
        put ("x = ");   put_line (x1);
        loop
           y := x1 + interval(0.2);
           --put ("y = ");   put_line (y);
           y := ln_x(y);
           y := interval (1.0/7.0) * y;
           x2 := exp(y);
           put ("x = ");   put_line (x2);
           exit when abs(Average_bound(x1)-Average_bound(x2))<=Float'Epsilon;
           x1 := x2;
        end loop;

        new_line;put_line("Second solution ********** close to 1.0");new_line;
        x1 := interval(-1.0);
        put ("x = ");   put_line (x1);
        loop
           y := x1 + interval(0.2);
           --put ("y = ");   put_line (y);
           y := interval(-1.0) * y;
           y := ln_x(y);
           y := interval (1.0/7.0) * y;
           x2 := exp(y);
```

```
        x2 := interval (-1.0) * x2;
        put ("x = ");   put_line (x2);
        exit when abs(Average_bound(x1)-Average_bound(x2))<=Float'Epsilon;
        x1 := x2;
     end loop;

     new_line;put_line ("Third solution ********** close to 0.0");new_line;
     x1 := interval (0.0);
     put ("x = ");   put_line (x1);
     loop
        y := x1 ** 7;
        --put ("y = ");   put_line (y);
        x2 := y - interval(0.2);
        put ("x = ");   put_line (x2);
        exit when abs(Average_bound(x1)-Average_bound(x2))<=Float'Epsilon;
        x1 := x2;
     end loop;

end iter_1;
```

This equation has three real roots; the approsimated solutions for these three roots are given in the following decimal and binary values:

```
First solution ********** close to 1.0

x = Lower_Bound : 9.99999999999999977795539507496 87E-01
                       (111111111111111111111111111111111111111111111111111)
    Upper_Bound : 1.00000000000000044408920985006 3E+00
                       (000000000000000000000000000000000000000000000000001)
x = Lower_Bound : 1.02638809625703908423588472942 4E+00
                       (000001101100000101011110110010100110110101111100101)
    Upper_Bound : 1.02638809625704019445890935458 0E+00
                       (000001101100000101011110110010100110110101111100111)
x = Lower_Bound : 1.02958245299760098312447098578 6E+00
                       (000001111001001010110111001101000010100011111111110)
    Upper_Bound : 1.02958245299760209334749561094 2E+00
                       (000001111001001010110111001101000010100100000000001)
x = Lower_Bound : 1.02996513169903280981998250354 1E+00
                       (000001111010101111001011011111001010101011101101000)
    Upper_Bound : 1.02996513169903392004300712869 7E+00
                       (000001111010101111001011011111001010101011101101010)
x = Lower_Bound : 1.03001091880528949928930160240 3E+00
                       (000001111010111011001011101010101100101010101000010)
    Upper_Bound : 1.03001091880529060951232622755 9E+00
                       (000001111010111011001011101010101100101010101000100)
x = Lower_Bound : 1.03001639636672259747740554303 2E+00
                       (000001111010111100100111100100001011110100100101000)
    Upper_Bound : 1.03001639636672370770043016818 8E+00
                       (000001111010111100100111100100001011110100100101011)
x = Lower_Bound : 1.03001705164166290629168543091 5E+00
                       (000001111010111100110010100011110001111110001111101)
    Upper_Bound : 1.03001705164166401651471005607 2E+00
                       (000001111010111100110010100011110001111110001111111)

Second solution ********** close to -1.0

x = Lower_Bound :-1.00000000000000044408920985006 3E+00
                       (000000000000000000000000000000000000000000000000001)
    Upper_Bound :-9.99999999999999977795539507496 87E-01
                       (111111111111111111111111111111111111111111111111111)
x = Lower_Bound :-9.68625085926998363738960051705 4E-01
                       (111011111110111110100000100101000100100010000110000)
    Upper_Bound :-9.68625085926996587382120651454 9E-01
```

```
                          (1110111111110111110100000100101000100100010000101000)
x = Lower_Bound :-9.631047123211229354922124912264E-01
                          (1110110100011100000011111001010000010101110111110010)
    Upper_Bound :-9.631047123211209370907681659446E-01
                          (1110110100011100000011111001010000010101110111101001)
x = Lower_Bound :-9.621134922871522610066108427418E-01
                          (1110110010011010001000111100000011010011110011100001)
    Upper_Bound :-9.621134922871503736274689799757E-01
                          (1110110010011010001000111100000011010011110011011001)
x = Lower_Bound :-9.619348615900636945141854994290E-01
                          (1110110010000010101110011110011011010010101111110100)
    Upper_Bound :-9.619348615900618071350436366629E-01
                          (1110110010000010101110011110011011010010101111101100)
x = Lower_Bound :-9.619026488478590319886052384390E-01
                          (1110110001111110100000010000010101111001100000010111)
    Upper_Bound :-9.619026488478571446094633756729E-01
                          (1110110001111110100000010000010101111001100000001110)
x = Lower_Bound :-9.618968391867195322220140951686E-01
                          (1110110001111101101111100001010011011101100011111000)
    Upper_Bound :-9.618968391867174227982673073711E-01
                          (1110110001111101101111100001010011011101100011101110)
x = Lower_Bound :-9.618957913751355892628680521739E-01
                          (1110110001111101100110101110110000111011100000100101)
    Upper_Bound :-9.618957913751335908614237268921E-01
                          (1110110001111101100110101110110000111011100000011100)
x = Lower_Bound :-9.618956023945179900636048841989E-01
                          (1110110001111101100101001001010011100110101111110110)
    Upper_Bound :-9.618956023945159916621605589171E-01
                          (1110110001111101100101001001010011100110101111101101)

Third solution ********** close to 0.0

x = Lower_Bound : 0.000000000000000000000000000000E+00
                          (0000000000000000000000000000000000000000000000000000)
    Upper_Bound : 9.881312916824930883531375857364E-324
                          (0000000000000000000000000000000000000000000000000001)
x = Lower_Bound :-2.000000000000000666133814775094E-01
                          (1001100110011001100110011001100110011001100110011110)
    Upper_Bound :-1.999999999999999555910790149937E-01
                          (1001100110011001100110011001100110011001100110011100)
x = Lower_Bound :-2.000128000000000738634042818376E-01
                          (1001100110100000010011111001011001000000001001000111)
    Upper_Bound :-2.000127999999999628411018193219E-01
                          (1001100110100000010011111001011001000000001001000101)
x = Lower_Bound :-2.000128057355011756968110603339E-01
                          (1001100110100000010100000101101101010010010010010010)
    Upper_Bound :-2.000128057355010646745085978182E-01
                          (1001100110100000010100000101101101010010010010001000)
```

7. Conclusion

In this paper, we have defined the concepts of *rough numbers*, *dyadic numbers*, *finite dyadic numbers*, *limited dyadic numbers* and *model numbers*. We also created a modified interval arithmetic so that each rough number can be easily fitted into a *model interval*. Therefore, the model of interval arithmetic can be used for the computation of *rough numbers. Interval computation* is a *self-validating computation* that provides users with a set of possible results and an absolute error for those results.

In addition, we have revised interval arithmetic from real number ending points to model numbers and from a closed interval to a closed and open interval. This will eliminate some possible rounding-errors, create the shortest initial intervals for each initial real number, and ensure the rounding-error for the computation is a minimum in any given machine. In this paper, we have successfully redefined overloading for +, – , ∗, and / on internal binary representations so that we can create the lower bound and the upper bound in model numbers for a given real number. Also, we have carefully developed bit-to-bit model interval computation. For each initial real number, we have created the shortest model interval to include the real number for computation with the required precision. This is a very important step because only with the shortest initial intervals can one obtain the final computational result with minimum error for the required precision.

Acknowlegment
I would like to thank Wenli Zhang who helped in developing of programs for computation.

References

[1] Aberth, O., *Precise Numerical Analysis*, Dubuque: Wm C Brown Publishers, 1988.
[2] Alefeld, G. and Herzberger, J., *Introduction to interval Computations*, New York: Academic Press, 1983.
[3] Barnes, J. G. P., *Programming Language in Ada*, 4th Ed. Addison-Wesley Publishing Company, 1992.
[4] Corliss, G. F. and Rall, L. B., "Adaptive, Self-Validating Numerical Quadrature," *SIAM Journal on Scientific and Statistical Computing* 8, 831-847, 1987.
[5] Herstein, I. N., *Topics in Algebra*, University of Chicago, 1971.
[6] IBM AIX *Ada/6000 User's Guide*, IBM Canada Ltd. Laboratory, North York, Ontario, Canada, M3C 1W3,1992.
[7] IEEE inc., *IEEE Standard for Binary Floating-point Arithmetic* (ANSI/IEEE std 754-1985), New York, 1985.
[8] Kelley, J. L., *General Topology*, D. Van Nostrand Company, Inc. 1955.
[9] Kennedy, W. J., Special Purpose Numerical Tools for Approximating Functions, *Statistical Computing and Statistical Graphics Newsletter*, pp. 3-6, May 1990.
[10] Mahler, Kurt, *P-adic Numbers and Their Functions*, Cambridge University Press, Cambridge, England, 1981.
[11] Moore, R. E., *Interval Analysis*, Prentice-Hall, Englewood Cliffs, New Jersey, 1966.
[12] Moore, R. E., *Methods and Applications of Interval Analysis*: SIAM Studies in Applied Mathematics, 2, Philadephia: Society for Industrial and Applied Mathematics, 1979.
[13] Pawlak, Z., Rough sets, *International Journal of Computer and Information Sciences*, V. 11, No. 5, 341-356, 1982.
[14] Ratschek, H. and Roken, J., *Computer Methods for the Range of Functions*, Chichester, England: Ellis Hoewood Limited, 1984.
[15] Wang, M. and Kennedy W. J., An Numerical Method for Accurately Approximating Multivariate Normal Probabilities, *Computational Statistics & Data Analysis* 13, 197-210, 1992.
[16] Watt, D. A., Wichmann, B. A., and Findlay, W., *Ada Language and Methodology*, Prentice-Hall, Englewood Cliffs, New Jersey, 1987.
[17] Wu, T, Ada Interval Mathematics Package: An unpublished software package.
[18] Wu, T., Rough Number Structure and Computation, *Proceedings of the Third International Workshop on Rough Sets and Soft Computing*, pp.360-367, 1994.
[19] Yao, Y. Y., Interval-set Algebra for Qualitative Knowledge Representation, *Proc. of the Fifth International Conference on Computing and Information*, 370-374, Sudbury, Ontario, Canada, 1993.

The Study of 3-D Surface Reconstruction in Digital Image Processing

Zhong Qu[a,b]

[a]*College of Computer Science and Technology, Chongqing University of Posts and Telecommunications, 400065,Chongqing, China*
[b]*College of Computer Science, Chongqing University, 400044,Chongqing, China*

Abstract

3-D surface reconstruction has been an important research topic in digital image processing for many years. Reconstruction from a few views is an ill-posed problem. To reduce uncertainty, the solution has to be regularized by incorporating some a priori information. This is the solution generally adopted for reconstruction methods described in the literature. Firstly, images are processed to remove air annulus. Then volume rendering is used to reconstruct the workpiece in Matlab environment. The cross section image resulted from the 2-D convolution projection reconstruction is the original image for 3-D reconstruction, which is image file. The noise caused by converting paper or film image into digital image is reduced in this way. Furthermore the additional step to match images is avoided by applying the fan-beam convolution projection algorithm that determines whether the images are matched. 3-D reconstruction and quantitative analysis can significantly improve the accuracy and reproducibility of the measurement of lesions and the estimates of luminal narrowing and geometry that characterize the hemodynamic extent and functional consequences of these lesions.

Keywords: 3-D surface reconstruction; Volume rendering; Edge detection

1. Introduction

Alternatively, it is possible to estimate the 3-D object shape from its already available various 2-D views so that storing 2-D image information from many different views becomes unnecessary. Rather, a 3-D structural model of the whole object is kept together with the texture information attached to it. Moreover, the object can be viewed from any angle regardless of the acquisition process, and displayed through standard rendering techniques [1]. Another source of information to capture 3-D object shape is the direct 3-D information acquired actively by laser range scanners or other coded light projecting systems. Such 3-D data acquisition systems can be very precise, but have several drawbacks. Most of them are very expensive, and they require special skill and know-how for the acquisition process itself. They often necessitate specific environmental conditions, and do not perform well with objects made of materials absorbing light such as fur or velvet, or when the object surface is very shiny. In addition, these scanners concern more with shape rather than with texture. Even regarding the shape only, they require very sophisticated approaches for 3-D matching and data fusion of different surface patches that are scanned separately, to reconstruct the object as a whole. Only few scanners are capable of recording 3-D shape information concurrently with color texture. For those with this capability, color information is incorporated either by an RGB camera or by using three different laser wavelengths during the 3-D

The corresponding author. E-mail address: quzhong@cqupt.edu.cn (Zhong Qu)

acquisition process. In the first case, the color acquisition device is usually different from the 3-D one [1]. The color texture and the 3-D data, being acquired with two different geometrical configurations, still have to be registered. In the second case, the color and the 3-D information are perfectly registered, but the obtained colors are not colorimetrically faithful since the spectral reflectance of the object surface is sampled only at three wavelengths [1]. The reconstruction of 3-D models involves four major steps: data acquisition, registration, surface integration, and texture mapping. People expect direct image of 3-D with the richer information. Therefore, 3-D reconstruction [2]-[3] is a hot topic. 3-D reconstruction can be divided into 2 groups: direct 3-D reconstruction and 3-D reconstruction from series of -D cross section images. The former gets the density distribution of 3-D object using 2-D projections acquired from horizontal detector. The latter gets the 3-D image by processing cross-sectional images [4] using computer graphics, image process and visualization methods [5]. Reconstruction from a few views is an ill-posed problem. To reduce uncertainty, the solution has to be regularized by incorporating some a priori information. This is the solution generally adopted for reconstruction methods described in the literature. The proposed methods differ either in their specific application or in the techniques they use [6]-[13].

2. Theory of reconstruction

3-D reconstruction with cross-sectional images is to reconstruct the 3-D shape of the object with a series of cross-sectional images. It mainly involves the acquisition of the cross section image, image interpolation, rendering and data compression.

It describes the complete reconstruction system that we have built by combining different techniques and addresses the details of the system components which determine the resulting reconstruction quality. The accuracy of the reconstructed object shape and the quality of the texture attached on it have been the main concerns during the development of the system presented here, for which the most limiting factor is the original image quality [1]. The precision of vertex positions is not limited to the resolution of the constructed surface model, rather it is determined by the resolution of available images. The octree representation and the marching cubes algorithm have been used in conjunction before for the general object reconstruction problem, however the use of these techniques in the case of the shape from silhouette problem necessitates specific considerations [1]. A robust background removal technique which is quite general and independent of the object type or the lighting conditions is proposed [1]. A particle-based texture mapping method is developed, that adequately addresses the photography related problems. In particular, the problem of highlights, which has been ignored in other works, is taken into account and handled during the texture mapping process [1].

There are 2 methods to acquire cross-sectional images. One is from film or the scanned printed image. The other is direct from image data. In this experiment, the second method is adopted. Our reconstruction method is passive, the only information needed being 2-D images. The cross section image resulted from the 2-D convolution projection reconstruction is the original image for 3-D reconstruction, which is image file. In this way the noise caused by converting paper or film image into digital image is reduced. Besides, fan-beam convolution projection algorithm determines that the images are matched, which avoids the step to match images. Figure 1 is mainly about the research of obtaining of 3-D vexel data.

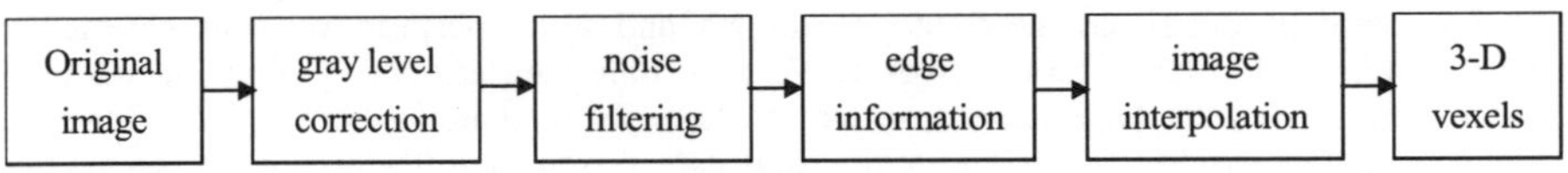

Fig. 1. The flow chart of 3-D data

The 3-D reconstruction of cross sectional images is implemented in Matlab environment. Matlab software has provided various functions for matrix computation, image operation and image display and image processing, such as image display function imshow, getimage, input and output function imread, imwrite, geometric operation function imcrop, imresize, imrotate and image analysis function edge, qtgetblk. Thus Matlab has been widely used in image processing area. Its powerful image processing functions make the image processing more convenient and effective.

3. Edge detection

Figure 2 shows the piecework consists of several different materials represented by different gray levels. Different algorithms are designed to peel the external material.

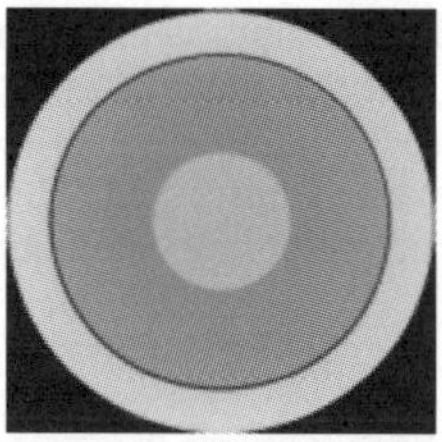

Fig. 2. The computed tomography of piecework 1

Algorithm 1: Measure the gray level range (a, b) of the external material from 2-D image. Then scan the image from the left to the right on horizontal line and read every pixel gray level value. If it is in the range of (a, b), set its value as background gray level value. Stop scan until the value is not in the range of (a, b), which means it is internal material pixel. Then scan from the right to the left along the same line until it is internal pixel. Then scan next line with the same method until all the external material is scanned. This algorithm skips the process for internal material.

The disadvantage of this algorithm is that it has low efficiency when calculating the gray level range (a, b) of external material. This inconsistence can cause big difference gray level of same material among different images. For example, in one image it can be in the range of (160, 180), but it can be (120, 140) for the same material in another image. Therefore, the values of a, b have to be determined manually for every image, which causes low efficiency and inaccuracy.

Algorithm 2 is designed to overcome the disadvantage of algorithm 1. It has the same process of algorithm 1 except it uses different method to choose gray level threshold, which increases the efficiency greatly.

Materials have discontinuity at boundary, and have uniformity at internal. Gray levels have an obvious hop at boundary for materials with different densities. As long as the coordination of the hop can be found the edge can be identified.

Fig. 3. Peeled images

For algorithm 2, even though the gray levels of same material have big difference because of inconsistency when peeling the cross section images, the gray levels have a hop at boundary between two different materials. Since the hop can be located, use the average value of gray level before and after the hop as the threshold value. Above this value is one material; below it is another material. The merit of this method is the average value for any cross section image is always effective no matter how big the difference of gray level at external material between cross section images. This way avoids keeping change value of (a, b) in algorithm 1 and efficiency is increased greatly. The gray level curve is shown in figure 3 by reading pixel values along with a line from cross section image using improfile in Matlab.These 2 algorithms are especially useful for circle shaped materials. The peeled images are shown in Figure 3.

When the internal structure of the material is complicated, algorithm 1 and 2 cannot work for this situation. Because for both algorithms, the threshold value is first calculated, then the material is peeled from left to right and the gray level of the peeled material is set as background value. This process stops at another material. Then do the same process from right to left side. Then next line is chosen and the same process is started until the whole image is processed. This method works for circle shaped objects. However, it has problem when peeling structure-complicated objects, as shown in Figure 4.

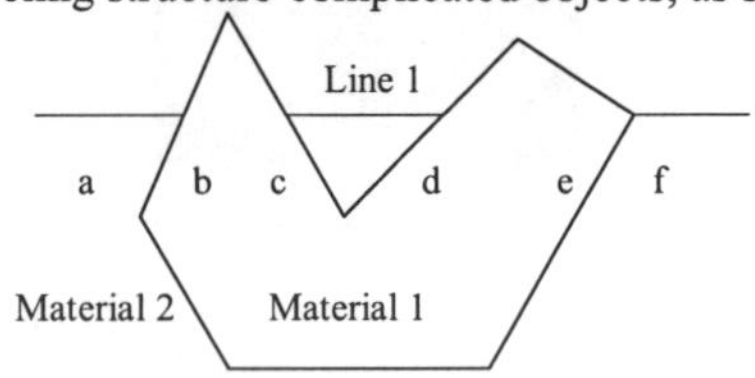

Fig. 4. An analysis image of the abuse

As shown in Figure 4, assume the inside of the irregular object is material 1, the outside is material 2. Conduct the edge detection from left to right side along line 1.

1) From point a to b, the same material is detected and set the gray level of all points between a and b as background gray level.

2) From point b to right, different material is detected, then stop the edge detection.

3) Along the same line 1 from right to left, same material between f and e is detected and set the gray level of these points as background gray level.

4) From point e to left, different material is detected, then stop.

5) Do the same process for next line.

According to the algorithm above, between points c and d is material 2 and is not set as background gray level. This way can't peel material 2 completely.

The cross section image in figure 4 is peeled with algorithm 2, as shown in Figure 5. The outer layers 1 and 2 are peeled by algorithm 2 to get wheel shaped object.

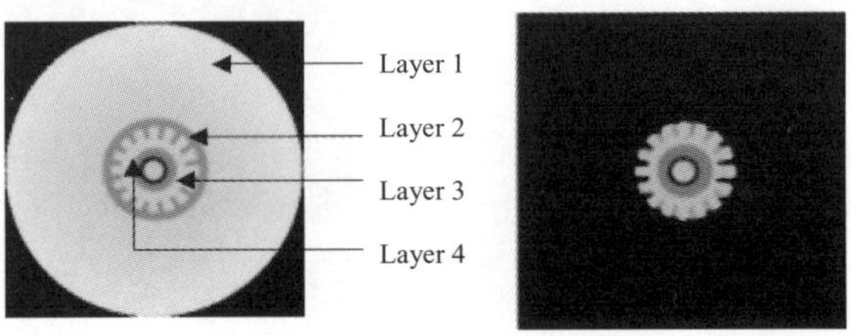

Fig. 5. The peeled image by algorithm 2

From the upper and lower parts of the image it tells that the material 2 between the wheels is not totally peeled. Thus another algorithm 3 is introduced for complicated shaped objects.

The steps of algorithm 3are described as following:

1) Filter the noise and get the threshold value by algorithm 2.

2) Image threshold processing to get binary image.

3) Process the binary image to minimize the effect of noise. Because the edge has some level of continuity, the 24 neighborhoods of points with value 1 can be checked; If the number of points with value 1 is small, it is noise.

4) Assume n layers materials are peeled, scan this binary image from left to right, set the first n points with value 1 as gray level 0; then scan the same line from right to left. Use the same way scan the whole image from top to bottom.

5) Use the resulted binary image as mask.

Set the gray level value of pixels outside of mask as background gray level value. The purpose of mask is: as show in figure 6, layer 2 and layer 4 are different layers, but are same material; When peel layer 2, don't want to peel layer 4.

Peel all cross section images by algorithm 3, then use 3-D reconstruction method to reconstruct 3-D image from these peeled images and get image as shown in Figure 5. The result is much better than that by algorithm 2.

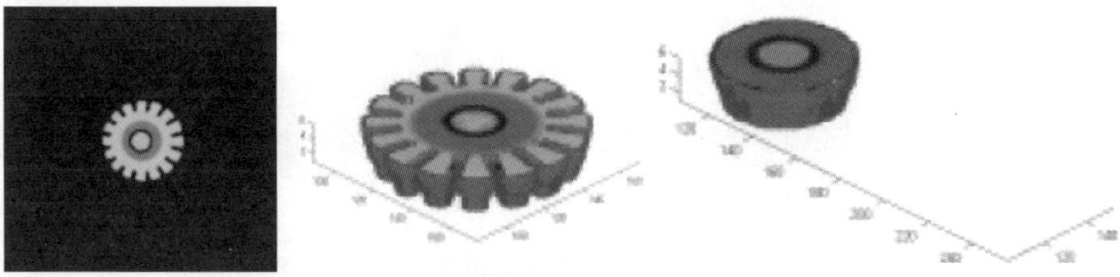

Fig 6. The peeled image by algorithm 3

4. The implementation of a series of cross-sectional images

In Matlab, edge function can be used to identify the image edge by changing parameters, and using different operators. But this kind of function operates on all of pixels, which takes longer time and is less efficient because internal cross section of object doesn't need edge detection. Besides, the image processed by edge function is binary image, which can't satisfy 3-D display requirement. This article designed a fast algorithm to remove air annulus using theory of edge detection, as show in Figure 7.

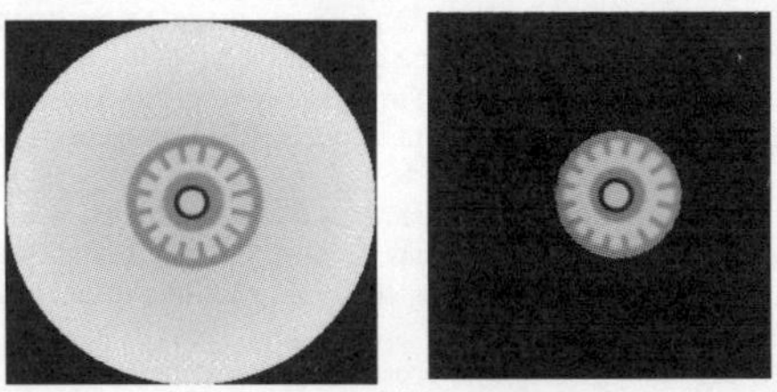

Fig. 7. Image after edge detecting

A pan bottom effect existed in air annulus can be found by imread, pixval functions. The closer to the center, the smaller the gray level is; the closer to edge, the bigger the gray level is. Gray level changes within the range (a, b), and the difference of gray level is very small between pixels in air annulus. The gray level outside of air annulus is different from the gray level between the point at cross section and the point close to air annulus, which is a quite big value n. Value n changes in the range of c and d. Values of a, b, c, d can be derived based on the available knowledge. This algorithm assumes a line scan the cross section image from left to right and checks the pixels on the line. Set gray level as 0 at the point if it is air. Keep check until reaching the edge of the object. Then do the same process from right to left. Thus the detection process is reduced at cross section. The process time is reduced and the efficiency is improved.

5. Conclusions

I have presented a complete 3-D reproduction system that reconstructs real objects. The various tasks have been considered one by one in detail throughout the paper. The experimental results concerning the different tasks of the whole system have already been presented throughout the text for demonstration purposes.

After the series of cross sectional images are processed by the method discussed above, 3-D data D is input, and a matrix $X \times Y \times S$ is generated, where X and Y are the coordination of cross section image, and S is the serial number. For example D(56,78,6) represents the pixel gray level at point (56,78) in the image. Use the function smooth 3(D) to smooth data, then use the function isosurface to calculate the surface date of D, finally use the function patch to display data in image format with illumination effect. Figure 8 shows the reconstruction result of 13 cross section images and breakout section images.

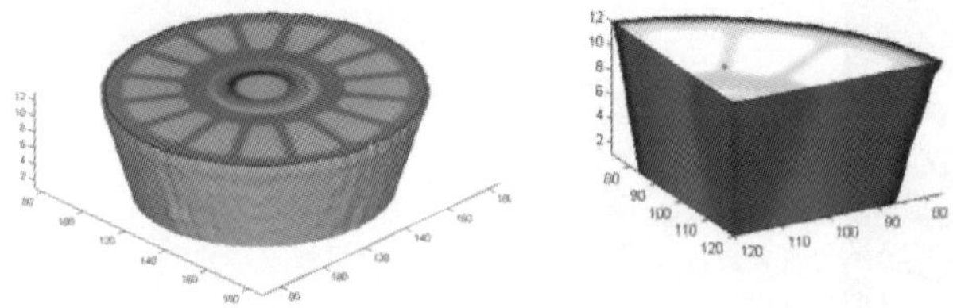

Fig. 8. The display image of 3-D reconstruction

References

[1] Y. Yemeza,*, F. Schmitt 3D reconstruction of real objects with high resolution shape and texture. Image and Vision Computing 22, 2004; 1137-1153.

[2] Roberto Marabini, Carlos O. S. Sorzano, Samuel Matej, José J. Fernández, José M. Carazo, and Gabor T. Herman. 3-D Reconstruction of 2-D Crystals in Real Space. IEEE Transactions on Image Processing, Vol.13, No.4, Apr. 2004; 549-561.

[3] Brian D. Rigling, and Randolph L. Moses. Three-Dimensional Surface Reconstruction From Multistatic SAR Images. IEEE Transactions on Image Processing, Vol.14, No.8, Aug. 2005; 1159-1171.

[4] W. C. Liu et. al. Dynamic elastic interpolation for 3-D medical image reconstruction from serial cross-sections. IEEE Transaction on Medical Imaging, Vol.7, No.3, Jul.1998; 225-232.

[5] Rick Archibald, Jiuxiang Hu, Anne Gelb, and Gerald Farin. Improving the Accuracy of Volumetric Segmentation Using Pre-Processing Boundary Detectionand Image Reconstruction. IEEE Transactions on Image Processing, Vol. 13, No. 4, Apr.2004; 459-466.

[6] D. J. Rossi and A. S. Willsky, Reconstruction from projections based on detection and estimation of objects parts I and II: Performance analysis and robustness analysis, IEEE Trans. Acoust. Speech Signal Process.ASSP-32, 1984; 886–906.

[7] L. V. Tran, R. c. Bahn, and J. Sklansky, Reconstructing the cross sections of coronary arteries from biplane angiograms, IEEE. Trans. Med. Imag. 11, 1992; 517–529.

[8] Y. Bresler, J. Fessler, and A. Macovski, Bayesian approach to reconstruction from incomplete projections of a multiple object 3-D domain, IEEE Trans. Pattern Anal. Mach. Intell. 11, 1989; 840–858.

[9] J. L. Prince and A. S. Willsky, Reconstructing convex sets from support line measurements, IEEE Trans. Pattern Anal. Mach. Intell. 12, 1990; 377–389.

[10] C. Pellot, A. Herment, M. Sigelle, P. Horain, and P. Perronneau, Segmentation, modelization, and reconstruction of arterial bifurcation in digital angiography, Med. Biol. Eng. Compt. 30, 1992; 567–583.

[11] C. Pellot, A. Herment, M. Sigelle, P. Horain, H. Maitre, and P. Peronneau, A 3-D reconstruction of vascular structures from two X-ray angiograms using an adapted simulated annealing algorithm, IEEE Trans. Med. Imaging 13, 1994; 48–60.

[12] P. Milanfar,W. C. Karl, and S.Willsky, Reconstructing binary polygonal objects from projections:Astatistical view, Comput. Vision Graphics Image Process. 56, 1994; 371–391.

[13] G. Ising, Beitrag zur theorie des ferromagnetismus, Zeit. Phys. 31, 1925; 253.